HISTORY
OF MATHEMATICS

David Eugene
Smith

VOLUME
II
SPECIAL TOPICS OF
ELEMENTARY MATHEMATICS

DOVER PUBLICATIONS, INC.
NEW YORK

Published in Canada by General Publishing Company, Ltd., 30 Lesmill Road, Don Mills, Toronto, Ontario.
Published in the United Kingdom by Constable and Company, Ltd.

This Dover edition, first published in 1958, is an unabridged and unaltered republication of the last edition. It is reprinted by special arrangement with Ginn and Company.

International Standard Book Number: 0-486-20430-8
Library of Congress Catalog Card Number: 58-11285

Manufactured in the United States of America
Dover Publications, Inc.
180 Varick Street
New York, N.Y. 10014

PREFACE

As stated in Volume I, this work has been written chiefly for the purpose of supplying teachers and students with a usable textbook on the history of elementary mathematics, that is, of mathematics through the first steps in the calculus. The subject has come to be recognized as an important one in the preparation of teachers and in the liberal education of students in colleges and high schools, showing, as it does, mathematics as constantly progressing instead of being a static mass of knowledge. Through a consideration of the history of the science the student comes to appreciate the fact that mathematics has continually adjusted itself to human needs, both material and intellectual; and thus he comes into sympathy with the effort to improve its status, either adding to its store through his own discoveries or bettering the methods of presenting it to those to whom it is taught in our schools.

In Volume I the reader found a general survey of the progress of elementary mathematics arranged by chronological periods with reference to racial and geographical conditions. In this volume he will find the subject treated by topics. The teacher of arithmetic will now see, in three or four chapters, a kind of moving picture of the growth of his subject,—how the world has counted, how it has performed the numerical operations, and what have been the leading lines of applications in which it has been interested. In geometry he will see how the subject arose, what intellectual needs established it so firmly, what influences led to its growth in various directions, and what human interest there is in certain of the great basal propositions. In algebra he will see, partly by means of facsimiles, how the symbolism has grown, how the equation looked three thousand years ago, the way its method of expression has changed from age to age, and how the science has so adjusted itself to world needs as now to be a necessity for the average citizen instead of a mental luxury for the selected few. He will learn how the number concept has enlarged as new needs have manifested themselves, and how the world struggled with fractions and with the mysteries of such artificial forms as the negative and the imaginary number, and will thus have a still clearer vision of mathematics as a growing science. The terminology of the subject will arouse interest; the common units of measure will mean something more than mere names; the minutes and seconds of time and of angles will take on a kind of human aspect; and the calendar will cease to be the mystery that it is to the youth. Trigonometry will have a new interest to the teacher who reads what Plutarch tells of the shadow-reckoning of Thales, and of the independent origin of the trigonometry of the sphere; and the calculus, which the freshman or sophomore burns in accordance with time-honored tradition, will be seen to have a history that is both interesting and illuminating. To see in its genetic aspect the subject that one is teaching or studying, and to see how the race has developed it, is oftentimes to see how it should be presented to the constantly arriving new generations, and how it can be made to satisfy their intellectual hunger.

While the footnote is frequently condemned as being merely an apology for obscurity or as an exhibition of pedantry, it would be difficult, in a work of this kind, to dispense with its aid. There are two principal justifications for such a device: first, it enables an author to place the responsibility for a statement that may be open to question; and, second, it encourages many students to undertake further study, either from secondary sources or, what is more important, from the original writings of the men who rank among the creators of mathematics. With these two points in mind, footnotes have been introduced in such a way as to be used by readers who wish further aid, and to be neglected by those who wish merely a summary of historical facts. For the student who seeks an opportunity to study original sources a slight introduction has been made to this field. The text of the book contains almost no quotations in foreign languages, the result being that the reader will not meet with linguistic difficulties in the general narrative. In the notes, however, it is frequently desirable to quote the precise words of an author, and this has been done with reference to such European languages as are more or less familiar. It is not necessary to translate literally all these extracts, since the text itself sets forth the general meaning. Students who have some general knowledge of Latin, French, or German will have little difficulty, and in many cases will have much interest, in seeing various statements in their original form. For special reasons a few notes have been given in Greek, but in every case the meaning is evident from the text.

In these footnotes and occasionally in the text there have, in this volume, been inserted a few names of minor importance which were purposely omitted in Volume I. These names refer to certain arithmeticians who contributed nothing to the advance of mathematics, but who, through popular textbooks, helped to establish the symbols and terms that are used in elementary instruction. In such cases all that has seemed necessary in the way of personal information is to give the approximate dates. In the case of names of particular importance further information may be found by referring to the Index.

The difficult question of the spelling and transliteration of proper names is always an annoying one for a writer of history. There is no precise rule that can be followed to the satisfaction of all readers. In general it may be said that in this work a man's name is given as he ordinarily spelled it, if this spelling has been ascertained. To this rule there is the exception that where a name has been definitely Anglicized, the English form has been adopted. For example, it would be mere pedantry to use, in a work in English, such forms as Platon and Strabon, although it is proper to speak of Antiphon and Bryson instead of Antipho and Bryso. When in doubt, as in the case of Heron, the preference has been given to the transliteration which most clearly represents the spelling used by the man himself.

In many cases this rule becomes a matter of compromise, and then the custom of a writer's modern compatriots is followed. An example is seen in the case of Leibniz. This spelling seems to be gaining ground in our language, and it has therefore been adopted instead of Leibnitz, even though the latter shows the English pronunciation better than the former. Leibniz himself wrote in Latin, and the family spelled the name variously in the vernacular. There seems, therefore, to be no better plan than to conform to the spelling of those recent German writers who appear to be setting the standard that is likely to be followed.

In connection with dates of events before the Christian era the letters B.C. are used; in connection with dates after the beginning of this era no distinguishing letters are added except in a few cases near the beginning of the period, in which the conventional letters A.D. have occasionally been inserted to avoid ambiguity. With some hesitation, but for a purpose which seems valid, dates are frequently given in parentheses after proper names. It is well recognized that a precise date, like 1202 after the name Fibonacci, is of no particular value in itself. It makes no difference, in ordinary cases, whether Fibonacci wrote his *Liber Abaci* in 1202, or in 1180, or in 1220, or whether *abacus* is spelled *abbacus*, as in some manuscripts, or in the more correct Latin form. On the other hand, two things are accomplished by a free use of such dates. In the first place, a reader is furnished with a convenient measuring instrument; he does not have to look in the Index or in a chronological table in order to see approximately where the particular writer belongs in the world's progress. The casual reader may well be pardoned if he does not recall where Bede, Alcuin, Gerbert, Jordanus, Fibonacci, and Roger Bacon stood chronologically with respect to one another, and in reading a technical history of this kind there is no reason why he should not be relieved of the trouble of consulting an index whenever he meets with such names as these. In the second place, it needs no psychologist to confirm the familiar principle that the mind comes, without conscious effort, to associate in memory those things which the eye has frequently associated in reading. At the risk, therefore, of disturbing the minds of those who are chiefly interested in the literary aspect of a general statement of the progress of mathematics, many important dates have been repeated, especially where they have not appeared in the pages immediately preceding.

The extent of a bibliography in a popular work of this kind is a matter of judgment. It can easily run to great length if the writer is a bibliophile, or it may receive but little attention. The purpose of giving lists of books for further study is that the student may have access to information which the author has himself used and which he believes will be of service to the reader. For this reason the secondary sources mentioned in this work are such as may be available, and in many cases are sure to be so, in the libraries connected with our universities, while the original sources are those which are of importance in the development of elementary mathematics or which may be of assistance in showing certain tendencies.

The first time a book was mentioned in Volume I, the title, date, and place of publication were given, together, whenever it seemed necessary, with the abbreviated title thereafter used. In general this plan has been followed in Volume II, at least in the case of important works. To find the complete title at any time, the reader has only to turn to the Index to find the first reference to the book. The abbreviation *loc. cit.* (for *loco citato*, in the place cited) is used only where the work has been cited a little distance back, since any more general use of the term would be confusing. The symbolism "I, 7" has been used for "Vol. I, p. 7," in order to conserve space, although exceptions have been made in certain ambiguous cases, as in the references to Heath's *Euclid*, references to Euclid being commonly by book and proposition, as in the case of Euclid, I, 47.

The standard works are referred to as given on pages xiv–xvi of Volume I.

In the selection of illustrations the general plan followed has been to include only such as will be helpful to the reader or likely to stimulate his interest; it would be undesirable to attempt to give, even if this were possible, illustrations from all the important sources, for this would tend to weary him. On the other hand, where the reader has no access to a classic that is being described, or even to a work which is mentioned as having contributed to the world's progress in some humbler manner, a page in facsimile is often of value. It is evident that space does not permit of the use of such bibliographical illustrations as those which comprise a large part of the facsimiles in the author's *Rara Arithmetica*.

In general the illustrations have been made from the original books or manuscripts in the well-known and extensive library of George A. Plimpton, Esq., who has been very generous in allowing this material to be used for this purpose, or from the author's collection of books, manuscripts, mathematical portraits and medals, and early mathematical instruments.

The scheme of transliteration and pronunciation of proper names is set forth fully on pages xvii–xxii of Volume I. Since Arabic, Persian,

Hindu, Chinese, and Japanese names are used less frequently in this volume, it will sufficiently meet the needs of the reader if he refers to the scheme there given.

As in Volume I, a few topics for discussion or for the personal consideration of students are suggested at the close of each chapter. Specific questions have been avoided, the purpose being not so much to examine the reader on the facts set forth as to encourage him to pursue his reading in other works upon the subject. In most cases this reading will be done in such encyclopedias as may be available, and, preferably, in other histories also; but in any case the reader will have his attention called to a number of general lines for further study, and he will have the consciousness that the present work is merely an introduction to the general subject, in which, it is hoped, his interest has increased.

On account of the extent of the index to Volume I it has not been combined with that of Volume II. It should therefore be consulted in connection with the index to this volume, particularly with respect to biographical and bibliographical references. Since, in many cases, textbooks are mentioned so frequently as to render a complete list so long as to be burdensome to the reader, thus defeating its purpose, such works are included only when the author is not mentioned in Volume I and when the work is of such importance as to make the reference valuable.

The author wishes to express his appreciation of the aid rendered by various friends in reading the proofsheets of both volumes, and especially by the late Herr Gustaf Eneström of Stockholm, by Professor R. C. Archibald, by Mr. Jekuthial Ginsburg of New York, and by Captain E. L. Morss of Boston.

DAVID EUGENE SMITH

CONTENTS

SPECIAL TOPICS OF ELEMENTARY MATHEMATICS

CHAPTER I

DEVELOPMENT OF THE ARITHMETICA

1. GENERAL SURVEY

Nature of Arithmetica. As stated in the Preface, it is the purpose of this volume to set forth in considerable detail the important steps in the historical development of the several branches of elementary mathematics. One of these branches is now known as arithmetic, a name which, as commonly understood in the English-speaking world, has little or no relation to the arithmetic of the ancients. In recent times the word has acquired the meaning given by the Greeks and Romans to logistic, or the art of computation, a much more humble discipline than that which they called arithmetic.

In order to make the distinction clear, the present chapter will set forth a sufficient number of simple details of the ancient arithmetic to enable the student to form an idea of its general nature, and the second chapter will consider the development of that elementary art which now bears the ancient name. It will be seen that the science which formerly appropriated the title was not related to ordinary calculation but was a philosophical study dealing with such properties as might now find place in a course in the theory of numbers if the latter had not outgrown most of these simple number relations and become a subject for the university student.

Modern Theory. The modern theory of numbers has so little direct relation to elementary mathematics that its history need only be referred to briefly in this volume.[1] Certain features like prime and composite numbers, polygonal numbers (such as squares), and solid numbers (such as cubes) are still found in elementary mathematics, however, and these features render essential a brief statement concerning the ancient arithmetic. In order to explain the position of this science in the ancient scheme of learning, it is desirable to speak first of the general range of knowledge according to the Greek schools of philosophy, and to distinguish between arithmetica, the classical theory of numbers, and arithmetic, the modern art of computation.

2. THE SEVEN LIBERAL ARTS

The Sevenfold Division. As stated in Volume I, three and seven have been the chief among mystic numbers in all times and among all peoples. Many reasons have been assigned for this universal habit of the race, most of them manifestly fanciful, and possibly no reason can be adduced that will command the general approval of scholars. If, however, we omit the number five, which was often used as a primitive radix and thus lost its element of mystery, a fairly satisfactory explanation is found in the fact that three and seven are the first prime numbers,— odd, unfactorable, unconnected with any common radix, possessed of various peculiar properties, and thus of a nature to attract attention in the period of superstition and mysticism.

One of the many results of this veneration for these numbers is seen in the fact that <u>the ancients numbered seven great branches of learning, just as they numbered the Seven Wonders of the World and the Seven Wise Men of Greece.</u> They separated these branches into two groups, four studies making up the domain of science as recognized by the Pythagoreans, and three constituting the nonscientific domain. Plato[2] spoke of

---

[1] The student will find it elaborately treated in L. E. Dickson, *History of the Theory of Numbers*, 3 vols., Washington, 1919–1923; hereafter referred to as Dickson, *Hist. Th. Numb.*

[2] *Republic*, IX. See also Aristotle's *Politics*, VIII, 1.

the liberal arts and separated them into two groups, but he did not limit them to any definite number. The scientific group, consisting of arithmetic, geometry, spherics, and music, constituted the ancient domain of mathematics.

The Seven Liberal Arts. It was probably in the work of Capella (*c.* 460), that the seven liberal arts were first distinctly specified.[1] These seven arts were thenceforth looked upon as necessary to the education of free men (*liberi*). They were then separated into the quadrivium,[2] constituting the Pythagorean group, and the trivium,[3] made up of grammar, dialectics, and rhetoric.[4]

The names of the seven arts are fairly descriptive of the subjects represented, with the exception of spherics, which related to mathematical astronomy; music, which related only to the theory of harmony;[5] and arithmetic, which had little in common with the subject known in English by this name.

[1] Varro (1st century B.C.) wrote a treatise on the "nine liberal disciplines," but the work is not extant. Capella introduced the liberal arts as the bridesmaids at the marriage of Philology and Mercury. Cassiodorus (*c.* 470–*c.* 564) placed the limit definitely at seven because of the seven pillars in the Temple of Wisdom (Proverbs, ix, 1).

[2] In medieval Latin also written *quadruvium*, the *quadruplex via*, as some writers have it. The term in its literal meaning is found as early as Juvenal. In its technical educational meaning it is used by Cassiodorus.

[3] Also written *truvium*.

[4] As "Hugnitio natione tuscus, civis pisanus, episcopus ferrariensis," to quote a medieval record, has it: "Et uero quia gramatica dialecta rethorica dicuntur triuuium quadam similitudine quasi triplex uia ad idem idest ad eloquentiam arismethica. musica. geometria. astronomia. quadam simili similitudine dicuntur quadriuuium quasi quadruplex uia ad idem idest ad sapientiam." See also the well-known verse quoted in Volume I, page 180.

[5] As an old Latin MS. has it:

Musicorum et cantorum magna est distantia:
Isti dicunt illi sciunt quae componit musica.

The distinction is well set forth in B. Veratti, *De' Matematici Italiani anteriori all' invenzione della stampa*, p. 4 (Modena, 1860). See also P. Tannery, "Du rôle de la musique grecque dans le développement de la mathématique pure," *Bibl. Math.*, III (3), 161; E. Narducci, "Di un codice . . . dell' opera di Giorgio Pachimere: περὶ τῶν τεσσάρων μαθημάτων," *Rendiconti della R. Accad. dei Lincei*, Rome, VII (1891), 191.

3. Early Writers on Number Theory

Origin of the Theory. There is no definite trace of the study of the theory of numbers before the time of Thales (*c.* 600 B.C.). Tradition says that this philosopher, filled with the lore of the Egyptians and probably well informed concerning the mysticism of the Babylonians, taught certain of the elementary properties of numbers in the Ionic School, of which he was the founder. Such meager knowledge as he had he imparted to his brilliant disciple, Pythagoras (*c.* 540 B.C.), who thereupon resorted to the priests of Egypt and probably of Babylon for further light. In the school which he established at Crotona, in southern Italy (Magna Græcia), he elaborated the doctrines of his teachers, including ideas which are distinctly Oriental, and made the first noteworthy beginning in the theory of arithmetica.

Little by little, first among the Pythagoreans and then in other schools of philosophy, the subject grew, a little being added here and a little there, until the time finally became ripe for the appearance of treatises in which the accumulated knowledge could be systematically arranged.

Books on the Theory. The first successful effort in the preparation of an expository treatise on the subject was made by Euclid (*c.* 300 B.C.), who is often known only as a geometer but who showed great genius in systematizing mathematical knowledge in other important lines as well. In his *Elements* he devotes Books II, V, VII, VIII, IX, X (in whole or in part) to the theory of numbers or to geometric propositions closely related thereto, and includes such propositions as the following:

If four numbers are proportional, they are also proportional alternately (VII, 13).

If two numbers are prime to two numbers, both to each, their products also will be prime to one another (VII, 26).

If a square number does not measure a square number, neither will the side measure the side; and if the side does not measure the side, neither will the square measure the square (VIII, 16).

If an odd number measures an even number, it will also measure the half of it (IX, 30).

The next worker in this field was that interesting dilettante in matters mathematical, Eratosthenes ($c.$ 230 B.C.), who worked on a method of finding prime numbers[1] by sifting out the composite numbers in the natural series, leaving only primes. This he did by canceling the even numbers except 2, every third odd number after 3, every fifth odd number after 5, and so on, the result being what the ancient writers called the sieve.[2]

His friend and sometime companion Archimedes ($c.$ 225 B.C.) did little with the theory of arithmetica, but made an effort to improve upon the Greek system of numbers,[3] his plan involving the counting by octads (10^8), in which he proceeded as far as 10^{52}, and making use of a law which would now be expressed by such a symbolism as $a^m a^n = a^{m+n}$, although he made no specific mention of this important theorem.

It was to the commentary on the *Timæus* of Plato, written by Poseidonius ($c.$ 77 B.C.), that the Greeks invariably went for their knowledge of the number theories of the Pythagoreans.[4] This is seen in the fact that the phraseology used by such writers as Theon of Smyrna ($c.$ 125) and Anatolius ($c.$ 280), in speaking of this subject, is simply a paraphrase of that used by Poseidonius.

Nicomachus, and Theon of Smyrna. The first noteworthy textbook devoted to arithmetica was written by Nicomachus ($c.$ 100), a Greek resident of Gerasa (probably the modern Jerash, a town situated about fifty-six miles northeast of Jerusalem). He was not an original mathematician, but he did for the theory of numbers what Euclid had done for elementary geometry and Apollonius ($c.$ 225 B.C.) for conic sections,—he summarized the accumulated knowledge in his subject. In his work are found such statements as the following: "Now fur-

[1]Vol. I, p. 109.   [2]Κόσκινον (*kos'kinon*), Latin, *cribrum.*
[3]Vol. I, p. 113. See also his Ψαμμίτης (*psammi'tes*, Latin *arenarius*, "sand reckoner"), *Archimedis Opera Omnia*, ed. Heiberg (Leipzig, 1880–1915), with revisions.
[4]F. E. Robbins, "Posidonius and the Sources of Pythagorean Arithmology," *Classical Philology*, XV (1920), 309. On Plato's appreciation of the value of this kind of work see F. Cajori, "Greek Philosophers on the Disciplinary Value of Mathematics," *The Mathematics Teacher* (December, 1920), p. 57.

thermore every square upon receiving its own side becomes heteromecic; or, by Zeus, on being deprived of its own side."[1]

The next writer of note was Theon of Smyrna (c. 125). He added several new propositions to the theory, two of them being of special interest: (1) If n be any number, n^2 or $n^2 - 1$ is divisible by 3, by 4, or by both 3 and 4; and if n^2 is divisible by 3 and not by 4, then $n^2 - 1$ is divisible by 4. (2) If we arrange two groups of numbers as follows:

$$n_1 = 1 + 0 \qquad\qquad d_1 = 1 + 0 = 1$$
$$n_2 = 1 + 1 \qquad\qquad d_2 = 2 + 1 = 3$$
$$n_3 = 2 + 3 \qquad\qquad d_3 = 4 + 3 = 7$$
$$n_4 = 5 + 7 \qquad\qquad d_4 = 10 + 7 = 17$$
$$\vdots \quad \vdots \qquad\qquad\qquad \vdots \quad \vdots$$
$$n_r = n_{r-1} + d_{r-1} \qquad d_r = 2\,n_{r-1} + d_{r-1}$$

then d^2 is of the form $2\,n^2 \pm 1$; for example, $d_1^2 = 1 = 2\,n_1^2 - 1$, $d_2^2 = 9 = 2\,n_2^2 + 1, \cdots$. The numbers d_r were called by Theon diameters. It is interesting to observe a fact unknown to him, namely, that the ratios $d_1 : n_1 = 1$, $d_2 : n_2 = \frac{3}{2}$, $d_3 : n_3 = \frac{7}{5}$, $\cdots$ are the successive convergents of the continued fraction

$$1 + \frac{1}{2} + \frac{1}{2} + \frac{1}{2} + \cdots,$$

and hence approach nearer and nearer the square root of 2.

Boethius. Boethius (c. 510) appropriated the knowledge of such writers as Euclid, Nicomachus, and Theon, incorporating it in his work *De institutione arithmetica libri duo* and producing a textbook that was used in all the important schools in the Middle Ages. It is the source with which a student may advantageously begin his study of this subject.

Later Writers. The most noteworthy writer on the subject in the medieval period is Fibonacci (1202), and with respect to him and subsequent writers, all of whom have been considered in Volume I, we shall later speak in detail as necessity arises.

---

[1] Introduction, XX. See G. Johnson, *The Arithmetical Philosophy of Nicomachus of Gerasa*, Lancaster, 1916, hereafter referred to as Johnson, *Nicomachus*. The meaning is that $x^2 \pm x$ is not a square but a heteromecic or oblong number.

4. NAMES FOR ARITHMETIC

Arithmetic and Logistic. The ancient Greeks distinguished between arithmetic,[1] which was the theory of numbers and was therefore even more abstract than geometry,[2] and logistic,[3] which was the art of calculating. These two branches of the study of numbers continued as generally separate subjects until the time of printing, although often with variations in their names; but about the beginning of the 16th century the more aristocratic name of "arithmetic" came to be applied to both disciplines. This use of the term was not universal, however, and even today the Germans reserve the word *Arithmetik* for the theoretical part of the science as seen in the operations in algebra, using the word *Rechnen* for the ancient logistic.[4] Various writers,[5] preserved the word "logistic" in the 16th century, but in the older sense it generally dropped out of use thereafter.

From the fact that computations were commonly performed on the abacus, the name of this instrument was used in the early Middle Ages as a synonym of logistic. Finally, however, the word "abacus" came to mean any kind of elementary arithmetic,[6] and this usage obtained long after printing was invented.[7]

In the Middle Ages the name "arithmetic" was apparently not in full favor, perhaps because it was not of Latin origin. Thus, in a manuscript attributed to Gerbert the word is spoken of as Greek, the Latin being "numerorum scientia."[8]

[1] Ἀριθμητική (*arithmetike'*), from ἀριθμός (*arithmos'*), number. It passed over into Latin as *arithmetica*.

[2] "Est enim Arithmetices subjectum purius quiddam & magis abstractum, quam subjectum Geometriae" (J. Wallis, *Opera Mathematica*, I, 18 (Oxford, 1695)).

[3] Λογιστική (*logistike'*), which passed over into Latin as *logistica*.

[4] Compare also the French *calcul*.

[5] J. Noviomagus, *De Numeris libri duo* (1539); Buteo (Lyons, 1559); Schoner edition of Ramus (1586); "Logistica quam uulgo uocant algoristicam et algorismum" (MS. notes in the 1558 edition of Gemma Frisius, in Mr. Plimpton's library). For biographical information relating to such writers as are of particular importance, see the Index of Volume I.

[6] As in Fibonacci's *Liber Abaci* (1202).

[7] See *Rara Arithmetica* for many works bearing such titles as *Libro d' abacho*.

[8] "Graece Arithmetica, latine dicitur numerorum scientia," from the colophon of the "Liber subtilissimus de arithmetica." See C. F. Hock, *Gerberto o sia Siivestro II Papc . . . trad. del . . . Stelzi*, p. 206 (Milan, 1846).

Vicissitudes of the Term. The word "arithmetic," like most other words, has undergone many vicissitudes. In the Middle Ages, through a mistaken idea of its etymology, it took an extra *r*, as if it had to do with "metric."[1] So we find Plato of Tivoli, in his translation (1116) of Abraham Savasorda, speaking of "Boetius in arismetricis."[2] The title of the work of Johannes Hispalensis, a few years later (*c.* 1140), is given as "Arismetrica," and fifty years later than this we find Fibonacci dropping the initial and using the form "Rismetrica."[3] The extra *r* is generally found in the Italian literature until the time of printing.[4] From Italy it passed over to Germany, where it is not uncommonly found in the books of the 16th century,[5] and to France, where it is found less frequently.[6] The ordinary variations in spelling have less significance, merely illustrating, as is the case with many other mathematical terms, the vagaries of pronunciation in the uncritical periods of the world's literatures.[7]

[1] Greek μέτρον, a measure, as in "metre" and "metrology."

[2] *Abhandlungen*, XII, 16. For such abridged forms see the Index of Volume I.

[3] This is in one of the MSS. formerly owned by Boncompagni. See the sale catalogue of his library, p. 104. Fibonacci (1202) commonly used "abacus."

[4] *E.g.*, see the "Brani degli Annali Decemvirali posseduti dall' archivo Decemvirale di Perugia," in Boncompagni's *Bullettino*, XII, 432; E. Narducci, *Catalogo di Manoscritti*, 2d ed. (Rome, 1892), No. 56, p. 26; hereafter referred to as Narducci, *Catalogo Manosc.*

[5] *E.g.*, "Die Kunst Arismetrica die aller edelst vnder den sybē freyen künsten," Köbel, 1514. A MS. in Scheubel's (*c.* 1550) handwriting in the Columbia University Library has "de Arrismetris." There is also a MS. copy made *c.* 1515 in Rome, by a Swedish savant, Peder Månsson, from the *Margarita phylosophica* of Gregorius Reisch (1503), in which the form "Arismetrice" is given. See *Bibl. Math.*, II (2), 17.

[6] So in a MS. written by Rollandus *c.* 1424 (see *Rara Arithmetica*, p. 446) the form "arismetica" is usually given, but the form "arismetrica" also appears. In an unpublished MS. entitled "traicte d'Arismetricque . . . faite et compillé A paris en lan mil 475" (for 1475) there is this curious etymology: "Arismeticve est vne des sept ars liberaulx & la premiere des quatre ars Mathematique En la quelle est la vertus de nombrer. Et est dicte de ares Nom grec qui est en latin Virtus Et de menos aussi nom grec qui est en Latin numerus parquoy est dicte Vertus de Nombre." E. Narducci, *Catalogo Manosc.*, No. 603, p. 395.

[7] Thus, we have "arimmetica" throughout Zuchetta's work of 1600 (see *Rara Arithmetica*, p. 425); "eritmeticha" in a 17th century MS. (see Narducci's *Catalogo Manosc.*, No. 446, 1, p. 267); "aristmeticque" in an anonymous French work, Paris, 1540; "Alchorismi de pratica Aricmetica," in a MS. of Sacrobosco, Boncompagni sale catalogue, No. 645.

Origin of Algorism. From the fact that the arithmetic of al-Khowârizmî (*c.* 825) was translated into Latin as *liber Algorismi* (*the book of al-Khowârizmî*), arithmetic based on the Hindu-Arabic numerals, more especially those that made use of the zero, came to be called algorism as distinct from the theoretical work with numbers which was still called arithmetic.[1] Since *al* often changes to *au* in French, we have "augrisme" and "augrime,"—forms which were carried over to England as "augrim,"[2] later reverting to "algorism"[3] or the less satisfactory form of "algorithm."[4]

The prefix *al* was dropped from this word by most Spanish writers, giving such forms as "guarisma"[5] and "guarismo,"[6] and in other countries there were many variations that were quite as curious.[7]

The word troubled many of the early Latin writers, and various fanciful etymologies were suggested, the best conjecture being that of Sacrobosco that it came from Algus or Argus,

---

[1] Thus, that part of the Rollandus MS. (*c.* 1424) relating to the theory is referred to in the phrase "Arismetrice pars primo tractanda est speculatiua," while the other part is called "algorismus." See also M. Chasles, *Comptes rendus*, XVI, 162.

[2] "¶ ouer the wiche degrees ther ben nowmbres of augrym; . . . & the nombres of the degres of tho signes ben writen in Augrim." Chaucer's *Astrolabe*, ed. Skeat, p. 5.
"Although a sypher in augrim have no might in significacion of it-selve, yet he yeveth power in significacion to other." Chaucer, *The Testament of Love*, ed. Skeat, Bk. II, chap. vii.

[3] Thus, Recorde (*c.* 1542) in his *Grovnd of Arts* (as spelled in the 1646 edition): "Some call it Arsemetrick, and some Augrime. . . . Both names are corruptly written: Arsemetrick for Arithmetick, as the Greeks call it, and Augrime for Algorisme, as the Arabians found it." 1646 ed., p. 8.

[4] One eccentric English writer, Daniel Fenning (1750), attempted to distinguish algorithm, as first principles, from algorism, as the practice of these principles.

[5] As in the Spanish *Suma de Arithmetica* of Gaspard de Texeda, Valladolid, 1546. The separate word *al* or *el* (the) was prefixed, however, and the form *algoritmo* is still preserved.

[6] ". . . de vn Filosofo llamado Algo, y por aquesta causa fue llamada el Guarismo" (Santa-Cruz, a Spanish writer, 1594); but see *Rara Arithmetica*, p. 407.

[7] "Arismethique qui vulgayrement est appellee argorisme" (E. de la Roche, a French writer, 1520). We also find such forms as *alkauresmus* and *alchocharithmus* in various MSS. of the same period.

a certain philosopher, this being merely a corruption of al-Khowârizmî.[1] It was not until 1849 that the true etymology was again discovered.[2]

The Etymology early Recognized and Forgotten. Very likely the etymology of the term "algorism" was known to such early translators or writers as Johannes Hispalensis[3] (c. 1140) and Adelard of Bath[4] (c. 1120). By the following century, however, al-Khowârizmî was quite forgotten by such Latin writers as Sacrobosco[5] (c. 1250) and Bacon (c. 1250). From that time on we have the word loosely used to represent any work related to computation by modern numerals[6] and also as synonymous

[1] So we have Chaucer's expression,

> Thogh Argus the noble covnter
> Sete to rekene in hys counter.
>
> *Dethe Blaunche* (c. 1369)

This derivation was followed by various writers, such as Santa-Cruz (1594), Cataldi (1602), and Tartaglia (1556 ed., I, fol. 3, r.). Of the other fanciful etymologies the following may be of interest: *argris* (Greek) + *mos* (custom); *algos* (Greek for "white sand") + *ritmos* (calculation); *algos* (art) + *rado* (number); *Algorus*, the name of a Hindu scholar; *Algor*, a king of Castile. See A. Favaro, Boncompagni's *Bullettino*, XII, 115; M. Cantor, *Mathematische Beiträge zum Kulturleben der Völker*, p. 267 (Halle, 1863); C. I. Gerhardt, *Ueber die Entstehung . . . des dekadischen Zahlensystems*, Prog., p. 26, n. (Salzwedel, 1853); K. Hunrath, "Zum Verständniss des Wortes Algorismus," *Bibl. Math.*, I (2), 70; and see VIII (2), 74. P. Ramus (*Scholarum Mathematicarum Libri XXXI*, p. 112 (1569)) derived it from *al* (Arabic for "the") + ἀριθμός (*arithmos'*, number), and J. Schoner (1534 edition of the *Algorithmvs Demonstratvs*, fol. A [iij], v.) did the same.

[2] By the orientalist J. Reinaud (1795–1867). See *Mém. de l'Institut national de France des inscriptions et belles-lettres*, XVIII, 303; Boncompagni's *Bullettino*, XII, 116. Even as late as 1861, however, L. N. Bescherelle's well-known French dictionary (Paris, 1861) gave *al* (the) + *ghor* (parchment), and the variants *algarthme, algarisme*. See also Boncompagni's *Bullettino*, XIII, 557.

"Incipit prologus in libro alghoarismi de pratica arismetrice. Qui editus est a magistro Johanne yspalensi." See F. Woepcke, *Journal Asiatique*, I (6), 519.

[4] Who uses such forms as *algoritmi* and *algorizmi*.

[5] "Hanc igitur scientiam numerandi compendiosam edidit philosophus nomine Algus, unde algorismus nuncupatur, vel ars numerandi, vel introductio in numerum." Halliwell ed., p. 1.

[6] "Ceste signifiance est appellée algorisme" (MS. of c. 1275); see C. Henry, Boncompagni's *Bullettino*, XV, 53. "Secondo Lalgorismo" (Ghaligai, 1521); ". . . calculandi artem, quam uulgus Algorithmum uocat" (Schoner, 1534). So the MS. of Scheubel (c. 1550), already mentioned as in the Columbia Uni-

with the fundamental operations themselves[1] and even with that form of arithmetic which makes use of the abacus.[2]

Names for Logistic. There have been various other names for logistic. The early Italian writers often spoke of a practical arithmetic as a *practica*,[3] *pratica*, or *pratiche*.[4] Many of the Latin writers of the Renaissance, particularly in the 16th century, spoke of it as the art of computing (*ars supputandi*).[5] The Dutch writers used the term "ciphering,"[6] particularly in the 16th and 17th centuries, and from this source, through New Amsterdam, came the common use of the word in the early schools of America.

In Italy, in the 15th century, logistic occasionally went by the name of the minor art,[7] and arithmetic and algebra by the name of the major art.[8]

5. ELEMENTARY CLASSIFICATIONS OF NUMBER

Abstract and Concrete. The distinction between abstract and concrete numbers is modern. The Greek arithmeticians were concerned only with the former, while the writers on logistic naturally paid no attention to such fine distinctions. It was not

versity Library, has such phrases as "Algebrae fundamenta seu algorismus," "Algorismus de surdis," and "Algorithmus quantitatis," showing the broader use of the term. Stifel (1544) used the term in the same way.

[1] Thus, Thierfelder (1587) uses "Der Algorithmus" and "Die Species" (p. 51) as synonymous. Similarly, "ALgorithmus ist ein lehr aus der man lernet Addiren/Subtrahiren/Multipliciren vnd Diuidiren" (Stifel, 1545).

[2] As in the *Algoritmus* of Kłos (1538), the first Polish arithmetic, which is purely a treatise on abacus reckoning. See S. Dickstein, *Bibl. Math.*, IV (2), 57. Similarly, there were several books entitled *Algorithmus linealis* published in Germany early in the 16th century, all dealing with the abacus.

[3] As in the Treviso arithmetic (1478).

[4] As in Cataneo's arithmetic (1546).

[5] Thus, Tonstall (1522) calls his work *De arte supputandi*, a title already used by Clichtoveus (1503) in the abridged form of *Ars supputādi*. Glareanus (1538) speaks of the "supputandi ars," and "supputation" (for computation) was a term in common use in England until the 19th century. For example, see W. Butler, *Arithmetical Questions*, London, 2d ed., 1795.

[6] Cyffering, cyffer-konst, cÿffer-boeck, and the like.

[7] *L'arte minore*.   [8] *L'arte maggiore*; or, in Latin, *Ars magna*.

until the two streams of ancient number joined to form our modern elementary arithmetic that it was thought worth while to make this classification, and then only in the elementary school. The terms "abstract" and "concrete" were slow in establishing themselves. The mathematicians did not need them, and the elementary teachers had not enough authority to standardize them. In the 16th century the textbook writers began to make the distinction between pure number and number to which some denomination attached, and so we find Trenchant (1566), for example, speaking of absolute and denominate number, the latter including not only 3 feet but also 3 fourths.[1]

From that time on the distinction is found with increasing frequency in elementary works. Such refinements, however, as required the product to be of the same denomination as the multiplicand are, in general, 19th century creations of the schools. Thus Hodder[2] asserts that "Pounds multiplied by 20, are shillings," and every scientist today recognizes such forms as "20 lb. × 10 ft. = 200 foot-pounds."

Digits, Articles, and Composites. One of the oldest classifications of numbers is based upon finger symbolism.[3] The late Roman writers seem to have divided the numbers below a hundred into fingers (*digiti*), joints (*articuli*), and composites

---

[1] "L'absolu est celuy qui n'a aucune denomination: comme 2, 7, 5, tel nombre est abstrét, & de forme nue se referant à la Theorique. Le denommé: est celuy qui si prononce auec quelque denomination . . . & se refére à la Pratique." The latter included "le vulgairement denommé, comme 8 auñ," and also "le rompu, comme ¾," although he says that in practice ¾ is considered as abstract unless some denomination is given to it: "lequel en pratiquant est entendu absolu s'il n'a quelque denomination de suget, comme disant ¾ d'aun" (1578 ed., p. 16).

Similarly, Stifel: "Numeri abstracti proprie dicuntur, q̄ nullā prorsus denominationē habet" (*Arithmetica Integra*, 1544, fol. 7, *v.*). Xylander (1577) used *ledige* and *benannte Zahlen*. [2] 1672 ed., p. 56.

[3] See page 196. Th. Martin, "Les signes numéraux," *Annali di mat. pura ed applic.*, V, 257, 337, and reprint (Rome, 1864); hereafter referred to as Martin, *Les Signes Num.* Suevus (*Arithmetica Historica*, 1593, p. 3) speaks of the finger origin: "Digitus heist ein Finger zal/die unter zehen bedeut"; M. Wilkens, a Dutch arithmetician (*Arithmetica*, Groningen, 1630; 1669 ed., p. 1), says: "Dese zijn Digiti, dat's Enckel ofte vingergetalen"; and many other early writers have similar statements.

(*compositi*) of fingers and joints, the joints being the tens, and the composites being numbers like 15, 27, and so on. In a passage attributed, but doubtfully, to Boethius it is said that this threefold division is due to the ancients.[1] While the terms were probably known in early times, they were not used commonly enough to appear in the places where finger symbolism is mentioned.[2] So far as extant works are concerned, the classification is medieval.

Meaning of "Digit." Since there are ten fingers, it is probable that the digits were originally the numbers from one to ten inclusive; but so far as appears from treatises now extant they were the numbers from one to nine inclusive, not the figures representing these numbers; that is, they were the numbers below the first "limit." The division of numbers into limits or differences (in which 10, 20, ···, 90 were of the first order; 100, 200, ···, 900, of the second order, and so on) is found in the works of such writers as Alcuin (*c.* 780), Jordanus Nemorarius (*c.* 1225), O'Creat (*c.* 1150), and Sacrobosco (*c.* 1250), and was evidently common.[3] Since unity was not considered a number until modern times, it was sometimes definitely omitted, leaving only eight digits.[4]

---

[1] Since this is the first time the division appears, so far as known, the passage is important enough to be quoted in the original: "Digitos vero, quoscunque infra primum limitem, id est omnes, quos ab unitate usque ad denariam summam numeramus, veteres appellare consueverunt. Articuli autem omnes a deceno in ordine positi et in infinitum progressi nuncupantur. Compositi quippe numeri sunt omnes a primo limite id est a decem usque ad secundum limitem id est viginti ceterique sese in ordine sequentes exceptis limitibus. Incompositi autem sunt digiti omnes annumeratis etiam omnibus limitibus." Boethius, ed. Friedlein, p. 395. See also G. Eneström, *Bibl. Math.*, XI (2), 116.

[2] Pliny, *Hist. Nat.*, 34, 7; 2, 23; Martin, *Les Signes Num.*, 51.

[3] G. Eneström, "Sur les neuf 'limites' mentionnés dans l''Algorismus' de Sacrobosco," *Bibl. Math.*, XI (2), 97. See also the 12th century MS. described by M. Chasles in the *Comptes rendus*, XVI (1843), 237; the *Compotus Reinheri*, p. 28; Boncompagni's *Bullettino*, X, 626; S. Günther, *Geschichte des math. Unterrichts*, p. 99 (Berlin, 1887) (for Bernelinus), hereafter referred to as Günther, *Math. Unterrichts*.

[4] *E.g.*, by Peletier (1549): "Le Nombre Entier se diuise en Simple, Article, & Composé. Le Simple est le Nombre plus bas que 10: ce sont les huict figures, 2, 3, 4, 5, 6, 7, 8, 9." He uses numbers and figures as synonymous, and uses "simple" for "digit."

Meaning of "Article" and "Composite." The articles were sometimes limited to nine in number (10, 20, $\cdots$, 90), but it was more common to take any multiple of ten. In the early printed books they were occasionally called decimal numbers,[1] and as such they finally disappeared.

The term "composite," originally referring to a number like 17, 56, or 237, ceased to be recognized by arithmeticians in this sense because Euclid had used it to mean a nonprime number.[2] This double meaning of the word led to the use of such terms as "mixed" and "compound" to signify numbers like 16 and 345.[3]

The oldest known French algorism (*c.* 1275) has the three-fold division[4] above mentioned, as does also the oldest one in the English language (*c.* 1300), already cited. The latter work is so important in the history of mathematics in this language as to justify a further brief quotation:

> Some numbur is called digitus latine, a digit in englys. Somme nombur is called articulus latine. An Articul in englys. Some nombur is called a composyt in englys. . . .
>
> ¶Sunt digiti numeri qui citra denarium sunt.[5]

---

[1] Thus, Pellos (1492, fol. 4) speaks of "numbre simple," "nūbre desenal," and "nūbre plus que desenal"; and Ortega (1512; 1515 ed., fols. 4, 5) has "lo numero simplice," "lo numero decenale," and "lo numero composto."

[2] *Elements*, II, def. 13. For other Greek usage see Heath's *Euclid*, Vol. II, p. 286.

Lazesio (1526), among others, pointed out this twofold usage "secūdo sacro busco ī suo algorismo" and "secōdo el senso di Euclide" (1545 ed., fol. 2). See also Pacioli, *Sūma* (1494 ed., fols. 9, 19); Tartaglia, *General Trattato* (1556, II, fol. 1, *v.*); Santa-Cruz (1594; 1643 ed., fol. 2).

Trenchant (1566; 1578 ed., p. 223) speaks of "Nombre premier, ou incomposé," and "Nombre second, ou composé," a natural use of "second" as related to "premier" (prime), and the same usage was doubtless common at that time.

[3] "Alius aūt mixte siue ppositus," in the *Questio haud indigna eiusq3 solutio ex aurelio Augustino, c.* 1507. So Hylles (1592; 1600 ed., fol. 7) says: "The third sort are numbers MIXT or compound"; Digges (1572; 1579 ed., p. 2) uses "compound" alone; and Hodder (10th ed., 1672, p. 5) has "A Mixt, or Compound." Dutch arithmeticians avoided the difficulty by using terms in the vernacular; thus, Mots (1640) gives "De enckel getallen" (digits), "Punct ofte leden-getallen" (articles), "t'samen-gevoeghde getallen" (composites).

[4] "Tu dois savoir ki sont .3. manieres de nombres car li .1. sont degit li autre article. li autre compost." See Ch. Henry in Boncompagni's *Bullettino*, XV, 53.

[5] The anonymous writer here quotes from the *Carmen de Algorismo* of Alexandre de Villa Dei (*c.* 1240). The translation follows.

¶Here he telles qwat is a digit, Expone versus sic. Nomburs digitus bene alle nomburs þat ben with-inne ten, as nyne, 8.7.6.5.4.3.2.1. . . . Articulis ben ben alle þat may be deuidyt into nomburs of ten & nothynge leue ouer, as twenty, thretty, fourty, a hundryth, a thousand, & such oþer. . . . Compositys ben) nomburs þat bene componyt of a digyt & of an articulle as fouretene, fyftene, sextene, & such oþer.

Recorde (*c.* 1542) sums the matter up by saying:

A diget is any numbre vnder 10. . . . And 10 with all other that may bee diuided into x. partes iuste, and nothyng remayne, are called articles, suche are 10, 20, 30, 40, 50, &c. 100, 200, &c. 1000. &c. And that numbre is called myxt, that contayneth articles, or at the least one article and a digette: as 12.[1]

At best such a classification is unwieldy, and many of the more thoughtful writers, like Fibonacci (1202), abandoned it entirely. Others, like Sacrobosco (*c.* 1250), struggled with it but were obscure in their statements;[2] while Ramus very wisely (1555) dismissed the whole thing as "puerile and fruitless."[3]

All that is left of the ancient discussion is now represented by the word "digit," which is variously used to represent the numbers from one to nine, the common figures for these numbers, the ten figures 0, 1, . . ., 9, or the first ten numbers corresponding to the fingers.

Significant Figures. After the advent of the Hindu-Arabic figures into Europe (say in the 10th century) the difference between the zero and the other characters became a subject of comment. The result was the coining of the name "significant figures" for 1, 2, 3, · · ·, 9. At the present time the meaning

---

[1] 1558 ed. of the *Grovnd of Artes*, fol. C iij. Similar classifications are found in most of the early printed books of a theoretical nature, but less frequently in the commercial books.

[2] Thus Petrus de Dacia (1291) confessed that he could not quite understand Sacrobosco, saying, "ita credo auctorem esse intelligendum."

[3] "Puerilis et sine ullo fructu." See also Boncompagni, *Trattati d'Aritmetica,* II, 27 (Rome, 1857); J. Havet, *Lettres de Gerbert,* p. 238 (Paris, 1889); Boncompagni's *Bullettino,* XIV, 91; *Abhandlungen,* III, 136.

has been changed, so that 0 is a significant figure in certain cases. For example, if we are told to give log 20 to four significant figures, we write 1.301. Similarly, we write 0.3010 for log 2, and 7.550 for $\sqrt{57}$. The term is doubtless to be found in medieval manuscripts; at any rate it appears in the early printed arithmetics[1] and has proved useful enough to be retained to the present time in spite of the uncertainty of its meaning.

Odd and Even Numbers. The distinction between odd and even numbers is one of the most ancient features in the science of arithmetic. The Pythagoreans knew it, and their founder may well have learned it in Egypt or in Babylon. It must have been common to a considerable part of the race, for the game of "even and odd" has been played in one form or another almost from time immemorial,[2] being ancient even in Plato's time.[3] The game consisted simply in guessing odd or even with respect to the number of coins or other objects held in the hand.

The odd number was also called by the geometric name of "gnomon," the primitive form of the sundial. If such a figure

---

[1] *E.g.*, Licht (1500); Grammateus (1518), "neun bedeutlich figuren"; Riese (1522), "Die ersten neun sind bedeutlich"; Gemma Frisius (1540); Stifel (1544), "Et nouem quidem priores, significatiuae uocantur"; Peletier (1549), "Chacune des neufs premieres (qui sont appellees significatiues)" . . . ; Recorde (*c.* 1542), "The other nyne are called Signifying figures"; Trenchant (1566).

[2] This is seen in such expressions as ἀρτιασμός, ἄρτια ἢ περιττὰ παίζειν, ζυγὰ ἢ ἄζυγα παίζειν. This ζυγὰ ἢ ἄζυγα, "yokes or not-yokes," is similar to the Sanskrit "yuj" and "ayug" for even and odd. Horace couples it with riding a hobby horse as a childish diversion:

Ludere par impar, equitare in harundine longa.

Satires, II, 3, 248

See also E. B. Tylor, "History of Games," in the *Fortnightly Review*, May, 1879, p. 735.

[3] In addition to the references to the Greek theory of numbers given in Volume I and in this chapter, consult Dickson, *Hist. Th. Numb.*; F. von Drieberg, *Die Arithmetik der Griechen*, Leipzig, 1819; G. Friedlein, *Die Zahlzeichen und das elementare Rechnen der Griechen und Römer*, Erlangen, 1869; Heath, *History*, I, 67–117. Heath mentions a fragment of Philolaus (*c.* 425 B.C.) which says that "numbers are of two special kinds, odd and even, with a third, even-odd, arising from a mixture of the two."

is turned to the east in the morning and to the west in the afternoon, the hours can be read on the horizontal arm as in the Egyptian sun clock mentioned in Volume I, page 50. Thus we have the origin of the right shadow, the *umbra recta*, used in early trigonometry. By such an instrument we come to "know"[1] the time, and by facing it to the south we also come to know the seasons, the solstices, and the length of the year.[2]

It is apparent that the gnomon here shown in the shaded part of the figure is of the form $2n + 1$ and hence, as stated above, is an odd number.[3] It is also apparent that $\sum_{0}^{n} (2n + 1)$ is a square, that is, that the sum of the first n odd numbers, including 1, is a square,—a fact well known to the Greeks, as is shown by the works of Theon of Smyrna[4] (*c.* 125).

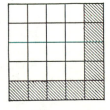

That there is luck in odd numbers is one of the oldest superstitions of the race, with such occasional exceptions as the case of the general fear of thirteen,—a fear that seems to have long preceded the explanation that it arose from the number present at the Last Supper.[5]

The general feeling that odd numbers are fortunate and even numbers unfortunate comes from the ancient belief that odd numbers were masculine and even numbers, always containing other numbers, were feminine. This led to the belief that odd numbers were divine and heavenly, while even numbers were human and earthly. The superstition was quite general among ancient peoples. Plato says: "The gods below . . . should receive everything in even numbers, and of the second choice, and

---

[1] Greek γνώμων (*gno'mon*), one who knows, from γιγνώσκειν, γνῶναι, know.

[2] Heath, *History*, I, 78.

[3] "Gnomon . . . quod Latini amussim seu normam vocant." J. C. Heilbronner, *Historia Matheseos Universae* (Leipzig, 1742), p. 173; see also page 193; hereafter referred to as Heilbronner, *Historia*.

[4] *Theonis Smyrnaei . . . expositio*, ed. Hiller, p. 31 (Leipzig, 1878). On this entire discussion see also Johnson, *Nicomachus*, and especially Heath, *History*, I, 77.

[5] Ernst Böklen, *Die Unglückszahl Dreizehn und ihre mythische Bedeutung*, Leipzig, 1913, with extensive bibliography.

ill omen; while the odd numbers, and of the first choice, and
the things of lucky omen, are given to the gods above,"[1] and
the phrase "Deus imparibus numeris gaudet" ("God delights in
odd numbers") probably goes back to the time of Pythagoras.[2]
The superstition runs through a wide range of literature.
Thus, Shakespeare, in the *Merry Wives of Windsor*, remarks
that "there is divinity in odd numbers, either in nativity, chance,
or death." Such beliefs naturally persist among the less advanced peoples and are common even today. For example, on
the island of Nicobar, India, an odd number of vessels of water
are dashed against the hut where a corpse is being laid out,
and the stretcher that bears it must always contain an odd
number of pegs.[3]

Further Classification. The Greeks not only recognized odd
and even numbers,[4] but they carried the classification much farther, including what Euclid calls "even-times-even numbers,"
"even-times-odd numbers," and "odd-times-odd numbers." His
definitions of the first two differ from those given by Nicomachus (*c.* 100) and other writers,[5] with whom an "even-times-
even number" is of the form 2^n; an "even-times-odd number"
is of the form $2(2n + 1)$; and an "odd-times-odd number" is of
the form $(2n + 1)(2m + 1)$. How far back these ideas go in
Greek arithmetic is unknown, for they were doubtless transmitted orally long before they were committed to writing.

Since the product of two equal numbers represents the numerical area of a square, this product was itself called a square,—
a word thus borrowed from geometry. The product of two
unequal numbers was called a heteromecic (different-sided)
number. Square and heteromecic numbers were called plane

---

[1] *Laws*, Jowett translation, V, 100.

[2] On the general number theory of Pythagoras, see Heath, *History*, I, 65.

[3] E. H. Man, "Notes on the Nicobarese," in the *Indian Antiquary*, 1899, p. 253.

[4] In Euclid's *Elements*, VII, 6, 7, ἄρτιοι and περισσοί.

[5] For particulars see Heath's *Euclid*, Vol. II, pp. 277, 281 seq. For the
"odd-times-even number," which Euclid seems to have taken as synonymous
with an "even-times-odd number," see *ibid.*, p. 283; on the general classifications by the Greeks, see K. G. Hunger, *Die arithmetische Terminologie der
Griechen*, Prog., Hildburghausen, 1874.

numbers, while the product of three numbers was called a solid number, the cube being a special case. These are

30 ARITHMETICAE

NICOLAVS.

Pariter im-
par.

Qvis eſt alter numerus par? IVST. Eſt pariter impar, uel à paribus impar, Græcis ἀρτιοπεριοσος, uel ἀρτιάκις πε ριοσος. Eſt autem, cum primum diuiditur, mox fit indiuiſibilis, ut 14. 18. 22. NICOL. Quomodo eum finit horum numerorum exquiſitus magiſter Euclides? IVST. Sic: ἀρτιάκις περιοσος ὅςιυ, ὁ ὑπό ἀρτίου ἀριθμῶ μετρύμενος ᴙⅼ περιοσον ἀριθμόν. Ita numerus diuidens par eſt, ſed diuiſorius mox impar exurget. NICOL. Cur id nominis illi inditum eſt? IVST. Ideo, quod qualibet eius ordinis numeri

Cur uocalur
pariter impar

pares, faĉti ſunt per imparem multiplicationem: ut bis ter, ſenarium, bis quinq; denarium conficiunt. Verum ſi cui altius contemplari libet, eundem uocabit imparem in ſua quantitate, ſed parem in deno minatione. Eſto exempli gratia, denarius, cuius altera pars eſt quinarius, qui quantitate, hoc eſt, moadum congregatione eſt impar, ſed quia à binario denominatur, par iudicabitur. Quæ ratio nominis ex Boethio colligitur: Alia autem Euclidi eſſe uidetur. NICOL. Sunt ne huic de illo aliquot theo

Symbola ex
pariter impa-
ri.

remata? IVST. Quid ni? Vnum eſt. Si numerus dimidium impar habuerit, pariter impar eſt tantum. Nam hic duntaxat extremum, quod maxi-

THEORY OF ODD AND EVEN NUMBERS

From the arithmetic of Willichius (1540). The page also illustrates the use of the catechism method in the 16th century

particular types of the figurate numbers mentioned later. The Boethian arithmetic made much of this classification;[1]

---

[1] *Boetii de institutione arithmetica libri duo*, ed. Friedlein, p. 17 (Leipzig, 1867); hereafter referred to as *Boethius*, ed. Friedlein.

For a full discussion see R. Bombelli, *L'antica numerazione Italica*, cap. x (Rome, 1876); hereafter referred to as Bombelli, *Antica numer.* For the status of the classification in the early printed books, see Pacioli, *Suma*, 1494 ed., fol. [1], v.(= A [1], v.).

the medieval writers, both Arab[1] and Latin,[2] did the same; and early writers in the vernaculars simply followed the custom.[3]

Prime Number. Aristotle, Euclid, and Theon of Smyrna defined a prime number as a number "measured by no number but by an unit alone," with slight variations of wording. Since unity was not considered as a number, it was frequently not mentioned. Iamblichus says that a prime number is also called "odd times odd," which of course is not our idea of such a number. Other names were used, such as "euthymetric" and "rectilinear," but they made little impression upon standard writers.[4]

The name "prime number" contested for supremacy with "incomposite number" in the Middle Ages, Fibonacci (1202) using the latter but saying that others preferred the former.[5]

Perfect Numbers. Conventionally we speak of the aliquot parts of an integral number as the integral and exact divisors of the number, including unity but not including the number itself. A number is said to be deficient, perfect, or abundant according as it is greater than, equal to, or less than the sum of its aliquot parts.[6]

---

[1] On Savasorda (c. 1100), or Abraham bar Chiia, and his classification, see *Abhandlungen*, XII, 16. On al-Ḥaṣṣâr (c. 12th century), see *Bibl. Math.*, II (3), 17.

[2] Thus Jordanus Nemorarius (c. 1225): "Par numerus est qui in duo equalia diuidi potest. Impar est in quo aliqua prima pars est absq3 pari: additq3 supra parem vnitatē. Parium numerorū alius pariter par: alius pariter īpar: et alius impariter par. Pariter par est quē nullus impar numerat. Pariter īpar est quē quicunq3 pares numerāt. Imparit⁹ par est quē quidā par scdm parē ī quidā scdm imparē numerāt." 1496 ed., fol. b (3).

[3] *E.g.*, Chuquet, *La Triparty* (1484); see Boncompagni's *Bullettino*, XV, 619. Curtze found an early German MS. at Munich (No. 14,908, Cod. lat. Monac.) with such terms as "gelich oder ungelich," "glich unglich," and the like. See *Bibl. Math.*, IX (2), 39.

[4] Heath, *Euclid*, Vol. I, p. 146; Vol. II, pp. 284, 285.

[5] "Nvmerorum quidam sunt incompositi, et sunt illi qui in arismetica et in geometria primi appellantur. . . . Arabes ipsos hasam appellant. Greci coris canon, nos autem sine regulis eos appellamus." *Liber Abaci*, I, 30.

[6] *E.g.*, 8 is a deficient number, since $8 > 1 + 2 + 4$; 6 is a perfect number, since $6 = 1 + 2 + 3$; 12 is an abundant number, since $12 < 1 + 2 + 3 + 4 + 6$. Various other names are given to abundant and deficient numbers, such as redundant or overperfect (ὑπερτελής, ὑπερτέλειος) and defective (ἐλλιπής). Heath, *History*, I, 10, 74.

This classification may have been known to the early Pythagoreans, but we have no direct evidence of the fact; indeed, their use of "perfect" was in another sense, 10 apparently being considered by them as a perfect number.

Euclid proved[1] that if $p = \sum_{0}^{n} 2^n$ and is prime, then $2^n p$ is perfect.[2] Nicomachus[3] separated even numbers into the classes above mentioned, and gave 6, 28, 496, and 8128 as perfect numbers, noting the fact that they ended in 6 or 8. Theon of Smyrna (c. 125) followed the classification of Nicomachus, but gave only two perfect numbers, 6 and 28. Iamblichus[4] (c. 325) did the same, but asserted that there was one and only one perfect number in each of the intervals 1 $\cdots$ 10, 10 $\cdots$ 100, 100 $\cdots$ 1000, 1000 $\cdots$ 10,000, and so on, and that the perfect numbers end alternately in 6 and 8,— statements which are untrue but which are found repeated in the arithmetic[5] of Boethius (c. 510). Subsequent writers in the Middle Ages and the Renaissance frequently followed Nicomachus or Iamblichus.

Fibonacci (1202) gave $\frac{1}{2} \cdot 2^2(2^2 - 1) = 6$, $\frac{1}{2} \cdot 2^3(2^3 - 1) = 28$, $\frac{1}{2} \cdot 2^5(2^5 - 1) = 496$ as perfect numbers, and so in general for $\frac{1}{2} \cdot 2^p(2^p - 1)$ where $2^p - 1$ is prime,— a rule which holds for all known perfect numbers.[6] Chuquet (1484) gave Euclid's rule and repeated the ancient error that perfect numbers end alternately in 6 and 8.

The fifth perfect number, 33,550,336, is first given, so far as known, in an anonymous manuscript[7] of 1456–1461. Pacioli (1494) incorrectly gave 9,007,199,187,632,128 as a perfect number.[8]

---

[1] *Elements*, IX, 36.

[2] On all this work see Dickson, *Hist. Th. Numb.*, with bibliography, I, i; R. C. Archibald, *Amer. Math. Month.*, XXVIII, 140, with valuable references to American contributions.

[3] *Arithmetica*, I, 14, 15.   [4] 1668 ed., p. 43.

[5] *Arithmetica*, I, cap. 20, "De generatione numeri perfecti."

[6] Dickson, *Hist. Th. Numb.*, I, 5.

[7] Codex lat. Monac. 14,908. Dickson, *Hist. Th. Numb.*, 1, 6.

[8] "Sia el nũero a noi pposto. 9007199187632128. q̄le cõmo e ditto: e pfec̣to." Fol. 7, v.

Charles de Bouelles (1509) wrote on perfect numbers[1] and asserted, without proof, that every perfect number is even. He stated that $2^{n-1}(2^n-1)$ is a perfect number when n is odd, which is substantially the incorrect rule of Fibonacci. This was also given by various other writers of the 16th century, including as good mathematicians as Stifel[2] (1544) and Tartaglia[3] (1556).

Robert Recorde[4] (1557) attempted to give the first eight perfect numbers, but three in his list were incorrect. Cataldi[5] showed that Pacioli's pretended fourteenth perfect number is in fact abundant, that the ancient belief that all perfect numbers end in 6 or 8 is unfounded, and that perfect numbers of the type given by Euclid's rule do actually end in 6 or 8.

Descartes thought that Euclid's rule covered all even perfect numbers and that the odd perfect numbers were all of the type ps^2, where p is a prime.[6]

Fermat (1636) and Mersenne (1634) paid much attention to the subject, and their investigations contributed to the theory of prime numbers.[7]

Euler at first (1739) asserted his belief that $2^{n-1}(2^n-1)$ is a perfect number for $n = 1, 2, 3, 5, 7, 13, 17, 19, 31, 41$, and 47, but afterward (1750) showed that he was in error with respect to 41 and 47. He proved that every even perfect number is of Euclid's type, $2^n \sum_0^n 2^n$, and that every odd perfect number is of the form $r^{4\lambda+1}p^2$, where r is a prime of the form $4n+1$.[8]

There are many references to perfect numbers in general literature,[9] in Hebrew and Christian writings on religious

---

[1] "De Numeris Perfectis," in his general work published at Paris in 1509–1510. See *Rara Arithmetica*, p. 89.

[2] *Arithmetica Integra*, fols. 10, 11 (Nürnberg, 1544); also *Die Coss Christoffs Rudolfs*, fols. 10, 11 (Königsberg, 1553).

[3] *La seconda Parte del General Trattato*, fol. 146, *v*. Venice, 1556.

[4] *The whetstone of witte*, fol. [4, *v*.]. London, 1557.

[5] *Trattato de' nvmeri perfetti*. Bologna, 1603.

[6] *Œuvres*, II, 429. Paris, 1898.   [7] Dickson, *Hist. Th. Numb.*, I, 11–13.

[8] *Ibid.*, p. 18. For the later theory, consult this work, I, i.

[9] Thus Macrobius, in his *Saturnalia*, says that six "plenus perfectus atque diuinus est." *Satvrnaliorvm Liber VII*, cap. xiii, ed. Eyssenhardt, 1868, p. 446.

doctrines,[1] including Isidorus of Seville and Rabbi ben Ezra, and in the works of medieval and Renaissance mystics.[2]

Amicable Numbers.[3] Two integral numbers are said to be amicable[4] if each, as in the case of 220 and 284, is equal to the sum of the aliquot parts of the other. These two numbers, probably known to the early Pythagoreans, are mentioned by Iamblichus. They occupied the attention of the Arabs, as in the works of Tâbit ibn Qorra (*c.* 870). It was asserted by certain Arab writers that talismans with the numbers 220 and 284 had the property of establishing a union or close friendship between the possessors, and this statement was repeated by later European writers, including Chuquet (1484) and Mersenne (1634).

For a long time the only amicable numbers known were the two given above, 220 and 284, but in 1636 Fermat[5] discovered a second pair, $17,296 = 2^4 \cdot 23 \cdot 47$ and $18,416 = 2^4 \cdot 1151$, and also found a rule for determining such numbers.[6] A third pair was discovered by Descartes[7] (1638), namely, $9,363,584 = 2^7 \cdot 191 \cdot 383$ and $9,437,056 = 2^7 \cdot 73,727$. Descartes gave a rule which he asserted[8] to be essentially the same as Fermat's, but which various later writers, apparently ignorant of this assertion, assigned to Descartes himself.

Euler[9] (1750) made a greater advance in this field than any of his predecessors, adding fifty-nine pairs of amicable numbers of the type am, an, in which a is relatively prime to m and n, and contributing extensively to the general theory. Dickson

---

[1] *E.g.,* J. J. Schmidt, *Biblischer Mathematicus*, p. 20 (Züllichau, 1736).

[2] See Curtze's mention of a Munich MS. (No. 14,908, Codex lat. Monac.) in the *Bibl. Math.*, IX (2), 39, with five perfect numbers.

Thierfelder (1587, fol. A₄, *r.*) says: "Den in sechsz tagen hat Gott Himmel vnd Erden/ vnd alles was dariñen ist/ gemacht/ das ist ein *Trigonal* oder dreyeckichte Zahl/ welche Zahlen für die heiligen Zahlen gehalten werden/ vnd ist darzu die erste perfect Zahl."   [3] Dickson, *Hist. Th. Numb.*, I, 36.

[4] The terms "amiable" and "agreeable" are also used.

[5] *Œuvres*, 1894, II, 72, 208.

[6] The rule is given in Dickson, *loc. cit.*, p. 37.

[7] *Œuvres*, 1898, II, 93.   [8] *Œuvres*, 1898, II, 148.

[9] *Opuscula varii argumenti*, 3 vols., II, 23 (Berlin, 1746–1751). See also *Bibl. Math.*, IX (3), 263; X (3), 80; XIV (3), 351; Cantor, *Geschichte*, III, 616.

(1911) has obtained two new pairs of amicable numbers and has also added to the general theory of the subject.[1]

Figurate Numbers. The Greeks were deeply interested in numbers which are connected with geometric forms and which therefore received the name of figurate numbers.[2] These are triangular if capable of being pictured thus:

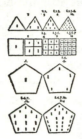

and are therefore of the form

$$\sum_{1}^{n} n = \tfrac{1}{2} n (n + 1).$$

FIGURATE NUMBERS

From the first printed edition of the arithmetic of Boethius (1488)

They are square if they can be represented by squares, such as ⦂⦂, and are then of the form n^2. They are pentagonal if in the form of a square with a triangle on top, thus:

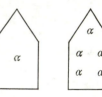

so that the form is $n^2 + \tfrac{1}{2} n (n - 1)$. Similarly, there are hexagonal numbers and other types of polygonal numbers.[3] In the Greek manuscripts they appeared in such forms as those here shown, the α's standing for 1's or possibly for ἀριθμός (*arithmos'*, number).[4]

Related to figurate numbers there are the linear numbers. Under this name Nicomachus (*c.* 100) included the natural numbers, beginning with 2; side and diagonal numbers[5]; area,

---

[1] See also his *Hist. Th. Numb.* and *Amer. Math. Month.*, XXVIII, 195.

[2] Boethius defined them as numbers "qui circa figuras geometricas et earum spatia demensionesque versantur." Ed. Friedlein, p. 86, l. 12. See also Heath, *History*, I, 76.   [3] Boethius, ed. Friedlein, p. 98 seq.

[4] These two forms are from a 10th century MS. of Nicomachus in Göttingen.

[5] The πλευρικοί καὶ διαμετρικοί ἀριθμοί of Theon of Smyrna (*c.* 125). See also Boethius, ed. Friedlein, p. 90.

or polygonal, numbers[1]; and solid numbers,[2] including cubic, pyramidal, and spherical numbers.[3] A relic of such numbers is

Pyramidum numeri hoc pacto digeruntur.

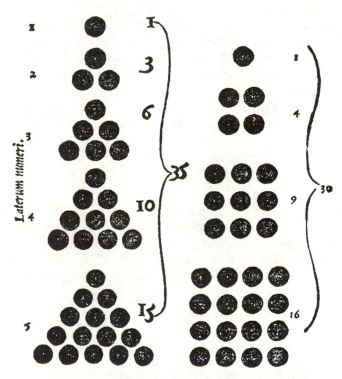

PYRAMIDAL NUMBERS

From Joachim Fortius Ringelbergius, *Opera*, 1531. The four layers of the two pyramidal numbers 35 and 30 are shown

seen in problems relating to the piling of round shot, still to be found in algebras. Indeed, it is not impossible that they may have been suggested to the ancients by the piling of spheres in

[1] See Nicomachus, *Introd.*, II, capp. 8–11.
[2] Στερεοί. Nicomachus, *Introd.*, II, 14.
[3] Boethius, *loc. cit.*, pp. 104, 121.

such games as the *Castellum nucum* to which Ovid refers in his poem *De Nuce*, where the pyramidal number is mentioned.[1] **Continuous and Discrete.** The distinction between continuous and discrete magnitude is commonly referred to the Pythagoreans or even to Pythagoras (*c.* 540 B.C.) himself,[2] the continuous magnitude being geometric and the discrete being arithmetic. The distinction was recognized by various Greek and Latin writers,[3] appearing in the works of such medieval authors as Fibonacci (1202) and Roger Bacon (*c.* 1250).[4]

Cardinals and Ordinals. The distinction between cardinal and ordinal numbers is ancient, but the names are relatively modern. A cardinal number is a number on which arithmetic turns[5] or depends, and hence is a number of importance,[6] while an ordinal number is one which denotes order.[7]

6. UNITY

Unity. Not until modern times was unity considered a number. Euclid defined number as a quantity made up of units,[8] and in this he is followed by Nicomachus.[9] Unity was defined by Euclid as that by which anything is called "one."[10] It was generally defined, however, as the source of number, as in the

---

[1] Quattuor in nucibus, non amplius, alea tota est,
    Cum sibi suppositis additur una tribus.
See also F. Lindemann, "Zur Geschichte der Polyeder und der Zahlzeichen," *Sitzungsberichte der math.-physik. Classe der K. Bayerischen Akad. der Wissensch. zu München,* XXVI, 625–757 (Munich, 1897).

[2] "Ogni quantità . . . secondo Pythagora, è o continua, ouer Discreta, la continua è detta Magnitudine, . . . & la discreta moltitudine." Tartaglia, *General Trattato,* I, fol. 1, *r.* (Venice, 1556).

[3] Boethius, ed. Friedlein, pp. 8, 16; Heath, *Euclid,* Vol. I, p. 234.

[4] *E.g.,* in the Sloane MS. fol. 94 of the *Communia.*   [5] Latin, *cardo,* a hinge.

[6] Compare cardinal, a prince of the Church. Glareanus recognized this metaphor: "Sunt enim quaedam, quae Cardinalia appellant, à cardine sumpta, ut opinor, metaphora, quòd ut in cardine ianua uertitur, ita huius artis primum ac praecipuum negocium in hisce consistat" (1538; 1543 ed., fol. 3, *r.*).

[7] On the history of these terms see E. Bortolotti, "Definizioni di Numero," *Esercitazioni Matematiche,* II, 253, and *Periodico di Matematiche,* II (4), 413.

[8] Ἀριθμὸς δὲ τὸ ἐκ μονάδων συγκείμενον πλῆθος. *Elements,* VII, def. 2. See also Heath, *History,* I, 69.   [9] *Introd.,* I, 7, 1.

[10] Μονάς ἐστιν καθ᾽ ἣν ἕκαστον τῶν ὄντων ἓν λέγεται. *Elements,* VII, def. 1; ed. Heath, Vol. II, p. 279, with references to other Greek writers.

anonymous *Theologumena*,[1] a Greek work of the early Middle Ages. The dispute goes back at least to the time of Plato, for the question is asked in the *Republic*, "To what class do unity and number belong?"—the two being thus put into separate categories.

It is not probable that Nicomachus (*c.* 100) intended to exclude unity from the number field in general, but only from the domain of polygonal numbers.[2] It may have been a misinterpretation of the passage from Nicomachus that led Boethius[3] to add the great authority of his name to the view that one is not a number. Even before his time the belief seems to have prevailed, as in the case of Victorius (457) and Capella (*c.* 460), although neither of these writers makes the direct assertion.[4] Following the lead of Boethius, the medieval writers in general, such as al-Khowârizmî[5] (*c.* 825), Psellus[6] (*c.* 1075), Savasorda[7] (*c.* 1100), Johannes Hispalensis[8] (*c.* 1140), and Rollandus[9] (*c.* 1424), excluded unity from the number field.[10] One writer, Rabbi ben Ezra (*c.* 1140), seems, however, to have

---

[1] Μονάς ἐστιν ἀρχὴ ἀριθμοῦ, θέσιν μὴ ἔχουσα. *Theologumena*, I, 1.

[2] Ἡ μὲν μονὰς σημείου τόπον ἐπέχουσα καὶ τρόπον. See also Johnson, *Nicomachus,* p. 7.

[3] "Numerus est unitatum collectio." Ed. Friedlein, p. 13, l. 10. In the Latin version of the so-called Boethian geometry it is asserted: "Primum autem numerum id est binarium, unitas enim, ut in arithmeticis est dictum, numerus non est, sed fons et origo numerorum. . . ." Ed. Friedlein, p. 397, l. 19. See also H. Weissenborn, *Gerbert*, p. 219 (Berlin, 1888).

[4] "Unitas illa, unde omnis numerorum multitudo procedit." From the *Calculus* of Victorius; see Boncompagni's *Bullettino*, IV, 443.

"Nec dissimulandum est ex eo quod monas retractantibus unum solum ipsam esse. ab eaque cetera procreari. Omniumque numerorum solam seminarium esse. solamque mensuram et incrementorum. causam. statumque detrimentorum." From a fragment of Capella; see E. Narducci in Boncompagni's *Bullettino*, XV, 566.

[5] "Quia unum est radix uniuersi numeri, et est extra numerum." From the supposed translation of Adelard of Bath.

[6] "Principium itaque omnis numeri est Μονὰς, non-numerus fons numerorum." See the 1532 edition, p. 13.

[7] "Numerus est ex unitatibus profusa collectio" (Plato of Tivoli's translation, 1145). See C. H. Haskins, *Bibl. Math.*, XI (3), 332.

[8] "Unitas est origo et prima pars numeri . . . sed ipsa extra omnem numerum intelligitur." See B. Boncompagni, *Trattati d' Aritmetica*, II, 25 (Rome, 1857); hereafter referred to as Boncompagni, *Trattati*.

[9] See Volume I, page 261. "Vnitas non est numerus sed principia numerorum" (Plimpton MS., Pt. I, cap. 1). [10] See also Boncompagni's *Bullettino,* XV, 126.

approached the modern idea. In his *Sefer ha-Echad* (*Book on Unity*) there are several passages in which he argues that one should be looked upon as a number.

Most of the authors of the early printed books excluded unity, as is seen in the works of Pacioli[1] (1494), Köbel[2] (1514), Tzwivel[3] (1505), and many others. Thus the English writer Baker (1568) remarks that "an vnitie is no number but the beginning and original of number."[4] In the 16th century, however, the more thoughtful writers began to raise the question as to whether this exclusion of unity from the number field was not like the trivial disputes of the schoolmen,[5] and by the end of the century it was recognized that the ancient definition was too narrow. Thus Hylles (1592), speaking of "an vnit or an integer (which sometimes I also cal an Ace)," is rather afraid to take a definite stand in the matter, but says that "the latter writers, as namely Ramus, and such as have written since his time, affirme not only that an vnite or one, is a number, but also that euery fraction or parte of an vnite, is a number. . . . I do accompt it after a sorte for the first or least number . . . euen as an egg, with[6] in power possibilitie containeth a bird though really and actually it is none." Stevin (1585), a much greater man, used the argument that a part is of the same nature as the whole, and hence that unity, which is part of a collection of units, is a number.[7] To this Antoine Arnauld, "le

---

[1] "Et essa vnita nō e numero: ma ben principio di ciascun numero" (1494 ed., fol. 9).

[2] "Daraus3 verstehstu das I. kein zal ist / sonder es ist ein gebererin / anfang / vnnd fundament aller anderer zalen" (*Zwey rechenbuchlin*, Frankfort ed., 1537, fol. 26). It is also in his *Rechenbüchlin*, 1531 ed. dedication, and 1549 ed., fol. 26.

[3] "Unitas em numerus non est. sed fons et origo numerorum" (fol. 2).

[4] 1580 ed., fol. 1.

[5] So Gemma Frisius (1540) makes it a matter of authority: "Nvmerum authores vocant multitudinem ex Vnitatibus conflatum. Itaque Vnitas ipsa licèt subinde pro numero habeatur, propriè tamen numerus non erit" (1563 ed., fol. 5). Also Trenchant (1566): " . . . l'vnité n'est pas nombre . . . Mais en la pratique, ou le nombre est tousiours adapté à quelque sugét . . . l'vnité est prinse pour nombre" (1578 ed., p. 9).

[6] *Sic*, for "which." From the 1600 edition.

[7] "La partie est de mesme nature que le tout. Unité est partie d'une multitude d' unitez . . . et par consequent nombre." See also the Girard edition, of 1634, p. 1, with slight change in wording.

grand Arnauld" (1612–1694), replied that the argument was worthless, for a semicircle is not a circle. Stevin also used the argument that if from a number there is subtracted no number, the given number remains; but if from 3 we take 1, 3 does not remain; hence 1 is not no number.[1] The school arithmetics kept the Boethian limitation until the close of the 18th century.[2]

Another common notion was that unity is, like a point, incapable of division,—an idea also due to the Greeks.[3]

7. Later Developments

Higher Domain of the Arithmetica. The later developments in the arithmetica do not belong to the domain of elementary mathematics. Their history has been treated with great erudition by Professor Dickson in his *History of the Theory of Numbers*.[4] As a matter of general information, however, a few of the theorems which have attracted wide attention will be stated.

Typical Theorems. In 1640 Fermat, in a letter to Bernard Frénicle de Bessy (*c.* 1602–1675), set forth the theorem that if p is any prime number and x is any integer not divisible by p, then $x^{p-1} - 1$ is divisible by p. The special case of $2 - 2$ being divisible by the prime p had long been known to Chinese scholars, but the general theorem is due to Fermat. Leibniz proved the proposition some time before 1683.

Euler stated Fermat's theorem in a communication to the Petrograd Academy[5] in the form: If $n + 1$ is a prime dividing neither a nor b, then $a^n - b^n$ is divisible by $n + 1$.

As stated in Volume I, page 459, Wilson discovered (*c.* 1760) that if p is prime, then $1 + (p - 1)!$ is a multiple of p. The manuscripts of Leibniz now preserved at Hannover show that he knew the theorem before 1683, but he published nothing upon

---

[1] *Abhandlungen*, XIV, 227.

[2] *E. g.*, Ward's *Young Mathematician's Guide*, p. 4 (London, 1771).

[3] "Il punto nella Geometria, & l'vnità nell' Arimmetica non è capace di partimento. Proclo sopra Euclide lib. 2.c.xi." Ciacchi, *Regole Generali d' Abbaco*, p. 352 (Florence, 1675).

[4] See also A. Natucci, *Il Concetto di Numero*, Turin, 1923.

[5] Presented in 1732, published in 1738.

the subject. Lagrange published a proof of the theorem in 1771, deduced it from Fermat's Theorem, and proved its converse. Fermat gave as his opinion that $2^{2^n}+1$ is always prime, but asserted that he was unable to prove it. Euler (1732)[1] showed that Fermat's opinion was not warranted, since

$$2^{2^5}+1 = 641 \cdot 6,700,417.$$

Fermat's connection with numbers of this form led to their being called "Fermat's Numbers."[2]

With respect to the sum and the number of divisors of a number there is an extensive literature.[3] For example, Cardan (1537) stated that a product P of k distinct primes has $1 + 2 + 2^2 + \cdots + 2^{k-1}$ aliquot parts; for example, that $3 \cdot 5 \cdot 7$ has $1 + 2 + 4$ aliquot parts. This rule was proved by Stifel in his *Arithmetica Integra* (1544). Frans van Schooten (1657) proved that a product of k distinct primes has $2^k - 1$ aliquot parts, which is only another expression for Cardan's rule.

Descartes (probably in 1638) showed that if p is a prime the sum of the aliquot parts of p^n is $(p^n - 1)/(p - 1)$, a law simply illustrated by the cases of 2^3 and 7^2.

Fermat proposed (1657) two problems: (1) Find a cube which, when increased by the sum of its aliquot parts, becomes a square, one example being $7^3 + (1 + 7 + 7^2) = 20^2$; (2) find a square which, when increased by the sum of its aliquot parts, becomes a cube. Problems of this general nature attracted the attention of men like Frénicle de Bessy, Lord Brouncker, Wallis, Frans van Schooten, Ozanam, and various later scholars.

Other Subjects of Investigation. Among other subjects investigated is that of the factors of numbers that can be expressed in the form of $a^n \pm b^n$; for example, to find all the prime factors of $2^{45} - 1$. There are also such questions as the infinitude of primes in general; the tests for primality; the number of primes between assigned limits; the curious properties connected with the digits of numbers; periodic fractions; and the general theory of congruent numbers.

---

[1] Published in 1738.   [2] R. C. Archibald, *Amer. Math. Month.*, XXI, 247.
[3] Dickson, *Hist. Th. Numb.*, I, 51.

TOPICS FOR DISCUSSION

1. The numbers three and seven in folklore and in literature.

2. The history of the seven liberal arts.

3. Distinction between arithmetic and logistic in ancient and medieval times.

4. History of the word "algorism" in various languages, particularly with reference to its forms and significance.

5. Various names given to what is now called arithmetic in the period known as the Renaissance.

6. History of the distinction between concrete and abstract numbers. The present status of the question, including that of operations with concrete numbers.

7. History of the finger names assigned to numbers, and the probable reason why such names attracted more attention in early times than at present.

8. Rise of the idea of significant figures and the present use of the term.

9. Probable reasons for the superstitions in regard to odd and even numbers and for the properties assigned to them at various times and by various peoples.

10. The gnomon and its relation to numbers and to other branches of mathematics.

11. Nature of and probable reason for certain other classifications of number in ancient times.

12. Probable cause for the special interest in prime numbers expressed by the ancients.

13. The historical development of the interest in amicable numbers and the present status of the theory.

14. The historical development of the theory of perfect numbers and the present status of the theory.

15. The interest in figurate numbers among the Greeks and the traces of such numbers in modern times.

16. The history of the concept of unity and of the controversy with respect to its being a number.

17. Traces of ancient arithmetic and logistic in modern textbooks in arithmetic and algebra.

18. Questions relating to the theory of numbers and attracting the attention of mathematicians during the 19th and 20th centuries.

CHAPTER II

LOGISTIC OF NATURAL NUMBERS

1. Fundamental Operations

Number of Operations. In America at the present time it is the custom to speak of four fundamental operations in arithmetic, that is, in what the ancients called logistic. This number is, however, purely arbitrary, and it is quite possible to argue that it should be increased to nine or more,[1] or even that it should be decreased to one.

The *Crafte of Nombrynge* (c. 1300) enumerates seven:

¶Here tells þat þer ben . 7. spices or partes of þis craft. The first is called addicioñ, þe secunde is called subtraccioñ. The thryd is called duplacioñ. The 4. is called dimydicioñ. The 5. is called multiplicacioñ. The 6. is called diuision. The 7. is called extraccioñ of þe Rote.[2]

Sacrobosco (*c.* 1250) had already spoken of nine of these operations,—numeration, addition, subtraction, duplation, mediation, multiplication, division, progression, and the extraction of roots,[3]—and Michael Scott had done the same.[4] This was a common number among medieval writers and, indeed, in the

---

[1] On the general question of the operations see J. Tropfke, *Geschichte der Elementar-Mathematik*, I (2) (Leipzig, 1921), hereafter referred to as Tropfke, *Geschichte*; Suzan R. Benedict, *A Comparative Study of the Early Treatises introducing into Europe the Hindu Art of Reckoning*, Dissertation, Univ. of Michigan, 1914.

[2] R. Steele, *The Earliest Arithmetics in English*, Oxford, 1923. The old letter þ is our *th*. Since it slightly resembles our letter *y*, the old word þe (the) is often ignorantly written as *ye*, as in "ye editor."

[3] In his *Algorismus*. See Volume I, page 222.

[4] Santa-Cruz (1594) refers to this, saying: ". . . las especies del qual, segũ Iuã de sacrovosco, y Michael Scoto, son nueue" (1643 ed., fol. 9).

early printed books.[1] Pacioli (1494), however, claimed credit
to himself for reducing the number to seven.[2] In due time a
further reduction was made to six,[3] then to five, as with most
16th-century writers, and then to four. When five operations
were taken, numeration was usually the first, the topic properly
including notation.[4] One of the first of the writers of any note
to reduce the number to four was Gemma Frisius (1540),[5] and
such was his influence that this number soon became common.
There have been those, indeed, who gave only three fundamental operations, multiplication being included in addition
as a special case. This number was given, for example, by
Elia Misrachi (c. 1500).[6]

Duplation and Mediation. The four operations generally
recognized at present will be considered later in this chapter;
the two operations of duplation (doubling) and mediation
(halving), with the reasons for their use, will be explained
briefly at this time.

[1] *E.g.*, Widman (1489), Peurbach (*c.* 1460; 1st ed., 1492), Huswirt (1501),
Tartaglia (1556), and Santa-Cruz (1594).

[2] He says that nine were given by "Giouā de sacro busco e Prodocimo de
beldemandis da padua dignissimo astronomo e molti altri in loro algorismi.
Ma noi le ditte noue reduremo a septe" (fol. 19, *r.*).
Since the names have some interest, the list is reproduced: "La prima sira
ditta numeratiõe ouer representatiõe: cioe sapere cognoscere e releuare le figure
e caratteri del nũero. La secõda sira ditta additiõe ouer recogliere: agiognere;
sũmare e acozare. La terza sira ditta subtractiõe ouer abattere: sotrare: cauare
e trare. La q̃rta fia ditta multiplicatiõe. La quinta sira ditta diuisiõe ouer
partire. La sexta sira ditta ℘gressiõe. La septima sira ditta delle radici
extractione." *Ibid.*

[3] *E.g.*, Glareanus (1538): "Eius sex, ut in epitome, prosequemur species, nu-
merationem, additionem, subtractionem, multiplicationem, diuisionem, ac pro-
gressionem" (1543 ed., fol. 9).

[4] Numeration (Latin *numeratio*, from *numerus*, a number) has lately been
used to mean the reading of numbers. Since medieval writers often called the
characters 1, 2, 3, . . . *notae* (compare the "notes" in music), the writing of
numbers has been called notation. The distinction is one chiefly of the school-
room. Ramus (1569; 1586 ed., p. 1) was one of the first prominent writers to
make it: "In numero spectatur primum notatio, deinde numeratio."

[5] "Qvatuor omnino sunt Arithmetices species" (1563 ed., fol. 6).

[6] G. Wertheim, *Die Arithmetik des Elia Misrachi*, Prog., Frankfort a. M.,
1893; hereafter referred to as Wertheim, *Elia Misrachi*. The first edition of
Misrachi appeared at Constantinople, 1532 or 1533; the second, at Basel, 1546.

The Egyptians often multiplied by continued doubling,[1] thus saving the trouble of learning a multiplication table. This was particularly convenient in working on the abacus. On this account duplation was generally recognized as a separate topic until the 16th century. Moreover, the Egyptian tables of measure were commonly arranged so as to make doubling and halving operations of great importance.[2] This method continued as long as the abacus was in use, and persisted for some time after that instrument was generally abandoned. An interesting illustration of mediation is seen, for example, in official papers of Russia prior to the time of Peter the Great (1672–1725), the word "half" being repeated as many as ten times to indicate a certain division.[3]

The use of duplation and mediation is seen in many of the Arab works,[4] and this fact influenced such medieval translators as Johannes Hispalensis[5] (c. 1140) and Adelard of Bath[6] (c. 1120). The processes were common in the theoretical works of the 15th century[7] but not in the commercial arithmetics, at least in Italy. The early printed books of Germany were less progressive in this respect than those of other countries, partly

---

[1] Thus, 7 × 15 = 2 × 2 × 15 + 2 × 15 + 15.

[2] See the common use of the fractions $\frac{1}{2}$, $\frac{1}{4}$, $\frac{1}{8}$, $\frac{1}{16}$, in the Edfu Survey, in H. Brugsch, *Thesaurus Inscriptionum Ægyptiacarum*, Leipzig, 1883–1891, Vol. III; H. Brugsch, *Numerorum apud Veteres Ægyptios Demoticorum Doctrina*, Berlin, 1849; T. E. Peet, *The Rhind Mathematical Papyrus*, London, 1923; hereafter referred to as Peet, *Rhind Papyrus*. For the survival of doubling and halving until the present time, see the tables in Mahmoud Bey, "Le système métrique actuel d'Égypte," *Journal Asiatique*, I (7), 69, 82.

[3] V. V. Bobynin, "Esquisse de l'histoire du calcul fractionnaire," *Bibl. Math.*, X (2), 97.

[4] E.g., al-Nasavî (c. 1025), on whose work see F. Woepcke, *Journal Asiatique*, I (6), 496; al-Ḥaṣṣar (12th century), on whose work see M. Steinschneider, *Abhandlungen*, III, 10; al-Khowârizmî (c. 825), on whose work and al-Ḥaṣṣar's see Suter, *Bibl. Math.*, II (3), 12.

[5] Who, however, speaks of them merely as special cases of multiplication and division. See Boncompagni, *Trattati*, II, 38. Similarly, as to Gernardus (13th century?), see G. Eneström, *Bibl. Math.*, XIII (3), 289, 292.

[6] Boncompagni, *Trattati*, I, 10.

[7] E.g., in the Rollandus MS. (1424) ; see *Rara Arithmetica*, p. 446. Rollandus gives: "addere. sbtᵣhere. mediare. duplare. diuidere. m̄ltiplicare. et radices invenirᵉ" (fol. 2).

because of the continued use of the abacus in that part of Europe, and so these two processes are found in the works of Tzwivel (1505), Köbel (1514), Grammateus (1518), Riese (1522), Rudolff (1526), and various other writers of that period, often with a statement that they are special forms of multiplication and division.[1] Stifel (1545) uses them only apologetically, and Scheubel (1545) omits them entirely.[2] They are rarely found in any of the printed arithmetics of Italy,[3] Spain,[4] France, or England. It is a curious fact, however, that Recorde (c. 1542) omits them with integers but includes them with fractions,—a vagary that endured at least as late as the 1668 edition of his *Ground of Artes.* His example was followed by Baker (1568), who had the notion that only fractions should be used with fractions, saying: "If you will double anye broken number you shall divide ye same by ½," and giving triplation and quadruplation in the same way. Gemma Frisius (1540) did as much as any other Continental writer to show the absurdity of following those "stupid people" who would include these operations.[5]

Names of the Operations. The awkward expression "the four fundamental operations" is modern. Several others used in the past possess the merit of greater brevity, and some of these are still found in various languages. A common name is "species," a term of the 13th century and made popular in the 16th century by the works of Riese (1522) and Gemma Frisius (1540).[6] Ramus (1569) used both "parts" and

---

[1] "Dupliren heist zwifeltigen/ ist nichts anders dañ ein zal mit 2 multiplicirn. Medijren heist halb machen od' halbiren/ ist nichts anders/ dañ ein zal in 2 abteilen." Köbel (1514).

[2] "De duplatione porro & Mediatione, cum illa multiplicationis, haec uero diuisionis pars sit, scribere quicq3, necesse non fuit." "Tractatus secundus" of the 1545 edition of his *De Numeris.*

[3] *E.g.,* Pacioli (1494): "Ma noi le ditte noue reduremo a septe. Peroche la duplatiõe īplicita in la multiplicatiõe: ela mediatiõe nella diuisiõe" (fol. 19, *r.*).

[4] *E.g.,* Santa-Cruz (1594): "Y porq el doblar no se distingue del multiplicar, ni el mediar del partir" (1643 ed., fol. 9).

[5] "Quid verò mouerit stupidos illos nescio" (1563 ed., fol. 12).

[6] In the Latin editions: "De speciebus Arithmetices"; "Vocamus autem species certas operandi . . ." In the Italian translation: "Delle Specie dell' Aritmetica."

"species,"[1] while most of the Spanish[2] and Dutch[3] arithmeticians of the 16th and 17th centuries used the latter only.

A common Italian name in the 16th century was "acts,"[4] although "passions" was also used.[5] When Clavius wrote his algebra (1608), he used the word *operationes*, and it is probable that this word worked down from algebra to arithmetic.

Sequence of Operations. Our present traditional sequence has by no means been generally recognized, particularly in relation to fractions. Although all writers place notation and a certain amount of addition first, there has been little further uniformity. Abraham bar Chiia (*c.* 1120), Rabbi ben Ezra (*c.* 1140), and Fibonacci (1202), for example, use this sequence: multiplication, division, addition, subtraction, fractions, proportion, and roots.[6] Grammateus (1518) used the order: addition, multiplication, subtraction, division,[7]—an order which has much to commend it.

2. READING AND WRITING NUMBERS

Babylonian Numerals. Since the early Babylonians were without papyrus or parchment, they doubtless followed the custom of most other early peoples and wrote upon leather. Living on an alluvial plain, they had no convenient access to stone for the purpose of permanent inscriptions, except in the northern region, and so they also resorted to the use of clay. They wrote by pressing into the clay with a stylus, the result being wedge-shaped (cuneiform) characters. These tablets were then baked in the sun or in a kiln of some kind, and thus they

---

[1]"Alii faciunt arithmeticae partes vel species . . .," *Arithmeticae libri dvo*, p. 111.

[2]*E.g.*, Santa-Cruz (1594): ". . . las especies del qual . . . son nueue."

[3]Thus Stockmans (1589), Houck (1676), and others speak of the "vier specien."

[4]So Sfortunati (1534) speaks of the "Cinque atti dell' arithmetica."

[5]Tartaglia (1556) prefers "atti" but says that "altri gli dicone Passioni del numero" (1592 ed., fol. 5). The word comes from the Latin *passio*, used by late Latin writers to mean "phenomenon"; originally, something endured.

[6]M. Steinschneider, *Abhandlungen*, III, 107.   [7]1535 ed., fol. A iii.

became fairly permanent records. For relatively small numbers the numerical system was simple, consisting of the following characters:

$$Y = 1, \qquad \blacktriangleleft = 10, \qquad Y\!\!\blacktriangleright = 100.$$

These symbols had different numerical meanings, however. The Y stood not only for 1 but also for 60, 3600, 12,960,000, and in general for 60". The $\blacktriangleleft$ stood for 10 · 60", and hence for 10, 600, 36,000, · · ·. In every case the context was depended upon to determine which value was to be taken. Furthermore, we often find the units represented by horizontal strokes, 10 represented by a vertical crossed by a horizontal stroke (like a plus sign), 20 represented by a vertical crossed by two horizontals, and so on. In certain tablets 71 is represented by 1 (for 60), the above symbol for 10, and a horizontal stroke for the unit.

BABYLONIAN NUMERALS FROM 1 TO 9

The forms vary in shape, but this gives an idea of the simpler numerals in common use. For the correct forms as seen in the clay tablets, see page 39

In writing their numerals the Babylonians made a slight use of the subtractive principle with which we are familiar in connection with the Roman notation. For example, the XIX of the Romans is equivalent to XX − I, a device that was anticipated some two thousand years by the Babylonians, who wrote $\blacktriangleleft\blacktriangleleft Y\!\!\vec{}$ for 19, the symbol $Y\!\!\blacktriangleright$ (*lal* or *lá*) meaning minus.[1] In this case, then, we have 20 − 1, or 19. It has been suggested that one reason for writing 19 as 20 − 1 instead of 10 + 9 is that

[1] There are numerous forms for this symbol, some of them very complex. See H. V. Hilprecht, *Mathematical, Metrological, and Chronological Tablets from the Temple Library of Nippur*, p. 23 (Philadelphia, 1906); hereafter referred to as Hilprecht, *Tablets*. See also G. Reisner, "Altbabylonische Maasse und Gewichte," *Sitzungsberichte der k. Preussischen Akad. der Wissensch.*, p. 417 (Berlin, 1896); G. Contenau, "Contribution à l'histoire économique d'Umma," *Bibliothèque de l'École des hautes études*, fascicule 219 (Paris, 1915), with excellent facsimiles of various numeral forms. See especially Plate XIV for the representation of 71 referred to in the text. The tablets date from *c.* 2300 to *c.* 2200 B.C.

it was an unlucky number. The nineteenth day of a lunar month was the forty-ninth day from the beginning of the preceding month, and this forty-ninth day was one to be specially avoided. To avoid writing 19, therefore, the Babylonians resorted to writing 20 − 1. This does not, however, account for such common forms as 60 − $\frac{1}{3}$ for 59$\frac{2}{3}$ and, as we shall presently see, the existence of the subtractive principle is easily explained on other and more rational grounds.

Since the larger numbers were used by relatively few scholars, there was no compelling force of custom to standardize them. The variants in these cases are not of importance for our purposes, and simply a few of the numerals will serve to show their nature. These illustrations[1] date from *c.* 2400 B.C.

⟨⟩	3600		
⟨⟩	36,000,	*i.e.,*	3600 × 10
⟨⟩	72,000,	*i.e.,*	3600 × (10 + 10)
⟨⟩𝔼	216,000		
𝖸𝖪𝖸𝖪𝖸𝖪𝖸𝖪	2400		
𝖸𝖸𝖸𝖸	171$\frac{1}{2}$,	*i.e.,*	2 × 60 + 50 + 1 + $\frac{1}{2}$
●	10		
●● 𝔹𝔹	36		
●●𝖸	19,	*i.e.,*	20 − 1
●●𝖸	18,	*i.e.,*	20 − 2
●●𝖸	17,	*i.e.,*	20 − 3
𝖣𝖣●𝖢	130$\frac{1}{2}$,	*i.e.,*	2 × 60 + 10 + $\frac{2}{4}$
●●●𝖣𝖣𝖣	53,	*i.e.,*	50 + 3

As mentioned later, the Babylonians also used a circle for zero, at least to the extent that they employed it to represent the absence of number, but it played little part in their system

---

[1]G. A. Barton, *Haverford College Library Collection of Cuneiform Tablets*, Part I, Philadelphia [1905]; Allotte de la Fuye, "En-e-tar-zi patési de Lagaš," in the *Hilprecht Anniversary Volume*, p. 121 (Chicago, 1909).

of notation. More commonly a circle simply stood for 10, particularly in the early inscriptions, as shown on page 38.

TABLET FROM NIPPUR

Contains divisors of 60^4, the quotients being in geometric progression. Date *c.* 2400 B.C. The top line reads $2(\cdot 60)$, $5(\cdot 60)$, and $12(\cdot 60)$. The left-hand figure is the original; the right-hand one is a drawing. Courtesy of the University of Pennsylvania

Chinese Numerals. The present forms of the Chinese numerals from 1 to 10 are as follows:

一 二 三 四 五
六 七 八 九 十

The number 789 may be written either from the top downwards or from left to right, as follows:

The second character in the number as written on the preceding line means hundred, and the fourth character ten.

It will be observed that the figure for 4, probably four vertical marks in its original form, resembles the figure for 8 inclosed in a rectangle. On account of this the Chinese have given it the fanciful name of "eight in the mouth."[1]

Chinese merchants also use the following forms for figures from 1 to 10:

| 丨 | 丨丨 | 丨丨丨 | Ⅹ | 𠂆 | 丄 | 圭 | 𡗜 | 夂 | 十 |

and they have special symbols for 100, 1000, and 10,000, besides those in which a circle is used for zero.

These symbols are not the same as the ancient forms, but our knowledge of the latter is imperfect.[2] There are many variants of each of the characters given above, as when Ch'in Kiu-shao (1247) used 𠃊 for 5, and both 𠂢 and 𠂤 for 9.[3] The numerals on the early coins also show the variations that are found from time to time. In the second century B.C., for example, we find the 5 given in the so-called seal characters in the form 乄,—a form which was used for hundreds of years.[4]

Rod Numerals. There were also numerals represented by rods placed on the counting board,—a device which will be described in Chapter III. These numerals appear in the *Wu-ts'ao Suan-king*, which may have been written about the beginning of our era, or possibly much earlier, and are found

---

[1] L. Vanhée, in *T'oung-Pao*, reprint. On the general subject of the Chinese numerals in their historical development the standard work is that of F. H. Chalfant in the *Memoirs of the Carnegie Museum*, Vol. IV, No. 1, and Plate XXIX.

[2] We have, however, various records going back to the early part of the Christian Era. For example, the Metropolitan Museum, New York, has a land grant of 403 in which the forms of the numerals are almost the same as those now in use.

[3] Chalfant, *loc. cit.*; Y. Mikami, *The Development of Mathematics in China and Japan*, p. 73 (Leipzig, 1913); hereafter referred to as Mikami, *China*. On the general topic see S. W. Williams, *The Middle Kingdom*, New York, 1882; 1895 ed., I, 619; hereafter referred to as Williams, *Middle Kingdom*. J. Hager, *An Explanation of the Elementary Characters of the Chinese*, London, 1801; J. Legge, *The Chinese Classics*, 2d ed., I, 449 (Oxford, 1893).

[4] H. B. Morse, "Currency in China," *Journal of the North China Branch of the R. Asiat. Soc.*, Shanghai, reprint (n.d.). Valuable on the history of Chinese money and weights.

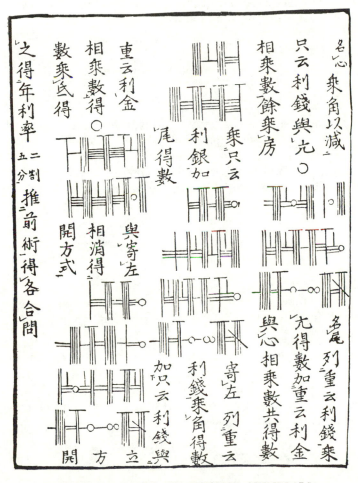

MONOGRAM FORMS OF CHINESE NUMERALS

The Japanese *sangi* were sticks used for representing numbers and were descended from the "bamboo rods" of the ancient Chinese. They gave rise to a *sangi* method of writing numbers. From a work by the Japanese mathematician Fujita Sadasuke (1779) in Chinese characters. The number in the first line at the top is 46,431

even as late as the 19th century. The oldest forms for the units are commonly arranged as follows:

I II III IIII IIIII T TT TTT TTTT

In the tens' column the symbols usually appear as follows:

_ = ≡ ≣ ≣ ⊥ ⊥ ⊥ ≛

the arrangement thereafter alternating, the hundreds being like the units, and so on. Sometimes the ⊤ was used instead of TT for 7 hundreds, and so on, and similarly for 8 and 9. By the common plan the number 7436 would appear as ⊥ IIII ≡ T. In this system the zero takes the form of a circle in the Sung Dynasty (950–1280), as is seen in a work of 1247,[1] where the subtraction 1,470,000 − 64,464 = 1,405,536 appears as

I ≡ O ≣ IIIII ≡ T   I ≡ TOOOO
                          T X IIII ⊥ X

with two forms for 4.

These numerals were frequently written in the monogram form;[2] for example, 123,456,789 appears as 彐彐川川筭.

Hindu Numerals. The history of those Hindu-Arabic numerals which may have developed into our modern European forms is considered later. It should be said, however, that there are various other systems in use in India and neighboring countries. Of these the most interesting is the modern Sanskrit, the numerals being as follows:

१ २ ३ ४ ५ ६ ७ ८ ९ ०

These characters are evidently related to the early Brāhmī forms which are mentioned later.

<hr>

[1] The *Su-shu Kiu-ch'ang* of Ch'in Kiu-shao. See A. Wylie, *Chinese Researches,* Pt. III, p. 159 seq. (Shanghai, 1897); L. Vanhée, in *T'oung-Pao,* reprint, thinks that the zero reached China from India somewhat earlier.

[2] A. Vissière, *Recherches sur l'origine de l'abaque chinois* (Paris, 1892), reprint from the *Bulletin de Géographie*; hereafter referred to as Vissière, *Abaque.*

Of the numerals of the same general character and in use in parts of Asia adjacent to India, the following are types:

	0	1	2	3	4	5	6	7	8	9	10
Siam											
Burma											
Malabar											
Thibet											
Ceylon											
Malayalam											

Their history has no particular significance, however, in a work of this nature, since the forms are local and are relatively modern.[1]

An American Place Value. When Francisco de Cordoba landed the first Spanish expedition on the coast of Yucatan, in 1517, he found the relics of a highly developed civilization, that of the Maya, which had received its deathblow in the wars of the preceding century.[2] Within a few years after the European invasion the independence of the Maya was completely lost. In 1565 Diego de Landa, bishop of Merida, in northern Yucatan, wrote a history of these people,[3] so that our knowledge of their achievements goes back to about the beginning of the period of European influence. They had an elaborate calendar before the Spaniards arrived, and capable investigators have asserted that the Maya cycle began as early as 3373 B.C.

[1] One of the best general works on Eastern notation is that of A. P. Pihan, *Exposé des Signes de Numération usités chez les Peuples Orientaux Anciens et Modernes*, Paris, 1860, with many tables.

[2] S. G. Morley, *An Introduction to the Study of the Maya Hieroglyphs*, p. 6 (Washington, 1915). This work should be consulted for details respecting this entire topic. Authorities vary as to the plural form of Maya, some giving Mayas and others Maya. See also C. Thomas, "Numeral Systems of Mexico and Central America," *Annual Report of the Bureau of American Ethnology*, XIX (1897–1898), 853 (Washington, 1900).

[3] In his *Relacion de las Cosas de Yucatan*, a work not printed until 1864.

The Maya counted essentially on a scale of 20, using for their basal numerals two elements, a dot (•) and a dash (—), the former representing one and the latter five. The first nineteen numerals were as follows, reading from left to right:

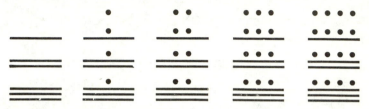

There were numerous variants of these forms,[1] but these offer no special peculiarities which we need consider.

The most important feature of their system was their zero, the character ⬭, which also had numerous variants. Since their scale was vigesimal, they wrote 20 as we write 10, using their characters for 1 and zero.[2] The following table shows the general plan that was used when they wrote on flexible material:

Numerals	• ⬭	•• ••• ≡	≡ ⬭	• ⬭ ⬭	•••• ≡ ••• •••
Our forms	1 0	1 17	15·20 0	360 0 0	19·360 13·20 13
Values	20	37	300	360	7113

We see here a fairly well developed place value, the lowest order being units from 1 to 19, the next being 20's from 1 · 20

[1] See Morley, *loc. cit.*, p. 89, which should be consulted for a description of the system.

[2] Their special hieroglyphic for 20, used for certain purposes, need not concern us. On the word "hieroglyphic" see page 45, note 2.

to $17 \cdot 20$; the next being 360's from $1 \cdot 360$ to $19 \cdot 360$; the next being 7200's, and so on,[1] representing a very satisfactory system. There is no evidence in any extant record that it was used for purposes of computation, its use in the texts being merely to express the time elapsing between dates. The fact, however, that the pebble and rod are apparently the basal elements in the writing of numbers leads us to feel that we have in these numerals clear evidence of the early use of an abacus. If, as many ethnologists believe, there is a connection between the Japanese and certain of our primitive Americans, the use of the rods may be traced back to Asia.

Egyptian Numerals. The Egyptians had four materials upon which they could conveniently record events. One of these was stone, a medium of these was stone, a medium supplied by the quarries along certain parts of the Nile. Another medium was papyrus, a

EARLY FORMS OF COMMON
EGYPTIAN NUMERALS

From a piece of pottery of the First Dynasty, c. 3400 B.C. The symbols for 10 and 100 are repeated several times

kind of paper made from strips of the pulp of a water reed which was apparently more common at one time than it is at present. The other two common materials were wood and pieces of pottery. Leather does not seem to have been so commonly used as in other countries.

In writing on stone the Egyptians took time for the work and made their characters with great care. These characters are called hieroglyphics.[2] The hieroglyphic characters were commonly written from right to left, but also from left to right.

---

[1] For complete description see Morley, *loc. cit.*, pp. 129–133.

[2] The sacred inscriptions; from the Greek ἱερός (*hieros'*), sacred + γλύφειν (*gly'phein*), to carve.

In the earlier inscriptions they are often written from the top down. This accounts for the various ways of writing the simple numerals, a character often being found facing in different directions. For our present purposes it suffices to give the ordinary form of hieroglyphic numerals,[1] as follows:

I	II	III	IIII	ᵚ	ᴵᴵᴵ	ᵚᵚ	IIII	ᴵᴵᴵ	∩
1	2	3	4	5	6	7	8	9	10

I∩	II∩	∩∩	ßß	∩∩∩∩	𝟗	𝟗𝟗	𝔦	(
11	12	20	40	70	100	200	1000	10,000

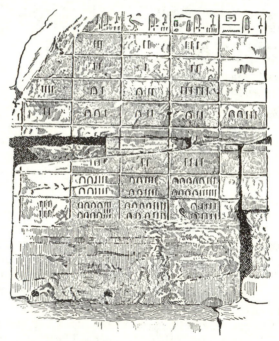

EGYPTIAN NUMERALS

Numerals reading from left to right. From the walls of a temple at Luxor

---

[1] These are as given in A. Eisenlohr, *Ein mathematisches Handbuch der alten Aegypter*, 2d. ed., table following p. 8 (Leipzig, 1877); hereafter referred to as Eisenlohr, *Ahmes Papyrus*. See also Peet, *Rhind Papyrus*; J. De Morgan, *L'Humanité Préhistorique*, p. 115 (Paris, 1921).

There were higher numerals, but the above will serve to show the general nature of the characters employed.

While the hieroglyphic forms were used in writing inscriptions on stone and in elaborate treatises on papyrus, other forms were early developed for rapid writing on papyrus, wood, and pieces of pottery. There were two forms of this writing, the hieratic (religious) and the demotic (popular). The former was a cursive script derived from the hieroglyphic, and the latter was a somewhat later form of the hieratic, beginning in the 7th century B.C. After the demotic forms came into general use, the hieratic was reserved for religious purposes.

HIEROGLYPHIC FOR 6000 (ABOUT 500 B.C.)

The meaning is, "The Falcon King led captive 6000 men of the Land of the Harpoon Lake," there being a harpoon just below this in the original inscription

The hieratic writing usually proceeded from right to left, although in early times it is occasionally found running from the top down. The numerals to 10 were of the following forms: [1]

$$ \text{I} \quad \text{ιI} \quad \text{ιΙI} \quad \text{ιΙιι} \quad \text{ᴗ} \quad \text{⋮⋮⋮} \quad \text{⌐} \quad \text{ʒ} \quad \text{ℓu} \quad \text{Λ} $$

1	2	3	4	5	6	7	8	9	10

The demotic forms offer no peculiarities of special interest. [2]

Greek Numerals. The first numeral forms of the Greeks seem to have been such upright strokes as were used in all Mediterranean countries, and perhaps represented the fingers. These strokes were repeated as far as the needs of the primitive inhabitants required. For example, in a stele from Corinth, of about the 5th century B.C., there is the numeral IIIIIIII, referring

[1] These are taken from the Ebers Papyrus as copied by Eisenlohr. Naturally the forms varied with different scribes.

[2] For a careful study of these forms, with numerous facsimiles, see H. Brugsch, *Numerorum apud Veteres Ægyptios Demoticorum Doctrina*, Berlin, 1849.

EARLY CYPRIOTE NUMERALS

From a fragment of a temple record found on the island of Cyprus. In the last two lines the numeral for 6 (||| |||) appears twice. Courtesy of the Metropolitan Museum of Art, New York

EARLY CYPRIOTE NUMERALS

The lower part of the fragment shown above. The numerals are the same as those on the tablets found at Knossos, Crete, where ◇ is used for 1000, ○ for 100, — for 10, and | for 1. The number 4 (||||) is in the first line and the number 14 (|||| —) in the line next to the last. The Phœnicians also used these symbols for ten and one. Courtesy of the Metropolitan Museum of Art, New York

EARLY CYPRIOTE NUMERALS

From a fragment of a receptacle in a sanctuary. The inscription reads, "Zeus's portion of wine is three measures." Courtesy of the Metropolitan Museum of Art

to a fine of eight obols for intruding on certain property.[1] Inscriptions illustrating this usage are found not only in Greece but in various islands of the eastern Mediterranean Sea, as shown in the illustrations of monumental records from Cyprus. By the time Greece had reached the period of her intellectual ascendancy there had developed a system of numerals formed from initial letters of number names. These forms appear in records of the third century B.C., and were probably in use much earlier, although the custom of writing large numbers in words seems to have been general. Many generations later the system was described so fully by Herodianus, a prominent grammarian of the latter part of the second century, that the symbols were thereafter known as Herodianic numerals, although this name has no worthy sanction. In recent times they have been known as Attic numerals, since they are the only pre-Christian number forms found in Attic inscriptions. The system is also known as the acrophonic (initial) system, the initials

GREEK NUMERALS OF THE PTOLEMAIC PERIOD

On an icosahedral die of the Ptolemaic Period in Alexandria, just before the Christian Era. Such dice are occasionally found, usually made of basalt or quartz. This one is basalt, whitened for the purpose of photographing. From the author's collection

[1] *American Journal of Archæology* (1919), p. 353.

of the several number names, as of πέντε (*pen'te*), five, being used, singly or in combination, in the following manner:

Γ, an old form for Π, the letter *pi*, initial of ΠΕΝΤΕ (*pen'te*), five, used as a numeral for 5;

Δ, the capital *del'ta*, initial of ΔΕΚΑ (*dek'a*), ten, used as a numeral for 10; it is often written like Ο in the Greek papyri, and an inscription at Argos has ☉;

H, the old Attic breathing, like our *h*, later represented by ʻ, initial of ΗΕΚΑΤΟΝ (*hekaton'*), hundred;

X, the capital *chi*, initial of ΧΙΛΙΟΙ(*chil'ioi*), thousand;

M, the capital *mu*, initial of ΜΥΡΙΟΙ (*myr'ioi*), ten thousand.

These numerals were frequently combined, thus:

Ͷ or Ͷ, *pente-deka*, was used for 50;
Ͷ, *pente-hekaton*, was used for 500;

and so on for other numbers.

The forms of the letters varied in different cities and states of Greece, but the variants need not concern us in this description.[1]

The following will show how the characters were used:

$$\Gamma I = 6 \qquad \qquad \Gamma HHH = 800$$
$$\Delta \Gamma IIII = 19 \qquad MMMM = 40,000$$
$$\Gamma \Delta \Delta \Delta = 90 \qquad \Gamma^m = 50,000$$

but in the manuscripts the forms vary so much as often to be exceedingly difficult to decipher.

[1] G. Friedlein, *Die Zahlzeichen und das elementare Rechnen der Griechen und Römer*, Erlangen, 1869; F. G. Kenyon, *Palæography of Greek Papyri*, Oxford, 1899; E. S. Roberts, *Greek Epigraphy*, p. 96 (Cambridge, 1887); J. Gow, "The Greek Numeral Alphabet," *Journal of Philology* (1884), p. 278; S. Reinach, *Traité d'Épigraphie Grecque*, pp. 216, 218 (Paris, 1885); J. P. Mahaffy, "On the Numerical Symbols used by the Greek Historians," *Trans. of the Royal Soc. of Literature*, XXVII (2), 160; Heath, *History*, I, 29. The best modern treatment is that of M. N. Tod, "Three Greek Numeral Systems," *Journal of Hellenic Studies*, XXXIII, 27, and "The Greek Numeral Notation," *Annual of the British School at Athens*, XVIII, 98. On the numerals of Crete see Sir A. J. Evans, *The Palace of Minos*, p. 279 (London, 1921), and *Scripta Minoa*, p. 258 (Oxford, 1909).

To these may be added the following characters related to numerical work:

T = talent and also $\frac{1}{4}$ obol

Ͱ = drachma

I = obol, with Ɔ or C for $\frac{1}{2}$ obol

Ϛ = stater, so that ϚϚϚϚϚϚϚ = 7 staters

and HHΔΔΔΔΓϚϚϚ = 248 staters

Ⲙ = 5 talents, ⋔ = 10 talents, ℍ = 100 talents

Contemporary with the development of the Ionic alphabet we find numerical values assigned to the letters, somewhat as we use letters to number the rows of seats in an assembly room. The oldest forms that we have are substantially as follows:

A = 1	H = 7	N = 13	T = 19
B = 2	Θ = 8	Ξ = 14	Y = 20
Γ = 3	I = 9	O = 15	Φ = 21
Δ = 4	K = 10	Π = 16	X = 22
E = 5	Λ = 11	P = 17	Ψ = 23
Ⲭ = 6	M = 12	Ϛ = 18	Ω = 24[1]

These were used very early, but the system was manifestly of no value for computation. A more refined alphabetic system appeared at least as early as the third century B.C., running parallel with the more primitive systems.

As seen above, the Greeks had twenty-four letters in their common Ionic alphabet, but for a more satisfactory system of numerals they needed twenty-seven letters. They therefore added the three forms F or Ϲ (the old digamma), ϙ or sometimes Ϙ (the Phœnician *koph*), and ⅶ[2] (perhaps the Phœnician

---

[1] S. Reinach, *loc. cit.*, p. 220.

[2] A modern name for the character is *sampi* (σαν+πι, *san'pi*), suggested because of its resemblance to π in its 15th century form. The form in the 2d century was ⅄, and it may go back to the Τ (s), which was used from the 5th to the 2d century B.C. See Roberts, *loc. cit.*, p. 10.

shin or tsadé), after which they arranged their system as follows:

Units	A	B	Γ	Δ	E	F	Z	H	Θ
	1	2	3	4	5	6	7	8	9
Tens	I	K	Λ	M	N	Ξ	O	Π	۹
	10	20	30	40	50	60	70	80	90
Hundreds	P	Ϛ	T	Y	Φ	X	Ψ	Ω	⅄
	100	200	300	400	500	600	700	800	900

To distinguish the numerals from letters, a bar was commonly written over each number, as in the case of $\bar{A}$, although in the Middle Ages the letter was occasionally written as if lying on its side, as in the case of ᐊ.[1]

The capital forms were used, the small letters being an invention of a much later period. In a manuscript of the 10th century in Göttingen the small letters are found, and there are no accents when these numerals appear in tables. When, however, they appear in the text, there are bars superscribed to distinguish the numerals from words, thus: $\bar{\alpha}$, $\bar{\epsilon}$, $\bar{\theta}$, $\bar{\iota\theta}$, etc. In modern books the forms usually appear as α', β', γ', δ', and so on, the accents being used to distinguish the numerals from letters. The thousands were often indicated by placing a bar to the left, thus:

/A,   /B,   /Γ,   · · ·   for   1000, 2000, 3000,   · · ·,

these appearing in modern Greek type as $_{\iota}\alpha$, $_{\iota}\beta$, $_{\iota}\gamma$, · · · ·.

The myriads, or ten thousands ($\mu\acute{\upsilon}\rho\iota o\iota$, *myr'ioi*), were represented by such forms as the following:

$\overset{\gamma}{M}$ or M, 10,000; $\overset{B}{M}$, 20,000; $\overset{\Gamma}{M}$, 30,000, and so on.

In late Greek manuscripts the symbol ⌒ was used for myriad, as in the case of $\overset{\frown\frown}{1\Delta}$ for 14 myriads (140,000). We also find such forms as $\overset{\epsilon}{\frown}$ for 5 myriads (50,000).[2]

---

[1] V. Gardthausen, *Die Schrift, Unterschriften und Chronologie im Byzantinischen Mittelalter,* 2d ed., p. 360 (Leipzig, 1913); hereafter referred to as Gardthausen, *Die Schrift.* See also F. E. Robbins, "A Greco-Egyptian Mathematical Papyrus," *Classical Philology,* XVIII, 328.

[2] Gardthausen, *Die Schrift,* p. 371.

In the early Christian period the three lines of letters representing units, tens, and hundreds respectively were called verses or rows, and the rectangular arrangement of the figures in these verses was probably of some value in computation.[1]

Hebrew Numerals. The Jewish scholars used the letters of their alphabet for numeral symbols in the same way as the Greeks did. We find this usage well established in the Maccabean period (2d century B.C.), but it is probably of an earlier date. In the Talmud the numbers above 400 are formed by composition, 500 being formed of the symbols for 400 and 100,[2] and 900 being a combination of the symbols for 400, 400, and 100.[3] Later writers, however, followed a plan introduced by the Massoretes,[4] in which certain final forms of letters were used for the hundreds above 400. These numeral forms as now recognized are as follows:

א	ב	ג	ד	ה	ו	ז	ח	ט
1	2	3	4	5	6	7	8	9

י	כ	ל	מ	נ	ס	ע	פ	צ
10	20	30	40	50	60	70	80	90

ק	ר	ש	ת	ך	ם	ן	ף	ץ
100	200	300	400	500	600	700	800	900

The thousands were represented by the same letters as the units. Since the number 15 would naturally be represented by 10 and 5, read from right to left, that is, by יה, and since these are the first two letters of the word יהוה (Jhvh, Jahveh, Jehovah), the Hebrews wrote 9 + 6 (טו) instead.

---

[1] "Primus igitur versus est a monade usque ad enneadem," etc. (Capella, VII, 745). Favonius Eulogius (c. 400) remarks: "Primi versus absolutio novenario numero continetur." See J. G. Smyly, in "Mélange Nicole," *Recueil de Mémoires de Philologie Class. et d'Archéol.*, p. 514 (Geneva, 1905).

[2] תק.   [3] התקק.

[4] The scholars engaged in the work of *Massorah*, the establishing of the traditional pronunciation and accents of the Hebrew scriptures. The work extended over a long period, closing in the 10th century. See *Jewish Encyclopedia*, IX, 348 (New York, 1905). For the zero, see Smith-Karpinski, p. 60.

u

Gematria. The fact that the letters of various ancient alphabets had numerical values, and hence were used in computation, led to the formation of a mystic pseudo-science known as gematria, which was very popular among the Hebrews as well as among other peoples.

Although it had many modifications, its general nature may be explained by saying that the numerical value of a name could be considered instead of the name itself. If two names had the same numerical value, this fact showed some relation between the individuals. It is probable that 666, "the number of the beast" in Revelations, was the numerical value of some name, this name being known to those who were in the secret, but being now lost. It is not improbable that it referred to "Nero Cæsar," which name has this value when written in Hebrew. For nearly two thousand years attempts have been made to relate the number to different individuals, particularly to those of a religious faith differing from that of the one suggesting the relationship. Thus, it has been assigned to various popes, to Luther, and to Mohammed; but it has also been related to statesmen, to the Latin Church, and to various other classes and organizations. In some cases a man's name and its gematria number have both appeared upon his tombstone. An interesting illustration of gematria is also found in our word "amen." Written in Greek, the numerical values of the letters are as follows: $A(\alpha) = 1$, $M(\mu) = 40$, $H(\eta) = 8$, $N(\nu) = 50$, the total being 99. On this account we find in certain Christian manuscripts the number 99 written at the end of a prayer to signify "amen."[1]

Roman Numerals. The theories of the origin of the Roman numerals are for the most part untenable. Priscian (6th century) believed that "I" was used for 1 because it was the initial of the Greek $\ddot{\iota}\alpha$, a dialectic Greek word for unity,[2] although long before the Greeks had any written language it was used for this purpose in Egypt, Babylon, and various

---

[1] Gardthausen, *Die Schrift*, p. 309.

[2] See the 1527 (Venice) edition of Priscian, fol. 271, *r*. For the feminine of *εἷς (heis)* the Æolic Greeks used *ἴα*; the other Greeks, *μία*. Homer used both forms.

other parts of the ancient world. His other theories were equally unscientific except in the cases of C and M. These symbols he took to be the initials of *centum* (hundred) and *mille* (thousand), and there was enough historical evidence for the late adoption of these letters as symbols for 100 and 1000 to justify him in making this statement. There are also various theories connected with stick-laying, but for these there is no historic sanction.

In the 16th century Mattheus Hostus[1] asserted that the theory of the early grammarians *frivolum est*; and while his own theories were generally about as frivolous, he made the plausible suggestion that the V was derived from the open hand, the fingers with the exception of the thumb being held together. This led naturally to taking the X as a double V,—a view held by various later writers and receiving powerful support from Mommsen (1850), the great German authority on Latin history and epigraphy. The theory is not inconsistent with the fact that the V is occasionally inverted (Λ), since this form, although an early one, may have developed relatively late with respect to X and may thus have represented half of that numeral.

Mommsen's most important suggestion, however, was that C and M are not primitive forms but are late modifications of such forms, influenced by the initials of *centum* and *mille*. The primitive forms for 50, 100, and 1000 he stated to be the Greek aspirates X (*chi*), from which L was derived; ϴ (*theta*), from which comes the C; and Φ (*phi*), which is the origin of the M. As to this theory there is positive evidence that one of the earliest forms for X (*chi*) was ↓, and this, with the later forms ⅃, ⊥, and L, was used for 50 in the inscriptions of about the beginning of our era.

As to the use of ϴ for 100, we have also the early forms ⊗, ⊕, ⊙, and ⊖. If the last of these were written rapidly with a stylus or a reed pen, the result might easily resemble C. We have not, however, any of these transition forms extant, although by analogy with L and M we might well accept this theory.

[1] *De numeratione emendata veteribus Latinis et Graecis usitata*, Antwerp, 1582.

The ф was also written ⊕, and the symbol for 1000 is very commonly given on the ancient monuments as CIↃ, ↀ, cIↄ, and the like, so that this part of the theory is reasonable. The M as a numeral is unusual on the older monuments, although an expression like $\overline{\text{II}}$M for 2000, where M evidently stands for the word *mille*, is not uncommon. Generally the Romans used one of the modifications of ф as stated above, or the symbol ∞, which is probably a cursive form of CIↃ, with numerous variants such as ⋈ and ∾.

As to the X for 10, there is the further theory that it may have come from the crossing off of ten single strokes for 1 by a *decussare* line, either as ⧺⧺⧺⧺⧺ or as ⧺⧺⧺⧺⧺, which was abbreviated as ⅄.[1] This is analogous to the possible Egyptian plan of grouping ten strokes by an arc and thus obtaining their symbol ∩. There is much to commend this *decussare* theory, for 20 was commonly written ⧺ or H, and similarly for 30 and 40. If this is the origin of the X for 10, then the V and Λ were naturally taken as halves of X. On the whole, this seems quite as probable as the hand theory. It has also been thought that X represents the crossed hands, thus giving two fives.

In 1887 Karl Zangemeister[2] advanced the theory that the entire system was based on the single *decussare* principle. Briefly, a crossing line multiplies any number by ten. Hence we have I and ⅄ for 1 and 10 respectively; X and ⳩ for 10 and 100, from the latter of which the X finally dropped out, leaving (, which became our C under the influence of the initial letter for *centum*; and ⳩ for 1000, which finally became the common ∞. Although the theory is interesting, it has never been generally accepted by Latin epigraphists, and so we at present fall back on the Mommsen theory as the most probable of any thus far suggested. It is quite as reasonable, however, to believe that the symbols were arbitrary inventions of the priests.

[1] *Decussare* is the verb form. The word also appears as *decussatio*, *decussatim*, and *decussis*, according to the sentence construction.

[2] "Entstehung der römischen Zahlzeichen," *Sitzungsberichte der Königl. Preuss. Akad. der Wissensch.*, XLIX, 1011, with a bibliography on page 1013.

An examination of the many thousand inscriptions collected in the *Corpus Inscriptionum Latinarum*[1] fails to solve the problem of origin, but it shows the change in forms from century to century. This change is even more marked in the medieval manuscripts. The following brief notes will serve to show how these numerals have varied.

The I is always a vertical stroke, or substantially so. Horizontal strokes are used in writing certain fractions. In late medieval manuscripts the stroke appears as i or, as a final letter, j.

The V also appears on the early monuments as U or Λ, and is frequently found in such contracted forms as X for 15. In the medieval manuscripts it varies with the style of writing, appearing as V, v, U, and u. In the late Roman times the character ꝙ, with numerous variants, was used for 6, possibly from the Greek numeral. To represent eight, for example, this character was combined with II.[2]

The X also appears on the monuments as ⋎ or ⋏. It is frequently combined with other letters in such forms as lxx for 70. In the medieval manuscripts it is often written as a small letter.

The L very frequently appears on the monuments of about the beginning of our era in the older forms of ↓, ↓, and ⊥. In the Middle Ages it often appears as a small letter, as in a case like Clxviij for 168.

The C has changed less than the other forms, appearing on the ancient monuments as a capital and frequently in the later manuscripts as a small letter.

The D is generally thought to be merely half of the CIƆ which stood for thousands. It is occasionally written CI and appears very commonly as IƆ even after the beginning of printing. In the Middle Ages it appears both as a capital and as a small letter. There is a possibility that the use of D to represent 500 is due to the fact that the Etruscans had no such letter in their early alphabet, and consequently took the Δ (*delta*) for this purpose, just as they took other

[1] Berlin, 1863 seq.
[2] L. A. Chassant, *Dictionnaire des Abréviations* . . . *du Moyen Age*, p. 114 (Paris, 1884).

Greek letters for numerical purposes. The *delta* was then changed, in the course of time, to the form with which we are now familiar.[1]

When the Romans used the M in representing numbers, it was commonly as the initial of *mille*, thousand. When written with other numerals, the thousand symbol was usually CIꓛ, Ⱈ, clɔ, Ⱈ, ∞, ▷◁, ᔕ, or some similar form, as in the case of ∞CIII for 1103. In the medieval manuscripts the M, usually a capital, replaced the earlier forms, as in the number Mcccclxxxxiiij for 1494.

The subtractive principle is found in certain cases like that of IV for 4, that is, 5 − 1. This principle was, as we have seen, used by the Babylonians in the 3d millennium B.C. It was also used by the Hebrews, at least in word forms, but apparently not before the Etruscans and Romans used it. The Etruscans[2] preceded the Romans in recognizing the principle and made a more extensive use of it. They commonly wrote their numerals from right to left, and so we have such forms as the following:[3]

XIIIXX, for 20 + (10 − 3), or 27;
XIIXXX, for 30 + (10 − 2), or 38;
↑III, for 50 − 3, or 47;
II↑X, for (50 − 10) + 2, or 42.

The Etruscans also used ↓ for X, and so we find such forms as ↓√ for XL; when read from right to left this means our LX (60), but when written √↓ it means our XL (40).[4] Such forms as XIꓶ[5] for the Roman LIX are also found.

Subtractive Principle Widespread. The subtractive principle was probably used by various other early peoples, for an immature mind finds it easier to count backwards by one or two from some fixed standard, like 5, 10, 15, 20, and so on, than to count forwards by three or four. Thus, the Romans found it easier to think of "two from twenty" (*duo de viginti*) than of

---

[1] B. Lefebvre, *Notes d'Histoire des Mathématiques*, p. 30 (Louvain, 1920).
[2] R. Brown, "The Etruscan Numerals," *Archæological Rev.*, July, 1889.
[3] W. Corssen, *Ueber die Sprache der Etrusker*, 2 vols., Leipzig, 1874, 1875.
Corpus Inscriptionum Etruscarum, I, Nos. 23, 27, 32, 38, *et passim* (Leipzig, 1893–).   [5] *Ibid.*, 4615.

"eight and ten" (*octodecim*), and of "one from twenty" than of "nine and ten." This is especially the case with numbers above five, since the difficulty is hardly experienced until nine or fourteen is reached.

As an indication of the tendency of primitive peoples to use the subtractive principle the fact may be mentioned that the Zuñi Indians, whose number names refer to the fingers, speak of four as "all the fingers almost complete," and of nine as "almost all are held up with the rest," each containing the idea of subtraction.[1] They had a system of knot numerals which involved the same principle. A medium knot indicated 5, and this with a small knot before it indicated 5 − 1, whereas if the small knot came after the medium one the number was 5 + 1. Similarly, a large knot indicated 10, and a small knot was used either before or after it so as to indicate 9 or 11 respectively.

Further Cases of the Subtractive Principle. It is because of the fact that the difficulty is not evident with so simple a number as 4 that the Romans did not commonly use the subtractive principle in this case, preferring the form IIII to the form IV. They used the principle more frequently in the case of 9, but even here they wrote VIIII oftener than IX. In the case of 400 they usually wrote CCCC, but occasionally they used CD. Even as late as the 16th century we often find a number like 1549 written in some such form as Mcccccxxxxviiij.

Relics of the subtractive principle are seen in our tendency to say "ten minutes of (or "to") six" instead of "fifty minutes past five," and to say "a quarter of (or "to") six" rather than "three quarters of an hour past five."

There is a possibility that the Romans avoided IV, the initials of IVPITER, just as the Hebrews avoided יה in writing 15, as the Babylonians avoided their natural form for 19, and as similar instances of reverence for or fear of deity occur in other languages.

---

[1] F. H. Cushing, "Manual Concepts," *American Anthropologist* (1892), p. 289; Th. W. Danzel, *Die Anfänge der Schrift*, p. 55 (Leipzig, 1912); L. L. Conant, *Number Concept*, p. 48 (New York, 1896).

Even when the subtractive principle was used, no fixed standard was recognized. The number 19 was commonly written XIX, but not infrequently IXX.[1] We also find IIX for 8 and IIXX for 18, but these were not so common. It is quite rare to find CD for 400 or CM for 900, and forms like MCM and DCD were never used in ancient or medieval times. In general, therefore, it may be said that the Romans recognized the subtractive principle but did not make much use of it.

Occasionally this principle was used with the fraction $\frac{1}{2}$, for which the Romans wrote the letter S, initial of *semis* (half). Thus we find SXC for $89\frac{1}{2}$ and SXXC for $79\frac{1}{2}$.

Large Numbers. The Romans had relatively little need for large numbers, and so they developed no general system for writing them. The current belief that they commonly used a bar, or *vinculum*, over a number to multiply it by 1000 is erroneous. What they ordinarily did, if they used numerical symbols at all, was to take some such forms as the following:

For 100,000:   CCCIƆƆƆ ⊓ ☥ ☥ ◎ ◉

For 10,000:   CCIƆƆ ⋔ ☩ cclɔɔ ⋔ ιιlιι

For 5,000:   IƆƆ ⋔ レ lɔɔ ⋀ l,,

To represent larger numbers, these forms were repeated. Thus, the symbol ⊂⊃, used for 100,000, is repeated twenty-three times on the *columna rostrata*,[2] making 2,300,000.

In the Middle Ages, however, we find such forms as ⌐X⌐ or IXI for million and ⌐M⌐ for hundred million, that is, for ten hundred thousand and one thousand hundred thousand.

Use of the Bar. The Romans commonly placed a bar over a number to distinguish it from a word, as in the case of IIVIR for *duumviri* (two men) and IIIVIR for the triumvirate. The

---

[1] In early inscriptions this form was sometimes used for 21, since the Romans occasionally wrote numbers from right to left, like the early Greeks.

[2] A Roman monument set up in the Forum to commemorate the victory of 260 B.C. over the Carthaginians. This is the earliest noteworthy example of the use of large numbers in a Roman inscription.

oldest example that we have of the bar to indicate thousands dates from about 50 B.C. Cicero (106–43 B.C.), or, more probably, some late copyist of his works, also used $\overline{XX}$.CD and CCIƆƆ CCIƆƆ CCCC as equivalent. The *vinculum* is found frequently in the works of Pliny (1st century), but it is not used

Quænam fuerunt notæ Romæ
norúm?

I.   1.
V.   5.
X.   10.
L.   50.
C.   100.

Ɔ. D. IƆ. 500. *Quingenta.*
CXƆ. ∞ . CIƆ. 1000. Χίλια. *Mille.*
   IƆƆ. 5000. *Quinque millia.*
CMƆ. . CCIƆƆ. 10000. Μύρια. *Decem millia.*
 . IƆƆƆ. 50000. *Quinquaginta millia.*
 . CCCIƆƆƆ. 100000.*Centum millia.*
IƆƆƆƆ.   500000. *Quingenta millia.*
CCCCIƆƆƆƆ. . CCCCIƆƆƆƆ. 1000000, *Decies*
   centena millia.

Romani numeri non progrediuntur ultra decies centena
millia illa et cú plura significare uolunt, duplicant notas; ut,

∞ . ∞ . 2000.
CIƆ. CIƆ. CIƆ. 3000.
CIƆ. IƆ. 1500. ∞ . D.

ROMAN NUMERALS

From the work of Freigius, a Swiss writer, published in 1582

with much uniformity and we are not sure how many of his numeral forms are due to later scribes. In the Middle Ages the *vinculum* was called a *titulus*,[1] but even then it was more commonly used to distinguish numerals from words than to indicate thousands.

[1] Thus Bernelinus: "Nam sicut prima unitas notatur per elementum I, ita millenarius primus per idem I, superaddito tantum titulo." A. Olleris, *Œuvres de Gerbert*, p. 360 (Paris, 1867)

The Romans did not use the double bar to indicate 1000 × 1000, as in $\overline{\overline{V}}$ for 5,000,000, but it is said to be occasionally seen in the late Middle Ages.[1]

Late Coefficient Method. In the later Roman times there arose a kind of coefficient method of representing large numbers. Thus, Pliny used XIIM for 12,000, and we have a relic of this method in our modern use of 10 M. In such cases, however, M was looked upon as abbreviation for *mille* rather than as a symbol for 1000, although the distinction is, of course, not noticeable. We find the same thing in the Middle Ages, as when O'Creat (*c.* 1150) writes X^{es}M. milia for ten thousand thousand. A somewhat similar usage appears in the *Compotus Reinheri* (13th century), where IIII^{or} milia. ccc^a.l.vi appears for 4356. Even as late as the 16th century the same plan was followed, as when Noviomagus (1539) wrote IIIM for 3000 and MM for 1,000,000, and when Robert Recorde (*c.* 1542) used vj.C for 600, ixM for 9000, CCC.M for 300,000, and 230 M, MM, M for 230 · 10^{12}. The coefficients were often written above the M, as in $\overset{lxxx}{MM}$ for 80,000,000 and $\overset{c}{MM}$ for 100,000,000, in a manu-

Numeratio.

CIƆ ccIƆɔ / ∞ ccIƆɔ	*9000.*
CCI·Oɔ / c·c·I·ɔ·ɔ	
$\overline{\text{X}}$ / $\overline{\overline{\text{X}}}$ / cc·I·cc / ɔMc / ɔMɔ / IMI	*10000.*
CCIƆƆ cIɔ / cc·I·ɔɔ ∞	*11000.*
CCIƆƆ cIɔ cIɔ / cc·I·ɔɔ ∾ ∾	*12000.*
CCIƆƆ cIɔ cIɔ cIɔ / ccIɔɔ ∾ ∾ ∾	*13000.*
CCIƆƆ cIɔ Iɔɔ / ccIɔɔ ∞ Iɔɔ	*14000.*
CCIƆƆ Iɔɔ	*15000..*

ROMAN NUMERALS

From Bongo's work on the mystery of numbers, Bergamo, 1584–1585

[1]A. Cappelli, *Dizionario di Abbreviature*, 2d ed., p. lii (Milan, 1912).

script of *c.* 1442.[1] They were also written below, as in $\underset{I}{C}$xxiij for
123 and $\underset{I}{C}$ $\underset{xxiij}{MM}$ $\underset{iiij}{C}$ $\underset{lvj}{M}$ $\underset{vij}{C}$ $_{lxxxix}$ for 123,456,789, in the arithmetic
of Bartjens.[2]

Epigraphical Difficulties. The Romans varied their numerals,
often according to the pleasure of the writers, and it takes a
skilled epigraphist to decipher many of those that appear upon
the amphoræ stating the amount or the price of wine. For example, the following numbers, which were taken from wine jugs
of about the 1st century, and which are by no means among
the most difficult, would certainly not be understood by the
casual observer:

for ∞ CCC = 1300

for IXCS = 89½

Such forms concern chiefly the student of epigraphy, however. The medieval numerals are more interesting, since they
involve new methods, and hence a few types will here be given:

$\overline{c}$ · $\overline{lxiiij}$ · ccc · l · i, for 164,351, Adelard of Bath (*c.* 1120).
$\overline{vi}$ · dclxvi, for 6666, Radulph of Laon (*c.* 1125).
II.DCCC.XIIII, for 2814, Jordanus Nemorarius (*c.* 1225).
MↃCLVI, for 1656, a monument in San Marco, Venice.
cIↃ. Iↄ. Iↄ, for 1599, edition of Capella, Leyden, 1599.
xxyiii, for 28, edition of Horace, Venice, 1520.
IIII$_{xx}$ et huit, for 88, a Paris treaty of 1388.[3]
ⅭDCXL, for 1640, edition of Petrus Servius, Rome,
1640.
four Cli.M, two Cxxxiiii, millions, sixe ClxxviiiM. fiue Clxvii,
for 451,234,678,567, Baker, 1568.

[1] A copy of Sacrobosco's arithmetic made *c.* 1442. See *Rara Arithmetica,*
p. 450.
[2] A Dutch work of the 18th century, 1792 ed., p. 8.
[3] This is simply the French *quatre vingt* (4 × 20) and is common in medieval
French MSS.

To the many other peculiarities of this system it is not possible to allow further space. The Roman forms persisted in use, especially outside of Italy, until printed arithmetics made our common numerals widely known. Even at the present time the fishermen of Chioggia, near Venice, use forms that closely resemble those of the early Etruscans, so persistent is custom in the humbler occupations of man.[1]

Our Common Notation. When we come to consider the origin of our common numerals, we are confronted by various theories, and the uncertainty is quite as marked as in the case of the Roman system. These symbols are generally believed to have originated in India, to have been carried to Bagdad in the 8th century, and thence to have found their way to Europe.[2] This is not certain, for various authors of scientific standing have attempted to show that these numerals did not originate in India at all,[3] but the evidence still seems much more favorable to the Hindu origin than to any other that has been suggested. The controversy has recently centered about the meaning of the word *hindasi*, which is often used by the Arabs in speaking of the numerals. It is asserted that the word does not mean Hindu, some claiming that it refers to Persia and others that it means that which is related to calculation. It is difficult, however, to explain away the following words of Severus Sebokht (*c.* 650), written in the 7th century and already quoted in Volume I:

I will omit all discussion of the science of the Hindus, a people not the same as the Syrians; their subtle discoveries in this science

---

[1] A. P. Ninni, "Sui segni prealfabetici usati . . . nella numerazione scritta dai pescatori Clodiensi," *Atti del R. Istituto Veneto delle sci. lett. ed arti*, VI (6), 679.

[2] Smith and Karpinski, *The Hindu-Arabic Numerals*, with bibliography, Boston, 1911 (hereafter referred to as Smith-Karpinski); G. F. Hill, *The Development of Arabic Numerals in Europe*, Oxford, 1915; J. A. Decourdemanche, "Sur la filiation des chiffres européens modernes et des chiffres modernes des Arabes," *Revue d'Ethnographie et de Sociologie*, Paris, 1912; G. Oppert, "Ueber d. Ursprung der Null," *Zeitschrift für Ethnographie*, XXXII, 122 (Berlin, 1900); G. N. Banerjee, *Hellenism in Ancient India*, p. 202 (Calcutta, 1919).

[3] *E.g.*, see Carra de Vaux, "Sur l'origine des chiffres," *Scientia*, XXI (1917), 273; but see F. Cajori, "The Controversy on the Origin of our Numerals," *The Scientific Monthly*, IX, 458.

of astronomy, discoveries that are more ingenious than those of the Greeks and the Babylonians; their valuable methods of calculation; and their computing that surpasses description. I wish only to say that this computation is done by means of nine signs.

Types of Early Hindu Numerals. The early numerals of India were of various types.[1] The earliest known forms are found in the inscriptions of King Aśoka, the great patron of Buddhism, who reigned over most of India in the 3d century B.C. These symbols are not uniform, the characters varying to meet the linguistic conditions in different parts of the country. The Karoṣṭhī forms, for example, are merely vertical marks, I II [III] IIII IIIII, and are not particularly significant. The Brāhmī characters found in some of these inscriptions are of greater interest. The only numerals thus far found in the Aśoka edicts are as follows:

I	II	+	ᵟᵉ	G	Ɔ	木	Ƴ	𝕴
1	2	4	6	50	50	200	200	200

The Nānā Ghāt Inscriptions. About a century after the Aśoka edicts certain records were inscribed on the walls of a cave on the top of the Nānā Ghāt hill, about seventy-five miles from the city of Poona. A portion of the inscriptions is as follows:

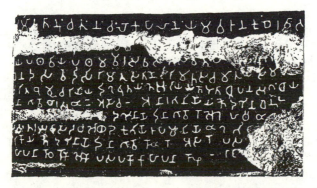

[1] For discussion and bibliography see Smith-Karpinski, p. 19.

The probable number forms contained in these inscriptions are as follows:

1	2	4		6	7	9	10	10	10

20	60	80	100	100	100	200	400

700	1000	4000	6000	10,000	20,000

The next important trace of the numerals is found in the caves at Nasik, India. These are of the 1st or 2d century and are as follows:

1	2	3	4	5	6	7	8	9

10	10	20	40	70	100	200	500

1000	2000	3000	4000	8000	70,000

The significant feature of these numerals is that they clearly resemble the Nānā Ghāt forms, and that in both we seem to have the progenitors of our present numerals.

It should be understood, however, that the interpretations of the inscriptions at Nānā Ghāt and Nasik are not universally accepted. All that we can say, in our present state of knowledge, is that these are the probable number forms as stated, that they resemble some of the numerals that were transmitted to Europe as of Hindu origin, that no zero appears in these early inscriptions, and hence that the place value, as we know it at present, had not yet been developed.

Variants of Hindu Forms. The variants of the Hindu forms preceding the invention of the zero may be seen in the table shown below.

NUMERALS	1	2	3	4	5	6	7	8	9	10	20	30	40	50	60	70	80	90	100	200	1000
1 Aśoka	∤ ||																			
2 Śaka	∤					×	×		×	××	? ?		??????						∕∣∕				
3 Aśoka	∤			+	⋌						ℂ						⋏			
4 Nāgarī	− =	Ŧ	↳↱		?∝∘				⊣	∾	ℋℋT									
5 Nasik	− = ≡	Ⴤ ⱴ	ⱬ ↱ ⱬ	∝ ⊖		×				⁊⁊⁊										
6 Kṣatrapa	− = ≡	Ⴤ ⱴ	ⱳ ⱬ	3 ⱬ ∝	⊖ ⱬ ×	⫣ ⱬ ⱶ	⊕ ⊕ ⱬ 𝔃														
7 Kuṣana	− = ≡	ⱶℲ	⧸ ⱴ	ⱬℙ	∝ ⊖	ⱴⱶⱳ	ℬ	× ∾ ⊕													
8 Gupta	− = ≡	ⱬⱴ	ⱴℙ	ⱴ 3	3ⱴ	∘⅄		⅄⅄∾ℭⱴℋ													
9 Valhabī	ⱱ = ≡	ⱬ ℱⱴ	ℱ ⱴ 𝔮	∾ ℭ ⱴ	ⱴ 𝔮 ⱬ	ℱ 𝔮 ∾	ℂ₂ℤ	T													
Nepal	ⱱ ≳ ⱬ	ⱬℭℛ	? ⅄ ⱶ ℤ ⱬ			∾	ⱨ														
Kalinga	ⱬ⅃	ℐ ⅄ ? ⅄	ℐ ∘	ℐ		∾	⁊														
Vākāṭaka	ⱨ	⅄ ⱴ ⊖																			

As to the original significance of these forms we are wholly ignorant except in the cases of the first three. As to I, II, III, or −, =, ≡, there is, of course, no question. The vertical forms may have represented fingers used in counting or they may have been the marks that one naturally makes with a stylus or brush in keeping a numerical record. The horizontal forms may be pictures of computing sticks, like those which the Chinese used in remote times, and which appear in the Chinese numerals. Such sticks are naturally laid horizontally with respect to the eye. The earliest Sumerian forms of the numerals

---

[1] Karoṣṭhī numerals, Aśoka inscriptions, *c.* 250 B.C. For sources of infor-mation with respect to this table see Smith-Karpinski, p. 25.

[2] Same, Śaka inscriptions, probably of the first century B.C.

[3] Brāhmī numerals, Aśoka inscriptions, *c.* 250 B.C.

[4] Same, Nānā Ghāt inscriptions, *c.* 150 B.C.

[5] Same, Nasik inscription, *c.* 100.

[6] Kṣatrapa coins, *c.* 200.

[7] Kuṣana inscriptions, *c.* 150.

[8] Gupta inscriptions, *c.* 300 to 450.　　　　[9] Valhabī, *c.* 600.

were horizontal,[1] and so the computing rod may have had its origin in Sumeria. The later Babylonian forms were vertical, and so the finger computation may have been in favor at that time. The Mediterranean lands adopted the vertical forms, and the Far East preferred the horizontal. From the vertical II came the following Egyptian forms:

Hieroglyphic II

Hieratic   U

Demotic   ɥ

From the symbol II came also the Arabic P, which is our 2 if turned on its side. Indeed, our 2 is merely a cursive form of = and our 3 is similarly derived from ≡, as is seen from a study of inscriptions and manuscripts.[2] These horizontal forms were early used by the Chinese and probably found their way from China to India.

Fanciful Theories. Numerous conjectures have been made as to the origin of the characters from four to nine, but no one of them has had any wide acceptance. We may dismiss at once all speculations as to their derivation from such combinations as ⊠ and ⊕ and from the number of sticks that might be laid down to make the figure 8. Such ideas are trivial and have no sanction from the study of paleography. There remain, however, various scientific theories, as that the forms are ancient initial letters of number words.[3] None of these theories, however, has stood the test of scholarly criticism, and today we have to confess that we are entirely ignorant as to the origin of the forms which began possibly in India in Aśoka's time and appear as the common numerals which we use.

---

[1] Sir H. H. Howard, "On the Earliest Inscriptions from Chaldea," *Proceedings of the Society of Biblical Archæology*, XXI, 301.

[2] An interesting fact in relation to the figure 2 is that the Romans often wrote $\frac{2}{12}$ as two lines, =, the twelfth being understood as we understand tenths when we write o.2. They also wrote this character cursively, ꝣ, which is the character used for our 2 in several early printed books.

[3] For details of this theory consult the bibliography given in Smith-Karpinski, p. 30.

Origin of the Zero. The origin of the zero is quite as uncertain as the origin of our other numerals. Without it the Hindu numerals would be no better than many others, since the distinguishing feature of our present system is its place value. The earliest undoubted occurrence of a zero in India is seen in an inscription of 876 at Gwalior. In this inscription 50 and 270 are both written with zeros.[1] We have evidence, however, that a place value was recognized at an earlier period, so that the zero had probably been known for a long time. The Babylonians, indeed, had used a character for the absence of number, and they made use of a primitive kind of place value;[2] but they did not create a system of numeration in which the zero played any such part as it does in the one which we now use. There is also a slight approach to a place value in some of the late Greek works. For example, Diophantus seems, judging by certain manuscripts, to have used $\cdot\dot{\mathrm{B}}\cdot\Lambda\phi\overline{\Pi}\mathrm{Z}$ for 23,587, the four points about the B being a late Greek symbol for myriads, and the position of Λ determining its value as 30 hundreds.[3]

The form of the zero may have been suggested by an empty circle, by the Greek use of omicron (O) to indicate a lacuna,[4] by the horned circle used in the Brāhmī symbols for ten, by the Hindu use of a small circle (o), as well as a dot, to indicate a negative, or in some other way long since forgotten. There is no probability that the origin will ever be known, and there is no particular reason why it should be. We simply know that the world felt the need of a better number system, and that the zero appeared in India as early as the 9th century, and probably some time before that, and was very likely a Hindu invention.

The Arabs represented 5 by a character that looked somewhat like the Hindu zero. In a manuscript of 1575 the numerals appear as ٩ ٨٧٤ ٥ ٣ ٢١. In other manuscripts we find such forms as ٮ, ∞, and ⊃. Because of the resemblance of their five to the circle the Arabs adopted a dot for their zero.

---

[1] See Datta, *Amer. Math. Month.*, XXXIII, 449.

[2] For particulars see *ibid.*, p. 51.

[3] For other cases see Gardthausen, *Die Schrift*, p. 372.

[4] Being the initial of οὐδέν (*ouden'*), nothing. Thus, Archimedes might have used O to indicate the absence of degrees or minutes. See Heath, *Archimedes*, lxxi.

II

For purposes of comparison the Sanskrit forms are here repeated and the modern Arabic forms are given:

Sanskrit,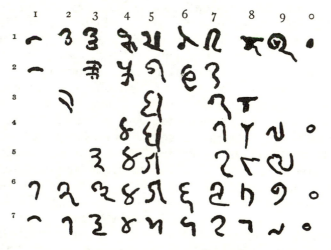

Arabic,

The various forms of the numerals used in India after the
zero appeared may be judged from the table here shown.

¹ Bakhshālī MS. See Volume I, page 164; Smith-Karpinski, pp. 49, 50.

² The 3, 4, 6, from H. H. Dhruva, "The Land-Grants from Sankheda," *Epi-graphia Indica*, Vol. II, pp. 19–24 with plates; date 595. The 7, 1, 5, from Bhan-darkar, "Daulatabad Plates," *Epigraphia Indica*, Vol. IX, Part V; date *c.* 798.

³ The 8, 7, 2, from "Buckhala Inscription of Nagabhatta," Bhandarkar, *Epi-graphia Indica*, Vol. IX, Part V; date 815. The 5 from "The Morbi Copper-Plate," Bhandarkar, *Indian Antiquary*, Vol. II, pp. 257–258, with plate; date 804.

⁴ The 8 from the above Morbi Copper-Plate. The 4, 5, 7, 9, and 0 from "Asni Inscription of Mahipala," *Indian Antiquary*, Vol. XVI, pp. 174–175; date 917.

⁵ The 8, 9, 4, from "Rashtrakuta Grant of Amoghavarsha," J. F. Fleet, *Indian Antiquary*, Vol. XII, pp. 263–272; date *c.* 972. See Bühler. The 7, 3, 5, from "Torkhede Copper-Plate Grant," Fleet, *Epigraphia Indica*, Vol. III, pp. 53–58.

⁶ From "A Copper-Plate Grant of King Tritochanapâla Chanlukya of Lata-deṣa," H. H. Dhruva, *Indian Antiquary*, Vol. XII, pp. 196–205; date 1050.

⁷ A. C. Burnell, *South Indian Palæography*, Plate XXIII, Telugu-Canarese numerals of the 11th century.

The following are later European and Oriental forms:

	1	2	3	4	5	6	7	8	9	0
1	ᒪ	2	3	4	ᔕ	ᖶ	7	ᖺ	𝟫	0
2	ᒋ	ᔓ	ᔕ	ᔑ	ᖲ	6	ᖺ	8	𝟫	0
3	I	ᔓ	3	4	ᖲ	6	7	8	𝟫	
4	I	ᔓ	ᔕ	ᖿ	ᖲ	6				0
5	ᔑ	ᔓ	ᔕ	ᙅᙈ	ᔕᐯ	ᐸ	ᒆ			0
6	ᒉ	ᔓ	ᔕ	ᙇᙈ	ᔕᐯ	ᐸ	ᔕ			0
7	0	𝟫	𝟛	ᙇ	ᖻ	ᔕ	ᔕ	ᔕ	𝟫	•

The Name for Zero. The name for zero is not settled even yet. Modern usage allows it to be called by the name of the letter *O*, an interesting return to the Greek name *omicron* used by Buteo in 1559. The older names are *zero, cipher,* and *naught.* The Hindus called it *śūnya,* "void," and this term passed over into Arabic as *aṣ-ṣifr* or *ṣifr.* When Fibonacci (1202) wrote his *Liber Abaci,* he spoke of the character as *zephirum.*[8] Maximus Planudes (*c.* 1340) called it *tziphra*[9] and this form was used by Fine (1530) in the 16th century. It passed over into Italian as *zeuero,*[10] *ceuero,*[11] and *zepiro,*[12] and in the medieval period it

---

[1] and [2] From a manuscript of the second half of the 13th century, reproduced in "Della vita e delle opere di Leonardo Pisano," Baldassare Boncompagni, Rome, 1852, in *Atti dell' Accademia Pontificia dei Nuovi Lincei,* anno V.

[3] and [4] From a 14th century manuscript.

[5] From a Thibetan MS. in the library of the author.

[6] From a specimen of Thibetan block printing in the library of the author.

[7] Śāradā numerals from *The Kashmirian Atharva-Veda, reproduced by chromophotography from the manuscript in the University Library at Tübingen,* M. Bloomfield and R. Garbe, Baltimore, 1901.

[8] ". . . quod arabice zephirum appellatur."

[9] From the Greek form τζίφρα (*tzi'phra*), used also by another writer, Neophytos, about the same time.

[10] Thus Jacopo da Firenze (1307), or Magister Jacobus de Florentia.

[11] As in the arithmetic of Giovanni de Danti of Arezzo (1370).

[12] As in a translation into Latin of the works of Avicenna.

had various other forms, including *sipos, tsiphron, zeron, cifra,* and *zero.* It was also known by such names as *rota, circulus, galgal, omicron, theca, null,* and *figura nihili.*[1]

Numerals outside of India. The first definite trace that we have of the Hindu numerals outside of India is in the passage already quoted from Severus Sebokht (*c.* 650). From this it seems clear that they had reached the monastic schools of Mesopotamia as early as 650.

The next fairly definite information as to their presence in this part of the world, and with a zero, is connected with the assertion that a set of astronomical tables was taken to Bagdad in 773 and translated from the Sanskrit into Arabic by the caliph's command. There is ground for doubt as to the assertion, but the translation is said to have been made by al-Fazârî (*c.* 773). It is probable that the numerals were made known in Bagdad at this time, and they were certainly known by the year 825. About that year al-Khowârizmî recognized their value and wrote a small book explaining their use. This book was translated into Latin, possibly by Adelard of Bath (*c.* 1120), under the title *Liber Algorismi de numero Indorum.*[2]

The Hindu forms described by al-Khowârizmî were not used by the Arabs, however. The Bagdad scholars evidently derived their forms from some other source, possibly from Kabul[3] in Afghanistan, where they may have been modified in transit from India. These numerals have been still further modified in some respects, and at present are often seen in the forms given on pages 70 and 71.

The Numerals move Westward. Owing to the fact that almost no records of a commercial nature have been preserved from

---

[1] For a full discussion see Smith-Karpinski, chap. iv.

[2] The Book of al-Khowârizmî on Hindu number. On this work see Smith-Karpinski, pp. 5 seq., 92 seq.

[3] It is curious that the old Biblical name of Cabul (1 Kings, ix, 13; Joshua, xix, 27) should be found in Afghanistan. Could it have been taken there by the ruling clan, the Duranis, who call themselves Beni Israel and who claim descent from the Israelites who fled to the Far East after the Assyrians devastated Samaria? If so, could these people, who also claim descent from Kish (1 Samuel, ix, 1), have taken the numerals from Egypt to Afghanistan?

the so-called Dark Ages of Europe, and that the number of scientific works that have come down to us is also very limited, we cannot say when the Hindu-Arabic numerals first found their way to the West. There are good reasons for believing that they reached Alexandria along the great pathway of trade from the East even before they reached Bagdad, possibly in the 5th century, but without the zero.[1] It would have been strange if the Alexandrian merchants of that time and later had not known the numeral marks on goods from India, China, and Persia. No system that did not contain a zero, however, would have attracted much attention, and so this one, if it was known at all, was probably looked upon only as a part of the necessary equipment of a trader with the East.

The Gobâr Numerals. At any rate, numerals are found in Spain as early as the 10th century, and some of these numerals differ so much from the rest that they evidently came through a different channel, although from the same source. These were called the dust numerals,[2] possibly because they were written on the dust abacus instead of being represented by counters. It is worthy of note that Albêrûnî (c. 1000) states that the Hindus often performed numerical computations in the sand.

If these numerals reached Alexandria in the 5th century, they probably spread along the coasts of the Mediterranean Sea, becoming known in all the leading ports. In this case they would have been familiar to the merchants for purposes of trade and to the inquisitive for reasons of curiosity. The soothsayer and astrologer would have adopted them as part of the mysticism of their profession, and the scholar would have investigated them as possibilities for the advancement of science. In that case a man like Boethius (c. 510) would have been apt to know of them and perhaps to mention them in his writings.

All this is speculative. The gobâr numerals exist as a fact, and this is their possible origin. In certain manuscripts of

---

[1] For bibliography and discussion see Smith-Karpinski.

[2] Ḥurûf al-ġobâr. The name appears in Tunis as early as the middle of the 10th century. There were also the ḥurûf al-jumal, or alphabetic numerals, used by the Jews and probably also by the Arabs.

Boethius there appear similar forms, but these manuscripts are not earlier than the 10th century and were written at a time when it was not considered improper to modernize a text. They do not appear in the arithmetic of Boethius, where we might expect to find them, if at all, but in his geometry, and their introduction breaks the continuity of the text. It therefore seems very doubtful that they were part of the original work of Boethius. Since any forms that reached Alexandria would probably have lacked the zero, and since a zero appears in the late Boethian manuscripts, there is the added reason for feeling that at least part and very likely all of the symbols were inserted by copyists.

These *gobâr* numerals varied considerably but were substantially as shown in the following table:

Gerbert and the Numerals. The first European scholar who is definitely known to have taught the new numerals is Gerbert (*c.* 980), who later became Pope Sylvester II (999). He went to Spain in 967 and may have learned about them in Barcelona.[6] He probably did not know of the zero, and at any rate he

---

[1] For sources of information with respect to these numerals see Smith-Karpinski, p. 69.

[2] Al-Ḥaṣṣār's forms, H. Suter, *Bibl. Math.*, II (3), 15.

[3] The manuscript from which these are taken is the oldest (970 A.D.) Arabic document known to contain all the numerals.

[4] and [5] Woepcke, "Introduction au calcul Gobārī et Hawaī," *Atti dell' Accademia Pontificia dei Nuovi Lincei*, Vol. XIX.

[6] On this question see Smith-Karpinski, p. 110.

did not know its real significance. He placed upon counters the nine *caracteres*, as they were called by his pupils Bernelinus (*c.* 1020) and Richer, and used these counters on the abacus. Such counters, probably in the form of flattened cones, were called *apices*, a term also used in connection with the numerals themselves. These numerals were severally called by the names *igin, andras, ormis, arbas, quimas, calctis, zenis, temenias, celentis,* and *sipos.*[1] The origin and meaning of these terms have never been satisfactorily explained, but the words seem to be Semitic.[2]

The oldest definitely dated European manuscript that contains these numerals was written in Spain in 976. A Spanish

EARLY EUROPEAN NUMERALS

Oldest example of our numerals known in any European manuscript. This manuscript was written in Spain in 976

copy of the *Origines* of Isidorus, dated 992, contains the numerals with the exception of zero. Dated manuscripts of the Arabs have been found which give some of these numerals a century earlier, that is, in 874 and 888. They also appear in a Shiraz manuscript of 970 and in an Arabic inscription in Egypt dated 961. The earliest occurrence of these numerals in a date on a coin is found on a piece struck in Sicily in 1138.

There is good reason for believing that Gerbert obtained his knowledge of the numerals from studying in the convent of Santa Maria de Ripoll, a well-known center of learning near Barcelona;[3] indeed, it is not improbable that he saw the very

---

[1] There were variants of these forms. The *sipos* does not appear in the works of the pupils of Gerbert, but is found in a MS. of Radulph of Laon (*c.* 1125).   [2] Smith-Karpinski, p. 118.

[3] J. M. Burnam, "A Group of Spanish Manuscripts," *Bulletin Hispanique,* XXII, 229 (Bordeaux, 1920). With respect to the library in this convent see R. Beer, *Die Handschriften des Klosters Santa Maria de Ripoll,* Vienna, 1907.

manuscript of 976 above mentioned. There is considerable evidence to support the belief that the monks in this cloister obtained their knowledge of these numerals through mercantile

sources which were in communication with the East, rather than through any Moorish channels in Mohammedan Spain.

The changes in the forms of the numerals may be seen in the table above.[1] The forms as they appeared just before the

[1] This is from a table prepared by Mr. G. F. Hill of the British Museum, and is reproduced by his permission. His noteworthy article on the subject appeared in *Archæologia*, LXII (1910). This was elaborated in book form under the title, *The Development of Arabic Numerals in Europe*, Oxford, 1915.

invention of European printing may be seen in the annexed facsimile from a Latin manuscript written by Rollandus at Paris, *c.* 1424. After Europe began to print books, the forms varied but little, most of the changes being due simply to the fashions set by designers of type. For example, the figures 4 and 5 were changed to their present forms in the 15th century and have since then remained fairly well standardized.

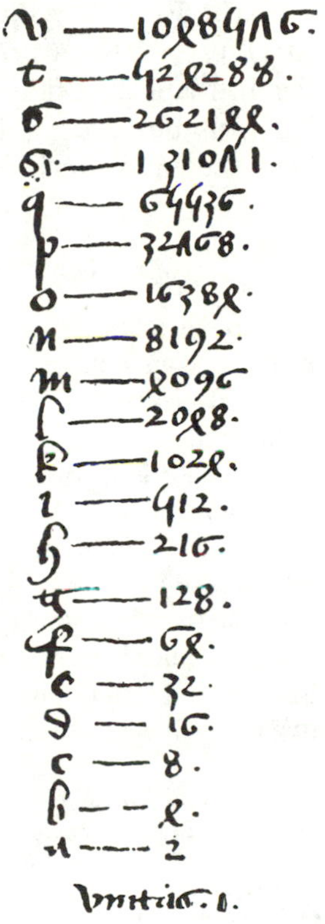

Not only did the forms of the numerals change considerably during the Middle Ages, but the method of writing the ordinary numbers also varied from century to century. Some scribes always placed a dot before and after each figure, as in the case of a number like .2. Others adopted a somewhat similar plan in the case of numbers having several figures. For example, one writer of *c.* 1400[1] gives 5. 7. 8. 2. for 5782, and one of 1384 gives 1000. 300. 80. 4 for 1384, as shown in the following illustration from an anonymous *computus* written in Italy:

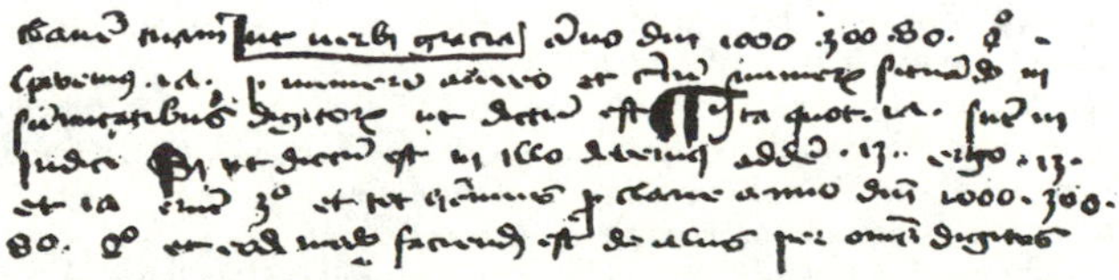

FROM THE ROLLANDUS
MANUSCRIPT OF *c.* 1424

NUMERALS FROM A COMPUTUS OF 1384

The method of writing the date, 1000. 300. 80. 4, illustrates the difficulties in using the numerals. From Mr. Plimpton's library

[1] F. J. Studnička, *Algorismus prosaycus magistri Christani*, p. 9 (Prag, 1893). This Magister Christanus was Christanus Prachaticensis, or Christian of Prag (born 1368; died 1439).

Early English Algorism. An interesting illustration of the early use of the word "algorism" (algorym, augrim) in the English language may be seen in a manuscript now in the British Museum, dating from *c.* 1300.[1] The first page, which is here shown in facsimile, reads as follows:

> Hec algorisms ars psens dicit[r] in qua
> Talibs indoꝫ fruim[r] bis quīq figuris.[2]

This boke is called þe boke of algorym or Augrym after lewder use. And þis boke tretys þe Craft of Nombryng, þe quych crafte is called also Algorym. Ther was a kyng of Inde þe quich heyth Algor, & he made þis craft. And aft his name he called hit algory. Or els anoth cause is quy it is called Algorym, for þe latyn word of hit s. Algorismus coms of Algos grece q ē ars, latine, craft on englis, and rides q ē nm̄s, latine, A nombr on englys. inde dr algorismus ꝑ addicōne huis sillabe ms & subtraccōnem d & E, qsi ars numāndi.[3]

¶ fforthermor[4] ye most undrstonde þt in þis craft ben usid teen figurys. as her ben writen for ensampul. φ.9.8.7.6.5.4.3.2.1. ¶ Expone þe too vsus a for;[5] þis psent craft is called Algorisms, in þe quych we use teen figurys of Inde. Questio. ¶ Why ten fyguris

---

[1] It was first privately printed by the Early English Text Society (transcription by Robert Steele), London, 1894. It has already been referred to in Volume I, page 238, and in this volume, page 32.

[2] These are the two opening lines of the *Carmen de Algorismo*, of Alexandre de Villedieu (*c.* 1240). They should read as follows:

> Haec algorismus ars praesens dicitur; in qua
> Talibus Indorum fruimur bis quinque figuris.

It is translated a few lines later: "This present craft is called Algorismus, in the which we use ten figures of India."

[3] "Inde dicitur Algorismus per addicionem huius sillabe *mus* & subtraccionem *d* & *e*, quasi ars numerandi (Whence it is called Algorismus by the addition of this syllable *mus*, and the taking away of *d* and *e*, as if the art of numbering)." This idea had considerable acceptance in the 13th century.

[4] "Furthermore," the *f* being doubled for a capital. "Furthermore you must understand that in this craft there are used ten figures." The forms of the numerals given in the original were the common ones of the 12th and 13th centuries. The zero was not usually our form, but frequently looked more like the Greek *phi*. The 7, 5, and 4 changed materially in the latter part of the 15th century, about the time of the first printed books. The sequence here shown is found in most of the very early manuscripts, the zero or nine being at the left.

[5] "Explain the two verses afore."

Folio 1r of the manuscript

FIRST PAGE OF THE CRAFT OF NOMBRYNG

Egerton MS. 2622 in the British Museum, one of the earliest manuscripts in
English which treat of any phase of mathematics

of Inde. Solucio.[1] for as I have sayd a fore þai wer fonde fyrst in
Inde of a kyng of þat Cuntre þt was called Algor. ¶ Pma sigt uno
duo vo scda[2] ¶ Tercia sigt tria sic pcede sinistre. ¶ Donc ad extmā
venias que cifra vocar. ¶ Capm pmū de significacoē figurarm ¶ In
þis verse is notifide þe significacōn of þese figuris.   And þus expose
þe verse the first signifiyth on. þe secūde signi[fiyth tweyn].[3]

Reading and Writing Large Numbers. One of the most striking features of ancient arithmetic is the rarity of large numbers.
There are exceptions, as in some of the Hindu traditions of
Buddha's skill with numbers,[4] in the records on some of the
Babylonian tablets,[5] and in the *Sand Reckoner*[6] of Archimedes,
with its number system extending to 10^{63}, but these are all
cases in which the élite of the mathematical world were concerned; the people, and indeed the substantial mathematicians
in most cases, had little need for or interest in numbers of any
considerable size.

The Million. The word "million," for example, is not found
before the 13th century, and seems to have come into use in
England even later. William Langland (*c.* 1334–*c.* 1400), in
Piers Plowman, says,

> Coueyte not his goodes
> For millions of moneye;

[1] "Answer."

[2] "The first means one, the second two, the third means three, and thus proceed to the left until you reach the last, which is called cifra." The author is
quoting from the *Carmen* of Alexandre de Villedieu:

> Prima significat unum; duo vero secunda;
> Tertia significat tria; sic procede sinistre
> Donec ad extremam venias, quæ cifra vocatur.

[3] "Capitulum primum de significacione figurarum (Chapter I, On the meaning
of the figures)."
"And thus explain the (Latin) verse: the first signifieth one."
"The second (secunde) signifieth twain."

[4] Sir Edwin Arnold speaks of this in *The Light of Asia*. See Smith-Karpinski,
p. 16.   [5] Hilprecht, *Tablets.*

[6] Ψαμμίτης (*Psammi'tes*), translated into Latin as *Arenarius.* For the text,
see *Archimedis opera omnia,* ed. Heiberg, with revisions, II, 242 (Leipzig, 1880–
1913) ; Heath, *Archimedes,* p. 221. See also M. Chasles, in the *Comptes rendus,*
April 11, 1842; the preface to the English translation of the work by Archimedes
made by G. Anderson, London, 1784; Heath, *History,* II, 81.

but Maximus Planudes (*c.* 1340) seems to have been among the first of the mathematicians to use the word.[1] By the 15th century it was known to the Italian arithmeticians, for Ghaligai

℀ Como si formano milion

ℒ million adoncha se die formar per sette figure in questo modo. 1000000 .perche laseptima figura tien elluogo demiara demiara:cpcr che millemiara fano vno million :et essendo in quel luogo lafigura che ripresenta vno pero bene edito vno milion.ⅅa.iquestp modo. 1 100000 diria vno milion e cento milia:perche oltra el million:in luogo de centenara de miara:sono lafigura che ripre senta vno si che bene edito vno million e cento milia.ⅅa in questo modo. 1 1 10000.diria vno million e cento e dicre milia perche oltra elmilion e cento milia : in luogo dele decene de miar:sono lafigura che ripresenta vno:si che bene edito vno milion e cento ediere milia.ⅅa in questo modo. 1 1 1 1000.diria vnmilion cento e vndere milia per che oltra elmilion cento ediere milia:in luogo de numeri demiar sono lafigura che rip senta vno:si che bene edito vnmilion cento e vndere milia.ⅅa in questo modo. 1 1 1 1 100.diria vnmilion cento e vndere mi lia e cento:perche oltra elmilion cento e vndere milia:in luogo de elsimplice centenar:sono lafigura che ripresenta vno:si che bene editovnmilion cento e vndere milia e cento.ⅅa in questo modo. 1 1 1 1 1 10.diria vnmilion. cento e vndere milla cento ediere:perche oltra elmilion cento e vndere milia e cento:i luogo dele simplice decene:sono lafigura che ripresenta vno. ⅅa in questo modo. 1 1 1 1 1 1.diria vnmilion cento e vnde re milia cento e vndere.per che ancbe in luogo dele simplice vnita .sono lafigura che ripresenta vno.si che bene edito vn milion cento e vndere milia cento e vndere.et cbosi procededo perfina .9999999.ponendo sempre alsuo luogi quelefigure reprezentante queli numeri ouero decene ocentenara.che si nomina et cetera.equesto basta cerc ba lo amaistramento del numerar.ben che in infinitum sipozia proceder .ma cbom vna zeneral figura mifozcero dicbiarir quanto potesse acbader.et faraito questo sotto posta

1000000

1 10000

1 1 10000

1 1 1 1000

1 1 1 1 100

1 1 1 1 1 10

1 1 1 1 1 1

9999999

THE WRITING OF LARGE NUMBERS IN 1484

From Pietro Borghi's *De Arte Mathematiche*, Venice, 1484. This illustration is from the 1488 edition

(1521; 1552 ed., fol. 3) relates that "Maestro Paulo da Pisa"[2] read the seventh order as millions. It first appeared in a printed

---

[1] H. Wäschke translation, p. 4 n. (Halle, 1878) ; hereafter referred to as Wäschke, *Planudes.* The word simply means "great thousand" (from *mille + on*), just as *salon* means "great hall" (from *salle + on*) and *balloon* means "great ball."

[2] "La settima dice numero di milione." This Paul of Pisa may have been the Paolo dell' Abbaco (Dagomari, *c.* 1340) mentioned in Volume I, page 232.

work in the Treviso arithmetic of 1478. Thereafter it found place in the works of most of the important popular Italian writers, such as Borghi[1] (1484), Pellos[2] (1492), and Pacioli[3] (1494), but outside of Italy and France it was for a long time used only sparingly. Thus, Gemma Frisius (1540) used "thousand thousand"[4] in his Latin editions, which were published in the North, while in the Italian translation (1567) the word *millioni* appears. Similarly, Clavius carried his German ideas along with him when he went to Rome, and when (1583) he wished to speak of a thousand thousand he almost apologized for using "million," referring to it as an Italian form which needed some explanation.[5]

In Spain the word *cuento*[6] was early used for 10^6, the word *millon* being reserved for 10^{12}. When the latter word was adopted by mathematicians, it was slow in coming into general use.[7]

[1] "Il miar de milliara ò vuol dir il million" (1540 ed., fol. 5).

[2] His names beyond units are *desena, centenal, millier, x^a de m^a, c^a de m^a,* and *million* (fol. 2).

[3] He uses *milioni* (fol. 9) but no higher special names, although he repeats this word, as in "Migliara de miliō de miliō" (fol. 19, *v.*), adding: "Et sic ī sequētib⁹ʼ ṗseqre."
The spelling varies in the early books, sometimes appearing as *miglioni* (Pagani, 1591).   [4] *Millena millia.*

[5] In the Latin edition (1583): "Iam vero si more Italorum millena millia appellare velimus Milliones, paucioribus verbis & fortasse significantius" (*Epitome,* cap. i).
In the Italian edition (1586) : "Hora se secōdo il costume d' Italia vorremo vn migliaio di migliaia chiamare millione, con manco parole, & forse piu significantemente" (p. 14).
In *De Cosmographia Libri IV* by Francesco Barozzi (*c.*1538–*c.*1587), a work published in Venice in 1585, it is stated that "septima (nota) pro Millenarii quem vulgus quidem Millionem appellant, Latini vero Milleno Millio."

[6] From *contar,* to count or reckon. Ciruelo, whose work was published in Paris in 1495, says: "Millies millena: quod vulgariter dicitur cuento: decies cuento/centies cuento/millies cuento/decies millies cuento/centies millies cuento [/millies millies cuento]/quod vulgariter dicitur millon." See 1513 ed., fol. a₂ and fol. A₃. He is not, however, uniform in the matter, using "millon" as synonymous with "cuento" in other places.

[7] As late as the 1643 edition of Santa-Cruz (1594) it was necessary to explain the word thus: "Millon que significa mil millares" (fol. 13, *r.*), the latter being the common form.

France early took the word "million" from Italy, as when Chuquet (1484) used it, being followed by De la Roche (1520), after which it became fairly common. The conservative Latin writers of the 16th century were very slow in adopting the word. Even Tonstall (1522), who followed such eminent Italian writers as Pacioli, did not commonly use it. He seems to have been influenced by the fact that the Romans had no use for large numbers;[1] or by the fact that, for common purposes, it sufficed to say "thousand thousand," as had been done for many generations.[2] He simply mentions the word as a piece of foreign slang to be avoided.[3] Other Latin writers were content to say "thousand thousand."[4]

The German writers were equally slow in abandoning "thousand thousand" for "million," most of the writers of the 16th century preferring the older form.[5] The Dutch were even more conservative, continuing the old form later than the writers in the neighboring countries.[6] Indeed, for the ordinary needs of business in the 16th century, the word "million" was a luxury rather than a necessity.

[1]"Non me latet Romanos ueteres prisco more, suos nūmos Sestertios computātes, numerum trāscendentem centum millia . . . Latine nō enunciasse . . . ."

[2]Even as late as 1501, Huswirt, a German scholar, writes "quadraginta quattuor mille millia. quingēta millia quinquaginta nouē millia. octingenta. octoginta sex" for 44,559,886.

[3]"Septimus millena millia: uulgus millionē barbare uocat."

[4]So Stifel uses "millia, millies" (*Arithmetica Integra*, 1544 ed., fol. 1); Ramus uses "millena millia" (*Libri II*, 1569, p. 1); Glareanus has "mille millia" (1538; 1543 ed., fol. 9).

Thus, "tausant mal tausant" is used by such writers as Köbel (*Zwey Rechenbüchlin*, 1514; 1537 ed., fol. 14), Grammateus (1518; 1535 ed., p. 5), Riese (1522; 1529 ed., p. 3), and Rudolff (1526; 1534 ed., fol. 3). Rudolff, however, uses it together with the older form in his *Rechenbuch* (1526), and in his *Exempelbüchlin* (1530; 1540 ed., exs. 62 and 137) he says: "Vnd wirt ein million mit ziffern geschriben 1000000," and "ist zehenmal hundert tausent."

[6]Thus, "duysent mael duysent" is used by such writers as Petri (1567, fol. 1), Raets (1580, fol. A₃, with "duysentich duysent"), Mots (1640, fol. B₂), Cardinael (1659 ed., fol. A₃), Willemsz (1708 ed., p. 5), and Bartjens (1792 ed., p. 8). There were exceptions, as when Wentsel (Wenceslaus, 1599) used both "millioenen" and "millions" (p. 2), Stockmans (1589; 1679 ed., p. 8) occasionally used "millioen," and Starcken (1714 ed., p. 2) used "million" rather apologetically.

England adopted the Italian word more readily than the other countries, probably owing to the influence of Recorde[1] (*c.* 1542). It is interesting to see that Poland was also among the first to recognize its value, the word appearing in the arithmetic of Klos in 1538.

The Billion. Until the World War of 1914–1918 taught the world to think in billions there was not much need for number names beyond millions. Numbers could be expressed in figures, and an astronomer could write a number like $9.15 \cdot 10^7$, or $2.5 \cdot 10^{20}$, without caring anything about the name. Because of this fact there was no uniformity in the use of the word "billion." It meant a thousand million (10^9) in the United States and a million million (10^{12}) in England, while France commonly used *milliard* for 10^9, with *billion* as an alternative term.

Historically the billion first appears as 10^{12}, as the English use the term. It is found in this sense in Chuquet's number scheme[2] (1484), and this scheme was used by De la Roche (1520), who simply copied parts of Chuquet's unpublished manuscript, but it was not common in France at this time, and it was not until the latter part of the 17th century that it found place in Germany.[3] Although Italy had been the first country to make use of the word "million," it was slow in adopting the word "billion." Even in the 1592 edition of Tartaglia's arithmetic the word does not appear. Cataldi (1602) was the first Italian writer of any prominence to use the term, but he sug-

---

[1] "203000000, that is, CCiii milliōs," "M. of millions," and "x.M. of millions" (1558 ed., fol. C$_8$).

[2] This plan is historically so important as to deserve being given in full. Chuquet gives the 6-figure periods, thus: 745324'8043000'700023'654321 (in which 8043000 should be 804300), and then says: "Ou qui veult le p̄mier point peult signiffier million Le second point byllion Le tiers poı̄t tryllion Le quart quadrillion Le cinq^e quyllion Le six^e sixlion Le sept^e septyllion Le huyt^e ottyllion Le neuf^e nonyllion et ansi des ault^{9s} se plus oultre on vouloit ⹀ceder ¶ Item lon doit sauoir que ung million vault mille milliers de unitez. et ung byllion vault mille milliers de millions. et tryllion vault mille milliers de byllions." From A. Marre's autograph copy of Chuquet. See also Boncompagni's *Bullettino*, Vol. XIII, p. 594.

[3] F. Unger, *Die Methodik der praktischen Arithmetik in historischer Entwickelung*, p. 71 (Leipzig, 1888), with the date of use of the word as 1681; hereafter referred to as Unger, *Die Methodik.*

gested it as a curiosity rather than a word of practical value.[1] About the same time the term appeared in Holland,[2] but it was not often recognized by writers there or elsewhere until the 18th century, and even then it was not used outside the schools. Even as good an arithmetician as Guido Grandi (1671–1742) preferred to speak of a million million rather than use the shorter term.[3]

The French use of *milliard*, for 10^9, with *billion* as an alternative, is relatively late. The word appears at least as early as the beginning of the 16th century as the equivalent both of 10^9 and of 10^{12}, the latter being the billion of England today.[4] By the 17th century, however, it was used in Holland[5] to mean 10^9, and no doubt it was about this time that the usage began to change in France.

As to the American usage, taking a billion to mean a thousand million and running the subsequent names by thousands, it should

---

[1] He generally used millions, thousand millions, million millions, and so on (p. 2); but he sometimes used *bilioni* for 10^9, although even then he preferred *duilioni*. His scheme of names is *millioni*, *bilioni* (or *duilioni*), *trilioni*, *quadrilioni* (or *quattrilioni*), *quintilioni*, for 10^6, 10^9, 10^{12}, 10^{15}, and 10^{18}. *Practica Aritmetica*, p. 5 (Bologna, 1602).

[2] Van der Schuere (1600) uses *millioen* (10^6), *bimillioen* (10^{12}), *trimillioen* (10^{18}), and *quadrimillioen* (10^{24}), but in a later edition (1634) of his arithmetic he gives *bimillion* and *billion* (10^9), *trimillion* and *trillion* (10^{12}), and so on to *nonemillion* and *nonilion*. Even as late as 1710 Leonhard Christoph Sturm (*Kurzer Begriff der gesamten Mathesis*, Frankfort a. d. Oder, 1710) used the words *trimillionen* and *bimillionen*.

[3] In his *Istituzioni di Aritmetica Practica*, p. 3 (Florence, 1740), he says: "millioni de' millioni (che possono dirsi Billioni) e li millioni di millioni di millioni (che si chiamano ancora Trillioni). . . . E così se fosse più lungo il numero, vi sarebbero ancora Quintillioni . . . Novillioni, ec. crescendosi ciascuno da ogni sei note."

[4] Thus, Trenchant (1566) uses *Miliars* (1578 ed., p. 14), and Peletier (1549) says: "Les François ont deux mots significatifs: l'un au septieme lieu, qui est Million, & l'autre au treizieme, qui est Milliart: c'est à dire, Million de Millions" (1607 ed., p. 15). Peletier states that the word was used by Budæus, and in the latter's *De Asse et partibus eius Libri quinq3* (1514; Paris edition of 1532, fol. 95, v.) the following appears: "hoc est denas myriadū myriadas, quod vno verbo nostrates abaci studiosi Milliartū appellāt, quasi millionū millione." In Boissière's arithmetic of 1554 there is a statement similar to the one in Peletier's work. In E. Develey, *Arithmétique d'Émile*, 2d ed., Paris, 1802, only "billion" is used for 1000 millions.

[5] "milliart/ofte duysent millioenen," as Houck's arithmetic (1676, p. 2) has it.

be said that this is due in part to French influence after the Revolutionary War, although our earliest native American arithmetic, the Greenwood book of 1729,[1] gave the billion as 10^9, the trillion as 10^{12}, and so on. Names for large numbers were the fashion in early days, Pike's well-known arithmetic (1788), for example, proceeding to duodecillions before taking up addition.

Writing Large Numbers. Although it is nearly a thousand years since our common numerals appeared in any European manuscripts now extant, we have not even yet decided on the method of writing large numbers.

Influenced by the crosses placed on the thousands' and millions' lines of the abacus (see page 181) to aid the eye, the medieval writers often placed a dot above the thousands and above every third place beyond, but sometimes they placed one or more dots below, and these customs also appear in the early printed books. Thus, we have such a form as $68\dot{5}4\dot{9}73$,[2] with the occasional variant of a dot over the units' figure also.[3] Recorde (*c.* 1542) gives the rule as follows:

Fyrst put a pricke ouer the fourth fygure, and so ouer the vij. And if you had so many ouer the x, xiij, xvj, and so forth, still leauing two fygurs betwene eche two pricks. And those roomes betwene the prickes are called Ternaries.[4]

Recorde also uses a bar (*virgula*) for separating the figures, saying:

[1] It was published anonymously, but, as is stated in the *Weekly News Letter* (Boston) of May 29, 1729, was written by Isaac Greenwood, sometime professor of mathematics at Harvard. As stated in Volume I, the first arithmetic printed in the New World appeared in Mexico in 1556; the first in what is now the United States was a reprint of Hodder's English arithmetic, Boston, 1719.

[2] Tonstall (1522, fol. C₁), Riese (1522; 1529 ed., p. 3), Rudolff (1526, fol. 3), Grammateus (1518; 1535 ed., p. 5), and many others. Widman (1489) recommends but does not use this plan: "Vnd setz vff ytlich tausent ain punct da by man mercken mag wie vil die letst figur mer tausent bedeut dann die vor ir" (1519 ed., fol. 5, *v.*).

[3] Thus Clavius (Italian edition of the *Epitome*, 1586, p. 14; Latin ed., 1583, p. 10). He recommends, however, the following: $4\overset{4}{2}3\overset{3}{2}0089\overset{2}{5}6\overset{1}{2}80\overset{0}{0}$.

[4] *Ground of Artes*, 1558 ed., fol. B₈. Similarly in Digges (1572; 1579 ed., p. 2), Baker (1568; 1580 ed., fol. 4), and Hodder (1672 ed.).

And some doo parte the nūbres with lynes after this forme
230|864|089|015|340, where you see as many lines as you made
pricks.[1]

Some writers used this symbolism in grouping by sixes.[2] Besides placing one dot above a figure, the medieval writers often
used such forms as 243756293842 1[3] and 2437562938421, and
these occasionally appear in the printed works. Fibonacci
(1202) used the arc, as in 678935784105296,[4] but this was not
a common form. A few of the other variations are given below :

7.538.275.136	Pellos (1492, fol. 4)
4.5.9.3.6.2.9.0.2.2 or ⎫	Reisch[5] (1503, Lib. IIII,
a c b a c b a c b a ⎬	
4593629022 ⎭	Tract. II, Cap. 4)
2578391062751234 6894352	Barozzi[6] (1585)
23.456.007.840.000.305.321	Santa-Cruz (1594, fol. 12, r.)
1,234,567 or 1.234.567	Greenwood (1729)
68|765|432|189|716|789|132	Blassière (1769)

The groups have been called by various names, such as
periods,[7] regions,[8] and ternaries,[9] and occasionally, as with
Trenchant (1566), there were four figures in the right-hand
group.[10]

Spanish Method of Writing Large Numbers. One of the most
interesting examples of the writing of large numbers found in
the books of the 16th century is seen in the work of Texeda,

---

[1] Similarly in Gemma Frisius (1540), Trenchant (1566), and various others.
[2] *E. g.*, the Dutch arithmetic of Wilkens, 1669, p. 8.
[3] As in a 14th century algorismus in the Columbia University Library. This
plan is also followed by Tartaglia (1556, I, fol. 7, r.).
[4] *Liber Abaci*, p. 1.
[5] These cases contain errors in printing in the first (1503) edition of the
Margarita phylosophica, but they are corrected in the later editions.
[6] Francesco Barozzi, *De Cosmographia Libri IV*, Venice, 1585.
[7] "Haec prima est periodus," etc., Ramus (1569; ed. Schoner, 1586, p. 2).
[8] Santa-Cruz (1594), fol. 12.
[9] Recorde, as quoted above.   [10] As in 10,500,340,8020 on page 16.

a Spanish writer of 1546. In seeking to explain algorism he writes numbers in the Spanish style (*en Castellano*) and also in algorism (*en guarismo*). The following cases are typical:

c. lx. U 462 q̃s . . ix U 621¹	160 U 462 q̃s 009 U 621
c. iij U. 75 q̃s c. ij U 300	103 U 075 q̃s 102 U 300
Dcccxcj Uccxxxiiij q̃s Dlx U.	891 U 234 q̃s 560/000
vij U . . . q̃s Dxlv Ucccclxijm̄	7 U 000 q̃s 545 U 462

It will be seen that in the Spanish forms, doubtless owing to the Arab influence, there is a tendency (not uniformly carried out) to use the dot for zero. Texeda also mixes his algoristic numerals with the Roman,—a custom not uncommon after the 12th century. The U stands for thousands, appearing in earlier times as ∪ with several variants, and being of uncertain origin. The q̃s stands for *quentos* (*cuentos*, millions). In the 16th century the Greek θ was also used instead of U, as in XXXVθCCCXXVI for 35,326, and 637θ500 for 637,500, and in the 18th century it often degenerated into a kind of inverted C.² In Portugal a symbol $ (*cifrão*) was used as early as the 16th century for the same purpose.

3. ADDITION

Terminology of Addition. The name of the operation which we call addition has had its vicissitudes. One writer of the 13th century, for example, used "aggregation" instead.³ Writ-

---

¹ In the original the ix is misprinted x.

The numbers at the left are *en castellano*; those at the right, *en guarismo*. The illustrations are from fols. iij, *v.*, to iiij, *v.*

The number is 160,462,009,621.

See F. Cajori, "Spanish and Portuguese symbols for 'Thousands,'" *Amer. Math. Month.*, XXIX, 201, who had not seen Texeda; he suggests that the U came from some variant of the Roman symbol for thousand.

² See the "Fragmentos del Archivo Particular de Antonio Perez, Secretario de Felipe II," *Revista de Archivos, Bibliotecas y Museos*, XXV (1920), 140 (Madrid, 1920). In the author's library are several Spanish manuscripts of c. 1725–1750 with the degenerate form of θ referred to in the text.

³ "Agregare est quoslibet duos numeros uel plures in unum colligere" (B. Boncompagni, *Trattati*, II, 30). We still preserve the phrase "in the aggregate." The word is merely the Latin for προστιθέναι (*prostithen'ai*), used by Euclid and Diophantus, or συντιθέναι (*syntithen'ai*), as used by Heron and Pappus.

ing about the year 1200, Fibonacci used "composition" and "collection" as well as "addition."[1] Nearly a century after Fibonacci the earliest French algorism (c. 1275) used "assemble"[2] for "add," and two centuries later the first printed arithmetic used "join."[3] In the early printed books the word "summation" was a rival of "addition,"[4] and we still speak of summing up, and of summing certain numbers. Addition being the operation most frequently used, the operation probably gave rise to the expression "to do a sum," meaning to solve a problem. Various other names for the process have been used,[5] but they have no special significance. With the English tendency to brevity, there is little prospect of change in this language in the words "add" and "addition."

In such of the early printed arithmetics as were intended for popular use there was ordinarily no word corresponding to our term "addend."[6] On the other hand, the theoretical books, generally printed in Latin, spoke of the *numeri addendi*, that is, the "numbers to be added,"[7] and from this came the word *addendi* alone, as used by Fine (1530),[8] Gemma Frisius (1540), and later writers. From this we have our English "addends." Those who seek for a change have occasionally used the less familiar "summands."

----

[1] In the Latin, *compositio, collectio,* and *additio.*

[2] "Se tu veus assambler .1. nombre a autre" (Boncompagni's *Bullettino,* XV, 53).   [3] *I.e., jongere.* Treviso arithmetic, 1478.

[4] "Addirn oder Summirn," in Rudolff's arithmetic of 1526 (1534 ed., fol. 3), Stifel's *Deutsche Arithmetica* (1545, fol. 1), Albert's arithmetic of 1534, and many others. Grammateus (1518) has "Additio oder Summierung," and Adam Riese (1522; 1550 ed.) has a chapter on "Addirn/Summirn/Zusamen legen," the last term derived from counter reckoning.

[5] *E.g.,* the German *Zusammenthuung,* the French *aiouster,* and the Italian *recogliere, summare,* and *acozzare.*

[6] *E.g.,* Recorde (c. 1542), Sfortunati (1534), Baker (1568), Digges (1572), Peletier (1549), Trenchant (1566), Pagani (1591), and Pacioli's *Sūma* (1494). The early Dutch arithmeticians rarely had such a word, and even the American Greenwood (1729) does not give one.

[7] As in Scheubel (1545, p. 13), Clavius (1583), Licht (1500, fol. 2), and many others, but curiously not in Tonstall (1522). There were also such terms as *termini addendi, numeri colligendi, numeri summandi,* and the like.

[8] Thus, he speaks of the *addendorum summa* as well as the *numeri addendi* (1555 ed., fol. 3).

The word "addend" was frequently used to refer only to the lower of two numbers to be added, as in the following case from the *Margarita phylosophica* (1503 ed.):

$$4'6'7'9' \text{ numerus cui debet fieri additio}$$
$$3'2'3'2' \text{ numerus addendus}$$
$$\overline{7\ 9\ i\ i} \text{ numerus ,pductus}$$

It was also used by many writers to refer to all the numbers to be added except the top one.[1]

The result obtained in addition has had a variety of names, although "sum" has been the favorite.[2] Next in order of popularity is "product," a term used for the result of any operation, but particularly in addition and multiplication. It was popular in Germany,[3] especially in early times,[4] and was also used in the Latin countries.[5]

Some of the Latin books of the 16th century also used *numerus collectus*, based upon the use of *collectio* for addition, and possibly we might now be using "collect" for "sum" if the Church had not appropriated the term.

The Operation of Addition. The operation of addition has not changed much since the Hindu-Arabic numerals began to be used. Even with the Roman numerals it was not a difficult process, and it is not probable that a Roman banker was com-

---

[1] George of Hungary (1499) calls only the lower of the two numbers the *numerus addendus*: "et numerus addendus, qui debet scribi in inferiori ordine" (Budapest reprint of 1894, p. 4). The same usage is found in an unpublished algorism of *c.* 1400 in the British Museum (Sl. 3281, fol. 4, *v.*).

[2] Thus, Chuquet (1484) uses *soɱe,* and similar forms appear in many early printed books, including those of Pacioli (1494), Fine (1530), Tonstall (1522), Klos (1538), Sfortunati (1534), and Riese (1522).

[3] Günther, *Math. Unterrichts,* p. 316.

[4] Joannes de Muris (*c.* 1350) says in his *Quadripartitum*: "Propositis namque numeris addicionis, supra figuras cuiuslibet numeri calculis situatis adde singulam singulis, arcubus obseruatis, et productum signa per calculos atque lege." *Abhandlungen,* V, 144. This is also interesting because it describes the use of counters on a Gerbert abacus.

[5] An interesting case sometimes occurs, as in Savonne's work of 1563, where "sum" is used for addend, and "product" for the result: "Adiouster est mettre plusieurs nombres ou sommes ensemble pour en sçauoir le produit."

pelled to resort to the abacus in ordinary addition. This will easily be seen by considering a case like the following:

$$\begin{array}{r} \text{DCCLXXVII} \\ \text{CC X VI} \\ \hline \text{DCCCCLXXXXIII} \end{array}$$

We might write this result CMXCIII, but a Roman would rarely if ever have done so. Even in the 16th century we find forms analogous to this, as in the work of Texeda (1546),[1] where we have the following parallel arrangement:

xxjUcxxvij	21U127
x vUccxviij	15U218
ijUcccliiij	2U454
jU.x.	1Uo10
xxxjUclxxxij	31U182
lxxUDccccxcj	70U991

By using their alphabetic numerals the Greeks were able to perform various operations without recourse to an abacus, although the work was somewhat more complicated than it is with our numerals.[2]

Hindu Method. Bhāskara (c. 1150) gives as the first problem in the *Lilāvati* the following:[3] "Dear intelligent Lilāvati, if thou be skilled in addition . . . , tell me the sum of two, five, thirty-two, a hundred and ninety-three, eighteen, ten, and a hundred, added together." In a commentary on this work, of unknown date, the following method is given:

Sum of the units, 2, 5, 2, 3, 8, 0, 0		20
Sum of the tens, 3, 9, 1, 1, 0		14
Sum of the hundreds, 1, 0, 0, 1		2
Sum of the sums, ...		360

[1] Fol. v, v.   [2] For details as to the Greek methods see Heath, *History*, I, 52.
[3] H. T. Colebrooke, *Algebra with Arithmetic and Mensuration from the Sanscrit*, p. 5 (London, 1817); hereafter referred to as Colebrooke, *loc. cit.*, or to special topics under the heads *Āryabhaṭa, Brahmagupta, Bhāskara, Vija Ganita*, with these spellings.

The Hindus seem generally to have written the sum below the addends, beginning with units' columns as we do. They had at one time another method, however, which they designated as inverse or retrograde, the operator beginning at the left and blotting out the numbers as they were corrected.[1]

$$
\begin{array}{r}
65391 \\
3279 \\
10420 \\
\hline
7\cancel{8}\cancel{9}\cancel{8}0 \\
909
\end{array}
$$

Arab Method and its Influence. The Arabs, on the other hand, often wrote the sum at the top, putting the figures of the check of casting out 9's at the side.[2] This plan was adopted by Maximus Planudes (*c.* 1340), the form used by him being here shown.[3]

8030	2
5687	8
2343	3

How the traces of the Oriental sand table, with its easily erased figures, and the traces of the old counter-reckoning, showed themselves in early English works is seen in the following passage in *The Crafte of Nombrynge*[4] (*c.* 1300):

lo an Ensampull of all

$$
\begin{array}{r}
326 \\
216
\end{array}
$$

Cast 6 to 6, & þere-of[5] wil arise twelue. do away þe hyer 6 & write þere 2, þat is þe digit of þis composit. And þen write þe articulle þat is ten ouer þe figuris hed of twene as þus

$$
\begin{array}{r}
1 \\
322 \\
216
\end{array}
$$

---

[1] This method is here indicated by canceling. The plan is one naturally adapted to the sand abacus. On the dispute as to whether the Hindus used this abacus, see Chapter III. See also C. I. Gerhardt, *Études historiques sur l'arithmétique de position*, Prog., p. 4 (Berlin, 1856) (hereafter referred to as Gerhardt, *Études*); J. Taylor, *Lilawati*, Introd., p. 7 (Bombay, 1816) (hereafter referred to as Taylor, *Lilawati*).

[2] H. Suter, "Das Rechenbuch des Abû Zakarîjâ al-Haṣṣar," *Bibl. Math.*, II (3), 15.

[3] Wäschke, *Planudes*, p. 6; Gerhardt, *Études*, p. 20. On such general early methods in the various operations see F. Woepcke, *Sur l'introduction de l'Arithmétique Indienne en Occident* (Rome, 1859).

[4] See pages 32 and 78.

[5] As stated on page 32, the old letter þ is our *th.*

Now cast þe articulle þat standus vpon þe figuris of twene hed to þe same figure, reken þat articul bot for one, and þan þere will arise thre. þan cast þat thre to þe neþer figure, þat is one, & þat wul be foure. do away þe figure of 3, and write þere a figure of foure. and let þe neþer figure stonde stil, & þan worch forth.

This is the oldest known satisfactory explanation of an example in addition in our language.[1]

Special Devices. In the way of special devices, Gemma Frisius (1540) gives one that is still used in adding long columns. It consists in adding each column separately, writing the several results, and then adding the partial sums as here shown. It will be observed that Gemma writes the largest number at the top, the object being to more easily place the various orders in their proper columns.[2]

In manuscripts of this period dots are sometimes used, as is the case today, to indicate the figure to be carried.

9279
389
479
27
22
9
9
10147

Carrying Process. The expression "to carry," as used in addition, is an old one and, although occasionally objected to by teachers, is likely to remain in use. It probably dates from the time when a counter was actually carried on the line abacus to the space or line above,[3] but it was not common in English works until the 17th century. Thus, we have Recorde (c. 1542) using "keepe in mynde," Baker (1568) saying "keepe the other in your minde," and Digges (1572) employing the same phraseology and also saying "keeping in memorie," and "keeping reposed in memorie." The later popularity of the word "carry" in English is largely due to Hodder (3d ed., 1664). In the 17th cen-

---

[1] On similar methods in the medieval manuscripts, not merely in addition but in the other operations, see L. C. Karpinski, "Two Twelfth Century Algorisms," *Isis*, III, 396.

[2] "Obseruandum igitur primò, vti maior numerus superiori loco scribatur, minores huic subscribantur" (1575 ed., fol. A 7).   [3] See Chapter III.

tury the expression "to carry" was often used in Italy.[1] Expressions like "retain," "keep in mind," and "hold" have, however, been quite as common.[2]

4. SUBTRACTION

Terminology of Subtraction. As with addition, so with subtraction, the name of the process and the names of the numbers used have varied greatly and are not settled even now. Outside the school the technical terms of arithmetic are seldom heard. When we hear a statement like "*Deduct* what I owe and pay me the *rest*," we hear two old and long-used terms instead of the less satisfactory words "subtract" and "difference."

Terms Meaning Subtract. While the word "subtract," meaning to draw away from under,[3] has been the favorite term by which to indicate the operation, it has by no means enjoyed a monopoly. When Fibonacci (1202), for example, wishes to say "I subtract," he uses some of the various words meaning "I take."[4] Instead of saying "to subtract" he says "to extract,"[5] and hence he speaks of "extraction."[6] These terms, as also "detract,"[7] which Cardan[8] (1539) used, are etymologically rather better than ours. "Subduction"[9] has also been used for "subtraction," both in Latin[10] and in English. Digges (1572),

---

[1] As in "summa senza portare," "portare decine," and the like. See, for example, the arithmetic of G. M. Figatelli Centese, fol. 21 (Bologna, 1664).

[2] *E.g.*, "die ander behalt" (Riese, 1522; 1533 ed.), "behalt die ander in sinn, welche ist zu geben der nechsten" (Grammateus, 1518), "et alterā mente reconda" (Clichtoveus, 1503; *c.* 1507 ed., fol. D2), "& secunda reservanda" (Ramus, ed. Schoner, 1586, p. 6), "ie . . . retien le nombre de diszeines" (Trenchant, 1566; 1578 ed., p. 24).

[3] *Sub* (under) + *trahere* (whence *tractum*) (to draw).

[4] *Tollo, aufero,* or *accipio.*   [5] *Extrahere* (to draw out or take away from).

[6] *Extra*[*c*]*tio.*   [8] *Practica,* 1539, capp. 7–14.

[7] *Detrahere* (to draw or take from).   [9] *Sub* (under) + *ducere* (to lead).

[10] *E.g.*, Tonstall (1522) devotes fifteen pages to *Subductio.* He also says: "Hanc autem eandem, uel deductionem uel subtractionem appellare Latine licet" (1538 ed., p. 23; 1522 ed., fol. E 2, *r.*). See also Ramus, *Libri duo,* 1569, 1580 ed., p. 3; *Schol. Math.,* 1569 ed., p. 115. Schoner, in his notes on Ramus (1586 ed., p. 8), uses both *subduco* and *tollo* for "I subtract." Gemma Frisius (1540) has a chapter *De Subductione siue Subtractione,* and Clavius (1585 ed., p. 26) says: "Subtractio est . . . subductio." In his arithmetic Boethius uses *subtrahere,* but in the geometry attributed to him he prefers *subducere.*

for example, says: "To subduce or subtray any sūme, is wittily to pull a lesse frō a bigger nūber." Our common expressions "to diminish" and "to deduct" have also had place in standard works, as in the translation of the *Liber algorismi*[1] and in the work of Hylles (1592).[2] Recorde (*c.* 1542) used "rebate" as a synonym for "subtract," and the word is used today in commercial matters in a somewhat similar sense.

In a manuscript written by Christian of Prag[3] (*c.* 1400) the word "subtraction" is at first limited to cases in which there is no "borrowing." Cases in which "borrowing" occurs he puts under the title *cautela* (caution), and gives this caption the same prominence as *subtractio*.[4]

The word "subtract" has itself had an interesting history. The Latin *sub* appears in French as *sub, soub, sou,* and *sous, subtrahere* becoming *soustraire* and *subtractio* becoming *soustraction*.[5] Partly because of this French usage, and partly no doubt for euphony, as in the case of "abstract," there crept into the Latin works of the Middle Ages, and particularly into the books printed in Paris early in the 16th century, the form *substractio*.[6] From France the usage spread to Holland[7] and England, and from each of these countries it came to America. Until the beginning of the 19th century "substract" was a common form in England and America,[8] and among those brought up in somewhat illiterate surroundings it is still to be found.

---

[1] Which uses both *diminuere* and *subtrahere*. See Boncompagni, *Trattati*, II, 32.
[2] He uses "abate," "subtract," "deduct," and "take away."   [3] See page 77.
[4] The passage begins: "*Cautela* . . . si figura inferioris ordinis non poterit subtrahi a sibi supraposita."
[5] With such variants as *soubstraction, soubstraire,* and the like.
[6] It appears in the *Geometria* of Gerbert, but the MSS. used are of *c.* 1200; ed. Olleris, p. 430. As to the early printed books, Clichtoveus (1503), for example, generally uses *substractio*, although *subtractio* is occasionally found. See also his edition of Boethius, and see the 1510 edition of Sacrobosco. The word also appears in the work of George of Hungary (1499), along with *subtractio*, so that the usage was unsettled.
[7] Thus Wentsel (1599), Van der Schuere (1600), Mots (1640), and, indeed, nearly all Dutch writers before 1800. Petri (1567; 1635 ed.), however, uses *subtractio* in the Latin form and *subtraheert* in the Dutch, and Adriaen Metius (1633; 1635 ed.) also omits the *s*.
[8] Our American Greenwood (1729), for example, always used "substract" and "substraction," but dropped the *s* in "subtrahend."

The incorrect form was never common in Germany,[1] probably because of the Teutonic exclusion of international terms.

Minuend and Subtrahend. The terms "minuend" and "subtrahend," still in use in elementary schools, are abbreviations of the Latin *numerus minuendus* (number to be diminished) and *numerus subtrahendus* (number to be subtracted).[3] The early manuscripts and printed books made no use of our abridged terms. The minuend and subtrahend were called the higher and the lower numbers respectively, as in *The Crafte of Nombrynge* (*c.* 1300), the upper and under numbers,[4] the number from which we subtract and the subduced,[5] the total and less,[6] the total and abatement, and the total and deduction.[7] Among the most popular terms have been "debt" and "payment,"[8] but better still are the terms "greater" and "less."[9]

---

[1] As witness Köbel (1514 ; 1549 ed., fol. 110), Stifel (*Arithmetica Integra*, 1544, fol. 2), Albert (1534; 1561 ed.), Thierfelder (1587, p. 11), and many others.

[2] Their early writers used such forms as *abzihung* and *abzyhung*, instead of "subtraction," just as the Dutch used such terms as *Af-trekkinge* (Van der Schuere, 1600; 1624 ed., fol. 10). While the Italians used *abattere* and *cavare*, they also used *sottrare* and *trarre* (as in Cataneo, 1546; 1567 ed., fol. 5).

[3] See Boncompagni, *Trattati*, II, 33, on Johannes Hispalensis (*c.* 1140) and his use of *numerus minuendus*.

[4] "Die vnder zal sol nit übertreffen die öbern" (Grammateus, 1518). Tonstall (1522) and other Latin writers have *numerus superior* and *numerus inferior*; the Italian edition of Clavius (1586) has *numero superiore* and *numero inferiore*.

[5] "Numerus ex quo subducitur" and "subducendus" (Gemma Frisius, 1540; 1563 ed., fol. 9).

[6] *Totalis*, *minor*, used by Tzwivel (1505), Clichtoveus (1503), and others.
[7] Hylles, 1600, fol. 19.
[8] Thus, the Dutch-French work of Wentsel (1599, p. 4) has:

Schult/Debte.	£. 15846
Betaelt/paye,	£. 5424
Reste	£. 10422

The Dutch names in the 16th and 17th centuries were generally *de Schult* and *de Betaelinghe*.

Similarly, we have the Italian *debito*, *pagato*, and *residuo* (as in the 1515 edition of Ortega), the French *dette* and *paye*, as well as *la superieure & inferieure* (Trenchant, 1566; 1578 ed., p. 30), and the Spanish *recibo* and *gasto* (Santa-Cruz, 1594; 1643 ed., fol. 20).

[9] Sfortunati (1534; 1544 ed., fol. 8), *il numero maggiore* and *il numero minore*; G. B. di S. Francesco (1689), *quantità maggiore* and *quantità minore*; Raets (1580), *Het meeste ghetal* and *Het minste ghetal*, with similar forms in other languages. See also Tartaglia, 1592 ed., fol. 9.

Name for Difference. The words "difference" and "remainder" have never been popular, in spite of the fact that they are commonly found in the textbooks of today. The popular term has been "rest," and in common parlance this is still the case, as when we say "Give me the rest," "Take the rest." It appears in the first printed arithmetic[1] and is found generally in the works of the Latin countries. Indeed, the verb "to rest" was not infrequently used to mean subtract.[2] In England, Tonstall (1522), writing in Latin, used sometimes *reliqua*[3] and sometimes an expression like "the number sought." Recorde (*c.* 1542) introduced "remayner" or "remainer," a term which Hylles (1592) also used, together with "remaynder," "remaynes," and "rest." The Latin writers commonly used *numerus residuus*,[4] *differentia, excessus*,[5] and *reliqua*. Of these terms we have relics in our language in the forms of "difference" and "excess," and another term commonly used by us is "balance."

An interesting illustration of the use of expressions which later resulted in technical terms is seen in the following from the *Margarita phylosophica* (1503 ed.):

$$9\,00\,i\,386 \quad \text{numerus a quo debet fieri subtractio}$$
$$\underline{7532436} \quad \text{numerus subtrahendus}$$
$$i468950 \quad \text{numerus relictus}$$

The Operation of Subtraction. The process of subtraction, unlike the processes of addition and multiplication, has never been standardized. There are four or five methods in common use today, the relative advantage of any one over the others not being decided enough to give it the precedence. A brief history of a few of the more prominent methods will be given.

---

[1] Treviso, 1478, p. 18.

[2] Thus, the Spanish writer Santa-Cruz (1594) uses *restar*, and Ortega (1512; 1515 ed.) begins a chapter *per sapere restare o / subtrahere*.

[3] Various other writers did the same. Thus, Glareanus (1538) has *relictum* and *reliquum*. Fibonacci (1202) used *residuum* and *reliquus*.

[4] *E.g.*, Fine (1530) and occasionally Clavius (1583). An unpublished algorism of *c.* 1400, now in the British Museum (Sl. 3281, fol. 4, *v.*), uses *a q″ sbtrahi, subtrahēd⁹*, and *residuū* for the three terms.

[5] Clavius speaks of *differentia siue excessus*. 1585 ed., p. 133.

1. The complementary plan is based upon the identity

$$a - b = a + (10 - b) - 10.$$

In particular, to find $13 - 8$ we may substitute the simpler process $13 + 2$ and then subtract 10. This plan is today used in the case of cologarithms and on certain types of calculating machines. It is not a modern device, however. Bhāskara (c. 1150) used it in the *Lilāvati*,[1] and no doubt it was even then an old one. It appears in *The Crafte of Nombrynge* (c. 1300), and the difficulty of the operation is apparent from the following extract:

lo an Ensampul.

```
2122
1134
```

take 4 out of 2. it wyl not be, þerfore borro one of þe next figure, þat is 2. and set þat ouer þe hed of þe fyrst 2. & releue it for ten. and þere[2] þe secunde stondes write 1. for þou tokest on[3] out of hym. þan take þe neþer figure, þat is 4, out of ten. And þen leues 6. cast[4] to 6 þe figure of þat 2 þat stode vnder þe hedde of 1. þat was borwed & rekened for 10, and þat wylle be 8. do away þat 6 & þat 2, & sette þere 8, & lette þe neþer figure stonde stille,

and so on with equal prolixity. The expression to "borro," used in this work, was already old. It was afterwards used by Maximus Planudes (c. 1340), acquired good standing in the works of Recorde (c. 1542) and Baker (1568), and has never lost its popularity.

The same method appears in the Treviso arithmetic[5] (1478),

---

[1] Taylor, *Lilawati*, Introd., p. 7.   [2] For "where."   [3] One.
[4] *I.e.*, add; a relic of the abacus. Compare our expression "cast accounts."
[5] The author adds 2 to 2, the result being 4, and then adds 1 to the next figure of the subtrahend, saying:

"al .4. tu die iongere .i. e levera .5. poi dira .5. da .5. che equale da equale: resta .o." (Treviso arithmetic, p. [19]). The *i* is used for 1, as on page 97 and as is the *j* in the following problem from Huswirt.

```
      452
      348
Lo resto 104
```

and Huswirt (1501) solved his first problem in subtraction by this means, saying:

5 from 4 I cannot. I take the distance[1] of the lower number, that is, 5 from 10, or 5, and this I add to the upper number, 4, and obtain 9, which I write directly under the bar and below the 5. I carry the j in mind or on the tablet,[2] first canceling the 4 and 5, and add it to the next number, that is, to 9. . . .

```
59j0j0j0j4
400j999j95
j908j0j8j9
```

Among other authors of early printed books who favored the plan there were such writers as Petzensteiner (1483), Pellos (1492), Ortega (1512), Fine (1530), Gemma Frisius (1540), Ramus (1555), Albert (1534), Baker (1568), and Digges (1572). Savonne (1563) also used it and indicated the borrowing of ten by means of a dot, as shown in the annexed example from his arithmetic.

The early American arithmeticians looked with some favor on the plan. Thus, Pike[3] says:

```
6·4·0·6·0
2 7 8 7 4
3 6 1 8 6
```

If the lower figure be greater than the upper, borrow ten and subtract the lower figure therefrom: To this difference add the upper figure.

2. The borrowing and repaying plan, in which the 1 that is borrowed is added to the next figure of the lower number, is one of the most rapid of the methods in use today and has for a long time been one of the most popular. It appears in Borghi's (1484) well-known work, the first great commercial arithmetic to be printed. Borghi takes the annexed example and says, in substance: "8 from 14, 6; 8 from 15, 7; 10 from 13, 3; 3 from 6, 3." The plan was already old in Europe, however. Fibonacci[4] (1202) used it, and so did Maximus Planudes (c. 1340). These writers seem to have inherited it from the Eastern Arabs, as did the Western Arab writer al-Qalasâdî

```
6354
2978
3376
```

---

[1] *Distantia*, for the complement.
[2] Very likely the wax tablet, still used in Germany at that time. See Chapter III.   [3] 1788; 1816 ed., p. 12.   [4] *Liber Abaci*, Boncompagni ed., I, 22.

(*c.* 1475). The arrangement of figures used by Maximus Planudes in the subtraction of 35843 from 54612 is here shown, the remainder being placed above the larger number, after the Arab and Hindu[1] custom. The top line was used only in the checking process. This method of borrowing and repaying was justly looked upon as one of the best plans by most of the 15th and 16th century writers, and we have none that is distinctly superior to it even at the present time.

$$\begin{array}{c} 54612 \\ 18769 \\ \hline 54612 \\ 35843 \\ \hline 1111 \end{array}$$

3. The plan of simple borrowing is the one in which the computer says: "7 from 12, 5; 2 from 3 (instead of 3 from 4), 1." This method is also very old. It appears in the writings of Rabbi ben Ezra[2] (*c.* 1140), the computer being advised to begin at the left and to look ahead to take care of the borrowing. This left-to-right feature is Oriental[3] and was in use in India a century ago.[4] It was the better plan when the sand table allowed for the easy erasure of figures, but it had few advocates in Europe.[5]

$$\begin{array}{c} 42 \\ 27 \\ \hline 15 \end{array}$$

When the computation began at the right, the borrowing plan was also advocated by such writers as Gernardus[6] (13th century?), Sacrobosco[7] (*c.* 1250), and Maximus Planudes[8] (*c.* 1340). The writers of the early printed arithmetics[9] were

---

[1] Taylor, *Lilawati*, Introd., p. 7.

[2] *Sefer ha-Mispar*, ed. Silberberg, p. 29 (Frankfort a. M., 1895); hereafter referred to as Silberberg, *Sefer ha-Mispar*.

[3] It is found in the works of al-Khowârizmî (*c.* 825), Behâ Eddîn (*c.* 1600), Albanna (*c.* 1300), and others. See H. Suter, *Bibl. Math.*, II (3), 15.

[4] See Taylor, *Lilawati*, Introduction.

[5] One of these was Ramus, who advocates "subductio fit á sinestra dextrorsum" (*Arith. Libri duo*, 1569; 1580 ed., p. 4; 1586 ed., p. 8).

[6] *Algorithmus demonstratus*, I, cap. ix. Formerly attributed to Jordanus Nemorarius (*c.* 1225). See G. Eneström, *Bibl. Math.*, XIII (3), 289, 292, 331.

[7] See J. O. Halliwell, *Rara Mathematica* (London, 1838–1839), 2d ed., 1841, p. 7; hereafter referred to as Halliwell, *Rara Math.*

[8] With one or two other methods.

[9] *E.g.*, such writers as Tzwivel (1505), Clichtoveus (1510 edition of his Boethius, fol. 39), Köbel (1514; 1549 ed., fol. 120), Stifel (1544), Ghaligai (1521), Raets (1580), and Clavius (1583). Some of the more pretentious writers, like Pacioli (1494) and Tartaglia (1556), gave all three methods.

not unfavorable to it, although they in general preferred the borrowing and repaying method.

4. The addition method, familiar in "making change," is possibly the most rapid method if taught from the first. To subtract 87 from 243 the computer says: "7 and 6 are 13; 9 and 5 are 14; 1 and 1 are 2"; or else he says: "7 and 6 are 13; 8 and 5 are 13; 0 and 1 are 1," the former being the better. The method was suggested by Buteo (1559) and probably by various other early writers, but it never found much favor among arithmeticians until the 19th century. It has been called the Austrian Method, because it was brought to the attention of German writers by Kuckuck (1874), who learned of it through the Austrian arithmetics of Mocnik (1848) and Josef Salomon (1849).

$$\begin{array}{r} 243 \\ 87 \\ \hline 156 \end{array}$$

5. MULTIPLICATION

General Idea of Multiplication. The development of the idea of multiplication and of the process itself is naturally more interesting than the evolution of the more primitive and less intellectual processes already described. Just as addition is a device for obtaining results that could be reached by the more laborious method of counting, so multiplication was developed as an abridgment of addition.[1] It was simply a folding together of many equal addends. This is expressed not merely in the Latin name[2] but in the corresponding names in various other

---

[1] Attention was called to this fact by various 16th century writers. Thus Ramus (1569) remarks: "Multiplicatio est qua multiplicandus toties additur, quoties unitas in multiplicante continetur, & habetur factus." Schoner, in his commentary, adds: "Ideoq3 multiplicatio est additio, sed ejusdem numeri secum, no diuersorum" (1586 edition of the *Libri duo*, p. 12). Even as early as *c.* 1341 Rhabdas mentioned the same fact. See P. Tannery, *Notices et extraits des manuscrits de la Bibl. nat.*, XXXII, 155.

[2] From *multus* (many) + *plicare* (to fold); compare also our word "manifold." The term is simply the Latin form of the Greek πολυπλασιάζειν (*polyplasia'zein*), as used by Euclid, Pappus, and Diophantus, or πολλαπλασιάζειν (*pollaplasia'zein*), as used by Heron and Pappus, the latter using both forms. Such words as "three-ply" and "four-ply" illustrate this use of *plicare*.

languages.[1] The Latin writers of the Middle Ages and the Renaissance speak of leading a number into this multiplicity,[2] which explains our use of the expression "*a* into *b*," still retained in algebra but discarded in arithmetic.

Definition of Multiplication. The definition of multiplication has often disturbed teachers of arithmetic because of their failure to recognize the evolution of such terms. It gave no trouble in the ancient *arithmetica,* for the numbers there involved, in speaking of such a process, were positive integers; whereas in the ancient *logistica* no attention, so far as we know, was paid to any definitions whatever. When, however, the notion of the necessity of exact definition entered the elementary school, teachers were naturally at a loss in adjusting the ancient limitations to the multiplication by a fraction or an irrational number, and by such later forms as a negative or a complex number.

One of the best of the elementary definitions referring to integers, and at the same time one of the oldest in our language, is found in *The Crafte of Nombrynge* (*c.* 1300): "multiplicacion is a bryngynge to-geder of 2 thynges in on nombur, þe quych on nombur contynes so mony tymes on, howe mony tymes þere ben vnytees in þe nowmbre of þat 2."[3] The same definition is found in the arithmetic of Maximus Planudes (*c.* 1340),[4] in the first printed arithmetic (1478),[5] in the first noteworthy com-

---

[1] Compare the German works of the 15th and 16th centuries, with their *mannigfaltigen* and *vervielfachen*. Grammateus (1518) speaks of "Multiplicatio oder Merung."

[2] "¶ Si aliquis numerus . . . ducatur," as Jordanus Nemorarius (*c.* 1225) says (1496 edition of the arithmetic, fol. C 3, *et passim*). Similarly Clichtoveus (1503), "Duco .4. in .3. et fit .jz."; Gemma Frisius (1540), "Mvltiplicare, est ex ductu vnius numeri in alterum numerum producere, qui toties habeat in se multiplicatum, quoties multiplicãs vnitatem"; and many others. In the Latin edition of his arithmetic (1583; 1585 ed., p. 36) Clavius has "Multiplicatio est ductus vnius numeri in alium . . . Vt numerus 6. in numerum 5. . . . duci dicitur . . . ," but in the Italian edition (1586, p. 35) he uses *per* for *in*, thus: "Moltiplicare vn numero per vn' altro."

[3] R. Steele's proof-sheet edition, p. 21 (London, 1894).

[4] Wäschke, *Planudes*, p. 13.

[5] "Che moltiplicare vno nũero per si ouero per vno altro: non e altro: che de do nnmeri ,ppositi: trouere vno terzo numero: el quale tante volte contien vno de quelli numeri: quante vnitade sono nel altro. Exempio .2. fia .4. fa .8. ecco che .8. cõtiẽ in se tante .4. quante vnitade sono nel .2." Treviso arithmetic, p. [27].

mercial arithmetic (1484),[1] and in numerous other works.[2]
Recorde (c. 1542) set the English standard by saying, "Multiplication is such an operaciō that by ij sumes producyth the
thyrde, whiche thyrde sūme so manye times shall cōtaine the
fyrst, as there are vnites in the second."[3]

A somewhat more refined definition, including the notion of
ratio, was necessary for fractional multipliers, and this appeared
occasionally in the early printed books, as in Huswirt (1501).[4]
Its use in English is largely due to the influence of Cocker's
popular arithmetic (1677), where it appears in these words:
"Multiplication is performed by two numbers of like kind, for
the production of a third, which shall have such reason [ratio]
to the one, as the other hath to unite." The idea is Oriental,[5]
appearing in various Arab and Russian works.[6]

The elementary teacher generally objects to such a form as
2 ft. × 3 ft. = 6 sq. ft., and on the ground of pedagogical theory
there is some reason for so doing, but not on logical or historical
grounds. With respect to logic, it all depends on how multiplication is defined; while with respect to history there is abundant
sanction for the form in the works of early and contemporary
writers. For example, Savasorda[7] (c. 1120) and Plato of Tivoli
(1116)[8] broaden the definition in such a way as to allow a line
to serve as a multiplier, and Baker (1568) remarks, "If you wil
multiply any number by shillinges and pence," an expression
commonly paralleled by children today. Few of our contemporary physicists would see anything to criticize in such an expression as 6 ft. × 10 lb. = 60 foot-pounds, and in due time such
forms will receive more recognition in elementary arithmetic.

[1]Borghi, 1540 ed., fol. 6.
[2]E. g., Tonstall, 1522, fol. G 1; Stifel, *Arithmetica Integra*, 1544, fol. 2; Sfortunati, 1534, 1544/5 ed., fol. 11; Tartaglia, 1556; Trenchant, 1566.
[3]1558 ed., fol. G1. Digges (1572; 1579 ed., p. 4) gives the same form.
[4]"Multiplicatio est numeri procreatio, proportionabiliter se habentis ad mul-
tiplicandū sicut multiplicans ad vnitatem se habet" (fol. 3).
[5]See Taylor, *Lilawati*, Introd., p. 15.
[6]E. g., Behâ Eddîn (c. 1600). It is still used in Russian textbooks on arithmetic.
[7]Abraham bar Chiia.
[8]In his translation of Savasorda: "Et primum quidem exponemus, quid significare velimus, cum dicimus: multiplicatio lineae in se ipsam."

Terminology of Multiplication. Of the terms employed, "multiplicand" is merely a contraction of *numerus multiplicandus*. In *The Crafte of Nombrynge* (*c.* 1300) it is explained as "Numerus multiplicandus, Anglice þe nombur þe quych to be multiplied." In most of the early printed Latin books it appears in the full form,[1] but occasionally the *numerus* was dropped, leaving only *multiplicandus*,[2] and this led the non-Latin writers to use the single term.[3] A few of the Latin writers suggested *multiplicatus*,[4] so that we had at one time a fair chance of adopting "multiplicate." In their vernacular, however, many writers tended to use no technical term at all, simply speaking of the number multiplied, as the Latin writers had done,[5] and to this custom we might profitably return. It is hardly probable that such terms as subtrahend, minuend, and multiplicand, signifying little to the youthful intelligence, can endure much longer.

The word "multiplier" has had a more varied career. *The Crafte of Nombrynge* (*c.* 1300) speaks of "numerus multiplicans. Anglice, þe nombur multipliynge," the former being the Latin name for "multiplying number." Since the word *numerus* was frequently dropped[6] by Latin writers, in the translations the technical term appeared as a single word, with such

---

[1] *E.g.*, Clichtoveus (1510 edition of Boethius, fol. 35), Tonstall (1522), Grammateus (1518), Scheubel (1545, I, cap. 4).

[2] *E.g.*, Pacioli (1494, fol. 26), Licht (1500, fol. 6), Huswirt (1501), Ciruelo (1495), Glareanus (1538), Fine (1530).

[3] Thus Trenchant used *multiplicande* (1566; 1578 ed., p. 35). On the Italian writers see B. Boncompagni, *Atti d. Accademia Pontificia di Nuovi Lincei*, XVI, 520; hereafter referred to as *Atti Pontif.*

[4] *E.g.*, Tzwivel (1505) used "nūerus multiplicādus siue multiplicatus," and Gemma Frisius (1540) used *multiplicandus* and *multiplicatus* interchangeably. So the Treviso arithmetic (1478) says: "Intendi bene. che ne la moltiplicatione sono prīcipalmente do numeri necessarii. zoe el nūero moltiplicatore: et el nūero de fir moltiplicato" (p. [27]), which is not quite the same usage.

[5] Thus Chuquet (1484) simply speaks of "le nombre multiplie," and similarly with Borghi (1484). Riese (1522; 1529 ed., p. 8), Sfortunati (1534; 1544/5 ed., fol. 11), and others. Digges (1572) speaks of it as "the other summe, or number to be multiplied."

[6] Thus Johannes Hispalensis (*c.* 1140); see Boncompagni, *Trattati*, II, 41. Clichtoveus (1510 edition of Boethius) uses *multiplicans* both as an adjective and as a noun. So also Huswirt (1501, fol. 3), Ciruelo (1495; 1513 ed., fol. A 6), Grammateus (1518; 1523 ed., fol. A 4), Gemma Frisius (1540), and many others.

variants as "multiplicans,"[1] "moltiplicante,"[2] "multiplicator,"[3] "multipliant,"[4] and "multiplier."[5]

The word "product" might with almost equal propriety be applied to the result of any other arithmetic operation as well as to multiplication. It means simply a result,[6] but it has some slightly stronger connection with multiplication on account of the use of the verb *ducere* in the late Latin texts.[7] It has, however, been used in the other operations by many writers, and its special application to the result of multiplication is comparatively recent. The tendency to simplify the language of the elementary school will naturally lead to employing some such term as "result" for the various operations.

The authors of the early printed books often took the sensible plan of having no special name for the result in multiplication. Certain of them used "sum"[8] or "sum produced,"[9] while *factus*, a natural term where *factor* is employed, had its advocates.[10] Finally, however, the *numerus* was dropped[11] from *numerus productus*, and "product" remained.[12]

[1] Pacioli, 1494, fol. 26.   [2] Cataldi, 1602, p. 21.

[3] *Multiplicador* (Pellos, 1492, fol. 8), *multiplicatore* (Ortega, 1512; 1515 ea., fol. 16), *multiplicatour* (Baker, 1568; 1580 ed., fol. 16). The word was common in English. Greenwood used it, with "multiplier," in the 1729 American arithmetic.

[4] Peletier, 1549; 1607 ed., p. 34.

[5] "The lesse is named the Multiplicator or Multiplyer." Digges, 1572; 1579 ed., p. 5.

[6] Latin *producere* (to lead forth); whence *productum* (that which is led forth).

[7] See pages 101, 102.

[8] *E.g.*, Pacioli (1494, fol. 26), Ortega (1512; 1515 ed., fol. 18), and Recorde (*c.* 1542; 1558 ed., fol. G 2). Fine (1530) uses it as well as *numerus productus*.

[9] Thus Hodder (10th ed., 1672, p. 25) speaks of "The Product, or sum produced." Similarly, Clichtoveus (1503) uses both *numerus productus* and *tota summa*, and Glareanus (1538) uses *summa producta*.

[10] Thus, Fibonacci (1202) uses "factus ex multiplicatione." He also speaks of the "contemptum sub duobus numeris." Ramus (1569) speaks of the *factus* in multiplication, and in his treatment of proportion he says: "Factus à medio aequat factum ab extremis."

[11] As in Licht (1500), Huswirt (1501), Gemma Frisius (1540), and Scheubel (1545).

[12] *Produit*, Trenchant (1566); *produtto*, Sfortunati (1534) and Cataldi (1602), or *prodotto* by later Italian writers. Unlike most of the Latin terms it found place in the early Teutonic vocabulary, as seen in Werner (1561) and such Dutch writers as Petri (1567), Raets (1580), and Coutereels (1599).

The Process of Multiplication. We know but little about the methods of multiplication used by the ancients. The Egyptians probably made some use of the duplation plan, 17 being multiplied by 15 as shown in the annexed scheme.[1] It is also probable that this plan was followed by other ancient peoples and by their successors for many generations, which accounts for the presence of the chapter on duplation in so many books of the Renaissance period. Indeed, even as good a mathematician as Stifel multiplied 42 by 31 by successive duplation, substantially as here shown.[2]

1	17
2	34
4	68
8	136
16	272
1	17
15	255

There is also a contemporary example of the use of duplation and mediation, found among the Russian peasants today. To multiply 49 by 28 they proceed to double 28 and to halve 49, thus:

$1 \cdot 42 =$	42
$2 \cdot 42 =$	84
$4 \cdot 42 =$	168
$8 \cdot 42 =$	336
$16 \cdot 42 =$	672
$31 \cdot 42 =$	1302

49	24	12	6	3	1
28	56	112	224	448	896

The fractions are neglected each time, and finally the figures in the lower row which stand under odd numbers are added, thus:

$$28 + 448 + 896 = 1372,$$

and this is the product of 28×49.[3]

Greek and Roman Methods. The abacus[4] was probably used so generally in ancient times that we need hardly speculate on the methods of multiplication used by the Greeks and Romans. It is quite possible, however, that the Greeks multiplied upon

---

[1] P. Tannery, *Notices et extraits des manuscrits de la Bibl. nat.*, XXXII, 125; *Pour l'histoire de la science hellène*, p. 82 (Paris, 1887); Heath, *History*, I, 52.

[2] See his *Rechenbuch* (1546), p. 12. He uses a similar process for division.

[3] In the above case we have

$\frac{49}{1}$	$\frac{49}{2}$	$\frac{49}{4}$	$\frac{49}{8}$	$\frac{49}{16}$	$\frac{49}{32}$
49	24	12	6	3	1
$1 \cdot 28$	$2 \cdot 28$	$4 \cdot 28$	$8 \cdot 28$	$16 \cdot 28$	$32 \cdot 28$

Here $49 \cdot 28 = (32 + 16 + 1) \cdot 28 = 896 + 448 + 28 = 1372$.

[4] See Chapter III.

their wax tablets about as we multiply, but beginning with their highest order;[1] and there is no good reason why the Romans should not have done very nearly the same. Indeed, in Texeda's arithmetic (1546) the "Spanish method" with Roman numerals is given side by side with the new method of "Guarism," thus:

ccclxxv————vijUD.vj	ƀɇƀ————7U506
ijqsccljUDccc..	2251U800
DxxvUccccxx.	525U420
xxxvijUDxxx.	37U530
ijqsDcccxiiijUDccl.	2q̊s814U750

It should be observed, however, that the small tradesman has never had much need for this kind of work.[2]

Pacioli's Eight Plans. Our first real interest in the methods of multiplication starts with Bhāskara's *Lilāvati* (c. 1150), although we have a few earlier sources. Bhāskara gives five plans and his commentators add two more. These plans had increased to eight when Pacioli published his *Sūma* (1494), and these will now be considered.

Our Common Method of Multiplying. Our common form was called by Pacioli "Multiplicatio bericocoli vel scachierij," and appears in his treatise (fol. 26, *r.*) in the following form:

Multiplicandus.
Producentes.   9 8 7 6
Multiplicans.   6 7 8 9

8	8	8	8	4

7	9	0	0	8

schachieri
Bericuocolo

summa   6 7 0 4 8 1 6 4

[1] P. Tannery, *Notices et extraits des manuscrits de la Bibl. nat.*, XXXII, 126; Heath, *History*, I, 54.
[2] An interesting witness to this fact is the first Bulgarian arithmetic (1833), described in *L'Enseignement Mathématique* (1905), p. 257.

He says that the Venetians called this the method "per scachieri" because of its resemblance to a chessboard,[1] while the Florentines called it "per bericuocolo" because it looked like the cakes called by this name and sold in the fairs of Tuscany.[2] In Verona it was called "per organetto,"[3] because of the resemblance of the lines to those of a pipe organ, and "a scaletta" was sometimes used because of the "little stairs" in the figure, as seen on page 107.[4]

```
 135
  12
 ───
  12
  36
  60
────
1620
```

```
 135
  12
 ───
 135
 270
────
1620
```

This method is not found directly in the *Lilāvati*, but two somewhat similar ones are given. In the first of these[5] the multiplier is treated as a one-figure number and the work begins at the left, as shown above; in the second,[6] which is shown in the above computation at the right, the multiplier is separated as with us, but the work begins at the left as in the preceding case.

---

[1] *Scacchero*, the modern *scacchiere*. Our word "exchequer" comes from the same root. See page 188. The spelling varies often in the same book.

[2] ". . . el primo e detto multiplicare ⵁ Scachieri in vinegia ouer per altro nome per bericuocolo in firença . . . el primo modo di multiplicare chiamano. Bericuocolo: perch' pare la figura de q̄sti bricuocoli: o cõfortini che se vendano ale feste" (fols. 26, *r.*, 28, *v.*). A MS. in Dresden, dated 1346, has "lo modo di moltiplicare per ischachiere." B. Boncompagni, *Atti Pontif.*, XVI, 436, 439. An undated MS. in the Biblioteca Magliabechiana (Florence, C. 7. No. 2645) gives the name as *iscacherio*, *scacherio*, and *ischacherio*. Cataneo (1546; 1567 ed., fol. 10), although printing his work in Venice, calls the method *biricvocolo*. Those who do not have access to Pacioli may find the methods in facsimile in Boncompagni, *Scritti inediti del P. D. Pietro Cossali*, p. 116 (Rome, 1857), a work more likely to be found in university libraries.

[3] So Feliciano da Lazesio (1526) says: "Del multiplicar per scachier vocabulo Venitiano, ouer baricocolo uocabolo Fiorentino, ouer multiplicar per organetto uocabolo Veronese" (1545 ed., fol. 12). Similarly, Tartaglia (1556) says: "Del secondo modo di multiplicare detto per Scachero, ouer per Baricocolo, ouer per Organetto" (*General Trattato*, I, 23 (Venice, 1556)).

[4] "Multiplicatione a scaletta .&. aggregatione a bericocolo," in a MS. at Paris, described by Boncompagni, *Atti Pontif.*, XVI, 331.

[5] The *Swarupa gunanam*, "the multiplier as a factor." It is Bhāskara's first method. For the method of Mahāvīra (*c.* 850) see his *Ganita-Sāra-Sangraha*, Madras, 1912, p. 9 of the translation (hereafter referred to as *Mahāvīra*).

[6] The *St'hana gunanam*, "multiplication by places." It is his fourth method.

Since multiplication on the abacus required no symbol for zero, the earlier attempts with the Hindu-Arabic numerals occasionally show the influence of the calculi. This is seen in a Paris MS. in which the multiplication of 4600 by 23 is described in a manner leading to the form here shown.[1] It is possibly in forms like this that the chessboard method had its origin.

CM	XM	M	C	X	I
		4	6		
		I	8		
	I	2			
	I	2			
	8				
I			5	8	
				2	3

The name *scachiero* was used for a century after the chessboard form had entirely disappeared. The Treviso arithmetic (1478) does not attempt to mark off the squares, but the author uses the name,[2] as did various other Italian writers.[3] It was also used occasionally in Germany,[4] England,[5] and Spain,[6] but less in other countries.

Even after this method was generally adopted, the relative position of the figures was for a long time unsettled. In the oldest known German algorism[7] the multiplier appears above the multiplicand. In the Rollandus MS. (Paris, *c.* 1424)[8] the arrangement is as here shown. In the Treviso arithmetic the multiplier is sometimes placed at the right, as seen in the

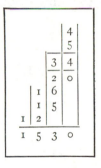

[1] M. Chasles, *Comptes rendus*, XVI, 234 (1843).

[2] " . . . attēdi al terzo modo. zoe al moltiplicare per scachiero " (fol. 19).

[3] Borghi (1484) gives only this method, designating it " per scachier."

[4] *E.g.*, Petzensteiner (1483) says: "Also ich wil multipliciren in Scachir."

[5] Recorde (*c.* 1542) speaks of "one way that is wrought by a checker table" (1558 ed., fol. G8).

[6] Thus Texeda (1546) describes multiplication "escaqr o berricolo."

[7] A. Nagl, "Ueber eine Algorismus-Schrift des XII. Jahrh.," *Zeitschrift für Mathematik und Physik*, Hl. Abt., XXXIV, 129. This journal is hereafter referred to as *Zeitschrift* (Hl. Abt.).

[8] *Rara Arithmetica*, p. 446.

annexed facsimile; Widman (1489) gives the same arrangement in his second method; and this lateral position of the
multiplier is preserved in our syn-
thetic multiplication in algebra.

$$\begin{array}{r} 9\ 3\ 4 \\ \hline 3\ 2\ 3\ 6\,/\,4 \\ 9\ 3\ 4\,/\,1 \\ 2\ 8\ 0\ 2\,/\,3 \\ \hline 2\ 9\ 3\ 2\ 7\ 6 \end{array}$$

SCACCHERO MULTIPLICATION

From the Treviso arithmetic, 1478.
In this case the multiplier is placed
at the right

The placing of the multiplier above
the multiplicand[1] is possibly due
to the fact that the writers of
that period did not greatly concern
themselves as to which of the
two numbers was the operator, although the smaller one was more
often chosen. The difficulty of
settling down to a definite arrange-
ment of figures is seen by a study of the various editions of the
Taglientes' popular *Libro dabaco*,[2] eight of which[3] give the
following examples:

(1515)	(1520)	(1541)	(1547)
456	456	456	456
23	23	23	23
1368	1368	136 8	1368
912	912	912	912
10488	10488	1048 8	10488

(1550)	(1561)	(1564)	(1567)
456	456	456	456
23	23	23	23
1368	1368	1368	1368
912	912	912	912
10488	10488	10488	10488

The various other editions give arrangements similar to the
above. Some of these forms are doubtless due to printers'
errors, but as a whole they go to show that a definite plan had

[1] *E.g.*, Sfortunati, 1534; 1545 ed., fol. 11.
[2] 1515, the work of two authors.   [3] The dates appear in parentheses.

not been agreed upon in the 16th century, although the general chessboard method was given the preference by most writers,[1] other methods being looked upon as mere curiosities. Thus Hylles (1600) says:

Also you shall vnderstand, that there are besides these sundrie other waies of Multiplication, aswell with squars as without which if you list to learne I referre you to M. Records ground of artes, where you may finde plentie of varietie.

The Castle Method. The second plan of multiplying laid down by Pacioli was, on account of the form of the work, known as "the castle" or, in Florence, "the little castle."[2] The significance of the name is best understood from the first example given by Pacioli.

	9876	6		Per .7
		6		
	6789	1		Proua
	61101000			
Castelucio	5431200			
	476230		[*sic*]	
	40734			
Sūma	67048164		.1.	

It will be observed that the figures are arranged somewhat like the wall and turret of a castle.[3] The scheme was merely a copy-book invention of the Italian schoolmasters and, although enduring until the close of the 17th century or later,[4] was always looked upon as a puerile method.[5]

---

[1] Thus Tartaglia (1556) calls it "vn modo generalissimo da nostri antichi pratici ritrouato, & piu di alcun' altro vsitato." Pagani (1591), although giving a list of methods like Pacioli, prefers this one, calling it "molto vago" (very pretty) and "molto sicuro" (very certain).

[2] "Del multiplicare per castello ouero castelluccio vocabulo Fiorentino" (Feliciano, 1526; 1545 ed., fol. 12). Pacioli (1494) says of it: "El secondo modo di multiplicare e detto castellucio" (fol. 26, *r*.), and a MS. of Benedetto da Firenze of *c*. 1460 calls it "elchasteluccio." In Spain (Texeda, 1546) it appears as "El .2. modo le dize castellucio."

[3] The figures at the right are the proof by casting out 7's.

[4] It appears in Ciacchi's *Regole Generali d' Abbaco*, fol. B3 (Florence, 1675).

[5] Thus Pagani (1591) says: "ma più tosto capriciose ch' vsitato, & vtile."

The Column Method. The third method given by Pacioli is known as the column or tablet plan.[1] By this method the computer refers to the elaborate tables, always in columns, like those used by the Babylonians, which are found in many of the 15th century manuscripts. It is essentially nothing but a step in the development of such elaborate and convenient multiplication tables as those of Crelle and others, which appeared in the 19th century.

Cross Multiplication. Pacioli's fourth method was that of cross multiplication, still preserved in our algebras and used, in simple cases, by many computers. To this he gave the name "crocetta" or "casella,"[2] adding that it is more fantastic and ingenious than the others.[3] His most elaborate illustration is that of 78 × 9876:

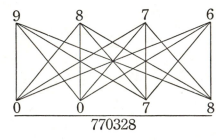

$$770328$$

---

[1] "El terço e detto multiplicare ρ colōna ouer atauoletta" (1494 ed., fol. 26). Cataneo (1546; 1567 ed., fol. 8) gives it another name, "Del mvltiplicar a la memoria detto uulgarmente Caselle o Librettine," and the Taglientes (1515) speak of it as " ρ cholonella." Texeda (1546), who follows Pacioli very closely, speaks of it as "colona o taboleta." As "per colona" it is the first method given in the Treviso (1478) arithmetic, and this name is also used by Borghi (1484). Tartaglia (1556) calls it the oral or mental plan ("per discorso, ouer di testa") as well as "per colona" and "per colonella."

[2] "De .4°. mō multiplicandi dicto crocetta siue casella." The Treviso arithmetic calls it the method of the simple little cross: "Attendi diligētamente a lo segondo modo: zoe moltiplicare per croxetta simplice." The name *crocetta* was the more common one, for Feliciano (1526) speaks of it as "per crocetta o voi dire per casela"; Cardan (1539) gives it only the name "modus multiplicādi ρ cruceta"; and Tartaglia (1556) and Cataneo (1546) call it merely "per crosetta." *Crocetta* means a little cross, and *Casella* (a little house) is often used for "pigeonhole."

[3] ". . . piu fantasia e ceruello che alcūo d' glialtri." He admires it, however, as "bella e sotii e fo bel trouato" (fol. 28, *r.*).

The filling of the vacant places by zeros, in 0078, was not unusual among the Arabs. Thus in a manuscript of one of the works of Qosṭâ ibn Lûqâ[1] (c. 900) the multiplication of 21,600 by 4 appears in this form:

The plan is ancient, appearing in the *Lilāvati* (c. 1150) as the *tatst'ha* method, or method of the stationary multiplier, in distinction from the advancing multiplier, where the multiplying figures were advanced one place to the right after each partial product was found. This method is shown on page 118. The method given in the *Lilāvati* is fully explained by Ganeśa (c. 1535) in his commentary on Bhāskara.[2] It also appears in the arithmetic of Planudes (c. 1340), but in the forms here shown:[3]

AN ITALIAN METHOD OF
MULTIPLICATION

From the 1541 edition of the Taglientes' *Opera* of 1515

$$\begin{array}{ccc} 840 & 114048 & 76842 \\ \hline 24 & 432 & 1423 \\ 35 & 264 & 0054 \end{array}$$

While the first of these cases is simple, it is doubtful if the other two were practically used.

Some of the work of the Taglientes (1515) is related to cross multiplication, as may be seen from the illustration here given.

[1] The MS. is dated 1106 A.H. (1695 A.D.).
[2] Colebrooke, *Lilāvati*, p. 6 n.

[3] Wäschke, *Planudes*, p. 14.

On account of the difficulty of setting the crossed lines, printers often used the letter X between the multiplicand and the multiplier, and this may have suggested to Wright (1618) the multiplication sign (x) used by him and his contemporaries. In Pagani's arithmetic of 1591, for example, the work in cross multiplication appears as follows:

$$\overset{3}{}\overset{2}{} \qquad \overset{4}{I}\overset{3}{X}\overset{2}{I}\overset{}{X}I$$

$$\underset{800}{\overset{2}{}\ 5} \qquad \underset{157248}{\overset{}{3}\ \overset{6}{}\ 4}$$

He recognized that the method is not very practical with numbers of more than two figures.[1]

The Method of the Quadrilateral. The fifth method given by Pacioli is that of the quadrilateral.[2] It was really nothing but the chessboard plan with the partial products slightly shifted, as is seen in the illustration from the Treviso arithmetic (1478) here shown.

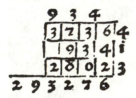

MULTIPLICATION PER
QUADRILATERO

From the Treviso arithmetic, 1478

Gelosia Method. There seems to be no good reason why Pacioli should have postponed to the sixth in order the so-called gelosia, or grating, method, also known by the name of the quadrilateral, the square,[3] or the method of the cells,[4]

[1] "Il moltiplicare a crocetta di tre figure, è assai piu dificile de primo. . . . Il moltiplicare per crocetta di 4. figure è più dificile delli sopra nominati, & quanto più sono figure, tanto più sono dificile" (p. 17).

[2] "El quinto mō e detto ꝑ q̄drilatero" (fol. 26, r.). It appears in Pagani's work (1591) as "per quadrato." It is often merged with the *gelosia* method next mentioned, as when Tartaglia (1556) calls it "Per Quadrilatero, ouero per gelosia."

[3] "El sexto modo e detto ꝑ gelosia ouer graticola." Tartaglia (1556) says: "Il quinto modo di multiplicare è detto Quadrilatero qual è assai bello, perche in quello nō vi occorre a tener a mente le decene." Various other names are used, such as "modo di quadrato" (Feliciano, 1526), "ꝑ quadro" (16th century MS. of Gio. Dom. Marchesi), "per squadrado" (undated Bologna MS.).

[4] ". . . per le figure de le camerete," "dala fugura dela camerella," the cells also being called "camere triangulate." This is in an undated Turin MS. See B. Boncompagni, *Atti Pontif.*, XVI, 448.

and to the Arabs after the 12th century by such names as the method of the sieve[1] or method of the net.[2]

The method is well illustrated by two examples from the Treviso (1478) book here shown. It will be observed that the diagonals separate the tens and units and render unnecessary

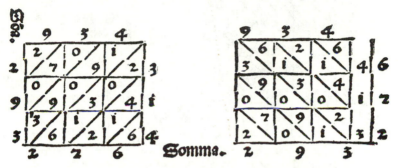

the carrying process except in adding the partial products. These diagonals sometimes slant one way and sometimes another, but in general the direction from the upper right-hand corner to the lower left-hand corner was the favorite.[3]

The method is very old and might have remained the popular one if it had not been difficult to print or even to write the net. It was very likely developed first in India, for it appears in Ganeśa's commentary on the *Lilāvati* and in other Hindu works.[4] From India it seems to have moved northward to

[1] In the writings of Albanna (*c.* 1300), possibly due to his commentator, al-Qalasâdî (*c.* 1475). See F. Woepcke, *Journal Asiatique*, I (6), 512.

[2] H. Suter, "Das Rechenbuch des Abû Zakarîjâ al-Ḥaṣṣâr," *Bibl. Math.*, II (3), 17.

[3] See the Marre translation of Behâ Eddîn's (*c.* 1600) *Kholâçat al Hissâb*, Paris, 1846 ; Rome, 1864, p. 13 ; hereafter referred to as Behâ Eddîn, *Cholâṣat*. Also see books as late as Giuseppe Cortese's *Aritmetica* (Naples, 1716).

The contrary direction is seen in MSS. of Ibn al-Hâ'im (*c.* 1400), dated 1132 A.H. (1720 A.D.), Albanna (*c.* 1300), and various other Arabic writers.

[4] Colebrooke, *Lilāvati*, p. 7. See also the introduction to Taylor, *Lilawati*, pp. 20, 33. It was called by the Hindus *Shabakh*.

China, appearing there in an arithmetic of 1593.[1] It also found its way into the Arab and Persian works, where it was the favorite method for many generations.

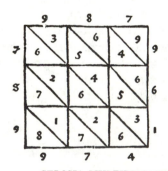

From the Arabs it passed over to Italy and is found in many manuscripts of the 14th and 15th centuries. In the printed books it appeared as late as the beginning of the 18th century,[2] but more as a curiosity than as a practical method.

As to the name *gelosia*, Pacioli's statement is more complete than that of any of his contemporaries:

GELOSIA METHOD OF MULTIPLYING

From an anonymous manuscript written in Florence c. 1430. The author describes it as *Multiplicha p modo de Quadrato*

The sixth method of multiplying is called *gelosia* or *gra-ticola* . . . because the arrangement of the work resembles a lattice or *gelosia*. By *gelosia* we understand the grating which it is the cus-

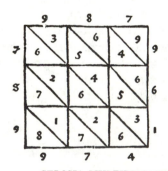
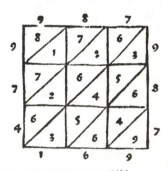

GELOSIA MULTIPLICATION AS GIVEN BY PACIOLI, 1494

Showing the same double arrangement of diagonals as in the Treviso book of 1478

---

[1] Libri, *Histoire*, I, 386, 389.

[2] *E.g.*, in the 1690 edition of Coutereels's *Cyffer-Boeck*; in Padre Alessandro's *Arimmetica* (Rome, 1714); and in Giuseppe Cortese's *Aritmetica* (Naples, 1716).

tom to place at the windows of houses where ladies or nuns reside, so they cannot easily be seen. Many such abound in the noble city of Venice.[1]

The Repiego Method. Another method that was popular enough to survive and to have a place in some of our modern textbooks was called by the early Italians the "modo per repiego," that is, the method by composition, or, more exactly, by decomposition, of factors. For example, to multiply by 72, multiply by 9 and then by 8, thus saving the addition of partial products.[2] It is one of several methods inherited by the Italians, through the Arabs, from Hindu sources.[3]

The Scapezzo Method. Pacioli's eighth method was commonly known among the Italians as *a scapezzo*, or multiplying by the parts, not the factors, of the multiplier.[4] Tartaglia (1556) gives as an illustration

$$26 \times 67 = (3 + 4 + 5 + 6 + 8) \times 67$$
$$= 201 + 268 + 335 + 402 + 536 = 1742 ;$$

but he could not have considered it as other than a curiosity, although it was recognized by such writers as Ramus and Schoner.[5] It goes back to Bhāskara at least.[6]

---

[1] "Gelosia intendiamo quelle graticelle c̄h si costumono mettere ale finestre de le case doue habitano dōne acio nō si possino facilmē e vedere o altri religiosi. Diche molto abonda la excelsa cita de uinegia" (fol. 28, *r.*). The word found its way into French as *jalousie*, meaning a blind, and thence passed into German and was carried even to the Far East, where it is met with today.

[2] Pacioli explains the term thus: "Repiego de vn numero se intende el producto de doi altri numeri che multiplicati vno nel laltro fanno quel tal numero aponto: del quale essi sonno ditti repieghi" (fol. 28, *v.*). It is Tartaglia's third method. See also Terquem's *Bulletin*, Vol. VI, and B. Boncompagni, *Atti Pontif.*, XVI, 404.

[3] It appears in Taylor's notes on the *Lilāvati* under the name *vibhaga gunanam* (submultiple multiplication). See translation, p. 8 n.

[4] "De octauo modo multiplicandi dicto aschapeçço." Pacioli, 1494, fol. 29, *r.* Tartaglia gives the name as "spezzato, ouer spezzatamente." *General Trattato*, 1556, I, 26. In Texeda's Spanish work of 1546 it appears as *escapeço*.

[5] 1586 ed., p. 16.

[6] See Taylor's translation, p. 8. The name in his notes is *khanda gunanam* (parts multiplication).

u

Minor Methods of Multiplying. Besides the leading methods given by Pacioli there are many variations to be found in other early works. One of the most valuable is the left-to-right method, the "allo adietro" plan of Tartaglia,[1] still used to advantage in some cases but hardly worth teaching. The Arabs occasionally used it,[2] and the Hindus varied it by beginning with the lowest order of the multiplier and the highest order of the multiplicand.[3]

Another variant is seen in a cancellation method which went by various names. The Arabs called it the Hindu plan,[4] and Taylor[5] found the Hindus using it early in the 19th century. Al-Nasavî (c. 1025)[6] and other Arab writers thought highly enough of the method to give it a place in their works. It may be illustrated by the case of 76 × 43. The figures were first written as shown in the upper rectangle. The multiplication began with 7 × 4 = 28. As soon as a figure had served its purpose it was erased on the abacus and its place was taken by another. This procedure was modified in India, probably long after 200 b.c., when ink came into use,[7] the figures being canceled as here shown.[8]

In the 15th and 16th centuries there were numerous vagaries of the copy-book makers, the extensive discussion of which is

---

[1] *General Trattato*, 1556, I, 25, *v.* Calandri (1491) gives a page (fol. 8) to the method.

[2] *E. g.*, al-Karkhî, *c.* 1020. See H. Hankel, *Geschichte der Mathematik*, Leipzig, 1874, pp. 56, 188; hereafter referred to as Hankel, *Geschichte*.

[3] An elaborate example of the late use of the method is in Marten Jellen's *Rekenkundige Byzonderheden* (1779), p. 13.

[4] This translation of the term *Hindasi*, used by many Arab writers, is, however, disputed. It also means numerical, a translation that would have little significance in a case like this. See page 64.   [5] *Lilawati*, Introd., p. 9.

[6] F. Woepcke, *Journal Asiatique*, I (6), 497.

[7] See G. Bühler, *Indische Palaeographie*, pp. 5, 91 (Strasburg, 1896).

[8] This example is from an Arab arithmetician, Mohammed ibn Abdallâh ibn 'Aiyash, Abû Zakarîyâ, commonly known as al-Ḥaṣṣâr (c. 1175?). See H. Suter, *Bibl. Math.*, II (3), 16. Substantially the same plan is used by al-Qalasâdî (c. 1475), *ibid.*, p. 17. For various arrangements followed by the Hindus see Colebrooke, *Lilāvati*, p. 7 n.

not worth while. Suffice it to say that certain teachers had their pupils arrange the partial products in the form of a rhombus,

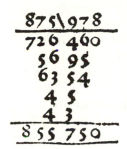

and even as good mathematicians as Tartaglia (1556) and Cataldi (1602) multiplied "per Rumbo" and "à Rombo." Others arranged the figures so that the outline of the work looked like a cup, chalice, or beaker.[1] The better writers, however, recognized that such work was time-consuming.[2]

THE FIRST EXAMPLE IN MULTIPLICATION PRINTED IN THE NEW WORLD

From the *Sumario cōpedioso* of Juan Diez, published in Mexico in 1556. The problem is to multiply 978 by 875 and the method is essentially *per copa*, that is, the method of the cup, so called because the figure resembles a drinking cup

Labor-Saving Short Methods. Besides the general methods already described there were many special devices for the saving of labor. Even when the multiplication table was learned, the medieval computers did not require it beyond 5 × 10, and various plans were developed for operating under this limitation when an abacus was not conveniently at hand. Of these methods, one of the best known is a complementary plan that is found in many of the 16th century books. To multiply 7 by 8, write the numbers with their complements to 10, as here shown.

Then either 8 − 3 or 7 − 2 is 5, and 2 × 3 = 6, so that 56 is the product, the operation not requiring the multiplication table beyond 5 × 10 in any case. It is given by such writers as Huswirt (1501), Fine (1530), Riese (1522), Rudolff (1526), Stifel (1544), Recorde (*c.* 1542),

---

[1] Thus Tartaglia calls the plan "Per Coppa, ouer per Calice," and Cataldi says: "dalla forma loro si possono chiamare à Calice, Coppa, Tazza, ò Bicchiere." In Spanish (Texeda, 1546) it appears as "per copa" and is incorrectly given as "a la frācesa" (a French method). It is essentially the method used by Juan Diez, Mexico, 1556, as shown in the illustration.

[2] So Tartaglia: "trouate piu per mostrar vn piu sapere, che per alcuna vtilità" (1592 ed., fol. 40). Other curious forms are given by Coutereels (1690 ed., p. 8) under the title "Vernakelijke Multiplicatie" (interesting multiplication).

Peletier (1549), and Baker (1568), while Peurbach (c. 1450; 1st ed., 1492) speaks of it as an ancient rule.[1] It was commonly used in connection with finger reckoning.[2] For example, to find 6×9, raise one finger on one hand and four fingers on the other hand, these representing the respective complements of 9 and 6. Then multiply the standing fingers for the units and add the closed fingers for the tens. The plan is still in use among the peasants in certain parts of Russia and Poland. As a variant of this method the following plan was until recently in use in certain towns in Russia: Number the fingers on each hand from 6 to 10, and give to each the value 10. Then the products of 8×8 and 7×9 are found as indicated below:

$$8 \times 8$$

```
10 ———————    ——————— 10
9  ———————    ——————— 9
8  ——— 10 —    —— 10 —— 8
7  ——— 10 —    —— 10 —— 7
6  ——— 10 —    —— 10 —— 6
```

$$7 \times 9$$

```
10 ———————         ——————— 10
                      10
9  ———————         —— 10 —— 9
8  ———————         —— 10 —— 8
7  ——— 10 —        —— 10 —— 7
6  ——— 10 —        —— 10 —— 6
```

$$8 \times 8 = 6 \times 10 + 2 \times 2 = 64$$

$$7 \times 9 = 4 \times 10 + 2 \times 10 + 1 \times 3 = 63$$

A few of the most common of the complementary methods will now be briefly indicated.

Various Arab and Persian writers[3] multiplied 8 by 7 by the relation $ab = 10b - (10 - a)b$, as here shown.[4]

In this case we have

$$7 \times 8 = 10 \times 8 - (10 - 7) \times 8$$
$$= 10 \times 8 - 3 \times 8$$
$$= 80 - 24$$
$$= 56.$$

$$\begin{array}{r} 10 \times 8 = 80 \\ 3 \times 8 = 24 \\ \hline 56 \end{array}$$

The method was considered valuable by writers like Widman (1489), Riese (1522), Rudolff (1526), and Scheubel (1545).

[1] ". . . regulam illam antiquam."
[2] See pages 196 and 201.
[3] E.g., Behâ Eddîn (c. 1600).
[4] G. Eneström, Bibl. Math., VII (3), 95.

Widman, Rudolff, and Grammateus (1518) used the relation

$$ab = 10\,b - 10(10 - a) + (10 - a)(10 - b)$$
$$= 10[b - (10 - a)] + (10 - a)(10 - b),$$

COMPLEMENTARY MULTIPLICATION

From Huswirt's *Enchiridion nouus Algorismi summopere visus De integris,*
Cologne, 1501. Much reduced

illustrated in the following multiplication of 8 by 7.
Here we have

$$7 \times 8 = 10 \times 8 - 10 \times (10 - 7)$$
$$+ (10 - 7)(10 - 8)$$
$$= 80 - 30 + 6$$
$$= 56.$$

$10 \times 8 =$	80
$- 10 \times 3 =$	$- 30$
$+ 3 \times 2 =$	6
	56

Further Algebraic Relations. The following relations also had their advocates:

$$ab = 10(a+b-10)+(10-a)(10-b);^{1}$$

$$ab = 10(a-b+2\cdot\overline{b-5})+(10-a)(10-b);$$

$$ab = (10-a)(10-b)+10(a+b)-100;^{2}$$

$$(10\,a+a)(10\,b+b) = [(10\,a+a)b+ab]\,10+ab\,;^{3}$$

$$(10\,a+b)(10\,a+c) = (10\,a+b+c)a\times10+bc\,;^{4}$$

$$(10\,a+b)(10\,a-c) = 100\,a^2+10\,ab-(10\,ac+bc)\,;^{5}$$

$$ab = (a+b-10)\times10-(a-10)(10-b),\ a>10,\ b<10\,;^{6}$$

$$(10+a)(10\,b+c) = (ab+10\,b+c)\times10+ac\,;$$

$$(3\,a)^2 = 10\,a^2-a^2;\quad (3\,a+1)^2 = (3\,a)^2+[(3\,a+1)+3\,a];$$

$$(3\,a-1)^2 = (3\,a)^2-[(3\,a-1)+3\,a];^{7}$$

$$a^2 = 10\,a-(10-a)\times a;$$

$$a^2 = (\tfrac{1}{3}\,a)^2\times10-(\tfrac{1}{3}\,a)^2,\ \text{where } a = 3\,m\,;$$

$$a^2 = (a-1)^2+[a+(a-1)],\ \text{where } a = 3\,m+1\,;$$

$$a^2 = (a+1)^2-[a+(a+1)],\ \text{where } a = 3\,m+2\,;^{8}$$

$$(a+b)(a-b) = a^2-b^2;^{9}$$

$$(5\,a)^2 = 10\,a^2\times2\tfrac{1}{2}\,;$$

---

[1] *E.g.*, Behâ Eddîn and Riese.

[2] The Petzensteiner arithmetic, Bamberg, 1483.

[3] Al-Karkhî (*c.* 1020), as in

$$22\times44 = (22\times4+8)\,10+2\times4 = 968.$$

[4] Al-Karkhî, Behâ Eddîn, Tartaglia. *E.g.*,

$$23\times27 = (23+7)\times2\times10+3\times7 = 600+21 = 621.$$

[5] Al-Karkhî and Tartaglia.   [6] An Arab writer, al-Kashî, *c.* 1430.

[7] Elia Misrachi (*c.* 1500) and Rabbi ben Ezra (*c.* 1140).

[8] Rabbi ben Ezra, who recognized that the limitations on *a* were unnecessary.

[9] Well known to the Greeks and given by Euclid.

$$(5\,a+1)^2=(5\,a)^2+[(5\,a+1)+5\,a];^1$$

$$(10+a)(10+b)=100+10(a+b)+ab;^2$$

$$ab=\left(\frac{a+b}{2}\right)^2-\left(\frac{a-b}{2}\right)^2.^3$$

The method of aliquot parts was also well known in the Middle Ages, both in Europe and among the Arabs, and the 16th century writers frequently gave our common rules of multiplying by numbers like 11 and 15. Beginning at least as early as the 14th century, multiplication by numbers ending in one or more zeros was commonly effected as at present.[4]

Contracted multiplication, the work being correct to a given number of significant figures, is a development intended to meet the needs of modern science. It began to assume some importance in the 18th century,[5] although a beginning had already been made by Bürgi (c. 1592) and Prætorius (c. 1599).[6]

The Multiplication Table. The oldest known arrangement of the multiplication table is by columns. This is the one always found on the Babylonian cylinders and the one commonly used by the Italian writers on mercantile arithmetic in the formative period of the subject. In general, no product appeared more than once; that is, after 2 × 3 = 6 was given, 3 × 2 was thought

---

[1] Elia Misrachi gives various rules of this kind.

[2] Huswirt (1501).

[3] This is the rule of quarter squares, which still has its advocates. It is probably due to the Hindus. See A. Hochheim, *Kâfî fîl Hisâb*, p. 7 (Halle a. S., 1878) (hereafter referred to as Hochheim, *Kâfî fîl Hisâb*); H. Weissenborn, *Gerbert*, p. 201 (Berlin, 1888). It is found in the *Talkhys* of Albanna (c. 1300), the work of al-Karkhî (c. 1020) mentioned above, and the works of Behâ Eddîn (c. 1600) and other Oriental writers. The preferred transliteration of the name of al-Karkhî's work is *al-Kâfî fi'l-Hisâb*, but the more familiar title as given in the European editions has been adopted in this work. See Volume I, page 283.

[4] E.g., Maestro Paolo dell' Abbaco (c. 1340); see G. Frizzo's edition, p. 42 (Verona, 1883). Bianchini's correspondence with Regiomontanus (1462) contains it; see M. Curtze, *Abhandlungen*, XII, 197, 270. It is also in the Treviso arithmetic (1478), Pellos (1492), and other early works. There are, however, various cases in which it was not recognized in the 16th century.

[5] Greenwood's American arithmetic (1729) gives the reversed multiplier.

[6] M. Curtze, *Zeitschrift* (Hl. Abt.), XL, 7.

to be unnecessary, a view still taken by Japanese arithmeticians[1] and having much to commend it. The early Italian mercantile arithmetics gave, for purpose of easy reference, tables with the products of all primes to 47 × 47, or often to 97 × 97. Computers turned to these columns for the simpler products needed in multiplication *per colonna*. The Italians obtained the idea from the East, Rhabdas (1341) giving the column tables "which the very wise Palamedes taught me."[2]

MEDIEVAL MULTIPLICATION TABLE

Part of a table from an anonymous Italian MS. of *c.* 1456, but apparently a copy of an earlier work of *c.* 1420

The second arrangement was the square form generally used by nonmercantile writers and known as the Pythagorean Table,[3] whereof, as Hylles (1600) remarks, "Some affirme Pythagoras to be the first author." This mistaken idea was held by various early writers,[4] although the better ones seem to have recognized

[1] Smith and Mikami, *History of Japanese Mathematics*, p. 37 (Chicago, 1914); hereafter referred to as Smith-Mikami.

[2] P. Tannery's translation in *Notices et extraits des manuscrits de la Bibl. nat.*, XXXII, 167. For their use by Benedetto da Firenze, Luca dell' Abacho, and others, see *Rara Arithmetica*, p. 464 and elsewhere. They are also found in Pacioli (1494), Pellos (1492), Borghi (1484), and the Treviso book (1478), and in many other works. See also D. E. Smith, "A Greek Multiplication Table," *Bibl. Math.*, IX (3), 193.

[3] Table de Pythagore, Tabula Pythagorica, Mensa Pythagorae, Mensula Pythagorae, Tavola Pitagorica, Mensa Pythagorica, and other similar names are common.

[4] Thus Köbel (1514) speaks of "Der Pythagorisch Tisch oder Tafel" as "von dem Fürstē Pythagora geordnet" (1518 ed., fol. 1ᵇ).

1	2	3	4	5	6	7	8	9	10
2	4	6	8	10	12	14	16	18	20
3	6	9	12	15	18	21	24	27	30
4	8	12	16	20	24	28	32	36	40
5	10	15	20	25	30	35	40	45	50
6	12	18	24	30	36	42	48	54	60
7	14	21	28	35	42	49	56	63	70
8	16	24	32	40	48	56	64	72	80
9	18	27	36	45	54	63	72	81	90
10	20	30	40	50	60	70	80	90	100

MULTIPLICATION TABLE (*c.* 1500)

The table as it appeared in an anonymous Latin MS. of *c.* 1500, being the same form as the one found in various MSS. of Boethius

that the later Pythagoreans were the inventors.[1] It is found in the arithmetic of Boethius[2] and in a work attributed to Bede (*c.* 710),[3] but the fact that Rhabdas (*c.* 1341) does not give it[4] suggests that the Greeks did not use it. It was common in the medieval works[5] and in the early printed books.[6] Some writers carelessly attributed it to Boethius,[7] while others arranged tables of addition, subtraction, and division on the same plan and gave to them the name of Pythagoras.[8]

The third standard form was the triangular array. It appears in a Prag manuscript[9] in the form here shown, but there are several variants. It is given in *The Crafte of Nombrynge*

1	2			
2	4	3		
3	6	9	4	
4	8	12	16	5

(*c.* 1300) as "a tabul of figures, where-by þou schalt se a-nonn) ryght what is þe nounbre þat comes of þe multiplicacion) of 2 digittes." Widman (1489) speaks of it as a Hebrew device,[10] and at any rate it is quite likely to be Arabic.[11] It was not so popular in the early textbooks as the columnar and square arrangements, although it was used by such writers as Widman (1489), Gemma Frisius (1540), Recorde (*c.* 1542), Baker (1568), and Trenchant (1566).

[1] Thus Boethius says: "Pythagorici . . . quam ob honorem sui praeceptoris mensam Pythagoream nominabant" (Friedlein ed., p. 396). See also A. Favaro in Boncompagni's *Bullettino*, XII, 148. Clavius (1583) says: "quòd Pythagoras eam vel primus excogitauerit, vel certe discipulos suos in ea mirifice exercuerit."

[2] Friedlein ed., p. 53. On the text see Boncompagni's *Bullettino*, XV, 139.

[3] *De arithmeticis numeris*, of doubtful authorship, where the "Pythagorica Mensa sive abacus numerandi" is given in full to 20 × 20, with the more important products as far as 1000².

[4] P. Tannery, *Notices et extraits des manuscrits de la Bibl. nat.*, XXXII, 121.

[5] *E.g.*, Jordanus Nemorarius (*c.* 1225), Rollandus (1424), and al-Kashî (*c.* 1430).

[6] *E.g.*, Tzwivel (1505) and such commentators as Faber Stapulensis and Clichtoveus.

[7] Thus Stifel (1545) says: "Dise tafel hat Boetius gesetzt."

[8] Possibly Ramus (1569) began this, for he gives these tables and says: "Hic Pythagoraeus additionis abacus est," and so for subtraction and division.

[9] See S. Günther, Boncompagni's *Bullettino*, XII, p. 149; very likely the MS. of Christian of Prag, already referred to on pages 77, 95.

[10] "Das erst ist eynn taffel geformiret auff den triangel geczogen ausʒ hebraischer zungen oder iudischer."

[11] Behâ Eddîn (*c.* 1600) gives it in his *Kholâṣat al-ḥisâb*.

The extent to which the tables were carried varied considerably from time to time. Tables used for reference, as we might use Crelle's tables today, go back to ancient times, one of the 5th century giving the important products to 50 × 1000.[1] The medieval writers were usually content to stop with 20 × 20,[2] however. For tables to be committed to memory it was sufficient, in the days of the medieval abacus, to go only to 5 × 10; even 4 × 9 was far enough for practical purposes.[3] Many of the 16th century writers outside of Italy found it necessary to urge their pupils

TRIANGULAR AND SQUARE FORMS OF THE
MULTIPLICATION TABLE

From Widman's arithmetic (Leipzig, 1489), the edition of 1500

1 "Victorii Calculus ex codice Vaticano editus a Godofredo Friedlein," in Boncompagni's *Bullettino*, IV, 443.

2 As in a MS. written before 1284 and copied in 1385, described by Steinschneider in the *Bibl. Math.*, XIII (2), 40. See also Beldamandi's work (1410), printed in 1483, where the products extend to 22 × 22.

3 See the devices for finding such products as 7 × 8, page 119. Thus Rudolff (1526; 1534 ed., fol. D 8) says: "Das ein mal eins . . . musten zum ersten wol in kopff fassen / doch nit weiter dañ bis auff 4 mal 9." Clavius, while recommending the learning of the table to 10 × 10, says: "Qvod si huiusmodi tabula in promptu nō sit, vtendum erit hac regula," namely, the one given on page 119.

very strongly to learn the table, showing that the custom was relatively recent in countries where the abacus had only just been abandoned or where its use was diminishing.[1] One Spanish writer says that it should be known as thoroughly as the Ave Maria,[2] and Digges (1572) encourages his pupils by saying: "This Table therefore first printe liuely in thy remembrance, and then boldly proceede farther, all difficultie I assure thee is past."

It may interest those teachers who feel that they must insist upon "two threes are six" instead of "two times three are six" to know that the former has at least some kind of remote sanction in a terse Latin form,[3] although in most languages the use of "times" has been general.

6. DIVISION

Definition of Division. Division has generally been considered as the fourth of the fundamental operations,[4] the fifth when numeration is included, or the seventh when duplation and mediation are considered separately. In general the operation

---

[1] So Chuquet (1484) says: "¶Item plus est necesze de sauoir tout de cueur la multiplication dune chascune des .10. figures par soy mesmes et aussi par une chascune des aultres La quelle chose est appelle le petit liuret de algorisme." (From A. Marre's MS. copy in the author's library.) This "livret de algorisme" was a common name for the small multiplication table, "gli libretti minori" of the Italians, "gli libretti maggiori" referring to the table beyond 10 × 10. (Spelling as in the Dagomari MS. described in *Rara Arithmetica*, p. 435.) The couplet often found in 16th century books,

¶ Lern wol mit fleisz das ein mal ein
So wirt dir alle rechnung gmein,

appeared first in print, so far as I have found, in Widman's work of 1489; 1508 ed., fol. 11.

[2] ". . . laquale tabula bisogna sapere ad memoria como la Aue Maria" (Ortega, 1512; 1515 ed., fol. 16). Thierfelder (1587) says: "Aber wer das ein mal eins nicht fertig lernet . . . wird nimermehr keinen fertigen Rechner geben" (p. 16). Metius is equally urgent: "Tabula Pitagorica, dieman wel vast in sijn memorie moet hebben" (1635 ed., p. 5).

[3] Thus Scheubel (1545, I, cap. 4) says: "Sixies septem sunt 42. septies quinq3 sunt 35," and so on, in which, however, the word "times" is concealed.

[4] The "quarto atto" of the Treviso arithmetic (1478).

has been known either as division[1] or as partition,[2] but many writers use both terms.[3] Thus Baker (1568) speaks of "Deuision or partition," and Digges (1572) says "To deuide or parte."[4] As in the case of multiplication, no satisfactory definition, adapted to the understanding of beginners, is possible, since the concept is constantly extended as the pupil proceeds. To say that "diuision sheweth onlely howe often the lesse summe is conteyned in the bigger,"[5] or that

> Diuision doth search how oft the diuisor
> In Diuidend may be quoted or found
> Whereof the quotient is the decidor,[6]

is to exclude cases like 6 ft. $\div$ 2 or $3 \div 4$, although the latter was intentionally barred out by many writers[7] for the reason that a result like $\frac{3}{4}$ could not be "times" in the primitive use of the word. The early idea was manifestly that of an integral divisor and an integral quotient.[8]

A second definition which has had some sanction is that of finding a number which is contained as many times in the dividend as unity is contained in the divisor. It has long been used, being found in Maximus Planudes (c. 1340)[9] and the Treviso arithmetic.[10] An improvement upon this definition, and quite

---

[1] E.g., with such medieval writers as Fibonacci (1202), *Liber Abaci*, p. 27. This is the idea of measuring, and so Euclid used the term μετρεῖν (*metrein'*) to mean both to measure and to divide.

[2] E.g., the Treviso arithmetic, Huswirt (1501), Ghaligai (1521), Stifel (1544), Scheubel (1545), Cataldi (1602), Ortega (1512), Savonne (1563), and Santa-Cruz (1594). This form was preferred by Heron, Pappus, and Diophantus, all of whom used μερίζειν (*meri'zein*, to part).

[3] E.g., Pacioli (1494), Tartaglia (1556), Trenchant (1566), Clavius (1583). [4] So with some of the Dutch writers. Thus in the Dutch-French work of Wentsel (1599): "Deuisio: dat is deelinge," "Diuisio: e' esta dire, partir."

[5] Digges, 1572; 1579 ed., p. 8.

[6] Hylles (1600), the word "quoted" being interesting as related to "quotient."

[7] Thus Tzwivel (1505) says: "Officiuȝ diuisionis est cognoscere quotiens minor nūerus in maiore re,piat," and Peletier (1549) says: "c'est sçauoir combien de fois vn moindre nombre est contenu en vn plus grand."

[8] Thus Clavius: "Divisio est distributio propositi numeri in partes ab altero numeró dato denominatas" (1583; 1585 ed., p. 48).

[9] Wäschke, *Planudes*, p. 23.

[10] "Trouare vno terzo nūero: el quale se troua tante volte nel mazore: quãte vnitade sono ꝑel menore."

sufficient for pure number, is one that is based upon ratio,—the finding of a number which has to unity the same ratio as the dividend has to the divisor. It is often found in the 16th century books.[1]

It was natural in the Middle Ages, when division as performed on the abacus was often based upon subtraction, to base the definition also upon the latter operation.[2] This plan was followed by such writers as Ramus,[3] Schoner,[4] and Peletier[5] and has not wholly died out even yet.

Of all the elementary definitions the one most generally approved describes the operation as seeking a number which, multiplied by the divisor, is equal to the dividend, and it serves the purpose fairly well. It is perhaps the oldest definition extant[6] and it has the sanction of many scholarly writers.[7]

Two-fold Nature of Division. The above definitions do not, in general, distinguish between the two notions of division illustrated by the cases 6 ft. ÷ 3 ft. = 2 and 6 ft. ÷ 2 = 3 ft., although the last definition includes both cases. Rudolff (1526) seems to have been the first to make this distinction perfectly clear,[8] and Stifel (1545) to have been the second.[9] Tartaglia[10] also gave it, and thereafter it was mentioned by various writers of the 16th and 17th centuries.

[1] "Diuisio est numeri ˌpcreatio ˌpportionabiliter se ad vnitatem habētis vt diuidēdus ad diuisorē" (Huswirt, 1501, fol. 5).

[2] "Numerum per numerum diuidere est maiorem secundum quantitatem minoris partiri, uidelicet minorem de maiore tociens subtrahi, quociens in eo poterit inueniri." Johannes Hispalensis, *Liber Algorismi* (*c.* 1140), in Boncompagni's *Trattati*, II, 41.

[3] *Arith. libri duo*, 1569.   [4] *Tabulae Astronomicae*, 1536, fol. A 3*a*.

[5] 1549; 1607 ed., p. 48, as a secondary definition.

[6] J. P. A. Erman, *Life in Ancient Egypt*, p. 364, English translation by Tirard, New York, 1894 (hereafter referred to as Erman, *Egypt*), attributes it to the Egyptians.

[7] *E.g.*, Cataldi: "Il partire è modo di trouare vna quantità, quale moltiplicata per vna quantità proposta (ouero con la quale moltiplicando vna quantità proposta) produca vna quantità data" (1602, p. 32).

[8] "Diuidirn heisst abteilen. Lernet ein zal in die ander teilen/auff das man sehe/wie offt eine in d'andern beschlossen werde/oder wieuil auff einen teil kome" (1534 ed., fol. 8).

[9] *Deutsche Arithmetica*, 1545, fol. 1, where it is more clearly stated.

[10] *General Trattato*, 1556, I, fol. 27, *r*.

Terminology of Division. Early writers commonly gave names to only two of the numbers used in division, the *numerus dividendus* (number to be divided) and the *numerus divisor*.[1] These are, of course, not technical terms, and they appear as mere colloquial expressions in various medieval works. Gradually, however, the *numerus* was dropped and *dividendus* and *divisor* came to be used as technical nouns, as at present.[2] Such names as "answer" or "result" were commonly used for quotient[3] and were quite as satisfactory.

The names of the terms have undergone various changes. The divisor has frequently been called the "parter"[4] or the "dividens,"[5] but our present term has been the one most commonly used. The dividend has generally been called by this name, although there have been terms equivalent to "partend," with the usual linguistic variants.[6] The quotient has frequently been called the product,[7] the part,[8] the exiens,[9] and the outcome,[10] but the term used by English writers has been the favorite in most of the leading European languages.[11]

---

[1] Thus Clichtoveus, in his commentary on Boethius (1503; 1510 ed., fol. 36), says: "In divisione tres requiruntur numeri. Primus est numerus diuidēd⁹ & maior/ex hypothesi dandus. Secundus/numerus diuisor siue diuidens: etiam assignandus ex hypothesi. Tertius est numerus ex diuisione proueniens: & hic est querendus," no name being given for this quotient and no mention being made of a remainder.

[2] *E.g.*, in the Rollandus MS. (1424), where *quotiens* is also used. Joannes de Muris (*c.*1350) used *dividendus* and *numerus quociens*, but not *divisor*. See *Abhandlungen*, V, 145.   [3] From *quoties*, how much.

[4] "L'autre qui le diuise, s'apele parteur, partisseur, ou diuiseur" (Trenchant, 1566; 1578 ed., p. 51). Chuquet (1484) calls it the *partiteur*, and Cataldi (1602) uses *il partitore*, following the Treviso book and other Italian works of the time. Pellos (1492), writing in a dialect mixture of French, Italian, and Spanish, called it the *partidor*. In the Teutonic languages it appeared in the 16th century as Theiler, Deyler, Teyler, Deeler, and deylder.

[5] ". . . nūer⁹ diuisor siue diuidēs," Tzwivel, 1505, fol. 6, and various other Latin works.

[6] Ortega (1512; 1515 ed.) calls it "la partitione," as he calls the multiplicand "la multiplicatione." Santa-Cruz (1594) calls it "suma partidera." Digges (1572) writes it "diuident."

[7] *E.g.*, Gemma Frisius (1540) and numerous other Latin writers.

[8] *E.g.*, the Treviso book (1478) gives "la parte."

[9] *E.g.*, Scheubel (1545).   [10] In Dutch, the *Uitkomst*.

[11] Of course with such variants as *quotiens* in the Latin books, *cociente* in the Spanish (Santa-Cruz, 1594), and so on.

For obvious reasons the name for the remainder has varied more than the others. The medieval Latin writers used *numerus residuus, residuus,* and *residua,* and various other related terms, and certain later authors employed the same word for the remainder as for the fraction in the quotient.[1]

The Process of Division. The operation of division was one of the most difficult in the ancient *logistica,* and even in the 15th century it was commonly looked upon in the commercial training of the Italian boy as a hard matter.[2] Pacioli (1494) remarked that "if a man can divide well, everything else is easy, for all the rest is involved therein." He consoles the learner, however, by a homily on the benefits of hard work.[3] So impressed was Gerbert (*c.* 980) by the difficulties to be overcome that he gave no less than ten cases in division, beginning with units by units, treated by continued subtraction.[4] Even as late as 1424 Rollandus gave only the simplest cases with small numbers, and nearly two centuries later Hylles (1600) recognized the difficulties when he said, "Diuision is esteemed one of the busiest operations of Arithmetick, and such as requireth a mynde not wandering, or setled vppon other matters."[5]

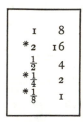

1	8
*2	16
$\frac{1}{2}$	4
*$\frac{1}{4}$	2
*$\frac{1}{8}$	1

Early Form of Division. Probably the oldest form of division is the one used by the Egyptians. This was based upon the processes of duplation and mediation. Thus, to divide 19 by 8 we may arrange the work as here shown. We take 2 × 8 = 16, $\frac{1}{2}$ of 8 = 4, and so on, and select the numbers in the right-hand column which have 19 for their sum; for example, 16 + 2 + 1 = 19. The quotient is therefore $2 + \frac{1}{4} + \frac{1}{8}$, the multipliers being marked here by asterisks.[6]

[1] *E.g.,* G. B. di S. Francesco (1689). In the case of 7 ÷ 3 = 2, and 1 remainder, or 2$\frac{1}{3}$, he uses *auanzo* for the 1 and also for the $\frac{1}{3}$. In a MS. of 1736 in the Woolwich Academy, England, "remainer" is used exclusively for "remainder."   [2] "Dura cosa e la partita" is a phrase often met.
  [3] "Peroche nulla virtus est sine labore. E questo aferma el phylosopho q̃ virtus cõsistit circa difficile" (fol. 32, *v.*).
  [4] M. Chasles, *Comptes rendus,* XVI, 284.
  [5] Fol. 37.   [6] Erman, *Egypt,* p. 365.

We are quite ignorant as to the way in which the Greeks and Romans performed the operation of division before the Christian Era. We have, however, a case described in the 4th century by Theon of Alexandria (c. 390), in which the literal numeral system of the Greeks is used and the work is not unlike our own, except that sexagesimal fractions are employed.[1] Since we know so little of the development of the operation among the ancients, we shall proceed at once to the history of the subject, showing particularly how long division was performed after the introduction of our modern numerals, say from about the year 1000.

Short Division. The simplest method, however, was the one which we call in English short division, which is based upon the recognition of the products in the columns of the multiplication table, and which has therefore been known as division by the column,[2] by rule, or by the table, as oral division, or as division in the head.[3] The method is illustrated in the Treviso book as follows:

$$\text{Lo partitore} \qquad .2. \qquad \left.\begin{matrix} 7624 \\ 3812 \end{matrix}\right| \text{o lauanzo,}$$
$$\text{La parte}$$

which means that $7624 \div 2 = 3812$, with o remainder. The arrangement used by Sfortunati (1534) for a similar case is seen in the following example taken from his arithmetic:

$$\begin{matrix} \text{P } 14 \\ \underline{1037382} \\ 74098\tfrac{10}{14} \end{matrix}$$

---

[1] The details are given in Heath, *History*, I, 58.

[2] *Per colona* (Treviso, 1478); *per cholona* (Borghi, 1484).

[3] "Partire a regolo: ouer a tauoletta" (Pacioli, 1494; 1523 ed., fol. 32). Pacioli advises: "E comenza a partire sempre da lultima (more arabū)," that is, to begin at the left as the Arabs do. If anyone claims that the method is difficult, says Pacioli, "Bonum est difficile: malum autem facile . . . Stultorum infinitus est numerus" (fol. 32, v.). "Del primo modo de partire detto per colona, ouer di testa, ouer per discorso, ouer per toletta . . . aregolo, ouer alla dritta, ouer tauoletta." Tartaglia, *General Trattato*, 1556, I, fol. 29; 1592 edition of *Arithmetica*, fol. 43.

meaning that $1{,}037{,}382 \div 14 = 74{,}098\frac{10}{14}$.[1] As with us, the method was generally used only with a divisor of one figure[2] and until recently has not been very popular with teachers,[3] requiring as it did some attention to a division table.[4]

Gerbert's Method. Of the methods which make use of our common numerals in long division, one of the oldest is often attributed to Gerbert (*c.* 980), although it is uncertain whether he originated it and although he did not use the zero.[5] It may be illustrated by the simple case of 900 ÷ 8. The process consists of dividing 900 by 10 − 2, 2 being the complement of the divisor, and was essentially as follows:

$$10 - 2\,)\,900\,(90 + 18 + 3 + 1 + \tfrac{1}{2} = 112\tfrac{1}{2}$$
$$\underline{900 - 180}$$
$$180$$
$$\underline{180 - 36}$$
$$36$$
$$30 - 6$$
$$\underline{6 + 6 = 12}$$
$$\underline{10 - 2}$$
$$2 + 2 = 4, \ \tfrac{4}{8} = \tfrac{1}{2}$$

The form actually used by certain of the successors of Gerbert may be seen from an example in an anonymous manuscript of the 12th century now in Paris,[6] no zero appearing in the computation. The combination of Roman and Hindu numerals is

---

[1] 1544/5 ed., fol. 15, under "Partire per testa."

[2] "Si chiama Partire à Colonna, quando il Partitore sarà d'vn Numero solo." Gio. Batt. di S. Francesco, 1689, p. 29.

[3] Pike's very widely used arithmetic employs long division in the cases of 175,817 ÷ 3 and 293 ÷ 8. See the 8th edition, New York, 1816, pp. 18, 60.

[4] Some 16th and 17th century writers in Italy gave a division table, and Onofrio (1670) speaks of his as "di grandissima vtilta." The Japanese learn a peculiar division table for their *soroban* and the Chinese for their *suan-pan.* See Smith-Mikami, p. 40.

[5] H. Weissenborn, *Zur Geschichte der Einführung der jetzigen Ziffern*, p. 14 (Berlin, 1892); *Gerbert*, p. 169 (Berlin, 1888).

[6] M. Chasles, *Comptes rendus*, XVI, 235, 243.

frequently seen in this period. The long explanation in the manuscript may be summarized in the following solution:

	C	X	I	
[10 − 8]			2	Differentia
			8	Divisor
	ϕ			Div^{dus}
[2 × 90]	1	8		
[2 × 10]		2		
[80 + 20]	1			
[2 × 10]		1		
[2 × 2]			4	
		1		⎫
		1		⎬ Denominaciones
		9	2	⎭
[Quotient]¹	1	1	2	

This same method is one of three given by Adelard of Bath (*Regulae abaci, c.* 1120), who attributes it to Gerbert. These three methods are the *divisio ferrea*, as above; the *divisio aurea*, somewhat like our long division; and the *divisio permixta*.²

Division by Factors. A third method of division that was common in the late Middle Ages consisted in using the factors of the divisor, and was known as "per repiego."³ By this method 216 ÷ 24 reduces to 216 ÷ 8 ÷ 3, the object being to

---

¹ The fraction ½ was neglected. The bracketed matter is not in the original.

² Of the "iron division" he says: "⟨De ferreis quidem diuisorib₃ [for "divisionibus," as in two MSS.] hec paucis dicta sufficiant. Tamen quia super his tractauit gibertus philosoph⁹ vir subtilis ingenij diligenter et compendiose quidam eciam quem discipulum eius predicant quē guichardum nominant/diligenter et prolixe." See Boncompagni's *Bullettino*, XIV, 67.

³ *Repiego* means "refolding." It appears with various spellings, often *ripiego*. In Texeda's Spanish arithmetic (1546) it appears as *repriego*. It was occasionally called "division by rule," a name also given to short division. Thus in a 14th century MS. in Mr. Plimpton's library: "Questo e partire per regola = cioe. Parti 9859 *p* 48 cioe *p*.6. & *p* .8. sua reghola. . . . fattiamo fino alpartm̄eo *p* Regholo." See also the repiego method of multiplication, page 117.

secure one-figure divisors that could be handled "per tavoletta."
The illustration given by Pacioli (1494) is that of 9876 ÷ 48.
He first divides 9876 by 6, the result being 1646. He then divides 1646 by "the other number of the repiego"[1] and obtains
205⅝ or 205¾. It is still used, although not commonly taught in
school.

Division by Parts. If the divisor was a multiple of ten, the
16th century writers frequently resorted to "Partire per il
scapezo," that is, "division by cutting up" the dividend. Thus,
to divide 84,789 by 20, the dividend was cut by a bar, 8478|9,
the first part being divided by 2 and the 9 being divided by 20,
—a plan that is found essentially in our modern books.[2]

The Galley Method. By far the most common plan in use
before 1600 is known as the galley, *batello*, or scratch, method
and seems to be of Hindu origin. It may be illustrated by the
case of 65,284 ÷ 594, as given in the Treviso arithmetic (1478).
To make the work clear, the first six steps are given separately
as follows:

(1)	(2)	(3)
65284 | 1	1	16
594	65284 | 1	65284 | 1
	594	594

(4)	(5)	(6)
5	5	5
168	168	168
65284 | 1	65284 | 10	65284 | 10
594	5944	5944
	59	599
		5

[1] " . . . dico che parta *p* laltro numero del repiego : cioe. *p* .8. neuen .205.
sani : e auāza .6." (1494 ed., fol. 33, *r.*).

[2] It is given in *Le Regoluzze di Maestro Paolo dell' Abbaco* (14th century),
ed. Frizzo, p. 43 (Verona, 1883). The relation of this to the decimal fraction
is discussed on page 238. The plan is given by many writers, including Borghi
(1484), Sfortunati (1534), Cataneo (1546), Baker (1568), Digges (1572), and
Pagani (1591).

The completed work, the explanation for which occupies two and one-half pages, is as follows:

$$\begin{array}{c} \not{1}5 \\ \not{5}\not{3}3 \\ \not{1}\not{6}\not{8}\not{7}8 \\ \not{6}\not{5}\not{7}\not{8}4 \\ \not{5}\not{9}444 \\ \not{5}\not{9}\not{9} \\ \not{5} \end{array} \Big| \ 109$$

That is, $65{,}284 \div 594 = 109$, with a remainder 538.

The method is by no means as difficult as it seems at first sight, and in general it uses fewer figures than our common plan. Maximus Planudes (c. 1340) throws some light upon its early history, saying that it is "very difficult to perform on paper, with ink, but it naturally lends itself to the sand abacus. The necessity for erasing certain numbers and writing others in their places gives rise to much confusion where ink is used, but on the sand table it is easy to erase numbers with the fingers and to write others in their places."[1] It thus appears that this method, which at first seems cumbersome, is a natural development of a satisfactory method used on the sand abacus.

$$\begin{array}{r} 6 \\ 149 \\ 18456 \\ 17 \\ 1085 \\ \tfrac{1}{7} \ 1085 \end{array}$$

It was adopted by Fibonacci (1202), as here shown for the case of $18{,}456 \div 17$.[2]

The names *galea* and *batello* referred to a boat which the outline of the work was thought to resemble.[3] An interesting

---

[1] From the French translation in the *Journal Asiatique*, I (6), 240. On the Hindu method, see Gerhardt, *Études*, p. 7.

[2] The Boncompagni edition (I, 32) gives no cancellation marks, and very likely Fibonacci made no use of them.

[3] As Pacioli says: "E q̄sto vocabulo li aduene a tale o̱pare p̱ certa similitudine materiale che li respōde del offitio e acto de la galea materiale q̄le e legno marittimo acto al nauigare" (1494 ed., fol. 34). Tartaglia remarks: "È detto in Vinetia per batello, ouer per galea per certe similitudini di figure" (1592 ed., fol. 48). The spelling varied, as usual, giving such forms as *battello, vatelo, galera,* and *galia.* There was occasionally a distinction between the *galea* and *batello* forms, as in Forestani, *Pratica d' Arithmetica*, Venice, 1603.

illustration of this resemblance is seen in a manuscript of
c. 1575, as here shown. Tartaglia[1] tells us that it was the custom of Venetian teachers to require such illustrations from their
pupils when they had finished the work.

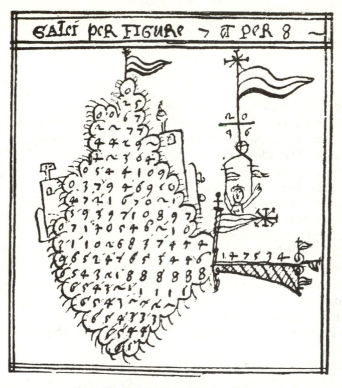

GALLEY DIVISION, 16TH CENTURY

From an unpublished manuscript of a Venetian monk. The title of the work
is "Opus Arithmeticā D. Honorati veneti monachj coenobij S. Laurētij." From
Mr. Plimpton's library

This method of dividing was used by the Arab writers
from the time of al-Khowârizmî (*c.* 825), of course with variations. For example, al-Nasavî (*c.* 1025), in finding that
$2852 \div 12 = 237\frac{8}{12}$, used the form on page 139. The advancing

[1] 1592 ed., fol. 53.

of the divisor one place to the right each time is here seen more clearly than in the usual Italian forms. The medieval Latin writers sometimes called this feature *anterioratio*.[1] This advancing of the divisor was not universal, however, Rudolff (1526) telling us that the French and other computers often set the divisor down but once.[2]

```
 12
 493
 237      237
2852        8
 12       12
 12
 12
```

As already stated, the galley method was the favorite one with arithmeticians before 1600, and it had many strong advocates up to the close of the 18th century.[3] It is found occasionally without cancel marks, probably owing in most cases to the lack of the necessary canceled types.[4] With or without this canceling, the method was preferred not merely by commercial computers but also by such scientists as Regiomontanus.[5] Even as good a mathematician as Heilbronner, in the middle of the 18th century, preferred it in all long examples.[6] One reason for this preference was, no doubt, that fewer figures were used; but even more important was the fact that the work was more compact,—an important item before

```
  OO,
  12
 0300
11 4400|4400
 26666
  222
```

FIRST EXAMPLE IN LONG DIVISION PRINTED IN THE NEW WORLD

From the *Sumario Compendioso* of Juan Diez, Mexico, 1556. It illustrates the galley method, without canceled figures, as applied to the case of $114{,}400 \div 26 = 4400$

---

[1] So Sacrobosco (*c.* 1250) uses this word and also the verb *anteriorare*. From this, no doubt, Chuquet (1484) was led to use *anteriorer*. See G. Eneström, *Bibl. Math.*, XIII (2), 54; Halliwell, *Rara Math.*, p. 17.

[2] "Frantzosen vnd etlich ander Nacion/welche den teyler nit mehr dann ein mal setzen/. . ." (1534 ed., fol. 11).

[3] Among those who preferred it to any other are Chuquet (1484), Widman (1489), Riese (1522), Tonstall (1522), Köbel (1514), Gemma Frisius (1540), Recorde (*c.* 1542), Baker (1568), Oughtred (1631), and certain Dutch writers even as late as Bartjens (1792).

[4] *E.g.*, Pellos (1492), Grammateus (1518), Albert (1534), and the Mexican work of 1556 as shown in the facsimile.

[5] See his correspondence with Bianchini in the *Abhandlungen*, XII, 197.

[6] *Historia*, pp. 776 *et passim.*

the days of cheap paper. Hodder, late in the 17th century, says that he "will leave it to the censure of the most experienced to judge, whether this manner of dividing be not plain, lineal, and to be wrought with fewer Figures than any which is commonly taught,"[1] and in this he follows the testimony of many of the best Italian writers for two centuries preceding.[2] The method is still taught in the Moorish schools of North Africa, and doubtless in other parts of the Mohammedan world.

1	4	4	(Quotient)
1	7	2	9	(Dividend)

Our Long Division. It is impossible to fix an exact date for the origin of our present arrangement of figures in long division, partly because it developed gradually. We find in various Arab and Persian works arrangements substantially like the one shown above for the case of 1729 ÷ 12 = 144, and 1 remainder.[3] This resembles our method, although it has several points in common with the galley plan.

In the 14th century Maximus Planudes gave what is called an Arab device. This is a step in advance of the one given above and yet is quite distinct from our method.[4] It appears in a form somewhat like the one here shown for the case of 625 ÷ 25.

```
25)625(25
   4
   22
   10
   125
   100
    25
    25
```

[1] 1672 ed., p. 54.

[2] Thus Pagani (1591) : "Il partire a Galera e molto sicuro & legiadro ch' ogn' altro partire," and Pacioli is even more pronounced in his opinion.

[3] This is a composite of solutions in various MSS. examined, including several of the 16th century. See also the work of al-Kashî (c. 1430) as referred to in Taylor, *Lilawati*, Introd., p. 22 ; Gerhardt. *Études*, p. 14.

[4] Gerhardt, *Études*, p. 22.

The 15th century saw the method brought into its present form under the name *a danda* ("by giving"). This name came

from the fact that when a partial product is subtracted we bring down the next figure and "give" it to the remainder.[1] An excellent illustration from a manuscript of *c.* 1460 is here shown, but it will be noticed that the remainder is repeated each time before the "giving." The name *danda*, or *dande* in parts of Tuscany, is still used to designate this method of dividing.[2] It has, however, been applied to forms quite different from the one shown above. For example, the case of 49,289 ÷ 23 = 2143 appears in the form shown below in a 14th century Italian manuscript,[3] and the author speaks of an analogous solution as *a danda*.[4] The earliest printed book to give the method is Calandri's work of 1491, and the first example of the kind is shown on page 142. It next appeared as the third method of Pacioli,[5] and was given with increasing

EARLY EXAMPLE OF LONG DIVISION

One of the earliest examples of the present method. From an Italian MS. of *c.* 1460

frequency in the following century, but rather as an interesting than as a particularly valuable device.[6] With the opening of the 17th century it began more effectively to replace the galley

---

[1] So Cataneo (1546) says: "È chiamato a danda il detto modo, perche à ogni sottration fatta nel operare se li da vna o piu figure dal lato destro" (1567 ed., fol. 15). [2] Boncompagni's *Bullettino*, XIII, 252 n.

[3] *Rara Arithmetica*, p 437. [4] "Questo sie' ilpartire adanda."

[5] "De tertio modo diuidendi dicto danda" (1494 ed., fol. 33).

[6] Pagani (1591) speaks of it thus: "Partire à danda è assai bello, & vago." Cognet (1573) mentions the advantage of not canceling: "Les Marchands Italiens, pour ne trencher aucune figure, divisent en la sorte qui s'ensuit"; and Trenchant (1566) remarks: "Il y a vne autre belle forme de partir, sans trencher aucune figure," or "sans rien couper."

method. Cataldi (1602) gives it as his first method, but with the quotient below the dividend, the first part of his work

FIRST PRINTED EXAMPLE OF MODERN LONG DIVISION

From Calandri's arithmetic, Florence, 1491. The problem is the division of 53,497 by 83

being as shown on page 143. In another example he places the quotient at the right, saying that this is the custom in

Milan.[1] In the galley method the most convenient place for the quotient was at the right; Cataldi's attempt at placing it below was awkward; the modern custom of placing it above the dividend in long division is the best of all, since it automatically locates the decimal point.

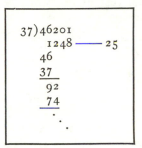

At the close of the 17th century the modern form of division was fairly well established, the galley method being looked upon more as a curiosity.[2]

There have been many variants of the *a danda* method, but the only one of any importance is that which omits the partial products as shown below. Cataldi (1602) calls it the abbreviated *a danda*.[3] It has had more or less vogue for three centuries, but it requires too much mental effort to become common. It was brought to the attention of American teachers by Greenwood (1729), who, speaking of the various methods, remarked that, as "most of the rest are at best an unnecessary Curiosity; I shall confine myself wholly to the Two ITALIAN Methods; which are the most usual," these two being *a danda* and the contracted form.

$$25)\overline{15625}\ (625$$

Of the various methods suggested in the 16th century one of the most interesting is that of Apianus (1527), particularly as it suggested the scheme of decimal fractions. To divide 11,664 by 48, Apianus first writes the aliquot parts of 48, with a corresponding series of numbers based on 48 as a unit, substantially as follows:

		48	corresponds to	1
$\frac{1}{2}$	of	48 = 24,		05
$\frac{1}{4}$	of	48 = 12,		025
$\frac{1}{8}$	of	48 = 6,		0125
$\frac{1}{16}$	of	48 = 3,		00625

[1] "Partire à Danda vsato in Milano."
[2] Thus Onofrio's *Aritmetica* (1670) gives it as "di poco ò nullo profitto."
[3] " . . . à Danda abbreuiato" (p. 88).

He then observes that $11 \div 48 \not> 1$, $11 \div 24 \not> 1$, $11 \div 12 \not> 1$, but $11 \div 6 > 1$. But $6 = \frac{1}{8}$ of 48, and hence the first part of the quotient is 0125. The rest of the work is substantially as follows:

$$11 - 6 = 5$$
$$5 \div 48 \not> 1$$
$$5 \div 24 \not> 1$$
$$\vdots$$
$$5 \div 3 > 1$$

```
        2
       ⁄22
      11664
      ────
      0125
      0062   5
      ○5
      ○5
      ────
      ○   5
Facit  243
```

hence we write $\frac{1}{16}$, or 00625, and so on. It is evident that Apianus had some idea of decimal fractions in his mind, although it was not developed in his treatise.

Clichtoveus (1503) gave a rule based upon the identity

$$\frac{10\left(a - \dfrac{c}{2}\right) + b}{c} + 5 = \frac{10\,a + b}{c}.$$

Thus, to find $29 \div 4$, take $a - \dfrac{c}{2} = 2 - \dfrac{4}{2} = 0$; then subtract 4 (or c) as often as possible from 9 (or b), thus finding that $9 \div 4 = 2\frac{1}{4}$. The final quotient[1] is then $2\frac{1}{4} + 5$, or $7\frac{1}{4}$.

Whatever method of dividing was used, a table of multiples of the divisor was early recognized as desirable. Such tables are found in many works, including those of Recorde ($c.$ 1542), Fine (1530), Ramus (1569), Hylles (1592), and Greenwood (1729).

7. Roots

Finding Square and Cube Roots of Numbers. The Greeks found the square root of a number by a method similar to the one commonly set forth in the elementary algebras and arithmetics of the present time. It was shown geometrically by Euclid[2] that $(a + b)^2 = a^2 + 2ab + b^2$ (a fact that was prob-

---

[1] Edition of $c.$ 1507, fol. D 4. He also gives a rule for the case of $a < \frac{1}{2}\,c$. For a few further notes on the history of division see E. Mathieu, "Méthodes de division en usage à la fin du siècle dernier," in *Journal de math. élément.,* V (4), 97.   [2] *Elements,* II, 4.

ably known long before his time), and by means of this relation Theon of Alexandria (*c.* 390), using sexagesimals, found the square root of a number by the following rule:

When we seek a square root, we take first the root of the nearest square number. We then double this and divide with it the remainder reduced to minutes, and subtract the square of the quotient; then we reduce the remainder to seconds and divide by twice the degrees and minutes [of the whole quotient]. We thus obtain nearly the root of the quadratic.[1]

By this rule he finds that $\sqrt{4500}$ = 67 4' 55" approximately.

From Greece the method passed over to the Arabs and Hindus, with no particular improvement. Thus Bhāskara (*c.* 1150) writes his number as follows:

$$| \quad - \quad | \quad - \quad |$$
$$8 \quad 8 \quad 2 \quad 0 \quad 9$$

and then proceeds much as Theon had done. He says:

Having deducted from the last of the odd digits[2] the square number, double its root; and by that dividing the subsequent even digit[3] and subtracting the square of the quotient from the uneven place,[4] note in a line the double of the quotient.[5]

2^1	3^8	5	
1	5		10
	2		10

One of the most interesting medieval examples of the finding of a square root is given by Maximus Planudes (*c.* 1340). To find $\sqrt{235}$ he arranges the work as here shown. This is quite

---

[1] For further details of the process see J. Gow, *History of Greek Mathematics,* pp 54–57 (Cambridge, 1884) (hereafter referred to as Gow, *Greek Math.*); K. Hunrath, *Ueber das Ausziehen der Quadratwurzel bei Griechen und Indern,* Prog., Hadersleben, 1883.

[2] That is, from 8, the third and last of the odd places denoted by a vertical line, counting from the right.

[3] Really, $(882 - 400) \div 40 = 9+$.

[4] Apparently meaning that 29^2 is subtracted from 882.

[5] For the rest of the rule see Colebrooke's *Lilāvati,* p. 9. For the work of the Arabs and Persians, see the Taylor translation, p. 23.

unintelligible without the accompanying explanation, which
may be condensed as follows:[1]

$$235(15$$

$$\frac{1}{13}$$

$$
\begin{array}{ll}
& 2 \\
2 \times 5 = & 10 \\
& \overline{30} \qquad = \text{twice the root.} \\
\text{Hence} \quad 15 \qquad = \text{the root.} \\
\text{But} \quad 15^2 \qquad = 225.
\end{array}
$$

Hence $\dfrac{235 - 225}{30} = \dfrac{1}{3}$, which must be added.

Hence the root is $15\frac{1}{3}$.

The early printed arithmetics generally used an arrangement
of figures similar to the one found in the galley method of division. Thus Pacioli (1494) gives the following:[2]

Extractio radicū

$$
\begin{array}{l}
00 \\
0\cancel{1}\cancel{8} \\
\cancel{1}\cancel{1}\cancel{1}0 \\
\cancel{2}\cancel{0}\cancel{8}\cancel{8}0 \\
0\cancel{9}\cancel{9}\cancel{6}\cancel{9}\cancel{8}0 \\
\cancel{1}\cancel{8}\cancel{7}\cancel{7}\cancel{8}\cancel{9}\cancel{8}0 \\
\cancel{9}\cancel{9}\cancel{9}\cancel{8}\cancel{0}\cancel{0}\cancel{0}\cancel{1}\ \underline{|9999} \\
\cdot\ \cdot\ \cdot\ \cdot \\
\cancel{9}8\cancel{9}8\cancel{9}8\cancel{9} \\
11999 \\
1
\end{array}
$$

that is, $\sqrt{99{,}980{,}001} = 9999$.

Gradually, in the 16th century, the galley method gave way
to our modern arrangement, although it was occasionally used
until the 18th century.[3] Among the early writers to take an

---

[1] For examples of a more elaborate nature see Wäschke, *Planudes.*

[2] Fol. 45, *r.* For those who do not have access to original works a good illustration of this method may be seen in the *Abhandlungen*, XII, 201, 269. A problem
of Chuquet's (1484) may also be seen in Boncompagni's *Bullettino*, XIII, 695.

[3] Among the better arithmeticians that used it in the 17th century was
Wilkens, a Dutch writer (1630).

important step toward our present method was Cataneo (1546),
who arranged the work substantially as follows:[1]

```
54756(234
  4          primo duplata  4
 14          secondo       46
 12
 27
  9
185
184
 16
 16
  0
```

Among the first of the well-known writers to use our method
in its entirety was Cataldi, in his *Trattato* of 1613.[2] Most
early writers gave directions for "pointing off" in periods of
two figures each, some placing dots above, as in $8\overset{.}{2}44\overset{.}{6}4$[3];
some placing dots below, as in $11\underset{.}{9}025$[4] or as in $21\,17\,\overset{.}{8}4\,\overset{.}{0}4$[5];
some using lines, as in $26\,00\overset{.}{0}\,00\overset{.}{0}\,000$[6]; some using colons,
as in $13:01:76:64$:[7]; and some using vertical bars, as in
$94|21|80|73|55$.[8] Many writers, however, did not separate the
figures into groups.[9]

[1] *Le Pratiche*, Venice, 1567 ed., fol. 72.

[2] For various forms used by other writers see P. Treutlein, *Abhandlungen*,
I, 64, 71.

[3] *E. g.*, Grammateus (1518), Scheubel (1545), Hartwell (1646 edition of Recorde's *Ground of Artes*), Wilkens (1630), and the American Greenwood (1729).

[4] *E.g.*, Gemma Frisius (1540), L. Schoner (1586), Peletier (1549), Santa-Cruz
(1594), and Metius (1625). Cardan sometimes placed them above and sometimes below.

[5] This from the *Epitome* of Clavius (1583 ; 1585 ed., p. 309), although generally (as on page 310) he places the dots immediately below the figures.

[6] This in cube root, from the Rollandus MS. (1424).

[7] From Ortega (1512 ; 1515 ed., fol. 99). He also writes 3 : 6 : 0 : 8 :
for the square root.

[8] This was very common and has much to commend it. It was given by
Chuquet (1484), Pellos (1492), Fine (1530), Trenchant (1566), and many
others.

[9] *E.g.*, the Arab al-Ḥaṣṣâr (*c.* 1175), Cataneo (see the example above), and
Feliciano da Lazesio (1526).

In finding the square root and the cube root most of the early writers[1] gave the rules without any explanation, or at the most with merely a reference to the fact that $(a + b)^2 = a^2 + 2ab + b^2$. Thus Buteo (1559) proceeded no farther with cube root than to find the first figure, saying that it is better to use a table of cubes;[2] and more than a century later de Lagny[3] asserted that it would take most computers more than a month to find the cube root of 696,536,483,318,640,035,073,641,037. Although the ponderous work of Tonstall (1522) naturally included roots, Recorde (c. 1542) did not think the subject worthy of a place in his *Ground of Artes*.[4]

A conviction of the value of the reasoning involved in the subject led various writers in the 16th century to give clear explanations based on the geometric diagram.[5] The use of the blocks for explaining cube root was found somewhat later, and became fairly common in the 17th century.[6] In the 17th and 18th centuries the blocks are even used in finding the fourth root, x cubes being taken, each composed of x^3 cubes.[7]

[1] *E.g.*, Brahmagupta (c. 628) and Bhāskara (c. 1150), ed. Colebrooke, pp. 10, 279; al-Karkhî (c. 1020), ed. Hochheim, II, 13. Fibonacci (1202) described cube root, and it also appears in Sacrobosco's *Algorismus* (c. 1250) and in the *Carmen de Algorismo* of Alexandre de Villedieu (c. 1240).

[2] He gives such a table up to 40^3, and a rule which we may express by the approximation formula

$$\sqrt[3]{a^3 + r} = a + \frac{r}{3a^2 + 3a}.$$

[3] *Nouveaux Élémens d'Arithmétique et d'Algèbre*, Paris, 1697; A. De Morgan, *Arithmetical Books*, p. 55 (London, 1847) (hereafter referred to as De Morgan, *Arith. Books*).

[4] In Hartwell's edition, however, there is "An Appendix concerning the Resolution of the Square and Cube in Numbers, to the finding of their side," in which he speaks of the "Quadrat root, or the side of any Quadrat number," and gives the geometric diagram (1646 ed., p. 573).

[5] *E.g.*, Tonstall (1522, fol. TV_3), Trenchant (1566), L. Schoner (1586, p. 255), and Gemma Frisius (1540).

[6] A good illustration is found in Hartwell's edition of Recorde's *Ground of Artes* (1646 ed., p. 587).

[7] *E.g.*, Cardinael (1644; 1659 ed., fol. E 8). When Bartjens (1633; 1752 ed., pp. 242, 243) wishes to know "Hoe veel is de $\sqrt{xx}$ van 576" or "de Radix xx uit 3136 is," such being his two symbols for square root, he uses the diagram. He then uses the blocks in "kubicq-wortel" when he "trekt de $\sqrt{x^3}$ uit 5832" (p. 251), and in fourth root when he "trekt de $\sqrt{x^4}$ van 81450625." See also the 1676 edition, p. 242.

Higher Roots. The conviction of the value of the subject as a mental exercise led various writers to include some work in higher roots, the work being based upon a knowledge of the binomial coefficients. These coefficients were occasionally arranged in the triangular form subsequently known as Pascal's Triangle.[1] This arrangement was known to the Chinese[2] as early as 1303, and also to the Arabs,[3] and in Europe it appeared in print on the title-page of a work by Apianus published in 1527 and in a work[4] by Scheubel that appeared in 1545.

This arrangement of the binomial coefficients was first seriously considered in a printed book, in connection with higher roots, simultaneously by Stifel (1544)[5] and Scheubel (1545).[6] The latter finds the tenth root of 1,152,921,504,606,846,976, for example, to be 64, and he carries the work as far as to the finding of a 24th root. A little later it was used in France by such writers as Trenchant (1566)[7] and Peletier (1549),[8] and it appeared also in the works of various Dutch writers.[9]

---

[1] *Traité du triangle arithmétique,* published posthumously in 1665. The form used by Pascal is given later (p. 510).

[2] It appears in the *Szu-yuen Yü-kien* of Chu Shï-kié (1303), but as something already known. See Mikami, *China,* p. 90.

[3] Cantor, *Geschichte,* I (2), 645. [4] *Rara Arithmetica,* pp. 156, 236.

[5] In the *Arithmetica Integra,* fol. 44. As to their use in his *Coss* (1554), see *Abhandlungen,* I, 77 ; II, 43.

[6] *De Numeris,* in the *tractatus quintus.* See *Rara Arithmetica,* p. 236.

[7] "Doctrine generale pour extrére toutes racines." He also says: "Pour fondement de la quelle, i'ay formé ce trigône semé de nombres, s'imbolisans & s'engendrans les vns les autres par vn ordre de grandissime consideration" (1578 ed., p. 249). It will be observed that, by placing 1 at each end of each row, the successive rows give the coefficients in the expansion of $(a + b)^n$ for n equal to 2, 3, 4, $\cdots$. This serves as a basis for the general rule for finding the nth root of any number. For example, to find the fifth root we observe that the arithmetic triangle gives the trial divisor as $5a^4$ and the complete divisor as $5a^4 + 10a^3b + 10a^2b^2 + 5ab^3 + b^4$, a principle well known to writers of the 16th century.

$$
\begin{array}{c}
2 \\
3 \cdot 3 \\
4 \cdot 6 \cdot 4 \\
5 \cdot 10 \cdot 10 \cdot 5 \\
\text{etc}
\end{array}
$$

[8] He speaks of it as a "Nouuelle maniere d'extraire les Racines, generale pour toutes extractions, jusques à infinité" (1607 ed., pp. 107, 178, 252).

[9] Thus Van der Schuere (1600) speaks of the "Drie-hoecks wijze" (triangle-like) arrangement. It is also used by Bartjens (1633), Cardinael (1644), and others.

Abbreviated Methods. Attempts at abbreviating the process are relatively late. One of the most popular rules for the abridgment of square root is attributed to Newton, and Greenwood (1729) gives it as follows:

> SIR Isaac Newton takes notice of a very useful *Contraction*, in these Cases, viz. *That when a Root is carried on half way or above, the Number of Figures you intend it shall consist of; the remaining Figures may be obtained by Dividing the remainder by the double of the Radical Figures.*[1]

The Meaning of the Term. It should be stated in this connection that the use of "root" to mean the square root, common in Europe today, has historic sanction. Indeed, all the world still recognizes it by taking the symbol $\sqrt{a}$ instead of $\sqrt[2]{a}$ to indicate the positive square root of a. The usage, however, was not entirely general, many early writers specifying the square root as carefully as the cube root.[2]

The Arab writers conceived a square number to grow out of a root, while the Latin writers thought of the side of a geometric square. Hence the works translated from the Arabic have *radix* for a common term, while those inherited from the Roman civilization have *latus*.[3] Hence the Latin writers "found" the *latus* and the Arab writers "extracted," or pulled out, the root. Our arithmetics, based largely upon Arab sources, still use "extract," although the older usage of "find" is better. The fact that from *radix* we have both "radical" and

---

[1] P. 77. See Newton's *Arithmetica Universalis*, p. 33 (Cambridge, 1707): "Ubi vero radix ad medietatem aut ultra extracta est, cæteræ figuræ per divisionem solam obtineri possunt."

[2] Thus Suevus (1593), under his "Regvla qvadrata," gives "Extractio Radicis Quadratae," and Digges (1572) speaks of "the square *Radix*," "quadrat roote," and "quadrate root."

Among the early writers who used "root" for "square root" were al-Nasavî (*c*. 1025), L. Schoner (1586), Rollandus (1424), and probably Bhāskara (*c*. 1150; Taylor, *Lilawati*, introduction, p. 6).

[3] Schoner speaks of this in *De numeris figuratis liber*, appended to his 1586 edition of Ramus: "Sic 9 est aequilaterus, & latus ejus est 3. Hoc latus aequilateri ab Arabibus etiam dicitur Radix" (p. 3). Fibonacci (*Liber Abaci*, p. 353) uses "find" instead of "extract" with the word "root," having used "extract" for "subtract."

"radish" makes the use of "extract" more easily understood.[1] This use is found in various modern languages,[2] but is by no means universal. Thus Digges (1572) says, "To find the square *Radix*, or Roote of any number" (p. 13), although he also says, "to search or pull out the *Radix*, or roote cubical" (p. 16).[3]

8. CHECKS ON OPERATIONS

Need for Checks. The fact that the intermediate steps in a long operation were erased on the various forms of the abacus rendered it impossible to review the work as may be done with our present methods. It was therefore necessary that some simple check should be used to determine the probable accuracy of a result. The inverse operation was generally too long to serve the purposes, and hence other methods were developed rather early.

Check of Nines. Of all these methods the check of nines is probably the best known. It is simple of application and serves to detect most of the errors that are likely to occur. The origin of the method is obscure. It is found in the works of various Arab writers, including al-Khowârizmî (*c.* 825), al-Karkhî (*c.* 1020), Behâ Eddîn (*c.* 1600), and others. Avicenna (*c.* 1020), however, in discussing the subject of roots, speaks of it as a Hindu method.[4] On the contrary, no Hindu writer is

---

[1] On the use of "root" see Wertheim's edition of Elia Misrachi (*c.* 1500), *Sefer-Hamispar*, p. 20 (Frankfort a. M., 1893), and Tartaglia's *General Trattato*, II, fol. 53, *v.* (1556).

[2] *E. g.,* "Uyttreckinge der wortelen" (Cardinael, 1659 ed., p. 2), and "cavere la radice qvadra" (Ciacchi, 1675 ed., p. 335).

[3] Of other forms of expression the following are types: "7097, cuius tetragonicū latus inquirens . . .," Buteo (1559; 1560 ed., p. 71); " . . . sacar rayz quadrada . . .," Iuan Perez de Moya (1562; 1615 ed., fol. 223), *sacar* meaning to extract; "Del trare la radice de numeri quadrati" (fol. 15), but "Del trouare la radice Cubica" (fol. 18, *v.*), in the Italian translation of Fine (Venice, 1587), showing both "extract" and "find"; "Del modo di trar la radice quadra . . ." (fol. 182, *v.*), but "La estrattione delle radici cube" (fol. 187, *v.*), Forestani, *Pratica d' Arithmetica . . .*, Venice, 1603.

[4] "Fâ' l-tharîk al-hindacî," an expression that has been variously interpreted. See F. Woepcke, *Journal Asiatique*, I (6), 500; Carra de Vaux, "Sur l'histoire de l'arithmétique arabe," *Bibl. Math.*, XIII (2), 33.

known to have used it before the 12th century,[1] while the Arabs certainly used it early in the 9th century. Nevertheless, as careful a writer as Paul Tannery is convinced that the evidence at present points to its invention in India but to its first considerable use in the School of Bagdad.[2]

There is some interesting evidence of the recognition of the excess of nines in the number mysticism of one of the late Greco-Roman writers, Hippolytus, who seems to have lived in the 3d century and who wrote several theological treatises as well as a *canon paschalis*. He made no use of the principle, however, in the verification of computations, and so far as we know he was ignorant of this application of the theory.[3] What he did was to make use of gematria, as in estimating the relative ability of individuals by means of the numerical values of the letters of their names. Instead, however, of simply stating this value in the usual way, he stated it with respect to the modulus nine. For example, the numerical value of Hector ("Εκτωρ) is 1225, but Hippolytus gave it as 1, which is the excess of nines in this number. He spoke of this plan as due to the Pythagoreans, meaning, no doubt, the Neo-Pythagoreans of a period much later than that of Pythagoras himself.

The check of nines seems to have come into general use in the 11th century, largely due to the influence of Avicenna (c. 1020) and his contemporary, al-Karkhî, and thereafter it is found in most of the other arithmetics of any importance for a period of about eight hundred years. Albanna (c. 1300) speaks of the Arab arithmeticians as giving proofs of their computations by the checks of 7, 8, 9, and 11, and as knowing of the checks by other numbers as well.

From the Arabs this method of checking passed over to the West, appearing in the works of the Hebrew-Arabic writer Kûshyâr ibn Lebbân (c. 1000), the Hebrew Rabbi ben Ezra

---

[1] G. R. Kaye, *Indian Mathematics*, p. 34 (Calcutta, 1915), hereafter referred to as Kaye, *Indian Math.*; Taylor's *Lilawati*, p. 7.

[2] P. Tannery, *Mémoires Scientifiques*, I, 185 (Paris, 1912). On the Arab writers see Boncompagni, *Trattati*, I, 12; *Bibl. Math.*, II (3), 17; XIII (2), 33; Hochheim, *Kâfî fîl Hisâb*, II, 10 n.

[3] P. Tannery, *Mémoires Scientifiques*, I, 185; Tropfke, I (2), 58.

($c.$ 1140),[1] the Hebrew-Christian Johannes Hispalensis ($c.$ 1140), and the Christian writers Fibonacci (1202), Maximus Planudes ($c.$ 1340), and their successors.

Fibonacci called the excess of nines the *pensa* or *portio*[2] of the number, and used it as a check in multiplication and division. Maximus Planudes arranged his work in multiplication as here shown, using 9 instead of 0 in the case of a zero excess, and apparently believing that the check was a complete one. Johannes Hispalensis and Fibonacci, however, recognized its limitations.

294849	9
543	3
543	3

In the early printed arithmetics the check is found quite generally. Pacioli (1494) speaks of it as "corrente mercatoria e presta,"[3] and Widman (1489) always concludes his operations by the query, "Wiltu probirn?" Scheubel (1545) considered the matter so important that he gave a table of multiples of nine for the convenience of computers.[4]

The failure of the check was considered at some length by Pacioli, but Clavius[5] was especially clear in his treatment of the case. So important was the whole matter considered that Santa-Cruz (1594) devoted twenty-two pages to the theory.[6] In the 17th century, owing to the general acceptance of the modern forms of computing, the revision of the operations became more simple, and hence some of the leading commercial arithmetics[7] discarded the check of nines. In England, however, the influence of Cocker[8] served to make it very popular, and such influence as Greenwood (1729) had in America was in the same direction. In the 19th century it dropped out of American arithmetics for the most part, but after 1900 it began to appear again.

So important did Tartaglia (1556) consider the check of nines, even in addition, that he gave a table of the excess of

[1]Silberberg, *Sefer ha-Mispar*, p. 94.   [2] *Liber Abaci*, I, 8.
[3]Fol. 20 [numbered 10], *r*.
[4]*De Nvmeris*, I, chap. 2, p. 12 (1545). He did the same for 7, 11, 13, and 19, using these numbers also for checking.   [5]*Epitome*, p. 22 (1583).
[6] 1643 ed., fol. 171. See also Sfortunati (1534; 1545 ed., fol. 8); Cataneo (1546; 1567 ed., fol. 18); and Pagani, p. 6 (1591).
[7] *E.g.*, Eversdyck's edition of Coutereels, p. 33 (1658); Mots (1640).
[8]*Arithmetick*, London, 1677, with later editions.

nines for each number from o to 90,[1] a waste of space that argues for the lack of appreciation of the ease with which one casts out the nines in any number, however large.

Checks with Other Numbers. Any other number besides nine may be used for checking, although nine is the most convenient. The use of other numbers is found in the works of various Arab writers, and Fibonacci[2] gives the checks for 7, 9, 11. Other medieval and Renaissance writers[3] also give such numbers as 2, 3, 5, 6, 13, and 19. Several of the early printed books show a preference for 7 on account of the diminished chance of error.[4] In general, however, they naturally give the proof by nines the preference.[5]

Inverse Operation. Although the check by the inverse operation took more time, it was more certain, and hence it found many advocates. It is so simple that its origin is probably remote, although it is not until the Middle Ages that we find it first stated definitely.[6] It appears frequently in the early printed books,—for example, in the works of Clichtoveus (1503), Albert (1534), and Thierfelder (1587). Tartaglia (1556) asserted that the method was illogical, since subtraction could not be used in checking addition, for the reason that it was taught after that subject,[7]—an objection that is of no practical significance.

---

[1] *General Trattato,* I, fols. 8, *v.,* and 9, *r.*

[2] *Liber Abaci,* pp. 8, 39, 45.

[3] B. Boncompagni, *Atti Pontif.,* XVI, 519. Rudolff (1526), Apianus (1527), Fischer (1549), Albert (1534), and Scheubel (1545) are particularly worth consulting.

[4] Thus Pellos (1492), comparing 7 with 2, says: "⫸Item sapias che ₚba de .7. es la plu segura ₚba che pusca esser car la ₚba de .2." (fol. 18). See also Borghi (1484).

[5] Thus Clavius (1583) prefers the proof "per abiectionem nouenarij" or, in the Italian edition, "col gettar via tutti li 9," to that "per abiectionem septenarij" or "col gettar via li 7"; and so with Chuquet, Pacioli, Buteo, Tartaglia, Cardan, and many others. The proof by other numbers than 9 and 11 is not often found after about 1600.

[6] For example, in the *Algorismus prosaycus magistri Christani* (c. 1400): "Et nota, quod subtraccio probat addicionem et addicio subtraccionem." Studnička ed., p. 9 (Prag, 1893).

[7] See the *General Trattato,* I, 8, *r.*

TOPICS FOR DISCUSSION

1. The number and the nature of the fundamental operations, and the reasons for the various classifications.

2. Significance of duplation and mediation in the development of logistic, particularly in early times.

3. Difficulties in adequately defining the fundamental operations as their nature expanded from time to time.

4. The leading principles determining systems of notation, with illustrations of each principle.

5. The leading systems of notation, with a study of their respective merits.

6. The significance and growth of the concept of place value in the writing of numbers.

7. The history of the Roman numerals, with a study of the variants from century to century.

8. The nature, history, and significance of the subtractive principle in the writing of numbers.

9. The history of our common numerals, with a study of the variants from century to century.

10. The reading and writing of large numbers at various periods and in various systems.

11. The terminology used from time to time in connection with the common operations.

12. Significant features of the work in addition and subtraction at different stages of the development of these operations and a study of the relative merits of the various methods.

13. A study of the different methods of multiplying, with a consideration of the relative merits of each and of the probable reason for the survival of the present common method.

14. A study of the different methods of division, with particular reference to the contest between our present plan (a modification of the *a danda* arrangement) and the galley method.

15. Traces of early methods of computations in our present operations with algebraic polynomials.

16. The historical development of the process of finding roots of numbers.

17. The historical development of the various methods of checking operations with integers.

CHAPTER III

MECHANICAL AIDS TO CALCULATION

1. THE ABACUS

Necessity for the Abacus. Since the numerals of the ancients were rather unsuited to the purposes of calculation, it is probable that some form of mechanical computation was everywhere necessary before the perfecting of the modern system. This probability becomes the stronger when we consider that all convenient writing materials were late developments in the history of civilization. Papyrus was unknown in Greece before the 7th century B.C., parchment was an invention of the 5th century B.C.,[1] and paper is a comparatively recent product,[2] while tablets of clay or wax were not suitable for calculation.

Meaning of the Term. In earliest times the word "abacus"[3] seems to have referred to a table covered with sand or with fine

---

[1] Pliny says, of the 2d century.

[2] It may have been brought into Europe in the 12th century by the Moors of Spain, but specimens dating from about the beginning of our era have been found on the eastern borders of China.

[3] The word comes from the Greek ἄβαξ (a'bax), probably from the Semitic אבק (abq), dust. Numerous other etymologies have been suggested. Among the most interesting is one given by Th. Martin (*Les Signes Num.*, p. 34) on the authority of Orion of Thebes, a lexicographer of the 5th century, and on that of several other scholars,—namely, that the word comes from α + βάσις (a + ba'sis, without base), referring to the fact that the computing tablet had no feet. A recent article by R. Soreau gives the improbable suggestion that ἄβαξ simply meant a numerical table, and came from α', β', + ἄξια (a, b, ax'ia, relating to value), meaning 1, 2, + αξ (indicating numerical values). See R. Soreau, "Sur l'origine et le sens du mot 'abaque,'" *Comptes rendus*, CLXVI, 67. The question was debated even in Pacioli's time, for he says (*Sūma*, fol. 19, *r.* (1494)): "e modo arabico e chiamase Abaco: ouer secōdo altri e dicta Abaco dal greco vocabulo." Of the various guesses, that of Joannes de Muris (*c.* 1350) is the most curious, that "abacus" is the name of the inventor: "Non est sub silencio transeundum de tabula numerorum, quam abacus adinuenit" (*Quadripartitum*, chap. xiv; in the *Abhandlungen*, V, 144).

dust, the figures being drawn with a stylus and the marks being erased with the finger when necessary. This at any rate is the testimony of etymology, and the dust tablet seems to have been the earliest form of the instrument.[1]

While all definite knowledge of the origin of the abacus is lost, there is some reason for attributing it to Semitic rather than to Aryan sources.[2]

The dust abacus finally gave place to a ruled table upon which small disks or counters were arranged on lines to indicate numbers. This form was in common use in Europe until the opening of the 17th century, and persisted in various localities until a much later date.

Meanwhile, and in rather remote times, a third form of abacus appeared in certain parts of the world. Instead of lines on which loose counters were placed there were grooves or rods for movable balls or disks, a form still found in Russia, China, Japan, and parts of Arabia.

We have, then, three standard types,—the ancient dust board, which probably gave the name to the abacus, the table with loose counters, and the table with counters fastened to the lines. These three, with their characteristic variants, will now be explained.[3]

The Dust Abacus. The dust abacus was merely a kind of writing medium of little greater significance in computation

---

[1] C. G. Knott, "The abacus in its historic and scientific aspects," *Transactions of the Asiatic Society of Japan*, Yokohama, XIV, 18 ; hereafter referred to as Knott, *Abacus.*   [2] Knott, *Abacus*, pp. 33, 44.

[3] The literature of the subject is extensive. The following are some of the general authorities consulted: Knott, *Abacus*; M. Chasles, *Comptes rendus*, XVI, 1409 ; F. Woepcke, *Journal Asiatique*, I (6), 516 ; Sir E. Clive Bayley, *Journal of the Royal Asiatic Society*, XV (N. S.) ; M. Hübner, "Die charakteristischen Formen des Rechenbretts," *Zeitschrift für Lehrmittelwesen und pädagogische Literatur*, II, 47 ; D. Martines, *Origine e progressi dell' aritmetica*, p. 19 (Messina, 1865) (hereafter referred to as Martines, *Origine aritmet.*) ; A. Terrien de Lacouperie, "The Old Numerals, the Counting Rods and the Swan-pan in China," *Numismatic Chronicle*, III (3), 297–340, reprinted in London in 1888 (hereafter referred to as Lacouperie, *The Old Numerals*). The most elaborate and scholarly work on the subject is F. P. Barnard, *The Casting-Counter and the Counting-Board*, Oxford, 1916 (hereafter referred to as Barnard, *Counters*).

than the clay tablet of the Babylonians, the wax tablet of the Romans, the slate of the Renaissance period, or the sheet of paper of today. In its use, however, is to be found the explanation of certain steps in the operations with numbers, and on this account it deserves mention.

The Hindus seem to have known this type in remote times but to have generally discontinued its use. Even in recent times, however, children have been instructed to write letters and figures in the dust or sand on the floor of the native school before being allowed to use the common materials for writing.[1] That the dust abacus was common a century ago is asserted by Taylor in the preface to his edition of the *Lilāvati*.

In the Greek and Roman civilizations the dust abacus was also well known. Figures were drawn upon it with a stylus, called by the Latin writers a *radius*,[2] much as they were drawn on the slate in recent times. The wax tablet, described later, was even more extensively used.

Nature of the Counter Abacus. As in the case of all such primitive instruments, the origin of the counter abacus is obscure.[3] We only know that in very early times there seems to have been a widespread knowledge of some kind of instrument in which objects (beads, disks, or counters) on one line indicated units, on the next line tens, on the next hundreds, and so on. Some general idea of this instrument may be obtained from the illustrations given on page 159. The first one shows the successive steps taken in the addition of numbers. The second illustration shows the use of the abacus in multiplication. Several variants of this type are given later.

---

[1] Sir E. Clive Bayley, *Journal of the Royal Asiatic Society*, XV (N.S.), 9 n., 15, and XIV (N.S.), Part 3 (in the reprint of the article "On the genealogy of ancient numerals" it appears in Part II, p. 71; see also Part I, p. 19, and Part II, pp. 50, 54); G. R. Kaye, "The use of the abacus in Ancient India," *Journ. and Proc. of Asiatic Soc. of Bengal*, IV (2), 293.

[2] "Ex eadem urbe humilem homunculum a pulvere et radio excitabo, qui multis annis post fuit Archimedes," Cicero, *Tusculan Disputations*, V, 23, 64; "Descripsit radio," Vergil, *Eclogues*, III, 41.

[3] On the history in general see A. Nagl, *Die Rechenpfennige und die operative Arithmetik*, Vienna. 1888.

There is some reason for believing that this form of the abacus originated in India, Mesopotamia, or Egypt. The whole

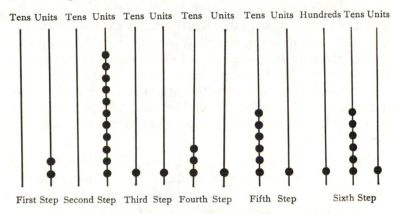

Tens Units	Tens Units	Tens Units	Tens Units	Tens Units	Hundreds Tens Units
First Step	Second Step	Third Step	Fourth Step	Fifth Step	Sixth Step

ADDITION ON THE ABACUS

An early computer, wishing to add 22 and 139, might have proceeded as follows: Place 2 pebbles on the units' line, as shown in the First Step. Then place 9 more, as shown in the Second Step. Then take away 10 of these pebbles and add one pebble to the tens' line, as shown in the Third Step. Then add 2 pebbles to the tens' line because of the 20 in 22, as shown in the Fourth Step. Then add 3 more because of the 30 in 139, as shown in the Fifth Step. Finally draw a line for hundreds, and on this place one pebble because of the 100 in 139. The answer is 161

matter is, however, purely speculative at the present time and it seems improbable that it will ever be definitely settled.

4132

2 × 4132 = 8264

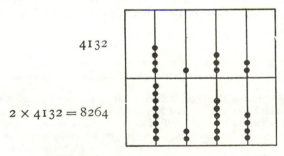

MULTIPLICATION ON THE ABACUS

Above the horizontal line in the middle it is easily seen that the number 4132 is represented. If we wish to multiply this by 2, we may simply double the objects (in this case the black dots) below the line, and the result is evidently 8264

The Abacus in Egypt. That the Egyptians used an abacus is known on the testimony of Herodotus, who says that they "write their characters and reckon with pebbles, bringing the hand from right to left, while the Greeks go from left to right." This right-to-left order was that of the Hieratic script, the writing of the priestly caste, and in this respect there is probably some relation between this script and the abacus.[1] No wall pictures thus far discovered give any evidence of the use of the abacus, but in any collection of Egyptian antiquities there may be found disks of various sizes which may have been used as counters.[2]

The Abacus in Babylonia. We have as yet no direct evidence of a Babylonian abacus. The probabilities are, however, that the Babylonians, like their neighbors, made use of it. Methods of computing were never chiefly confined to the learned class whose written records have survived. It was the trader first of all who used the abacus, and it was he who carried the customs and manners from country to country. Tradition not infrequently assigns the origin of the abacus to the Middle East, as in the writings of Iamblichus (*c.* 325), who not only states that Pythagoras introduced the instrument into Greece, but hints that he may have brought knowledge of this kind from Babylon.[3] The tradition that the primitive home of the abacus was in or near Babylon is also recorded by Radulph of Laon (*c.* 1125)[4] and other writers who had no special knowledge of the subject.

---

[1] On the Egyptian abacus see M. Cantor, *Geschichte*, I, chap. i; J. P. Mahaffy, *Old Greek Education*, p. 56 (New York, 1882), derives the Greek abacus from Egypt.

[2] In a papyrus of the time of Menephtah I (1341–1321 B.C., Lepsius) is a drawing which looks at first sight like an abacus (Cantor, *Geschichte*, I (1), 51), but which is more likely a record of the delivery of grain. Numerous similar illustrations are to be found in collections of Egyptian antiquities, as in the Archeological Museum at Florence (Egyptian coll., 2631 and 2652).

[3] *De Vita Pythagorae*, cap. v, § 22. "Primo itaque illum in arithmeticam et geometriam introduxit, demonstrationibus in abaco propositis. . . ." For further evidence as to the Babylonian origin see Volume I, page 40.

[4] "Et quum instrumenti hujus Assirii inventores fuisse perhibeantur." From a MS. in Paris, transcribed by F. Woepcke, *Journal Asiatique*, I (6), 48 n.

The Abacus in Greece. The abacus[1] and the counters[2] are mentioned several times in Greek literature. It is possible that one of the pictures on the so-called Darius vase in the Museum at Naples is intended to represent such an instrument, although what various writers have stated to be an abacus may be merely the table of the receiver of tribute. In the lowest line of figures in the illustration the king's treasurer may be seen as the figure next to the last one on the left. The other figures represent the bearers of tribute. On the table itself are the letters MΨHΔΓO<T, which are the ordinary numerals representing ten thousands, thousands, hundreds, tens, and fives, together with the symbols for the obol, half obol, and quarter obol. These symbols resemble those on the Salamis abacus mentioned below. The receiver of tribute holds a diptych, or two-

THE DARIUS VASE

The collector of tribute mentioned in the text is the figure next to the left-hand one in the lowest row. He has a tablet in one hand, and there is a table in front. From the Museum at Naples

leaved wax tablet, in his hand. Upon this tablet are the letters TΛΛATA:H, which seem to stand for $τάλα(ν)τα$ $έ(κατόν)$ (*tal'anta hekaton'*, hundred talents). The receiver of tribute seems to be casting something on the table, the picture refers to the Persian wars of the time of Darius, these wars took place about 500 B.C., and coins were then known; hence he may have

---

[1] Ἄβαξ, ἀβάκιον (*a'bax, aba'kion*).   [2] Ψῆφοι (*pse'foi*).

been casting either coins or counters. The one thing that leads to the belief that the table is an abacus is the numerals, but there are no lines such as are found on the Salamis specimen. The date of the vase itself is unknown, but the style shows it to be of the best Greek period. It was found in 1851.[1]

Salamis Abacus. While there is some question as to the figure on the Darius vase, there seems to be little respecting an abacus found on the island of Salamis. It is of white marble, 1.49 m. long and 0.75 m. wide, and is broken into two unequal parts, but is otherwise well preserved and is now in the Epigraphical Museum at Athens.[2] Of the history of this specimen but little is known. It was found before the days of the careful keeping of records, and we are ignorant of its date and of the exact place in which it was discovered. It may have been the computing table in the counting house of some dealer in exchange, and in some of its features it is not unlike the tables used by bankers in the Middle Ages; or, as Kubitschek thinks, it may have been used in some school. The theory that it may have been used in scoring games of some kind seems to have no substantial foundation. In any case it was apparently used for the mechanical representation of numbers by means of counters. It should be observed that, although the crosses are at intervals of three spaces, the first is not on the fourth line as in the medieval European abacus.

[1] The vase is unusually large, being 1.3 meters high. For a good description see A. Baumeister, *Denkmäler des klassischen Altertums*, I, 408 (Munich, 1885). I have slightly changed the inscription from a personal examination of the vase. See Heath, *History*, I, 48; M. N. Tod, "Greek Numeral Notation," *Annual of the British School at Athens*, XVIII, 124.

[2] A description was first published by Rangabé in the *Revue Archéologique*, III, 295 seq., with a comment by A. J. H. Vincent, p. 401. Until 1899 all reproductions of the stone seem to have been derived from the drawing in Rangabé's article. In that year Dr. Nagl (*Zeitschrift* (Hl. Abt.), IX, 337–357, and plate) published an illustration of the abacus under the mistaken impression that it was different from Rangabé's specimen. In the same year W. Kubitschek set forth the facts and gave a satisfactory photograph in the *Wiener Numismatische Zeitschrift*, XXXI, 393–398, Plate XXIV. The author has had a cast taken from the original, and from this the above description is made. See also Harper's *Dictionary of Classical Literature and Antiquities*, p. 2 (New York, 1897); hereafter referred to as Harper's *Dict. Class. Lit.*; Heath, *History*, I, 49–51; Tod, *loc. cit.*, p. 116.

THE SALAMIS ABACUS

It will be seen that the marble slab is ruled as usual, so that counters could be placed on the lines. On three sides are Greek characters substantially as follows:

Ⱶ	1, drachma, a mutilated form of E, for ἕν
Γ	5, old form of Π, for πέντε
Δ	10, for δέκα
ⱀ	50, for Π and Δ, five tens
H	100, for HEKATON, old form for ἑκατόν
ⱀ	500, for Π and H, five hundreds
X	1000, for χίλιοι
I	the obol
C	the half obol
T	the quarter obol
χ	for χαλκοῦς, the eighth of an obol
ⱀ	5000, for Π and X, five thousands
T	the talent of 6000 drachmas.

The lines at the top were for fractions. In the illustration the lines and symbols have been accentuated for the sake of clearness.

As to whether the Greeks commonly used loose counters or not we can only infer from this single extant specimen of an abacus, and possibly from the Darius vase. The former and possibly the latter lead us to believe that the loose counters were preferred to those sliding on wires or rods. We do not know any details as to the actual methods of computing, and in spite of the effort of Herodotus to be clear on the subject[1] we are uncertain whether the rows were horizontal or vertical with respect to the computer.[2] It seems probable that the Greeks made less use of the abacus than the Romans, the Greek numerals being better adapted to the purposes of computation, particularly of multiplication and division.[3]

[1] Liber II, cap. 36.

[2] H. Weissenborn, *Zur Geschichte der Einführung der jetzigen Ziffern*, p. 2 (Berlin, 1892), and authorities cited.

[3] J. G. Smyly, "The employment of the alphabet in Greek logistic," in the *Mélanges Jules Nicole*, p. 521 (Geneva, 1905); H. Suter, *Geschichte der math. Wissenschaften*, 2d ed., I, 11 (Zürich, 1873); Heath, *History*, I, 52.

The Abacus in Rome. There were at least three forms of abacus used by the Romans,—a grooved table with beads a
marked table for counters, and the primitive dust board.[1] In
respect to each of these forms Latin writers give us considerable information. Horace, for example, speaks of the schoolboy with his bag and table hung upon his left arm, the table
referring to the abacus or the wax tablet;[2] and Juvenal mentions both the table and the counters.[3] Cicero refers to counters
when he speaks of the *aera* (bronzes), the computing pieces being then made of bronze,[4] and Lucilius the satirist, who lived a
generation earlier, does the same.[5] The common name for these
counters was, however, *calculi* or *abaculi*, and the material from
which they were made was originally stone and later ivory and
colored glass.[6] The word *calculus* means "pebble" and is the

[1] S. Hoüel, in his review of Friedlein's "Die Zahlzeichen," in Boncompagni's
Bullettino, III, 78; Günther, *Math. Unterrichts*, p. 95 n.; A. J. H. Vincent, *Revue Archéologique*, III, 401; A. Kuckuck, *Die Rechenkunst im sechzehnten
Jahrhundert*, p. 6 (Berlin, 1874). Although we have numerous references to the
use of loose counters, it is curious that no ancient writer speaks definitely of the
ruled table on which they are used ; see Gerhardt, *Études*, p. 16. On the abacus
as a gaming table, particularly for dice, see G. Oppert, *On the original inhabitants of Bharatavarṣa or India*, p. 329 (London, 1893); W. Ramsay and R. Lanciani, *Manual of Roman Antiquities*, 17th ed., p. 497 (London, 1901) (hereafter
referred to as Ramsay and Lanciani). For a bibliography and description of the
κύβοι (*ku'boi*) and *tesserae* see J. Marquardt, *La vie privée des Romains*,
French translation, II, 522 (Paris, 1893).

[2] Laevo suspensi loculos tabulamque lacerto.
Sat., I, 6, 74

[3] Computat . . . ponatur calculus, adsint
Cum tabula pueri ; numera sestertia quinque
Omnibus in rebus, numerentur deinde labores.
Satire IX, 40

[4] "Si aera singula probasti." *Philosoph. Fragmenta*, V, 59.
[5] Hoc est ratio ? perversa aera, summa est subducta improbe !
L. 886, ed. Marx ; l. 740, ed. Lachmann
[6] Adeo nulla uncia nobis
Est eboris, nec tessellae nec calculus ex hac
Materia. Juvenal, XI, 131
Fragmenta teporata . . . fundi non queunt praeterquam abrupta sibimet in
guttas, veluti cum calculi fiunt, quos quidem abaculos appellant aliquos et pluri-
bus modis versicolores.—Pliny, *Hist. Nat.*, XXXVI, 26, 67
Capitolinus (*Pertinax*, I, 4), speaking of the boyhood of Pertinax (126–193),
says: "Puer litteris elementariis et calculo imbutus."
Martial (II, 48) includes among his modest wants "tabulamque calculosque."
II

diminutive of *calx*, a piece of limestone (often referring to the special form of chalk, the name of which comes from the same root). It is therefore our word "marble" as applied to the small spheres with which children play games. From it came the late Latin *calculare*,[1] to calculate. Teachers of calculation were known as *calculones* if slaves, but *calculatores* or *numerarii* if of good family.[2] To calculate means literally, therefore, to pebble, and a calculator is a pebbler. The word *calculi* was transmitted by the Romans to medieval Europe and was in common use until the 16th century.[3]

We are not sure whether the small disks found in Roman remains were counters for purposes of calculation, counters for games (like American poker chips), or draughts. The games of backgammon and draughts are both very old,[4] and the former is our nearest approach, aside from such abaci as we still use, to the Roman and medieval abacus.[5]

The abacus in which the beads were allowed to slide in grooves or on rods is not mentioned by any early writer and seems to have been of relatively late invention. Indeed, in the 15th and 16th centuries it was commonly asserted that Ap-

---

[1] The Romans used *calculos subducere* instead of *calculare*. This word, in the sense of "to calculate," is first found in the works of the poet Aurelius Prudentius, who lived in Spain *c.* 400; see *Nouvelles Annales de Math.*, XVII, supplementary bulletin, p. 33.

[2] Tertullian, evidently with reference to the dust abacus, calls them "primi numerorum arenarii."

[3] Thus Clichtoveus, in his arithmetic of 1503 (1507 ed., fol. b, iij, *v.*), says: "Numeratio calcularis est cuiusqȝ numeri suo loco et limite apta per calculos dispositio"; and Noviomagus (1539, fol. 9, *r.*) says: "Ut detur autem hac forma in calculis seu ut nunc fit nummis."

[4] They appear in various Egyptian, Greek, and Roman remains. For example, in the British Museum is an ancient model of an Egyptian barge on which a game of draughts is in progress, and A. Baumeister (*Denkmäler des klassischen Altertums*, I, 354 (Munich, 1885)) has reproduced an illustration of a similar game from an old Greek terracotta. There were two Roman games, the *ludus latrunculorum* and *ludus duodecim scriptorum*, on which pieces called *calculi* were used, but their exact nature is unknown. See Ramsay and Lanciani, p. 498; J. Marquardt, *La vie privée des Romains*, French translation, II, 530 (Paris, 1893); Harper's *Dict. Class. Lit.*, p. 562; J. Bowring, *The Decimal System*, p. 198 (London, 1854).

[5] It was probably the *ludus duodecim scriptorum* already mentioned, or the διαγραμμισμός (*diagrammismos'*), the late τάβλα (*ta'bla*), of the Greeks.

puleius invented this form of the instrument in the 2d century,[1]—a statement for which there is no standard authority.

Our knowledge of the grooved abacus is derived from a few specimens of uncertain date which have come down to modern times. One of these, formerly owned by Marcus Welser of Augsburg, was made of metal, is said to have been 4.2 cm. long

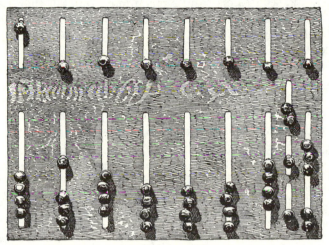

ROMAN ABACUS

Ancient bronze abacus of uncertain date, now in the British Museum

and 3.5 cm. wide, and had nineteen grooves and forty-five counters or buttons (*calculi*)[2]. Another was once owned by the reformer Ursinus (*c.* 1575), but is now lost. A third specimen, of bronze, is now in the Kircherian Museum at Rome. The general plan of the Roman abacus may be seen from the illustration here given, representing a specimen in the British Museum.

The symbols found on such specimens as are extant are usually the common Roman numerals from 1,000,000 down to 1,

---

[1] Unger, *Die Methodik*, p. 69.

[2] This was twice described before it was lost, once in Amsterdam in 1674 and once in Nürnberg in 1682. The measurements are questionable. See G. A. Saalfeld, "Der griechische Einfluss auf Erziehung und Unterricht in Rom," *Neue Jahrbücher für Philologie*, CXXVI, 371.

together witn o (or θ) for *uncia*, or $\frac{1}{12}$ of the *as*; S for *semiuncia*; ɔ for the *sicilicus*, or $\frac{1}{4}$ *uncia*; and Z for the *duella*, or $\frac{1}{3}$ *uncia*.

The Abacus in China. At the present time the use of the abacus is universal in China. In banks, shops, and counting houses of all kinds the computations are performed on the *suan-pan*.[1] The computer works very rapidly, like an expert typist or pianist, and secures his results much more quickly than can be done by our common Western methods.[2] He learns its use by practical experience in business, probably as the Romans and Greeks learned it, and not in the village schools.[3]

The *suan-pan* is, however, a relatively late development of the abacus in China, appearing first, so far as we know at the present time, in the 12th century.[4] It is true that many writers[5] have placed its introduction much earlier, but there is no definite description of the instrument in Chinese before about 1175.[6]

[1] The term means computing plate or computing tray, often incorrectly translated as computing board. It is also called the *su-pan*, and there are other variants. It is called *suinbon* in Calcutta, where it is used by all the Chinese shroffs (computers, accountants, cashiers) in the counting houses. The common spelling is suan-pan, swan p'an, or swan pan. The instrument is also in common use in Siam and wherever Chinese merchants have determined business customs.

[2] See Knott, *Abacus,* p. 44; J. D. Bell, *Things Chinese,* p. 1 (New York, 1904); J. Goschkewitsch, "Ueber das chinesische Rechnenbrett," *Arbeiten der kaiserlich Russischen Gesandtschaft zu Peking,* I, 293 (Berlin, 1858); Smith-Mikami; R. van Name, "On the Abacus of China and Japan," *Journal of the Amer. Orient. Soc.,* X (*Proceedings*), p. cx; J. Bowring, *The Decimal System,* p. 193 (London, 1854).

[3] A. H. Smith, *Village Life in China,* p. 105 (New York, 1899).

[4] One of the most scholarly articles on the history of the *suan-pan* is the one already cited, by Lacouperie, *The Old Numerals,* pp. 297–340. It contains an excellent bibliography of the subject up to 1883.

[5] J. Hager, *An Explanation of the Elementary Characters of the Chinese,* p. x (London, 1801); H. Cordier, *Bibliotheca Sinica,* col. 509 (Paris, 1881–1895); L. Rodet, "Le Souan-pan et la Banque des Argentiers," *Bulletin de la Société Mathématique de France,* Vol. VIII (Paris, 1880). Chinese writers record that a work, *Su-shuh ki-i,* "Anecdotes of mathematics," written about 200, mentions various methods of computing, including "bead computation" and "hand computation," but the work gives no description of any process. See also an interesting early essay, Smethurst, "Account of the *shwan pan,*" *Phil. Trans.,* XLVI (1749), 22.

[6] This occurs in two works, the *Pan chu tsih* and the *Tseu pan tsih,* which appeared in the Shun-hi dynasty, 1174–1190. They describe the *pan,* or tray, the word *suan-pan* not being then in use. Indeed, as late as the 16th century the name *pan shih* (board to measure) was used. See Lacouperie, p. 38.

As to the origin of the Chinese abacus, the evidence seems to point to Central or Western Asia. At the time of its appearance, China was largely under the domination of the Tangut or Ho-si state and of the Liao and Kin Tartars. The Tangutans were a mercantile race, and the Tartars were favorable to learning. Moreover, Arab and Persian traders are known to

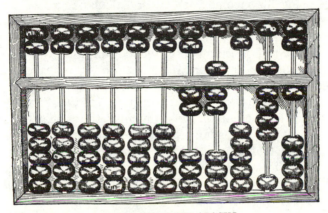

MODERN CHINESE ABACUS

The *suan-pan*, known to have been used as early as the 12th century

have been in Canton in the 8th century, and the Nestorians were in contact with the northwest, so there was plenty of opportunity for such a simple device to make its way into China from Khorasan or some neighboring province. The fact that it seems to have reached Russia from Central Asia[1] adds to the belief that China may have received it from the same source.[2]

Before the time of the *suan-pan* the counting rods, often called the bamboo rods, had been used for more than a thousand years. They were known *c.* 542 B.C.[3] and are referred to as counting stalks in a statement of Hiao-tze, the ruler of Ts'in from 361 to 337 B.C. They are mentioned again about 215 B.C., and some specimens of this period were displayed in a museum

---

[1] Larousse, *Grand Dictionnaire Universel,* I, 636; Vissière, *Abacque*; A. Wylie, *Notes on Chinese Literature,* p. 91 (Shanghai, 1902).

[2] Lacouperie. *The Old Numerals,* p. 41.   [3] See Volume I, page 96.

of the Emperor Ngan (397–419). These were about 18 inches long, some made of bone and others of horn. In the reign of Wu-tí (140–87 B.C.) of the Han dynasty, it is related that an astronomer Sang Hung (about 118 B.C.) was very skillful in his use of the rods. In the third century of our era it is recorded that Wang Jung, a minister of state, spent his nights in reckoning his income with ivory calculating rods, and the expression "to reckon with ivory rods" is still used as an allusion to wealth. In the time of the Emperor Ch'eng (326–343) the counting rods were made of wood, ivory, or iron, and two centuries later the Emperor Siuen Wu (500–516) had counting rods cast in iron for the use of his people.[1]

The Chinese historian Mei Wen-ting (1633–1721), in his work on ancient calculating instruments,[2] states that about the beginning of the Christian era 271 rods constituted a set, or handful, and that they formed a hexagon that had nine rods on a side. This means that they were arranged in six groups of which the ends of each formed a triangular number of $1 + 2 + \cdots + 9$ units, or 45 in all.   Six of these make 6×45,

or 270, and these six were grouped about one central rod, making 271, thus affording an illustration of the use of figurate numbers in the East.

It seems from Mei Wen-ting's work that the rods were in general use until the 13th century. With respect to the suan-pan, he places the date somewhat later than other writers, saying, "If in my ignorance I may be allowed to hazard a guess, I should say that it began with the first years of the Ming dynasty," which would make the date about 1368. Subsequent writers are probably correct, however, in placing it a century or two earlier.

The Abacus in Japan. The primitive method of computing in Japan is quite unknown, but from the time of the Empress Suiko (593–628) the bamboo rods (*chikusaku*) were used.[3]

---

[1] Lacouperie, *The Old Numerals*, pp. 34–36 of reprint.
[2] *Ku-suan-k'i-k'ao.*   [3] Smith-Mikami, chap. iii, with bibliography.

These were round sticks about 2 mm. in diameter and 12 cm. in length, but because of their liability to roll they were in due time replaced by the *sanchu* or *sangi*, rectangular prisms about 7 mm. thick and 5 cm. long. The *soroban*, the name being probably the Japanese rendering of the Chinese word *suan-pan*, was developed but not generally adopted in the 16th century.[1]

THE SANGI BOARD IN JAPAN

Intended for computation with the *sangi* (rods). From Satō Shigeharu's *Tengen Shinan*, 1698

What may prove to be a relic of a very early Japanese system is seen in the tally sticks used in the Luchu (Liu Kiu, Riu Kiu) Islands, near Formosa, and known as *Shō-Chū-Mā*.[2]

The Abacus in Korea. The bamboo rods of China passed over to Japan by way of Korea, and in the latter country they re-

[1] Smith-Mikami, p. 19.
[2] B. H. Chamberlain, *Journal of the Anthropological Institute of Great Britain and Ireland*, XXVII, 383. For a brief mention of these tallies see the *Geographical Journal*, June, 1895.

mained in use long after they were abandoned elsewhere. The
commercial class was acquainted with the *suan-pan* for a long

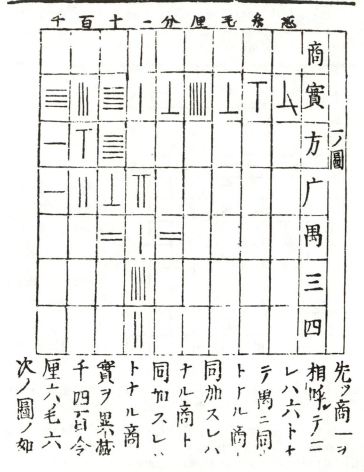

SANGI BOARD WITH NUMBERS INDICATED

From Nishiwaki Richyū's *Sampō Tengen Roku,* 1714. The *sangi* board was a
board ruled as shown, the *sangi* being placed in the rectangles

time before the Japanese conquest, and now the *soroban* is
common among the officials. But most of those who were edu-

cated in the native schools used the counting sticks until recent times, while those with but little education performed their

THE SANGI BOARD IN USE

From Miyake Kenryū's *Shōjutsu Sangaku Zuye*, 1716 (1795 ed.)

simple computations mentally or on their fingers. The counting sticks (*Ka-tji san*) were of bone, as in the illustration on page 174, or of bamboo split into long prisms. About a hun-

JAPANESE ABACUS

The *soroban*, known to have been used in Japan as early as the 16th century, and in universal use there at present

dred fifty were used in ordinary calculation, and these were kept in a bamboo case on the computer's desk. The sticks were laid as follows to represent the first twelve numbers:

I	II	III	IIII	X	XI	XII	XIII	XIIII	—	T	TT
1	2	3	4	5	6	7	8	9	10	11	12

In computing, the Koreans used the rods in substantially the same way as the Chinese and Japanese had used theirs. The process was so cumbersome that it has recently given way to the Chinese and Japanese methods with the *suan-pan* and the *soroban*.[1] The Koreans also used pebbles and coins for the same purpose.[2]

The Abacus among the Mohammedans. The Arabs, Persians, Armenians, and Turks have a form of abacus which differs from that of the Far East and from the one used by the Romans, having ten beads on each line. Its early history is unknown, but since it resembles neither the abacus of China nor that of Western Europe, it probably originated among the Arab or Persian computers. The Turks call it the *coulba* and the Armenians the *choreb*.[3]

This form of the abacus does not seem to have been generally used by the Saracens in the Middle Ages. In

KOREAN COMPUTING RODS

Computing rods made of bone. Until quite recently these were used in the schools of Korea. The numbers were represented as shown on page 173

[1] On the mathematics of Korea in general, see P. Lowell, *The Land of the Morning Calm*, p. 250 (Boston, 1886). For the *Song yang hoei soan fa* or *Song yang houi san pep* (*Treatise on Arithmetic of Yang Hoei of the Song Dynasty*), which was for a long time a classic, see M. Courant, *Bibliographie Coréenne*, III, 1 (Paris, 1896). See also the *Grammaire Coréenne*, p. 44 (Yokohama, 1881), in which the description of the laying of the sticks recalls the Japanese method and differs from the one shown on page 173, which was given to the author by an educated Korean in Peking.   [2] *Grammaire Coréenne, loc. cit.*

[3] Pacioli speaks of this form of the abacus when he says that the orders of numbers increase from right to left "more arabū de simil arte pratica primi inuētori secōdo alcuni vnde ꝑ ignorātia et vulgo a corropto el vocabulo dicēdo la Abaco: cioe modo arabico. Che loperare suo e modo arabico e chiamase Abaco: ouer secōdo altri e dicta Abaco dal greco vocabulo." *Sūma*, fol. 19, *r.* (Venice, 1494).

that period the dust board was common and the numeral forms derived from being written on such a tablet were therefore, as already stated, called in the schools of the western Arabs the *ġobâr* (dust) numerals.[1] Thus the Moorish writer al-Qalasâdî (*c.* 1475), in his commentary on the *Talchîs* of Albanna (*c.* 1300), speaks of "a man of the Indian nation who took fine powder and sprinkled it on a table and marked on it the multiplications, divisions, or other operations, and this is the origin of the term *ġobâr*" (dust).[2] Further evidence of the rarity of any other form of the abacus among the Saracens in the Middle Ages is to be found in the silence of Maximus Planudes (*c.* 1340) upon the subject; for the contact with the East of one writing upon arithmetic in Constantinople would almost certainly have led him to speak of the bead abacus if it had been in common use among the Arabs of his time. It may be, however, that the dust abacus was used in some parts of the Mohammedan domain, and the bead abacus in other parts, the latter giving to Christian nations the line abacus. Some reason for this belief is found in the fact that certain medieval writers derived the word "abacus" from the Arabic,[3] while William of Malmesbury, although by no means a reliable chronicler, writing in the 12th century, says that Gerbert (*c.* 1000) obtained his idea of the instrument from the Saracens.[4] There is also a possible reference to the line abacus by Alchindi[5] (*c.* 860).

The Abacus in Russia. From the Mohammedan countries the bead abacus worked its way northward, and in comparatively

---

[1] H. Weissenborn, *Gerbert*, p. 235 (Berlin, 1888); *Zur Geschichte der Einführung der jetzigen Ziffern*, p. 7 (Berlin, 1892); Smith-Karpinski, p. 65.

[2] From an Arabic MS. in Paris, described by F. Woepcke, *Journal Asiatique*, I (6), 60.

[3] Thus a 12th century MS., *Regulae abaci*, published by M. Chasles in the *Comptes rendus* for 1843 (XVI, 218), asserts, "Ars ista vocatur abacus: hoc nomen vero arabicum est et sonat mensa."

[4] "Abacam certe primus a Sarazenis rapiens, regulas dedit, quae a sudantibus abacistis vix intelligentur." On the unreliability of this chronicler, see H. Weissenborn, *Gerbert*, p. 236 (Berlin, 1888).

[5] The reference is in the chapter "De numeris per lineas & grana hordeacea multiplicandis Liber I" of the Latin translation of his arithmetic. See H. Weissenborn, *Zur Geschichte der Einführung der jetzigen Ziffern,* p. 7.

recent times was adopted in Russia. It is still found in every school, shop, and bank of Russia proper, although in the former provinces of Finland and Poland it is seldom used. The

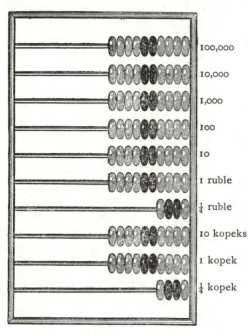

computers handle it with much the same ease as the Chinese show in their use of the *suan-pan*, and there seems to be no reason why they should not continue to use it until it is replaced by more elaborate calculating machines. The Russians call their abacus the *s'choty*,[1] and the form is the same as that of the Armenian *choreb* or the Turkish *coulba*. They occasionally speak of it as the Chinese abacus, so that there is this ground for the claim that it was introduced from China by way of Siberia, although the form of the instrument would go to show that it came from the South.

RUSSIAN ABACUS

The *s'choty* of the Russians. It is of the same form as the Armenian *choreb* and the Turkish *coulba*

In the 16th century the German form of line reckoning was used in Poland,[2] and when this disappeared it was not replaced by the Russian abacus but by the algorism of Western Europe.

----

[1] Variously transliterated, but this form gives the pronunciation more nearly than the others.

[2] The third arithmetic printed in Poland, but the first in the Polish language, is that of Kłos (1538). It is devoted almost entirely to this kind of computation. See the Baraniecki reprint (Cracow, 1889) and Dickstein's article in *Bibl. Math.*, IV (2), 57.

The Abacus in Western Europe. In medieval times in Western Europe the abacus had various names and forms. The followers of Boethius (*c.* 510) called it the Pythagorean table (*mensa Pythagorica*), a name also given to the square array of the multiplication table.[1] It was also known as the geometric table (*tabula geometricalis, mensa geometricalis*), table of the abacus (*tabula abaci*), and Pythagorean arc (*arcus Pythagoreus*), although *abax* or *abacus* was the common medieval name.[2] So common was this name that the verb "to abacus" became recognized,[3] and the arithmeticians of about the 11th century and later were occasionally called abacists.

The Dust Table. Of the various forms of abacus used in Europe, the dust table, already described as known in the Orient and in classical times, was one. We have evidence of its use at the close of the 9th century, when Remigius of Auxerre (*c.* 900), in his commentary on Capella's arithmetic, speaks of the table as being sprinkled with blue or green sand and the

---

[1] Cantor called attention to this distinction in his *Mathematische Beiträge zum Kulturleben der Völker,* Halle, 1863, p. 204 (hereafter referred to as Cantor, *Beiträge*) ; and Eneström did the same in the *Bibl. Math.,* I (2), 90. Adelard of Bath (*c.* 1120), in his *Regulae Abaci,* says: "... quidem mensam pithagoream ob magistri sui reuerentiam. sed postī tamē abacum dixerunt" (Boncompagni's *Bullettino,* XIV, 68). In this he apparently had in mind a passage in the *Ars Geometria* of Boethius (ed. Friedlein, p. 396): "Pythagorici vero ... descripserunt sibi quandam formulam, quam ob honorem sui praeceptoris mensam Pythagoream nominabant ... a posterioribus appellabatur abacus." Adelard even goes so far as to assert, with no foundation except tradition, that the abacus is due to Pythagoras himself: "Pythagorici vero hoc opus [abacum] composuerunt/ut ea que magistro suo pitagora docente audierant. ocul' subiecta retinerent: et firmius custodirent."

[2] Turchillus, writing on the abacus about 1200, says: "Ab antiquis mensa pytagorica, a modernis autem uel abax vel abacus nuncupatur" (Boncompagni's *Bullettino,* XV, 135). An anonymous MS. of the 12th century, in the Vatican, says: "(T)abula abaci quę pytagorea m̄sa uocatur" (*ibid.,* pp. 132, 154). See also M. Chasles, "Développements et détails historiques sur divers points du système de l'abacus," *Comptes rendus,* XVI, 1393, with references to other MSS.

[3] In a MS. in the Bibliothèque nationale in Paris there is some correspondence between one Radulph of Liège and Rogimbold of Cologne in the early part of the 11th century in which the writer says: "Hoc si abacizando probaveris." In the same MS. there is a letter addressed to Hermannus Contractus (*c.* 1050) in which the statement is made: "Ut meam abicizandi notem inscitiam." See Chasles, *loc. cit.,* p. 1417.

figures as being drawn with a *radius*.[1] A certain Papias, who wrote a *Vocabularium* in 1053, and who may be considered as representing the knowledge of his time, also speaks of the abacus as a table covered with green sand.[2]

The Wax Tablet. Allied to the dust table is the old wax tablet of classical times. This consisted of a tablet of wood or bone on which a thin coat of black wax was smeared, the figures being written with an iron stylus of which one end was pointed and the other was somewhat spoon-shaped, the latter being used for erasing by smoothing the wax down again.[3] This tablet passed into the medieval schools and counting houses, and specimens are extant which were in use as late as the 16th century.[4]

[1] "Abacus tabula est geometricalis super quam spargebatur puluis uitreus siue glaucus. Ibique cum radio uirgę formabantur figurę geometrię." See Boncompagni's *Bullettino*, XV, 572, and III, 84. In the same journal (X, 625) there is a description of a medieval MS. of unknown date in which the following passage appears: "Abacus vocatur mensa geometricalis que et in numeris et formis numerorum diuisa . . . ."

[2] "Abacus vel abax tabula: in qua uiridi puluē formae depinguntur." From the first printed edition of the *Vocabularium*, Milan, 1476, fol. 2, *v*. See also Boncompagni's *Bullettino*, XIV, 69.

It would seem that Adelard of Bath (*c.* 1120) referred to this form of abacus when he wrote: "Vocatur (Abacus) etiam radius geometricus, quia cum ad multa pertineat, maxime per hoc geometricae subtilitates nobis illuminantur." Radulph of Laon (*c.* 1125) had the same form in mind in writing the following: ". . . ad arithmeticae speculationis investigandas rationes, et ad eos qui musices modulationibus deserviunt numeros, necnon et ad ea quae astrologorum sollerti industria de variis errantium siderum cursibus . . . Abacus valde necessarius inveniatur." See Chasles, *loc. cit.*, p. 1414.

[3] One of the best specimens of this kind seen by the author is a 6th century Roman piece in the Rylands Library at Manchester, England. This is made of bone and is a diptych bound with iron and having an iron hinge. It has three iron styli, one end being pointed for writing and the other end being spoon-shaped for erasing.

[4] There is an elaborate set of Comptes de l'hôtel Saint Louis, written in 1256–1257, in the Musée des Archives, Paris. It consists of between ten and twenty placques. There is a 15th century specimen in the Germanic Museum at Nürnberg, from southern Germany; a piece apparently of the 16th century, from a church in Switzerland, may be seen in the British Museum; but the most interesting one of the medieval period that has come to the author's attention is in the Rathhaus at Goslar a. Harz,—a book of eight tablets bound together and making sixteen pages, two compartments to a page. This Goslar specimen is a Bürgerrolle of the 14th or 15th century, the numerals being Roman and the original stylus being annexed.

The Slate. In the later Middle Ages the slate replaced the wax tablet and sand table, and continued in use until the manufacture of cheap paper rendered it nearly obsolete at the close of the 19th century. The earliest printed reference to it is

Prosdocimi de beldamandis algo-
rimi tractatus perutilis z neceffarius
foeliciter incipit.qui de generibus cal-
culationum fpecié pretcri.t nullaȝ.q̃ falté
neceffaria ad b⁹ art̃ ꝛgnitóȝ fuerat

Iueni inq̃ꝫ pluribus libris algoꝛifmi nūcupa
tis.mós circa numeros opandi fatis narics:
atqȝ diuerfos.q licet boni exiftcrét.atqȝ veri
erát.tu faftidiofi:tu ꝓpf ipaꝛ regulaꝛ mul-
titudiné:tu ꝓpf eaꝛ deleatóes.tu etiá ꝛpter
ipaꝛ operationu̅ꝑbaȝóes:vtꝛ.f.bone fuerint uel ue.Æ rát z eti
am ifti modi inu̅m faftidiofi: q̃ fi in aliq? calculo aftroloico erics
ꝛtiȝifȝ:calculatoꝛé opató̃ȝ fuam a capite incipere oportebat: da
to q̃ erroꝛ fuus adhuc fatis ꝓpiquus exifteret. z hoc ꝓpf figu
ras in fua opatóe deletas.Ȝndigebat etiá calculatoꝛ femp aliq?
lapide uel fibi ꝛfoꝛmi.fup quo fcribere atqȝ faaliter delere pofȝ
figuras cu̅ qbuf opabat̃ in calculo fuo.Ɫt qa bec oia fans fa:.i

FIRST PRINTED REFERENCE TO A SLATE

From Beldamandi's work, 1483. See the word *lapide* in next to the last line

probably in the *Algorismus* of Prosdocimo de' Beldamandi, a work written in 1410 and first printed in 1483,[1] in which the author speaks of the necessity which a computer has for a slate from which he can easily erase what he has written.[2] The gain to the art of computation which resulted from this invention can hardly be realized at the present time.

[1] "Anno domini .1410. die .10. Iunij compilata." See *Rara Arithmetica*, p. 13.
[2] "Indigebat etiã calculator semₚ aliq? lapide uel sibi Ɔformi, suₚ quo scribere atqȝ faciliter delere possȝ figuras cũ qbus oₚabat' in calculo suo" (1540 ed., fol. 2, *v.*). Compare also J. T. Freigius, *Pædagogvs*, Basel, 1582: "Numeri in abaco scribendi."

In this period, but we do not know precisely when, there came into general use the blackboard, arranged for hanging on a wall. This is frequently shown in the illustrations in the early printed books, as in the case of Böschensteyn's work of 1514.

EARLY ILLUSTRATION OF A BLACKBOARD

From Johann Böschensteyn's *Rechenbiechlin*, Augsburg, 1514

Gerbert's Abacus. To Gerbert (*c.* 1000) there is attributed the arc or column abacus.[1] If we could put one counter marked "4" on a line instead of putting four counters upon it, there would seem at first thought to be some gain. This apparent gain is offset, however, by the loss of time in selecting the counters and still more by the necessity for learning certain tables. The plan was followed by Gerbert, and possibly some of his succes-

---

[1] *Arcus Pythagoreus, tableau à colonnes.*

sors, the counters being the apices already mentioned on page 75. They represented the number 2,056,708, for example, as follows:

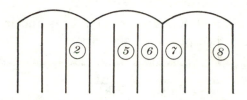

Over each triad of columns an arc was drawn to aid the eye, whence the name "arc abacus"; and in each column in which a number was to be represented a counter bearing that number was placed. As already stated, however, the gain over the older form was more apparent than real, for the computer was under the necessity of picking out the right counter each time. If Gerbert had understood the significance of the zero, he would not have used this device.

The Line Abacus. The most popular abacus of Western Europe consisted of a table ruled horizontally to represent different decimal orders, counters being placed upon the lines and in the spaces.

| | Ten thousands |
| Five thousands |
| | Thousands |
| Five hundreds |
| | Hundreds |
| Fifties |
| | Tens |
| Fives |
| | Units |

GENERAL PLAN OF A MEDIEVAL COMPUTING TABLE

This plan shows the arrangement of lines on the kind of computing table used in most parts of Europe in the Middle Ages

It was often called a calculating table or simply a table,[1] and in England it received the name "counting table" or "counter."

The illustration of the line abacus, from Köbel's work of 1514, shows the form which was common in all Western Europe

---

[1] So Hudalrich Regius, in his *Epitome* (1536), says, "Abacus vulgo mensa dicitur calculatoria quibusdam distincta lineis"; and Radulph of Laon (*c.* 1125) asserts that "Greci enim Mensam abacum dicunt." See *Abhandlungen*, V, 96.

II

for several hundred years. The line nearest the computer represents units, the space above it, fives; the second line, tens;
the second space, fifties; and so on.[1]

Representation of Numbers. On the lines and in the spaces,
counters were then placed as shown in this illustration in the

COUNTER RECKONING IN 1514

From the title-page of Köbel's *Rechenbiechlin*, Augsburg, 1514

two columns on the table. In the right-hand column, which is
the left-hand column of the computer who sits by the window,
the number 26 is represented, 2 being on the tens line, 1 in the

---

[1] One of the best of the older authorities on this type is the work of T. Snelling, *A view of the origin, nature, and use of Jettons or Counters*, London, 1769.
See also D. E. Smith, *Computing Jetons*, New York, 1921. The standard authority, however, is Barnard's work already mentioned.

fives space, and 1 on the units line. There should never be more than one counter left in a space or more than four counters on a line; for if there are five on a line, one is "carried" to the space above, and if there are two in a space, one is "carried" to the line above; whence our expression "to carry" in addition.[1] The thousands and millions lines were each marked by a small cross. This aided the eye in reading the numbers and is the origin of our system of separating the figures in groups of three by means of a comma.[2]

The intervals between the horizontal lines were commonly called "spaces" (*spacia*), and the divisions made by the vertical lines were called *cambien*, from the Italian *cambio* (exchange).[3]

Reckoning on the Lines. Computation on this form of abacus was called reckoning "on the lines," and many of the early German arithmetics include the expression "auf den Linien."[4] As a result, a boy who knew his abacus was said to "know the lines."[5] When he represented a number by means of counters on the lines, he was said to "lay" the sum;[6] and when he also knew the modern form of computing which had developed in Italy, he was said to be able to reckon "on the lines and with the pen."[7] He was often advised to "lay and seize" correctly, meaning that he must be careful to place the counter prop-

---

[1] Other illustrations from early printed books will be found in *Rara Arithmetica*. A good description of the common German abacus is given by A. Kuckuck, *Die Rechenkunst im sechzehnten Jahrhundert*, p. 7 (Berlin, 1874).

[2] "Dieselbe verzeichne mit einem Creutzlin," as J. Albert says in his arithmetic of 1534, mentioned below in note 4.

[3] Also "Cambien oder Bankir" by various other writers; e.g., Köbel, *Zwey rechenbüchlin* (1514; title of edition of 1517). Hudalrich Regius (1536) used the term *viculi*, and J. Albert (1534) speaks of the divisions as *Feldungen* as well as *Cambien* and *Cambiere*.

[4] For example, J. Albert, *Rechenbüchlein Auff der Federn*, Wittemberg, 1534 (title from 1561 ed.); A. Riese, *Rechenung auff der linihen vnd federn*, Erfurt, 1522.

[5] "Die Linien zu erkennen, ist zu mercken, das die underste Linie (welche die erste genent wird) bedeut uns, die ander hinauff zehen, die dritte hundert," etc. J. Albert, *loc. cit.*

[6] "Leg zum ersten die fl." J. Albert, *loc. cit.* This may be connected with our expression "to lay a wager."

[7] As in Riese's work of 1522.

erly and pick it up with the same care. Thus Albert (1534)
tells him:

> Write right, lay right, seize right, speak right,
> And you will always get the answer right.[1]

Addition and Subtraction. The operation of addition can be
understood by studying the illustration on page 185 from Recorde's work and by considering the following figure suggested
by Albert's arithmetic:

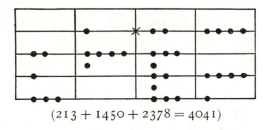

$$(213 + 1450 + 2378 = 4041)$$

Subtraction was merely the reverse of the above operation,
and the word "borrowing" had a more definite meaning than
with us. The following figure is also suggested by Albert's work:

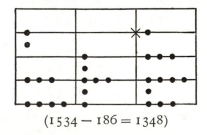

$$(1534 - 186 = 1348)$$

Multiplication and division were more complicated and are
not of enough importance to warrant a description in this work.

Extent of Use of the Line Abacus. During the 15th century
the line abacus furnished almost the only means of commercial
computation throughout most of Western Europe north of the

[1] "Schreib recht/leg recht/greiff recht/sprich recht/
So kômpt allzeit dein Facit recht."

Alps. In the 16th century we find it given prominently in the printed arithmetics of Germany, Holland, Poland, and Austria, somewhat less prominently in England, and still less in France. We may say that those countries which were chiefly influenced

ADDITION.
Maſter.

The eaſieſt way in this arte, is to adde but two ſummes at ones together: how be it, you maye adde moꝛe, as I wil tel you anone. therefoꝛe whenne you wylle adde two ſummes, you ſhall fyꝛſte ſet downe one of them, it foꝛceth not whiche, and then by it dꝛaw a lyne croſſe the other lynes. And afterwarde ſette doune the other ſumme, ſo that that lyne maye be betwene them: as if you woulde adde 1659 to 8342, you muſt ſet your ſumes as you ſee here.

And then if you lyſt, you maye adde the one to the other in the ſame place, oꝛ els you may adde th:m bothe togither in a new place: which way, bycauſe it is moſt plyneſt

3

PAGE FROM ROBERT RECORDE'S *GROUND OF ARTES*, *c.* 1542

This page shows the treatment of addition. It is from the 1558 edition

by the customs of the Italian merchants tended to abandon the abacus, while those which were in closer contact with the German counting houses continued to use it.

The popularity of the method may be seen in the fact that abacus-reckoning was a favorite subject of illustration in the title-pages of the arithmetics of Adam Riese, Gemma Frisius, and Robert Recorde, which were among the most widely circulated textbooks of the 16th century.

Origin of this Form of Abacus. It is not known when this form of the abacus first appeared. Indeed, there is a break of several centuries in the use of counters in any manner. We are ignorant as to how the Western world computed at the beginning of the Middle Ages, or what method Bede (*c.* 710) and Alcuin (*c.* 775) used in their calculations. In the 13th century counters were used for practical business computation, as they had been in the Roman days, but in the long interval the ancient scheme had changed, the vertical lines giving place to the horizontal. When or where this change took place there is at present no means of knowing.[1] Fibonacci (1202) names three methods of computing in use in his day,—finger reckoning, algorism (Hindu numerals), and the Gerbert abacus.[2] Of the use of ordinary counters he has nothing to say. Certain it is that counters were generally unknown in the 15th century in Italy, for we have the positive assertion of the Venetian patrician Ermolao Barbaro (*d.* 1495) that they were used only in foreign countries.[3]

The Counters in England. The most common of the English names for the small disks used on the line abacus was "counter," a word derived from the Latin *computare* through the French forms *conteor* and *compteur* and the Middle English *countere*, *cowntere*, and *countour*.[4] So in a work entitled *Know Thyself*, written about 1310, we are told to "sitte doun and take countures rounde.... And for vche a synne lay thou doun on Til thou thi synnes haue sought vp and founde," and in a work of 1496 mention is made of "A nest of cowntouris to the King." In the laws of Henry VIII (Act 32, cap. 14, 1540) we read, "Item for euery nest of compters .xviii.*s*," so that the expression was a common one and referred to the box or bag full of computing

---

[1] A. Nagl, *Die Rechenpfennige und die operative Arithmetik*, p. 8 (Vienna, 1888); A. Kuckuck, *loc. cit.*, p. 15.

[2] "Computatio manibus, algorismus, arcus pictagore."

[3] "Calculos sive abaculos . . . eos esse intelligo . . . qui mos hodie apud barbaros fere omnes servatur." Nagl, *loc. cit.*, p. 40.

[4] By a false etymology we have "comptroller," although this word is properly "controller," one who controls an account, from the Middle Latin *contrarotulum* (*contra* + *rotulus*), a counter roll, a check list.

pieces. Such a nest is probably referred to by Alexander Barclay (*c.* 1475–1552), in his *Egloges*, when he speaks of "The kitchin clarke . . . Jangling his counters."

In the Middle Ages in England it seems to have been the custom of merchants, accountants, and judges who had to consider financial questions to sit on benches (banks) with checkered boards and counters placed before them. Hence the checkered board came to represent a money changer's office, finally becoming a symbol for an inn, probably because innkeepers followed the trade of the money changer.

In Shakespeare's time abacus computation was in low repute, for the poet speaks contemptuously of a shopkeeper

THE GROVND OF
A R T E S:
Teaching the woorke and practise of
Arithmetike, both in whole numbies
and Fractions , after a moie easper
and eracter forte, then anye lyke
hath hytherto beene
set toith: with di=
uers new ad=
ditions.
Made by M. R O B E R T E
R E C O R D E
Doctor of Physike.

COUNTER RECKONING
From the 1558 edition of Recorde's *Ground of Artes*

as a "counter caster." Counters apparently lost their standing only in the last half of the 16th century, for Robert Recorde, writing *c.* 1542, says: "Nowe that you haue learned the common kyndes of Arithmetike with the penne, you shall see the same arte in counters,"[1] and an anonymous arithmetic of 1546 has

[1] From the 1558 edition of the *Ground of Artes*, in which Recorde devotes forty pages to this phase of the subject.

"An introduction for to lerne to reken with the pen, or with the counters accordying to the trewe cast of Algorisme." A century later Hartwell, in his appendix to the *Ground of Artes* (1646 ed.), speaks of ignorant people as "any that can but cast with Counters." Even in the first half of the 16th century people had begun to doubt the value of line reckoning, for Palegrave writes in 1530: "I shall reken it syxe tymes by aulgorisme or you can caste it ones by counters." That the abacus died out here before it did in Germany is also evident from the fact that German counters of the 15th and 16th centuries are very common in numismatical collections, while most of those used in England at this time were imported.[1]

From the use of "counter" in the sense described, the word came to mean an arithmetician. Thus we find in one of the manuscripts in the Cotton library the statement, "Ther is no countere nor clerke Con hem recken alle," and Hoccleve (1420) writes: "In my purs so grete sommes be, That there nys counter in alle cristente Whiche that kan at ony nombre sette."[2] The word also came to mean the abacus itself. Thus, in his *Dethe Blaunche* (*c.* 1369) Chaucer says: "Thogh Argus[3] the noble covnter Sete to rekene in hys counter."

Court of the Exchequer. Aside from the mere history of computation an interest attaches to the abacus in England because of its relation to the Court of the Exchequer, the *Chambre de l'échiquier* of the French.[4] In the *Dialogus de Scaccario*

---

[1] Barnard, *Counters*, p. 63.

[2] See Murray's *New English Dictionary*, II, 1057.

[3] The passage comes from the *Roman de la Rose*, in which this name, with also the spelling Algus, is given for al-Khowârizmî. Chaucer also speaks of the counters as "augrim (*i.e.*, algorism, from al-Khowârizmî) stones." On this subject see L. C. Karpinski, "Augrim Stones," *Modern Language Notes*, November, 1912 (Baltimore).

[4] The best original source of information as to the exchequer is the *Dialogus de Scaccario*, a work written by one Fitz-Neal in 1178–1179 (1181 according to Stubbs) and first edited by Madox in 1711. See also F. Liebermann, *Einleitung in den Dialogus de Scaccario*, Göttingen, 1875, and the Oxford edition of 1902. It is published in E. F. Henderson, *Select Historical Documents of the Middle Ages*, p. 20 (London, 1892). Consult also H. Hall, *The Antiquities and Curiosities of the Exchequer*, London, 1891 (reviewed somewhat adversely in *The Nation*, New

a disciple and his master discuss the nature of the exchequer as follows:

Disciple. What is the exchequer?

Master. The exchequer[1] is a quadrangular surface about ten feet in length, five in breadth, placed before those who sit around it in the manner of a table, and all around it has an edge about the height of one's four fingers, lest anything placed upon it should fall off. There is placed over the top of the exchequer, moreover, a cloth[2] bought at the Easter term, not an ordinary one but a black one marked with stripes, the stripes being distant from each other the space of a foot or the breadth of a hand. In the spaces moreover are counters placed according to their values.

The rest of the description is too long to be given, but it shows that a kind of abacus, although not the one above described, characterized this ancient court.

The counters finally came to be used to keep the scores in games,[3] as in the American game of poker and in the use of markers in billiards. They also remained in the schools for the purpose of teaching the pupils the significance of our number system, sometimes in the form of an abacus with ten beads on a

York, February 25, 1892, p. 157); R. L. Poole's Ford Lectures at Oxford in 1911, published under the title *The Exchequer in the Twelfth Century*; J. H. Ramsay, *The Foundations of England*, II, 323 (London, 1898); Martin, *Les Signes Num.*, p. 32; J. H. Round, *The Commune of London, and other Studies*, p. 62 (London, 1899); C. H. Haskins, "The abacus and the king's curia," *English Historical Review* (1912), p. 101. On the *Dialogus* as the earliest work on English government, consult J. R. Green, *Short History of the English People.*

[1] The word is a corrupt form of the Old French *eschequier* and Middle English *escheker*, based on the mistaken idea that the Latin *ex-* is taken with *scaccarium*. The term *scaccarium* for exchequer first appears under Henry I, about 1100. Before him, under William the Conqueror and William Rufus, we find the terms *fiscus* and *thesaurus*. "Exchequer" was later used to mean a chessboard, as in a work of 1300: "And bidde the pleie at the escheker"; and in Caxton's work on chess (*c.* 1475), p. 135, where it appears as *eschequer*.

[2] There is in the National Museum at Munich a green baize cloth embroidered in yellow with the ordinary arrangement of the medieval German abacus, intended to be laid on the computing table in the manner here described. For illustrations of such pieces see Barnard, *loc. cit.*, plates.

[3] "They were marking their game with Counters." Steele, in *The Tatler*, No. 15 (1709).

line, as seen in primary classes today, and sometimes for the purpose of teaching fractional parts. A specimen possibly intended chiefly for this purpose is seen in an abacus formerly used in the Blue Coat School in London, and here shown.

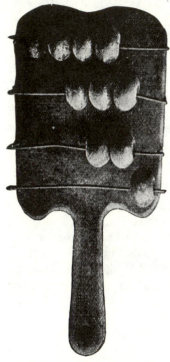

ABACUS FOR TEACHING
FRACTIONS

Formerly used in the Blue Coat School, London. From Mr. Plimpton's collection

Counters in Germany. In no country was the line abacus more highly esteemed in the 15th and 16th centuries than in Germany. Its use had died out in Italy, the great commercial center of the world, but in the counting houses of Germany it was almost universal until the era of printed arithmetics. Indeed, even in the 16th century its superiority was stoutly maintained by various German Rechenmeisters,[1] and as late as the middle of the 18th century its use had not died out in a number of the towns.[2] Even after the leading merchants had learned the method of algorism the common people continued to do their simple sums by the aid of counters purchased from the itinerant peddler,[3] and the most popular of the

---

[1] Thus Apianus (*Eyn Newe . . . Kauffmansz Rechnung*, Ingolstadt, 1527) asserted: ". . . die Summirung der Register in gewicht mass vnd müntz durch die rechenpfenning auf der linie brauchsamer ist vnd vil schneller vnd füglicher geschicht dann durch die federn oder kreide."

[2] Heilbronner, in his *Historia*, p. 890, says that in his time the counters were still used "in pluribus Germaniae atque Galliae provinciis a mercatoribus," and speaks of computing on the line as "arithmetica calculatoria sive linearis est Scientia numerandi per calculos vel nummos metallicos."

[3] The priest Geiler of Kaiserberg (1445–1510) tells of peddlers' selling them in his day, and the custom doubtless continued. See *Bibl. Math.*, IV (3), 284.

early German arithmetics were based on "Rechnung auf Linien."[1] As late as 1587 Thierfelder testified to the fact that the use of counters was still common in Germany,[2] and in 1591 two arithmetics based on line computation were published.[3] Even as late as 1621 a textbook[4] on the subject appeared at Hildesheim. One of the last writers to describe the process fully was Leonhard Christoph Sturm, whose work was published in 1701.[5]

The common German name for the counter was *Rechenpfennig*,[6] although *Zahlpfennig*[7] and *Raitpfennig* were also used.

Counters in France. In the later Middle Ages France, like all the Latin countries, made less of counter reckoning than the Teutonic lands. The first printed description of counter reckoning in that country dates from about 1500,[8] and although

[1] So Köbel's well-known arithmetics, which went through various editions beginning with 1514, gave only the counter reckoning; Adam Riese's famous textbooks, beginning in 1518, favored it; and various other popular textbooks gave it prominent place. Even as good a mathematician as Rudolff introduced it immediately after the treatment of algorism in his *Kunstliche rechnung* of 1526.

The first arithmetics printed in Germany appeared at Bamberg in 1482 and 1483. Neither seems to have had anything to say about counters, although we have only a fragment of the earlier one, and perhaps this failure explains the lack of popularity of these books. The subject was so popular that Stifel (1544) calls it *Haussrechnung*. On the general subject, see H. Schubert, "Die Rechenkunst im 16. Jahrhundert," *Deutsche Blätter für Erziehenden Unterricht*, III, 69, 105.

[2] On the first page of his *Rechenbuch* he says: "Wie vil sind Arten oder Weisen/die im Rechnen am meysten gebraucht werden? Fürnehmlich zwo/ Die erst mit der Feder/oder Kreyden/durch die Ziffern. Die ander mit den Zahlpfenning auff den Linien."

[3] One (Swiss) by Mewrer in Zürich and the other by Kauder in Regensburg.

[4] The work was anonymous. The title is *Ein new Rechenbüchlein auf Linien und Ziffern*. See Nagl, *Die Rechenpfennige*, p. 27, for other cases.

[5] *Kurtzer Begriff . . . Mathesis.*

[6] Reckoning penny. The spelling varies. Stifel, for example, in his *Deutsche Arithmetica*, 1545, fol. 1, calls them *Rechenpfenning* (both singular and plural). *Rechenpfennig* was merely a translation of the medieval Latin name, as is seen in the arithmetic of Clichtoveus (1503): "quos denarios supputarios vocant," and in his commentary on Boethius (1510, fol. 33).

[7] Number penny. Rudolff, in his edition of 1534, uses this form, while Thierfelder, in his *Rechenbuch* of 1587, uses the form *Zahlpfenning*.

[8] In this anonymous and undated work, *De arte numerandi siue arismetice (perfectionis) summa quadripartita*, the author treats of the operations "per proiectiles," and says: "hec licet breuiter de proiectilibus sint dicta. negotianti tamen atque se exercenti per eos frequenter. abundantissime hec pauca sufficient." See Treutlein in the *Abhandlungen*, I, 24.

arithmetics on the subject are not so common as in Germany, they are sufficiently numerous to show that the system was well known.[1] The subject dropped out of the business textbooks in the last quarter of the 16th century, although the counter was used by women long after men had come to use algorism, writing not being so common among the former as among the latter.[2]

Because the counters are thrown upon the table the medieval Latin writers often called them *projectiles*.[3] The French translated this word, omitting the prefix, as *jetons*,[4] a word which still survives in France to mean a game counter, a small medal, or a token.

The older French jetons frequently bear such inscriptions as "pour les Comtes" and "pour les Finances,"[5] showing their use. Sometimes the legends are admonitory, thus: "Gectez, Entendez au Compte," "Gardez vous de Mescomptes," and "Jettez bien, que vous ne perdre Rien."[6]

The Tally and Related Forms. The subject of the abacus should not be dismissed without mention of the tally and certain other related forms. The tally was originally a piece of wood on which notches or scores[7] were cut to designate num-

---

[1] Among these books may be mentioned the following: Clichtoveus, *Ars supputādi tam per calculos q3 notas arithmeticas*, Paris, 1507; Clichtoveus, *De Mystica numerorum* (Paris, 1513, fol. 33, *r*.), subdividing *supputatio* into *calcularis* and *figuralis*, giving five common operations under the former and eight under the latter; Blasius, *Liber Arithmetice Practice* (1513); an anonymous *Le livre des Getz* (about 1500), in which is taught "la pratique de bien sçavoir conter aux getz comme à la plume"; Cathalan, *Arithmetiqve*, Lyons, 1555, in which the author explains "a Chiffrer & compter par la plume & par les gestz"; an anonymous *Arithmétique par les jects*, Paris, 1559; Trenchant, *Arithmetiqve*, 1566, the 1578 edition of which gives thirteen pages to "L'art et moyen de calcvler avec les Getons," the 1602 edition dropping the subject.

[2] Thus F. Legendre, in his arithmetic of 1729, says: "Cette manière de calculer est plus pratiquée par les femmes que par les hommes. Cependant plusieurs personnes qui sont employées dans les finances et dans toutes les jurisdictions s'en servent avec beaucoup de succès." [3] *Pro* (forward) + *jacere* (to throw).

[4] Also found in the following forms: *jettons, gects, gectz, getoers, getoirs, gettoirs, getteurs, jectoers, jectoirs, jetoirs, giets, gietons*, and *gitones*. Consult Snelling, *loc. cit.*, p. 2.

[5] Also "Getoirs de la chambre des comptes, Le Roi," "Ce sont les getoirs des ?tes [Comptes]. La Reinne."

[6] See also Snelling, *loc. cit.*, p. 3; Nagl, *loc. cit.*, p. 11; and Barnard, where many photographic plates may be consulted. [7] Scars; related to "shear."

bers. The word comes from the French *tailler* (to cut), whence our word "tailor." The root is also seen in the Italian word *intaglio* and is from the Latin *talea* (a slender stick).[1] The word has even been connected with the German *Zahl* (number) through the primitive root *tal*.[2]

The idea of keeping numerical records on a stick is very ancient, and in a bas-relief on the temple of Seti I (*c.* 1350 B.C.), at Abydos, Thot is represented as indicating by means of notches on a long frond of palm the duration of the reign of Pharaoh as decreed by the gods.[3]

In the Middle Ages the tally formed the standard means of keeping accounts. It was commonly split so as to allow each party to have a record, whence the expression "our accounts tally."[4]

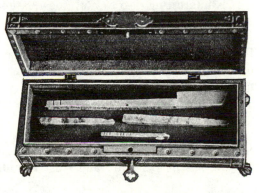

TALLY STICKS OF 1296

Fragments found at Westminster in 1904. In the author's collection

The root also appears in "tail," as when Piers Plowman, speaking of his gold, says that he "toke it by taille," meaning by count;[5] and in "tailage" (or "tallage") for toll or tax,[6]—a relic of the days when "our forefathers had no other books but the score and the tally."[7]

---

[1] Consult also Greenough and Kittredge, *Words and their Ways*, pp. 45, 266 (New York, 1901).

[2] G. Rosenhagen, "Was bedeutet Zahl ursprünglich?" *Zeitschrift für deutschen Altertum und deutsche Literatur*, LVII, 189.

[3] It is reproduced in G. Maspero, *Dawn of Civilization*, 3d ed. by Sayce, p. 221 (New York, 1897).

[4] Other expressions, like "keeping tally" at a game, "to tally up," and "stocks," are traced to this device.

[5] See also Chaucer, *General Prologue to the Canterbury Tales*, l. 570.

[6] The first poll tax is said to have been the "tailage of groats" levied by Parliament in 1377.   [7] Shakespeare, *2 Henry VI*, IV, vii, 38.

Usually a hazel stick was prepared by the "tally cutter," and the notches were cut before it was split, a large notch meaning M (£1000), a smaller one C (£100), a still smaller one X (£10), and so on down to pence. In earlier times the twentieth mark was a larger scar than the others, and the number was therefore called a score.[1] The system was used in the English Exchequer as late as 1812.

It is interesting also to note the relation of the tally stick to modern forms of investment. Formerly, if a man lent money to the Bank of England, the amount was cut on a tally stick.[2] This was then split, the bank keeping the "foil" (*folium*, leaf) and the lender receiving the "stock" (*stipes*), thereby becoming a "stock" holder and owning "bank stock."[3]

The tally was used in Germany for keeping accounts in the 13th and 14th centuries. Even at the beginning of the 15th century, and in as progressive a city as Frankfort am Main, the so-called *Kerbenrechnung* was common,[4] nor did the custom die out in Germany and Austria until the 19th century.[5]

---

[1] Teutonic *Stiege*, a word often used for twenty.

[2] The British Museum has a number of tally sticks, from 1348 to the "hop tally" still used in Kent and Worcestershire. There are six very perfect but not very old specimens in the Museum of Folklore at Antwerp. In cleaning the Chapel of the Pyx at Westminster in 1904 several specimens were found. These dated, as the inscriptions show, from 1296, and some of them came into the author's possession and are shown on page 193. Scotch "nick sticks" and Scandinavian calendar sticks belong to the same general class.

On the method of cutting and using tallies in England see the *Publications of the Pipe Roll Society*, Vol. III (London, 1884). On tally charts used by sailors, see *Zeitschrift für Ethnologie*, XXXV, 672 (Berlin, 1903). On their use as a means of communication see W. von Schulenberg, *Verhandlungen d. Berliner Gesellsch. für Anthropologie, Ethnol., und Urgeschichte*, XVIII, 384; *Zeitschrift für Ethnologie*, XIV, 370. On their other uses see Gyula v. Sebestyén, "Ursprung der Bustrophedonschrift," *Zeitschrift für Ethnologie*, XXXV, 755.

[3] Poole, *loc. cit.* See also C. H. Jenkinson, "Early wooden tallies," *Surrey Archæological Collection*, XXIII, 203.   [4] Günther, *Math. Unterrichts*, p. 287.

[5] F. Villicus, *Geschichte der Rechenkunst*, 3d ed., p. 15 (Vienna, 1897) (hereafter referred to as Villicus, *Geschichte*) ; R. Andree, *Braunschweiger Volkskunde*, p. 247 (Braunschweig, 1901), with several bibliographical notes of value; *Verhandlungen d. Berliner Gesellsch. für Anthropologie, Ethnol., und Urgeschichte*, XI, 763. With our "keeping tally" and baseball "scores" compare the German "Er hat viel auf dem Kerbholze." For interesting accounts of the use of the tally in Bohemia, see W. Wattenbach, *Das Schriftwesen im Mittelalter*, 3d ed., p. 95 (Leipzig, 1896).

In Italy the tally was evidently common in the 16th century, for Tartaglia (1556) gives a picture of one in his arithmetic, saying that one of the two parts was called by the Latin name *tessera*, a word often used to mean a counter.[1]

Knotted Cords. Related to the tally, in that they were used for recording numbers but not for purposes of calculation, are the knotted cords. These are used in various parts of the world and have such an extended history that only a passing reference can be given to them. Lao-tze, "the old philosopher," as the Chinese call him, in his *Tao-teh-king* of the 6th century B.C., referring to the earlier use of this device, says, "Let the people return to knotted cords (*chieh shing*) and use them."[2] Herodotus (IV, 98) tells us that the king of Persia handed the Ionians a thong with sixty knots as a calendar for two months, and a similar device of modern India may be seen in the museum at Madras. Indeed, in taking the census in India in 1872, the Santáls in the wilder parts of Santál Parganas used knots on four colors of cords, the black signifying an adult man, the red an adult woman, the white a boy, and the yellow a girl. The census was taken by the headmen, who, being unable to write, simply followed the popular method of keeping a numerical record.[3]

In the New World the knotted cord is best illustrated in the Peruvian *quipu*.[4] In each city of Peru there was, at the time

---

[1] "E laltro di questi dui pezzi lo chiamauano Tessera." *General Trattato*, I, fol. 3, *v*. (Venice, 1556).

[2] See Carus's English edition of the *Tao-teh-king*, pp. 137, 272, 323 (Chicago, 1898).

[3] *Proceedings of the Asiatic Society of Bengal*, p. 192 (Calcutta, 1872).

[4] The leading work on the general question of the *quipu*, with analyses of about fifty specimens, and with an extensive bibliography, is that of L. L. Locke, *The Quipu*, New York, 1923, published by the American Museum of Natural History. There is an article by the same author, entitled "The Ancient Quipu," in the *American Anthropologist* for 1912, p. 325. See also E. Clodd, *Storia dell' Alfabeto*, trad. del Nobili, cap. iii (Turin, 1903), or the English original; E. B. Tylor, *Early History of Mankind*, p. 160; *Westminster Review*, London, XI, 246; A. Treichel, *Verhandlungen d. Berliner Gesellsch. für Anthropologie, Ethnol., und Urgeschichte*, XVIII, 251; W. von Schulenberg, "Die Knotenzeichen der Müller," *Zeitschrift für Ethnologie*, XXIX, 491; H. G. Fegencz, "Kinderkunst und Kinderspiele," *Anzeiger d. Ethnolog. Abteilung d. Ungarischen National-Museums*, Budapest, XI, 103.

of the European invasion, a *quipucamayocuna* ("official of the knots") who may have performed duties not unlike those of a

SPECIMEN OF QUIPU

The knotted cords of the ancient Peruvians

city treasurer today. At any rate, we have no evidence that the knots were used for any other purpose than the recording of numerical results, just as the Peruvian shepherd today uses them for keeping account of his herds.[1]

The knotted cords found in various forms of religious regalia may originally have recorded the number of prayers, pilgrimages, or sacrifices of the devotee. Examples of these are seen in the Lama rosary (*prenba*) and the rosaries of the Mohammedans, the Buddhists of Burma, and the Catholic Christians. Somewhat similar in use is the notched praying stick of the pilgrim, such as may be seen at the shrine of St. Fin Barr at Gouganebarra, Ireland.

2. FINGER RECKONING

Finger Notation. The absence or rarity of suitable writing material led most early peoples to represent numbers by positions of the fingers,—a system not unlike the digital language of the deaf mutes of today. While this is manual rather than mechanical, it may properly be explained in this chapter. It is not improbable that the idea developed from the primitive method of counting on the fingers, usually beginning by point-

[1] They are also related to the wampum of the American Indian and possibly to the *lo-shu* and *ho-t'u* symbols of the ancient Chinese *I-king*.

ing at the little finger of the left hand with the second finger of the right, this being the result of holding the hands in a natural position for such a purpose. The person counting would thus proceed from right to left, and this may have influenced some of the early systems of writing numbers.[1]

The general purposes of digital notation were to aid in bargaining at the great international fairs with one whose language was not understood, to remember numbers in computing on an abacus, and to perform simple calculations.[2]

For the mere representing of the small numbers of everyday life the left hand sufficed. In this way it became the custom to represent numbers below 100 on the left hand and the hundreds on the right hand. Juvenal refers to this custom in his tenth satire, saying: "Happy is he indeed who has postponed the hour of his death so long and finally numbers his years upon his right hand."[3]

FINGER SYMBOLISM ABOUT THE YEAR 1140

One of a large number of drawings in a manuscript copy of Bede's works in the Biblioteca Nacional at Madrid, dating from *c.* 1140. The number 2000 is indicated. From a photograph by Professor J. M. Burnam

[1] On the general relation of the finger numbers to systems of counting and writing there is an extensive literature. See, for example, F. W. Eastlake, *The China Review*, Hongkong, IX, 251, 319, with a statement that the Chinese place the system before the time of Confucius; S. W. Koelle, "Etymology of the Turkish Numerals," *Journal of the Royal Asiat. Soc.*, London, XVI (N. S.), 141; Sir E. Clive Bayley, *ibid.*, XIV (reprint, part 2, p. 45 n.); M. Barbieri, *Notizie istoriche dei Mat. e Filosofi . . . di Napoli*, p. 10 (Naples, 1778); Villicus, *Geschichte*, p. 6; Bombelli, *Antica Numer.*, I, 108, 115 n., with bibliography and plates.

[2] P. Treutlein, *Abhandlungen*, I, 21; H. Stoy, *Zur Geschichte des Rechenunterrichts*, I. Theil, Diss., p. 31 (Jena, 1876).

[3] Felix nimirum, qui tot per saecula mortem
   Distulit atque suos iam dextra computat annos.

11

That the system was familiar to the people is evident from a remark of Pliny[1] to the effect that King Numa dedicated a statue of two-faced Janus, the fingers being put in a position to indicate the number of days in a common year, and Macrobius testifies that the hundreds were indicated on the right hand.

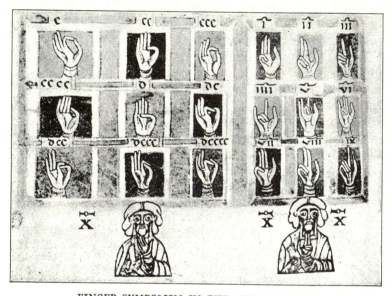

FINGER SYMBOLISM IN THE 13TH CENTURY

From the *Codex Alcobatiensis* in the Biblioteca Nacional at Madrid, dating from *c.* 1200. From a photograph by Professor J. M. Burnam

The system was in use among the Greeks in the 5th century B.C., for Herodotus tells us that his countrymen knew of it. Among the Latin writers it is mentioned by Plautus, Seneca, Ovid, and various others.[2]

[1] See I. Sillig, edition of Pliny's works, V, 140 (Hamburg and Gotha, 1851), and the *Hist. Nat.*, XXXIV, vii, 16, 33.

[2] L. J. Richardson, "Digital Reckoning among the Ancients," *Amer. Math. Month.*, XXIII, 7; Bombelli, *Antica Numer.*, p. 102; A. Dragoni, *Sul Metodo aritmetico degli antichi Romani*, p. 10 (Cremona, 1811); Günther, *Math. Unterrichts*, p. 12. Possibly Juan Perez de Moya (1573) was correct in saying that the Egyptians used the system because they were "friends of few words,"—"los Egipcianos eran amigos de pocas palabras . . . destos deuio salir."

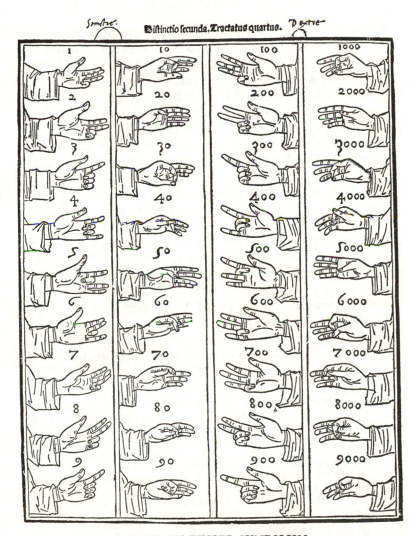

PACIOLI ON FINGER SYMBOLISM

From the *Süma* of Pacioli, Venice, 1494. The two columns at the left represent the left hand, the other two representing the right hand

Finger symbolism was evidently widely spread during many centuries, for there are also numerous references to it in both the Hebrew and the Arabic literature. Our precise knowledge of the subject is due chiefly, however, to a few writers,—to the Venerable Bede (*c.* 710), Nicholas Rhabdas (*c.* 1341), and a Bavarian writer, Aventinus (1522).[1] In the works of these writers the system is fully described, but brief summaries, often with illustrations, may be found in various books of the 16th century, including those of Andrés[2] (1515), Recorde[3] (*c.* 1542), Moya[4] (1562), Valerianus[5] (1556), and Noviomagus[6] (1539). The later literature of the subject is also extensive.[7]

The general scheme of number representations may be sufficiently understood from the illustrations given from the works of Pacioli and Aventinus. Bede gives a description of upwards of fifty finger symbols, the numbers extending through one million. No other such extended description has been given except the one of Rhabdas, but the works of Pacioli and Aventinus contain what are probably the best-known pictorial illustrations of the process.

---

[1] Bede, "De loquela per gestum digitorum," in his *Opera Omnia*, I, 686 (Paris, 1850); Nicholas Rhabdas, Ἔκφρασις τοῦ δακτυλικοῦ μέτρου; J. Aventinus, *Abacvs atqve vetvstissima, vetervm latinorum per digitos manusq̃ numerandi . . . cõsuetudo*, Nürnberg, 1522 (title from Regensburg edition, 1532). See also St. Augustine, *Enarrationes in Psalmos*, xlix, 9, 1; *Sermones*, ccxlviii, ccxlix, cclii; and *Contra Iulianum*, iii, 11, 22; and M. Capella, *De Nuptiis Philologiae et Mercurii*, ii, 102, and vii, 729 and 746. On the Rhabdas symbolism see Heath, *History*, II, 551.

[2] Mossen Juan Andrés, *Sumario breve d' la prática de la Arithmetica*, Valencia, 1515. [3] *Ground of Artes*, London, *c.* 1542.

[4] *Arithmetica Practica*, p. 627 (Salamanca, 1562).

[5] Joannes Pierius Valerianus Bellunensis, *Hieroglyphica*, p. 454 (Frankfort a. M., 1556; 1614 ed.).

[6] Cap. XIII of the 1544 edition of his *De nvmeris libri II*, Cologne.

[7] *E. g.*, see *Abhandlungen*, V, 91, 100; the Basel edition of St. Jerome's works, IX, 8 (1516); L. A. Muratori, *Anecdota*, Naples, 1776, with the "Liber de computo S. Cyrilli Alexandrini"; V. Requeno, *Scoperta della Chironomia*, Parma, 1797, with illustrations; M. Steinschneider, *Bibl. Math.*, X (2), 81; A. Marre, "Manière de compter des anciens avec les doigts . . .," Boncompagni's *Bullettino*, I, 309; Bombelli, *Antica Numer.*, cap. xiv, especially p. 109 n.; A. F. Pott, *Die quinäre und vigesimale Zählmethode bei Völkern aller Welttheile*, Halle, 1847, with an "Anhang über Fingernamen," p. 225; F. T. Elworthy, *The Evil Eye*, p. 237 (London, 1895), with illustrations; E. A. Bechtel, "Finger-Counting among the Romans in the Fourth Century," *Classical Philology*, IV, 25.

The representation of numbers below 100 was naturally more uniform, since they were in international use by the masses, while the representation of the higher numbers was not so well standardized.

Finger Computation. From finger notation there developed an extensive use of finger computation. This began, of course, with simple counting on the fingers, but it was extended to include particularly the simpler cases of multiplication needed

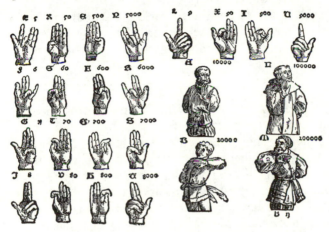

AVENTINUS ON FINGER SYMBOLS

From the *Abacvs* of Johannes Aventinus, Nürnberg, 1522 (Regensburg edition of 1532)

by the illiterate. For example, to multiply 7 by 8, raise two fingers on one hand and three on the other, since $5 + 2 = 7$ and $5 + 3 = 8$. Then add the numbers denoted by the raised fingers, $2 + 3 = 5$, and multiply those denoted by the others, $3 \cdot 2 = 6$, and the former result is the tens, 50, and the latter is the units, the product being 56. This depends, of course, upon the fact that $(10 - a)(10 - b) = 10 \ (5 - a + 5 - b) + ab.$[1] The same principle is frequently seen in written work in arithmetic in the Middle Ages, since by its use it was unnecessary to learn the multiplication table above $5 \cdot 5$. In a somewhat

[1] See also pages 119 and 120.

similar way we may find the product of numbers from 10 to 15. For example, to find the product of 14 and 13, raise four fingers on one hand and three on the other, since $14 = 10 + 4$ and $13 = 10 + 3$. Then to 100 add ten times the sum of the number of fingers raised, and the product of the same numbers, the result being $100 + 10 (4 + 3) + 4 \cdot 3 = 182$. The method is evidently general, since

$$(10 + a)(10 + b) = 100 + 10(a + b) + ab.$$

Such work is still to be seen in various parts of the world.

In the time of Fibonacci (1202) finger symbols were still used,[1] especially in remembering certain numbers in division.[2]

3. MODERN CALCULATING MACHINES

Napier's Rods. The first important improvement on the ancient counter computation was made by Napier (1617). In his *Rabdologia*[3] he explains a system of rods arranged to represent the *gelosia* method of multiplication as seen in the illustration on page 203. The plan shows how crude were the methods of calculating even as late as the 17th century, although it would have had some value in connection with trigonometric functions if logarithms had not been invented. These rods were commonly known as Napier's Bones, as in Leybourn's *The Art of Numbring By Speaking-Rods: Vulgarly termed Nepeir's Bones,*[4] London, 1667. They attracted considerable attention, not merely in Europe but also in China and Japan.

---

[1] ". . . opportet eos qui arte abbaci uti uoluerint, ut subtiliores et ingeniores appareant scire computum per figuram manuum, secundum magistrorum abbaci usum antiquitus sapientissime inuentam." *Liber Abaci,* I, 5.

[2] So Fibonacci has a chapter, "De diuisione numerorum cordetenus in manibus per eosdem numeros," with such expressions as "ponens semper in manibus numeros ex diuisione exeuntes." *Ibid.,* I, 30.

[3] *Rabdologiae, Sev Nvmerationis Per Virgulas Libri Dvo,* Edinburgh, 1617; Leyden, 1626. Translations, Verona, 1623; Berlin, 1623. *Rabdologia* = late Greek ῥαβδολογία (*rhabdologi'a*), a collection of rods, from ῥάβδος (*rhab'dos,* rod) + λογία (*logi'a,* collection). Probably Napier took the word from the *Glossaria H. Stephani,* Paris, 1573, where the above meaning is given.

[4] W. Leybourn (c. 1670) derived *Rabdologia* from ῥάβδος(*rhab'dos,* rod)+λόγος (*log'os,* speech), and this etymology is still accepted by some writers.

Modern Machines. The essential superiority of the modern calculating machine over an instrument like the *suan-pan* is that the carrying of the tens is done mechanically instead of being done by the operator. For this purpose a disk is used which engages a second disk, turning the latter one unit after nine units have been turned on the former.[1]

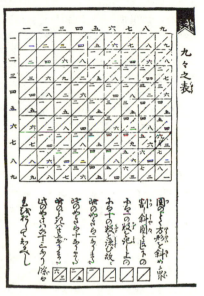

The first of these instruments seems to have been suggested by a Jesuit named Johann Ciermans, in 1640,[2] but apparently nothing was done by him in the way of actually constructing such a machine.

The real invention may properly be attributed to Pascal (1642), who, at the age of nineteen and after many attempts, made an instrument of this kind, receiving (1649) a royal privilege for its manufacture,[3] and one particularly interesting specimen is still preserved in the Conservatoire National des Arts et Métiers at Paris. It is an adding machine adapted to numbers

NAPIER'S RODS IN JAPAN

The Napier Rods found their way into China at least as early as the beginning of the 18th century, and into Japan in the century following. This illustration is from Hanai Kenkichi's *Seisan Sokuchi*, of the middle of the 19th century

---

[1] For a succinct description of modern machines see F. J. W. Whipple, "Calculating Machines," in E. M. Horsburgh, *Handbook of the Exhibition at the Napier Tercentenary Celebration*, p. 69 (Edinburgh, 1914), hereafter referred to as Horsburgh, *Handbook*. For slide rules, *ibid.*, pp. 155 and 163.

[2] Kästner, *Geschichte*, III, 438; Cantor, *Geschichte*, II (1), 657.

[3] Cantor, *Geschichte*, II (1), 661; M. D'Ocagne, *Le Calcul simplifié par les procédés mécaniques et graphiques*, Paris, 1894; 2d ed., 1905; "Histoire des machines à calculer," *Bulletin de la Société d'Encouragement pour l'Industrie Nationale*, tome 132, p. 554, and other articles in the same number, with an extensive bibliography on pages 739–759.

of six figures and is one of the later attempts of Pascal. On the inside of the box is this inscription:

Esto probati instrumenti symbolum hoc; Blasius Pascal; arvernus, inventor. 20 mai 1652.[1]

In the same museum there are two other machines, apparently also of Pascal's make, one of which was verified and presented by a collateral descendant.

In 1673 Sir Samuel Morland (1625–1695), an English diplomat, mathematician, and inventor, made a machine for multiplying, and about the same time (1671) Leibniz constructed one in Germany. In 1709 the Marchese Giovanni Poleni (1683–1761), then professor of astronomy at Padua, made a similar attempt in Italy; and in 1727 there was described in Germany a machine constructed just before his death by Jacob Leupold (1674–1727), a Leipzig mechanic. These various attempts were recorded in 1735[2] by Christian Ludwig Gersten (1701–1762), then professor of mathematics at Giessen, in connection with a description of a machine invented by himself. It was not, however, until the 19th century that any great advance was made. In 1820 Charles Babbage began the construction of a machine for calculating mathematical tables, and in 1823 the Royal Society secured aid from the British government to enable him to continue his work. Babbage's progress not being satisfactory, this aid was soon withdrawn, but the work continued until 1856, when it was abandoned.[3] From the time when Babbage began to the present, however, the modern calculating machine has been constantly improved, first by Thomas de Colmar (1820), and various types are now in extensive use.[4]

---

[1] "Let this signature be the sign of an approved instrument. Blaise Pascal, of Auvergne, inventor. May 20, 1652."

[2] *Phil. Trans.*, Abridgment, 1747, VIII, 16.

[3] One of the best descriptions of this machine is given in *Babbage's Calculating Machine; or Difference Engine*, printed by the Victoria and Albert Museum, London, 1872; reprinted in 1907.

[4] There is a large collection in the Conservatoire National des Arts et Métiers at Paris.

Slide Rule. In 1620 Edmund Gunter designed the logarithmic "line of numbers," on which the distances were proportional to the logarithms of the numbers indicated. This was known as Gunter's Scale, and by adding or subtracting distances by the aid of compasses it was possible to perform multiplications and divisions. Thus the inventor worked out the principle of the slide rule, but instead of having the sliding attachment he used a pair of compasses.[1] This instrument was subsequently used in navigation.

In 1628 Edmund Wingate published at London his *Construction and Use of the Line of Proportion*, but this, like Gunter's Scale, was merely a rule in which the spaces on one side indicated numbers, while those on the other indicated the mantissas of these numbers.

About 1622 William Oughtred invented the slide rule,[2] but descriptions of his instrument did not appear in print until 1632. A pupil of Oughtred's, Richard Delamain, published at London in 1630 a small pamphlet entitled *Grammelogia; or the Mathematicall Ring*, in which he described a circular slide rule, apparently of his own invention. Oughtred, however, seems unquestionably to have invented the rectilinear logarithmic slide rule, and also, independently of Delamain, to have invented a circular one.

In the year 1654 a slide rule was made in which the slide moved between parts of a rigid stock, and a specimen of this type, now in the Science Museum at South Kensington, is inscribed "Made by Robert Bissaker, 1654, for T. W."[3] Who this T. W. was we do not know, but the invention was a notable step in the development of the modern type.

From that time on there were numerous inventors who improved upon the instrument. Among them were various obscure artisans, but there was also Newton, who devised a system of

---

[1] On this entire topic see F. Cajori, *A History of the Logarithmic Slide Rule*, New York, 1909; hereafter referred to as Cajori, *Slide Rule.*

[2] F. Cajori, *William Oughtred*, p. 47 (Chicago, 1916); hereafter referred to as Cajori, *Oughtred.*

[3] Horsburgh, *Handbook*, p. 163.

concentric circles for the solution of equations. In the 17th century the slide rule of the type now used attracted little attention, either in England or on the continent. In the following century, however, its value began to be recognized, and the instruments in use at that time resemble in several particulars those with which we are familiar. About 1748 George Adams made spiral slide rules that were carefully engraved and probably were of a higher degree of accuracy than those of his predecessors.

The first one to make a decided step in advance, however, was William Nicholson (1753–1815). He described (1787) the various types of rules then known, and suggested noteworthy improvements, particularly in the way of a rule which, through the device of a system of parallels, gave the effect of an instrument more than 20 feet in length. He also designed a spiral slide rule, apparently ignorant of the work done in this field by various predecessors. At about the same time various French and German writers contributed to the perfecting of the instrument, notably Jean Baptiste Clairaut (1720), who designed a new circular slide rule.

The most marked advance in the middle of the 19th century was made by Amédée Mannheim[1] (1831–1906), who (c. 1850) designed the Mannheim Slide Rule, which is still a standard, although modified in various particulars.[2] These modifications related (1) to increasing the length of the scales without increasing the size of the instrument; (2) to adapting the rule to specialized branches of science; and (3) to increasing the mechanical efficiency of the device.[3] Few such instruments have gained so much popularity in such a short time.

---

[1] *L'Enseignement Mathématique*, IX (1907), 169.

[2] For details as to other inventors see Cajori, *Slide Rule*. On the history of the planimeter, which may be classified among instruments relating to the calculus or among those having to do with calculation, see A. Favaro, "Beiträge zur Geschichte der Planimeter," Separat-Abdruck aus der *Allgemeinen Bauzeitung*, Vienna, 1873. The instrument seems to have been first designed *c.* 1814 by J. M. Hermann, but it attracted little attention. The first published description was that of an Italian inventor, Gonella; it appeared in 1825.

[3] Horsburgh, *Handbook*, p. 156.

TOPICS FOR DISCUSSION

1. Consider the difficulties of multiplying a number like 4275 by a number like 876, using only the Roman, Greek, Egyptian, or Babylonian numerals.

2. Reason for the persistence of the abacus in business calculations until the 17th century.

3. Reason for abandoning the abacus in Italy before it was abandoned in northern Europe.

4. The various etymologies of the term "abacus" as given in such dictionaries as are accessible.

5. Reason for believing that the origin of the line abacus may be Semitic.

6. Etymologies of the terms used in connection with the abacus, and the relation of the word "calculus" to other words in our language.

7. Words used in various languages to mean computing disks, with their etymologies.

8. Comparison of the various types of line abacus, with a discussion of their respective merits.

9. The various forms of the abacus used in the Far East, with a comparison of their merits.

10. A study of the evolution of the paper tablet used for computation by pupils at the present time, beginning possibly with the dust table or the wax tablet.

11. Gerbert's abacus and its chief defects.

12. History of the British Court of the Exchequer.

13. General use of counters in the countries of Western Europe.

14. History of the tally.

15. The history of finger symbols and finger computation, with special reference to the international character of the symbols themselves.

16. The general character of the quipu and of similar knot-tying devices in various parts of the world.

17. Relation of the "cat's cradle" to the knotted cords and possibly to the tying together of the stars to make the constellations of ancient astronomy.

18. Rise of the modern calculating machine.

19. Types and history of modern calculating machines and planimeters as described in current encyclopedias.

CHAPTER IV

ARTIFICIAL NUMBERS

1. Common Fractions

Origin of Artificial Numbers. The natural numbers seem to have served the purposes of the world until about the beginning of the historic period. Men broke articles and spoke of the broken parts, but even after weights came into use it was not the custom to speak of such a fraction as $\frac{3}{4}$ of a pound. The world avoided difficulties of this kind by creating such smaller units as the ounce and then speaking of the particular number of ounces. For example, the commercial fractions of Rome were referred to the *as*,[1] 16 *asses* making a *denarius*.[2] A twelfth part of the *as* was the *uncia*, whence the modern "ounce" and "inch." Hence the Romans used this scheme:

Multiples of the as

$$1 \;\; as = \tfrac{1}{16} \; denarius = \tfrac{1}{24} + \tfrac{1}{48} = Denarii \; semuncia \; sicilicus$$
$$2 \;\; asses = \tfrac{1}{8} \; denarius = \tfrac{1}{12} + \tfrac{1}{24} = Denarii \; uncia \; semuncia$$
$$3 \;\; asses = \tfrac{3}{16} \; denarius = \tfrac{1}{6} + \tfrac{1}{48} = Denarii \; sextans \; sicilicus$$
$$\cdots \cdots \cdots \cdots \cdots \cdots \cdots \cdots \cdots \cdots$$
$$15 \;\; asses = \tfrac{15}{16} \; denarius = \tfrac{11}{12} + \tfrac{1}{48} = Denarii \; deunx \; sicilicus$$

Submultiples of the as

$\frac{11}{12}$, *deunx*, i.e., $1 - \frac{1}{12}$, *de uncia*, $\frac{1}{12}$ taken away. The symbol is $S = = -$, meaning *semis* $+ \frac{5}{12}$.

$\frac{10}{12}$, *dextans*, i.e., $1 - \frac{1}{6}$, *de sextans*, $\frac{1}{6}$ taken away. The symbol is $S = =$, *semis* $+ \frac{4}{12}$.

---

[1] Originally a pound of copper, but reduced by successive depreciations of coin until (191 B.C.) it weighed half an ounce.

[2] Originally a coin of 10 asses, but later of 16 asses, about 16 American cents.

$\frac{9}{12}$, *dodrans*, *i.e.*, $1 - \frac{3}{12}$, *de quadrans*, $\frac{1}{4}$ taken away. The symbol is $S = -$, *semis* $+ \frac{3}{12}$.

$\frac{8}{12}$, *bes*, *i.e.*, *bi as* for *duae partes*, $\frac{2}{3}$. The symbol is $S =$, *semis* $+ \frac{2}{12}$.

$\frac{7}{12}$, *septunx*, *i.e.*, *septem unciae*. The symbol is $S -$, *semis* $+ \frac{1}{12}$.

$\frac{6}{12}$, *semis*, half. The symbol is S, Σ, or (.

$\frac{5}{12}$, *quincunx*, *i.e.*, *quinque unciae*. The symbol is $= = -$.

$\frac{4}{12}$, *triens*, one third. The symbol is $= =$.

$\frac{3}{12}$, *quadrans*, one fourth. The symbol is $= -$.

$\frac{2}{12}$, *sextans*, one sixth. The symbol is $=$.

$\frac{1}{12}$, *uncia*, ounce, inch. The symbol is $-$.

There were similar special names and symbols for $\frac{1}{24}$ (*semuncia*, Σ, ( ), $\frac{1}{48}$ (*sicilicus*, ꟼ), $\frac{1}{72}$, $\frac{1}{144}$, $\frac{1}{288}$ (*scriptulum, scripulum, scrupulum*, ꝶ, surviving in our "scruple"), and other fractions.[1]

It will be seen that the Roman merchant could speak of $\frac{3}{8}$ of a *denarius* as 6 *asses*, of $\frac{1}{24}$ of the *as* as a *semuncia*, and so on, without considering fractions at all, and this was the case with all ancient peoples. In fact, the origin of such compound numbers as 3 yd. 2 ft. 8 in. is to be sought in the effort of the world to avoid the use of fractions.

Gradually, however, the notion of a unit fraction developed; then came the idea of a general fraction; then the surd appeared; and so on through various types of fractions, irrational numbers, transcendental numbers, complex numbers, and other kinds of artificial numbers. Each was created to satisfy an intellectual need, and in due time each, excepting the latest creations, has satisfied important practical needs as well.

First Steps in Fractions. The first satisfactory treatment of fractions as such is found in the Ahmes Papyrus (*c.* 1550 B.C.).[2]

---

[1] For a brief discussion of Roman fractions, with bibliography, see Pauly-Wissowa, *Real-Encyclopädie*, II, 1114 (Stuttgart, 1896); hereafter referred to as Pauly-Wissowa. See also Ch. Daremberg and E. Saglio, *Dictionnaire des Antiquités Grecques et Romaines*, Paris, 1877. Of the early printed works on the subject the classical one is G. Budé (or Budaeus), *De asse et partibus ejus, Libri V*, Paris, 1516, with several later editions.

[2] F. Hultsch, *Die Elemente der ägyptischen Theilungsrechnung*, reprint from the *Abhandl. d. k. Sächs. Gesellsch. d. Wissensch.*, Bd. XXXIX; Eisenlohr, *Ahmes Papyrus*; Peet, *Rhind Papyrus*, where the date is put somewhat earlier.

Artificial numbers of this kind had already been used by the Babylonians,[1] but we have no noteworthy treatment of fractions prior to the work of Ahmes. The notion of the unit fraction was already old in Egypt, however, for the tables given by Ahmes bear evidence of a development through a long period. The essential feature of the early Egyptian treatment is the unit fraction. The arithmeticians had long been able to conceive of $\frac{1}{10}$,[2] but they had no plural for it either verbally or mentally. By the time of Ahmes, however, an idea akin to that of ratio had developed. The number 2 was divided, say into 43 equal parts, and what is essentially the ratio of 2 to 43, or twice $\frac{1}{43}$, was expressed, using modern symbols, as

$$2 : 43 = \tfrac{1}{42} + \tfrac{1}{86} + \tfrac{1}{129} + \tfrac{1}{301}.$$

Indeed, most of the ancient theory of fractions centered about the concept of ratio, and in such theoretical works as that of Boethius it lasted until the 16th century.

In the Ahmes Papyrus the fraction $\frac{1}{42}$, for example, is written $\iota\iota\dot{\frown}$, where the dot is the unit-fraction symbol, $\frown$ is the Ahmes hieratic symbol for 40, and $\iota\iota$ is used to denote 2.[3] It is a curious fact that the dot is occasionally found in modern times as a fraction symbol, as in the case of $\dot{2}$ and $\dot{4}$ for $\frac{1}{2}$ and $\frac{1}{4}$ in English copy-books of the 18th century.

How these unit fractions were derived we do not know. It is evident that more than one solution is possible, but it is not always evident why any given one should be preferred to any other. For example,

$$\begin{aligned}
2 : 43 &= \tfrac{1}{24} + \tfrac{1}{258} + \tfrac{1}{1032} \\
&= \tfrac{1}{30} + \tfrac{1}{86} + \tfrac{1}{645} \\
&= \tfrac{1}{36} + \tfrac{1}{86} + \tfrac{1}{645} + \tfrac{1}{172} + \tfrac{1}{774} \\
&= \tfrac{1}{40} + \tfrac{1}{860} + \tfrac{1}{1720} \\
&= \tfrac{1}{42} + \tfrac{1}{86} + \tfrac{1}{129} + \tfrac{1}{301},[4]
\end{aligned}$$

[1] D. E. Smith, "The Mathematical Tablets of Nippur," in the *Bulletin of the Amer. Math. Soc.*, XIII (2), 392; H. F. Lutz, "A Mathematical Cuneiform Tablet," *American Journal of Semitic Languages*, XXXVI, 249.

[2] *Re-met*, "mouth of ten." Erman, *Egypt*, p. 365. Compare the Hebrew *pe-esr*.

[3] For the complete work in facsimile, see the British Museum edition.

[4] Eisenlohr, *Ahmes Papyrus*, p. 12; Peet, *Rhind Papyrus*, p. 42.

and so on, to which, of course, may be added $\frac{1}{43} + \frac{1}{43}$. Of all these possibilities Ahmes and his predecessors took the form

$$2 : 43 = \frac{1}{42} + \frac{1}{86} + \frac{1}{129} + \frac{1}{301},$$

although $\frac{1}{24} + \frac{1}{258} + \frac{1}{1032}$ has the advantage that the first fraction is nearer the value of $\frac{2}{43}$ than it is in the others. Although there are numerous rules for forming the unit fractions, no one of them applies to all the cases. This shows that the treatise combined the results of earlier computers, each working by a secret rule of his own, or else that each solution was worked out laboriously by repeated trials.[1]

The Egyptians indicated a unit fraction by a fraction symbol with the denomination underneath. In hieroglyphics this symbol was ◯, but in the cursive hieratic writing it was merely a dot. Thus, $\frac{1}{5}$, $\frac{1}{10}$, and $\frac{1}{30}$ appear respectively as ⌒ ⌒ ⌒, and $\frac{1}{42}$ appears as either ⌒ or ⌒ in hieroglyphic, but in the hieratic it appears as shown on page 210. For $\frac{1}{2}$, $\frac{1}{3}$, $\frac{1}{4}$, and $\frac{2}{3}$ there were special symbols, this having been rendered necessary by the frequent use of these fractions. Thus the symbol for $\frac{2}{3}$ was ⊕.

The symbol ◯ was also used with a different meaning. For example, in the Archeological Museum at Florence there is a marble cubit divided into parts marked with such characters as ⌒ ⌒ ⌒ ··· ‖‖⌒, representing 3, 4, 5, ··· 16 fractional parts, not $\frac{1}{3}$, $\frac{1}{4}$, $\frac{1}{5}$, ··· $\frac{1}{16}$. In the Louvre there is a similar measure made of wood with the symbols ‖‖◯ ··· ∩◯,··· ‖‖∩◯. It should be said, however, that the first of these symbols may be looked upon as meaning $\frac{1}{3}$ if we consider it as applying to $\frac{1}{3}$ of the subdivision of the cubit, say to $\frac{1}{3}$ of an inch, and similarly for the other fractions.

---

[1] E. g., when $b + c = ka$, we have

$$\frac{a}{bc} = \frac{1}{b \cdot \dfrac{b + c}{a}} + \frac{1}{c \cdot \dfrac{b + c}{a}},$$

and this gives the Ahmes result in certain cases but not in others. Thus, $\dfrac{2}{15} = \dfrac{2}{3 \cdot 5}$, and $3 + 5 = 4 \cdot 2$. This fraction, therefore, is equal to $\frac{1}{12} + \frac{1}{20}$, but Ahmes gives $\frac{1}{10} + \frac{1}{30}$. See Eisenlohr, *Ahmes Papyrus*, p. 28; G. Loria, *Bibl. Math.*, VI (2), 97; VII (2), 84; Peet, *Rhind Papyrus*, p. 34.

Later Development of Unit Fractions. The separation of a fraction into partial fractions is an illustration of the force of tradition. The predecessors of Ahmes decomposed their quotients in this way, and so Ahmes did the same. Although the Greeks had meanwhile developed a fairly good system of fractions, Heron (*c.* 50?) followed the Egyptian tradition, adopting the standard set by Ahmes nearly two thousand years earlier.[1] Some six or seven centuries later, so the Akhmim Papyrus (*c.* 8th century) informs us, the identical method of Ahmes was still in vogue in the temple schools of Egypt. Even as late as the 10th century Rabbi Sa'adia ben Joseph al-Fayyumî[2] (died 941), a Hebrew writer living in Egypt, made much use of unit fractions in his computations relating to the division of inheritances.

Not all tables of fractions made by the Egyptians followed precisely the Ahmes type, as may be seen in one dating from about the 4th century and recently acquired by the University of Michigan.[3] This table gives the unit fractional parts up to tenths of the units from 1 to 9, of the tens from 10 to 90, of the hundreds to 900, and of the thousands to 9000. It then gives the elevenths, twelfths, and so on to the seventeenths of the units up to 11, 12, and so on to 17 respectively. For example, $\frac{1}{9}$ of 50 is given as $5\frac{1}{2}\frac{1}{18}$, and $\frac{1}{11}$ of 9 as $\frac{1}{2}\frac{1}{4}\frac{1}{22}\frac{1}{44}$.

Upwards of two centuries after Rabbi Sa'adia, Fibonacci gave a rule for separating fractions into partial fractions,[4] of which the separation into unit fractions is a special case. Until recently our textbooks in algebra have given similar directions, although the subject had no immediate application that the pupil could then understand.

In the Middle Ages unit fractions were sometimes called "simple fractions," the more general form being known as "composite fractions."[5] These "simple fractions" were not

---

[1] Professor Loria has called attention to the fact that Heron was not very accurate about it, for he gives $\frac{3}{5}\frac{2}{1}\frac{7}{2}$ as $\frac{1}{2} + \frac{1}{8} + \frac{1}{73}$, while the $\frac{1}{73}$ should be $1 : (73 + \frac{1}{7})$. *Bibl. Math.*, VII (2), 88.   [2] *Traité des successions*, ed. Joël Muller. Paris, 1897.

[3] L. C. Karpinski, "Michigan Mathematical Papyrus, No. 621," in *Isis*, vol. iv.
[4] He called it a "regula uniuersalis in disgregatione partium numerorum." See *Liber Abaci*, p. 82.

[5] So the Rollandus MS. (*c.* 1424) says: "sicut $\frac{1}{2}$. $\frac{1}{3}$. $\frac{1}{4}$ que simplices fractões dicũtᵉ. sic $\frac{2}{3}$. $\frac{3}{5}$. $\frac{4}{5}$. que cõposite siue pregnãtes dicuntᵉ."

infrequently favored even by Renaissance mathematicians of some prominence, Buteo, for example, giving $1350534\frac{1}{2}\frac{1}{64}$ as the square of $1162\frac{1}{8}$,[1] even though he knew the other forms. As late as the 17th century Russian manuscripts on surveying speak of a "half-half-half-half-half-third" of a certain measure instead of $\frac{1}{96}$ of the measure,[2] and even today the unit fraction is used to some extent in the diamond trade in speaking of parts of a carat.

The difficulty met in early times in solving problems involving fractions is illustrated by an example from Ahmes: "A number together with its fifth makes 21 [; find the number]." Our solution would be $\frac{6}{5}x = 21$, whence $x = \frac{5}{6} \times 21 = 17\frac{1}{2}$; but Ahmes went through substantially this process: Multiplying 1 and $\frac{1}{5}$ by 5, we have 5 and 1, which make 6, and this is too small. To find how many times too small we divide 21 by 6, the result being $\frac{1}{2}$ and 3. Multiplying 5 by this result, the answer is $17\frac{1}{2}$.[3]

Development of the General Fraction. It seems probable that, except in very simple cases, the idea of a fraction with numerator greater than unity arose in Babylon. Although the unit fraction and possibly some idea of the sexagesimal fraction appear in the cuneiform records of *c.* 2000 B.C., the fraction forms also include special symbols for $\frac{2}{3}$, $\frac{2}{18}$, $\frac{4}{18}$, $\frac{5}{6}$, and other cases of a like degree of difficulty. No such elaborate treatment of the subject as that given by Ahmes, however, has been found as yet among the Babylonian remains.[4] In spite of this early use of the general fraction, our present forms are not due to Babylonian influences, at least not directly, but apparently to the Hindu arithmeticians.[5]

---

[1] *Ioan Bvteonis De Qvadratvra circuli Libri duo*, p. 39. Lyons, 1559.

[2] V. V. Bobynin, "Quelques mots sur l'histoire des connaissances mathématiques," *Bibl. Math.*, III (2), 104.

[3] Peet, *Rhind Papyrus*, p. 62, No. 27.

[4] For examples of the non-unit fractions among the Babylonians see Contenau, *loc. cit.* (cited on page 37), plates 3, 35, and 100; Peet, *Rhind Papyrus*, p. 28.

[5] R. C. Dutt, *History of Civilization in Ancient India*, I, 273 (London, 1893); hereafter referred to as Dutt, *History Civ. in Anc. India*. See also V. V. Bobynin, "Esquisse de l'histoire du calcul fractionnaire," *Bibl. Math.*, X (2), 100.

Greek Fractions. The Greeks followed the ancient plan of avoiding, by the use of submultiples, the difficulty of computing with fractions; but in due time the need for a fraction symbolism became so apparent that they developed a system that served their purposes fairly well. They designated $\frac{1}{3}$ ($\tau\rho\iota\tau\sigma\nu$, *tri'ton*) by the symbol $\overset{o\nu}{\Gamma}$, the Γ being the symbol for three. This was further abbreviated to Γ. Similarly, for $\frac{1}{4}$ they used Δ (for four) with two accent marks, thus: $''\Delta$. In the same way they accented their other numerals, a method represented in modern typography by γ'', δ'', ϵ'', $\cdots$.

The more common fractional unit, one half ($\overset{\prime}{\eta}\mu\iota\sigma\nu$, *he'misy*), had a special symbol, (, which was often written in a form resembling the Greek Σ or the Latin S. Two thirds ($\delta\iota\mu\sigma\iota\rho\sigma\nu$, *di'moiron*) had various abbreviations, such as (ς'', that is, $\frac{1}{2}+\frac{1}{6}$.[1]

Aristarchus (*c.* 260 B.C.) wrote the word for the numerator and the numeral for the denominator, as we might write "ten 71sts."[2] Various methods were afterwards used, such as writing the numeral for each term but doubling it for the denominator, as in the case of $2'5''5''$ ($\beta'\epsilon''\epsilon''$) for $\frac{2}{5}$; or writing the numerator, then the words "in part,"[3] and finally the denominator, as in the case of "3,069,000 in part 331,776," for $\frac{3069000}{331776}$.[4] Heron (*c.* 50?) and Diophantus (*c.* 275) used a symbol that naturally seems strange to the modern reader, namely, our common fraction reversed; that is, they wrote the equivalent of $\frac{1}{4}9$ or $\frac{19}{4}$ for four nineteenths,[5] and similarly in other cases. Ordinarily, however, the unit fraction was preferred, $\frac{15}{16}$ being written as $\frac{1}{2}+\frac{1}{4}+\frac{1}{8}+\frac{1}{16}$.[6]

Roman Fractions. As already stated, the Romans, like their predecessors, avoided fractions to a great extent by the device

---

[1] For bibliographical references and for this general topic see Pauly-Wissowa, II, 1077; Heath, *History*, I, 42.

[2] $\Delta\epsilon\kappa\alpha$ $o\alpha''$. All the Greek symbols used hereafter in this section are modern.

[3] '$E\nu$ $\mu\sigma\rho\iota\psi$, from $\mu\sigma\rho\iota\sigma\nu$, a piece, portion, or section, much as we should say "divided by."

[4] In modern Greek symbols, $\overline{\tau\varsigma}$. $\overline{\iota\theta}$ $\mu\sigma\rho$. $\lambda\gamma$. $\overline{\iota\alpha\psi\sigma\varsigma}$.

[5] *I.e.*, $\frac{\iota\theta}{\delta}$ ($\frac{1}{4}9$), or possibly $\frac{\iota\theta}{\delta}$ ($\frac{19}{4}$), for $\frac{4}{19}$.

[6] *I.e.*, ($\delta''\eta''\iota\varsigma''$. For a further discussion, see Pauly-Wissowa and Heath.

of compound numbers, although using a few convenient symbols. Even such names as *semuncia* (half-twelfth) were not numerous. Marcus Terentius Varro[1] (116–28 B.C.) mentions twelve such fractions, and Volusius Maecianus[2] (2d century) gives only two more. Of the later Latin writers, Isidorus (*c.* 610) mentions only eight and Papias (11th century) has eighteen.[3] Adelard of Bath (*c.* 1120) mentions twenty-four.

Chinese Fractions. The Chinese seem to have made use of fractions of considerable difficulty at a very early date.[4] The *Chóu-peï*, probably of about 1105 B.C. but possibly much earlier, has various problems involving such numbers as $247\frac{933}{1460}$, not stated, however, in numerical symbols but given in words. The work includes such divisions as that of 119,000 by $182\frac{5}{8}$, both of these expressions being multiplied by 8 before dividing. The unit fraction also entered into their work, as it did in all earlier civilizations. For example, in the *Nine Sections*, a work of very uncertain date but probably of the second millennium B.C.,[5] there is given the problem:

There is a field whose length is one *pu* and a half, one-third *pu*, one-fourth *pu*, and one-fifth *pu*. If the area is 240 square *pu*, what is its breadth?

Present Writing of Common Fractions. It is probable that our method of writing common fractions is due essentially to the Hindus, although they did not use the bar. Brahmagupta (*c.* 628) and Bhāskara (*c.* 1150), for example, wrote $\frac{2}{3}$ for $\frac{2}{3}$.[6] The Arabs introduced the bar, but it was not used by all their writers, and when Rabbi ben Ezra (*c.* 1140) adopted the Moorish forms he generally omitted it.[7] It is ordinarily found

---

[1] *De Lingua Latina*, 1st ed. s.l.a., but Rome. Hain mentions six editions s.l.a. before 1501, and one dated 1474 and another 1498.

[2] *Assis Distributio*, 1st ed., Paris, 1565.

[3] *Vocabularium*, 1st ed., Milan, 1476. For a full discussion of these fractions see Boncompagni's *Bullettino*, XIV, 71, 109. Not the Papias of the 2d century.

[4] Y. Mikami, "Arithmetic with Fractions in Old China," *Archiv for Mathematik og Naturvidenskab*, Christiania, XXXII, No. 3.

[5] See Volume I, page 32.

[6] Taylor, *Lilawati*, Introd., p. 12; text, p. 24 n.; Villicus, *Geschichte*, p. 54.

[7] Silberberg, *Sefer ha-Mispar*, p. 104.

in the Latin manuscripts of the late Middle Ages, but when printing was introduced it was frequently omitted, doubtless owing to typographical difficulties. This inference is confirmed by such books as Rudolff's *Kunstliche rechnung* (1526), where the bar is omitted in all ordinary fractions like $\frac{2}{3}$ and $\frac{8}{12}$ but is inserted in all fractions printed in larger type and in those having large numbers.[1] The same inference is drawn from his *Exempel-Büchlin* (1530), $\frac{1}{2}$ having the bar because that fraction was in the font, and the other fractions not having it because of the necessity for piecing them up. One of the interesting evidences of the troubles of early printers is seen in Ciacchi's *Regole generali d' abbaco* (Florence, 1675), where, in order to secure better alignment, every fraction in the book is set up like $\frac{12}{3}$, for $\frac{2}{3}$. The difficulties of the early printers probably account also for such forms as "Z3& septe octaui" for $23\frac{7}{8}$, and "Z3&V octaui" for $23\frac{5}{8}$, in Chiarino's work of 1481.

The omission of the bar was not, however, entirely a matter of typography. Hylles (1592), for example, omitted it after the first fraction in a case like $\frac{1}{2}$ of $\frac{2}{3}$ of $\frac{3}{4}$, writing this expression $\frac{1}{2} \cdot \frac{2}{3} \cdot \frac{3}{4}$ and saying:

And here you see the first fractions to wit $\frac{1}{2}$ being a true fraction, written with his lyne as it ought to be. and the other two that is to say $\frac{2}{3}$ and $\frac{3}{4}$ to be written without any lyne as their vse and order is.[2]

Recorde (*c.* 1542) tells us that "some . . . expresse them thus in slope forme,"[3] as here shown: $\frac{\frac{3}{4}}{\frac{2}{3}}{\frac{1}{2}}$

The common use of 2/3 for $\frac{2}{3}$ is the result of a desire to simplify written and printed forms.[4]

---

[1] Edition of 1534 examined.

[2] Fol. 11, *v*. Even as good a writer as Paolo Casati, *Fabrica et vso del compasso di proportione*, Bologna, 2d ed., 1685, however, omits the bar entirely.

[3] *Ground of Artes*, 1558 ed., fol. R iij, *v*.

[4] The questionable statement that 2/3 comes from 2 f 3, the f meaning *fratto* (fraction), is made by G. Frizzo, *Le Regoluzze del M. Paolo dell' Abbaco*, Bologna, 1857; 2d ed., enlarged, Verona, 1883, p. 45; but the manuscript was first published in G. Libri, *Histoire des Mathématiques*, III, 295.

Since the bar is an Oriental device, it was never used by the Greeks or Romans to indicate a fraction, at least in the way that we use it today. In Renaissance times, however, when Arab devices mingled with classical forms, we find the Roman numerals occasionally used[1] in cases like $\frac{IX}{XI}$, and the Greek numerals employed[2] in a similar manner.

KÖBEL'S USE OF COMMON FRACTIONS

From Köbel's *Rechen biechlin* (1514; 1518 ed.), showing the attempt to use Roman numerals with common fractions

The Name "Fraction." The word "fraction" is from the Latin *frangere* (to break). It is a broken number and was often so called. Baker (1568), for example, speaks of "fractions or broken numbers," calling a fraction of a fraction a "broken of broken," and various other English writers did the same. The word "fragment" is from the same root and was not infrequently used for "fraction."[3]

---

[1] Köbel, *Ain New geordnet Rechen biechlin* (1514; 1518 ed.), fol. xxxiii, *r.*
[2] V. Strigelius (Strigel), *Arithmeticus Libellvs*, Leipzig, 1563.
[3] Thus, in the Italian edition (1586) of Clavius (p. 75) the word appears as *fragmēto*. The idea goes back to the Egyptians. See Peet, *Rhind Papyrus*, p. 15.

The use of this root has not, however, been universal. Boethius (c. 510) does not speak of fractions as such in his arithmetic, introducing instead an elaborate system of ratios; but in the geometry attributed to him there is a chapter *De Minutiis*,[1] so that if he spoke of fractions at all, other than as ratios, he called them minutes, and in this he was followed by various medieval writers.[2] In the 12th century, for example, Adelard of Bath used *minuciae*,[3] while about the same time Johannes Hispalensis preferred *fractiones*.[4] In the translation of al-Khowârizmî attributed to Adelard, however, *fraciones* is used.[5] There are many instances in the early printed books of the use of the two terms interchangeably, each signifying a common fraction.[6] Several reputable writers used "parts" as a synonym of "fractions."[7] In English the word "fraction" appeared early,[8] however, and has been the general favorite.

Since *ruptus*, like *fractus*, means broken, this has been the root of a name for fraction. In Italian it appears as *rotto* (plural, *rotti*),[9] in Spanish as *rocto*,[10] and in French in various forms.[11]

---

[1] Friedlein ed., pp. v, 425.

[2] We shall see that the term was also applied specifically to sexagesimals, although by no means generally.

[3] In his *Regulae abaci*. See Boncompagni's *Bullettino*, XIV, 109.

[4] In his *Liber Algorismi de pratica arismetrice*. See Boncompagni, *Trattati*, II, 49; *Abhandlungen*, III, 111. Fibonacci (1202) generally used *fractio*.

[5] For *fractiones*. *Minuta* is used (for *minutae*) to mean sixtieths. See Boncompagni, *Trattati*, I, 17.

[6] Thus Huswirt (1501): "Minutia siue fractio nihil aliud est q̄3 pars integri" (fol. 11, *r.*); and Clavius (1583) expresses certain of his quantities "in numeris fractis, qui alio nomine Mintutiae, fractionesve dici solent vulgares" (p. 81).

[7] Thus Fine (1530): "De minutis, siue quotis eorundem integrorū partibus (quas uulgares appellant fractiones)"; and Gemma Frisius (1540): "Fractiones minutias aut partes." Gosselin, in his translation (Paris, 1578) of Tartaglia's arithmetic, uses *parties* more commonly than any other term for fractions. Hylles (1592) has the expression: "fractions of fractions (or as some men call them particles, that is as you would say parcels of parts)." Ramus (1555) speaks of "fractio sive pars."

[8] Thus Chaucer, in his *Astrolabe* (c. 1391), uses *fraccion*.

[9] Pacioli (1494) ordinarily speaks of *rotti*, although he also uses *fractioni* and *fracti* (fol. 48). Most of the 16th century Italian writers use *rotti*.

[10] Ortega (1512).

[11] Chuquet (1484), "nombres routz"; Savonne (1563), "roupt"; and later writers, "nombre rompu."

In the Teutonic languages the custom was followed of using vernacular expressions, and so the Latin *fractio* appeared as "broken number."[1]

Common Fraction. The expression "common fraction" was originally used to distinguish the fractions employed in trade from the sexagesimal fractions found in astronomy. It refers merely to the form of writing a fraction, $\frac{5}{10}$ being a common fraction, o.5 being a decimal, and 30' being a sexagesimal, although the values of the three are the same. In Latin the expression was *fractiones vulgares*, whence the "vulgar fractions" of the English. The adjective "common" is used at present in America, although this has not always been the case,[2] nor have the English uniformly followed their present usage.[3]

Definition of Fraction. In general a fraction has been defined as one or more parts, or equal parts, of a unit,[4] sometimes with the limitation that the numerator must be less than the denominator.[5] Occasionally the more scientific writers based the definition upon division, usually of a smaller number by a larger.[6] The idea of an improper fraction, like $\frac{3}{2}$, is a late development. Occasionally a 16th century writer like Recorde[7] (*c.* 1542), Gemma Frisius[8] (1540), or Tartaglia[9] (1556) mentioned this type of fraction as an expression of division, but little was done with it. Complex fractions, those in which a fraction appears

---

[1] Thus Riese (1522) speaks of "Ein gebrochene zal" (1550 ed., fol. 14, *v.*), and Grammateus (1518) has a chapter "Von Prüchen," speaking of a fraction as "ein iglicher pruch (welchen man in latein fraction nennet)." So in Dutch we find Raets (1580) speaking of "Die ghebroken ghetalen," and Mots (1640) and others speaking of "Ghebroken."

[2] Similarly in France, instead of *fraction ordinaire* Trenchant (1566) used *fraction vulgaire.* Our colonial arithmeticians usually followed the English use of "vulgar."

[3] Thus Digges (1572) speaks of "the vulgare or common Fractions."

[4] *E.g.*, Pacioli (1494): "Rotto e vno o vero piu parti de vno ītegro" (fol. 48, *r.*); Santa-Cruz (1594): "Quebrados es vna parte, ò partes dela cosa entera."

[5] *E. g.*, Pagani's arithmetic (1591).

[6] *E.g.*, Ramus (1555). The Dutch arithmetic of Raets (1580) defines a fraction as "een ghetal diuideert met een grooter." On the fusion of the notions of fraction and quotient, see V. V. Bobynin, *Bibl. Math.*, XIII (2), 81.

[7] See the 1558 edition of the *Ground of Artes*, fol. S vi, *v.*

[8] "Fractiones quae plus Integro valent."   [9] 1556 ed., I, fol. 107, *r.*

in either numerator or denominator, or in both, are older than might be expected. Rabbi ben Ezra, for example, has a problem involving the product of two such forms.[1]

Terms of a Fraction. The medieval Latin writers found it convenient to devise names for the terms of a fraction written after the Arab manner, and so they called the upper number by such names as *numerator* (numberer) and *numerus* (number),[2] while the lower number was called the *denominator* (namer). These terms are hardly destined to endure, but no others have been generally accepted. Among the medieval and Renaissance writers the numerator was often designated by such words as *nominator*,[3] "topterme," "top,"[4] *superior*,[5] and *denominato*,[6] and the denominator by such names as *base, inferior*, and *denominante*. Both the numerator and denominator took on vernacular forms with later Teutonic writers.[7] In the Latin languages, however, the favorite names were *numerator* and *denominator*, the former of which Tartaglia (1556) speaks of as being written above a *virgoletta* (little bar), and the latter as being written below it.[8]

[1] *Sefer ha-Mispar*, 39.

[2] So the Rollandus MS. (*c.* 1424) speaks of the "nūator et denom̃tor." See also the correspondence of Regiomontanus and Bianchini, *Abhandlungen*, XII, 287. In the 16th century Ramus (1555) speaks of the *superior terminus* as the *numerus sive numerator*.

[3] *E. g.*, Digges (1572), although he also used *numerator*. See pages 20, 24, 27 of the 1579 edition.

[4] Thus Hylles (1592): "Numerator which also for more shortnesse is sometimes called the Topterme or top onely: and that the lower term is vsually called the Denominator or Base."

[5] As in Gemma Frisius (1540), although he also uses *numerator*.

[6] Paolo dell' Abaco (*c.* 1340): "Sappi che ogni rotto si scrive con due numeri: il minore sta sopra la verga e chiamasi denominato; e il maggiore sotto la verga e chiamasi denominante" (ed. Frizzo, 1883, p. 45). The name was used by various 16th century writers, such as Sfortunati, *Nuovo Lume* (1534).

[7] *E. g.*, Widman (1489) has the Latin forms, but a little later the words *Zeler* and *Nenner*, with variants, came into general use. Occasionally a Dutch writer like Wentsel (1599) used the Latin forms, but most arithmeticians preferred "teller" and "noemer," with such variants as "telder" and "nommer," and similarly with the Scandinavian writers.

[8] ". . . l' uno di quali è detto numeratore (& questo si scriue sempre sopra vna virgoletta) l' altro è chiamato denominatore, e questo si scriue sempre sotto a quella tal virgoletta." *General Trattato*, I, 106, *v.*; 107, *r.*

Reduction of Fractions. Until recently the reduction of a fraction to lower or lowest terms was commonly known as abbreviation. Thus Digges (1572) says:

To abbreuiate any Fragment, is to bring a Fraction to his lest denomination. To make this abbreuiation, yee must diuide the Numerator of the Fraction, and so in the like maner the Denominator by the biggest number, that is some common part of them both.[1]

The word "depression" was also used, and, like "abbreviation," is more suggestive than "reduction,"[2] which sometimes had the special meaning of bringing fractions to a common denominator.[3]

Before the invention of decimals such fractions as $\frac{3345312}{4320864}$ were not uncommon,[4] and it was necessary to reduce them to lowest terms in order to operate with them. In general the cancellation of all common factors was not convenient, and hence the long form of greatest common divisor was essential. First, however, factors were canceled. A factor thus eliminated was called by the Italians a *schisatore*.[5] On account of the necessity for recognizing common factors, many of the early manuscripts and printed works gave the ordinary tests for divisibility by 2, 3, and 5, and even some kind of test for

[1] 1579 ed., p. 24. So Hodder (1672 ed.) says: "I would abreviate $\frac{24}{120}$," and J. Ward (1771 ed.) has a caption "To Abbreviare or Reduce Fractions into their Lowest or Least Denomination." The expression is much older than this, however, for Chuquet (1484) says: "Abreuier est poser ou escripre vng nombre rout par moins de figures . . ." (fol. 12, *r.*, of his MS.). Early Spanish writers used the same expression, as in Santa-Cruz (1594), "De abreuiar quebrados." The Dutch writers of the same period used various terms, including *abbreviatio*, *verminderinge, and vercontinghe*, and Van der Schuere (1600) says: "om ghebroken ghetallen te vercorten ofte minderen."

[2] Thus Pacioli (1494): "De vltima depͤssione fractorum siue modo schisandi dicto," adding "Che ī frãçese si chiama Abreuier" (fol. 48, *v.*).

[3] *E. g.,* Pellos (1492, fol. 21, *r.*), Chuquet (1484, fol. 10, *r.*), and others. This special meaning was not general, for Tartaglia uses it in the broader sense (1592 ed., I, fol. 169, *r.*).

[4] This and similar fractions are in the Treviso arithmetic (1478). Fractions like $\frac{9432500}{14406000}$ and $\frac{100207693}{300800000}$ are given in the *Epitome* of Clavius (1583; 1585 ed., pp. 77, 124).

[5] A word suggesting canceling "across" (*schisa*), whence *schisare,* to reduce a fraction. So Pacioli (1494) speaks "De diuersis modis in ueniendi schisatorem" (fol. 49, *r.*), and Cataneo (1546) tells "Come si schisino i rotti" and speaks of "Lo schisamento" and "di schisare."

divisibility by 7.[1] When common factors were not readily seen, the greatest common divisor was resorted to at once, being found by the Euclidean method.[2] This is given in al-Karkhî's *Kâfî fîl Hisâb* (*c.* 1015),[3] and in various other Oriental works, manifestly all derived from Greek sources.

Greatest Common Divisor. The greatest common divisor went by various names in the early printed books.[4] The theoretical works usually gave a rule for finding it, although the mercantile works often omitted the subject entirely, the former making use of long fractions and the latter ignoring them. One of the earliest printed rules is stated by Pacioli (1494) and is credited to Boethius (*c.* 510).[5] In this the smaller term is continually subtracted from the larger, a smaller remainder from that, and so on, an evident modification of the Euclidean method. Several early writers used the latter method for "abbreviating," without mentioning the greatest common divisor as such.[6]

Sequence of Operations. By analogy to the sequence of operations in the case of integers, the sequence in fractions has generally begun with addition. Medieval[7] and Renaissance writers, however, often took the more sensible course of beginning with multiplication,—a course to which the primary schools have now returned. Recorde (*c.* 1542) was earnest in his advocacy of this method, saying:

---

[1] A good illustration of the use of these tests in the later works of the 16th century may be found in Van der Schuere's *Arithmetica* (1600).

[2] That is, the one given in the *Elements*. See Heath's *Euclid*, Vol. II, pp. 118, 299.

[3] Hochheim ed., p. 10.

[4] In the Latin books it usually appears as *maximus communis divisor*, and in the Italian works as *il maggior comune ripiego* (Cataneo's spelling, 1546) or *massima cõmune misura* (Cataldi's spelling, 1606).

[5] "Vn altro modo se elice da Boetio nel secondo della sua Arithmetica per trouare ditto schisatore" (fol. 49, *v.*). See Friedlein's *Boethius*, p. 77.

[6] *E.g.*, Chuquet (1484), under "Aultre stile de abreuir," and the Dutch writer Petri (1567). Somewhat similar treatments are given by Baker (1568), Raets (1580), Rudolff (1526), and others. The phraseology used by Grammateus (1518) is interesting ("¶Prüch kleyner zumachen"), and that of Rudolff is analogous to it ("Wie man gewiszlich erkennen mag/ob ein bruch müg noch kleiner gemacht werdē od' nit").

[7] For example, Abraham ben Ezra (*c.* 1140).

There is an other ordre to be folowed in fractions then there was in whole numbres. for in whole numbres this was the ordre, Numeration, Addition, Subtraction, Multiplyplication, Diuision and Reduction. but in fractions (to folowe the same aptnesse in procedyng from the easyest woorkes to the harder) we muste vse this ordre of the woorkes, Numeration, Multiplication, Diuision, Reduction, Addition, and Subtractiō.

The book is in the form of a dialogue, and upon the pupil's saying, "I desyre to vnderstond ye reason," the master says:

As in the arte of whole numbres ordre woulde reasonablye begyn with the easiest, and so go forwarde by degrees to the hardest, even so reason teacheth in Fractions the lyke ordre.[1]

Addition and Subtraction. In adding or subtracting, early writers usually took for a new denominator the product of the given denominators, reducing the final result to lowest terms.[2] Because of the size of the common denominator thus found,[3] the early Rechenmeisters in Germany ordinarily added but two fractions at a time. Although the plan of reducing to the least common denominator before adding or subtracting was occasionally used by 15th and 16th century arithmeticians,[4] it was not until the 17th century that it began to be generally recognized,[5] and even then the name was slow of acceptance.[6]

---

[1] For further discussion see the 1558 edition of the *Ground of Artes*, fol. R iiii, *v*. The same order is followed by Pacioli (1494, fol. 51), Pagani (1591, pp. 34, 41), and others. Giovanni Battista di San Francesco, *Elementi Aritmetici* (Rome, 1689), even begins with division "that it may be better understood."

[2] Thus, $\frac{3}{4} + \frac{5}{6} = \frac{18}{24} + \frac{20}{24} = \frac{38}{24} = 1\frac{8}{2} = 1\frac{7}{12}$. The method is given by Bhāskara (*c.* 1150); see Taylor's translation, p. 24. It appears in many medieval MSS. and in such early arithmetics as those of Petzensteiner (1483), Pellos (1492), Riese (1522), Recorde (*c.* 1542), and Baker (1568).

[3] *E.g.*, the Dutch arithmetician Wilkens (1630) reduces $\frac{3}{4}$, $\frac{4}{5}$, $\frac{5}{6}$, and $\frac{7}{8}$ to 960ths before adding.

[4] Chuquet (1484) gives it (fols. 13 and 14), and it is found in such works of higher class as those of Tartaglia (1556) and Clavius (1583).

[5] So Cataldi (1606) reduces to the "minor commune denominatore"; the well-known Coutereels (1599), to "het minste ghetal"; and Wilkens (1630), to "Kleynste gemeyne Noemer."

[6] The shorter name of "general denominator" was used by some writers. See Starcken's Dutch work of 1714, with "General Nenner." Ramus (1569) suggested "cognomen" for common denominator,—not a bad term.

The arrangement of an example in addition was somewhat uniform before the 17th century, and it may be understood from the following case of $\frac{5}{6} + \frac{3}{8}$ as given by Pacioli[1]:

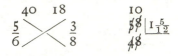

Multiplication of Fractions. Although our present interest in the multiplication of fractions relates to such simple cases as $\frac{2}{3} \times \frac{4}{5}$, it is desirable to set forth some of the difficulties met by ancient writers. These difficulties appear in the works of the Egyptians and Greeks, but they are sufficiently evident in a single example given by Rabbi Sa'adia ben Joseph al-Fayyumî, a Hebrew scholar of the 10th century already mentioned on page 212. The problem, which shows the difficulties met with in the use of unit fractions, is to find the product of $61\frac{1}{3}\frac{1}{9}$ by $61\frac{1}{3}\frac{1}{9}$. The solution is substantially as follows: $61 \times 10 = 610$, $61 \times 20 = 1220$, $61 \times 40 = 2440$, $61 \times 60 = 3660$, the last three being found by doubling or by adding. Then $61 \times 61 = 3721$, evidently found by adding 61 to 3660. Then $\frac{1}{3} \times 61 = 20\frac{1}{3}$, and $20\frac{1}{3} + \frac{1}{9}$ of $61 = 26\frac{2}{3}$ (*sic*), the double of which is $53\frac{1}{3}$. Adding this to 3721, he obtains $3774\frac{1}{3}$, and this increased by $\frac{2}{3}$ gives the result, 3775. What he tries to do is to square $61\frac{1}{3}\frac{1}{9}$ by taking $61^2 + 2 \times 61 \times (\frac{1}{3} + \frac{1}{9}) + (\frac{1}{3} + \frac{1}{9})^2$ (a rule which was, of course, well known), but he fails in his computation.

With respect to the ordinary operation with simple fractions, the process of multiplication has not changed materially during the last few centuries except that cancellation was not generally used by early writers,[2] although a few of the better arithmeticians saw its advantages.[3]

---

[1] 1494 ed., fol. 51, with an error in the quotient as printed.

[2] *E. g.*, Calandri (1491) multiplied $\frac{4}{5}$ by $\frac{3}{4}$ thus:

Multiplica $\frac{3}{4}$ uie $\frac{4}{5}$

$$\frac{3}{4} \underline{\hspace{1cm} \frac{12}{20} \hspace{1cm}} \frac{4}{5}$$

Fanno $\frac{3}{5}$

Even in the Greenwood American arithmetic (1729) this method is followed.

[3] Thus Rudolff (*Kunstliche rechnung*, 1526) says "das man ein ober vnd ein vnter gegen einander mag auffheben oder kleyner machen" during the operation.

In the matter of language, the schools have usually protested against the broadening of the meaning of any technical term. A teacher will object to saying " $\frac{3}{2}$ times 4" but will say " $1\frac{1}{2}$ times 4." The contest is an old one; thus Ortega (1512) would not write "$3\frac{1}{4}$ ducats," preferring the awkward expression "3 ducats and one fourth of a ducat";[1] but Rudolff (1526), at about the same time, did not hesitate to speak of " $\frac{1}{2}$ times" a number.[2]

Of the various special rules, most of which came from the Arabs, a single one may serve as a type. Expressed in modern symbols,

$$\frac{a}{b} \cdot \frac{c}{d} = \frac{\frac{ac}{b}}{d} = \frac{\frac{ac}{d}}{b},$$

that is,

$$\frac{2}{3} \cdot \frac{5}{7} = \frac{\frac{10}{3}}{7} = \frac{\frac{10}{7}}{3},$$

thus reducing the work to dividing a fraction by an integer and suggesting cancellation more strongly.[3]

Many of the early writers expressed concern over the fact that the product of a number by a proper fraction was less than the multiplicand. Borghi (1484) seems to have been the first author of a printed book to discuss the matter, and various 16th century writers had much to say about it.[4]

Few writers before the 17th century made any attempt at explaining the process, although Trenchant (1566) devoted some attention to it, using the illustration of a square cut into smaller squares.[5]

[1] "¶Se 3 ducati e vn quarto de ducato . . . guadagnano 5 fiorini e vn terzo," much as we say "a dollar and a half."

[2] " . . . dañ ich hab die sechs nur ein halbs mal haben wóllen" (1534 ed., fol. C iiij, v.).

[3] This is given by al-Karkhî (c. 1020).

[4] Among them was Ramus (1569). In the 1586 edition (p. 73) his commentator, Schoner, gives a whole page to it. Cataneo (1546) also devotes a page to it, seeking particularly to combat Borghi and Pacioli. (See 1567 edition, fol. 21, v.) Even Tartaglia did not see the point of the controversy (see 1592 ed., I, fol. 187, r.).

[5] Cardinael's School Boecken (1650; 1674 ed.) goes into the matter more fully, using several diagrams.

Division of Fractions. Naturally the most difficult óperation was division. Multiplication by the inverted divisor is so simple that we hardly realize that it has come into general use only recently, although it was known in the early Middle Ages by both the Hindus and the Arabs.[1] Influenced by the notion that only fractions could deal with fractions, medieval writers often substituted for the division of a fraction by an integer the process of multiplying by the reciprocal of the integer;[2] that is, $\frac{2}{3} \div 4 = \frac{2}{3} \times \frac{1}{4}$.

The early printed books gave two leading methods. The first of these reduced the fractions to a common denominator and took the quotient of the numerators,[3] as in the case of

$$\frac{2}{3} \div \frac{3}{4} = \frac{8}{12} \div \frac{9}{12} = \frac{8}{9}.$$

The second method is one of cross multiplication. Thus, in the case of $\frac{2}{3} \div \frac{3}{4}$ we have

$$\frac{2}{3} \diagdown\!\!\!\!\diagup \frac{3}{4} = \frac{8}{9},$$

which involves the same operations that enter with the inverted divisor. This was the favorite method in the early printed books,[4] and the name "cross multiplication" or its equivalent[5] was common, the divisor being usually placed on the left, but sometimes on the right.[6] One writer expresses the opinion that the divisor was placed at the left because the process may have come from the Hebrews, who write toward the left.[7]

---

[1] Brahmagupta (c. 628) and Bhāskara (c. 1150) both gave it (Colebrooke translation, pp. 17, 278), and al-Ḥaṣṣâr (c. 1175?) recognized it, at least with integral dividends (*Bibl. Math.*, II (3), p. 36).

[2] Thus Rollandus (c. 1424).

[3] *E.g.*, Chuquet (1484, fol. 16, whose manuscript was so extensively appropriated by De la Roche and in part printed in 1520), Trenchant (1566), and Ramus (1555).

[4] *E.g.*, Widman (1489): "Nū wiltu teilē $\frac{9}{13}$ in $\frac{5}{6}$ sprich 6 mal 9 ist 54 dy setz für den zeler vnd sprich darnach 5 mal 13 ist 65 die setz für den nēner also $\frac{54}{65}$ " (1508 ed., fol. 30, *v.*).

[5] Thus Hodder (1672 ed.) says "multiply cross wise"; Riese (1522), "so multiplicir im creutz"; Peletier (1549), "multiplier en croix"; Pagani (1591), "moltiplica in croce."

[6] Thus Hudalrich Regius (1536), Pagani (1591), Mots (1640), and others.

[7] Wentsel, 1599, p. 88.

The idea would have been more reasonable, so far as immediate origin is concerned, if he had spoken of the Arabs.

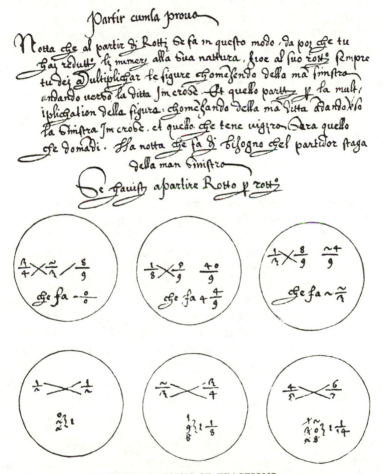

EARLY DIVISION OF FRACTIONS

From an anonymous Italian MS. of 1545 in Mr. Plimpton's library

The Inverted Divisor. As already said, the method of multiplying by the inverted divisor was known to certain Hindu
and Arab writers. It seems, however, to have dropped out of

sight for three or four hundred years, reappearing in Stifel's works in 1544.[1] It was not at once accepted, only a few of the 16th century writers making any use of it,[2] but in the 17th century it became fairly common.

Before the inverted divisor came into general use there were several special rules that met with some favor. One of these, given by Gemma Frisius (1540), may be expressed in modern symbols thus:

$$\frac{a}{b} \div \frac{ka}{c} = \frac{c}{kb},$$

as in

$$\frac{3}{5} + \frac{12}{13} = \frac{3}{5} \div \frac{4 \times 3}{13} = \frac{13}{4 \times 5} = \frac{13}{20}.$$

2. SEXAGESIMAL FRACTIONS

Nature of Sexagesimals. For scientific purposes the medieval writers usually followed the late Alexandrian astronomers in the use of fractions written on the scale of sixty.[3] This custom has continued until now in the measures of time, angles, and arcs, as when we write 2 hr. 20 min. 45 sec., that is, $(2 + \frac{20}{60} + \frac{45}{3600})$ hours, instead of $(2 + \frac{1}{3} + \frac{1}{80})$ hours. The measure of time meets a popular need, and so the sexagesimal fraction gives no present evidence of being abandoned for this purpose, but for circular measure it is losing its hold as decimals become better known, and seems destined soon to disappear.

In the Middle Ages the scientific workers carried the sexagesimal divisions still farther than the Greeks, as if we were to write 2 10′ 30″ 45‴ 5^{iv} 7^{v}, meaning thereby

$$2 + \frac{10}{60} + \frac{30}{60^2} + \frac{45}{60^3} + \frac{5}{60^4} + \frac{7}{60^5}.$$

---

[1] "Ego Diuisionis regulam reduco ad regulam Multiplicationis Minutiarum, hoc modo: Diuisoris terminos commuto," etc. *Arithmetica Integra*, 1544, fol. 6, *r*.

"Thu im also. Den Teyler . . . kere vmb/also ausz dem Zeler werde der nenner/vnd ausz dem nenner der Zeler. So steht denn das exemplum mit vmbgekereten Teyler also" (*Deutsche Arithmetica*, 1545, fol. 13, *v*.).

[2] Among them were Thierfelder (1587) and Clavius (1583). The latter says: "ac si termini diuisoris commutentur, & regula multiplicationis seruetur" (1585 ed., p. 118, and similarly the Italian edition of 1586, p. 106).

[3] Latin *sexagesimus* or *sexagensumus*, sixtieth, from *sexaginta*, sixty.

Thus Sibt al-Mâridînî,[1] an astronomer at the mosque of al-Azhar in Cairo in the middle of the 15th century,[2] gave $45° 50' \div 1° 25' = 33° 45' 52'' 56''' 28^{iv} 14^{v} 7^{vi} 3^{vii} 31^{viii} 45^{ix} 52^{x} \cdots$, and similar cases occur in many medieval works.

Names of Sexagesimals. Sexagesimals were usually known as physical fractions in the Middle Ages.[3] The name may possibly have come from their use in physics, this word (more frequently "physic"), as applied to natural philosophy, not being so recent as is sometimes thought. On the other hand, it may come from the fact that the denominators were understood to proceed in the natural[4] order of the powers of 60, somewhat as we speak of "natural numbers" at present, this being an opinion expressed in the 16th century.[5]

They were also called astronomical fractions,[6] the reason being quite apparent.[7]

Origin of Sexagesimals. There is a common idea that sexagesimal fractions came from Babylon,—an idea which arose from the fact that 60 plays an important part in the number

[1]Mohammed ibn Mohammed ibn Aḥmed, Abû 'Abdallâh, Bedr ed-dîn al-Miṣri, born in 1423, died in 1494/95. He wrote a number of works on arithmetic and astronomy.

[2]He gives the result only to 31^{viii}, the fraction then repeating,—an interesting case of a circulating sexagesimal. See Carra de Vaux, "Sur l'histoire de l'arithmétique arabe," *Bibl. Math.*, XIII (2), 33. The above symbols are, of course, modern. The problem is substantially that of $45\frac{5}{6} \div 1\frac{5}{12} = 33\frac{3}{4} +$.

[3]Thus we find in the MSS. such expressions as "Modum representationis minuciarum vulgarium et physicarum" (anonymous MS. of 1466) and "Minucie duplices sunt scilicet phisice et vulgares" (anonymous MS. of 15th century). In the early printed books they are called "fraciones phisice" (Ciruelo, 1495), "fraciones fisicas" (Texeda, 1546), "Minucciamenti Fisici" (Italian edition of Gemma Frisius, 1567), and by other similar names. [4]Φύσις (*phy'sis*, nature).

[5]Thus Trenchant (1566): "S'apele phisic, c'est à dire, naturel: pour ce que ses denominateurs, & caracteres, sont selon l'ordre naturel du nombre commençant à l'vnité (1578 ed., p. 19).

[6]"De Fractionibus Astronomicis, siue de minutiis Physicis," as Gemma Frisius (1540) says in his Latin editions, the Italian having "Rotti Astronomici." Trenchant has "Du nombre phisic, ou fractions astronomiques." Peletier (Peletarius) in his notes on Gemma Frisius (1563 ed.) speaks of "Fractiones Astronomicae, quas vulgò Physicis vocat." The name was used by Abraham ben Ezra (*c.* 1140) and probably by the late Greek writers.

[7]As Peletier (1549) says, because they "seruent aux supputations des mouuements celestes."

ν

system of that country. The assertion of this origin was first made, so far as we know, by Achilles Tatius, an Alexandrian rhetorician of the 5th or 6th century. It has also been assumed that the Babylonians divided the circle into 360 equal parts, because of the early notion that a year consisted of 360 days, and because their scientists knew that the radius employed in stepping around a circle divided it into six equal arcs, thus making 60 a mystic number. This reason may possibly be valid, but there is no authority for asserting that it is historical. The Babylonians divided the circle into 8, 12, 120, 240, and 480 equal parts, but not into 360 such parts.[1] It is true that six-spoked wheels are found represented on the Babylonian monuments, but no more frequently than the eight-spoked wheels, and the six-spoked type is more common in Egypt where the number 60 was not used to any great extent. It would seem, therefore, that the number 60 was not derived from the division of the circle into six equal arcs.

It is true, however, that the Babylonians wrote the equivalent of 11 for 60 + 1, 111 for $60^2 + 60 + 1$, and 44 26 for 44 × 60 + 26, although there is no reason for believing that this is a proof of their use of sexagesimal fractions. In a certain tablet of c. 2000 B.C., for example, the equivalent of the square of 44 26 40 is given as 32 55 18 31 6 4. This may be interpreted to mean the square of either 44 × 60^2 + 26 × 60 + 40 or $44 + \dfrac{26}{60} + \dfrac{40}{60^2}$. In the latter case we have sexagesimal fractions; in the former, numbers written on the scale of sixty, an interpretation more in harmony with the system of compound num-

---

[1] On this entire discussion see A. H. Sayce and R. H. M. Bosanquet, "The Babylonian Astronomy," in *Monthly Notices of the Royal Astron. Society*, XL, 108; E. Hoppe, *Archiv der Math.*, XV (3), 304; E. Löffler, *ibid.*, XVII (3), 135; and Hochheim, *Kâfî fîl Hisâb*, p. 23. The claim that the Chinese used a sexagesimal system in the third millennium B.C. (Vol. I, p. 24) is not supported by sufficient evidence to be considered at present. It is very improbable that it involved anything more than a recognition of 60 as a convenient unit for subdivision. On the Greek development of sexagesimals see Heath, *History*, I, 44. See also a discussion by E. Hincks, "Cuneiform Inscriptions in the British Museum," *The Literary Gazette*, XXXVIII, 707 ; C. Bezold, *Catalogue of the Cuneiform Inscriptions . . . British Museum*, I, 24.

bers used by all ancient peoples. Similarly, we find the case of
$1 \div 81$, but whether this is to be interpreted as having the
dividend 60 or some power of 60 is uncertain. In any case we
have no evidence of any such general use of sexagesimal fractions as is found among the Greek astronomers.[1]

The division of the circle into 360 parts as practiced by such
Greek astronomers as Ptolemy (c. 150) was probably the out-
growth rather than the origin of the sexagesimal system. The
Babylonians counted decimally by preference, although the
base of 60 played a considerable part in their system. They
counted decimally to 60, that is, to a *soss*; then by *sosses* and
the number over to the *ner*, which was 10 *sosses*, or 600; then
by *ners*, *sosses*, and the number over to the *saru*, which was
6 *ners*, or 3600; but they never counted 60, 360, 3600, so that
360 was not a natural step in their sexagesimal system.[2]

Greek Use of Sexagesimals. We do not know why the Greek
astronomers should have developed a scale of 60 in such a complete form, although we càn readily surmise the cause. There
seems to be no reason to doubt that the number 60 was suggested to them from Babylon, but the system of sexagesimal
fractions, as we know and use it, was, so far as now appears,
their own invention. Ptolemy used these fractions to represent
his chords in terms of a radius 60;[3] that is, the chord of 24°
would then be 24.9494, or, in sexagesimals, 24 56′ 58″. It seems
clear, however, that the Greeks needed for their astronomical
work a better type of fraction than the unit type of the Egyptians; that their habit of using such submultiples, as in feet and
inches, naturally led them to a similar usage in fractions, as
would be the case with degrees and minutes; and that the 60
of Babylon was a convenient radix, since it has as factors 2, 3,
4, 5, 6, 10, 12, 15, 20, and 30, and so permits of the ready use
of halves, thirds, fourths, fifths, sixths, tenths, twelfths, and so

---

[1] For arguments in favor of the fraction interpretation see F. Cajori, "Sexagesimal Fractions among the Babylonians," *Amer. Math. Month.*, XXIX, 8. See
also Heath, *History*, I, 29.   [2] Hilprecht, *Tablets*.
[3] For a discussion of this point see A. Schülke, "Zur Dezimalteilung des Win-
kels," *Zeitsch. für math. und naturw. Unterr.*, XXVII, 339; Heath, *History*, I, 45.

on. The Greeks may thus have been led to divide the radius into 60 equal parts and the diameter into 120 of these parts. Since the common value of π was 3 in ancient times, the circumference was naturally taken as 3 × 120, or 360.

Such was the influence of the Greek scholars that all the medieval astronomers, Christian, Jewish, and Mohammedan, used the sexagesimal system;[1] but some of the mathematical writers referred the system to India instead of Greece, influenced therein by the belief that our numerals came from the Hindus.[2]

Terms Used. When the Greeks decided to take $\frac{1}{360}$ of a circle as a unit of arc measure, they called this unit a degree.[3] They called $\frac{1}{60}$ of a degree a first part,[4] $\frac{1}{3600}$ a second part,[5] and so on.

Multiplication involving Sexagesimals. The operations of addition and subtraction with sexagesimals involved no difficulties, but multiplication and division were not so simple. It is meaningless to us to multiply 4° 7′ 38″ by 5° 6′ 29″, or even 4 7′ 38″ by 5 6′ 29″, but to the medieval scientist it meant

---

[1] As a noteworthy illustration, see the *Libros del saber de Astronomia del Rey Alfonso X*, Madrid, 1863. The Alfonsine astronomical tables date from *c.* 1254, but for argument as to a later date, see A. Wegener, "Die Astronomischen Werke Alfons X," *Bibl. Math.*, VI (3), 138.

[2] Thus Johannes Hispalensis (*c.* 1140): "placuit tamen Indis, denominationem suarum fractionum facere a sexaginta. Diuiserunt enim gradum unum in sexaginta partes, quas uocauerunt minuta" (B. Boncompagni, *Trattati*, II, 49). He may have had his idea from al-Khowârizmî (*c.* 825): "Set indi posuerunt exitum partium suarum ex sexaginta: diuiserunt enim unum in .LX. partes, quas nominauerunt minuta" (from a Cambridge MS. of the *Algoritmi de Numero Indorum,* in the *Trattati*, I, 17).

[3] Μοῖρα (*moi'ra*); medieval Latin, *de + gradus* (step). The Arabs translated μοῖρα by *daraja* (ladder, scale, step), which led G. H. F. Nesselmann (*Die Algebra der Griechen*, p. 137 (Berlin, 1842), hereafter referred to as Nesselmann, *Alg. Griechen*) to think that this word was the original form of the word "degree." It may have influenced the final form.

[4] Πρῶτα ἑξηκοστά (*pro'ta hexekosta'*); Latin, *pars minuta prima* (first small or fractional part). From this came our "minute." The Greeks also used λεπτά (*lepta'*, minute, the adjective). In the 12th century Walcherus (see Volume I, page 205) spoke of the minutes as *puncta*, and the same term is so used in an algorism of *c.* 1200. See L. C. Karpinski, "Two Twelfth Century Algorisms," *Isis*, III, 396.

[5] Δεύτερα ἑξηκοστά (*deu'tera hexekosta'*); Latin, *pars minuta secunda*, from which our "second." See also Wertheim, *Elia Misrachi*, p. 19 n.

simply the finding of $(4 + \frac{7}{60} + \frac{38}{3600}) \times (5 + \frac{6}{60} + \frac{29}{3600})$. In the operation there is, for example, $7' \times 6' = 42''$, which means simply that $\frac{7}{60} \times \frac{6}{60} = \frac{42}{3600}$.[1] In such work it became conven-ient to have multiplication and division tables, and these are found in various medieval manuscripts.[2] Some idea of the difficulty of operating with these fractions may be inferred from a prob-lem in the work of Maximus Planudes

3	28	54	8
	14	23	
	8	16	

(c. 1340).[3] His multiplication of $14°$ $23'$ by $8°$ $16'$, giving the product 3 signs[4] $28°$ $54'$ $8''$, is here shown.

Division involving Sexagesimals. In division Maximus Planudes reduced all the terms to the same denomination. For example, the operation

$$\frac{3° \ 23' \ 54''}{2° \ 34' \ 24''} = 1° \ 19' \ 14'',$$

or $\dfrac{3 + \frac{23}{60} + \frac{54}{3600}}{2 + \frac{34}{60} + \frac{24}{3600}} = 1 + \frac{19}{60} + \frac{14}{3600},$

is worked out as illustrated here.[5]

The finding of roots by the aid of sexagesimals appears in the works of

3	23	54
2	34	24
12234		1
9264		
2970		
178200		19
9264		
2184		
131040		14
9264		

[1] Thus in the translation of al-Khowârizmî (c. 825) attributed to Adelard of Bath (c. 1120): "Sex minuta multiplicata in VII. minuciis, erunt XLII. secunda" (Boncompagni, *Trattati*, I, 18).

[2] In the adaptation of the *Liber Algorismi* by Johannes Hispalensis (c. 1140) the multiplication table is given up to *nona* times *nona*, that is, up to $\frac{1}{60^9} \times \frac{1}{60^9} = \frac{1}{60^{18}}$ (see Boncompagni, *Trattati*, II, 103). The printed arithmetics occasionally gave such tables; e.g., those of Cardan (1539, cap. 38), Fine (1530; 1555 ed., fol. 38, r.), Trenchant (1566), and Peletier (1549). Schoner, in his *De logistica sexagenaria* (1569; 1586 ed., p. 370), calls it the "abacus logis-ticus," and a table of products up to 60 × 60, for use with sexagesimals, is called by Fine (1530) a "tabvla proportionalis." Division tables are also given by various writers; e.g., Fine (1530) and Trenchant (1566).

[3] Wäschke, *Planudes*, p. 34.

[4] The "sign" was 30°, and the 12 signs of the zodiac gave 360°.

[5] He says that the division may be continued farther.

Theon of Alexandria[1] (*c.* 390) and Maximus Planudes[2] (*c.* 1340), and in several of the 16th century arithmetics.[3] Its nature may be inferred from the work in division.

Symbols. The symbols ($°$ $'$ $''$) are modern. In medieval and Renaissance times there were several methods used for designating the sexagesimal orders. Thus in a manuscript of Leonardo of Cremona[4] (15th century) we have

$$.5. \quad \frac{aj''}{19} \cdot \frac{2''}{57} \cdot \frac{3''}{38} \quad \text{for} \quad 5° \ 19' \ 57'' \ 38''',$$

and

$$\frac{aj}{.46.} \quad \frac{2}{39} \cdot \frac{3}{12} \cdot \frac{4}{36} \quad \text{for} \quad 46' \ 39'' \ 12''' \ 36^{iv}.$$

Gemma Frisius (1540) wrote[5]

$$\text{S.} \quad \widetilde{g}. \quad \widetilde{m}. \quad \widetilde{2}. \quad \widetilde{3}. \quad \widetilde{4}$$
$$1. \quad 16. \quad 25. \quad 17. \quad 21. \quad 27$$

for

$$1^s \quad 16° \quad 25' \quad 17'' \quad 21''' \quad 27^{iv}.$$

Peletier (1571 edition of Gemma Frisius) used m˜ or ı˜ for minutes, 2˜ for seconds, 3˜ for thirds, and so on. Jean de Lineriis (*c.* 1340)[6] used the symbols s̈, ḡ, m̄, 2̄, 3̄, 4̄, and these, with slight modifications, are the ones most commonly seen up to the close of the 16th century. About that time there came into use such forms as

$$\text{IIae} \quad \text{Iae} \quad \overset{o}{.} \quad \text{I} \quad \text{II} \quad \text{III}$$
$$3 \cdot 15. \quad 7. \quad 50. \quad 34. \quad 23.$$

---

[1] The process is given in Heath, *History*, I, 60, and in Gow, *Greek Math.*, p. 55.

[2] Heath, *History*, II, 547, where the date of Planudes is given a little earlier.

[3] *E.g.*, the Peletier (Peletarius) revision (1545) of Gemma Frisius.

[4] In Mr. Plimpton's library; *Rara Arithmetica*, p. 474. The title is *Artis metrice pratice compilatio.*

[5] "Circulus 12 Signis constat: Signum, 30 Gradibus." The relation to the zodiac is apparent.

[6] In the *Algorismvs de Minutijs* appended to Beldamandi's work (1483 ed.).

in which $\frac{o}{7}$ stands for units,[1] and in which the symbols are evidently the forerunners of the ones now in common use.[2] One curious example of symbolism is seen in the multiplication table given by Fine (1530) for use in sexagesimal computation, the product 8 × 42 being given as 5.36, that is, 5 × 60 + 36, the period being essentially a sexagesimal point.

3. DECIMAL FRACTIONS

Need for Decimal Fractions. Before the beginning of printing, operations with common fractions having large terms are not frequently found. In mercantile affairs they were not needed, and in astronomical work the sexagesimal fraction served the purpose fairly well. The elaborate Rollandus manuscript of c. 1424 contains the addition of no common fractions more difficult than $\frac{207}{220}$ and $\frac{197}{280}$, and the work in multiplication involves no fraction more elaborate than $\frac{29}{36}$. There are exceptions to the general rule,[3] but they are not numerous. In the recording of results in division, however, elaborate common fractions were frequently used.[4] By the advent of printing, writers were led into various excesses. Widman (1489), for example, used in business computations fractions far beyond any commercial needs, his successors were even more reckless,[5] and the theorists naturally went still farther.[6]

[1] This example is from Schoner, *De logistica sexagenaria*, 1569; 1586 ed., p. 366. Of the 7 he says, "qui & partium numerus dicitur, circulus," and he speaks of "7 unitates."

[2] Peletier (1549) remarks: "Les Degrés dont seront au milieu de la numeration Astronomique: & seront representés par °: les Minutes par ï les Secondes par $\ddot{z}$: . . . Et ansi des autres" (*L'Arithmetiqve*, p. 107 (1607 ed.)).

[3] E. g., in a Dutch MS. of the 15th century (Boncompagni sale cat., No. 477) the square root of $2520\frac{2426735209}{3197947740 1 6}$ is required.

[4] E.g., in the *Svmme Arismetice* of Stephano di Baptista delli Stephani da Mercatello (MS. of c. 1522), a pupil of Pacioli's, there are results like $34\frac{29872}{55592}$ (fol. 74, r.).

[5] Thus Widman uses $\frac{365157}{697857}$, Trenchant (1566) has $1080\frac{1739251 4 1}{6001020016}$ (1578 ed., p. 286), and Wentsel (1599) has several fractions as difficult as $\frac{481079}{9215214}$; and all of these were commercial writers.

[6] As when Scheubel, in his *De nvmeris, tractatus quintus* (1545), gives $31\frac{2227971490}{84477154757}$, and Coutereels (Eversdyck's edition of 1658) gives a result like $2\frac{17249122 4 2}{53774608 7 8}$ days.

Forerunners in the Invention. As usual in the case of an important invention there were various scholars who had some intuition of the need for such a device as the decimal fraction long before it was finally brought to light. Such a man was Joannes de Muris, or Jean de Meurs, who wrote early in the 14th century.[1] The most interesting of the early influences tending to the invention, however, was a certain rule for the extraction of $\sqrt[n]{a}$, expressed in modern symbols by $\dfrac{\sqrt[n]{a \cdot 10^{kn}}}{10^{k}}$.

In particular, $\sqrt{3} = \dfrac{\sqrt{30000}}{100}$, or $\dfrac{\sqrt{3000000}}{1000}$, the actual process of extracting the root being quite like our present one with decimals. It was known to the Hindus, to the Arabs, and to Johannes Hispalensis ($c.$ 1140), and is found in the works of Johann von Gmünden ($c.$ 1430), Peurbach ($c.$ 1460), and their successors until the close of the 16th century.[2] The most interesting step from this rule in the direction of the decimal fraction appears in certain tables of square roots, in connection with which the statement is made that, the numbers having been multiplied by 1,000,000, the roots are 1000 times too large. Such a table, from Adam Riese's *Rechnung auff der Linien vnd Federn* (Erfurt, 1522), is reproduced on page 237. The same plan is given by such later writers as Trenchant (1566) and Bartjens (1633). Even after the decimal fraction was well known, the analogous plan of using a radius of 10,000,000, in order to express the trigonometric functions as whole numbers, remained in use for more than two centuries. It even extended to the reckoning of interest "to the Radius 100,000," as Thomas Willsford says in his appendix to the 1662 edition of Recorde's *Grovnd of Arts*,[3] so as to avoid decimals.

[1] L. C. Karpinski, *Science* (N. Y.), XLV, 663.

[2] Buckley, for example, an English arithmetician, who died $c.$ 1570, gave the rule in Latin verse as follows:

> Quadrato numero senas præfigito ciphras
> Producti quadri radix per mille secetur.
> Integra dat Quotiens, et pars ita recta manebit
> Radici ut veræ, ne pars millesima desit.
> *Arithmetica memorativa. c.* 1550

[3] **As** spelled in this edition.

fo Den erſten Pünct ſetz 1. vnd ſetze daſür dꝛei
nulla / Ziehe dañ Radicem quadꝛatam darvon
ſo kommen 1000. Dann preponir dem anderen
Puncten/das iſt der Ziffern 2. auch ſechs o/vnd
ziehe Radicem quadꝛatã davon/ſo komen 414.

Den dritten Punct mach auch alſo. Setz 3. vñ
darnach ſechs o. Extrahir dann Radicem quadꝛatam davon/kommen 832. Alſo thů mit allen
Puncten/ſo machſtu die Tafel ſelber. Es iſt a-
ber groß mühe vnd verdꝛoſſen arbeyt/ Darum
hab ich dir hie ein Tafel außgezogen / die gehet
biß vff 240. Punct der tieffe/ der mañ gnůg hat
vff groß oder kleyne vaß.

Tabula Radicum quadratarum.

1	1000	17	123	33	747
2	414	18	242	34	833
3	732	19	358	35	917
4	2000	20	472	36	1000
5	234	21	584	37	82
6	449	22	692	38	163
7	645	23	767	39	244
8	828	24	900	40	324
9	1000	25	1000	41	403
10	162	26	98	42	481
11	316	27	195	43	558
12	446	28	290	44	634
13	606	29	384	45	709
14	742	30	477	46	783
15	873	31	567	47	856
16	1000	32	659	48	928

EARLY STEPS TOWARD DECIMALS (1522)

From Adam Riese's arithmetic, showing a table of square roots in which the
figures of the decimal fractions appear, but without any form of decimal point.
From a later edition of the work

Another influence leading to the invention of the decimal fraction was the rule for dividing by numbers of the form $a \cdot 10^n$, attributed by Cardan (1539) to Regiomontanus. This appears in several manuscripts of the 15th century,[1] as in the case of $470 \div 10 = 47$ and $503 \div 10 = 50\frac{3}{10}$. Borghi (1484) elaborates this rule, but it appears in its most interesting form in the rare arithmetic of Pellos (1492), who unwittingly made use of the decimal point for the first time in a printed work (p. 239). The use of the dot before and after integers had been common in the medieval manuscripts, as in the case of Chuquet's work already mentioned, but its use to separate the integer from what is practically a decimal fraction is first seen here. Later writers commonly used a bar for this purpose, as was the case with Rudolff (1530; see page 241), Cardan (1539), Cataneo (1546), and various other writers. Even as late as the 1816 edition of Pike's *Arithmetick* (New York, 1816) 46,464 is divided by 7000 thus:

$$7 | 000) \, 46 | 464 \, (6 \frac{4\,4\,6\,4}{7\,0\,0\,0}$$
$$\underline{42|}$$
$$4 | 464$$

Pellos, however, did not recognize the significance of the decimal point, as is evident from the facsimile on page 239, and no more did Cardan appreciate the significance of the bar that he used for the same purpose.[2]

The initial steps in the invention of the decimal fraction were not confined to the West, however; indeed, the credit for first recognizing the principle of this type of fraction may well be given to al-Kashî,[3] the assistant of the prince astronomer Ulugh Beg and the first director of the latter's observatory at Samarkand. In his *al-Risâli al-moḥiṭîje* (*Treatise on the circum-*

---

[1] Thus Chuquet (1484): "Comme qui vouldroit partir .470. par .10. fault oster .o. qui est la p̄me' figure de .470. et demeurent .47. et tant monte la part. Ou que vouldroit partir 503. par .10. fault oster .3. et les mettre dessus .10. et lon aura .50. $\frac{3}{10}$. pour quotiens." Fol. 8, *v.*

[2] See his *Practica* (1539), cap. 38.

[3] See Volume I, page 289, n. 5. He died *c.* 1436, or possibly, as some writers assert, *c.* 1424.

Partir per   2 0|

    7 9 6 5 4 8 3 9 .7
                              1 7
quocient 3 9 8 2 7 4 1 9
                              2 0

Partir per   3 0

    5 8 3 6 0 4 .3
                           2 3
quocient 1 9 4 5 6 4
                           3 0

Partir per   7 0

    9 5 3 7 9 1 .9
                          6 9
quocient 1 3 6 2 5 5
                          7 0

Partir per   1 0 0

    6 9 7 6 5 .8 7
quocient

Partir per   4 0 0|

    7 8 9 6 5 .7 3
                          1 7 3
quocient 1 9 7 4 1
                          4 0 0

Partir per   3 0 0 0

    8 7 6 5 8 .7 9 1
                            1 7 9 1
quocient 2 9 2 1 9
                            3 0 0 0

FROM THE PELLOS ARITHMETIC (1492)

ference) he not only gives the value of π to a higher degree of accuracy than any of his predecessors, but he writes it (using Arabic characters) as follows:

sah-hah

3   1415926535898732,

the word *sah-hah* meaning complete, correct, integral.[1] We have, therefore, a fraction which we may express as follows:

Integer

3   14159 · · · ,

the part at the right being the decimal. Manifestly it is, therefore, a clear case of a decimal fraction, and it seems to be earlier than any similar one to be found in Europe.

The Invention. The first man who gave evidence of having fully comprehended the significance of all this preliminary work seems to have been Christoff Rudolff, whose *Exempel-Büchlin* appeared at Augsburg in 1530. In this work he solved an example in compound interest, and used the bar[2] precisely as we should use a decimal point today (see page 241). If any particular individual were to be named as having the best reason to be called the inventor of decimal fractions, Rudolff would seem to be the man, because he apparently knew how to operate with these forms as well as merely to write them, as various predecessors had done. His work, however, was not appreciated, and apparently was not understood, and it was not until 1585 that a book upon the subject appeared.

The first to show by a special treatise that he understood the significance of the decimal fraction was Stevin, who published a work[3] upon the subject in Flemish, followed in the same year

---

[1] The modern Turkish form is *sahih*. I am indebted for these facts to Professor Salih Mourad of Constantinople.

[2] On the general question of notation see Gravelaar, "De Notatie der decimale Breuken," *Nieuw Archief voor Wiskunde*. IV (2).

[3] *De Thiende*. A copy of this rare pamphlet was fortunately saved at the time of the destruction of the Louvain library, having been borrowed a few days before by the Reverend H. Bosmans. S. J. See the *Revue des Questions Scientifiques*, January, 1920. There was an English translation by Robert Norton, London, 1608.

375.   1875.

fl. 393|75 hauptgůt vñ gewin des erstē jars.
196875
413|4375 Andern
20 671875
434|109375 Dritten
21705 46875
455|8148 4375 Vierdten
22 790742 1875
478|605585 9375 Fünfften
23 9302792968 75
502|5358652343 75 Sechsten
25 12679326 171875
527|662658496093 75 Sibenden
26 38313292480468 75
554|0457914208084375 Achteten
277022895710449 21875
581|748080991943359375 Nēündtē
29 087404090597167968 75
fl.610|8354850415405273437 5 Zehētē
ß.   6|6878803323242 1875000
♃ 20|6164099697265625 0000.

72 Die 120 ff̄ tragē 2 jar p̄ hauptgůt zins vnd
zinszins 132 ff̄ 2 ß 12 ♃. Bringt zinß vñ zinszins
12 ff̄ 2 ß 12 ♃. Darnach die 250 ff̄ tragē 3 jar
Haupteg. zins vñ zinszinß 289 ff̄ 3 ß 7 ♃ ½. Vnd
ist halber zins des vierdtē jars 7 ff̄ 1 ß 26 ♃ 1 7/16.
                                              b ij

EARLY APPROACH TO DECIMAL FRACTIONS

From the 1540 edition of Rudolff's *Exempel-Büchlin* (1530), showing the use of
decimal fractions in compound interest

(1585) by a French translation. This work, entitled in French *La Disme*, set forth the method by which all business calculations involving fractions can be performed as readily as if they involved only integers.[1] Stevin even went so far as to say that the government should adopt and enforce the use of the decimal system, thus anticipating the modern metric system.[2] He was the first to lay down definite rules for operating with decimal fractions, and his treatment of the subject left little further to be done except to improve the symbolism. Some idea of his treatment of the subject and of his symbols may be obtained from the facsimile shown on page 243.

The Symbolism. The decimal fraction had now reached the stage in its progress when the symbolism had to be settled. As already stated, Pellos (1492) had used a period to separate the decimal from the integral part, but he had not comprehended the nature of the fraction. This, however, was hardly more strange than that as good a computer as Vlacq[3] (1628) should use decimals in his calculations and tables and yet give a result in the form $1295\frac{65032}{100000}$. Several writers had used the bar to mark off the decimal part, and Rudolff had probably grasped the significance of the new fraction. Stevin had fully comprehended and clearly expounded the theory, but his symbols were not adapted to use. The improvement in the symbolism was due largely to Bürgi, Kepler, and Beyer, and to the English followers of Napier.

Jobst Bürgi (1552–1632) dropped the plan used by Regiomontanus—that of taking 10,000,000 as the *sinus totus* in trigonometry—and took 1 instead, the functions therefore becoming decimal fractions. He was not clear as to the best method of representing these fractions, however, and in his manuscript of 1592 he used both a period and a comma for the

---

[1] ". . . facilement expedier par nombres entiers sans rompuz toutes comptes se rencontrans aux affaires des Hommes."

[2] Adriaen van Roomen (1609) tells us that Bishop Ernst of Bavaria had similar ideas as to measures.

[3] *Arithmetica Logarithmica*, pp. 35 *et passim* (Gouda, 1628), evidently thinking that the decimal form of the result would not be understood by most readers.

SECONDE PARTIE DE LA DISME DE L'OPE-

RATION.

PROPOSITION I, DE L'ADDITION.

Estant donnez nombres de Disme à ajouster : Trouver leur somme :

Explication du donné. Il y a trois ordres de nombres de Disme, desquels le premier 27 ⓪ 8 ① 4 ② 7 ③, le deuxiesme 37 ⓪ 8 ① 7 ② 5 ③, le troisiesme 875 ⓪ 7 ① 8 ② 2 ③.

Explication du requis. Il nous faut trouver leur somme. *Construction.* On mettra les nombres donnez en ordre comme ci joignant, les aioustant selon la vulgaire maniere d'aiouster nombres entiers, en ceste sorte :

⓪	①	②	③
2 7	8	4	7
3 7	6	7	5
8 7 5	7	8	2
9 4 1	3	0	4

Donne somme (par le 1 probleme de l'Arithmetique) 941304, qui sont (ce que demoustrent les signes dessus les nombres) 941 ⓪ 3 ① 0 ② 4 ③. Ie di, que les mesmes sont la somme requise. *Demonstration.* Les 27 ⓪ 8 ① 4 ② 7 ③ donnez, font (par la 3e definition) 27 $\frac{8}{10}$, $\frac{4}{100}$, $\frac{7}{1000}$, ensemble 27 $\frac{847}{1000}$, & par mesme raison les 37 ⓪ 6 ① 7 ② 5 ③ vallent 37 $\frac{675}{1000}$, & les 8,75 ⓪ 7 ① 8 ② 4 ③ feront 875 $\frac{782}{1000}$, lesquels trois nombres, comme 27 $\frac{847}{1000}$, 37 $\frac{675}{1000}$, 875 $\frac{782}{1000}$, font ensemble (par le 10e probleme de l'Arith.) 941 $\frac{304}{1000}$, mais autant vaut aussi la somme 941 ⓪ 3 ① 0 ② 4 ③, c'est

decimal point,[1] and also wrote 14$\underset{\circ}{1}$4 for 141.4. In his use of these fractions he was followed by Prætorius, in a manuscript of 1599.[2]

In his tables of 1612[3] Pitiscus assumed the radius to be 100,000 and gave sin 10″ as 4.85. Since this sine for the radius 1 is 0.00004848, the point after the 4 is possibly intended as a decimal point. Occasionally he used several points, as when he gave sin 89° 59′ 30″ as 99999. 99894. 23. In his trigonometry, of which the tables are a part, he used a vertical line to mark off the decimal.[4] In the 1600 edition both the point and the vertical line are used for other purposes, the former to separate sexagesimals[5] and the latter to separate (as above) a large number into periods, usually of five figures each.

It is unquestionably true that the invention of logarithms had more to do with the use of decimal fractions than any other single influence. When Napier published his tables in 1614 he made no explicit use of decimal fractions, the sine and the logarithm each being a line of so many units. In the 1616 translation of this work, however, the translator, Edward Wright, made use of the decimal point. One line will serve to show the appearance of the table:

| Deg. o. | | | +\|− | | | |
Min.	Sines.	Logarith.	Differen	Logarith.	Sines	
30	8726	4741385	4741347	38 · 1	999961 · 9	30

In his *Rabdologiae . . . Libri Dvo* of 1617 Napier made some observations upon the subject and wrote both 1993,273 and 1993, 2′ 7″ 3‴ for the number which we now, in America, write 1993.273. In the Leyden edition of this work (1626) the

---

[1] Cantor, *Geschichte*, II (2), 617.   [2] *Ibid.*, p. 619.

[3] *Canon Triangulorum Emendatissimus et ad usum accommodatissimus. Pertinens ad Trigonometriam Bartholomaei Pitisci . . .*, Frankfort, 1612.

[4] "Deinde pro latere AC nuper invento 13|00024 assumo 13 fractione scilicet $\frac{24}{100000}$ · · ·." *Trig. Problematum Geod. Liber Unus*, p. 12.

[5] As in this subtraction (p. 67):

$$
\begin{array}{r}
70°.\ 0′ \\
46.\ 8 \\
\hline
23.\ 52′
\end{array}
$$

Stevin notation is used, by which the above number would appear as 1993, ⓞ2,①7,③3.[1] In any case it is evident that Napier understood something about the decimal fraction, that he did not invent our modern symbolism, and that the practical use of logarithms soon made a knowledge of decimals essential.

In 1616 Kepler wrote a work on mensuration[2] in which he distinctly took up the decimal fraction, using both a decimal point (comma) and the parentheses to separate the fractional part.[3] He stated it as his opinion that these fractions were due to Bürgi,[4] although it seems strange that he was not familiar with the work of Stevin. In his edition of Tycho Brahe's *Tabulae Rudolphinae*, published at Ulm in 1627, he uses (p. 25) the period for a decimal point, thus: "29.032 valet $29\frac{32}{1000}$."

In the year 1616 Johann Hartmann Beyer (1563–1625) wrote to Kepler concerning his work, and in the letter he used both the decimal comma and the sexagesimal symbolism for the decimal, writing 314, 1′ 5″ 9‴ 2⁗ 6″‴ 5⁗‴ + for 314.15926 +. Beyer had before this (1603) published a work on these fractions, *Logistica decimalis*, and on this account had laid claim to their invention, although he had long been preceded by Stevin. Adriaen Metius (1571–1635) took about the same step in symbolism when he wrote both 47852°:8′0″4‴ and 47852/8′0″4‴ for 47852.804. He also spoke[5] of " $\frac{481481}{1000000}$ ofte 4′8″1‴4⁗8″‴1⁗‴."

---

[1] C. G. Knott, *Napier Memorial Volume*, pp. 77, 182, 188, 190, 191. Edinburgh, 1914.

[2] *Ausszug auss der uralten Messe-Kunst Archimedis.* It appears in Volume V of the Frisch edition of Kepler's works, 1864.

[3] "Fürs ander, weil ich kurtze Zahlen brauche, derohalben es offt Brüche geben wirdt, so mercke, dass alle Ziffer, welche nach dem Zeichen (,) folgen, die gehören zu dem Bruch, als der zehler, der nenner dazu wirt nicht gesetzt." He then gives an example in interest:

3,65
6 mal

facit 21 (90)

[4] "Dise Art der Bruchrechnung ist von Jost Bürgen zu der sinusrechnung erdacht."

[5] *Opera Omnia*, 1633, pp. 19, 31. 49, 50. When De Morgan (*Arithmetical Books*, p. 41) said of a 1640 edition of Metius that "sexagesimal fractions are taught, but not decimal ones," he may have confused the symbols.

II

There are numerous examples of writers of the same period who used these awkward symbols. Girard, the editor of Stevin, whose first edition of the latter's works appeared in 1625, did much to make known the works of his master, but he apparently added nothing to the theory or the symbolism. Even as late as 1655 we find the period used to separate an integer and common fraction,[1] as in the case of 198.$\frac{1}{2}$, and in the 1685 edition of Casati's work[2] we have 0.00438 represented by

$$\frac{438}{100000}$$

In 1657 Frans van Schooten[3] used the symbol 17579625 . . . ③ for 17579.625. It had the advantage that, in finding the product of two decimals, the indices in the circles need only be added in order to determine the proper index in the result.

The use of smaller type for the decimal part was not uncommon,[4] and it is still seen on the continent of Europe. As to the development of these fractions in England, Professor Cajori has suggested that Oughtred's (1631) use of the symbol o|56 for 0.56 was one of the causes for delay in the general adoption of the decimal point.

It should also be said that the symbolism is by no means settled even yet. In England 23$\frac{45}{100}$ is written[5] 23.45, in the United States it appears as 23.45, and on the Continent such forms as 23,45 and 23$_{45}$ are common. Indeed, in America we commonly write \$23.$\underline{45}$ or \$23$\frac{45}{100}$ instead of \$23.45, to avoid forgery.

---

[1] B. Capra, *Vsvs et fabrica circini* . . ., p. 25 (Bologna, 1655). The first edition, however, seems to have been 1607.

[2] P. Casati, *Fabrica et vso Del Compasso di Proportione*, p. 123. Bologna, 1685. In the first edition (1664), however, he writes such a number (p. 86) as a common fraction, with the bar between the terms.

[3] *Exercitationum Mathematicarum Libri quinque*, liber primus, p. 33. Leyden, 1657.

[4] *E.g.*, in some editions of Vieta's tables; also in R. Butler, *The Scale of Interest* (London, 1630), where 1$_{125}$ is used for 1.125.

[5] Not always, however. In a MS. at the Woolwich Academy, of date 1736, the decimal point is always a comma. Hodder wrote a *Decimal Arithmetick* in 1668, in which he used both the comma and the dot.

Summary. The historical steps in the invention of the decimal fraction may be summed up as follows: Pellos (1492) used a decimal point where others had used a bar, but the idea of the decimal fraction was not developed by him. Rudolff (1530) worked intelligently with decimal fractions, using a bar for the separatrix, but he did not write upon the theory. Stevin (1585) wrote upon the theory but had a poor symbolism. About 1600, several writers attempted to improve the symbolism, and Bürgi, in 1592, actually used a comma for the decimal point, without the common sexagesimal marks, and comprehended the nature and advantages of these fractions. Napier knew something of the theory of decimals and rendered their use essential, but did not himself contribute to the symbolism.[1] In the mere writing of the decimal fraction, at least, all these efforts had been anticipated by al-Kashî (c. 1430), whose symbolism was quite as good as that of any European writer for the next century and a half.

It is thus difficult to pick out the actual inventor, although Rudolff and Stevin are entitled to the most credit for bringing the new system to the attention of the world. It should be added that these fractions were mentioned by Richard Witt in his *Arithmeticall Questions* in 1613, and that Henry Lyte (1619) wrote *The Art of Tens, or Decimall Arithmeticke,*—a work which did for England what the work of Stevin had done for the Continent.

Percentage. Long before the decimal fraction was invented the need for it was felt in computations by tenths, twentieths, and hundredths, and this need gave rise to a peculiar notation which took the place of the decimal forms and which has persisted to the present time in the symbol %.

The computations of the Romans that led up to the subject of percentage may be illustrated by the *vicesima libertatis*, a tax of $\frac{1}{20}$ on every manumitted slave; by the *centesima rerum venalium*, a tax of $\frac{1}{100}$ levied under Augustus on goods sold at

---

[1] How imperfect his knowledge was may be seen by examining his *De arte logistica*, pp. 60, 65, 75, *et passim*.

auction; and by the *quinta et vicesima mancipiorum*, a tax of $\frac{1}{25}$ on every slave sold.[1] Without recognizing per cents as such, the Romans thus made use of fractions which easily reduce to hundredths.

In the Middle Ages, both in the East[2] and in the West, there was a gradual recognition of larger denominations of money than the ancients had commonly known, and this led to the use of 100 as a base in computation. In the Italian manuscripts of the 15th century it is common to find examples involving such expressions as 20 ꝑ 100, xꝑ cento, and vi ꝑ c°, for our 20%, 10%, and 6%.

When commercial arithmetics began to be printed, this custom was well established, and so in Chiarino's work of 1481 there are numerous expressions like "xx. per .c." for 20%, and "viii in x percēto" for 8 to 10%. Borghi (1484) and Pellos (1492) made less use of per cents than one would expect of such commercial authorities, although each recognized their value.[3] The demand was growing, however, and Pacioli (1494), familiar with the large commerce of the *giudecca* at Venice, had much to say of it.[4] Beginning early in the 16th century the commercial arithmetics made considerable use of per cents in connection with interest and with profit and loss, sometimes in relation to the Rule of Three,[5] so popular with merchants of that period, but more frequently in relation to isolated problems.[6]

---

[1] Harper's *Dict. Class. Lit.*, p. 1634.

[2] *E.g.*, Bhāskara (*c.* 1150) uses per cents in the interest problems in the *Lilāvati*. See Taylor, *Lilawati*, p. 47; Colebrooke, *Lilāvati*, p. 39.

[3] *E.g.*, Berghi: ". . . guadagno a rason de .20. per cento"; and Pellos: ".12. ꝑ .100.," "ꝑ .3. ans a rason de .16. ꝑ cent."

[4] His printed forms include "ꝑ cēto" (fol. 65), ".10. ꝑ cento" (fol. 66), and "per cēto" (fol. 66).

[5] Thus Ortega (1512; 1515 ed.) has a chapter on "Regvla de tre de centenare" (fol. 51).

[6] Thus Wå'ckl (1536): "Itē einer leihet einem 200 fl. 3 iar vnd eines ieden iars nimbt er 10fl võ 100 ist die frag wieuil die 3 iar thůt gwiñ vnd gwiñszwinn" (fol. B 7). So, also, Rudolff (1530; 1540 ed.): "Weñ man vom hundert zu jårlichem zins geben sol 5 flo . . ." (Ex. 71 in the *Exempel-Büchlin*). He shows that the Italian "pro cento" was not yet well known in Germany, for he says (Ex. 156): ". . . vnd wieuil pro cento (verstee an Hundert floren)."

In America at present the expression 6% is identical in meaning with 0.06, per cent having come to signify merely hundredths. This was not the original meaning, nor does it conform to the present usage in England and certain other countries, where expressions like "£6 per cent" are in common use. This usage is historically correct, the 15th and 16th century writers, with whom percentage begins in any large way, having always employed it.[1]

EARLY PER CENT SIGN

From an Italian MS. of c. 1490. Notice also the old symbol for pounds, which may have suggested the dollar sign

Chief Use for Per Cents. The chief use for per cents in the 16th century was in relation to the computation of interest, and by the beginning of the 17th century the rate was usually quoted in hundredths.[2] It also appears in computing profit and loss, at first indirectly, as in the following addition to Recorde by John Mellis (1594): "If one yard cost 6s̄——8 pence:

[1] Thus Sfortunati (1534; 1545 ed.) uses "libre .30. per 100"; Riese (1522) uses "10 lb. von 100," "10 fl zum 100," "10 fl am 100," and other similar forms; and Albert (1534) has "10 fl mit 100 fl."

[2] E.g., Trenchant (1566) has "à raison de 12 pour 100" with a 12% interest table; Petri (1567) speaks of "8 ten hondert" and "12 ten 100"; Raets (1580) gives the rate as "15 ten hondert," and Wentsel (1599) as "10. ten 100 tsjaers,"—all of which shows the high rates of interest prevailing.

and the same is sold againe for 8 s̄—— 6 pence: the question is, what is gayned in 100 pounds laying out on such commoditie." Many books, however, stated the problems substantially as at present.

The Per Cent Sign. In its primitive form the per cent sign (%) is found in the 15th century manuscripts on commercial arithmetic, where it appears as "per c̊" or "p c̊," a contraction

THE PER CENT SIGN IN THE 17TH CENTURY

From an anonymous Italian MS. of 1684

for "per cento."[1] As early as the middle of the 17th century it had developed into the form "per -°∘-," after which the "per" finally dropped out. The solidus form (%) is modern.

Permillage. It is natural to expect that percentage will develop into permillage, and indeed this has not only begun, but it has historic sanction. Bonds are quoted in New York "per M," and so in various other commercial lines. This was already common in the 16th century.[2] At present, indeed, the symbol °/∘∘ is used in certain parts of the world, notably by German merchants, to mean per mill, a curious analogue to % developed without regard to the historic meaning of the latter symbol.

---

[1] *Rara Arithmetica*, pp. 439, 441, 458, with facsimiles.

[2] Thus Cardan (1539) says that "tara cōputāda est ad 100. vel ad 1000." *Arithmetica*, 1539, capp. 57, 59.

4. Surd Numbers

Ancient Idea of Irrationals. Proclus (*c.* 460) tells us that the Pythagoreans discovered the incommensurability of the diagonal and the side of a square,[1] which is only a geometric view of the irrationality of $\sqrt{2}$. Proclus also states that they were led to study the subject of commensurability through their work with numbers. Plato says[2] that Theodorus of Cyrene (*c.* 425 B.C.) discovered that "oblong numbers, 3, 5, 6, 7, ⋯, are composed of unequal sides." He also states that "Theodorus was writing out for us something about roots, such as the roots of three or five, showing that they are incommensurable by the unit: he selected other numbers up to seventeen—there he stopped."[3]

With respect to other writers on incommensurable lines, Diogenes Laertius (2d century) tells us that Democritus (*c.* 400 B.C.) composed a treatise upon the subject.[4]

Summary of Greek Ideas on Irrationals. Summarizing the work of the Greeks, there seems to be good reason for believing that the immediate followers of Pythagoras knew and demonstrated the incommensurability of the diagonal and the side of a square, but that they looked upon this case of irrationality as a peculiarity of the square. Theodorus seems to have carried the investigation farther, recognizing that irrationality of square roots was not confined to $\sqrt{2}$. Theætetus (*c.* 375 B.C.) appears to have laid the foundations for a general theory of quadratic irrationals[5] and to have established their leading

---

[1] Heath's *Euclid*, Vol. III, p. 1 seq., to the notes of which the reader is referred. See also Cantor, *Beiträge*, p. 108. The proof is given in Euclid's *Elements*, numbered X, 117 in early editions, but is now relegated to an appendix. See also H. Vogt, "Die Entdeckungsgeschichte des Irrationalen nach Plato . . .," *Bibl. Math.*, X (3), 97; Heath, *History*, I, 65, 90, 154.

[2] *Theætetus*, 147 D; Jowett translation, IV, 123; Heath, *History*, I, 203.

[3] It should be observed that the method of proof for $\sqrt{3}$ is quite different from that for $\sqrt{2}$, and so for other surds. See Heath, *Euclid*, Vol. III, p. 2, and *History*, I, 155.

[4] Περὶ ἀλόγων γραμμῶν καὶ ναστῶν β΄. See F. Hultsch, *Neue Jahrbücher für Philologie und Pädagogik*, CXXIII, 578; Heath, *History*, I, 156.

[5] Heath, *History*, I, 209.

properties. Euclid (*c.* 300 B.C.) took the final important step due to the Greek geometers, classifying square roots and introducing the idea of biquadratic irrationals.[1]

This discovery, then, was the second noteworthy step in the creation of types of artificial numbers. The Greeks showed that all magnitudes are either rational (ῥητά, *rheta'*) or irrational (ἄλογα, *a'loga*), their idea of an irrational number being such a number as cannot be expressed as the ratio of two integers.

The geometric treatment of incommensurables naturally led to the arithmetic and algebraic treatment of irrationals, the subject of the present discussion.[2]

The Name "Surd." Al-Khowârizmî (*c.* 825) spoke of rational numbers as "audible" and of surds as "inaudible,"[3] and it is the latter that gave rise to the word "surd" (deaf, mute). So far as now known, the European use of this word begins with Gherardo of Cremona (*c.* 1150).[4] The term was also used by Fibonacci (1202), but to represent a number that has no root.[5] The Arabs and Hebrews often called surds "nonexpressible numbers,"[6]—a name which may have suggested the "inexplicable sides" of the Renaissance writers.[7] It is simply a translation from the Euclidean term ἄλογος (*a'logos*, without ratio, irrational, incommensurable).

As to what constitutes a surd, however, there has never been a general agreement. It is admitted that a number like $\sqrt{2}$

---

[1] Vogt, *loc. cit.* On Professor Zeuthen's discussion of Vogt's conclusions, see H. Bosmans, in the *Revue des Questions Scientifiques*, July, 1911. See also Heath, *History*, I, 402.

[2] On the history of transcendental numbers see the statement on page 268.

[3] Rosen ed., p. 192.   [4] *Bibl. Math.*, I (3), 516.

[5] "Nam quidem numeri habent radices, et uocatur [*sic*] quadrati; et quidam non; quorum radices, que surde dicuntur, cum inpossibile sit eas in numeris inuenire . . . " (*Liber Abaci*, p. 353). By "root" he refers here, as usual, to square root.

[6] *E.g.*, al-Karkhî (*c.* 1020); see Hochheim, *Kâfî fîl Hisâb*, II, 12.

[7] *E. g.*, Schoner, in his *De numeris figuratis* (1569; 1586 ed., p. 213), says: "Explicabile latus est, cujus ad 1. ratio explicari potest. Ut latus 4 est 2, & dicitur explicabile. . . . Inexplicabile latus contra est, cujus ad 1. ratio explicari non potest. Ut 3. . . ."
Stevin (1585) speaks of "nombres, comme $\sqrt{8}$, & semblables, qu'ils appellent absurds, irrationels, irreguliers, inexplicables, sourds, &c" (1634 ed., p. 9).

is a surd, but there have been prominent writers who have not included $\sqrt{6}$, since $\sqrt{6} = \sqrt{2} \times \sqrt{3}$;[1] and $\sqrt{2+\sqrt{3}}$ is commonly excluded.[2]

Approximate Values. An interest in the irrational showed itself strongly among the ancients. Here was a mystery to be fathomed, and from the time of Pythagoras to that of Weierstrass the nature of irrationals and the ability to work with them occupied the attention of a considerable part of the mathematical world. Among the noteworthy efforts was the one which sought to find an approximate value for an expression like $\sqrt{2}$. As already said (p. 144), the Greeks found the square root of a number in much the same way as that which is commonly taught in school today, but their ignorance of the decimal fraction made the process of approximation very difficult in the case of surds. For this reason the ancient and medieval writers resorted to various rules which can best be appreciated by first considering the principle involved.

Let a be an approximation to $\sqrt{A}$ by defect. Then A/a must be an approximation by excess, and the arithmetic mean, $a_1 = \frac{1}{2}\left(a + \frac{A}{a}\right)$, is an approximation of the second order by excess,[3] and the harmonic mean, A/a_1, is an approximation of the second order by defect. This process may evidently be carried on indefinitely. If $A = a^2 + r$, we have in particular,

$$a_1 = a + \frac{r}{2a}, \qquad \frac{A}{a_1} = a + \frac{r}{2a + \frac{r}{a}},$$

---

[1] *E.g.*, Behâ Eddîn (*c.*1600), al-Karkhî (*c.*1020), and other Arab writers included only non-squares not divisible by the digits 2, 3, . . ., 9. See Hochheim, *Kâfî fîl Hisâb*, II, 13 n.

[2] G. Chrystal, *Algebra*, 2d ed., I, 203 (Edinburgh, 1889): ". . . a surd number is the incommensurable root of a commensurable number. . . . For example . . . $\sqrt{e}$ is not a surd. . . . Neither is $\sqrt{(\sqrt{2}+1)}$."

[3] For $\dfrac{x^2}{x-n} = x + n + \dfrac{n^2}{x-n}$, so that if we divide by a number that is n less than the square root, we shall have a result that is more than n in excess of the square root.

and so on. Recognizing that

$$a_1 = \tfrac{1}{2}\left(a + \frac{A}{a}\right) = a + \frac{r}{2a}$$

is an approximation by excess, the Arabs took

$$a_1' = a + \frac{r}{2a+1}$$

as an approximation by defect, but this rule is not found among the Greeks.[1]

Of the various rules for approximation to $\sqrt{A}$, the one most commonly used in the past may be expressed as

$$\sqrt{A} = \sqrt{a^2 + r} = a + \frac{r}{2a},$$

as in
$$\sqrt{10} = \sqrt{9+1} = 3 + \frac{1}{2 \times 3} = 3\tfrac{1}{6},$$

an approximation by excess. The corresponding approximation by defect is:

$$\sqrt{A} = \sqrt{a^2 + r} = a + \frac{r}{2a+1},$$

as in
$$\sqrt{10} = \sqrt{9+1} = 3 + \tfrac{1}{7} = 3\tfrac{1}{7},$$

which probably explains why $\sqrt{10}$ was so often used for π by early writers. Of these approximations, the one by excess is found in the works of Heron ($c.$ 50?).[2] The medieval writers used both of these approximations,[3] often with variations. For

---

[1] P. Tannery, "L'extraction des racines carrées d'après Nicolas Chuquet," *Bibl. Math.*, I (2), 17; "Du rôle de la musique grecque dans le développement de la mathématique pure," *ibid.*, III (3), 171.

[2] A fact noted by Clavius, *Epitome*, 1583; 1585 ed., p. 318. It should occasionally be repeated that, as stated in Volume I, page 125, this date is uncertain. Heron may have lived as late as the 3d century.

[3] *E.g.*, al-Ḥaṣṣâr ($c.$ 1175?) made use of both the one by defect and the one by excess, together with $a + \dfrac{r+1}{2a+2}$ and $a + \dfrac{r}{2a} - \dfrac{\left(\dfrac{r}{2a}\right)^2}{2\left(a + \dfrac{r}{2a}\right)}$ for closer work.

For other cases see Hochheim, *Kâfî fi' Hisâb*. II, 14; Wertheim, *Elia Misrachi*, p. 21; B. Boncompagni, *Atti Pontif.*, XII, 402.

example, Rhabdas ($c.$ 1341), following an Arabic method,[1] obtained a first approximation to $\sqrt{10}$ by using a rule equivalent
to
$$\sqrt{A} = a + \frac{A - a^2}{2\,a},$$
giving
$$\sqrt{10} = 3 + \frac{10 - 9}{6} = 3\tfrac{1}{6} = \frac{19}{6}.$$

Then, since $10 \div \tfrac{19}{6} = 3\tfrac{3}{19}$, he takes the mean of $3\tfrac{1}{6}$ and $3\tfrac{3}{19}$, which is $3\tfrac{37}{228}$. This he shows to be a close approximation, since $(3\tfrac{37}{228})^2 = 10\tfrac{1}{51984}$.

A somewhat different method, also involving averages, is given by Chuquet (1484). Let two approximate values of $\sqrt{A}$ be $\frac{p_0}{q_0}$ and $\frac{p_1}{q_1}$, the first being too great and the second being too small, and let $p' = p_0 + p_1$ and $q' = q_0 + q_1$. Then p'/q' is a new approximation intermediate in value, and whether it is by excess or defect is found by squaring. In the same way an approximation is found between p'/q' and one of the others, and so on. This rule was employed by several later writers.[2] De la Roche (1520), who plagiarized Chuquet, asserted that any study of "imperfect roots" was useless, although custom required it.[3] Such approximations are common in the works of the 16th century, together with similar rules for cube root.[4] With all this there naturally developed many evidences of ignorance, as when Peter Halliman (1688) gave substantially the rule
$$\sqrt[n]{a^n + r} = a + \frac{r}{(a+1)^n - r^n},$$

[1] P. Tannery, *Notices et extraits des manuscrits de la Bibl. nat.*, XXXII, 185.

[2] *E. g.*, Ortega (1512); substantially by Clavius (*Epitome*, 1583; 1585 ed., p. 318, where he gives $\sqrt{20} = 4\tfrac{5473}{1159}2$, approximately); and substantially by Metius when he found the value of π by interpolating between $3\tfrac{17}{120}$ and $3\tfrac{15}{106}$, the result being the Chinese value, $\tfrac{355}{113}$. See P. Tannery, *Bibl. Math.*, I (2), 17.

[3] The study of "racines imparfaites" was "labeur sans vtile," but "pour la perfection de ce liure" he gave a method "par la regle de mediation entre le plus et le moins,"—an elementary method of interpolation. This is described by Treutlein in the *Abhandlungen*, I, 66.

[4] Thus Stevin gives substantially the rule that
$$\sqrt[3]{A} = \sqrt[3]{a^3 + r} = a + \frac{r}{3\,a(a+1)+1}.$$
Œuvres, 1634 ed., p. 30.

celebrating his discovery by the doggerel verse,

> Now Logarithms lowre your sail,
> And Algebra give place,
> For here is found, that ne'er doth fail,
> A nearer way, to your disgrace.[1]

It should also be understood that such rules for roots are ancient. For example, Heron ($c.$ 50?) gives what may possibly be the equivalent of the formula

$$\sqrt[3]{n} = b + \frac{a(n - b^3)}{n + a(n - b^3)},$$

where $a > \sqrt[3]{n} > b$, and $a - b = 1$. By means of this rule we should find that $\sqrt[3]{109} = 4.7785$ instead of $4.7769.$[2]

Criteria for Squares. In order to determine whether or not $\sqrt{a}$ is a surd, those who were interested in number theory developed from time to time criteria for ascertaining whether a number is a square. Such criteria are found in various ancient and medieval works, both Arab[3] and European. A Munich manuscript[4] of the 15th century, for example, states that if a square ends in an even number, it is divisible by 4;[5] that if it ends in zeros, it ends in an even number of zeros;[6] that it cannot end in 2, 3, 7, or 8;[7] and that every square is of the form $3n$ or $3n + 1$.[8] Such rules, often extended, are found in various works of the classical and Renaissance periods.[9]

---

[1] From *The Square and Cube Root compleated and made easie* (London, 1688), quoted by A. De Morgan, *Arithmetical Books*, p. 52.

[2] J. G. Smyly, "Heron's formula for cube root," *Hermathena*, XLII, 64, correcting M. Curtze, *Zeitschrift*, Hl. Abt. (1897), p. 119, and referring to Heron's *Metrica*, III, 20. See also the interpretations in Heath, *History*, II, 341.

[3] *E.g.*, al-Karkhî ($c.$ 1020) and al-Qalasâdî ($c.$ 1475). See Hochheim, *Kâfi fîl Hisâb*, II, 13.

[4] No. 14,908, described by Curtze in *Bibl. Math.*, IX (2), 38.

[5] "Omnis quadratus, cuius prima est par, est per 4 divisibilis."

[6] "Omnis quadratus in primis locis habet parem numerum ciffrarum."

[7] "Nullus quadratus recipit in primo loco 2, 3, 7 vel 8, sed alios bene." This is a very old rule.

[8] "Omnis quadratus est simpliciter vel subtracte unitate per 3 divisibilis."

[9] Thus Buteo (1559) adds that a square number cannot end in 5 unless it ends in 25.

Surds in Algebra. The placing of the study of surds in the books on algebra is a tradition which began with the Renaissance. The books on logistic, used in commercial schools, had no need for the subject; it properly belonged in the books on the theory of numbers, the ancient *arithmetica*. Since, however, algebra took over a considerable part of the latter in the Renaissance period, surds found a place in this science. Furthermore, since these forms are needed in connection with irrational equations, they were usually considered before that topic in the study of algebra. In the 15th century, however, they are often found in the theoretical arithmetics.[1]

5. NEGATIVE NUMBERS

Early Use of Negative Numbers. No trace of the recognition of negative numbers, as distinct from simple subtrahends, has yet been found in the writings of the ancient Egyptians, Babylonians, Hindus, Chinese, or Greeks. Nevertheless the law of signs was established, with the aid of such operations as $(10 - 4) \cdot (8 - 2)$, and was known long before the negative number was considered by itself.

The Chinese made use of such numbers as subtrahends at a very early date. They indicated positive coefficients by red computing rods, and negative ones by black, and this color scheme is also found in their written works.[2] The negative number is mentioned, at least as a subtrahend, in the *K'iu-ch'ang Suan-shu* (*c.* 200 B.C.),[3] and in various later works, but the law of signs is not known to have been definitely stated in any Chinese mathematical treatise before 1299, when Chu Shï-kié gave it in his elementary algebra, the *Suan-hio-ki-möng* (*Introduction to Mathematical Studies*).

---

[1] Thus the Rollandus MS. (*c.* 1424) has surds in the arithmetic just before algebra is begun, and similarly in Pacioli's *Sūma* (1494). On the modern problem consult the *Encyklopädie der mathematischen Wissenschaften*, I, 49 (Leipzig, 1898–); hereafter referred to as *Encyklopädie*.

[2] Mikami, *China*, pp. 18, 20, 21, 27, 89, *et passim*; Cantor, *Geschichte*, I (2), 642.

[3] See Volume I, page 31, for discussion as to earlier date for the original work. It may have been written before 1000 B.C.

The first mention of these numbers in an occidental work is in the *Arithmetica* of Diophantus (*c*. 275),[1] where the equation $4x + 20 = 4$ is spoken of as absurd (ἄτοπος), since it would give $x = -4$. Of the negative number in the abstract, Diophantus had apparently no conception. On the other hand, the Greeks knew the geometric equivalent of $(a - b)^2$ and of $(a + b)(a - b)$; and hence, without recognizing negative numbers, they knew the results of the operations $(-b) \cdot (-b)$ and $(+b) \cdot (-b)$.

In India the negative number is first definitely mentioned in the works of Brahmagupta (*c*. 628). He speaks of "negative and affirmative quantities,"[2] using them always as subtrahends but giving the usual rules of signs. The next writer to treat of these rules is Mahāvīra (*c*. 850), and after that time they are found in all Hindu works on the subject.

The Arabs contributed nothing new to the theory, but al-Khowârizmî (*c*. 825) states the usual rules,[3] and the same is true of his successors.

When Fibonacci wrote his *Liber Abaci* (1202) he followed the Arab custom of paying no attention to negative numbers, but in his *Flos* (*c*. 1225) he interpreted a negative root in a financial problem to mean a loss instead of a gain.[4] Little further was done with the subject by medieval writers, but as we approach the Renaissance period we find the negative number as such receiving more and more recognition. For example, among the problems set by Chuquet (1484) is one[5] which leads to an equation with roots "$\tilde{m}.7.\frac{3}{11}$" and "$27.\frac{3}{11}$," that is, $-7\frac{3}{11}$ and $27\frac{3}{11}$.

Modern Usage. The first of the 16th century writers to give noteworthy treatment to the negative number was Cardan. In

---

[1] Nesselmann, *Alg. Griechen*, p. 311; Heath, *Diophantus*, 2d. ed., pp. 52, 200 (Cambridge, 1910); Cantor, *Geschichte*, I (2), 441.

[2] Colebrooke translation, pp. 325, 339.

[3] Rosen translation, p. 26.

[4] "Hanc quidem quaestionem insolubilem esse monstrabo, nisi concedatur, primum hominem habere debitum." *Scritti*, II, 238.

[5] Boncompagni's *Bullettino*, XIV, 419, Ex. xiv. Chuquet adds, "Ainsi ce calcule est vray que aulcuns tiennent Impo.[ble]

his *Ars Magna* (1545) he recognized negative roots of equations and gave a clear statement of the simple laws of negative numbers.[1]

Stifel (1544) distinctly mentioned negative numbers as less than zero,[2] and showed some knowledge of their use. By this time the rules of operation with numbers involving negative signs were well understood, even though the precise nature of the negative number was not always clear. Thus Bombelli (1572) gave these rules and applied them intelligently to such cases as $(+15) + (-20) = -5$.[3] It was due to the influence of men like Vieta, Harriot, Fermat, Descartes, and Hudde, however, that the negative number came to be fully recognized and understood. The idea of allowing a letter, with no sign prefixed, to represent either a positive or a negative number seems due to Hudde (1659).

Names and Symbols for Negative Numbers. The Hindu writers who mentioned negative numbers, or numbers used as subtrahends, placed a dot or a small circle over or beside each, as stated on page 396. The names used were the equivalent of our word "negative." The early European usage has already been mentioned, but it remains to speak of the establishing of our modern terminology.

As already stated, the Chinese wrote positive numbers in red and negative numbers in black, and so indicated them by their stick symbolism. They also had another method for indicating negative coefficients, one that may have been due to Li Yeh (1259). This consisted in drawing a diagonal stroke through the right-hand digit figure of a negative number, as in the case of IO╫╫ for −10,724, and of IOⅩOO for −10,200.[4]

In the 15th century the names "positive" and "affirmative" were used to indicate positive numbers, as also "privative"

---

[1] Thus on fol. 3, *v.*, speaking of squares, he says: "At uero quòd tam ex 3, quàm ex m : 3, fit 9, quoniam minus in minus ductū ₚducit plus."

[2] "Finguntur numeri infra o. id est, infra nihil." *Arithmetica Integra*, fol. 249, *r.*

[3] "Più via più fà più. Meno via meno fà più" etc. (p. 70). Also: "E p.15 con m.20 fà m.5. perche se io mi trouassi scudi 15, e ne fossi debitore 20, pagati li 15 restarei debitorₑ ₅" (p. 72).

[4] Mikami. *China*, p. 82.

and "negative" for negative numbers,[1]—a usage followed by Scheubel (1551).[2]

Cardan (1545) spoke of "minus in minus" as being plus, but in general he called positive numbers *numeri ueri*[3] and negative numbers *numeri ficti*.[4] His symbol for a negative number is simply m:, as in the case of m:3 for -3.[5]

Stifel (1544) called these numbers "absurd" and wrote $0 - 3$ as an illustration.[6]

Tartaglia (1556) spoke of a negative number as "the term called minus,"[7] laying down the usual rules.[8]

Bombelli (1572) used the word "minus" (*meno*) as we do in such rules as "minus times minus gives plus," his symbol for -5 being m.5. Unlike Cardan, he had a definite sign for $+5$ also, writing it p.5.

Tycho Brahe, the astronomer (1598), spoke of the negative number as "privative" and indicated it by the minus sign.[9]

Napier (*c.* 1600) used the adjectives *abundantes* and *defectivi* to designate positive and negative numbers, Sturm (1707) spoke of *Sache* and *Mangel*, and various other names and symbols have been suggested.[10]

---

[1] H. E. Wappler, *Zur Geschichte der deutschen Algebra im XV. Jahrhundert*, Prog., p. 31 (Zwickau, 1887).   [2] Tropfke, *Geschichte*, II (2), 79.

[3] Or *ueri numeri*. He used both forms, as on fol. 3, *v.*, of the *Ars Magna*.

[4] So in speaking of the roots of an equation he says "una semper est rei uera aestimatio, altera ei aequalis, ficta."

[5] So he gives the roots of $x^2 = 16$ thus: "res est 4, uel m: 4."

[6] "Finguntur numeri minores nihilo ut sunt $0 - 3$" (*Arithmetica Integra*, fol. 48, *r.*). Later (fol. 249, *v.*) he speaks of zero as "quod mediat inter numeros veros et numeros absurdos."

[7] ". . . il termine chiamato men" (*General Trattato*, II, fol. 83, *r.*).

[8] *E.g.,* "Terzo regola, a multiplicare men fia men fa sempre piu" (*ibid.,* fol. 85, *v.*). His illustration is as follows:

a multiplicar	$9 \tilde{m} 2$
per ———————	——————— $8 \tilde{m} 3$
fa	72 men 43 piu 6
che sara	35 a ponto

[9] ". . . quod eorum alii positivi sunt alii privativi; positivi ii quibus vel nullum signum est additum vel praefigi debet hoc $+$; privativi verò qui praefixum habere debent signum hoc $-$" (*Tabulae Rudolphinae,* p. 9 (Ulm, 1627)).

[10] On the later treatment of the negative number see Cantor, *Geschichte*, IV, 79–88.

6. COMPLEX NUMBERS

Early Steps in Complex Numbers. The first trace of the square root of a negative number to be found in extant works is in the *Stereometria* of Heron of Alexandria (*c.* 50?), where $\sqrt{81-144}$ is taken to be $\sqrt{144-81}$, or $8-\frac{1}{16}$. The problem involved is impossible of solution, and this step should have been left $\sqrt{-63}$ or $3\sqrt{-7}$,[1] but whether the error is due to Heron or to some copyist is uncertain.

The next known recognition of the difficulty is found in the *Arithmetica* of Diophantus[2] (*c.* 275). In attempting to compute the sides of a right-angled triangle of perimeter 12 and area 7, Diophantus found it necessary to solve the equation $336x^2 + 24 = 172x$. He stated that the equation cannot be solved unless the square of half the coefficient of x diminished by 24×336 is a square, not otherwise seeming to notice that this equation has complex roots.

Mahāvīra (*c.* 850) was the first to state the difficulty clearly, saying, in his treatment of negative numbers, that, "as in the nature of things a negative [quantity] is not a square [quantity], it has therefore no square root."[3]

Bhāskara (*c.* 1150) used about the same language in his *Bija Ganita*:

The square of an affirmative or of a negative quantity is affirmative; and the square root of an affirmative quantity is two-fold, positive and negative. There is no square-root of a negative quantity: for it is not a square.[4]

The Jewish scholar Abraham bar Chiia (*c.* 1120) set forth the same difficulty in discussing the equations $xy = 48$ and $x + y = 14$.[5]

---

[1] On this general topic see W. W. Beman, "A Chapter in the History of Mathematics," vice-presidential address in Section A, *Proc. of the American Assoc. for the Adv. of Sci.*, 1897; E. Study, in the *Encyklopädie der Math. Wissensch.*, I, A, 4 (Leipzig, 1898); H. Hankel, *Vorlesungen über die complexen Zahlen*, Leipzig, 1867; G. Loria, *Scientia*, XXI, 101; F. Cajori, *Amer. Math. Month.*, XIX, 167. On Heron, see the Schmidt edition (Leipzig, 1914), V, 35.

[2] Heath, *Diophantus*, 2d ed., p. 244 (Cambridge, 1910).

[3] *Ganita-Sāra-Sangraha*, p. 7.

[4] Colebrooke's translation, p. 135.   [5] *Abhandlungen*, XII, 46.

II

The Arabs and Persians seem to have paid no special attention to the subject, and the next step was taken in Italy and after the invention of printing.

Early European Efforts. Pacioli (1494) stated in his *Sŭma* that the quadratic equation $x^2 + c = bx$ cannot be solved unless $\frac{1}{4} b^2 \geqq c$,[1] so that he recognized the impossibility of finding the value of $\sqrt{-a}$. About the same time Chuquet (1484) seems to have found that $\sqrt{-a}$ represents an impossible case.[2]

Cardan (1545) spoke of the equation $x^4 + 12 = 6x^2$ as being impossible,[3] referring to the roots of such equations as *ficta* or *per m̃*. He was the first to use the square root of a negative number in computation, the problem being to divide 10 into two parts whose product is 40.[4] He found the number to be $5 + \sqrt{-15}$ and $5 - \sqrt{-15}$, spoke of the solution "by the minus root,"[5] and proved by multiplication that his results were correct.

The next to attack the problem was Bombelli (1572). In his algebra[6] he speaks of such quantities as $+\sqrt{-a}$ and $-\sqrt{-a}$, but he made no advance upon Cardan's theory.

Stevin (1585) noted the difficulty of working with imaginaries, but could only remark that the subject was not yet mastered.[7] Girard (1629) found it necessary to recognize complex roots in order to establish the law as to the number of roots of an equation.[8]

---

[1] Fol. 147, r. E.g., $x^2 + 7 = x + 5$: "Dico q̄sto essere impossibile." As in all such cases, the symbolism here shown is modern.

[2] Boncompagni's *Bullettino*, XIV, 444.

[3] "Quòd si caruerit estimatiõe uera, carebit etiam ea, quę est per m: uelut 1 q̄d^ti q̄d^m p: 12, aeq̄tur 6 q̄d^tis, quia non potest aequationẽ ueram habere, carebit etiam ficta, sic em̃ uocamus eam, quae debiti est seu minoris" (*Ars Magna*, fol. 3, v.).

[4] *Opera*, IV, 287. Lyons, 1663.   [6] *L'Algebra*, p. 294 seq. Bologna, 1572.

[5] "Per radicem m̃ "   [7] *Œuvres*, 1634 ed., pp. 71, 72.

[8] "On pourroit dire à quoy sert ces solutions qui sont impossibles, je respond pour trois choises, pour la certitude de la reigle generale, et qu'il ny a point d'autre solutions, et pour son utilité" (*Invention nouvelle en l'algebre*, fol F 1 (Amsterdam, 1629)). The solution of "1 (2) est esgale à 6 (1) − 25," that is, $x^2 = 6x − 25$, which gives $x = 3 \pm \sqrt{-16}$, he calls "inexplicable" (p 114). He places in the same category (p. 130) numbers like $\sqrt{-3}$. From this point on, the reader may profitably consult the *Encyklopädie*, I, 148.

Approach to a Graphic Representation. Wallis (1673) seems to have been the first to have any idea of the graphic representation of these quantities. He stated[1] that the square root of a negative number was thought to imply the impossible, but that the same might also be said of a negative number, although we can easily explain the latter in a physical application:

These *Imaginary* Quantities (as they are commonly called), arising from the *Supposed* Root of a Negative Square (when they happen,) are reputed to imply that the Case proposed is Impossible.

And so indeed it is, as to the first and strict notion of what is proposed. For it is not possible that any Number (Negative or Affirmative) Multiplied into itself can produce (for instance) − 4. Since that Like Signs (whether + or −) will produce + ; and therefore not − 4.

But it is also Impossible that any Quantity (though not a Supposed Square) can be *Negative*. Since that it is not possible that any *Magnitude* can be *Less than Nothing* or any *Number Fewer than None*.

Yet is not that Supposition (of Negative Quantities,) either Unuseful or Absurd; when rightly understood. And though, as to the bare Algebraick Notation, it import a Quantity less than nothing: Yet, when it comes to a Physical Application, it denotes as Real a Quantity as if the Sign were + ; but to be interpreted in a contrary sense.

Having shown that we may have negative lines, he asserts that we may also have negative areas and that a negative square must have a side, thus:

Now what is admitted in Lines must, on the same Reason, be allowed in Plains also...

But now (supposing this Negative Plain, − 1600 Perches, to be in the form of a Square;) must not this Supposed Square be supposed to have a Side? And if so, what shall this Side be?

We cannot say it is 40, nor that it is − 40···

But thus rather that it is $\sqrt{-1600}$, or ···$10\sqrt{-16}$, or $20\sqrt{-4}$, or $40\sqrt{-1}$.

[1] *Algebra*, cap. LXVI; Vol. II, p. 286, of the Latin edition; but for his 1673 statement, differing somewhat from that in his algebra, see Cajori in *Amer. Math. Month.*, XIX, 167. See also G. Eneström, *Bibl. Math.*, VII (3), 263.

Where $\sqrt{}$ implies a Mean Proportional between a Positive and a Negative Quantity. For like as $\sqrt{bc}$ signifies a Mean Proportional between $+b$ and $+c$; or between $-b$ and $-c$; $\cdots$ So doth $\sqrt{-bc}$ signify a Mean Proportional between $+b$ and $-c$, or between $-b$ and $+c$.

He therefore reached the position where he would be supposed to draw a line perpendicular to the real axis and say that this might be taken as an imaginary axis, but although he touched lightly upon this possibility, he did nothing of consequence with the idea.

Leibniz on Complex Numbers. Leibniz was the next to take up the study of imaginaries. He showed (1676) that $\sqrt{1+\sqrt{-3}}+\sqrt{1-\sqrt{-3}}=\sqrt{6}$, and (1702) that x^4+a^4 is equal to $\left(x+a\sqrt{-\sqrt{-1}}\right)\left(x-a\sqrt{-\sqrt{-1}}\right)\left(x+a\sqrt{\sqrt{-1}}\right)\left(x-a\sqrt{\sqrt{-1}}\right).$[1] He was much impressed by the possibilities of the imaginary, but he seems never to have grasped the idea of its graphic representation.[2]

Modern Analytic Treatment. In 1702 Jean Bernoulli brought the imaginary to the aid of higher analysis by showing the relation between the $\tan^{-1}x$ and the logarithm of an imaginary number.[3] Newton's work with imaginaries (1685) was confined to the question of the number of roots of an equation,[4] a subject that was continued by Maclaurin[5] and other English algebraists.

---

[1] *Werke*, Gerhardt ed. (Berlin, 1850), II (3), 12; (Halle, 1858), V (3), 218, 360. See Tropfke, *Geschichte*, I (1), 171.

[2] "Itaque elegans et mirabile effugium reperit in illo Analyseos miraculo, idealis mundi monstro, pene inter Ens et non-Ens Amphibio, quod radicem imaginariam appellamus" (*Werke*, V, 357).

"Ex irrationalibus oriuntur quantitates impossibiles seu imaginariae, quarum mira est natura, et tamen non contemnenda utilitas; etsi enim ipsae per se aliquid impossibile significent, tamen non tantum ostendunt fontem impossibilitatis, et quomodo quaestio corrigi potuerit, ne esset impossibilis, sed etiam interventu ipsarum exprimi possunt quantitates reales" (*ibid.*, VII, 69).

[3] *Opera*, I, 393 (Lausanne, 1472); Tropfke, *Geschichte*, II (2), 83; Cantor, *Geschichte*, III, 348.

[4] *Arithmetica Universalis*, p. 242 (Cambridge, 1707), the imaginaries being called "radices impossibiles."

[5] *Phil. Trans.*, XXIV (1726), 104; XXXVI, 59; *Algebra*, 1748 ed., p. 275.

The first important step in the new theory to be taken by a British mathematician was made by Cotes (*c.* 1710) when he stated[1] that log $(\cos\phi + i\sin\phi) = i\phi$, corollaries to which are the important formula

$$e^{\phi i} = \cos\phi + i\sin\phi,$$

which bears Euler's name, and the well-known relation

$$(\cos\phi + i\sin\phi)^n = \cos n\phi + i\sin n\phi,$$

suggested by De Moivre in 1730 but possibly known by him as early as 1707.[2] Euler (1743, 1748) was the first to prove that this relation holds for all values of *n*, and also that

$$\sin\phi = \frac{e^{\phi i} - e^{-\phi i}}{2\,i}$$

and
$$\cos\phi = \frac{e^{\phi i} + e^{-\phi i}}{2}.$$

Graphic Representation. Although some approach to the graphic representation of the complex number had been made by Wallis, and although the goal had been more nearly attained by H. Kühn, of Danzig, it was a Norwegian surveyor, Caspar Wessel,[3] who first gave the modern geometric theory. In 1797 he read a paper on the subject before the Royal Academy of Denmark. This was printed in 1798 and appeared in the memoirs of the Academy in 1799.[4] In this he says:

Let us designate by $+1$ the positive rectilinear unit, by $+\epsilon$ another unit perpendicular to the first and having the same origin; then the angle of direction of $+1$ will be equal to $0°$, that of -1 to $180°$, that of ϵ to $90°$, and that of $-\epsilon$ to $-90°$ or to $270°$.

---

[1] *Harmonia mensurarum* (posthumous), p. 28 (Cambridge, 1722): "Si quadrantis circuli quilibet arcus, radio *CE* descriptus, sinum habeat *CX*, sinumque complementi ad quadrantem *XE:* sumendo radium *CE* pro Modulo, arcus erit rationis inter $EX + XC\ \sqrt{-1}$ & *CE* mensura ducta in $\sqrt{-1}$." See also *Bibl. Math.*, II (3), 442.

[2] *Bibl. Math.*, II (3), 97–102.

[3] Born at Jonsrud, June 8, 1745; died 1818.

[4] See the French translation, *Essai sur la représentation analytique de la direction*, Copenhagen, 1897.

The plan was therefore the same as the one now used and until recently attributed to various other writers, including Henri Dominique Truel[1] (1786), A. Q. Buée[2] (1805), J. R. Argand[3] (1806), Gauss,[4] J. F. Français[5] (1813), and John Warren.[6] W. J. G. Karsten (1732–1787), at one time professor at Halle, gave a method (1768) of representing imaginary logarithms.[7] Gauss took as his four units the roots of the equation $x^4 - 1 = 0$. Eisenstein[8] developed a theory based on the roots of the equation $x^3 - 1 = 0$, and Kummer[9] based his on the roots of the equation $x^n - 1 = 0$.[10]

Terms and Symbols. As already mentioned, Cardan (1545) spoke of a solution like $5 + \sqrt{-15}$ as "per radicem m̃," or "sophistic quantities," and Bombelli called numbers like $+\sqrt{-n}$ and $-\sqrt{-n}$ *piu di meno* and *meno di meno*,[11] abbreviated to *p. di m.* and *m. di m.* Descartes (1637) contributed the terms "real" and "imaginary."[12]

Most of the 17th and 18th century writers spoke of $a + b\sqrt{-1}$ as an imaginary quantity.[13] Gauss (1832) saw the desirability of having different names for $a\sqrt{-1}$ and $a + b\sqrt{-1}$,

---

[1] Cauchy mentions this fact, but nothing is known of the man. His MSS. are lost.

[2] In a paper *Sur les Quantités Imaginaires*, read before the Royal Society, London, 1805. See the *Philosophical Transactions*, London, 1806.

[3] *Essai sur une manière de représenter les quantités imaginaires dans les constructions géométriques*, Paris, 1806; 2d ed., Paris, 1874.

[4] He refers to the subject in his *Demonstratio nova* (1798), but does nothing with it. In his *Theoria residuorum biquadraticorum, Commentatio secunda* (1831) he presents the theory in its present form, evidently ignorant of Wessel's work.

[5] See Gergonne's *Annales*, IV, 61.

[6] *A Treatise on the Geometrical Representation of the Square Roots of Negative Quantities*, Cambridge, 1828. The general plan is that of Wessel, but the treatment of the subject is very abstract.

[7] F. Cajori, *Amer. Math. Month.*, XX, 76.

[8] On Gauss's estimate of him see Volume I, page 509.

[9] See Volume I, pages 507, 508.

[10] Tropfke, *Geschichte*, II (2), 88.

[11] "Plus of minus" and "minus of minus." *L'Algebra*, p. 294 seq. (Bologna, 1572).

[12] "Au reste tant les vrayes racines que les fausses ne sont pas toujours réelles, mais quelquefois seulement imaginaires." *La Geometrie*, 1705 ed., p. 117.

[13] Thus d'Alembert (1746): "Une fonction quelconque de tant et de telles grandeurs imaginaires, qu'on voudra, peut toujours être supposée égale à $p + q\sqrt{-1}$" (*Hist. de l'Acad. d. Berlin*, II, 195).

and so he gave to the latter the name "complex number."[1]   The use of i for $\sqrt{-1}$ is due to Euler (1748).[2]   Cauchy[3] (1821) suggested the name "conjugates" (*conjuguées*) for $a+bi$ and $a-bi$ and the name "modulus" for $\sqrt{a^2+b^2}$. Weierstrass called the latter the "absolute value" of the complex number and represented it by $|a+bi|$. Gauss had already given to a^2+b^2 the name of "norm."[4]

Quaternions and Ausdehnungslehre. The development of complex numbers, with their graphic representation in a plane, naturally led to the consideration of numbers of this type that might be graphically represented in a space of three dimensions. Argand (1806) attempted to take this step but found himself unable to do so, and Servois (1813) also made the attempt and failed.

In 1843 Sir William Rowan Hamilton[5] discovered the principle of quaternions, and presented his first paper on the subject before the Royal Irish Academy. His first complete treatment was set forth in his *Lectures on Quaternions* (1853). His discovery necessitated the withdrawal of the commutative law of multiplication, the adherence to which had proved to be a bar to earlier progress in this field.

The most active of the British scholars who first recognized the power of quaternions was Peter Guthrie Tait. Becoming acquainted with Hamilton soon after the latter's *Lectures on Quaternions* appeared, Tait began with him a correspondence that was carried on until his death.[6]   Tait had been a classmate

---

[1] "Tales numeros vocabimus numeros integros complexos" (*Werke*, II, 102 (Göttingen, 1876)).

[2] ". . . formulam $\sqrt{-1}$ littera i in posterum designabo, ita vt sit $ii = -1$" (*Institutionum calculi integralis volumen IV*, 184 (Petrograd, 1794)). In his *Introductio in Analysin Infinitorum* (Lausanne, 1748), he first used the symbol: "Cum enim numerorum negativorum Logarithmi sint imaginarii . . . erit l. − n quantitas imaginaria, quae sit = i." See also W. W. Beman in *Bulletin of the Amer. Math. Soc.*, IV, 274, 551.

[3] *Cours d'Analyse algébrique*, p. 180 (Paris, 1821).

[4] Tropfke, *Geschichte*, II (2), 90. For other attempts at explaining the imaginary number see Cantor, *Geschichte*, IV, 88–91, 303–318, 573, 712–715, 729–731.

[5] A. Macfarlane, *Ten Brit. Math.*, p. 43 (New York, 1916). See also P. G. Tait's article on "Quaternions" in the *Encyc. Britan.*, 9th ed., XX, 160.

[6] A. Macfarlane, "Peter Guthrie Tait," *Physical Review*, XV, 51.

of Clerk Maxwell's at the Edinburgh Academy and, like him, was deeply interested in physical studies. Partly as a result of this early training he soon began to apply the theory of quaternions to problems in this field, his results appearing in the *Messenger of Mathematics* and the *Quarterly Journal of Mathematics*. Recognizing Hamilton's wish, if not at his request, he delayed the publication of his work on the theory until after the former's *Elements* appeared. His own *Elementary Treatise on Quaternions* was therefore not issued until 1866, after which he continued to write upon the subject until his death. In 1873 he published with Professor Kelland a work entitled *An Introduction to Quaternions*, which did much to make the subject known to physicists. The theory has not, however, been as favorably received by scientists as had been anticipated by its advocates. It should be added that Gauss, about the year 1820, gave some attention to the subject but without developing any theory of importance.[1]

At about the same time that Hamilton published his discovery of quaternions Hermann Günther Grassmann published his great work, *Die lineale Ausdehnungslehre* (1844), although he seems to have developed the theory as early as 1840.[2]

7. TRANSCENDENTAL NUMBERS

Transcendental Numbers Considered Elsewhere. Among the artificial numbers should, of course, be included not only such types as surds but also all such nonalgebraic types as are found in connection with the trigonometric functions, logarithms, the study of the circle, and the theory of transcendental numbers in general. These may, however, be more conveniently considered in connection with algebra, geometry, and trigonometry, as they are commonly found in the teaching of these subjects.

---

[1] Tropfke, *Geschichte*, II (2), 88.

[2] V. Schlegel, "Die Grassmann'sche Ausdehnungslehre," Schlömilch's *Zeitschrift*, XLI. For A. Macfarlane's digest of the views of various writers see *Proceedings of the American Assoc. for the Adv. of Sci.*, 1891. See also E. Jahnke, in *L'Enseignement Mathématique*, XI (1909), 417; F. Engel, in Grassmann's *Gesammelte math. und physikal. Werke*, III (Leipzig, 1911).

TOPICS FOR DISCUSSION

1. The sequence of development of artificial numbers, with the causes leading to the successive steps.

2. General nature of compound numbers at various periods in their development.

3. Nature of fractions among the Egyptians, and the reasons for the persistence of the unit fraction.

4. Greek symbolism for fractions compared with that which the Romans used.

5. Origin and development of our common fractions.

6. The etymology of terms used in fractions, with the change in these terms from time to time.

7. The sequence of operations with fractions in the early printed works on arithmetic.

8. The development of methods of performing the operations with common fractions.

9. The origin, development, symbolism, and present status of sexagesimal fractions.

10. The origin and development of decimal fractions, including the question of symbolism.

11. The human needs that led to the development of the various types of fraction.

12. The origin and development of the idea of per cent, including the question of symbolism.

13. The reason for the interest of the Greek mathematicians in incommensurable numbers.

14. The development of surd numbers, particularly among the Greek and Arab writers.

15. History of the various methods of approximating the value of a surd number.

16. The origin and development of negative numbers, including the question of symbolism.

17. The origin and development of the idea of complex numbers, including the question of symbolism.

18. The origin and development of the graphic representation of complex numbers.

19. The origin and development of the idea of complex exponents in algebra.

CHAPTER V

GEOMETRY

1. General Progress of Elementary Geometry

Intuitive Geometry. All early geometry was intuitive in its nature; that is, it sought facts relating to mensuration without attempting to demonstrate these facts by any process of deductive reasoning. The prehistoric geometry sought merely agreeable forms, as in the plaiting of symmetric figures in a mat. The

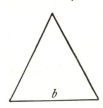

next stage was that of the mensuration of rectangles and triangles, and geometry was in this stage when the Ahmes Papyrus (*c.* 1550 B.C.)[1] was written. In this work the area of an isosceles triangle of base *b*, height *h*, is given as $\frac{1}{2} bh$. For the area of a circle of diameter *d* Ahmes used a rule which may be expressed in modern symbols as $(d - \frac{1}{9}d)^2$, which shows that he took 3.1605– as the value of π,—a value based on experiment.[2]

In Babylonia the same conditions existed. The tablets which have come down to us contain numerous cases of mensuration,[3] but the rules seem based chiefly on experiment.

[1] It should be recalled that Professor Peet (*Rhind Papyrus*, p. 3) puts this date as probably before 1580 B.C.

[2] On the general history of the development of geometry see G. Loria, *Il Passato ed il Presente delle Principali Teorie Geometriche*, Turin, 3d ed., 1907; hereafter referred to as Loria, *Passato-Presente Geom.* This work first appeared in the Turin *Memorie della R. Accad.*, XXXVIII (2), and was translated into German by F. Schütte, Leipzig, 1888. See also the *Encyklopädie,* Vol. III; R. Klimpert, *Geschichte der Geometrie*, Stuttgart, 1888; E. F. August, *Zur Kenntniss der geometrischen Methode der Alten*, Berlin, 1843; and the various general histories of mathematics. On the early Egyptian geometry see E. Weyr, *Ueber die Geometrie der alten Aegypter*, Vienna, 1884; Eisenlohr, *Ahmes Papyrus.* On a papyrus which may be slightly earlier than that of Ahmes, see page 293. [3] Hilprecht, *Tablets.*

The native mathematics of China was also of this type. The *Nine Sections*, written perhaps *c.* 1100 b.c., contains statements which show that the author knew the relations of the sides of certain right-angled triangles,[1] but there is no evidence of any proof of such relations.

In the later Chinese mathematics there are many ingenious examples involving mensuration, but nowhere does there appear any further idea of geometric demonstration, as we understand the term, than is found in the earliest works.

In India the same conditions existed, the native geometry giving us no evidence of any approach to a sequence of deductive proofs. There was a large amount of mensuration,[2] and considerable ability was shown in the formulation of rules, but the basis of the work was wholly empirical.

The Romans were interested in mathematics only for its immediately practical value. The measurement of land, the laying out of cities, and the engineering of warfare appealed to them, but for demonstrative geometry they had no use. Indeed, it may be said that, outside of those lands which were affected by the Greek influence, the ancient world knew geometry only on its intuitive side. Demonstrative geometry was Greek in its origin, and in the Greek civilization it received its only encouragement for more than a thousand years.

Demonstrative Geometry. The idea of demonstrating the truth of a proposition which had been discovered intuitively appears first in the teachings of Thales (*c.* 600 b.c.). It is probable that this pioneer knew and proved about six theorems,[3] each of which would have been perfectly obvious to anyone without any demonstration whatever. The contribution of Thales did not lie in the discovery of the theorems, but in their proofs. These proofs are lost, but without them his work in geometry would have attracted no attention, either among his contemporaries or in the history of thought.

---

[1] See Volume I, pages 30, 33.

[2] *E.g.*, consider Bhāskara's *Lilāvati*, with special sections on ponds, walls, timber, heaps, shadows, and excavations.

[3] See Volume I, page 67; Heath, *History*, I, 130.

From the time of Thales until the decay of their ancient civilization demonstrative geometry was the central feature of the mathematics of the Greeks. The history of the general progress of the science has been sufficiently outlined elsewhere in this work.[1]

The Arabs recognized the Greek culture more completely than any other people until the period of the awakening of Western Europe. They translated the Greek classics in geometry as they did also in philosophy and natural science, but they never made any additions of real significance to the works of Euclid and Apollonius.

It was chiefly through the paraphrase of Boethius (c. 510) that Euclid's *Elements* (c. 300 B.C.) was known in the Dark Ages of Europe. The study of geometry received some encouragement at the hands of Gerbert (c. 1000), Fibonacci (1220), and a few other medieval scholars, but no progress was made in the advance of the great discipline which had been so nearly perfected in Alexandria more than a thousand years earlier.

With the invention of European printing the work of Euclid became widely known, the first printed edition appearing in 1482.[2] Little by little new propositions began to be suggested, but the invention of analytic geometry early in the 17th century took away, for a considerable period, much of the interest in improving upon the ancient theory.

The next advance in the pure field was made in the 17th century, when Desargues[3] (1639) published a work which treated of certain phases of projective geometry. The new analytic treatment of the subject, however, was so novel and powerful as to take the attention of mathematicians from the work of Desargues, and it was not until the 19th century that pure geometry again began to make great progress.

The Greek theory of conics has already been considered sufficiently for our purposes (Vol. I, Chap. IV). The analytic and modern synthetic geometries are considered later.

---

[1] See Volume I, pages 59–146; Heath, *History*, I, chap. iv.
[2] Venice, Erhard Ratdolt; the Campanus translation. See Volume I, page 251.
[3] See Volume I, page 383.

2. Name for Geometry

Reason for Uniformity. When we consider that our elementary geometry is essentially the *Elements* of Euclid, and that the subject never flourished in ancient times outside the Greek sphere of influence, it is apparent that the Greek name would be the one generally used to designate the science. It is derived from the words for "earth" and "measure"[1] and therefore was originally, as it is in some languages today, synonymous with the English word "surveying." Since the latter science was well developed in Egypt before the Greeks founded Alexandria, the name is probably a translation of an Egyptian term. It was in use in the time of Plato and Aristotle, and doubtless goes back at least to Thales.

Euclid did not call his treatise a geometry, probably because the term still related to land measure, but spoke of it merely as the *Elements*.[2] Indeed, he did not employ the word "geometry" at all, although it was in common use among Greek writers.[3]

When *Euclid* was translated into Latin in the 12th century, the Greek title was changed to the Latin form *Elementa*,[4] but the word "geometry" is often found in the title-page, first page, or last page of the early printed editions.[5]

There have been, as would naturally be expected, various fanciful names for textbooks on geometry. In the 16th century such names were common in all branches of learning. Among the best-known of these titles is the one seen in Robert Recorde's *The pathway to Knowledg* (London, 1551 and 1574).

---

[1] Γῆ (*ge*), earth, and μετρεῖν (*metrein'*), to measure.

[2] In Greek, στοιχεῖα (*stoichei'a*). So in the *editio princeps* of the Greek text (Basel, 1533) the title appears as ΕΥΚΛΕΙΔΟΥ ΣΤΟΙΧΕΙΩΝ ΒΙΒΛ⟩ ΙΕ⟩.

[3] Thus Plato (*Theœtetus*, 173 E; *Meno*, 76 A; *Republic*, 546 C, 511 D), Xenophon (*Symposium*, 6, 8, etc.), and Herodotus (II, 109) use the word in some of its forms, but always to indicate surveying.

[4] So in the *editio princeps* (1482) the first page begins: "Preclarissimus liber elementorum Euclidis perspicacissimi: in artem geometrie incipit quā foelicissime." The colophon also has the name *geometria*.

[5] E.g., the first English edition (London, 1570) has the title *The Elements of Geometrie of the most auncient Philosopher Evclide of Megara.*

3. Technical Terms of Euclidean Geometry

Point. The history of a few typical terms of elementary geometry will now be considered.[1] The Pythagoreans defined a point as "a monad having position,"[2] and this definition was adopted by Aristotle (*c.* 340 b.c.). Plato (*c.* 380 b.c.) called a point "the beginning of a line,"[3] and Simplicius (6th century) called it "the beginning of magnitudes and that from which they grow," adding that it is "the only thing which, having position, is not divisible." Euclid (*c.* 300 b.c.) gave the definition: "A point is that which has no part." Heron (*c.* 50?) used the same words, but added "or a limit without dimension or a limit of a line." When Capella (*c.* 460) translated the definition into Latin, he made it read, "A point is that of which a part is nothing,"[4] which is a different matter.

Modern writers usually resort to analogy and give only a quasi definition, or else they make use of the idea of limit.

Line. The Platonists defined a line as length without breadth, and Euclid[5] did the same. Aristotle objected to such a negative definition, although Proclus (*c.* 460) observes that it is positive to the extent that it affirms that a line has length. An unknown Greek writer[6] defined it as "magnitude extended one way,"[7] a phrase not unlike one used by Aristotle. The latter defined it as a magnitude "divisible in one way only,"[8] in contrast to a surface, which is divisible in two ways, and to a solid, which is divisible in three ways. Proclus suggested defining a line as the "flux of a point,"[9] an idea also going back to

---

[1] For further discussion see J. H. T. Müller, *Beiträge zur Terminologie der griechischen Mathematiker*, 1860; Heath, *Euclid*, Vol. I, p. 155; H. G. Zeuthen, "Sur les définitions d'Euclide," *Scientia*, XXIV, 257, on the general nature of Euclid's definitions. See also Heath, *History*, on all such details.

[2] Μονὰς προσλαβοῦσα θέσιν (monad with position added).

[3] Ἀρχὴ γραμμῆς. On this and other definitions consult Heath, *Euclid*, Vol. I, p. 155; H. Schotten, *Inhalt und Methode des planimetrischen Unterrichts*, Vol. I (Leipzig, 1890); Vol. II (Leipzig, 1893); hereafter referred to as Schotten, *Inhalt*.

[4] "Punctum est cuius pars nihil est."

[5] Γραμμὴ δὲ μῆκος ἀπλατές.

[6] Alluded to by al-Nairîzî (*c.* 910) as one Heromides or Herundes.

[7] Μέγεθος ἐφ' ἓν διάστατον.   [8] Μοναχῇ διαιρετόν.   [9] Ῥύσις σημείου.

Aristotle, who remarked that "a line by its motion produces a surface, and a point by its motion a line." This occasionally appears as "A line is the path of a moving point."

Straight Line. It is evident that certain terms are so elementary that no simpler terms exist by which to define them. This is true of "point" and "line," but it is more evidently true of terms like "straight line" and "angle." Plato defined a straight line as "that of which the middle covers the ends," that is, relatively to an eye placed at either end and looking along the line. Euclid endeavored to give up the appeal to sight and defined it as "a line which lies evenly with the points on itself." Proclus explains that Euclid "shows by means of this that the straight line alone [of all lines] occupies a distance equal to that between the points on it," adding that the distance between two points on a circumference or any other line, and measured on this line, is greater than the interval between them. Archimedes (c. 225 B.C.) stated this idea more tersely by saying that "of all lines having the same extremities the straight line is the shortest," which is the source of the definition often found in textbooks, "a straight line is the shortest distance between two points," although "line" and "distance" are two radically different concepts. "The shortest path between two points" is an expression that is less objectionable, but it merely shifts the difficulty.

Heron (c. 50?) defined a straight line as "a line stretched to the utmost toward the ends," and Proclus adopted this phrase with the exception of "toward the ends." It is evidently objectionable, however, because it appeals to the eye and relates to a physical object. Heron also suggested the idea that "all its parts fit on all [other parts] in all ways," a definition substantially adopted by Proclus. Still another definition due to Heron is "that line which, when its ends remain fixed, itself remains fixed when it is, as it were, turned round in the same plane." This too was used with slight change by Proclus, and it appears in various modern works as "that which does not change its position when it is turned about its extremities (or any two points in it) as poles."

Surface. The Pythagoreans used a word[1] meaning "skin" or "color" to designate a surface. Aristotle, like Plato, used other words,[2] and spoke of a surface as extended or continuous or divisible in two ways, and as the extremity or the section of a solid. Aristotle recognized as common the idea that a line by its motion produces a surface.[3] Euclid defined a surface as "that which has only length and breadth."

Plane Surface. The same difficulties that the ancients had in defining a straight line were met when they attempted to define a plane. Euclid stated that "a plane surface is a surface which lies evenly with the straight lines on itself." Heron (c. 50?) added that it is "the surface which is stretched to its utmost," this being analogous to his definition of a straight line. He also defined it as "a surface all the parts of which have the property of fitting on" [each other], and as "such that if a straight line passes through two points on it, the line coincides wholly with it at every spot, all ways." Proclus (c. 460), adopting an assumption stated by Archimedes, defined it as "the least surface among all those which have the same extremities," and also used a modification of Euclid's definition, "a surface such that a straight line fits on all parts of it," or "such that the straight line fits on it all ways." There was no material improvement on these statements until the 18th century, when Robert Simson (1758) suggested the definition that "a plane superficies is that in which any two points being taken, the straight line between them lies wholly in that superficies,"[4] a statement which Gauss (c. 1800) characterized as redundant. Fourier (c. 1810) gave the definition that a plane is formed by the aggregate of all the straight lines which, passing through one point on a straight line in space, are perpendicular to that

---

[1] Χροιά (chroia').

[2] Ἐπιφάνεια (epipha'neia) and ἐπίπεδον (epi'pedon). From the former, a word meaning "appearance," we have our word "epiphany." The latter word, meaning a plane surface, occurs in our word "parallelepiped." Later Greek writers also used ἐπιφάνεια to indicate any kind of surface, and Plato used ἐπίπεδον in the same way.

[3] On the different kinds of lines and surfaces, consult Heath, *Euclid*.

[4] Compare one of Heron's definitions above.

straight line. This is, of course, merely putting into another form a well-known theorem of Euclid.[1] Crelle (1834) suggested that a plane is the surface containing throughout their entire lengths all the straight lines passing through a fixed point and also intersecting a straight line in space.[2]

Angle. Euclid's definitions of an angle are as follows:

A plane angle is the inclination to one another of two lines in a plane which meet one another and do not lie in a straight line.
And when the lines containing the angle are straight, the angle is called rectilineal.

This excludes the zero angle, straight angle, and in general the angle $n\pi$, and defines angle by the substitution of the idea of inclination,—in modern form, the difference in direction. Even less satisfactory is the definition of Apollonius (*c.* 225 B.C.) which asserts that an angle is "a contracting of a surface or a solid at one point under a broken line or surface." Plutarch (1st century) and various other writers defined it as "the first distance under the point," which Heath[3] interprets as "an attempt (though partial and imperfect) to get at the rate of divergence between the lines at their point of meeting." Perhaps this idea was also in the mind of Carpus of Antioch (1st century) when he said that the angle is "a quantity, namely a distance between the lines or surfaces containing it."

Later writers often return to the qualitative idea of Aristotle, as in the definition that an angle is a figure formed by two lines which meet. This was refined by Professor Hilbert of Göttingen[4] as follows:

Let α be any arbitrary plane and h, k any two distinct half-rays lying in α and emanating from the point O so as to form a part of two different straight lines. We shall call the system formed by these two half-rays h, k an *angle*.

---

[1] *Elements*, XI, 5.
[2] For further consideration of modern definitions see Heath, *Euclid*, Vol. I, p. 174, and Schotten, *Inhalt*, II.   [3] Heath, *Euclid*, Vol. I, p. 177.
[4] *Foundations of Geometry*, translated by E. J. Townsend, p. 13 (Chicago, 1902).
II

Circle. The ancient writers defined a circle substantially as Euclid did:

A circle is a plane figure contained by one line such that all the straight lines falling upon it from one point among those lying within the figure are equal to one another;

And the point is called the center of the circle.

Euclid had already defined a figure as "that which is contained by any boundary or boundaries," so that a circle is, in his view, the portion of a plane included in the bounding line. This bounding line Euclid usually calls the periphery ($\pi\epsilon\rho\iota\phi\epsilon\rho\epsilon\iota\alpha$), a word translated into the Latin as *circumferentia,* whence our "circumference." Euclid is not consistent however, for he speaks of a circle as not cutting a circle in more than two points,[1] the word "circle" here referring to the bounding line.

This uncertain use of the term has been maintained until recent times. The influence of analytic geometry has led to defining the circle as a line, but there is still no uniformity in the matter.

Diameter and Radius. Euclid used the word "diameter"[2] in relation to the line bisecting a circle and also to mean the diagonal[3] of a square, the latter term being also found in the works of Heron.

The term "radius" was not used by Euclid, the term "distance" being thought sufficient. Boethius (*c.* 510) seems to have been the first to use the equivalent of our "semidiameter."[4] A similar use also appears in India, in the writings of Āryabhaṭa[5] (*c.* 510). Ramus[6] (1569) used the term, saying: "Radius est recta a centro ad perimetrum."

---

[1] *Elements,* III, 10, where the Greek κύκλος (*ky'klos,* circle) is used.

[2] Διά (*dia',* through) + μετρεῖν(*metrein',* to measure).

[3] Διαγώνιος, from διά (*dia',* through) + γωνία (*goni'a,* angle).

[4] Thus in the *Ars geometriae,* ed. Friedlein, p. 424: "Conscribitur age emicyclus XXVIII in basi et in semidiametro XIIII pedes habens" (MS. of the 11th century).

[5] L. Rodet, "Leçons de Calcul d'Āryabhaṭa," *Journ. Asiatique,* XIII (7), 398; reprint (Paris, 1879), p. 10: "The chord of the sixth part of the circumference is equal to a semidiameter."

[6] On the general question see *Bibl. Math.,* II (3), 361, and P. Ramus, *Scholarvm Mathematicarvm, Libri vnvs et triginta,* p. 155 (Basel, 1569).

From India it seems to have passed over to Arabia and thence to Europe. So Plato of Tivoli (*c.* 1120) used *mediatas diametri* and *dimidium diametri*, Fibonacci (1220) used *semidyameter*,[1] and Jordanus Nemorarius (*c.* 1225) preferred the form *semidiameter*.[2]

The early printed books,—such as those of Maurolico (1558), Tartaglia (1560), and Pedro Nunes (1564),—commonly used the word *semidiameter*.[3]

The word "radius" as used in this sense is modern. It appears, as above stated, in the *Scholarvm Mathematicarvm, Libri vnvs et triginta* of Ramus (1569), and a little later was used by Thomas Fincke[4] (1583) in his *Geometria Rotundi*. It was then adopted by Vieta[5] (*c.* 1590), and after that time it became common.

Parallel Lines. The word "parallel"[6] means "alongside one another." Euclid defined parallel straight lines as "straight lines which, being in the same plane and being produced indefinitely in both directions, do not meet one another in either direction." Rather less satisfactory is the definition of Poseidonius (*c.* 100 B.C.) as those lines "which, in one plane, neither converge nor diverge, but have all the perpendiculars equal which are drawn from the points of one line to the other." This definition is substantially that ascribed to Simplicius (6th century), that two straight lines are parallel "if, when they are produced indefinitely both ways, the distance between them, or the perpendicular drawn from either of them to the other, is always equal and not different." The direction theory, one of the least satisfactory of all, is due to Leibniz.[7]

---

[1] *Scritti*, II, 85. See also *dimidium dyametri* on page 86.

[2] See his "De Triangulis" in the *Mitteilungen des Coppernicus-Vereins* . . . *zu Thorn*, VI (1887).

[3] Tropfke, *Geschichte*, IV (2), 108.

[4] Variously spelled. A Danish mathematician (1561–1646). See Volume I, page 348.

[5] "Posito X radio seu semidiametro circuli." See Tropfke, *Geschichte*, IV (2), 108.

[6] Παράλληλος (*paral'lelos*).

[7] For further discussion, including the various bases for a definition, see Heath, *Euclid*, Vol. I, p. 192; Schotten, *Inhalt*, II, 188.

4. Axioms and Postulates

Distinction between Axioms and Postulates. The Greek writers recognized the existence of first principles "the truth of which," as Aristotle affirmed, "it is not possible to prove." These, he stated, were of two kinds: (1) those which are common to all sciences, and for which the name "axiom" was used by the Stoic philosophers and by Aristotle himself; (2) those which relate to the particular science, and to which the name "postulate" was given by later writers.[1] The distinction was not completely recognized, even by Euclid, for his fourth axiom (see page 281) is rather a geometric postulate or a definition.

Aristotle had other names for axioms, speaking of them as "the common [things]"[2] or "common opinions."[3]

Proclus (c. 460) states that Geminus (c. 77 B.C.) taught that axioms and postulates "differ from one another in the same way as theorems are also distinguished from problems,"—an opinion which is quite at variance with that of Aristotle.

Euclid seems to have used the term "common notion" to designate an axiom, although he may have used the term "axiom" also.[4]

The word "postulate" is from the Latin *postulare*, a verb meaning "to demand." The master demanded of his pupils that they agree to certain statements upon which he could build. It appears in the early Latin translations of Euclid[5] and was commonly used by the medieval Latin writers.

As to the number of these assumptions, Aristotle set forth the opinion which has been generally followed ever since, that "other things being equal that proof is the better which proceeds from the fewer postulates or hypotheses or propositions."

---

[1] Heath, *Euclid*, Vol. I, p. 117. On the general question of foundation principles in geometry, see the *Encyklopädie*, II, 1.

[2] Τὰ κοινά (*ta koina'*). [3] Κοιναὶ δόξαι (*koinai' dox'ai*).

[4] But not in his extant writings. On the doubt that has been raised as to his giving a list of axioms at all, see Heath, *Euclid*, Vol. I, p. 221.

[5] "Postulata. I. Postuletur, ut a quouis puncto ad quoduis punctum recta linea ducatur." The Greek word for postulates used by Euclid was αἰτήματα (*aite'mata*) (*Euclid*, ed. Heiberg, I, 8, 9).

Axioms. Euclid laid down certain axioms, or "common notions," probably five in number, as follows:

1. Things which are equal to the same thing are equal to one another.
2. If equals be added to equals, the wholes are equal.
3. If equals be subtracted from equals, the remainders are equal.
4. Things which coincide with one another are equal to one another.[1]
5. The whole is greater than the part.

The axioms of inequality, of doubling, and of halving may have been given by Euclid, but we are not certain.[2]

It will be observed that Euclid built up his geometry on a smaller number of axioms than many subsequent writers have thought to be necessary.

Postulates. Euclid does not use a noun equivalent to the Latin *postulatum*, but says:

Let the following be postulated:

1. To draw a straight line from any point to any point.
2. To produce a finite straight line continuously in a straight line.
3. To describe a circle with any center and [any] distance.
4. That all right angles are equal to one another.
5. That, if a straight line falling on two straight lines makes the interior angles on the same side less than two right angles, the two straight lines, if produced indefinitely, meet on that side on which are the angles less than the two right angles.[3]

Considerable criticism of these postulates developed among the later Greeks. Zeno of Sidon (1st century B.C.) asserted that it was necessary to postulate that two [distinct] straight lines cannot have a segment in common. If this is not done, he claimed, one or more of the proofs in Book I are fallacious.

Others asserted that postulates 4 and 5 are theorems capable of proof. Proclus (*c.* 460) attempted a proof of postulate 4,

---

[1] Essentially a postulate or a definition.
[2] For the evidence *pro* and *contra*, see Heath, *Euclid*, Vol. I, p. 223. Heiberg's edition of Euclid (I, 10, 11) numbers these 1, 2, 3, 7, 8, giving in Greek the doubtful axioms. [3] Heath, *Euclid*, Vol. I, pp. 154, 195.

but it was fallacious. He also claimed that the converse is not necessarily true which asserts that an angle which is equal to a right angle is also a right angle, for he said that, in the figure given, $\angle ABC = \angle XBY$ and yet $\angle XBY$ is not a right angle. Sac-cheri[1] (1733) gave a proof of the postulate, but he assumed other statements equally fundamental upon which to base his argu-

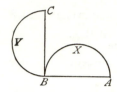

ment. Modern writers often adduce a simple proof based upon the postulate of the equality of straight angles, but this simply substitutes one postulate for another.

Postulate 5, the "Postulate of Parallels," has been frequently attacked.[2] Ptolemy attempted to prove it, one of his arguments being that if $a + b$ is a straight angle, then $c + d$ must be a straight angle. Hence if the lines meet at

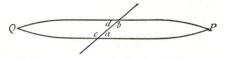

P they also must meet at Q, and in that case the two straight lines inclose space. Proclus gave a more seductive argument relating to the meeting of lines in general, thus:

Draw the lines AK and CL so that $\angle A + \angle C <$ two right angles.

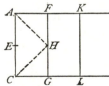

Bisect AC at E and lay off $AF = AE$ and $CG = EC$.

Then AF and CG cannot meet on FG, as at H; for if they did we should have $AH = AE$ and $CH = CE$, and so the sum of two sides of a triangle would be equal to the third side.

Bisect FG at H, make $FK = FH = HG = GL$.

[1] *Euclides ab omni naevo vindicatus*, p. x (Milan, 1733).

[2] Heath, *Euclid*, Vol. I, p. 202; Engel and Stäckel, *Die Theorie der Parallel-linien*, Leipzig, 1895; G. B. Halsted, *Saccheri's Euclides Vindicatus* (translation), p. 7 (Chicago, 1920); G. Boccardini, *L'Euclide emendato del P. Gerolamo Saccheri*, Milan, 1904 (incomplete translation); R. Bonola, "Sulla teoria delle parallele e sulle geometrie non-euclidee," in F. Enriques, *Questioni riguardanti le matematiche elementari*, p. 248 (Bologna, 1900), with an English translation of his *La Geometria non-euclidea*, by H. S. Carslaw, 1912. On the modern theory and treatment of postulates see C. J. Keyser, *Mathematical Philosophy*, Lecture II (New York, 1922).

Then *FK* and *GL* cannot meet on *KL*, for the same reason, and so on however far we go.

Hence the lines described in the postulate cannot meet at all, even though $\angle A + \angle C < 180°$.[1]

Further attempts at a proof of Postulate 5 were made by al-Ṭûsî (*c.* 1200), Wallis (*c.* 1660), Saccheri (1733), Lambert (*c.* 1766), Legendre (1794 and later), and many others.[2]

As an alternative postulate Proclus stated in substance, and Playfair (1795) made well known to the modern world, the following:

Through a given point only one parallel can be drawn to a given straight line; [or]
Two straight lines which intersect one another cannot both be parallel to one and the same straight line.

Playfair's form of the postulate was practically given, however, somewhat earlier than 1795. Joseph Fenn, in his edition of Euclid's *Elements*, published at Dublin in 1769,[3] stated it as follows: "A straight line which cuts one of two parallel lines will necessarily cut the other, provided this cutting line is sufficiently produced." Substantially the same assumption was also given by William Ludlam[4] in 1785, and, indeed, was given by Proclus (*c.* 460), as asserted above, in a note to Euclid, I, 31.

It has been observed by various writers that Euclid tacitly assumed other postulates, such as one relating to the intersection of plane figures and one which asserts that space is homogeneous or that a figure may be transposed without deformation.

---

[1] For the rest of his treatment of Postulate 5, see Heath, *Euclid*, Vol. I, p. 207.

[2] On the general question of the validity of the postulate, see page 335.

[3] *First Volume of the Instructions given in the Drawing School established by the Dublin-Society ... under the direction of Joseph Fenn, heretofore Professor of Philosophy in the University of Nants*, Dublin, 1769. F. Cajori, "On the history of Playfair's parallel-postulate," *School Science and Mathematics*, XVIII, 778.

[4] Born *c.* 1718; died in Leicestershire, March 16, 1788. He wrote various works on astronomy. His *Rudiments of Mathematics* first appeared in 1785.

5. Typical Propositions of Plane Geometry

Pons Asinorum. Most of the basic theorems of elementary plane geometry are found in Euclid's *Elements*. Of these a relatively small number have any interesting history, and only a few typical ones need be considered, the first being Proposition 5 of Book I. As given by Euclid, this reads as follows:

In isosceles triangles the angles at the base are equal to one another, and, if the equal straight lines be produced further, the angles under the base will be equal to one another.

Proclus states that Thales (*c.* 600 B.C.) was the first to prove this proposition. At any rate it was well known to Aristotle (*c.* 340 B.C.), who discusses one of the proofs then possibly

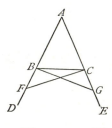

current. Proclus (*c.* 460) says that Pappus (*c.* 300) proved the theorem without using any auxiliary lines, simply taking the triangle up, turning it over, and laying it down upon itself. The question as to how he could lay the triangle itself down upon itself has caused a change in the phraseology on the part of modern writers.

The proposition represented substantially the limit of instruction in many courses in the Middle Ages. It formed a bridge across which fools could not hope to pass, and was therefore known as the *pons asinorum,* or bridge of fools.[1] It has also been suggested that the figure given by Euclid resembles the simplest form of a truss bridge, one that even a fool could make. The name seems to be medieval.

The proposition was also called *elefuga,* a term which Roger Bacon (*c.* 1250) explains as meaning the flight of the miserable ones, because at this point they usually abandoned geometry.[2]

---

[1] The term is sometimes applied to the Pythagorean Theorem.

[2] "Sic est hic quod isti qui ignorant alicujus scientiae, ut sit geometriae, nisi sint pueri qui coguntur per virgam, resiliunt et tepescunt, ut vix volunt tres vel quatuor propositiones scire. Unde ex hoc accidit quod quinta propositio geometriae Euclidis dicitur *Elefuga,* id est, fuga miserorum; *elegia* enim Graece dicitur, Latine *miseria;* et *elegi* sunt *miseri.*" *Opus Tertium,* cap. vi.

Congruence Theorems. The second of the usual congruence theorems relates to the case of two angles and the included side of a triangle. Proclus (*c.* 460) says of this:

Eudemus (*c.* 335 B.C.) in his geometrical history refers this theorem to Thales (*c.* 600 B.C.). For he says that, in the method by which they say that Thales proved the distance of ships in the sea, it was necessary to make use of this theorem.

How Thales could have used this theorem for the purpose is purely a matter of conjecture. He might have stood at *T*, the

top of a cliff *TF*, and sighted to the ship *S*, using two hinged rods to hold the angle *STF*. He could then have turned and sighted along the same rod to a point *P* along the shore. If he kept the angle constant, he would then merely have to measure *FP* to find the unknown distance *FS*. Since in those days the ships were small and remained near the shore in good weather, this plan would have been quite

EARLY METHODS OF MEASURING DISTANCES

From Belli's *Libro del Misvrar con la vista*, Venice, 1569, but representing essentially the method probably used by Thales

feasible. Thales probably had some simple instrument like the astrolabe by which he could measure angles when observing the stars, and he could have used this. We shall presently see that such instruments, in primitive form, were known to the Babylonians before his time.

Euclid stated the proposition in a more complicated form than the one now in use, but his proof had the advantage that it did not employ superposition. The latter form of proof is given by al-Nairîzî (c. 910), who ascribes it to some unknown predecessor.

Renaissance writers often used the theorem in practical mensuration, as Thales is thought to have done. The illustration on page 285, from Belli's work[1] of 1569, shows two methods of using it, and there is a story that one of Napoleon's engineers gained imperial favor by quickly applying it on an occasion when the army was held up by a river.

Areas. The sources of such propositions as those relating to the area of the triangle, the rectangle, the trapezium (trapezoid), and other rectilinear figures are, of course, unknown. The theorems relating to these areas are found in Euclid and apparently were common property long before his time. It is interesting to know, however, that the Egyptian surveyors, even after the time of Euclid, were in the habit of finding the area of a field by taking the product of the half-sums of the opposite sides. This is correct in the case of a rectangle, but in the case of a general convex quadrilateral it gives a result that is too large. The error was corrected to a certain extent by omitting in the calculation all fractions less than $\frac{1}{64}$ of the large unit of length.[2]

The rule for the area of an inscribed convex quadrilateral, expressed by the formula

$$A = \sqrt{(s-a)(s-b)(s-c)(s-d)},$$

was given by both Brahmagupta (c. 628) and Mahāvīra (c. 850), but without the limitation that it holds only for an

---

[1] S. Belli, *Libro del Misvrar con la vista*, Venice, 1569.

[2] This unit was the σχοῖνος (*schoi'nos*), equal to 100 cubits. Our information comes from a papyrus in the British Museum. See H. Maspero, *Les Finances de l'Égypte sous les Lagides*, p. 135 (Paris, 1905).

inscribed figure. If $d = 0$, the figure becomes a triangle and the formula reduces to

$$A = \sqrt{s(s-a)(s-b)(s-c)},$$

the rule being given by Heron ($c.$ 50?), but being ascribed on Arabic authority to Archimedes. Both Brahmagupta and Mahāvīra[1] gave the equivalents of the following formulas, without limitations, for the lengths of the diagonals of a quadrilateral:

$$m = \sqrt{\frac{(ac+bd)(ab+cd)}{ad+bc}},$$

$$n = \sqrt{\frac{(ac+bd)(ad+bc)}{ab+cd}}.$$

The first of these formulas was rediscovered by W. Snell, who gave it in his edition (1619) of Van Ceulen's works.

Among the other interesting formulas related to those concerning areas is one discovered by Lhuilier and published in 1782. It gives the radius of the circle circumscribing a quadrilateral and reduces to the following:[2]

$$r = \tfrac{1}{4}\sqrt{\frac{(ab+cd)(ac+bd)(ad+bc)}{(s-a)(s-b)(s-c)(s-d)}}.$$

Angle Sum. The fact that the sum of the angles of a triangle is equal to two right angles has long been recognized as one of the most important propositions of plane geometry. Eutocius ($c.$ 560) tells us that Geminus ($c.$ 77 B.C.) stated that "the ancients investigated the theorem of the two right angles in each individual species of triangle,—first in the equilateral, again in the isosceles, and afterwards in the scalene triangle." Proclus ($c.$ 460) says that Eudemus ($c.$ 335 B.C.) ascribed the theorem to the Pythagoreans.

There is also a possibility that Thales knew this property of the triangle, for Diogenes Laertius (2d century) quotes Pamphile (1st century) as saying that he was the first to inscribe

---

[1] See Volume I, page 163.

[2] I am indebted to Professor R. C. Archibald for this information, as for many other valuable suggestions.

a right-angled triangle in a circle, the proof of which solution requires this proposition, at least in a special case. The theorem was certainly well known before Euclid, for Aristotle refers to it several times.

Pythagorean Theorem. The relation of the sides of a triangle when these sides are 3, 4, and 5 (that is, $3^2 + 4^2 = 5^2$) was well known long before the time of Pythagoras. We find in the *Nine Sections* of the Chinese, perhaps written before 1100 B.C., this statement: "Square the first side and the second side and add them together; then the square root is the hypotenuse."

The Egyptians knew a similar relation for numerical cases, for a papyrus of the 12th dynasty (c. 2000 B.C.), discovered at Kahun, refers to four of these relations, one being $1^2 + (\frac{3}{4})^2 = (1\frac{1}{4})^2$. It was among these people that we first hear of the "rope stretchers,"[1] those surveyors who, it is usually thought, were able by the aid of this property to stretch a rope so as to draw a line perpendicular to another line, a method still in use at the present time.

Pythagorean Numbers in India. The Hindus knew the property long before the beginning of the Christian era, for it is mentioned in the *Śulvasūtras*,[2] the sacred poems of the Brahmans. The *Śulvasūtra* of Āpastamba gives rules for constructing right angles by stretching cords of the following lengths: 3, 4, 5; 12, 16, 20; 15, 20, 25; 5, 12, 13; 15, 36, 39; 8, 15, 17; and 12, 35, 37. Although the date of these writings is uncertain,[3] it is evident that the relations were known rather early in India.[4]

Did Pythagoras prove the Theorem? The proof of the proposition is attributed to Pythagoras (c. 540 B.C.) by various writers, including Proclus (c. 460), Plutarch (1st century), Cicero

---

[1] Ἁρπεδονάπται (*harpedonap'tae*) (Vol. I, p. 81). See Peet, *Rhind Papyrus*, p. 32.
[2] Curiously the word is sometimes interpreted to mean rope-stretching.
[3] Perhaps the 4th or 5th century B.C.
[4] A. Bürk, "Das Āpastamba-Sulba-Sūtra," in the *Zeitschrift der deutschen morgenländischen Gesellschaft*, LV, 543, and LVI, 327; G. Thibaut, *Journal of the Royal Asiatic Society of Bengal*, XLIV, reprint 1875, and his articles in *The Pandit*, Benares, 1875/6 and 1880; Heath, *Euclid*, Vol. I, p. 360.

(*c.* 50 B.C.), Diogenes Laertius (2d century), and Athenæus (*c.* 300). No one of these lived within, say, five centuries of Pythagoras, so that we have only a weak tradition on which to rest the general belief that Pythagoras was the first to prove the theorem.[1] It would seem as if such an important piece of history would have some mention in the works of a man like Aristotle; but, on the other hand, it is difficult to see how such a tradition should be so generally received unless it were well founded. Not only are we not positive that the proof is due to Pythagoras at all, but we are still more in doubt as to the line of demonstration that he may have followed.

Hundreds of proofs have been suggested for the proposition, but only two are significant enough to be mentioned at this time. Of these the first is the one in the *Elements*, a proof which Proclus tells us was due to Euclid himself. Although Schopenhauer, the philosopher, calls it "a proof walking on stilts" and "a mousetrap proof," it has stood the test of time better than any other.

The second noteworthy proof is that of Pappus (*c.* 300). In this figure we have any triangle *ABC*, with *CM* and *CN* any parallelograms on *AC* and *BC*, and with *QR* equal to *PC*. Then $AT = CM + CN$, a re-lation that reduces to the Pythagorean Theorem when *ABC* is a right-angled triangle and when the parallelograms are squares.[2]

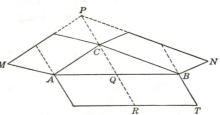

The Pythagorean The-orem is not uncommonly called the *pons asinorum* by modern French writers. The Arabs called it the "Figure of the Bride," possibly because it represents two joined in one. It is also called the "Bride's Chair," possibly because the Euclid figure is not

---

[1] For arguments against this belief see H. Vogt, *Bibl. Math.*, VII (3), 6, and IX (3), 15; G. Junge, *Wann haben die Griechen das Irrationale entdeckt*, Halle, 1907.

[2] *Mathematicae Collectiones*, ed. Commandinus, Bologna, 1660, liber quartus, p. 57; Hultsch ed., IV, 177.

unlike the chair which a slave carries on his back and in which the Eastern bride is sometimes transported to the ceremony.[1]

Of the rules for forming rational right-angled triangles the following related ones are among the most important:[2]

$$n^2 + \left(\frac{n^2-1}{2}\right)^2 = \left(\frac{n^2+1}{2}\right)^2 \qquad \text{Pythagoras } (c.\ 540\ \text{B.C.})$$

$$(2n)^2 + (n^2-1)^2 = (n^2+1)^2 \qquad \text{Plato } (c.\ 380\ \text{B.C.})$$

$$(2n+1)^2 + \left[\frac{(2n+1)^2-1}{2}\right]^2 = \left[\frac{(2n+1)^2-1}{2}+1\right]^2$$

$$\text{Proclus } (c.\ 460)$$

Recent Geometry of the Triangle. In the 19th century the geometry of the triangle made noteworthy progress. Crelle (1816) made various discoveries in this field, Feuerbach (1822) soon after found the properties of the nine-point circle, and Steiner set forth some of the properties of the triangle, but it was many years before the subject attracted much attention. Lemoine[3] (1873) was the first to take up the subject in a systematic way and to contribute extensively to its development. His theory of "transformation continue" and his "géométrographie" should also be mentioned. Brocard's contributions to the geometry of the triangle began in 1877, and certain critical points of the triangle bear his name.

The Pentagon and Decagon. The tenth proposition of Book IV of Euclid is the problem: "To construct an isosceles triangle having each of the angles at the base double the remaining one." This makes the vertical angle 36° and each of the others 72°, and therefore permits of the construction of

---

[1] The Greeks are said to have called it the "theorem of the married women," and Bhāskara to have spoken of it as the "chaise of the little married women." E. Lucas, *Récréations Mathématiques*, II, 130.

[2] H. A. Naber, *Das Theorem des Pythagoras*, Haarlem, 1908; Heath, *Euclid*, Vol. I, p. 350; the names are those of the authors of rules approximately represented by these formulas. The assertion as to Pythagoras is open to doubt.

[3] D. E. Smith, "Émile-Michel-Hyacinthe Lemoine," *Amer. Math. Month.*, III, 29; J. S. Mackay, various articles on modern geometry in the *Proceedings of the Edinburgh Mathematical Society*; E. Vigarié, "La bibliographie de la géométrie du triangle," *Mathesis*, XVI, suppl., p. 14.

a regular decagon and of a regular pentagon. The problem seems to have been known to the Pythagoreans, for Proclus (*c.* 460) tells us that they discovered "the construction of the cosmic figures,"—a statement anticipated by Philolaus (*c.* 425 B.C.) and Iamblichus (*c.* 325), and this construction requires using the problem. Lucian (2d century) and the scholiast to the *Clouds* of Aristophanes tell us that the pentagram, the star pentagon, was the badge of the Pythagorean brotherhood, and the construction of such a figure depends upon this proposition.[1]

The solution is related to that of the division of a line in extreme and mean ratio.[2] This was referred to by Proclus when he said that Eudoxus (*c.* 370 B.C.) "greatly added to the number of the theorems which Plato originated regarding the *section.*" This is the first trace that we have of this name for such a cutting of the line.

In comparatively modern times the section appears first as "divine proportion,"[3] and then, in the 19th century,[4] as the "golden section."

6. Typical Propositions of Solid Geometry

Prism. Since the Greeks were so much more interested in the logic of geometry than in its applications to mensuration, and since they found a sufficient field for their activities in the work with plane figures, they did not develop the science of solid geometry to any great extent, as witness the *Elements* of Euclid. This is one reason why even the technical terms were not so completely standardized as those of plane geometry. Of the special types of prism the right parallelepiped is naturally

---

[1] Heath, *Euclid*, Vol. II, p. 97.   [2] Euclid, II, 11, and VI, 30.

[3] So Pacioli gave to his work of 1509 the title *De diuina proportione*. Ramus (*Scholarvm Mathematicarvm, Libri vnvs et triginta*, Basel, 1569; *ibid.*, 1578; Frankfort, 1599, p. 191) referred to it in these words: "Christianis quibusdam divina quaedam proportio hic animadversa est . . ." ; and Kepler (Frisch ed. of his *Opera*, I, 377 (Frankfort, 1858)) spoke of it in the following terms: "Inter continuas proportiones unum singulare genus est proportionis divinae . . . ."

[4] The term seems to have come into general use in the 19th century. It is found in the *Archiv der Math. und Physik* (IV, 15–22) as early as 1844.

the most important. The word "parallelepiped"[1] means parallel surfaces. Although it is a word that would naturally be used by Greek writers, it is not found before the time of Euclid. It appears in the *Elements* (XI, 25) without definition, in the form of "parallelepipidal solid," the meaning being left to be inferred from that of the word "parallelogrammic" as given in Book I.

We have as yet no generally accepted name for a rectangular solid or right parallelepiped, nor had the Greeks. The word "cuboid" is significant and in modern times has had some sanction.

Euclid used *cubus*[2] for cube, and Heron (*c*. 50?) did the same. Heron also used "hexahedron"[3] for this purpose and then applied *cubus* to any right parallelepiped.

Although the Greeks knew that the diagonal of a right parallelepiped of edges a, b, c was $\sqrt{a^2 + b^2 + c^2}$, strangely enough the statement is not found in any of their works. It first appears, so far as now known, in the *Practica geometriae* of Fibonacci (1220).[4]

The word "prism" is Greek.[5] Euclid defines and treats it as we do at the present time.[6]

Pyramid. The Greeks probably obtained the word "pyramid"[7] from the Egyptian. It appears, for example, in the Ahmes Papyrus (*c*. 1550 B.C.). Because of the pyramidal form of a flame the word was thought by medieval and Renaissance writers to come from the Greek word for fire,[8] and so a pyramid was occasionally called a "fire-shaped body."[9]

---

[1] From παράλληλος (*paral'lelos*, parallel) + ἐπίπεδον (*epi'pedon*, plane surface).

[2] Κύβος, Latin *cu'bus*. See *Elements*, XI, def. 25.

[3] Ἐξάεδρον, from ἔξ (*hex*, six) + ἔδρα(*hed'ra*, seat).

[4] ". . . ut in Solido .aei. cuius dyameter sit linea .tb." (*Scritti*, II, 163).

[5] Πρίσμα (*pris'ma*), from πρίζειν (*pri'zein*), to saw; hence something sawed off.

[6] *Elements*, XI, def. 13.

[7] Πυραμίς (*pyramis'*), pl. πυραμίδες (*pyrami'des*), perhaps from the Egyptian *piromi*, but also thought to come from πυρός (*pyros'*, grain), as if a granary. On the uncertainty of the origin see Peet, *Rhind Papyrus*, p. 98.

[8] Πῦρ (*pyr*), as in "pyrotechnic."

[9] Thus the 16th century writer W. Schmid (*Das erste Buch der Geometrie*, Nürnberg, 1539) speaks of the "feuerförmige Cörper."

Euclid's treatment of the pyramid has remained substantially unchanged, except as to the proposition relating to the equivalence of pyramids of the same height and of equivalent bases. Cavalieri (1635) applied to this proposition his method of indivisibles, and Legendre[1] (1794) gave a simple proof that is now in common use. Āryabhata (c. 510) gave the volume as half the product of the base and height, or at least it so appears in the extant manuscripts.[2]

Frustum of a Pyramid. The method of finding the volume of the frustum of a square or a triangular pyramid is set forth in Heron's *Stereometry*.[3] The modern formula $V=\frac{1}{3}h(b+b'+\sqrt{bb'})$ first appears, in the form of a rule, in Fibonacci's geometry (1220),[4] unless we accept a gloss upon an Arabic manuscript of the 12th century as evidence that it was known at that time.[5] The method itself was probably known to the Egyptians, at least for a special case, long before the time of Heron; for in a hieratic papyrus, apparently a little earlier than that of Ahmes, there is a statement which seems to show familiarity with the method for the case of a square pyramid.[6]

Brahmagupta (c. 628) also gave a rule[7] for the volume of a frustum of a pyramid with square base of sides s_1 and s_2, substantially as follows: $V=\frac{1}{3}h(s_1^2+s_2^2+s_1s_2)$.

---

[1] *Elémens*, 1st ed., VI, 17 (Paris, 1794). The order differs slightly in the different editions.

[2] Rodet, "Leçons de Calcul d'Āryabhata," *Journal Asiatique*, XIII (7), 393; reprint, pp. 9, 10, 20.

[3] I, capp. 33, 34. See Tropfke, *Geschichte*, II (1), 383, with reference to the MS. of the Μετρικά (*metrika'*) discovered recently and published by Schöne, Leipzig, 1903. In his *Stereometry* he considers the pyramid with a square base; in the *Metrica*, with a triangular base. [4] *Scritti*, II, 174.

[5] See the *Sitzungsberichte der physikalisch-medizinischen Societät zu Erlangen*, 50.–51. Band, p. 270, hereafter referred to as *Erlangen Sitzungsberichte*.

[6] This was first published by B. A. Touraeff (Turajev) in 1917, my attention being called to the fact by Professor R. C. Archibald. The manuscript is now in Moscow. See *Ancient Egypt*, Part III, p. 100 (London, 1917).

[7] Colebrooke ed., p. 312. On Āryabhata see Rodet, *loc. cit.*, pp. 9, 10. On Bhāskara, see Colebrooke ed., p. 97. On Mahāvīra, see his work, p. 259. On the later treatment of the tetrahedron see Tropfke, *Geschichte*, II (1), 385. For the cylinder and cone, which are closely related to the prism, pyramid, and circle, see *ibid.*, p. 387.

11

Frustum of a Cone. The late Greeks knew how to find the volume of the frustum of a cone, deriving it from the rule that Heron used for the frustum of a pyramid, and thereafter the same method appeared in various mathematical treatises. Heron, however, used an approximation method which is probably of Egyptian origin, namely, that of taking the product of the altitude and the area of the circle midway between the bases.[1] An interesting example of the use of this approximation method has been found among the Greek papyri on arithmetic, being probably the work of a schoolboy of about the 4th century. The problems in this papyrus resemble those found in the Akhmîm Papyrus, which was written somewhat later. The first one of these problems relates to finding the contents of a circular pit of which the circumference is 20 cubits at the top and 12 cubits at the bottom, and of which the altitude is 12 cubits. The writer makes an error in his methods as well as his calculations, but endeavors to use Heron's approximation.[2]

Sphere. The word "sphere" comes to us from the Greek through the Latin.[3] The pure geometers of Greece had little interest in its measurement, although Archimedes tells us that "earlier geometers . . . have shown . . . that spheres are to one another in the triplicate ratio of their diameters."[4]

Archimedes also states that the volume of any sphere is four times that of the cone with base equal to a great circle of the sphere and with height equal to the radius of the sphere,—a statement that amounts to saying that $V = \frac{4}{3}\pi r^3$. He also stated that a cylinder with base equal to a great circle of the sphere and with height equal to the diameter of the sphere is equal to $1\frac{1}{2}$ times the sphere,—a statement that amounts to the same thing.[5]

---

[1] *Stereometrie*, ed. Hultsch, p. 157.

[2] J. G. Smyly, "Some Examples of Greek Arithmetic," *Hermathena*, XLII (1920), 105.

[3] Σφαῖρα (*sphaî'ra*), Latin *sphaera*.

[4] Allman, *Greek Geom.*, p. 96; Archimedes, ed. Heiberg, II, 265; Heath, *Archimedes*, 234.

[5] Heiberg ed., Vol. I, *De sphaera et cylindro*.

Polyhedrons. The word "polyhedron"[1] is not found in the *Elements* of Euclid; he uses "solid," "octahedron," and "dodecahedron," but does not mention the general solid bounded by planes. The chief interest of the Greeks in figures of this type related to the five regular polyhedrons. It seems probable that Pythagoras (*c*. 540 B.C.)[2] brought his knowledge of the cube, tetrahedron, and octahedron from Egypt, but the icosahedron and the dodecahedron seem to have been developed in his own school. The Pythagoreans assigned the tetrahedron to fire, the octahedron to air, the icosahedron to water, the cube to earth, and the dodecahedron, apparently the last one discovered, to the universe. They seem to have known that all five polyhedrons can be inscribed in a sphere. They passed the study of these solids on to the school of Plato (*c*. 380 B.C.), where they attracted so much attention as to be known to later writers as "Platonic bodies" or "cosmic figures." It is not probable, however, that the early Pythagoreans actually constructed the figures in the sense that Euclid (*c*. 300 B.C.)[3] and Pappus (*c*. 300)[4] constructed them.

We have specimens extant of icosahedral dice that date from about the Ptolemaic period in Egypt.[5] There are also a number of interesting ancient Celtic bronze models of the regular dodecahedron still extant in various museums. There was probably some mystic or religious significance attached to these forms. Since a stone dodecahedron found in northern Italy dates back to a prehistoric period, it is possible that the Celtic people received their idea from the region south of the Alps,[6] and it is also possible that this form was already known in Italy when the Pythagoreans began their teaching in Crotona.

---

[1] From πολύς (*polys'*, many) + ἕδρα (*hed'ra*, seat).

[2] Heath, *Euclid*, Vol. III, p. 525.

[3] *Elements*, XIII, props. 13–17. See Heath, *Euclid*, Vol. III, pp. 467–503, with notes on the solutions suggested by Pappus.

[4] Pappus, ed. Commandino, p. 45 seq. (Bologna, 1660); ed. Hultsch, III, 142 seq.

[5] See the illustration on page 49.

[6] F. Lindemann, "Zur Geschichte der Polyeder und der Zahlzeichen," *Sitzungsberichte der math.-physik. Classe der K. Bayerischen Akad. der Wissensch. zu München*, Munich, XXVI, 625.

The five regular polyhedrons attracted attention in the Middle Ages chiefly on the part of astrologers. At the close of this period, however, they were carefully studied by various mathematicians. Prominent among the latter was Pietro Franceschi, whose work *De corporibus regularibus* (*c.* 1475) was the first to treat the subject with any degree of thoroughness. Following the custom of the time, Pacioli (1509) made free use of the works of his contemporaries, and as part of his literary plunder he took considerable material from this work and embodied it in his *De diuina proportione*.[1]

Albrecht Dürer, the Nürnberg artist, showed how to construct the figures from a net[2] in the way commonly set forth in modern works. The subject of stellar polyhedrons begins with Kepler[3] (1619) and has attracted considerable attention since his time.

Polyhedron Theorem. Among the most interesting of the modern formulas relating to a polyhedron is the one connecting the faces, vertices, and edges. This formula, often known as Euler's Theorem, may be stated as $f + v = e + 2$. It was possibly known to Archimedes (*c.* 225 B.C.),[4] but not until the 17th century was it put into writing in a form still extant. Descartes (*c.* 1635) was the first to state it.[5] Euler seems to have come upon it independently. He announced it in Petrograd in 1752, but with simply an inductive proof. General proofs have been given by various writers.

Pappus-Guldin Theorem. Pappus of Alexandria (*c.* 300) stated, in substance, the basis of the theorem that the volume of a figure formed by the revolution of a plane figure about an axis is equal to the area of the figure multiplied by the length of the line generated by the center of gravity.[6] He also con-

---

[1] Venice, 1509.

[2] *Underweyssung der messung mit dem zirckel uñ richtscheyt, in Linien, ebnen unnd gantzen corporen*, Nürnberg, 1525.

[3] Frisch edition of his works, V, 126, where he speaks of such a figure as the "stella pentagonica" and carries the discussion "in solido."

[4] Tropfke, *Geschichte*, II (1), 398.

[5] But the fact was not made known until his *Œuvres inédites* appeared in 1860, where it appears on page 218.   [6] Hultsch edition of his works, p. 682.

sidered the case in which there was not a complete revolution. The proposition was soon forgotten and so remained until revived by Kepler (1615),[1] who extended the theory to include the study of the revolution of various plane figures. He gave special attention to the torus formed by the revolution of a circle or an ellipse. When the axis was tangent to the circle or the ellipse, he called the resulting solid an *annulus strictus*. Although he treats of only special cases, he doubtless knew the general theorem.

Various writers made use of the general principle in the 17th century, but it was brought into special prominence by Habakuk Guldin[2] (1577–1643), a Swiss scholar. It appeared in Book II of his *Centrobaryca* (1641). He added nothing to the theory except as he stated the general proposition.

In 1695 Leibniz suggested that the proposition could be extended to include the case of a plane revolving about an axis on any other path than a circle, provided it is always perpendicular to this path,[3] an idea that was considered also by Euler in 1778.

7. The Three Famous Problems

Nature of the Problems. The Greeks very early found themselves confronted by three problems which they could not solve, at least by the use of the unmarked ruler and the compasses alone.[4]

The first was the trisection of any angle. The trisection of the right angle was found to be simple, but the trisection of any arbitrary angle whatever attracted the attention and baffled the efforts of many of their mathematicians. To this problem may be added the related ones of dividing any given angle into any required number of equal parts and of inscribing in a circle a regular polygon of a given number of sides.

[1] In his *Stereometria Doliorum*. See the *Opera Omnia*, ed. Frisch, IV, 551–670.
[2] Known as Paul Guldin after he entered the Catholic Church.
[3] *Acta Eruditorum*, 1695, p. 493.
[4] See any of the histories of Greek mathematics. A good summary is given in H. G. Zeuthen, *Histoire des Mathématiques dans l'Antiquité et le Moyen Age*, French translation by J. Mascart, p. 57 (Paris, 1902).

The second problem was the quadrature of the circle, that is, the finding of a square whose area is the same as that of a given circle. The solution would be simple if we could find a straight line that is equal in length to the circumference of the circle; that is, if we could rectify the circumference. This is easily accomplished by rolling the circle along a straight line, but such a proceeding makes use of an instrument other than the ruler and compasses, namely, of a cylinder with a marked surface.

The third problem was the duplication of a cube,[1] that is, the finding of an edge of the cube whose volume is twice the volume of a given cube. This was known as the Delian Problem, one story of its origin being that the Athenians appealed to the oracle at Delos to know how to stay the plague which visited their city in 430 B.C. It is said that the oracle replied that they must double in size the altar of Apollo. This altar being a cube, the problem was that of its duplication. Since problems about the size and shape of altars appear in the early Hindu literature, it is not improbable that this one may have found its way, perhaps through Pythagoras, from the East. It was already familiar to the Greeks in the 5th century B.C., for we are told by Eratosthenes[2] that Euripides (c. 485–406 B.C.) refers to it in one of his tragedies which is no longer extant.[3]

Trisection Problem: the Conchoid. There are various ways of trisecting any plane angle, but it will suffice at this time to give only a single one. Probably the best known of the Greek attempts is the one made by Nicomedes (c. 180 B.C.). He used a curve known as the conchoid.[4] We take a fixed point O which

---

[1] N. T. Reimer, *Historia problematis de cubi duplicatione* (Göttingen, 1798); C. H. Biering, *Historia problematis cubi duplicandi* (Copenhagen, 1844); Archimedes, *Opera*, ed. Heiberg, III, 102; A. Sturm, *Das Delische Problem* (Linz, 3 parts, 1895, 1896, 1897), a critical historical study with extensive bibliography.

[2] See Archimedes, *Opera*, ed. Heiberg, III, 102.

[3] Too small hast thou designed the royal tomb.
Double it; but preserve the cubic form.

[4] Thus Proclus: "Nicomedes trisected every rectilineal angle by means of the conchoidal lines, the inventor of whose particular nature he is, and the origin, construction, and properties of which he has explained. Others have solved the same problem by means of the quadratrices of Hippias and Nicomedes

is d distant from a fixed line AB, and we draw OX parallel to AB and OY perpendicular to OX. We then take any line OA through O, and on OA produced lay off $AP = AP' = k$, a con-

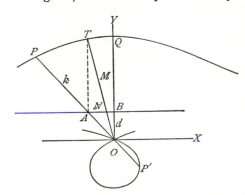

stant. Then the locus of points P and P' is a conchoid. According as $k \gtreqless d$ we have O a node, a cusp, or a conjugate point. The equation of the curve is $(x^2+y^2)(x-d)^2 - k^2x^2 = 0$.

In order to trisect a given angle we proceed as follows:

Let YOA be the angle to be trisected. From A construct AB perpendicular to OY. From O as pole, with AB as a fixed straight line and $2\,AO$ as a

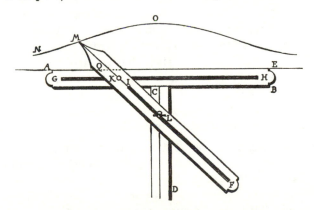

TRAMMEL FOR CONSTRUCTING THE CONCHOID

From Bettini's *Apiaria Universae Philosophiae Mathematicae*, Bologna, 1641

constant distance, describe a conchoid to meet OA produced at P and to cut OY at Q. At A construct a perpendicular to AB

. . .; others, again, starting from the spirals of Archimedes" (Proclus, ed. Friedlein, p. 272 (translation by Allman); see also Gow, *Greek Geom.*, p. 266; Heath, *History*, I, 235, 238)

meeting the curve at T. Draw OT and let it cut AB at N. Let M be the mid-point of NT.

Then $MT = MN = MA$.

But $\qquad NT = 2\,OA$ by construction of the conchoid.

Hence $\qquad MA = OA$.

Hence $\quad \angle AOM = \angle AMO = 2\angle ATM = 2\angle TOQ$.

That is, $\angle AOM = \frac{2}{3}\angle YOA$, and $\angle TOQ = \frac{1}{3}\angle YOA$.

The Quadratrix. A Greek geometer named Hippias, probably Hippias of Elis[1] (*c.* 425 B.C.), invented a curve which he used in the trisection of an angle. In this figure, X is any point on the quadrant AC. As the radius OX revolves at a uniform rate from the position OC to the position OA, the line MN moves at a uniform rate from the position CB to the position OA, always remaining parallel to OA. Then the locus of P, the intersection of OX and MN, is a curve CQ. Manifestly, when OX is one nth of the way from OC around to OA, MN is one nth of the way from CB down to OA. If, therefore, we make $CM = \frac{1}{3}CO$, MN will cut CQ at a point P such that OP will trisect the right angle. In the same way, by trisecting OM we can find a point P' on CQ such that OP' will trisect angle AOX, and so for any other angle. The method evidently applies to the multisection as well as to the trisection of an angle.

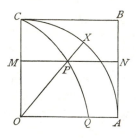

Other Methods of Trisection. The next prominent investigation of the problem is one that is attributed to Archimedes[2] (*c.* 225 B.C.), although probably not due to him in the form that has come down to us. The plan is as follows:

Produce any chord AB of a circle until the part produced, BC, is equal to r, the radius. Join C to the center O and produce CO to the circle at D.

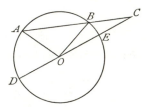

[1] But this has been questioned. For the arguments, see Allman, *Greek Geom.*, p. 93. See also Volume I, page 82, n., and Heath, *History*, I, 225.

See Heath, *Archimedes*; Allman, *Greek Geom.*, p. 90, with references.

Then arc AD is three times arc EB; that is, $\angle EOB$ is $\frac{1}{3}\angle AOD$. This, however, is manifestly no solution of the problem.

Vieta ($c.$ 1590) was led by this to suggest the following:

Let $\angle AOB$ be the angle to be trisected. Describe any circle with center O. Suppose the problem solved and that $\angle AOP=\frac{1}{3}\angle AOB$. Through B suppose BRQ drawn parallel to PO.
Then $\angle RBO=\angle ORB=2\angle AOP=$
$2\angle Q$. But $\angle ORB=\angle Q+\angle ROQ$,
and hence $\angle Q=\angle ROQ$, and so
$OR = QR$. The problem is therefore reduced to the following:

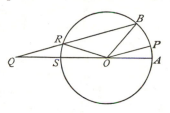

From B draw BRQ so that the part RQ, intercepted between the circle and the diameter AS produced, shall be equal to the radius, a construction involving the use of a marked straightedge.

Regular Polygons. If we can trisect an angle of 360° we can inscribe a regular polygon of three sides in a circle, and similarly for the inscription of other regular polygons. The trisection problem therefore naturally suggests the larger problem of the inscription of a polygon of any given number of sides. It was for a long time believed that the Greeks had exhausted all the possibilities in this line. In 1796, however, Gauss showed that it was possible, by the use of the straightedge and compasses alone, to inscribe a polygon of 17 sides. He even extended the solution to include polygons of 257 and 65,537 sides. The general proposition, as it now appears, is as follows: A regular polygon of p sides, where p is a prime number greater than 2, can be constructed by ruler and compasses if and only if p is of the form $2^{2^t} + 1$. For $t = 0$, 1, 2, 3, and 4 the values of p are respectively 3, 5, 17, 257, and 65,537; but if $t = 5$, $p = 641 \times 6700417$ and hence is not prime.[1] Gauss has left the following interesting record of the discovery:

The day was March 29, 1796, and chance had nothing to do with it. Before this, indeed during the winter of 1796 (my first semester in

[1] For the theory see J. W. A. Young, *Monographs on Modern Mathematics* (New York, 1911), article by L. E. Dickson, p. 378, with references.

Göttingen), I had already discovered everything related to the separation of the roots of the equation

$$\frac{x^p - 1}{x - 1} = 0$$

into two groups. . . . After intensive consideration of the relation of all the roots to one another on arithmetical grounds, I succeeded, during a holiday in Braunschweig, on the morning of the day alluded to (before I had got out of bed), in viewing the relation in the clearest way, so that I could immediately make special application to the 17-side and to the numerical verification. . . . I announced this discovery in the *Literaturzeitung* of Jena, where my advertisement was published in May or June, 1796.[1]

Squaring the Circle. The second famous problem of antiquity was that of squaring the circle. The first attempts were of course empirical. They were made long before the scientific period of the Greek civilization, and they naturally resulted in rude approximations.

The first definite trace that we have of an approximate value of π is in the Ahmes Papyrus[2] (*c.* 1550 B.C.). There is given in that work a problem requiring the finding of the area of a circle, the method, expressed in modern symbols, being as follows:

$$A = (d - \tfrac{1}{9} d)^2,$$

which amounts to saying that $\pi = 3.1605 -$, a result apparently arrived at empirically, as already stated on page 270.

It is probable that 3 is a much older value of π, although we have no extant literature to prove this fact. We find such a value in early Chinese works,[3] in the Bible,[4] in the Talmud,[5]

[1] R. C. Archibald, "Gauss and the Regular Polygon of Seventeen Sides," *Amer. Math. Month.*, XXVII, 323, with bibliography.

[2] Peet, *Rhind Papyrus*, p. 90; Eisenlohr, *Ahmes Papyrus*, p. 117; G. Vacca, "Sulla quadratura del circolo secondo l' Egiziano Ahmes," *Bollettino di bibliografia e storia delle scienze matematiche*, XI, 65.

[3] *E.g.*, in the *Chóu-pei*. See Mikami, *China*, pp. 8, 46, 135. It should be mentioned again that there are doubts as to the reliability of ancient Chinese texts.

[4] "And he made a molten sea, ten cubits from the one brim to the other: it was round all about . . . and a line of thirty cubits did compass it round about" (1 Kings, vii, 23. See also 2 Chronicles, iv, 2).

[5] In both Mishna and Talmud the value is always 3, the reason being traditional, based upon Solomon's "molten sea."

in the early Hindu works,[1] and in the medieval manuscripts, so that it was generally accepted in all countries and until relatively modern times.

The Greeks were not content with results that were merely empirical, however, and so the rectification of the circumference or the related problem of the squaring of the circle attracted the attention of their philosophers. For example, Anaxagoras (c. 440 B.C.) is said by Plutarch[2] to have been put in prison in Athens, and while there to have first attempted the solution. The results of his work are, however, unknown.

Methods of Attacking the Quadrature. There are three methods of attacking the problem: first, by the use of the ruler and compasses only; second, by the use of higher plane curves; third, by such devices as infinite series, leading to close approximations. The leading Greek mathematicians seem to have found the futility of the first method, although they did not prove that it is impossible; with the second method they were successful; with the third method they were less skillful.[3]

Method of Exhaustion. Antiphon (c. 430 B.C.) attempted the quadrature by inscribing a polygon (some early writers say a square and others a triangle), and then doubling the number of sides successively until he approximately exhausted the area between the polygon and the circle. By finding the area of each polygon he was thus able to approximate the area of the circle.[4]

Attempts of Hippocrates. Hippocrates of Chios (c. 460 B.C.) attempted the solution and was the first to actually square a curvilinear figure. He constructed semicircles on the three

---

[1] "The diameter and the square of the semidiameter, being severally multiplied by three, are the practical circumference and area. The square-roots extracted from ten times the squares of the same are the neat values." Colebrooke, *Brahmagupta*, p. 308. See also Mahāvīra, p. 189; Colebrooke, *Bhāskara*, p. 87.

[2] *De exilio*, cap. 17, ed. Dübner-Didot of the *Moralia*, I, 734 (Paris, 1885). See also the Leipzig edition of 1891, III, 573.   [3] Heath, *History*, I, 220.

[4] F. Rudio, "Der Bericht des Simplicius über die Quadraturen des Antiphon und des Hippokrates," in *Bibl. Math.*, III (3), 7, and also in book form, with Greek text (Leipzig, 1907); Allman, *Greek Geom.*, pp. 64, 81. With respect to Bryson see *ibid.*, pp. 77, 82; but compare also Volume I, page 84.

sides of an isosceles right-angled triangle and showed that the sum of the two lunes thus formed is equal to the area of the

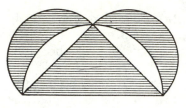

triangle itself. Having a triangle equal in area to a lune, he had only to construct a square equal to the triangle. His proof involves the proposition that the areas of circles are proportional to the squares of their diameters, —a proposition which Eudemus (*c.* 335 B.C.) tells us that Hippocrates proved.[1] To the quadrature problem as such, however, his contribution was not important. His method of attack was substantially as follows:

In a semicircle *ABCD*, center *O*, he inscribed half of a regular hexagon, *h*. On the three sides and on *OB* he described semicircles as here shown. Then the four small semicircles are together equal to the large semicircle. Subtracting the common shaded parts, the three lunes together with the semicircle on *OB* are equal to *h*, the half of the regular hexagon. Now take from *h* a surface equal to the sum of the lunes, which can be found by the method already given (and here is the fallacy), and there remains a rectilinear figure equal to the semicircle on *OB*. It will be observed that Hippocrates as-

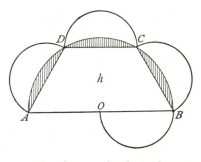

sumed that every lune can be squared, whereas he has shown, as we have seen, that this is possible only in the special case of a right triangle.[2]

---

[1] *Eudemi fragmenta*, ed. Spengel, p. 128. The proposition is equally true for any right-angled triangle, but Hippocrates proved it only for the isosceles case. See Allman, *Greek Geom.*, p. 66; W. Lietzmann, *Der Pythagoreische Lehrsatz*, p. 32 (Leipzig, 1912); E. W. Hobson, *Squaring the Circle*, p. 16 (Cambridge, 1913); Heath, *History*, I, 183, with a summary of recent literature on the subject.

[2] For details relating to the work of Hippocrates on lunes in general, see Heath, *History*, I, 183–201; Allman, *Greek Geom.*, p. 69.

The Quadratrix. The next noteworthy attempt was made by Deinostratus (*c.* 350 B.C.). Pappus[1] makes this statement:

> For the quadrature of the circle a certain curve was employed by Deinostratus, Nicomedes, and some other more recent geometers, which has received its name from the property that belongs to it; for it is called by them the quadratrix.[2]

This is the curve used by Hippias in the trisection of an angle. In the figure given below it can be shown that

$$\frac{CXA}{CO} = \frac{CO}{OQ};$$

and since these terms are all straight lines except the quadrant *CXA*, it is possible to construct a straight line equal in length to the quadrant, and hence to rectify the circumference.

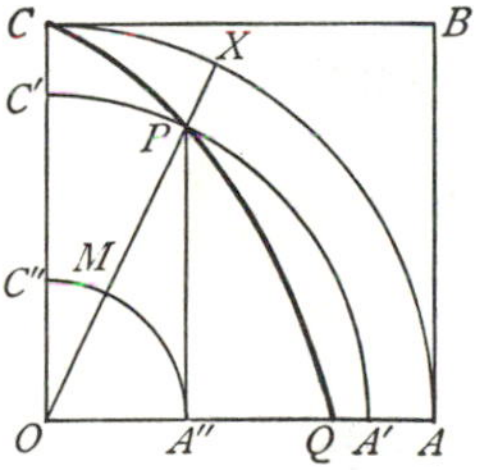

To prove the proposition Pappus states that the *reductio ad absurdum* was employed. If

$$\frac{CXA}{CO} \neq \frac{CO}{OQ},$$

then the proportion can be made true by increasing or by decreasing OQ; but it will be shown that this cannot be done without leading to an absurdity.

First, suppose that $\quad \dfrac{CXA}{CO} = \dfrac{CO}{OA'}.$ $\qquad\qquad OA' > OQ$

Then, since $\qquad \dfrac{CXA}{C'PA'} = \dfrac{CO}{C'O} = \dfrac{CO}{OA'},$

it follows that $\qquad C'PA' = CO.$

And since, from the property of the curve,

$$\frac{CXA}{XA} = \frac{CO}{PA''},$$

we have $\qquad \dfrac{CXA}{XA} = \dfrac{C'PA'}{PA'} = \dfrac{CO}{PA'};$

[1] Pappus, *Collectiones*, IV, cap. xxx (ed. Hultsch, I, 253); Hankel, *Geschichte*, p. 151; Cantor, *Geschichte*, I (2), 233; Heath, *History*, I, 226.
[2] Τετραγωνίζουσα (*tetragoni'zousa*).

whence arc $PA' = PA''$, which is impossible, since an arc cannot be equal to its chord, and since their halves cannot be equal.

Next, suppose that $\dfrac{CXA}{CO} = \dfrac{CO}{OA''}.$ $\qquad OA'' < OQ$

Draw the quadrant $C''MA''$.

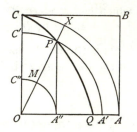

Then $\qquad \dfrac{C''MA''}{MA''} = \dfrac{CXA}{XA} = \dfrac{CO}{PA''}.$

Therefore $\quad \dfrac{CXA}{C''MA''} = \dfrac{CO}{C''O} = \dfrac{CO}{OA''}.$

But $\qquad \dfrac{CXA}{CO} = \dfrac{CO}{OA''}.$ By hyp.

Hence CO must be equal to $C''MA''$ if the hypothesis is correct.

But $\qquad \dfrac{C''MA''}{MA''} = \dfrac{CXA}{XA},$

and because CPQ is a quadratrix we have

$$\dfrac{CXA}{XA} = \dfrac{CO}{PA''};$$

whence $\qquad \dfrac{C''MA''}{MA''} = \dfrac{CO}{PA''}.$

But in the same way as it was shown in the first part of the proof that

$$C'PA' = CO$$

on the hypothesis there made, so it may be shown here that

$$C''MA'' = CO$$

on this hypothesis. It then follows from this proportion that

$$MA'' = PA'',$$

which leads to the absurdity that the circumference of a circle is equal to the perimeter of a circumscribed polygon.

Hence the second hypothesis is also untenable.

Hence $\qquad \dfrac{CXA}{CO} = \dfrac{CO}{OQ},$

and CXA, a quadrant, can be constructed.

Method of Archimedes. The next noteworthy contribution was that of Archimedes (*c.* 225 B.C.), who asserted that:

1. The area of a circle is equal to the area of a right-angled triangle one of whose sides forming the right angle is equal to the circumference of the circle and the other to the radius.[1]

2. The ratio of the area of a circle to the square on the diameter is approximately 11 : 14.

3. The ratio of the circumference of a circle to the diameter is less than $3\frac{1}{7}$ and greater than $3\frac{10}{71}$.

To prove the third proposition Archimedes inscribed and circumscribed regular polygons, found their areas up to polygons of 96 sides, and showed that the area of the circle lies between these results. These limits, expressed in modern decimal form, are 3.14285714 ⋯ and 3.14084507 ⋯. If our present notation and our methods of finding a square root had been known, the result would have been closer, since the geometric method permitted of any desired degree of approximation.

The Romans were little concerned with accurate results in such matters as this, and so it is not surprising that Vitruvius (*c.* 20 B.C.) speaks of the circumference of a wheel of diameter 4 feet as being $12\frac{1}{2}$ feet, thus taking π as $3\frac{1}{8}$.[2]

Other Greek Approximations of π. After the time of Archimedes the value $3\frac{1}{7}$ became recognized as a satisfactory approximation and appeared in the works of Heron (*c.* 50?), Dominicus Parisiensis (1378), Albert of Saxony (*c.* 1365), Nicholas Cusa (*c.* 1450), and others. Since one of the common approximations for a square root in the Middle Ages was

$$\sqrt{n} = \sqrt{a^2 + r} = a + \frac{r}{2a+1},$$

and since this gives $\sqrt{10} = 3 + \dfrac{1}{2 \times 3 + 1} = 3\frac{1}{7}$, it is natural to expect that $\sqrt{10}$, which is 3.1623 ⋯, would often have been given as the value of π, and this was in fact the case.

---

[1] Heath, *Archimedes*, p. 231–233; ed. Heiberg, I, 258; Heath, *History*, II, 50–56.

[2] *De Architectura*, X, cap. 14. Rose's edition (1889) gives the diameter as $4\frac{1}{6}$ feet, which would make π equal to 3. See *Bibl. Math.*, I (3), 298.

Ptolemy (c. 150) seems to have taken the Archimedean limits and to have expressed them in sexagesimals, obtaining substantially $3\frac{1}{7} = 3\ 8'\ 34.28''$ and $3\frac{10}{71} = 3\ 8'\ 27.04''$. He then improved upon the mean between these results by taking $3\ 8'\ 30''$ as the approximate value of π, although a still closer approximation is $3\ 8'\ 29.73355''$. Since $3\ 8'\ 30'' = 3.1416$, his result was very satisfactory.[1]

Hindu Values of π. The Hindu mathematicians took various values of π, and no writer among them seems to have been uniform in his usage.

Āryabhaṭa (c. 510), or possibly Āryabhaṭa the Younger, gave the equivalent of 3.1416, his rule being:

Add 4 to 100, multiply by 8, add 62,000, and you have for a diameter of two *ayutâs* the approximate value of the circumference.[2]

Brahmagupta (c. 628) criticized Āryabhaṭa for taking the circumference as 3393 for both diameters 1080 and 1050, which would make π either $3\frac{17}{120}$ or $3\frac{81}{350}$; that is, 3.1416 or 3.2314.

A certain astronomer, Puliśa,[3] to whom Brahmagupta refers, gives $3\frac{177}{1250}$, which is $3.18 +$, and Ya'qûb ibn Ṭâriq (c. 775) mentions certain Hindu astronomical measurements which give the same value. He also states that Puliśa used a value equivalent to 3.14183, and Brahmagupta a value equivalent to 3.162. For himself Ya'qûb ibn Ṭâriq used in one case a value equivalent to 3.1411.[4]

In case the value 3.1416 is due to either of the Āryabhaṭas, it may have been obtained from the Alexandrian scholars,

[1] Heath, *History*, I, 233.

[2] L. Rodet, "Leçons de Calcul d'Āryabhaṭa," *Journal Asiatique* as cited, reprint, p. 11. There is some doubt as to whether this rule is due to either of the Āryabhaṭas; see G. R. Kaye, "Notes on Indian Mathematics, No. 2, Āryabhaṭa," in *Journ. and Proc. of the Asiatic Soc. of Bengal*, IV, reprint. The word *ayutâs* means myriads, that is, 10,000's. The rule is translated more simply in Volume I, page 156.

[3] The name appears as Paulisa, Pulisa, and Pauliśa. For a discussion, see Sachau's translation of Albêrûnî's *India*, II, 304 (London, 1910); hereafter referred to as Albêrûnî's *India*. On Brahmagupta's criticism of Āryabhaṭa, see I, 168. Nothing is known concerning the life of Puliśa.

[4] Albêrûnî's *India*, II, 67.

by whom it was then known and whose works may well have reached India, or it may have been found independently.

Brahmagupta[1] (c. 628) used 3 as the "practical" value and $\sqrt{10}$ as the "exact" value, and these values are also given by Mahāvīra[2] (c. 850) and Srīdhara[3] (c. 1020).

Bhāskara (c. 1150) used $\frac{3927}{1250}$ for the "near" value and $\frac{22}{7}$ in finding the "gross circumference adapted to practice,"[4] the former being the same as the value $3\frac{177}{1250}$ of Puliśa.[5]

Chinese Values of π. The Chinese found various values of π, but the methods employed by the early calculators are unknown. The value 3 was used probably as early as the 12th century B.C.[6] and is given in the *Chóu-peï* and the *Nine Sections*.[7] Ch'ang Höng (c. 125) used $\sqrt{10}$, and Wang Fan (c. 265) used $\frac{142}{45}$, which is equivalent to 3.1555 ···. Liu Hui (263) gives us the first intimation of the method used by the Chinese in finding the value. He begins with a regular inscribed hexagon, doubles the number of sides repeatedly, and asserts that "if we proceed until we can no more continue the process of doubling, the perimeter ultimately comes to coincide with the circle."[8]

Among other early Chinese values of no high degree of accuracy are those of Men (c. 575), who gave 3.14, and Wu (c. 450), whose value was 3.1432 +.

Tsu Ch'ung-chih (c. 470) was able, by starting with a circle of diameter 10 feet, to obtain 3.1415927 and 3.145926 for the limits of π, and from these, by interpolation, he obtained the "accurate and inaccurate" values $\frac{355}{113}$ and $\frac{22}{7}$. No closer approximations were made in China until modern times.[9]

---

[1] Colebrooke's translation, p. 308.   [2] Mahāvīra, pp. 189, 200.
[3] Colebrooke's translation, p. 87.
[4] Colebrooke's translation, p. 87.
[5] On the general subject of the Hindu quadratures see C. M. Whish, *On the Hindù Quadrature of the Circle*, a paper read before the Madras Literary Society, December 15, 1832.   [6] Mikami, *China*, p. 46.
[7] For discussion of the dates of these works, see Volume I, page 31.
[8] For his computations, see Mikami, *China*, p. 48.
[9] On the work of the later writers, after the introduction of European mathematics, see Mikami, *China*, p. 135. In none of these early approximations was the decimal fraction used.

The Japanese did no noteworthy work in this field until the 17th century. They then developed a kind of native calculus and also made use of European methods which gave them fair approximations to the required ratio.[1]

Later Approximations of π. The following is a brief summary of some of the later European approximations of π, with the names of those who used them:

Franco of Liège[2] (c. 1066), $\pi = \frac{22}{7} = 3.142857+$.

Fibonacci[3] (1220), $\pi = \frac{864}{274} = 3.141818$. He also gave the limits as 3.1427 and 3.1410.

Al-Kashî[4] (c. 1430), 3.14159265358987₃₂.

Al-Kashî[4] (c. 1430), 3.1415926535898732.

Tycho Brahe[5] (c. 1580), $\pi = \dfrac{88}{\sqrt{785}} = 3.1409$.

Simon Duchesne[6] (c. 1583), $\pi = 3\frac{69}{484} = (\frac{39}{22})^2 = 3.14256198$.

Vieta (c. 1593), $3.1415926535 < \pi < 3.1415926537$.

Adriaen van Roomen (1561–1615) gave π to 17 decimal places.

Ludolf van Ceulen (1540–1610) gave π to 35 decimal places, and German textbooks still speak of π as the "Ludolphische Zahl."

Adriaen Anthoniszoon (c. 1600) and his son Adriaen Metius (1571–1635), $\pi = \frac{355}{113}$, the Chinese value.[7]

J. H. Lambert[8] (c. 1770),

$$\pi = (\tfrac{7}{4})^2, \ (\tfrac{16}{9})^2, \ (\tfrac{62}{35})^2, \ (\tfrac{39}{22})^2, \ (\tfrac{218}{123})^2, \ (\tfrac{296}{167})^2, \ \cdots .$$

---

[1] Smith-Mikami, pp. 60, 63, et passim.

[2] Abhandlungen, IV, 139. The decimal equivalents are modern in all these cases.

[3] "Practica geometriae," in his Scritti, II, 90. See also H. Weissenborn, "Die Berechnung des Kreisumfanges bei Archimedes und Leonardo Pisano," Berliner Studien für klassische Philologie und Archäologie, XIV.

[4] See Volume I, pages 289, 290; Volume II, pages 238, 240.

[5] Original name, Tyge Ottesen. It is not known how he came to give this curious value. See F. J. Studnička, Bericht über die Astrologischen Studien des . . . Tycho Brahe, p. 49 (Prag, 1901).   [6] Cantor, Geschichte, II (2), 592.

[7] They took the approximation $3\frac{15}{106} < \pi < 3\frac{17}{120}$, added the numerators $(15 + 17 = 32)$ and the denominators $(106 + 120 = 226)$, took the means (16 and 113), and gave $\pi = 3\frac{16}{113} = \frac{355}{113} = 3.1415929$, a very close approximation for the time. Priority for this is claimed for Valentinus Otto (c. 1550–1605).

[8] Vorläufige Kenntnisse für die, so die Quadratur und Rektifikation des Circuls suchen, II, 140 (Berlin, 1772).

The value of π was carried to 140 decimal places (136 correct) by Georg Vega (1756–1802), to 200 by Zacharias Dase (1824–1861), to 500 by Richter (died in 1854), and to 707 by William Shanks (*c.* 1853).

Continued Products and Series.   Vieta (*c.* 1593) gave another interesting approximation for π, using continued products for the purpose.   His value may be obtained from the following equation:

$$\frac{2}{\pi} = \sqrt{\tfrac{1}{2}} \cdot \sqrt{\tfrac{1}{2} + \tfrac{1}{2}\sqrt{\tfrac{1}{2}}} \cdot \sqrt{\tfrac{1}{2} + \tfrac{1}{2}\sqrt{\tfrac{1}{2} + \tfrac{1}{2}\sqrt{\tfrac{1}{2}}}} \cdots.$$

John Wallis (1655)[1] gave the form

$$\frac{4}{\pi} = \frac{3 \cdot 3 \cdot 5 \cdot 5 \cdot 7 \cdot 7 \cdot 9 \cdot 9 \cdot 11 \cdot 11 \cdots}{2 \cdot 4 \cdot 4 \cdot 6 \cdot 6 \cdot 8 \cdot 8 \cdot 10 \cdot 10 \cdot 12 \cdots}.$$

This is related to Lord Brouncker's value (*c.* 1658) in which use is made of continued fractions, as follows:

$$\frac{4}{\pi} = 1 + \cfrac{1}{2 + \cfrac{9}{2 + \cfrac{25}{2 + \cfrac{49}{2 + \cdots}}}}.^{2}$$

Leibniz[3] (1673),

$$\frac{\pi}{4} = 1 - \frac{1}{3} + \frac{1}{5} - \frac{1}{7} + \frac{1}{9} - \cdots.$$

Abraham Sharp (*c.* 1717),

$$\frac{\pi}{6} = \sqrt{\tfrac{1}{3}} \cdot \left(1 - \frac{1}{3 \cdot 3} + \frac{1}{3^2 \cdot 5} - \frac{1}{3^3 \cdot 7} + \frac{1}{3^4 \cdot 9} - \cdots \right),$$

from which he found the value of π to 72 decimal places.

---

[1] *Arithmetica Infinitorum* (1655), included in his *Opera*, I, 469.

[2] See L. Euler, *Opuscula analytica*, Vol. I (Petrograd, 1783); also (1785), II, 149.

[3] A special case of Gregory's (1671) series. De Lagny (1682) discovered it independently,

John Machin[1] (*c.* 1706),

$$\frac{\pi}{4} = 4 \cdot \left(\frac{1}{5} - \frac{1}{3 \cdot 5^3} + \frac{1}{5 \cdot 5^5} - \frac{1}{7 \cdot 5^7} + \cdots\right)$$
$$- \left(\frac{1}{239} - \frac{1}{3 \cdot 239^3} + \frac{1}{5 \cdot 239^5} - \frac{1}{7 \cdot 239^7} + \cdots\right).$$

Matsunaga Ryōhitsu[2] (1739), a Japanese writer,

$$\frac{\pi}{3} = 1 + \frac{1^2}{4 \cdot 6} + \frac{1^2 \cdot 3^2}{4 \cdot 6 \cdot 8 \cdot 10} + \frac{1^2 \cdot 3^2 \cdot 5^2}{4 \cdot 6 \cdot 8 \cdot 10 \cdot 12 \cdot 14} + \cdots.$$

The Symbol π. The symbol $\dfrac{\delta}{\pi}$ was used by Oughtred (1647) to represent the ratio of the diameter to the circumference.[3] Isaac Barrow (from *c.* 1664) used the same symbolism, and David Gregory (1697) used $\dfrac{\pi}{\rho}$ for the ratio of circumference to radius.

The first to use π definitely to stand for the ratio of c to d was an English writer, William Jones. In his *Synopsis Palmariorum Matheseos* (1706) he speaks (p. 243) of "Periphery (π)"; but on p. 263 he is more definite, giving

$$\text{" } 3.14159, \&c. = \pi,\text{''}$$
$$\text{" } d = c \div \pi = \overline{a \div \tfrac{1}{4}\pi}\big|^{\frac{1}{2}},\text{''}$$
$$\text{" } c = d \times \pi = \overline{a \times 4\pi}\big|^{\frac{1}{2}},\text{''}$$

and

$$\text{" } a = \tfrac{1}{4}\pi \times d^2 = c^2 \div 4\pi.\text{''}$$

Euler adopted the symbol in 1737, and since that time it has been in general use.

---

[1] W. Jones, *Synopsis Palmariorum Matheseos*, p. 243 (London, 1706), gives π "True to above a 100 Places; as Computed by the Accurate and Ready Pen of the Truly Ingenious Mr. John Machin."

[2] For a further list of values of π consult D. E. Smith, "The History and Transcendence of π," on page 396 of J. W. A. Young, *Monographs on . . . Modern Mathematics* (New York, 1911); Tropfke, *Geschichte*, IV (2), 195; F. Rudio, *Archimedes, Huygens, Lambert, Legendre. Vier Abhandlungen über die Kreismessung*, Leipzig, 1892.

[3] "Si in circulo sit 7.22 :: $d \cdot \pi$ :: 113.355: erit $\delta \cdot \pi$ :: 2R.P: periph." This symbolism appears first in the 1647 edition of the *Clavis Mathematicae* (1631). This quotation is from the 1652 edition. See Cajori, *Oughtred*, p. 32.

The proof of the transcendence of π was first given by F. Lindemann (1882), thus showing the impossibility of squaring the circle by the use of ruler and compasses alone.[1]

Duplication of the Cube.[2] Hippocrates of Chios (*c.* 460 B.C.) showed that the problem of duplicating the cube resolves itself into the finding of two mean proportionals between two given lines. If $a : x = x : y = y : b$, then $x^2 = ay$, $y^2 = bx$, and hence $x^4 = a^2y^2 = a^2bx$, or $x^3 = a^2b$. If $b = 2a$, then $x^3 = 2a^3$. That is, the cube of edge x will then have double the volume of a given cube with edge a. Since we have the three equations $x^2 = ay$ (parabola), $y^2 = bx$ (parabola), and $ab = xy$ (hyperbola), we can evidently solve the problem by finding the intersection of two parabolas or of a parabola and a hyperbola. These methods are credited to Menæchmus (*c.* 350 B.C.).[3]

Archytas (*c.* 400 B.C.) had already found the two mean proportionals, solving the problem by means of three surfaces of revolution. Eratosthenes (*c.* 230 B.C.) tells us that he "is said to have discovered them by means of his semicylinders."[4] It is possible that Archytas led Menæchmus to discover a solution by means of conics.

Eratosthenes also tells us that Eudoxus (*c.* 370 B.C.) solved the problem "by means of the so-called curved lines," but what these lines were we do not know.[5] The two statements here attributed to Eratosthenes are contained in a letter formerly (but incorrectly) credited to him. In the main, however, this letter sets forth facts with which he was familiar, as is shown from other sources.

[1] D. E. Smith, "The History and Transcendence of π," *loc. cit.*, p. 387; consult this work also with respect to transcendental numbers in general.

[2] J. S. Mackay, *Proc. of the Edinburgh Math. Soc.*, IV, 2; F. G. Teixeira, *Obras sobre Mathematica*, VII, 283–415 (Coimbra, 1915); C. H. Biering, *Historia Problematis Cubi Duplicandi*, Copenhagen, 1844; Heath, *History*, I, 244; F. Enriques, *Fragen der Elementar-Geometrie*, II. Teil (Leipzig, 1907).

[3] On this point see Heath, *History*, I, 251–255; Allman, *Greek Geom.*, p. 160.

[4] In his letter to Ptolemy III. See Archimedes, *Opera*, ed. Heiberg, III, 104, 106. On the solution, as Eudemus relates it, see Allman, *Greek Geom.*, p. 111. See also P. Tannery, "Sur les solutions du problème de Délos par Archytas et par Eudoxe," in his *Mémoires Scientifiques*, I, 53 (Paris, 1912); Heath, *History*, I, 246. [5] Archimedes, *Opera*, ed. Heiberg, III, 66.

Plato (*c.* 380 B.C.) is said to have solved the problem by means of a mechanical instrument[1] but to have rejected this method as not being geometric.

We are told by Joannes Philop'onus[2] that Apollonius (*c.* 225 B.C.) had a method of finding the two mean proportionals. The construction, however, assumes a postulate which begs the whole question.[3]

Cissoid of Diocles. One of the best-known of the ancient solutions was that of Diocles (*c.* 180 B.C.), who used a curve known as the cissoid (from κισσοειδής (*kissoeides'*), ivylike). In this figure, if $OL = OT$ and AQ is drawn, then P is a point

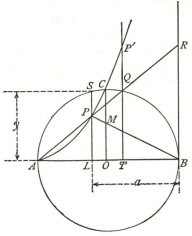

on the curve. Similarly, AS will determine on TQ produced a point P' on the curve. The cissoid evidently passes through A, BR is an asymptote, and the curve is symmetric with respect to AB.

By the aid of the curve two mean proportionals can be found in the following manner:[4]

Let $OM = \frac{1}{2} r$, determine P by producing BM to the curve, and draw AP and produce it to R, letting it cut the circle at Q.

Through P and Q respectively draw SL and QT perpendicular to AB.

Let $BL = a$, $SL = y$, $AL = x$.

Then $\dfrac{a}{PL} = \dfrac{BO}{OM} = \dfrac{r}{\frac{1}{2}r} = 2$; whence $\frac{1}{2}a = PL$.

---

[1] For figure and description, based on a statement of Eutocius (*c.* 560), see Tropfke, *Geschichte*, II (1), 42; Allman, *Greek Geom.*, p. 173; Archimedes, *Opera*, ed. Heiberg, III, 66; Heath, *History*, I, 255.

[2] Ἰωάννης ὁ Φιλόπονος (*Ioan'nes ho Philop'onos*), also known as ὁ Γραμματικός (*ho Grammatikos'*), an Alexandrian scholar of the 7th century; not a very reliable source.   [3] The proof appears in Heath, *Apollonius*, p. cxxv.

[4] For a somewhat different proof see Heath, *History*, I, 264.

Because $\triangle ALP$ is similar to $\triangle ATQ$, which is congruent to $\triangle BLS$, which is similar to $\triangle ALS$, we have

$$\frac{a}{y} = \frac{y}{x} = \frac{x}{PL} = \frac{x}{\frac{1}{2}a}.$$

We therefore have two mean proportionals between a and $\frac{1}{2}a$.

Hence $\qquad ax = y^2, \; a^2 x^2 = y^4,$

and $\qquad x^2 = \dfrac{y^4}{a^2}.$

Also, $\qquad \frac{1}{2}ay = x^2 = \dfrac{y^4}{a^2};$

whence $\qquad \frac{1}{2}a^3 = y^3$

and $\qquad a^3 = 2y^3.$

Therefore in the above figure we simply have to make $PL = \frac{1}{2}a$, which we get by having made $OM = \frac{1}{2}r$. Then SL is the side of a cube equal to $\frac{1}{2}a^3$, or $a^3 = 2y^3$. Hence

$$\frac{SL}{\text{side of given cube}} = \frac{a}{\text{side of required cube}}.$$

Later Methods. Several modern writers have suggested methods for duplicating the cube. Among these are Vieta,[1] Descartes,[2] Fermat,[3] de Sluze,[4] and Newton.[5] Descartes considered not only the question of finding two mean proportionals, as required in solving the problem, but also that of finding four; and Fermat[6] went so far as to consider certain cases involving n mean proportionals, a line of work which was later followed by Clairaut.

Viviani[7] solved the problem by the aid of a hyperbola of the second order ($xy^2 = a^3$). Huygens (1654) gave three methods

---

[1] *Opera Mathematica*, ed. Van Schooten, p. 242. Leyden, 1646.

[2] Suggested in *La Géométrie*, Book III.

[3] In his memoir *Ad locos planos et solidos Isagoge*, written before Descartes published his work, but not made known until after Fermat's death.

[4] In his *Mesolabium*, 1668.

[5] *Arithmetica Universalis*, 1707, p. 309.   [6] *Œuvres*, I, 118.

[7] *Quinto libro di Euclide o Scienze universale delle proposizioni spiegate colla dottrina del Galileo* (Florence, 1647).

of solving. Newton (1707) suggested several methods but preferred one which made use of the limaçon of Pascal. One of the comparatively recent methods is that employed by Montucci,[1] who made use of the curve defined by the equation

$$y = \sqrt{ax} + \sqrt{ax - x^2}.$$

8. ANALYTIC GEOMETRY

Three Principal Steps. There are three principal steps in the development of analytic geometry: (1) the invention of a system of coordinates; (2) the recognition of a one-to-one correspondence between algebra and geometry; and (3) the graphic representation of the expression $y = f(x)$. Of these, the first is ancient, the second is medieval, and the third is modern.

Ancient Idea of Coordinates. The idea of coördinates in the laying out of towns and lands seems unquestionably to have occurred to the Egyptian surveyors. It is to them that Heron was apparently indebted for his fundamental principles, and from them the Roman surveyors acquired their first knowledge of the science. Indeed, the districts (*hesp*) into which Egypt was divided[2] were designated in hieroglyphics by a symbol derived from a grid, as here shown, quite as we designate a survey today.[3]

Latitude and Longitude. The first definite literary references to the subject appear, however, in the works of the early Greek geographers and astronomers. Hipparchus (*c.* 140 B.C.) located points in the heavens and on the earth's surface by means of their longitude[4] and latitude,[5] the former being reck-

---

[1] *Résolution de l'équation du cinquième degré*, Paris, 1869.

[2] Known to the Greeks as νομοί (*nomoi'*, nomes) and to Pliny as *praefecturae oppidorum*.

[3] E. W. Budge, *The Mummy*, p. 8 (Cambridge, 1893). See also Cantor, *Geschichte*, I (2), 67.

[4] Μῆκος (*me'kos*, length; Latin, *longitudo*), *i.e.*, distance from east to west; so called because the length of the known world was along the Mediterranean Sea.

[5] Πλάτος (*pla'tos*, width; Latin, *latitudo*), *i.e.*, distance from north to south; so called because the width of the known world was north and south.

oned from the meridian of Rhodes, where Hipparchus took his observations. He also located the stars by means of coordinates. In the second century Marinus of Tyre (*c.* 150) took his prime meridian through the *Fortunatae Insulae*,[1] and perhaps through the most western point, this being the end of the earth as then known, and Ptolemy (*c.* 150) used the same line.[2]

Ancient Surveyors. The ancient surveyors located points in much the same way as the geographers. Heron (*c.* 50?), apparently following the Egyptian surveyors, laid out a field with respect to one axis quite as we do at present, although, strictly speaking, two coordinates are used.[3]

The Romans brought the science of surveying to the highest point attained in ancient times.[4] They laid out their towns with respect to two axes, the *decimanus*, which was usually from east to west, and the *cardo*, an axis perpendicular to the *decimanus*. They then arranged the streets on a rectangular coordinate system, much as in most American cities laid out in the 19th century.[5]

Rectangular Axes in Greek Geometry. In their treatment of geometric figures the Greeks made use of what were substantially two rectangular axes. Menæchmus (*c.* 350 B.C.), for example, may have used that property of the parabola expressed by the equation $y^2 = px$, and also that property of the rectangular hyperbola expressed by the equation $xy = c^2$. Archimedes (*c.* 225 B.C.), who no doubt was indebted to the lost work of Euclid on conics in general, used the same relation for the parabola, his results being expressed as usual in the form of a proportion.[6]

---

[1] Probably, as stated in Volume I, including the Canary, Madeira, and Azores groups.

[2] Halma's edition of Ptolemy, VI, 17 (Paris, 1828).

[3] See his *Opera quae supersunt omnia*, V, 5 (Leipzig, 1899–1914), on his stereometry and mensuration.

[4] M. Cantor, *Die römischen Agrimensoren*, Leipzig, 1875.

[5] As Frontinus states it in his Liber I, "ager . . . decimanis et cardinibus continetur"; "Ager per strigas [rows] et per scamna [steps] diuisus." He also used oblique coordinates. See the Lachmann and Rudorff edition for diagrams.

[6] On all this discussion see Heath, *Apollonius*, p. cxv seq.; *History*, II, 122.

Apollonius carried the method much farther, as may be seen by the following statement:

If straight lines are drawn from a point so as to meet at given angles two straight lines given in position, and if the former lines are in a given ratio, or if the sum of one of them and of such a line as bears a given ratio to the second is given, then the point will be on a given straight line.[1]

This is only a nonsymbolic method of stating that the equation $x = ay - b$ represents a straight line, a and b being positive.

Sir Thomas Heath calls attention to another essential difference between the Apollonian and Cartesian points of view:

The essential difference between the Greek and the modern method is that the Greeks did not direct their efforts to making the fixed lines of the figure as few as possible, but rather to expressing their equations between areas in as short and simple a form as possible. Accordingly they did not hesitate to use a number of auxiliary fixed lines, provided only that by that means the areas corresponding to the various terms in x^2, xy, . . . forming the Cartesian equation could be brought together and combined into a smaller number of terms. . . . In the case, then, where two auxiliary lines are used in addition to the original axes of coordinates, and it appears that the properties of the conic (in the form of equations between areas) can be equally well expressed relatively to the two auxiliary lines and to the two original axes of reference, we have clearly what amounts to a transformation of coordinates.[2]

Ordinate and Abscissa. As to technical terms, the Greeks used an equivalent of "ordinate."[3] For "abscissa"[4] they used such expressions as "the [portion] cut off by it from the diameter towards the vertex." Apollonius uses the word "asymptote,"[5] but the word had a broader meaning than with us, referring to any lines which do not meet, in whatever direction they are

---

[1] On the work of Apollonius see Volume I, page 116; Heath, *Apollonius*.

[2] Heath, *Apollonius*, p. cxviii. See also his *History*, II, 126-196.

[3] That is, τεταγμένως (*tetagmen'os*, ordinate-wise). The same term is used for the tangent at the extremity of a diameter. *Ibid.*, p. clxii.

[4] Latin *ab* + *scissa*, from *ab* (off) + *scindere* (to cut).

[5] Ἀσύμπτωτος (*asym'ptotos*, from ά privative + σύν, together, + πτωτός, falling).

produced. The names "ellipse," "parabola," and "hyperbola" are probably due to Apollonius,[1] although two of them are found in late manuscripts of the works of Archimedes.

Oresme's Contribution. In the Middle Ages Nicole Oresme (c. 1360) wrote two works[2] in which he took a decided step in advance. He considered a series of points which have uniformly changing *longitudines* and *latitudines*, the first being our abscissas and the second our ordinates.[3] The series of points determined by the ends of the *latitudines* was called a *forma*, and the difference between two successive *latitudines* was called a *gradus*. If the *latitudines* are constant, the series of points was described as *uniformis eiusdem gradus*; but if the *latitudines* varied, the *forma* of the series of points was *difformis per oppositum*. The difference between two successive *latitudines* was the *excessus graduum*, and this might or might not be constant. In the former case the *forma* was *uniformiter difformis*; in the latter case, *difformiter difformis*. The *formae* considered were series of points arranged in rectilinear, circular, or parabolic order. Of course only positive *latitudines* were considered. Here, then, we find the first decided step in the development of a coordinate system, apart from the locating of points on a map of some kind; but we also find a lack of any idea of continuity in the point systems. The method was the subject of university lectures at Cologne as early as 1398,[4] as witness the statutes of that period.[5] Kepler and Galileo recognized its value, and the former was influenced to

---

[1] Heath, *Apollonius*, p. clxiii; *History*, II, 138.

[2] *Tractatus de latitudinibus formarum* and *Tractatus de uniformitate et difformitate intensionum*. See the *Zeitschrift* (Hl. Abt.), XIII, 92. The first of these tractates was printed at Padua in 1482 and again in 1486, in Venice in 1505, and in Vienna in 1515.

[3] Tropfke, *Geschichte*, II (1), 409; H. Wieleitner, *Bibl. Math.*, XIV (3), 210

[4] Hankel, *Geschichte*, p. 351.

[5] "Item statuimus quod Bacalarius temptandus debet audivisse libros infrascriptos . . . aliquem tractatum de latitudinibus formarum." See F. J. von Bianco, *Die alte Universität Köln*, I, Anlagen, p. 68 (Cologne, 1885); S. Günther, "Die Anfänge und Entwickelungsstadien des Coordinatenprincipes," *Abhandlungen d. naturf. Gesellsch. zu Nürnberg*, VI (1877); reprint, p. 16; hereafter referred to as Günther, *Die Anfänge*.

make much use of it in his astronomical work.[1] Indeed, the use of a kind of coordinate paper for the graphic representation of the course of the planets is found much earlier even than Oresme, for Günther[2] has called attention to a manuscript of the 10th century[3] in which the graphs closely resemble similar forms of the present day.

Relation of Algebra to Geometry. The second step in the development of analytic geometry has to do with the relation of algebra to geometry. If we consider such a proposition of Euclid (*c.* 300 B.C.) as the one relating to the square on the sum of two lines,[4] we see that it is the analogue of the algebraic identity $(a + b)^2 = a^2 + 2ab + b^2$. Euclid, however, had no algebraic symbolism, and while of course he recognized the analogy to the square of the sum of two numbers, it cannot be said that he related algebra to geometry in the way that we do with our modern symbols.

When Archimedes (*c.* 225 B.C.), Heron (*c.* 50?), and Theon of Alexandria (*c.* 390) found square roots, they used this proposition of Euclid; but, again, they can hardly be said to have grasped the relation that is so familiar to us today.

It is among the Arab and Persian writers that we first find geometric figures used in works devoted solely to algebra. Thus al-Khowârizmî (*c.* 825) considered numerous cases such as the

following: "A square and ten Roots are equal to thirty-nine Dirhems."[5] Here he uses the annexed figure, the square AB having for its side one of the roots of the given equation

$$x^2 + 10x = 39,$$

a favorite equation with subsequent writers.

Omar Khayyam (*c.* 1100) made continued use of geometric figures in his work on algebra,[6] thus recognizing the one-to-one

[1] *Opera omnia*, Frisch ed., IV, 610 seq. (Frankfort a. M., 1863).

[2] *Die Anfänge*, p. 19 and Fig. 2 in the plates.

[3] Munich Cod. Lat. 14,436: "Macrobius Boetius in Isagog. Saec. X."

[4] *Elements*, II, 4.   [5] Rosen's translation, p. 13.

[6] *L'Algèbre d'Omar Alkhayyâmî*, translated by F. Woepcke, Paris, 1851. So Woepcke remarks: "Il est une particularité de cette algèbre qui mérite d'être

correspondence between algebra and geometry even more than al-Khowârizmî had done before him.[1] The Hindu algebraists also used geometric figures in their work. For example, Bhāskara (c. 1150)[2] has such problems as this: "Tell two numbers, such, that the sum of them, multiplied by four and three, may, added to two, be equal to the product." In such cases he gives two solutions, one algebraic and the other geometric.

Europeans relate Algebra to Geometry. Among the Europeans, Fibonacci was the first mathematician of prominence who recognized the value of relating algebra to geometry. In his *Practica geometriae* (1220) he uses algebra in solving geometric problems relating to the area of a triangle.[3]

In the early printed books there was more or less use of geometric figures in connection with algebraic work. Thus Vieta tells us that Regiomontanus solved algebraically problems which he could not solve by geometry. Pacioli (1494) continually uses geometric figures in his solution of quadratics,[4] and Cardan (1545) does the same.[5] After the publication of Cardan's work the recognition of the relationship became common. Vieta, for example, generalized the idea of the ancients as to representing points on a line, although adhering to the use of proportion in most of his geometric work instead of using the equation form.

The first textbook on algebraic geometry was that of Marino Ghetaldi (1630),[6] who may have been influenced by Vieta.[7] In his solution of geometric problems he freely brought algebra

---

remarquée et discutée d'abord. C'est que l'auteur se fait une loi, pour toutes les équations dont il s'occupe, de joindre la résolution numérique ou arithmétique à la construction géométrique" (*Préface*, p. vij).

[1] Rosen ed., p. 13 and elsewhere.   [2] Colebrooke translation, p. 270.

[3] "Quare quadratum lateris .cg. erit $\frac{1}{3}$ rei; et multiplicabo .cf. in dimidium .cg., hoc est radicem rei in radicem $\frac{1}{12}$ rei, ueniet radix $\frac{1}{12}$ census. . . ." *Scritti*, II, 223.

[4] *E.g., Sūma*, 1494 ed., fol. 146, *v., et passim.*

[5] For the use of geometric figures in his first solution of a cubic, see *Ars Magna*, fol. 29, *v*.

[6] See E. Gelcich, "Eine Studie über die Entdeckung der analytischen Geometrie mit Berücksichtigung eines Werkes des Marino Ghetaldi. . . ." *Abhandlungen*, IV, 191.   [7] Tropfke, *Geschichte*, II (1), 414.

to his aid, but it cannot be said that he in any way anticipated the work of the makers of analytic geometry.

Invention of Analytic Geometry.[1] The invention of analytic geometry is commonly attributed to Descartes, he having published (1637) the first treatise on the subject. There seems to be no doubt, however, that the idea occurred to Fermat at about the same time as to Descartes. In the British Museum there are eight volumes of Harriot's manuscripts, and among them was supposed to be "a well-formed analytical geometry, with rectangular coordinates and a recognition of the equivalence of equations and curves,"[2] but it is an interpolation by a later hand.

Fermat on Analytic Geometry. In a letter to Roberval, written September 22, 1636, and hence in the year before Descartes published *La Geometrie*, Fermat shows that he had the idea of analytic geometry some seven years earlier;[3] that is, in 1629. The details of this work appear in his *Isagoge ad locos planos et solidos*, which was published posthumously.[4] He used rectangular axes and followed Vieta in representing the unknowns by vowels (in this case only A and E) and the knowns by consonants. A general point on the curve was represented by I, and the foot of the ordinate from I to the axis of abscissas was represented by Z. The equation of a straight line through the origin was indicated by

$$D \text{ in } A \text{ aequetur } B \text{ in } E,[5]$$

---

[1] G. Loria, *Passato-Presente Geom.*; M. Chasles, *Aperçu historique sur l'origine et le développement des méthodes en géométrie*, 3d ed., Paris, 1889 (hereafter referred to as Chasles, *Aperçu*) ; M. Chasles, *Rapport sur les Progrès de la Géométrie*, Paris, 1870; A. Cayley and E. B. Elliott, "Curve," *Encyc. Britannica*, 11th ed.; Günther, *Die Anfänge*; E. Picard, *Bulletin of the Amer. Math. Soc.*, XI, 404; H. Wieleitner, *Zeitschrift für math. und naturw. Unterr.*, XLVII, 414; W. Dieck, *Mathematisches Lesebuch*, 4. Band, Sterkrade, 1920.

[2] F. V. Morley, "Thomas Hariot," *The Scientific Monthly*, XIV, 63. These manuscripts should be carefully studied. The spelling "Hariot" was used by some of his contemporaries.   [3] ". . . il y a environ sept ans étant à Bourdeaux."

[4] In his *Varia Opera*, p. 2 (Toulouse, 1679) ; Tropfke, *Geschichte*, II (1), 418; Günther, *Die Anfänge*, p. 43; *Œuvres de Fermat*, ed. P. Tannery and Ch. Henry, Vol. I (Paris, 1891), Vol. II (Paris, 1894).

[5] *I.e.*, $D \cdot A = B \cdot E$, which we should write as $ax = by$.

and that of a general straight line was given by the proportion

$$\text{ut } B \text{ ad } D, \text{ita } R - A \text{ ad } E.[1]$$

The equation of a circle appears as

$$Bq. - Aq. \text{ aequetur } Eq.[2]$$

If the ratio of $Bq. - Aq.$ to $Eq.$ is constant, Fermat asserted that the resulting figure is an ellipse;[3] and if the ratio of $Bq. + Aq.$ to $Eq.$ is constant, the figure is a hyperbola.[4] He also knew that $xy = a^2$ is the asymptotic equation of a hyperbola[5] and that $x^2 = ay$ is the equation of a parabola.[6]

Descartes publishes La Géométrie. Descartes published his Geometry in 1637, although he had been working upon it for some years,—even as early as 1619.[7] The treatise formed an appendix to his *Discours de la Methode* and was divided into three books. The first book treats of the meaning of the product of lines.[8] The second book defines two classes of curves, the geometric and the mechanic. We might now define the former as curves in which dy/dx is an algebraic function, and the latter as curves in which it is a transcendental function. In this book there is also much attention given to tangents and normals to a curve. The third book is largely algebraic, being entitled, "On the construction of solid or hypersolid problems." It treats particularly of such topics as the number of roots of an equation, "false roots," the increasing or decreasing of the roots, and the transformation of equations.

[1] *I.e.*, $a : b = c - x : y$, or $ay = b \ (c - x)$.
[2] *I.e.*, $B^2 - A^2 = E^2$, or $r^2 - x^2 = y^2$.
[3] "$Bq. - Aq.$ ad $Eq.$ habeat rationem datam, punctum I erit ad ellipsin." *I.e.*, $r^2 - x^2 = ky^2$ is the equation of an ellipse.
[4] "Si $Bq. + Aq.$ est ad $Eq.$ in data ratione, punctum I est ad hyperbolen."
[5] "A in E aeq. Z pl., quo casu punctum I est ad hyperbolen."
[6] "Si $Aq.$ aequatur D in E, punctum I est ad parabolen."
[7] J. Millet, *Descartes. Sa vie, ses travaux, ses découvertes, avant 1637*, p. 100 (Paris, 1867); E. S. Haldane, *Descartes*, p. 59 (London, 1905); C. Rabuel, *Commentaires sur la Geometrie de M. Descartes*, Lyons, 1730; *La Geometrie*, various editions from 1637. See the author's facsimile edition with translation (Chicago, 1925).
[8] "Des Problêmes qu'on peut construire n'y employant que des cercles & des lignes droites."

Instead of using the name "coordinates," Descartes spoke of roots or unknowns. The name "coordinate" is due to Leibniz,[1] as are also the terms "abscissa" and "ordinate," although, as we have seen, the Greeks used terms that were similar to them. Newton, Euler, Cramer, and various other writers used "applicate" to represent an ordinate.

Descartes had an idea of oblique coordinates, but he used only the x-axis and positive perpendicular ordinates in common practice.

Later Writers. In 1658 Jan (Johan) de Witt[2] wrote a work on curve lines[3] in which he set forth a number .of typical equations and gave the geometric character of each.

Further work was done by Lahire,[4] after which the elements of plane analytic geometry may be considered as having become established. The most noteworthy single contributor to the elements of the subject thereafter was Newton.[5] In his work on cubic curves he showed that a cubic has at least one real point at infinity, that any cubic belongs to one of four characteristic types, and that there are seventy-two possible forms of a cubic, a number since increased by six.[6] The discussion of the subject was nearly exhaustive, and was the most elaborate one of the kind that had been made up to that time.

The idea of polar coordinates seems due to Gregorio Fontana (1735–1803), and the name was used by various Italian writers of the 18th century.[7]

---

[1] *Acta Eruditorum* (1692), p. 170.

[2] Born at Dordrecht, September 12/24, 1625; died at The Hague, August 20, 1672. For biography see *The Insurance Cyclopædia*, Vol. II (London, 1873).

[3] *Elementa Cvrvarvm Linearvm*, Leyden, 1659; Amsterdam, 1683.

[4] *Les Lieux Géométriques*, Paris, 1679; *Construction des Equations Analytiques*, Paris, 1679.

[5] In his *Principia* (London, 1687) and his *Arithmetica Universalis* (Cambridge, 1707), but chiefly in his *Enumeratio linearum tertii ordinis*, which probably dates from 1668 or 1669 and which was published as an appendix to his *Optics* in 1704. See also W. W. R. Ball, "On Newton's classification of cubic curves," *Transactions of the London Mathematical Society* (1891), p. 104.

[6] G. Loria, *Ebene Kurven, Theorie und Geschichte*, p. 20 (Leipzig, 1902; 2d ed., 1910–1911), hereafter referred to as Loria, *Kurven*.

[7] For a discussion of the later types of coordinates see the *Encyklopädie*, III, 596, 656; Cantor, *Geschichte*, IV, 513.

Solid Analytic Geometry. Descartes clearly mentioned solid analytic geometry, but he did not elaborate it. Frans van Schooten the younger suggested the use of coordinates in three-dimensional space (1657), and Lahire (1679) also had it in mind. Jean Bernoulli (1698) thought of equations of surfaces in terms of three coordinates, but published nothing upon the theory at that time.

The first work on analytic geometry of three dimensions was written by Antoine Parent and was presented to the French Académie in 1700.[1] A. C. Clairaut (1729) was the first to write on curves of double curvature.[2] The third great contributor to the theory was Euler (1748), with whose work the subject advanced beyond the elementary stage.

Euler also laid the foundations for the analytic theory of curvature of surfaces, attempting to do for the classification of surfaces of the second degree what the ancients had done for curves of the second order. Monge introduced the notion of families of surfaces and discovered the relation between the theory of surfaces and the integration of partial differential equations, enabling each to be advantageously viewed from the standpoint of the other.

Modern Theory. Möbius began his contributions to geometry in 1823, and four years later published his *Barycentrische Calcul*. In this great work he introduced homogeneous coordinates. Of modern contributors to analytic geometry, however, Plücker stands easily foremost. In 1828 he published the first volume of his *Analytisch-geometrische Entwickelungen*, in which there appeared the modern abridged notation. In the second volume (1831) he set forth the present analytic form of the principle of duality. To him is due (1833) the general treatment of foci for curves of higher degree, and the complete classification of plane cubic curves (1835) which had been so frequently attempted before him. He also gave (1839) an

---

[1] "Des effections des superficies." This appears in his *Essais et Recherches*, Paris, 1705 and 1713.

[2] *Recherches sur les courbes à double courbure*, printed in 1731. It was presented to the Académie when Clairaut was only sixteen years old.

enumeration of plane curves of the fourth order. In 1842 he gave his celebrated "six equations," by which he showed that the characteristics of a curve (order, class, number of double points, number of cusps, number of double tangents, and number of inflections) are known when any three are given. To him is also due the first scientific dual definition of a curve, a system of tangential coordinates, and an investigation of the question of double tangents. The theory of ruled surfaces, begun by Monge, was also extended by him. Possibly the greatest service rendered by Plücker was the introduction of the straight line as a space element, his first contribution (1865) being followed by his well-known treatise on the subject (1868–1869).[1]

Certain Well-Known Curves. There are certain curves that are so frequently met in textbooks on analytic geometry as to deserve mention in an elementary history. Several of these have been considered elsewhere in this work, and a few others, with additional notes on those already given, will now be mentioned,[2] and for convenience will be given alphabetically.

Brachistochrone,[3] the curve of quickest descent, was studied by Galileo, Leibniz, Newton, and the Bernoullis, and was shown to be the cycloid. The name is due to the Bernoullis.[4]

Cardioid, the epicycloid $(x^2 + y^2 - 2ax)^2 = 4a^2(x^2 + y^2)$. The name is due to Giovanni Francesco M. M. Salvemini, called from his birthplace de Castillon (1708–1791), *De curva cardioide* (1741). It had already been studied by Ozanam.[5]

---

[1] On the various coordinate systems, see the *Encyklopädie*, III, 221, 596.

[2] H. Brocard, *Notes de Bibliographie des Courbes*, lith. autog., Bar-le-Duc, 1897; Partie complémentaire, 1899. See also Chasles, *Aperçu*, and the *Encyklopädie*, III, 185, 457; E. Pascal, *Repertorium der höheren Mathematik*, German translation by A. Schepp, Leipzig, 1902, especially Vol. II, chap. xvii; Loria, *Kurven*; Joaquin de Vargas y Aguirre, "Catálogo General de curvas," *Memorias de la Real Acad. de Ciencias exactas*, XXVI, Madrid, 1908; F. G. Teixeira, *Traité des courbes spéciales remarquables planes et gauches*, 3 vols., Coimbra, 1908, 1909, 1915.

[3] From βράχιστος (*brach'istos*, shortest) and χρόνος (*chron'os*, time). Formerly spelled *brachystochrone* by a confusion of the superlative βράχιστος with its positive βραχύς.

[4] Cantor, *Geschichte*, III, chap. 92.   [5] It is a special case of the limaçon.

Catenary, the French *chaînette*, $y = \frac{1}{2} a \left(e^{\frac{x}{a}} + e^{-\frac{x}{a}} \right)$. The name of the curve (*catenaria*) and the discovery of the equation and its properties are due to Leibniz.[1]

Cissoid of Diocles, the "ivy-shaped" curve, $y^2 = x^3/(2a - x)$, due to Diocles (*c.* 180 B.C.).[2]

Cochlioid,[3] $r = (a \sin \theta)/\theta$, a spiral curve discussed by J. Perk, *Phil. Trans.*, 1700, this form being a late one, due to J. Neuberg, a Belgian geometer. The name originated (1884) with two recent writers, Bentham and Falkenburg.

Cocked hat, the French *bicorne*, $x^2 + \left(\dfrac{a^2 - x^2}{y} - 2a \right)^2 - a^2 = 0$, a name also applied to another curve.[4]

Conchoid of Nicomedes (*c.* 180 B.C.), the "shell-shaped" curve.[5] The Cartesian equation is $(x - a)^2(x^2 + y^2) - b^2x^2 = 0$ and the polar equation is $r = a/\cos\theta + b$.

Conchoid of de Sluze, the cubic curve $a \ (r \cos\theta - a) = k^2 \cos^2 \theta$ or $a \ (x - a) \ (x^2 + y^2) - k^2x^2 = 0$, first constructed by René de Sluze (1662).

Curve of Pursuit, French *courbe du chien* as a special case. The name *ligne de poursuite* seems due to Pierre Bouguer[6] (1732), although the curve had been noticed by Leonardo da Vinci.

Cycloid, the transcendental curve

$$x = a \text{ arc } \cos \frac{a - y}{a} - \sqrt{2 \ ay - y^2}.$$

This curve, sometimes incorrectly attributed to Nicholas Cusa (*c.* 1450), was first studied by Charles de Bouelles (1501). It then attracted the attention of Galileo (1599), Mersenne

---

[1] On the history of this curve see C. A. Laisant, *Association Française pour l'avancement des Sciences, Congrès de Toulouse*, p. 64 (1887); Loria, *Kurven*, II, 204.

[2] See Volume I, page 118; Volume II, page 314.

[3] From κοχλίας (*cochli'as*, snail) and εἶδος (*ei'dos*, form).

[4] On this and the other curves see Archibald, "Curves," *Encyc. Brit.*, 14th ed.

[5] See Volume I, page 118.

[6] Born at Croisic, Brittany, February 16, 1698; died in Paris, August 15, 1758. He was one of the French geodesists sent to Peru in 1735 to measure an arc of a meridian. See articles by F. V. Morley, R. C. Archibald, H. P. Manning, and W. W. Rouse Ball, *Amer. Math. Month.*, XXVIII.

(1628), and Roberval (1634). Pascal (1659) called it the "roulette," completely solved the problem of its quadrature, and found the center of gravity of a segment cut off by a line parallel to the base.[1] Jean and Jacques Bernoulli showed that it is the brachistochrone curve, and Huygens (1673) showed how its property of tautochronism might be applied to the pendulum.

Devil's Curve, French *courbe du diable*, in general represented by the equation $y^4 - x^4 + ay^2 + bx^2 = 0$, and in particular by $y^4 - x^4 - 96\,a^2y^2 + 100\,a^2x^2 = 0$. The polar equation is

$$r = 2\,a\sqrt{(25 - 24\tan^2\theta)/(1 - \tan^2\theta)}.$$

It was studied by G. Cramer (1750)[2] and Lacroix (1810)[3] and is given in the *Nouvelles Annales* (1858), p. 317.

Elastic Curve, French *courbe élastique*, the differential equation of which is

$$dy = x^2 dx / \sqrt{a^4 - x^4}.$$

It was first studied by Jacques Bernoulli (1703).

Epicycloid, literally "epicycle-shaped," a curve traced by a point on a circle which rolls on the convex side of a given circle. The equation is $(x^2 + y^2 - a^2)^2 = 4a\,[(x - a)^2 + y^2]$. The curve was recognized by Hipparchus (*c.* 140 B.C.) in his astronomical theory of epicycles. Albrecht Dürer (1525) was the first to describe it in a printed work. It was next studied by Desargues (1639), but it first received noteworthy consideration by Lahire (1694) and Euler (1781).

Folium of Descartes, a curve represented by the equation $x^3 + y^3 = 3axy$. The problem was proposed to Roberval to determine the tangent to this curve, and through an error he was led to believe that the curve had the form of a jasmine flower, and hence he gave it the name *fleur de jasmin*, which was afterwards changed. It is also known as the *nœud de ruban*.

---

[1] H. Bosmans, "Pascal et les premières pages de l'"Histoire de la Roulette,'" *Archives de Philosophie*, I (1923), cah. 3.

[2] *Introduction a l'analyse des lignes courbes algébriques*, p. 19 (Geneva, 1750).

[3] *Traité du calcul différentiel et . . . intégral*, I, 391 (Paris, 1797; 1810 ed.).

Helix, the name given by Archimedes (*c.* 225 B.C.) to a spiral already studied by his friend Conon.[1] It is now known as the spiral of Archimedes. The equation is $r = a\theta$, or $\tan \phi = \theta$. It is one of the class of which the general equation is $r = a\theta^n$. The name is now usually applied to a curve traced upon a cylinder and cutting the generatrices under a constant angle. There are also the conical helix, the spherical helix (or loxodrome), and other types.

Lemniscate,[2] a curve first mentioned by Jacques Bernoulli (1694).[3] Its principal properties were discovered by Fagnano (1750). The analytic theory of the curve is due to Euler (1751, 1752). The general lemniscate has for its equation $(x^2 + y^2)^2 = 2a^2(x^2 - y^2) + b^4 - a^4$, while that of Bernoulli is represented algebraically by $(x^2 + y^2)^2 = 2a^2(x^2 - y^2)$, and is called the hyperbolic lemniscate. The general lemniscate is also known as Cassini's oval, after Giovanni Domenico (Jean Dominique) Cassini, who described it in 1680.

Limaçon, French *limaçon* (a snail), Italian *lumaca*, from Latin *limax*, called also by the French the *conchoïde du cercle*. The curve is

$$(x^2 + y^2 - ax)^2 = b^2(x^2 + y^2), \quad \text{or} \quad \rho = a + b\cos\phi.$$

Roberval called it the limaçon of Pascal, Étienne Pascal (father of Blaise) having discovered it. German writers speak of it as the *Pascal'sche Schnecke*.

Lituus (the Latin word for an augur's staff), the curve $r^2\theta = a^2$. The name is due to Cotes (*c.* 1710).

Logarithmic or Equiangular Spiral, the curve $r = ae^{k\theta}$, or $k\theta = \log\dfrac{r}{a}$, studied by Jacques Bernoulli (1692), who spoke of it as a *spira mirabilis*. It is still to be seen, in rude form, upon his tomb in Basel. The logarithmic spiral was the first non-algebraic plane curve to be rectified.

---

[1] But see Volume I, page 107.

[2] From λημνίσκος (*lemnis'kos*, Latin *lemniscus*), a ribbon on which a pendant is hung.

[3] "... formam refert jacentis notae octonarii ∞, seu complicitae in nodum fasciae, sive lemnisci." See F. Cajori, *Hist. of Math.*, 2d ed., p. 221.

Pearls, a name given by Pascal and de Sluze to the curves whose equation is

$$a^{p+q-r}y^r = x^p(a-x)^q,$$

or, in particular,

$$a^2y = x^2(a-x).$$

De Sluze proposed their consideration to Huygens (1658), and the latter made a careful study of them.

Roseate Curve, Rosace, or *Rhodonea,* the curve whose general polar equation is $r = a \cos m\theta$. The name *Rhodonea* is due to Guido Grandi (1713). The *Rosace à quatre feuilles,* or *Quadrifolium,* has for its polar equation $r = a \sin 2\theta$, and for the Cartesian form $(x^2 + y^2)^3 = 4a^2x^2y^2$.

Semicubic Parabola, or *Neile's Parabola,* $y^3 = ax^2$. It was the second curve to be rectified. William Neile discovered the curve in 1657. The method of rectification was published by Wallis in 1659, credit being given by him to his pupil, Neile, although there is still some dispute as to whether it was due to him, to Fermat, or to the Dutch writer Van Heuraet.[1]

Serpentine Curve, a name proposed by Newton for the curve $y = abx/(a^2 + x^2)$.

Spiral of Archimedes, the curve $r = a\theta$, mentioned under *Helix.*

Spiral of Fermat, the curve $r^2 = \theta$, proposed by Fermat in a letter to Mersenne, June 3, 1636.

Strophoid, French *strophoïde,*[2] a name proposed by a modern writer, Montucci (1846), for the curve $y = x\sqrt{(a-x)/(a+x)}$. Lehmus had already proposed (1842) the name *kukumaeïde* (cucumber seed), and various other names have been used. The curve has been studied by Barrow, Jean Bernoulli, Agnesi (1748), James Booth (1858), and various others.

Tractrix, the *tractoria* of Huygens (1693). The differential equation is

$$\frac{dy}{dx} = \frac{y}{\sqrt{a^2 - y^2}}.$$

---

[1] Hendrik van Heuraet, born at Haarlem in 1633. His brief *Epistolae de curvarum linearum in rectas transmutatione* was published by Van Schooten in 1659.

[2] From στρόφος (*stroph'os,* a twisted band, a cord) + εἶδος (*ei'dos,* form).

Witch of Agnesi, Versiera, Cubique d'Agnesi, or *Agnésienne,* the curve $y^2x + r^2(x - r) = 0$, discussed by Maria Gaetana Agnesi in 1748 in her *Istituzioni Analitiche.*[1]

9. MODERN GEOMETRY

Four Periods of Geometry. In order to appreciate the historical setting of modern geometry it is well to remember that the history of geometry in general may be roughly divided into four periods: (1) the synthetic geometry of the Greeks, including not merely the geometry of Euclid but the work on conics by Apollonius and the less formal contributions of numerous other writers; (2) the birth of analytic geometry, in which the synthetic geometry of Desargues, Kepler, Roberval, and other writers of the 17th century merged into the coordinate geometry already set forth by Descartes and Fermat; (3) the application of the calculus to geometry,—a period extending from about 1650 to 1800, and including the names of Cavalieri, Newton, Leibniz, the Bernoullis, l'Hospital, Clairaut, Euler, Lagrange, and d'Alembert, each one, especially after Cavalieri, being primarily an analyst rather than a geometer; (4) the renaissance of pure geometry, beginning with the 19th century and characterized by the descriptive geometry of Monge, the projective geometry of Poncelet, the modern synthetic geometry of Steiner and Von Staudt, the modern analytic geometry of Plücker, the non-Euclidean hypotheses of Lobachevsky, Bolyai, and Riemann, and the foundations of geometry.

Descriptive Geometry. Descriptive geometry as a separate science begins with Monge. He had been in possession of the theory for over thirty years before the publication of the *Géométrie Descriptive* (1794),[2]—a delay due to the jealous

---

[1] G. Loria, *Bibl. Math.*, XI (2), 7. See the English translation by J. Hellins, I, 222 (London, 1801). See also Volume I, page 519.

[2] G. Loria, *Storia della Geometria Descrittiva* (Milan, 1921), the leading authority on the subject; Chr. Wiener, *Lehrbuch der darstellenden Geometrie,* Leipzig, 1884–1887; *Geschichte der darstellenden Geometrie, ibid.,* 1884. See *Encyklopädie,* III, 517; F. J. Obenrauch, *Geschichte der darstellenden und projectiven Geometrie,* Brünn, 1897.

desire of the military authorities to keep the valuable secret. Certain of its features can be traced back to Frézier, Desargues, Lambert, and other writers of the preceding century, but it was Monge who worked it out in detail as a science, although Lacroix (1795), inspired by Monge's lectures in the École Polytechnique, published the first work on the subject. After Monge's work[1] appeared, Hachette (1812, 1818, 1821) added materially to the theory.

Period of Projective Geometry. It is also in this period that projective geometry has had its development, even if its origin is more remote. The origin of any branch of science can always be traced far back in human history, and this fact is patent in the case of this phase of geometry. The idea of the projection of a line upon a plane is very old. It is involved in the treatment of the intersection of certain surfaces, due to Archytas (*c.* 400 B.C.), and appears in various later works by Greek writers. Similarly, the invariant property of the anharmonic ratio was essentially recognized both by Menelaus (*c.* 100) and by Pappus (*c.* 300). The notion of infinity was also familiar to several Greek geometers and to the Latin writer Lucretius (*c.* 100), so that various concepts that enter into the study of projective geometry were common property long before the science was really founded.

Desargues, Pascal, Newton, and Carnot. One of the first important steps to be taken in modern times, in the development of this form of geometry, was due to Desargues. In a work published in 1639 Desargues set forth the foundation of the theory of four harmonic points, not as done today but based on the fact that the product of the distances of two conjugate points from the center is constant. He also treated of the theory of poles and polars, although not using these terms. In the following year (1640) Pascal, then only a youth of sixteen or seventeen, published a brief essay on conics in which he set forth the well-known theorem that bears his name.

---

[1] *Essais sur les plans et les surfaces*, Paris, 1795; *Complément des Élémens de Géométrie ou Élémens de Géométrie descriptive*, Paris, 1796; *Essais de Géométrie sur les plans et les surfaces courbes*, Paris, 1812.

In the latter part of the 17th century Newton investigated the subject of curves of the third order and showed that all such curves can be derived by central projection from five fundamental types. In the 18th century relatively little attention was given to the subject, but at the close of this period, as already stated, the descriptive geometry of Monge was brought into prominence,—itself a kind of projective geometry, although not what is technically known by this name.

Inspired by the general activity manifest in the 18th century, and following in the footsteps of Desargues and Pascal, Carnot treated chiefly of the metric relations of figures. In particular he investigated these relations as connected with the theory of transversals,—a theory whose fundamental property of a four-rayed pencil goes back to Menelaus and Pappus, and which, though revived by Desargues, was set forth for the first time in its general form by Carnot in his *Géométrie de Position* (1803), and supplemented in his *Théorie des Transversales* (1806). In these works Carnot introduced negative magnitudes, the general quadrilateral, the general quadrangle, and numerous other similar features of value to elementary geometry.

Poncelet on Projective Geometry. The origin of projective geometry as we know it today is generally ascribed to Poncelet.[1] A prisoner (1813–1814) in the Russian campaign, confined at Saratoff on the Volga, with no books at hand,[2] he was able in spite of all such discouragement to plan the great work[3] which he published in 1822. In this work he made prominent for the first time the power of central projection in demonstration and the power of the principle of continuity in research. His leading idea was the study of projective properties, and as a foundation principle he introduced the anharmonic ratio, a

---

[1] On the whole question consult the *Encyklopädie*, III, 389.

[2] "Privé de toute espèce de livres et de secours, surtout distrait par les malheurs de ma patrie et les miens propres."

[3] J. V. Poncelet, *Traité des propriétés projectives des figures*, Paris, 1822; *ibid.*, 1865–1866; *Applications d'analyse et de géométrie*, ed. Mannheim and Moutard, 2 vols., Paris, 1862, 1864. On the general subject of the development of modern geometric methods see J. G. Darboux, *Bulletin of the Amer. Math. Soc.*, XI, 517. See also Volume I, page 496.

concept which possibly dates back to the lost porisms of Euclid, and which Desargues (1639) had used. The anharmonic point-and-line properties of conics have since then been further elaborated by Brianchon, Chasles, Steiner, Plücker, Von Staudt, and other investigators. To Poncelet is also due the theory of "figures homologiques," the perspective axis and perspective center (called by Chasles the axis and center of homology), an extension of Carnot's theory of transversals, and the "cordes idéales" of conics which Plücker applied to curves of all orders. Poncelet also considered the circular points at infinity and completed the first great principle of modern geometry, the principle of continuity. Following upon the work of Poncelet, Möbius made much use of the anharmonic ratio in his *Barycentrische Calcul* (1827), but he gave it the name "Doppelschnitt-Verhältniss (ratio bisectionalis)," a term now in common use under Steiner's abbreviated form "Doppelverhältniss." The name "anharmonic ratio" or "anharmonic function" ("rapport anharmonique" or "fonction anharmonique") is due to Chasles, and "cross-ratio" was suggested by Clifford.

Gergonne, Steiner, and Von Staudt. Joseph-Diez Gergonne[1] (1813) introduced the term "polar" in its modern geometric sense, although Servois (1811) had used the expression "pole." Gergonne was the first (1825-1826) to grasp completely the principle to which he gave the name of "Principle of Duality," the most important principle, after that of continuity, in modern geometry. He used the word "class" in describing a curve, explicitly defining class and degree (order) and showing the duality between them. He and Chasles were among the first to study surfaces of higher order by modern methods.

Jacob Steiner, the most noted of the Swiss geometers of the 19th century, gave the first complete discussion[2] of the projective relations between rows, pencils, etc. and laid the foundation for the subsequent development of pure geometry. For the present, at least, he may be said to have closed the theory of conic sections, of the corresponding figures in three-dimensional

---

[1] See Volume I, page 495.
[2] *Systematische Entwickelungen* . . . , Berlin, 1832. See Volume I, page 524.

space, and of surfaces of the second order, and hence there opens with him the period of the special study of curves and surfaces of higher order.

Between 1847 and 1860 Karl Georg Christian von Staudt set forth a complete system of a pure geometry[1] that is independent of metrical considerations. All projective properties are here established independently of number relations, number being drawn from geometry instead of conversely, and imaginary elements being systematically introduced from the geometric side. A projective geometry, based on the group containing all the real projective and dualistic transformations, is developed, and imaginary transformations are introduced.

Non-Euclidean Geometry. The question of Euclid's fifth postulate, relating to parallel lines, has occupied the attention of geometers ever since the *Elements* was written.[2] The first scientific investigation of this part of the foundation of geometry was made by Girolamo Saccheri[3] (1733), a work which was

[1] *Geometrie der Lage*, Nürnberg, 1847; *Beiträge zur Geometrie der Lage*, 3 parts, Nürnberg, 1856, 1857. See M. Noether, *Zur Erinnerung an K. G. C. von Staudt*, Erlangen, 1901, and Volume I, page 505.

[2] F. Engel and P. Stäckel, *Die Theorie der Parallellinien von Euklid bis auf Gauss*, Leipzig, 1895; G. B. Halsted, various contributions, including "Bibliography of Hyperspace and Non-Euclidean Geometry," *American Journal of Mathematics*, Vols. I, II; *Amer. Math. Month.*, Vol. I; translations of Lobachevsky's Geometry, Vassilief's address on Lobachevsky, Saccheri's Geometry, Bolyai's work and his life; "Non-Euclidean and Hyperspaces," *Mathematical Papers of Chicago Congress*, p. 92; G. Loria, *Die hauptsächlichsten Theorien der Geometrie*, Leipzig, p. 106; A. Karagiannides, *Die Nichteuklidische Geometrie vom Alterthum bis zur Gegenwart*, Berlin, 1893; E. McClintock, "On the Early History of Non-Euclidean Geometry," *Bulletin of New York Mathematical Society*, II, 144; W. B. Frankland, *Theories of Parallelism*, Cambridge, 1910 (particularly valuable); H. Poincaré, "Non-Euclidean Geometry," *Nature*, XLV, 404; P. Stäckel, *Wolfgang und Johann Bolyai, Geometrische Untersuchungen*, 2 vols., Leipzig, 1913. See also Volume I, Chapter X, under the several names mentioned. On the general question of the modern synthetic treatment of elementary geometry, see the *Encyklopädie*, III, 859; for the analytic treatment, *ibid.*, 771. See also C. J. Keyser, *Mathematical Philosophy*, p. 342 (New York, 1922). For an excellent bibliography up to the time it was printed see D. M. Y. Sommerville, *Bibliography of Non-Euclidean Geometry*, London, 1911.

[3] Born at San Remo, September 4 or 5, 1667; died at Milan, October 25, 1733. The work was *Euclides ab omni naevo vindicatus*, Milan, 1733; English translation by G. B. Halsted, Chicago, 1920. Saccheri was a Jesuit and taught mathematics in Turin, Pavia, and Milan.

not looked upon as a precursor of Lobachevsky, however, until Beltrami (1889) called attention to the fact. Johann Heinrich Lambert (1728–1777)[1] was the next to question the validity of Euclid's postulate, in his *Theorie der Parallellinien* (posthumous, Leipzig, 1786), the most important treatise on the subject between the publication of Saccheri's work and the works of Lobachevsky and Bolyai. Legendre (1794) also contributed to the theory, but failed to make any noteworthy advance.

During the closing years of the 18th century Kant's[2] doctrine of absolute space, and his assertion of the necessary postulates of geometry, were the object of much scrutiny and attack. At the same time Gauss was giving attention to the fifth postulate, although at first on the side of proving it. It was at one time surmised that Gauss was the real founder of the non-Euclidean geometry, his influence being exerted on Lobachevsky through his friend Bartels,[3] and on János Bolyai through the father Farkas, who was a fellow student of Gauss, and it will presently be seen that he had some clear ideas of the subject before either Lobachevsky or Bolyai committed their theories to print.

Lobachevsky. Bartels went to Kasan in 1807, and Lobachevsky was his pupil. The latter's lecture notes fail to show that Bartels ever mentioned the subject of the fifth postulate to him, so that his investigations, begun even before 1823, seem to have been made on his own motion, and his results to have been wholly original. Early in 1826 he set forth the principles of his famous doctrine of parallels, based on the assumption that through a given point more than one straight line can be drawn which shall never meet a given straight line coplanar with it. The theory was published in full in 1829–1830, and he contributed to the subject, as well as to other branches of mathematics, until his death.

---

[1] D. Huber, *Lambert nach seinem Leben und Wirken*, Basel, 1829. See Volume I, page 480.   [2] E. Fink, *Kant als Mathematiker*, Leipzig, 1889.

[3] Johann Martin Christian Bartels, born at Braunschweig, August 12, 1769; died at Dorpat, December 19, 1836. He was professor of mathematics at Kasan and later at Dorpat.

The Bolyais and Gauss. János Bolyai received, through his father, Farkas, some of the inspiration to original research which the latter had received from Gauss. When only twenty-one he discovered, at about the same time as Lobachevsky, the principles of non-Euclidean geometry, and he refers to them in a letter of November, 1823. They were committed to writing in 1825 and were published in 1832. Gauss asserts in his correspondence with Schumacher[1] (1831–1832) that he had thought out a theory along the same lines as Lobachevsky and Bolyai, but the publication of their works seems to have put an end to his investigations. His statement on the subject is as follows:

I will add that I have recently received from Hungary a little paper on non-Euclidean geometry in which I rediscover all my own ideas and results worked out with great elegance. . . . The writer is a very young Austrian officer, the son of one of my early friends, with whom I often discussed the subject in 1798, although my ideas were at that time far removed from the development and maturity which they have received from the original reflections of this young man. I consider the young geometer Von Bolyai a genius of the first rank.[2]

This was not, however, the first statement of Gauss upon the subject, for in a letter written on November 8, 1824, he remarked:

The assumption that the sum of the 3 angles is smaller than 180° leads to a new geometry entirely different from ours [the Euclidean] —a geometry which is throughout consistent with itself, and which I have elaborated in a manner entirely satisfactory to myself, so that I can solve every problem in it with the exception of the determining of a constant which is not *a priori* obtainable.[3]

[1] Heinrich Christian Schumacher (1780–1850), the astronomer.

[2] Sedgwick and Tyler, *A Short History of Science*, p. 338 (New York, 1917).

[3] P. Stäckel, *Wolfgang und Johann Bolyai*, I, 95 (Leipzig, 1913). The letter was written to one Taurinus, who, two years later, published a *Geometriae prima elementa* (1826), in which he gives evidence of having thought upon a non-Euclidean trigonometry. See Volume I, page 527.

Riemann's Theory. Of all the contributions which appeared after Bolyai's publication the most noteworthy, from the scientific standpoint, is that of Georg Friedrich Bernhard Riemann. In his Habilitationsschrift (1854) he applied the methods of analytic geometry to the theory and suggested a surface of negative curvature, which Beltrami called "pseudo-spherical," thus leaving Euclid's geometry on a surface of zero curvature midway between his own and Lobachevsky's. He thus set forth three kinds of geometry, Bolyai having noted only two. These Klein (1871) called the elliptic (Riemann's), parabolic (Euclid's), and hyperbolic (Lobachevsky's) geometry.

10. PERSPECTIVE AND OPTICS

Relation of Perspective to Mathematics. While all painters seek to secure proper perspective in their pictures, the most successful of the painters of the Renaissance made an effort to base their treatment of the subject on mathematical principles. Of late these principles have interested architects more than painters, but in any case the subject is largely a mathematical one.[1]

The Greeks included perspective in their science of optics, and the Arabs in their science of appearances, their title being translated into Medieval Latin as *De aspectibus*.[2] Therefore, while there is a manifest difference between perspective and optics as we consider these terms today, it is necessary to treat of them as closely related.

Ancient Works. While several Greek writers wrote on the subject of perspective, the earliest mathematical work that has come down to us is the *Optics* of Euclid.[3] In this work Euclid

---

[1] On the history of the subject a beginning can be made with N. G. Poudra, *Histoire de la Perspective*, Paris, 1864, a rather poorly arranged work with no index.

[2] The first translation (1505) of Euclid's *Optics*, however, used the term *perspectiva*.

[3] The latest Latin edition of the *Optics* is that of J. L. Heiberg, in *Euclidis opera omnia*, Vol. VII, Leipzig, 1895. There are various translations of the text from Greek into Latin. The first is that of Zamberto (Venice, 1505), in the collected works of Euclid; the second is that of J. Pena (Péna, de la Pêne), Paris, 1557, or

lays down a series of axioms,[1] quite as he does in his *Elements*, the first being: "Therefore it is assumed that [visual] rays emitted from the eye are carried in a straight line, whatever may be the distance."[2]

On the axioms Euclid bases his propositions, sixty-one in number, proving them geometrically after the plan used by him in the *Elements*.

There is also a work on catoptrics containing thirty-one propositions and attributed to Euclid, but it is doubtful if the text published by Gregory[3] and Heiberg is his.

Some idea of the nature of Euclid's work may be obtained from a single proposition in his *Optics*: "If from the center of a circle a line be drawn at right angles to the plane of the circle, and the eye be placed at any point on this line, the diameters of the circle will all appear equal."[4]

Later Classical Writers. The only Roman writer who paid any attention to the subject is Vitruvius (*c.* 20 B.C.), who, in his work on architecture, has something to say on the plans and elevations of buildings. He seems to have had the idea of two projections, these being on two planes perpendicular to each other and arranged as in descriptive geometry.

Heron of Alexandria (*c.* 50?) is known to have written on dioptrics, but only a fragment of the work exists.[5] His theory of light involved the usual error of most of the Greek scientists, that the rays of light proceed from the eye to the object instead of from the object to the eye.

---

that of Dasypodius which appeared at Strasburg in the same year. See also G. Ovio, *L'ottica di Euclide*, Milan, 1918; D. Gregory, *Euclidis quae supersunt omnia*, Oxford, 1703, p. [599], with parallel Greek and Latin texts; *La prospettiva di Evclide*, . . . *tradotta dal R. P. M. Egnatio Danti*, Florence, 1573; *La perspective d'Euclide, traduite en français* . . . *par R. Fréart de Chanteloup*, Mans, 1663.

[1] In the Gregory edition (1703, p. 604) θέσεις (*the'seis*) and *positiones*.

[2] That the eye emitted the visual rays was Plato's idea. Aristotle held a view more in accordance with our own, asking why, if the older idea were correct, we cannot see in the dark.   [3] *Loc. cit.*, p. 643.   [4] Prop. XXXV.

[5] *Opera quae supersunt omnia*, Leipzig, 1899–1914; *Traité de la dioptre*, ed. A. J. H. Vincent, Paris, 1858, in the *Bibliothèque Nationale, Notices et extraits*, XIX, Pt. 2, pp. 157–347.

Ptolemy (*c.* 150) is said to have written upon the subject, but it is not certain that he did so.[1] The work attributed to him contains five books, the first dealing with the properties of light, the second with the nature of vision, the third with reflection, the fourth with concave mirrors and with two or more mirrors, and the fifth with refraction.

The next Greek writer on the subject was Heliodorus of Larissa,[2] whose date is uncertain but who lived after Ptolemy. His work is little more than a commentary on Euclid.[3]

Medieval Writers. One of the greatest of the medieval writers on perspective was the Arab scholar Alhazen (*c.* 1000).[4] His work was the basis of Peckham's *Perspectiva* mentioned below. The following well-known problem relating to optics bears his name: "From two given points within a circle to draw to a point on the circle two lines which shall make equal angles with the tangent at that point."[5]

Of the European writers the first one of importance was Roger Bacon (*c.* 1250). In his *Opus Majus* he devotes Part V (*De scientia perspectiva*) to perspective,[6] dividing it into three parts. Part I explains the general principles of vision, Part II deals with direct vision, and Part III discusses reflection and refraction. In the *Opus Tertium* there is also a brief *tractatus*

---

[1] There is a MS. in Paris beginning: *Incipit Liber Ptholemaei de Opticis sive Aspectibus translatus ab Ammiraco* [or *Ammirato*] *Eugenio Siculo*, consisting originally of five books. For a discussion, see W. Smith, *Dict. of Greek and Roman Biog.*, III, 573 (London, 1864). See also N. G. Poudra, *Histoire de la Perspective*, p. 28 (Paris, 1864).

[2] Possibly his name was Damianus. At any rate some of the MSS. bear the title Δαμιανοῦ φιλοσόφου τοῦ Ἡλιοδώρου Λαρισσαίου περὶ ὀπτικῶν ὑποθέσεων βιβλία β'.

[3] *La Prospettiva di Eliodoro Larisseo, Tradotto Dal Reverendo Padre M. Egnatio Danti*, Florence, 1573, bound with *La Prospettiva di Evclide*. There are other translations.

[4] Al-Ḥasan . . . ibn al-Ḥaitam. See Volume I, page 175. A Latin translation, under the title *Opticae Thesauri Libri VII*, was published at Basel in 1572.

[5] For a discussion of the problem see *American Journal of Mathematics*, IV, 327.

[6] *Rogerii Baconis angli, viri eminentissimi, Perspectiva*, Frankfort, 1614. This is best found, however, in the editions of the *Opus Majus* by S. Jebb (London, 1733; Venice, 1750) and J. H. Bridges (2 vols., Oxford, 1897; suppl. vol., London, 1900). See also E. Wiedemann, "Roger Bacon und seine Verdienste um die Optik," in A. G. Little, *Roger Bacon Essays*, p. 185 (Oxford, 1914).

de perspectiva.[1] Besides this, Bacon wrote two other brief treatises[2] on the subject, and still others are attributed to him without historic sanction.[3]

The work that had the greatest influence upon the subject of perspective in the Middle Ages was the *Perspectiva communis* of John Peckham[4] (*c.* 1280). This work was the recognized standard for three hundred years. It was edited and published by Cardan's father and went through various editions. As already stated, Peckham drew largely upon Alhazen's work. The work is divided into three parts, the second containing fifty-six propositions on reflection, and the third containing twenty-two on refraction.

About the same time as Peckham, the German (or possibly Polish) scholar Witelo (*c.* 1270)[5] was called to Rome and there became conversant with the works of the ancients as well as those of the Arabs in the science of perspective. Georg Tanstetter von Thannau[6] (1480–1530) and Apianus[7] prepared editions of his work which were published at Nürnberg in 1533 and 1551. The treatise is divided into ten books, the first four being a summary of the works of earlier writers; the fifth, a treatment of reflection; the sixth, reflection by convex spheric mirrors; the seventh, cylindric and compound mirrors; the eighth, concave spheric mirrors; the ninth, concave conic mirrors and irregular mirrors; and the tenth, refraction.

Among the other medieval writers on perspective were William of Moerbecke[8] (*c.* 1250) and Campanus[9] (*c.* 1260).

Renaissance Writers. The first writers of the Renaissance to take up the subject were the painters and engravers. Pietro

---

[1] This in a Paris MS., formerly attributed to Alpetragius, discovered by Duhem and not yet printed. See Little, *loc. cit.*, p. 390.

[2] *De speculis comburentibus* and *Notulae de speculis*, both published at Frankfort (1614) in Combach's *Specula mathematica*, pp. 168–207.

[3] See Little, *loc. cit.*, p. 409 seq.

[4] See Volume I, page 224. It was often printed. For editions, see Kästner, *Geschichte*, II, 264; for Kästner's history of optics in general, *ibid.*, p. 237.

[5] See Volume I, page 228. On his work at Padua see A. Birkenmajer, *Witelo e lo Studio di Padova*, reprint, Padua, 1922.

[6] Professor of astronomy at Vienna.   [7] See Volume I, page 333.

[8] William Fleming. See Volume I, page 229.   [9] See Volume I, page 218.

II

Franceschi (or Della Francesca), for example, who died in 1492, wrote the work *De corporibus regularibus* and a work *De perspectiva pingendi*,[1] which is still extant in manuscript, and in which he takes up the theory of perspective.[2] There were also such artists as Leonardo da Vinci[3] (*c.* 1500), many of whose ideas on perspective, and particularly on the nature of vision and the camera obscura, were a distinct advance in knowledge; Benvenuto Cellini,[4] whose work on perspective was largely taken from Leonardo;[5] and Albrecht Dürer, whose work on drawing[6] includes some treatment of perspective.

One of the first men in this period to write a work of any note, devoted solely to optics, was Ramus[7] (*c.* 1550). This work was published by his pupil, Friedrich Risner[8] (died 1580), who also published the works of Alhazen (*c.* 1000) and Witelo (*c.* 1270).[9] The work of Ramus is in four books, but it contains little that Witelo did not give.

Optics in the 17th Century. In the 17th century the science of optics took a great step forward, notably through the efforts of Kepler. These efforts first appear in his unpretentious work of 1604, the *Paralipomena ad Vitellionem*, this Vitello (Witelo) being the German or Polish scholar already mentioned. In this little work Kepler explained the mechanism of the eye, comparing the retina to the canvas on which images were depicted. He showed that imperfect vision is caused by the failure of the rays of light to converge properly on the retina. In 1611 he published a work on dioptrics in which he set forth his ideas,

---

[1] G. Pittarelli, "Intorno al libro 'de perspectiva pingendi' di Pier dei Franceschi," *Atti del Congresso internazionale di scienze storiche*, XII (Rome, 1904), 262.

[2] H. Wieleitner, "Zur Erfindung der verschiedenen Distanzkonstruktionen in der malerischen Perspektive," *Repertorium für Kunstwissenschaft*, XLII (1920), 249.

[3] *Trattato della pittura*, Paris, 1651. See Volume I, page 294.

[4] Born 1500; died *c.* 1571. Various dates of his death are given, ranging from December 13, 1569, to February 25, 1571.

[5] P. Duhem, *Études sur Léonard de Vinci*, sér. I, p. 225 (Paris, 1906); G. P. Carpani, *Memoirs of B. Cellini*, English translation by Roscoe, London, 1878; *Œuvres complètes de Benvenuto Cellini*, 2d ed., 2 vols., Paris, 1847.

[6] *Underweysung der messung*, Nürnberg, 1525; see Volume I, page 326.

[7] See Volume I, page 309.

[8] *Opticae libri quatuor, ex voto Petri Rami novissimo, per Fr. Risnerum . . .*, Cassel, 1606 (posthumous). [9] Basel, 1572. See page 341.

imperfect though they were, upon the law of refraction. He also gave a scientific explanation of the telescope, then recently invented. In the same year (1611) Antonio de Dominis, archbishop of Spalato, published his *De Radiis Lucis in Vitris Perspectiva et Iride*, in which he explained more fully than his

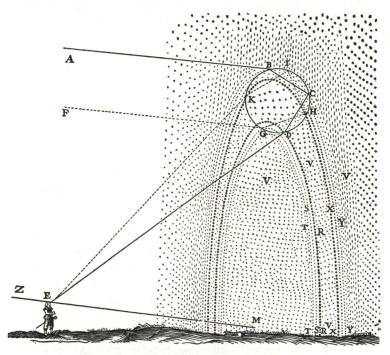

DESCARTES'S EXPLANATION OF THE RAINBOW

From his *Meteora*, 1656 ed., p. 214

predecessors the phenomena of the rainbow, basing them upon principles of refraction. It was Descartes, however, who in his *Dioptrica* (1637) gave the law that the sine of the angle of incidence has a constant ratio to the sine of the angle of refraction, the ratio being a function of the medium. The law was, in fact, known to Snell twenty years earlier, but he had failed to set it forth in print, although he had taught it. Nevertheless,

Descartes was living in Holland at that time, and there is some suspicion that he had there heard of Snell's discovery. In his *Dioptrica*, Descartes completed the theory of the rainbow by giving an explanation of the outer bow.

Just before Descartes produced his work, François Aguillon (1566–1617), a Belgian Jesuit, published a treatise[1] of some importance. In this he used the term "stereographic projection," although the idea was known to the Greeks.

Frans van Schooten the Younger published in 1656–1657 a book of mathematical exercises[2] in which he treated of perspective,[3] but it contained little that was original.

Newton's Work. Newton began to work seriously on optics about 1666. In his treatise[4] of 1704 he states that part of the treatise was written in 1675; and in his posthumous work the editor states that Newton "first found out his Theory of Light and Colours" as early as 1666, lecturing upon it in 1669. By this time the elementary theory of optics was well established.

11. INSTRUMENTS IN GEOMETRY

Early Instruments. Before the invention of the telescope, microscope, and vernier there can hardly be said to have been any instruments of precision. For practical land measure, however, for leveling, and for the measuring of heights, the world developed several interesting instruments worthy of mention.

In general, the ancient surveyors measured distances by the use of a rope or a wooden rod, the units of measure varying in different localities. They laid off right angles by the use of an

---

[1] *Francesci Aguilonii e societate Jesu Opticorum libri VI*, Antwerp, 1613.

[2] *Exercitationum Mathematicarum libri V*, Amsterdam, 1656–1657; Dutch edition, *ibid.*, 1659.

[3] "Een korte verhandeling van de Fondementen der Perspective." It was also separately printed, Amsterdam, 1660.

[4] *Opticks : or, a Treatise of the Reflexions, Refractions, Inflexions, and Colours of Light* . . . , London, 1704, with various later editions and translations. His second work in point of publication, but not of composition, was his *Optical Lectures Read in . . . 1669*, published posthumously, London, 1728.

instrument resembling the carpenter's square of the present time, by a kind of cross placed horizontally on a staff, or by the 3–4–5 relation applied to a stretched cord. For finding a level they ordinarily used a right-angled isosceles triangle with a plumb line. Illustrations of such instruments are found on monuments to certain ancient surveyors.[1]

Early Printed Books. The early printed books give us much information as to the nature of the instruments inherited from the Middle Ages. Of these there may be mentioned the mirror for the measuring of heights by the forming of similar triangles, the geometric square (*quadratum geometricum*), the quadrant,

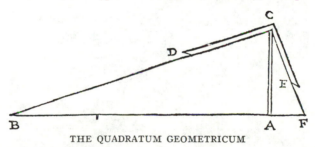

THE QUADRATUM GEOMETRICUM

From Oronce Fine's *De re & praxi geometrica*, Paris, 1556. The two triangles being similar, *AB* is easily found from the distances *AC* and *AF*

the astrolabe, and the cross-staff (*baculum*, also called the *baculus*). The method of using most of these instruments is evident, but a brief description of some of them will be helpful.[2]

The Square. The simplest of all the instruments of this class was the ordinary carpenter's square, known in some of the works on mensuration as the geometric square. Its use in finding short distances by means of the principle of similar triangles will be easily understood from the above illustration.

[1] See page 357.
[2] W. E. Stark, "Measuring Instruments of Long Ago," *School Science and Math.*, X, 48, 126; M. Curtze, "Ueber die im Mittelalter zur Feldmessung benutzten Instrumente," *Bibl. Math.*, X (2), 65; M. Cantor, *Die Römischen Agrimensoren*, Leipzig, 1875; E. N. Legnazzi, *Del Catasto Romano*, Verona, 1887; G. Rossi, *Groma e squadro ovvero Storia dell' Agrimensura Italiana dai tempi antichi al secolo XVII°*, Turin, 1877.

The Baculum. In its simplest form the baculum, arbalète (crossbow), geometric cross, cross-staff, or Jacob's staff[1] was a rod about 4 feet long, of rectangular cross section, and having a crosspiece that could slide upon it and always remain perpendicular to it. The staff was marked off in sections each equal in length to the crosspiece. In actual use the crosspiece was

PRACTICAL MATHEMATICS IN THE 17TH CENTURY

From Simon Jacob's work of 1560 (1565 ed.)

first placed at one of the division marks of the staff, the observer then facing approximately the mid-point of a line that he wished to measure and standing at a distance such that, when he sighted along the staff, the crosspiece should be parallel to the line and just cover it. The crosspiece was then moved to the next division on the staff, the observer taking a position where the first process of covering the line with the crosspiece could be repeated, as shown in the illustration on page 347.

---

[1] This name had various other uses, however.

The length of the line to be measured was then the same as the distance between the two positions of the observer. There were also various other methods of using the instrument.[1]

Sector Compasses. About the year 1597 Galileo invented the proportional compasses,[2] or sector compasses, an ingenious device for solving a variety of problems often met by architects, engineers, and others who have much to do with applied mathematics. The instrument consists ordinarily of two brass rules

THE BACULUM, OR CROSS-STAFF

From Oronce Fine's *De re & praxi geometrica*, Paris, 1556, showing the methods of measuring distances

hinged at one end. There are usually six pairs of lines, three on each face, radiating from the pivot. One pair might, for example, represent equal parts; another, squares; and the third, lines of polygons; but this varied according to the purpose of the particular instrument.

To give a single illustration of its use, suppose that each line of equal parts is divided into 200 equal segments, numbered by tens, beginning at the pivot. Then, to divide any given line

[1] For a brief résumé see G. Bigourdan, *L'Astronomie, évolution des idées et des méthodes*, p. 116 (Paris, 1911; 1920 ed.), hereafter referred to as Bigourdan, *Astronomie.*   [2] *Le operazioni del compasso geometrico e militare*, Padua, 1606.

segment into any number of equal parts, say nine, open a pair of ordinary dividers to the length of the segment, then open the sector compasses so that one point of the dividers rests on

ASTROLABE OF CHAUCER'S TIME

Fine piece of medieval workmanship now in the British Museum. It may well be that Chaucer himself made use of this in preparing his treatise on the astrolabe

90 on one face and the other point rests on 90 on the other face; then the distance from the 10 on one face to the 10 on the other is one ninth of the length of the given line segment.

Astrolabe. Of all the early astronomico-mathematical instruments none was better known than the astrolabe. The name

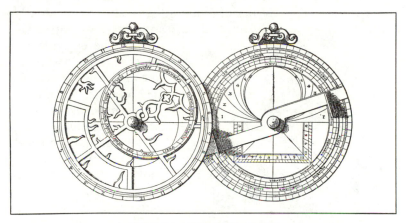

ITALIAN ASTROLABE OF 1558

It bears the inscription "Patavii Bernardinvs Sabevs faciebat MDLVIII."
From the author's collection

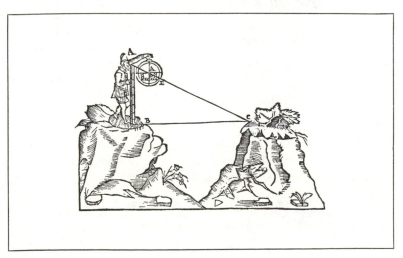

THE ASTROLABE IN SIMPLE MENSURATION

From Bartoli's *Del Modo di Misvrare*, Venice, 1589, showing simple work in a
crude kind of trigonometry

is Greek and means the taking of the stars.[1] Hence any instrument for measuring the angles by which a star was "taken" (as a sailor today speaks of "taking" the sun) was, strictly speaking, an astrolabe. One of the early forms was the armillary sphere, so called from the *armillae*,[2] or rings, which were so arranged as to form two, or sometimes three, circles, ordinarily placed at right angles to one another. One ring usually corresponded to the plane of the equator and the other to the plane of the meridian. By these two circles the ancients determined the two coordinates of a star. The astrolabe described by Ptolemy the astronomer is a kind of armillary sphere,[3] and furthermore these spheres are first heard of in connection with the school with which he was associated. It is asserted by early writers that Eratosthenes, through his interest in geodesy and astronomy, induced King Ptolemy III to have such instruments made and placed in the museum at Alexandria.

CHAMPLAIN'S ASTROLABE

Found near the Ottawa River about 1870. It was made in Paris in 1603. This is the type of astrolabe known as the planisphere. From the collection of Samuel V. Hoffman, New York

---

[1] From ἄστρον (*as'tron*, a heavenly body) + λαμβάνειν, λαβεῖν (*lamba'nein, labein'*, take). Ptolemy spoke of the two circles that he used in locating a star as ἀστρολάβοι κύκλοι (*astrola'boi ky'kloi*) and spoke of the whole instrument as ἀστρολάβον ὄργανον (*astrola'bon or'ganon*) or, commonly, as ὁ ἀστρολάβος (*ho astrola'bos*). See J. Frank, "Zur Geschichte des Astrolabs," *Erlangen Sitzungsberichte*, 50–51. Band, p. 275; R. T. Gunther, *Early Science in Oxford*, II, 181.

[2] *Armilla* means an armlet, bracelet, hoop, or ring. It is probably a diminutive of *armus*, the shoulder or upper arm.

[3] *Almagest*, VII, 2, 4.

Planisphere. Another ancient and common form of the astrolabe consisted simply of a disk upon the rim of which were marked the units of angle measure. Such instruments were probably well known in ancient times among all who made any scientific study of the stars. That they were familiar in ancient Babylon we have definite proof.[1] Fragments of several such instruments have been found and the inscriptions interpreted. They go back to the 2d millennium B.C., which goes to show that the early Greeks undoubtedly knew of their value and made

THE QUADRANT

From the *Protomathesis* of Oronce Fine, Paris, 1530–1532

use of them in angle measure. These astrolabes are in the form of planispheres and are made of clay, baked like the tablets.[2] A planisphere may be defined as a stereographic projection of the celestial sphere either upon the plane of the equator or upon the plane of the meridian.

Such instruments were used in various practical ways in which angle measure was the chief purpose, and this use continued until recent times. Even now they are seen in the Orient in the hands of the astrologers.

[1] E. F. Weidner, *Handbuch der Babylonischen Astronomie*, Lieferung I, 62; with bibliography, Leipzig, 1915.

[2] For a photographic reproduction, see Weidner, *loc. cit.*, p. 107, from A. Jeremias, *Handbuch der altorientalischen Geisteskultur*, Leipzig, 1913. There is a good specimen in the British Museum.

The planisphere in common use in later times represents the stereographic projection of the celestial sphere upon the plane of the equator, the eye being at the pole. Planispheres of

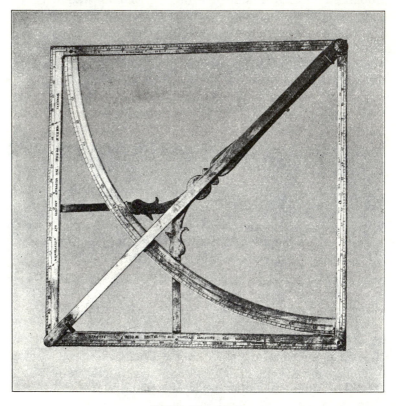

BRASS QUADRANT

Austrian work of the 18th century. The original is 29.5 cm. square. From the author's collection

various types were used by early navigators for the purpose of finding the elevation of the north star, or for other angle measurements, and were often furnished with several plates which could be so adjusted as to allow the instrument to be used in different latitudes.

The Astrolabe in the East. From Babylon[1] the astrolabe may have passed to China and India, or vice versa. At any rate, Mesopotamia seems to have been the source from which the Greeks derived their knowledge of the instrument. It is probable that Thales used it in measuring the distances of ships, since the Babylonian astronomy was already becoming known in the Greek civilization. It may be inferred from Plato's *Timæus* that some such instrument was in use in his day, but in any case an astrolabe of some type was known to Eratosthenes, Hipparchus, and other Greek astronomers even before Ptolemy described the armillary sphere.[2]

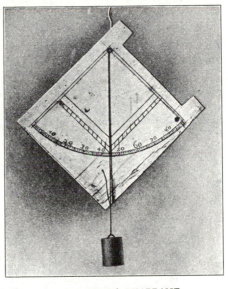

SMALL IVORY QUADRANT

Italian work of the 18th century. If we sight through holes in the two projections on the upper right-hand edge, the angle of elevation is indicated by the plumb line and the arc. The original is 5 cm. by 5.9 cm. From the author's collection

Arab Treatises on the Astrolabe. Led by their study of Greek astronomy, the Arabs, beginning in the 9th century, wrote numerous works upon the astrolabe, and these, in turn, influenced the medieval scholars of Europe. Thus we find Messahala (*c.* 800) composing a work upon the subject, which formed the basis of two manuscripts by Rabbi ben Ezra (*c.* 1140). From one of these manuscripts Chaucer (*c.* 1380) seems to have drawn his information for his treatise upon the astrolabe.

---

[1] A. H. Sayce and R. H. M. Bosanquet, "Babylonian Astronomy," *Monthly Notices of the Royal Astron. Society*, XL, No. 3, with illustrations.

[2] R. Wolf, *Geschichte der Astronomie*, p. 160 (Munich, 1877).

INSTRVCTIO
CAPVT XII.
QVA RATIONE ELICIENDA SIT AL-
TITVDO ALICIVS REI, QVAE ACCESSVM NON
admittit : Vt funt Mænium, arcium, &c.

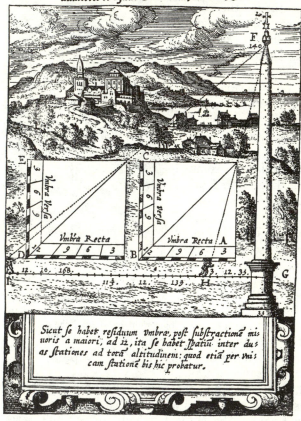

Sicut ſe habet reſiduum vmbræ, poſt ſubſtractionē mis
noris a maiori, ad 12, ita ſe habet ſpatiū inter dus
as ſtationes ad totā altitudinem: quod etiā per vnis
cam ſtationē bis hic probatur.

EXPLANATION OF THE QUADRANT

From *De Quadrante Geometrico*, usually referred to Cornelius de Judeis, Nürnberg, 1594, but in fact written by Levinus Hulsius. Cornelius made the drawings
with the help of Martin Geet

Quadrant. Closely related to the astrolabe is the quadrant, an instrument in which only a quarter of a circle is used. It

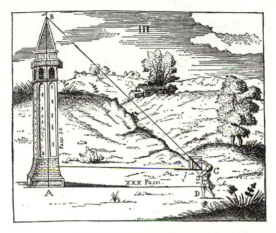

USE OF THE QUADRANT

From Ottavio Fabri's *L' Uso della Squadra Mobile*, Trent, 1752

appears in various forms, sometimes without an arc, the angles being read on the sides of a square. The earliest description that we have is given in the *Almagest*, and on this account the

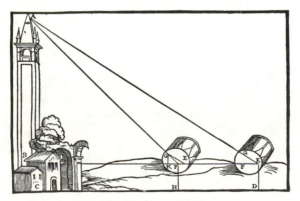

DRUMHEAD TRIGONOMETRY

A common method of triangulating in the 16th century. From Belli's *Libro del Misvrar*, Venice, 1569

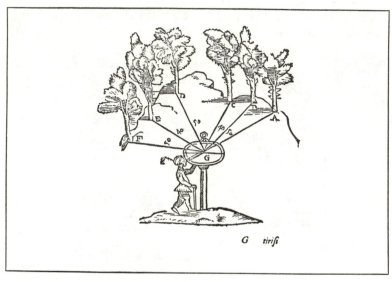

G   *tirifi*

EARLY APPROACH TO THE PLANE TABLE

The plane table in various forms was probably developed from such an instrument as the one here shown. In this case the table was used merely for taking horizontal angles. From Cosimo Bartoli, *Del Modo di Misvrare*, Venice, 1589

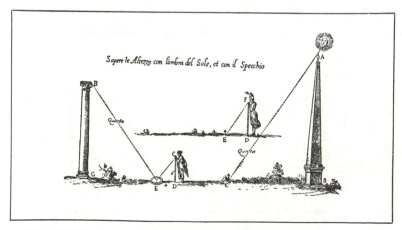

Sapere le Altezze con l'ombra del Sole, et con il Specchio

USE OF THE SHADOW AND THE MIRROR IN MEASURING HEIGHTS
From Giovanni Pomodoro's *La Geometria Prattica*, Rome, 1624

honor of its invention is usually awarded to Ptolemy. He used a stone cube, on one of the faces of which the quadrant was cut. On this was mounted a small cylindric pipe, as we should mount a telescope, and by this device he was able to take the height of the sun, evidently by means of the ray of light which shone through the cylinder. There is no indication of the size of Ptolemy's quadrant, but if we judge by the later specimens in use in the East, and by the incomplete records, it was probably a large one. He says that he used his quadrant in taking many astronomical observations; but he gives no results, and it is rather doubtful whether he did more than suggest the instrument, depending upon the results secured by Eratosthenes and others of his predecessors.

The quadrant is described in many works of the 16th, 17th, and 18th centuries, but with the invention of the telescope all devices of this kind gradually gave way to the transit in astronomical work. The sextant was invented by Thomas Godfrey, of Philadelphia, in 1730.

ANCIENT LEVELS AND SQUARE

From the tomb of Marcus Æbutius Macedo. Like the tomb of Lucius Æbutius Faustus (page 361), it is of uncertain date

Drumhead Trigonometry. The continual warfare of the Renaissance period shows itself in many ways in the history of mathematics. Some of these manifestations are mentioned from time to time in this work, and one of them is related to the subject now under consideration. Several writers of the 16th century give illustrations of the use of the drumhead as a simple means of measuring angles of elevation in computing distances to a castle or in finding the height of a tower. Such an illustration is shown on page 355 and is self-explanatory.

II

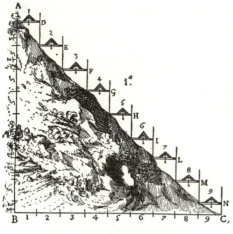

EARLY METHODS OF LEVELING

From Pomodoro's *La Geometria Prattica*, Rome, 1624. This was an early Egyptian method and was transmitted through the Greek and Roman surveyors

Somewhat related to this crude instrument is one for taking horizontal angles, as illustrated on page 356. From this device the plane table was probably developed.

The Mirror. In the early printed works on applied geometry there are frequent references to the speculum, a horizontal mirror used in measuring heights by the aid of similar triangles. The method is still in use for certain purposes, but in the 16th and 17th centuries it seems to have been extensively employed. On account of the difficulty of obtaining a satisfactory level, and the fact that one triangle was

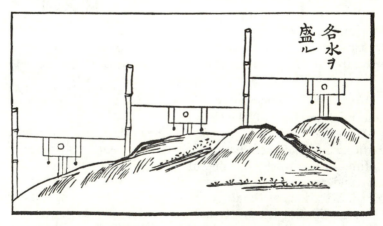

JAPANESE LEVELING INSTRUMENT

From Murai Masahiro's *Ryōchi Shinan*, a work on surveying, about 1732

small and not easily measured with accuracy, the method was not of much value. The plan of using the speculum was based upon the principle of similar triangles and is illustrated on page 356.

Leveling Instruments. The common leveling instrument of ancient times was the isosceles triangle with a plumb line from the vertex. This is found in Egyptian remains, is represented on the monuments of Roman surveyors,[1] is referred to by medieval writers, and is still in general use in various parts of the world. Until the invention of the telescope, and the consequent increase in accuracy of observation, it satisfied all ordinary needs. There are many reasons for believing that the early Egyptian surveyors who laid out the pyramids made use of this instrument for establishing their levels. An interesting variant of this instrument is seen in the quadrilateral which the Japanese scholars developed before the free influx of Western mathematics. Such a device is shown on page 358.

The principles underlying the later forms of leveling instruments were not numerous; in fact, the fundamental ones were only two in number,[2] the older one depending upon the plumb line, and the later one upon the state of

EARLY JAPANESE SUR-
VEYING INSTRUMENT

From a drawing in a manu-
script of a work (see
page 358) by Murai Masa-
hiro, about 1732

---

[1] On the leveling instruments of the Romans see C. G. de Montauzan, *Essai sur la science et l'art de l'ingénieur aux premiers siècles de l'Empire Romain*, pp. 46, 62, 74 (Paris, 1908). On Greek and Roman engineering instruments see Pauly-Wissowa; R. C. Skyring Walters, *Transactions of the Newcomen Society*, II, 45, and T. East Jones, *ibid.*, p. 61; E. N. Legnazzi, *Del Catasto Romano*, Verona, 1887.

[2] For a discussion of the subject see N. Bion, *Traité de la Construction et des principaux Usages des Instrumens de Mathematique*, p. 285 (Paris, 1713; ed. of The Hague, 1723). This is the best of the early classical treatises upon the subject of mathematical instruments, and is profusely illustrated.

equilibrium of some kind of liquid. The former was used in the ancient triangle illustrated on page 358, and in related types, while the latter is still seen in the ordinary level used by carpenters and in the leveling instrument used by engineers,—the *niveau à l'eau* described by Bion (1713). The triangle level had various special forms, such as an inverted T (that is, ⊥) with a plumb line along the vertical arm. In this form it was called by French writers the *niveau d'air*. The horizontal part was usually a tube through which the observer could sight when running a level line.

After the telescope was invented the tube was fitted with lenses,[1] and the instrument became, either with the plumb line or with the water level attached, not unlike the instrument in common use today. Sometimes the plumb line and the level were attached to the same instrument. Huygens invented a level in the form of a cross on which an inverted T was hung by a ring at the top, the telescope being kept horizontal by means of a weight.[2]

Until the advent of a new type of engineering, made possible by the commercial use of structural steel, the level was used chiefly for two purposes. The first of these was the construction of canals for purposes of irrigation, particularly in Mesopotamia and Egypt, and of aqueducts as a result of the Roman demand for pure water. The second use was seen in the building of fortifications, particularly during and as a result of the wars of the 17th century. The textbooks of that century on applied geometry (mensuration) gave much attention to the subject. The general practice in leveling was not unlike that of the present time, the chief difference being in the degree of precision of the instruments used. It is evident that in construction work of any extent the level was always necessary, but its elaborate use in modern engineering—as in railway gradients, tunnels, and bridges, and as in the erecting of modern office buildings of great height—surpasses anything conceived of in ancient times.

[1] The *niveau d'air à lunette* of Bion's treatise of 1713.
[2] Bion, *loc. cit.*, p. 110.

Other Surveying Instruments. It is not possible, in the limited space that should be allowed the subject in a work of this kind, to mention all the simple surveying instruments and devices that have come down to us from the Egyptian, Greek, and Roman civilizations. The simple staff, with a crude diopter through which to sight in running a line, is found in all parts of the world and is probably very ancient.

From such a humble origin sprang the *groma* used by the Roman surveyors and here illustrated from the tomb of Lucius Æbutius Faustus. He is mentioned in the third line as a *mensor*, but the term more commonly used in the case of a land surveyor was *agrimensor* (field measurer) or *gromaticus* (one who used the *groma*). The *groma* (*cruma, gruma*) consisted of the *stella* (the star-shaped part) and the *pondera* (the plumb lines). Surveyors in the time of the empire often spoke of the *machina* or *machinula* which they used and which consisted of two parts, the *groma* or *stella* and the *ferramentum* (the iron standard). Hence Hyginus (*c.* 120) says "ferramento groma superponatur" (let the groma be placed upon the iron standard).

ANCIENT SURVEYING
INSTRUMENT

From the tomb of Lucius Æbutius Faustus, a *mensor* (*agrimensor*, surveyor, or perhaps a measurer for architects and builders). The entire inscription was as follows: Tribv Clavdia Lvcivs Aebvtivs Lvcii libertvs Favstvs mensor sevir sibi et Arriae Qvinti libertae avctae vxori et svis et Zepyre libertae vivvs fecit. The instrument shown is the *groma*

In the 17th and 18th centuries, stimulated by the metal work of the Italian artists, the instrument-makers of France and Italy produced many beautiful pieces of workmanship designed with much ingenuity. These pieces are occasionally seen in museums, and one is shown in the following illustration.[1]

ELABORATE MATHEMATICAL INSTRUMENT

Showing artistic metal work of the 18th century. Now in the Metropolitan Museum of Art, New York

Such instruments were often elaborately engraved and some of those apparently made for the noble patrons of the sciences were even gold-plated. One of the elaborate forerunners of the range finder is shown on page 363.

[1] R. T. Gunther, *Early Science in Oxford* (Vols. I and II, Oxford, 1922, 1923), with a catalogue of the early mathematical instruments belonging to the University and colleges of Oxford. See particularly II, 192–233.

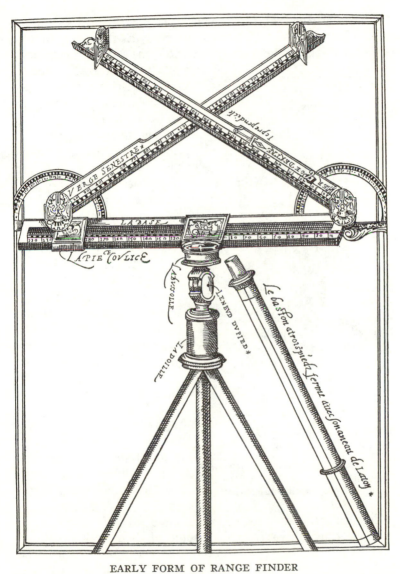

EARLY FORM OF RANGE FINDER

From Danfrie's *Declaration de l'Vsage du Graphometre*, Paris, 1597; appendix on trigonometry, p. 11

Other Astronomical Instruments. While this work is only indirectly concerned with astronomy, many astronomical instruments are distinctly mathematical, and some of them are particularly interesting as works of art. Such are the elaborate bronze pieces on the walls of the city of Peking, mostly due to the Jesuit influence which began about 1600, but partly native in their design and general plan. It was through the devising and use of instruments like these that such missionaries as F. Verbiest and J. A. Schall von Bell (*c.* 1660) were able to make observations that demonstrated, even to the hostile critics, the superiority of European astronomy over that of the Chinese. One of these pieces is shown in the illustration.[1]

BRONZE QUADRANT ON THE WALLS OF PEKING

One of several elaborate bronze instruments, most of them made under the influence of Jesuit missionaries

In Persia and India there are still to be found celestial spheres of great beauty, generally dating from the 17th century. These are usually of bronze, sometimes with silver stars.

The Hindus, Persians, and Arabs have also left many astrolabes of beautiful workmanship, some of them with constellations or particular stars represented in silver. Until the invention of the telescope their smaller types of astronomical instruments were unsurpassed both in beauty and in accuracy.

[1] See Volume I, page 272.

The most interesting of the Hindu instruments are found in the five observatories built by the Maharajah Jai Singh between 1728 and 1734.[1] These observatories were located at Delhi, Jaipur, Benares, Ujjain, and Mathurā, and represent the Arab astronomico-astrological science instead of the native Hindu or the European. Jai Singh was a Sikh by birth and was so interested in astronomy that he translated Ulugh Beg's

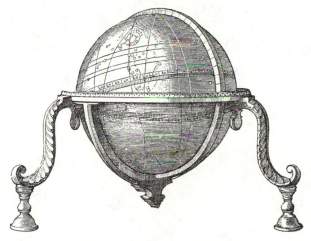

HINDU CELESTIAL SPHERE

This piece is of bronze, the stars being inlaid in silver. It was made *c.* 1600. From the author's collection

catalogue of the stars (*c.* 1435). He was of the opinion that the small brass instruments used in Samarkand were not accurate enough, and hence he determined to construct pieces so large and substantial as to leave no doubt about the validity of the observations. The results were monumental and are still the object of admiration to those interested in the science of India. An illustration showing one of the most elaborate and carefully preserved of these observatories (the one at Jaipur) will be found on page 366.

[1] G. R. Kaye, *The Astronomical Observatories of Jai Singh*, Calcutta, 1918; see also the review in the *Journal of the Royal Asiatic Soc.*, July, 1919, p. 427.

The Jaipur observatory was constructed by the Maharajah Jai Singh about 1734. The Jesuit missionary Joseph Tieffenthaler, in a work published in 1785, speaks of it as follows:

OBSERVATORY AT JAIPUR, INDIA

Showing the kinds of instruments generally used before the days of the telescope. This observatory, although relatively modern, is based upon ancient models

It is such a work as is never seen in this part of the world and, by the novelty and grandeur of the instruments, strikes one with astonishment. . . . What attracts most attention is a gnomon (*axis mundi*), remarkable for its height of 70 Paris feet. . . . There are three very large astrolabes, cast in copper, suspended by iron rings.

This is all quite as impressive to the visitor now as it was then. The instruments, which had become damaged through age and neglect, were restored in 1902.

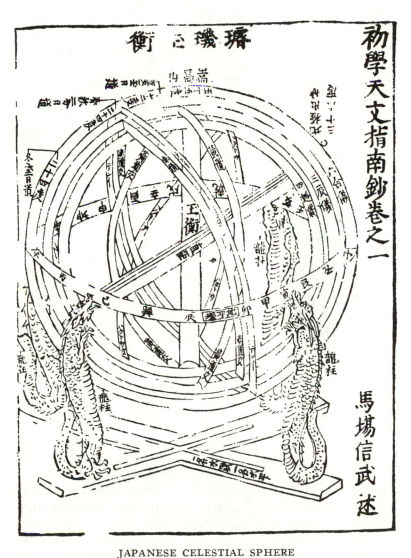

JAPANESE CELESTIAL SPHERE

From a wood engraving in Baba Nobutake's *Shogaku Tenmon* (1706)

The Chinese influence shows itself in the Japanese works of the 17th and 18th centuries, as is seen in the illustration from Baba Nobutake's work of 1706. We also find in Japan in this same period the use of the pierced sphere in astronomical observations and in the work of the astrologers. This device was common in Europe in the latter part of the Middle Ages and is found in various printed works of the 16th century.

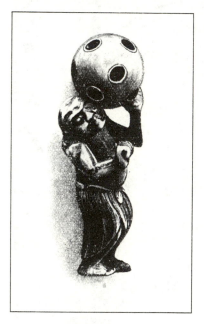

JAPANESE FIGURE OF AN
ASTRONOMER

Caricature in ivory. From the author's
collection

12. THE PROBLEM OF EARTH MEASURE

Need for Instruments of Precision. The need for instruments of a high degree of precision was first felt in connection with astronomy and the measure of the earth. The subject is too extensive to be considered at length in a work of this kind, but its general nature will be understood by a brief reference to the history of the measure of the earth's circumference and density.

Circumference determined from Arc. It should first be understood that the solution of this problem did not involve the ratio of the circumference to the diameter; it required the finding of the circumference when the diameter was unknown. When first undertaken it had nothing to do with navigation, economics, or military conquest; it developed as a purely abstract contribution to human knowledge. The plan adopted by the Greeks was the same, in basic principle, as the one used today, namely, that of measuring the amplitude and the length of an arc of a great circle (generally a meridian) and from these data computing the circumference. This led to one of the many branches of geodesy, a subject into the history of which we cannot enter at length in this work.

Application of Circle Measure to Geodesy. Pythagoras (*c.* 540 B.C.) was the first, so tradition asserts, to teach that the earth is a sphere and that it is situated in the center of the universe.[1] This idea was accepted by various Greek philosophers, and Aristotle (*c.* 340 B.C.) states that "the mathematicians who have attempted to calculate the circumference of the earth say that it may be forty myriads of stadia,"[2] that is, 400,000 stadia. The stadium varied so much with ancient writers that this does not give us any very satisfactory information.[3] Taking a rough approximation, however, say ten stadia to an Anglo-American mile, this makes the circumference 40,000 miles. Aristotle gives us no information as to the names of the mathematicians who made the calculations, and none as to the method employed, but it has been thought that the approximation is due to Eudoxus (*c.* 370 B.C.). It is evident, however, that the circumference was found by multiplying the length of a known arc, and not by using the ratio of the circumference to the diameter.

Four Greek Computations of the Earth's Circumference. From the time of Aristotle to that of Ptolemy (*c.* 150) there were four noteworthy attempts at measuring the earth's circumference. Of these the first is referred to by Archimedes (*c.* 225 B.C.), who speaks of certain writers as having stated that the circumference is 30 myriads of stadia, say about 30,000 miles. He does not mention the writers, and it is possible that he may have referred to some of the earlier attempts made by his friend Eratosthenes (*c* 230 B.C.). In his computation of the number of grains of sand in the universe, however, he takes the circumference as ten times this distance, so as to be on the safe side.[4]

---

[1] On this entire subject see Bigourdan, *Astronomie*, p. 144 seq.

[2] *De Cælo*, II, 16.

[3] One of the stadia was 125 paces (double steps), or, say, 625 Roman feet, equal to 606¾ Anglo-American feet.   [4] *Archimedes*, ed. Heiberg, I, 221.

The third important attempt at the measure of the earth's circumference is definitely known to have been made by Eratosthenes, and the fourth by Poseidonius (*c.* 100 B.C.).

Eratosthenes on the Measure of the Earth. The first attempt of which we have any details is this third one,—the one briefly described as due to Eratosthenes. Supplementing the description given in Volume I, page 110, it may be said that Eratosthenes used the arc of a great circle extending from Syene (the modern Assouan) to Alexandria. He took the length of this arc as 5000 stadia, but how this length was ascertained is not stated in any ancient writings. It is probable that the official pacers,[1] employed by Alexander and other military leaders in planning their campaigns, had made reports of all such standard distances, and Eratosthenes, as librarian at Alexandria, doubtless had access to their records.

It was well known that on the day of the summer solstice the sun's rays lighted up completely the wells of Syene at noontime, and that a body like an obelisk cast no shadow. On the other hand, Eratosthenes found that the zenith distance of the sun on this day, as measured at Alexandria, was $\frac{1}{50}$ of the circumference. It is not known how this angle was found, but it has been thought that Eratosthenes used certain armillary spheres which tradition says were furnished by the king, Ptolemy Euergetes. We are told by Cleomedes (*c.* 40 B.C.), however, who wrote on the *Circular Theory of the Heavenly Bodies,* that he used the *sca'phe,*[2] a concave sundial, and this may easily have been the case. Since the zenith distance of the sun changed $\frac{1}{50}$ of the circumference in 5000 stadia, Eratosthenes concluded that the circumference was 50 × 5000 stadia, or 250,000 stadia, roughly equivalent to 25,000 miles, or 40,000 kilometers. This would make $694\frac{4}{9}$ stadia to a degree; and since Eratosthenes was naturally aware that his measurements were merely approximate, he felt it allowable to take 700

---

[1] Βηματισταί (*bematistai'*), singular βηματιστής (*bematistes'*), from βῆμα (*be'ma*), a step.

[2] Σκάφη, originally anything dug out; hence a bowl, and then the bowl of a hemispherical sundial.

stadia as a more convenient measure for 1°. He had crude instruments with which to work, he did not take into consideration the difference of longitude of his two stations, and the stadium was a varying unit at best, so that his assumption of 700 stadia was not an unreasonable one.

Poseidonius on the Measure of the Earth. Poseidonius (*c.* 100 B.C.) was, as we have seen,[1] a Stoic philosopher, well known as an astronomer, a geographer, a historian, and a statesman. After having traveled extensively in all the Mediterranean countries, he opened a school at Rhodes and had among his pupils both Cicero and Pompey. Although his works are lost, Cleomedes (*c.* 40 B.C.) has given us a certain amount of information as to his method of measuring the circumference of the earth. Like Eratosthenes he took a known arc, selecting the one from Rhodes to Alexandria and estimating its length as 5000 stadia. He then observed that when the star Canopus was on the horizon at Rhodes, it was $\frac{1}{4}$ of a sign (that is, $\frac{1}{4}$ of 30°, or $\frac{1}{48}$ of 360°) above the horizon at Alexandria. He concluded that the circumference is 48 × 5000 stadia, or 240,000 stadia. This made the length of the degree $666\frac{2}{3}$ stadia.

It is hardly probable that Poseidonius considered these results as close approximations, since neither the length of his arc nor the elevation of the star could be measured with any approach to accuracy by instruments then available.

Ptolemy on the Measure of the Earth. The last of the noteworthy attempts of the ancient Greeks to find the circumference of the earth was made by Ptolemy (*c.* 150). He took a degree as 500 stadia, thus finding the circumference to be 180,000 stadia. He asserted that it was unnecessary to take the arc of a meridian, an arc of any other great circle being sufficient. We are without information, however, as to his method of measuring the arc selected. It will be noticed that his result is to that of Poseidonius as 3 :4; and since this is the ratio between two of the stadia employed by the ancients, it is possible that he simply used the latter's computations.

---

[1] Vol. I, p. 118. See O. Viedebantt, "Poseidonius," *Klio*, XVI, 94.

The theory has been advanced that all these results set forth by the Greeks were due to Egyptian or other measurements which are no longer extant, but there is no scientific basis for the conjecture.

Arab Measure of the Earth. It was some seven centuries after the last of the Greek geodesists that the Arabs engaged in the work of measuring the circumference of the earth. By order of al-Mâmûn certain mathematicians of Bagdad undertook the necessary surveys on the plain of Sinjar in Mesopotamia. They formed two groups, one party going to the north and one to the south, each proceeding to a point at which the elevation of the pole changed $1°$ from that of the base station. They then measured the respective distances, one being found to be 57 miles and the other $56\frac{1}{4}$ miles; this mile was given as 4000 "black cubits," but the length of this cubit is now unknown. The difference in the two measurements illustrates the lack of the necessary instruments of precision, even among a people who had brought the construction of such instruments to the highest degree of perfection known at that time.

Invention of the Telescope. With respect to instruments used in astronomy and geodesy the greatest improvement is due to the invention of the telescope, although much is also due to modern technique in manufacture and to the use of such devices as the vernier and the micrometer.

Roger Bacon (*c.* 1250) stated that it was possible to construct tubes by means of which distant objects could be seen as if they were near at hand,[1] but we have no evidence that this was other than a prophetic statement by a man who seemed peculiarly gifted in this respect. Possibly he was led to this prophecy by a knowledge of spectacles, which appeared sometime in the 13th century, for they were certainly known in 1299, and a certain Salvino degli Armati, a Florentine (died in 1317), is mentioned as their inventor.

---

[1] "Ita ut in incredibili distancia videremus arenas et litteras minias minutas, et ut altissima videantur infima et e contrario." Sloane Ms., fol. 84, a, 2. Cf. *Opus Majus* (ed. Bridges), ii, 164; *Opus Tertium* (ed. Little), 41.

The possibility of the telescope is also mentioned by Fracastorius, who, in his *Homocentricorum seu de stellis Liber Unus* (1538), speaks of using two superposed lenses in looking at a distant object. Somewhat similar statements were made by Giambattista della Porta in his *Magia Naturalis* (Naples, 1558) and by Kepler in his *Paralipomena* (Frankfort, 1604).

The invention seems due, however, not to the work of scientists like those mentioned, but largely to chance. It is uncertain who was the actual inventor, the claims of three artisans being about equal. These men are Zacharias Janszoon (Jansen), Johann Lippersheim (Lippershey, Lipperseim, Laprey, *c.* 1608), and Jacob Metius Adriaenszoon[1] (*c.* 1608).

Janszoon lived at Middelburg, was a grinder of lenses, and, apparently with the aid of his father, improved the microscope in 1590 and is known to have had a telescope in 1610.

Lippersheim was also a lens grinder of Middelburg. He is known to have asked for a patent in 1608 for an instrument intended to see distant objects, the lenses being of rock crystal.

Descartes and others attributed the invention to Jacob Metius Adriaenszoon (*c.* 1608), who happened to make the necessary combination of certain lenses and burning mirrors. He also asked for a patent in 1608, a few days after Lippersheim had made his request.[2]

The invention is known to have been made public in October, 1608, and the knowledge of the instrument spread all through Europe with astonishing rapidity. Even in 1608, and still more in 1609, instruments were made in France, England, Italy, and Germany. Hearing of the new device, Galileo, in 1609, invented an instrument of his own and by its aid at once made remarkable discoveries in astronomy; and from this time on a precision of observation unknown to earlier scientists, although the instrument was not accepted by all astronomers, became possible. It was only after the invention of achromatic lenses, however, that satisfactory results were obtained.

---

[1] He was a brother of the Adriaen Metius (1571–1635) mentioned in Volume I, page 340. He was born at Alkmaar and was interested in the grinding of lenses.

[2] Bigourdan, *Astronomie*, p. 124.

Modern Measures of the Earth. The first noteworthy modern attempt at measuring the earth was made by Jean Fernel (*c.* 1528), physician to Henri II of France. Fernel took the arc determined by Paris and Amiens, two stations being taken on the same meridian. Knowing the altitude of the sun at Paris, he proceeded northward to a point where the altitude was 30′ less than that at Paris. He then measured the arc by taking the number of revolutions of a wheel of known circumference. No record is available as to his method of allowing for errors, but certain compensations were made and the conclusion was reached that $1° = 57,099$ toises, 1000 toises being a little more than a geographic mile. This result is remarkable, the mean afterwards obtained by Lacaille[1] and Delambre in the latter part of the 18th century being 57,068 toises.

In 1617 Snell undertook the measurement of an arc by an elaborate system of triangulation, and although his results were satisfactory as to length of arc, they were not so as to amplitude.

Further French Attempts. In 1669 and 1670 Jean Picard[2] carried on an elaborate system of triangulation, measured an arc from a point near Corbeil to one near Amiens, and found that $1°21′54″$ corresponded to 68,347 toises 3 pieds, which gave 57,060 toises to $1°$. He estimated that the amplitude was correct to within $2″$ or $3″$.

In 1686 Newton proved that the earth is an oblate spheroid, a result not generally accepted by French scientists, chiefly owing to the conclusions reached by Jacques Cassini (Cassini II) as mentioned below. It was therefore decided that France should undertake a more elaborate and careful survey, not confined to that country alone, but including arcs nearer to and more remote from the equator.

---

[1] Nicolas Louis de Lacaille (La Caille); born at Rumigny, May 15, 1713; died in Paris, March 21, 1762. He wrote upon mathematics and physics, but chiefly upon astronomy.

[2] Born at La Flèche, Anjou, July 21, 1620; died in Paris, July 12, 1682. He wrote upon physics and astronomy and was particularly well known for his work on the measure of the earth.

The degree of accuracy reached by Picard was increased through the efforts of Giovanni Domenico (Jean Dominique) Cassini (Cassini I), who extended Picard's meridian in 1701 southward to the Pyrenees. It was afterwards extended northward to Dunkirk (Dunkerque), although the results, published by Jacques Cassini (Cassini II) in 1720, provoked great opposition because of their lack of precision and the incorrect conclusions reached with respect to the elongated form of the earth. In 1735 France sent a mission to Peru, and an elaborate survey was made for the purpose of measuring an arc. This work was carried on under the direction of Bouguer,[1] Condamine,[2] and Godin.[3] By 1745 they had completed the measurement of an arc of 3°. D'Alembert spoke of the work as the greatest scientific enterprise that had thus far been undertaken. In the following year another mission, including such French scientists as Maupertuis and Clairaut, and the Swedish scientist Celsius,[4] began a similar work in Lapland. The result of this survey was the measurement of an arc of 1°. The conclusions reached in Peru and Lapland confirmed Newton's assertion of the flattening of the earth at the poles and led to Voltaire's reference to Maupertuis, against whom he had a personal grudge, as the "great flattener" (*grand aplatisseur*). The form of the earth appears from the fact that degrees of latitude increase in length as we approach the poles.

In 1739 and 1740, owing chiefly to the work of Lacaille, an arc of the meridian was again measured in France, the result being a correction of the errors published in 1720 and a new confirmation of Newton's theory of the shape of the earth.

Toward the close of the 18th century France undertook a third great survey, this time for the purpose of determining the

---

[1] See page 327.

[2] Charles Marie de la Condamine, born in Paris, January 28, 1701; died in Paris, February 4, 1774. He wrote extensively on geodesy.

[3] Louis Godin, born in Paris, February 28, 1704; died at Cadiz, September 11, 1760. He wrote chiefly on astronomy.

[4] Anders Celsius; born at Upsala, November 27, 1701; died at Upsala, April 25, 1744. He was professor of astronomy at Upsala, but spent some years in France, Germany, and Italy.

length of the standard meter. In this undertaking a number of the greatest French scientists were engaged, but for the geodetic work Delambre and Méchain were chiefly responsible.

In the 19th and 20th centuries extensive triangulations have been made, and with the methods employed there have been connected such prominent names as those of Biot, Arago, Schumacher, Legendre, Laplace, Gauss, and Bessel. The Ordnance Survey of Great Britain, begun in 1783 and completed in 1858, resulted in the measurement of an arc of 10° 13', extending from the Isle of Wight to one of the Shetland group; the triangulation of India (1790–1884) gave an arc of about 24°; and the Russo-Scandinavian measurements, begun in 1817, resulted in an arc of 25° 20'. The arc recently measured in Africa, extending over a distance of about 65°, will, joined to the Russo-Scandinavian arc, give an arc of about 106°.

Mass of the Earth. The determination of the earth's density depends on the law of gravitation, and so it began in the work of Newton, who estimated it as five or six times that of water.

The first of the later methods depends upon the deflection of a plumb line due to the attraction of a mountain. This was first used by Pierre Bouguer, in Peru (*c.* 1740). By this plan Maskeleyne[1] (1774–1776) placed the density between 4.5 and 5.

The second method is based upon a comparison of the vibrations of a pendulum at sea level with those at the top of a high mountain. Francesco Carlini, the Italian astronomer, used the method in 1821 and obtained a density of 4.84.

The third method is due to Henry Cavendish[2] (1798) and is based upon the mutual attraction of known masses. Francis Baily[3] (1843) obtained the result of 5.67 by this method.

The fourth method uses a finely graduated balance to determine the attraction of known masses. By its use results of 5.69 were obtained by Von Jolly in 1881, and 5.49 by Poynting in 1891. The latest experiments give the result as about 5.53.

---

[1] Nevil Maskeleyne, born in London, October 5 (O. S.), 1732; died at Greenwich, February 9, 1811. He became astronomer royal in 1765.

[2] Born at Nice, October 10, 1731; died in London, February 24, 1810.

[3] Born at Newbury, Berkshire, April 28, 1774; died in London, August 30, 1844.

TOPICS FOR DISCUSSION

1. Intuitive geometry as it shows itself in the primitive decoration used by various peoples.

2. Intuitive geometry as it shows itself in the early stages of mathematics in various countries.

3. The rise of demonstrative geometry and the six most important contributors to the science in ancient Greece.

4. The various names used for geometry and the special significance of each.

5. The development of the terminology of elementary geometry, especially among the Greeks.

6. The development of the postulates and axioms of elementary geometry before the 19th century.

7. Propositions of elementary geometry of which the origin is known or which have any history of special interest.

8. The various methods of solving each of the Three Famous Problems of antiquity.

9. The historical development of methods for finding the approximate value of π.

10. The principal steps taken by the Greeks in the development of geometric conics.

11. The principal steps in the development of plane analytic geometry, with special reference to the 17th century.

12. A discussion of the history of solid analytic geometry.

13. The history of the most important higher plane curves commonly found in the study of elementary analytic geometry, together with the applications of these curves.

14. The nature, purpose, and history of descriptive geometry.

15. The relation of the fine arts to geometry in the 15th century.

16. The development of projective geometry.

17. The development of the non-Euclidean geometries, with special reference to the work of Bolyai, Lobachevsky, and Riemann.

18. The development of perspective and optics considered as mathematical subjects.

19. A study of the most interesting of the primitive instruments.

20. The general development of geodesy, particularly among the Greeks and in modern times, and with reference to the measure of the circumference of the earth.

CHAPTER VI

ALGEBRA

1. General Progress of Algebra

Nature of Algebra. When we speak of the early history of algebra it is necessary to consider first of all the meaning of the term. If by algebra we mean the science which allows us to solve the equation $ax^2 + bx + c = 0$, expressed in these symbols, then the history begins in the 17th century; if we remove the restriction as to these particular signs, and allow for other and less convenient symbols, we might properly begin the history in the 3d century; if we allow for the solution of the above equation by geometric methods, without algebraic symbols of any kind, we might say that algebra begins with the Alexandrian School or a little earlier; and if we say that we should class as algebra any problem that we should now solve by algebra (even though it was at first solved by mere guessing or by some cumbersome arithmetic process), then the science was known about 1800 B.C., and probably still earlier.[1]

A Brief Survey proposed. It is first proposed to give a brief survey of the development of algebra, recalling the names of those who helped to set the problems that were later solved by the aid of equations, as well as those who assisted in establishing the science itself. These names have been mentioned in Volume I and some of them will be referred to when we consider the development of the special topics of algebra and their application to the solution of elementary problems.

[1] For a brief study of the early history see H. G. Zeuthen, "Sur l'origine de l'Algèbre," in the *Kgl. Danske Videnskabernes Selskab, Math.-fysiske Meddelelser,* II, 4, Copenhagen, 1919; M. Chasles, "Histoire de l'Algèbre," *Comptes rendus,* September 6, 1841; but the subject is treated of in any general history of mathematics and in the leading encyclopedias.

It should also be stated as a preliminary to this discussion that Nesselmann[1] (1842) has divided the history of algebra into three periods: the rhetorical, in which the words were written out in full; the syncopated, in which abbreviations were used; and the symbolic, in which the abbreviations gave place to such symbols as occur in statements like $\sqrt{x} - x^2 = a^{\frac{2}{3}}$. There are no exact lines of demarcation by which to establish these divisions, Diophantus, for example, having made use of certain features of all three; but the classification has some advantages and the student will occasionally find the terms convenient.

It should be borne in mind that most ancient writers outside of Greece included in their mathematical works a wide range of subjects. Ahmes ($c.$ 1550 B.C.),[2] for example, combines his algebra with arithmetic and mensuration, and even shows some evidence that trigonometry was making a feeble start. There was no distinct treatise on algebra before the time of Diophantus ($c.$ 275).

Algebra in Egypt. The first writer on algebra whose works have come down to us is Ahmes. He has certain problems in linear equations and in series, and these form the essentially new feature in his work. His treatment of the subject is largely rhetorical, although, as we shall see later, he made use of a small number of symbols.

There are several other references to what may be called algebra in the Egyptian papyri, these references consisting merely of problems involving linear or quadratic equations. There is no good symbolism in any of this work and no evidence that algebra existed as a science.

Algebra in India. There are only four Hindu writers on algebra whose names are particularly noteworthy. These are Āryabhaṭa,[3] whose *Āryabhaṭiyam* ($c.$ 510) included problems in

---

[1] G. H. F. Nesselmann, *Alg. Griechen*, p. 302.

[2] As already stated, the period may have been $c.$ 1600 B.C. or earlier.

[3] See Volume I, page 153, and remember that there were two Āryabhaṭas and that we are not certain which one of them is entitled to the credit for various contributions.

series, permutations, and linear and quadratic equations; Brahmagupta, whose *Brahmasiddhānta* (*c.* 628) contains a satisfactory rule for solving the quadratic, and whose problems include the subjects treated by Āryabhaṭa; Mahāvīra, whose *Gaṇita-Sāra Sangraha* (*c.* 850) contains a large number of problems involving series, radicals, and equations; and Bhāskara, whose *Bija Gaṇita* (*c.* 1150) contains nine chapters and extends the work through quadratic equations.[1]

Algebra in China. It is difficult to say when algebra as a science began in China. Problems which we should solve by equations appear in works as early as the *Nine Sections*[2] and so may have been known by the year 1000 B.C. In Liu Hui's commentary on this work (*c.* 250) there are problems of pursuit, the Rule of False Position, explained later in this chapter, and an arrangement of terms in a kind of determinant notation.[3] The rules given by Liu Hui form a kind of rhetorical algebra.

The work of Sun-tzï[4] (perhaps of the 1st century, but the date is very uncertain and may be several centuries earlier) contains various problems which would today be considered algebraic. These include questions involving indeterminate equations of which the following is a type:

There are certain things whose number is unknown. If they are divided by 3 the remainder is 2; by 5, the remainder is 3; and by 7, the remainder is 2. Find the number.

Sun-tzï solved such problems by analysis and was content with a single result, even where several results are admissible.

The Chinese certainly knew how to solve quadratics as early as the 1st century B.C., and rules given even as early as the *K'iu-ch'ang Suan-shu* above mentioned involve the solution of such equations.

---

[1] H. T. Colebrooke, *Algebra with Arithmetic and Mensuration, from the Sanscrit*, pp. 129–276 (London, 1817). For the various spellings of *Bija Ganita* see Volume I, page 278.

[2] *K'iu-ch'ang Suan-shu.*

[3] Mikami, *China*, pp. 19, 23.

[4] *Sun-tzï Suan-king.*

Liu Hui (*c.* 250) gave various rules which would now be stated as algebraic formulas and seems to have deduced these from other rules in much the same way as we should deduce formulas at the present time.[1]

By the 7th century the cubic equation had begun to attract attention, as is evident from the *Ch'i-ku Suan-king* of Wang Hs'iao-t'ung (*c.* 625).

The culmination of Chinese algebra is found in the 13th century. At this time numerical higher equations attracted the special attention of scholars like Ch'in Kiu-shao (*c.* 1250), Li Yeh, (*c.* 1250), and Chu Shï-kié (*c.* 1300),[2] the result being the perfecting of an ancient method which resembles the one later developed by W. G. Horner (1819).

With the coming of the Jesuits in the 16th century, and the consequent introduction of Western science, China lost interest in her native algebra and never fully regained it.

Algebra in Greece. Algebra in the modern sense can hardly be said to have existed in the golden age of Greek mathematics.[3] The Greeks of the classical period could solve many algebraic problems of considerable difficulty, but the solutions were all geometric. Hippocrates (*c.* 460 B.C.), for example, assumed a construction which is equivalent to solving the equation $x^2 + \sqrt{\frac{3}{2}} \cdot ax = a^2$, and Euclid (*c.* 300 B.C.), in his *Data*, solved problems equivalent to the following:

1. $xy = k^2$, $x - y = a$ (Prob. 84).
2. $xy = k^2$, $x + y = a$ (Prob. 85).
3. $xy = k^2$, $x^2 - y^2 = a^2$ (Prob. 86).

In his *Elements* (II, 11) Euclid solved the equivalent of $x^2 + ax = a$, and even of $x^2 + ax = b^2$, substantially by completing the geometric square and neglecting negative roots.

After Euclid there came a transition period from the geometric to the analytic method. Heron (*c.* 50?), who certainly

---

[1] Mikami, *China*, pp. 35, 36.   [2] Mikami, *China*, pp. 63, 79, 89.
[3] Nesselmann, *Alg. Griechen*; Heath, *Diophantus*. On the "application of areas" see Heath, *History*, and R. W. Livingstone, *The Legacy of Greece*, p. 111 (Oxford, 1922).

solved the equation $144 x (14 - x) = 6720$, may possibly have used the analytic method for the purpose of finding the roots of

$$\tfrac{11}{14} x^2 + \tfrac{29}{7} x = 212.$$

With Diophantus ($c.$ 275) there first. enters an algebraic symbolism worthy of the name, and also a series of purely algebraic problems treated by analytic methods. Many of his equations being indeterminate, equations of this type are often called Diophantine Equations. His was the first work devoted chiefly to algebra, and on this account he is often, and with much justice, called the father of the science.

Algebra among the Arabs and Persians. The algebraists of special prominence among the Arabs and Persians were Mohammed ibn Musa, al-Khowârizmî, whose *al-jabr w'al muqâbalah* ($c.$ 825) gave the name to the science and contained the first systematic treatment of the general subject as distinct from the theory of numbers; Almâhânî ($c.$ 860), whose name will be mentioned in connection with the cubic; Abû Kâmil ($c.$ 900), who drew extensively from al-Khowârizmî, and from whom Fibonacci (1202) drew in turn; al-Karkhî ($c.$ 1020), whose *Fakhrî* contains various problems which still form part of the general stock material of algebra; and Omar Khayyam ($c.$ 1100), whose algebra was the best that the Persian writers produced.

Medieval Writers. Most of the medieval Western scholars who helped in the progress of algebra were translators from the Arabic. Among these were Johannes Hispalensis ($c.$ 1140), who may have translated al-Khowârizmî's algebra; Gherardo of Cremona ($c.$ 1150), to whom is also attributed a translation of the same work; Adelard of Bath ($c.$ 1120), who probably translated an astronomical work of al-Khowârizmî, and who certainly helped to make this writer known; and Robert of Chester, whose translation of al-Khowârizmî's algebra is now available in English.[1]

[1] L. C. Karpinski, *Robert of Chester's Latin Translation . . . of al-Khowarizmi.* New York, 1915.

FIRST PAGE OF AL-KHOWÂRIZMÎ'S ALGEBRA

From a MS. of 1456. It begins, "Liber mahucmeti de Algebra et almuchabala."
In Mr. Plimpton's library

The greatest writer on algebra in the Middle Ages was Fibonacci, whose *Liber Quadratorum* (*c.* 1225) and *Flos* both relate to the subject. The former work includes the treatment of such problems as $x^2 + y^2 = z^2$ and other well-known types, and shows great ingenuity in the solution of equations. Of the German algebraists in the Middle Ages the leading writer was Jordanus Nemorarius (*c.* 1225). His *De Numeris Datis*, already described (Vol. I, p. 227), contains a number of problems in linear and quadratic equations of the type still familiar in our textbooks. In general, however, the medieval writers were more interested in mathematics as related to astronomy than in mathematics for its own sake.[1]

The Renaissance. Algebra in the Renaissance period received its first serious consideration in Pacioli's *Sŭma* (1494), a work which summarized in a careless way the knowledge of the subject thus far accumulated. By the aid of the crude symbolism then in use it gave a considerable amount of work in equations.

The next noteworthy work on algebra, and the first to be devoted entirely to the subject, was Rudolff's *Coss* (1525). This work made no decided advance in the theory, but it improved the symbolism for radicals and made the science better known in Germany. Stifel's edition of this work (1553–1554) gave the subject still more prominence.

The first epoch-making algebra to appear in print was the *Ars Magna* of Cardan (1545). This was devoted primarily to the solution of algebraic equations. It contained the solution of the cubic and biquadratic equations,[2] made use of complex numbers, and in general may be said to have been the first step toward modern algebra.

The next great work on algebra to appear in print was the *General Trattato* of Tartaglia (1556–1560), although his side of the controversy with Cardan over the solution of the cubic equation had already been given in his *Qvesiti ed invenzioni diverse* (1546).

---

[1] On the general topic see P. Cossali, *Origine, trasporto in Italia, primi progressi in essa dell' Algebra*, 2 vols., Parma, 1797–1799.

[2] In chapters xi and xxxix et seq.

Algebra in the New World. As already stated,[1] the first mathematical work published in the New World was the *Sumario Compendioso* of Juan Diez. This appeared in the City of Mexico in 1556 and contains six pages on algebra. Some idea of its general nature may be obtained from two of the problems relating to the subject. Of these the first, literally translated and requiring the solution of the quadratic equation $x^2 - 15\frac{3}{4} = x$, is as follows:

Find a square from which if $15\frac{3}{4}$ is subtracted the result is its own root.

Let the number be *cosa* [x]. The square of half a *cosa* is equal to $\frac{1}{4}$ of a *zenso* [x^2]. Adding 15 and $\frac{3}{4}$ to $\frac{1}{4}$ makes 16, of which the root is 4, and this plus $\frac{1}{2}$ is the root of the required number.

Proof: Square the square root of 16 plus half a *cosa*, which is four and a half, giving 20 and $\frac{1}{4}$, which is the square number required. From $20\frac{1}{4}$ subtract 15 and $\frac{3}{4}$ and you have 4 and $\frac{1}{2}$, which is the root of the number itself.

The second problem, also literally translated, requires the solution of the quadratic equation $x^2 + x = 1260$:

A man takes passage in a ship and asks the master what he has to pay. The master says that it will not be any more than for the others. The passenger on again asking how much it would be, the master replies: "It will be the number of pesos which, multiplied by itself and added to the number, gives 1260." Required to know how much the master asked.

Let the cost be a *cosa* of pesos. Then half of a *cosa* squared makes $\frac{1}{4}$ of a *zenso*, and this added to 1260 makes 1260 and a quarter, the root of which less $\frac{1}{2}$ of a *cosa* is the number required. Reduce 1260 and $\frac{1}{4}$ to fourths; this is equal to $\frac{5041}{4}$, the root of which is 71 halves; subtract from it half a *cosa* and there remains 70 halves, which is equal to 35 pesos, and this is what was asked for the passage.

Proof: Multiply 35 by itself and you have 1225; adding to it 35, you have 1260, the required number.

[1] See Volume I, page 353. D. E. Smith, *The Sumario Compendioso of Brother Juan Diez*, Boston, 1921.

First Teachable Textbooks in Algebra. The first noteworthy attempt to write an algebra in England was made by Robert Recorde, whose *Whetstone of witte* (1557) was an excellent textbook for its time. The next important contribution was Masterson's incomplete treatise of 1592–1595, but the work was not up to the standard set by Recorde.

The first Italian textbook to bear the title of algebra was Bombelli's work of 1572. In this book the material is arranged with some attention to the teaching of the subject.[1]

By this time elementary algebra was fairly well perfected, and it only remained to develop a good symbolism. As will be shown later, this symbolism was worked out largely by Vieta (*c.* 1590), Harriot (*c.* 1610), Oughtred (*c.* 1628), Descartes (1637), and the British school of Newton's time (*c.* 1675).

So far as the great body of elementary algebra is concerned, therefore, it was completed in the 17th century.

2. Name for Algebra

Early Names. The history of a few of the most familiar terms of algebra not elsewhere discussed will now be considered, and of these the first is naturally the name of the science itself.

Ahmes (*c.* 1550 B.C.) called his treatise "Rules for inquiring into nature, and for knowing all that exists, [every] mystery, . . . every secret,"[2] and this idea is not infrequently expressed by later writers. Thus Seki (*c.* 1680) called a certain part of algebra the *kigen seihō*, meaning a method for revealing the true and buried origin of things, and we find the same idea in the titles of algebras by Follinus (1622)[3] and Gosselin (1577),[4] and in a note on Ramus written by Schoner in 1586.[5]

---

[1] *L'Algebra parte maggiore dell' arimetica* . . . , Bologna, 1572. There is a second edition, differing only in the title-page, Bologna, 1579.

[2] He adds: "Behold, this roll was written . . . [under . . . the King of Upper] and Lower Egypt, Aauserrē. . . . It was the scribe Aḥmōse who wrote this copy." Peet, *Rhind Papyrus*, p. 33. Professor Peet gives the probable date as between 1788 and 1580 B.C.

[3] *Algebra sive liber de rebus occultis.*

[4] . . . *de occulta parte numerorum.*

[5] " . . . Almucabalam, hoc est, librum de rebus occultis" (p. 322).

Since the Greeks gave the name "arithmetic" to all the theory of numbers, they naturally included their algebra under that title,[1] and this explains why the algebra of Diophantus went by the name of arithmetic. The Hindu writers had no uniform name for the science. Āryabhaṭa (c. 510) included algebra in his general treatise, the Āryabhaṭiyam; Brahmagupta (c. 628) placed it in his large treatise, giving a special name (kuṭaka, the pulverizer)[2] to his chapter on indeterminate equations. Mahāvīra (c. 850) included it in his Gaṇita-Sāra-Sangraha, a title meaning a brief exposition of the compendium of calculation. Bhāskara (c. 1150) had a name for general arithmetic, Bija Gaṇita,[3] meaning the calculation of seeds, that is, of original or primary elements,[4] and a special name for algebra, Avyakta gaṇita,[5] or Avyakta-kriyā, the former referring to the calculation with knowns and the latter to that with unknowns.

The Chinese used various fanciful titles for their books containing algebra and spoke of the method of the t'ien-yuen (celestial element),[6] meaning the algebra that made use of calculating rods (the Japanese sangi), to indicate coefficients.[7] Similar fanciful names were used in Japan, as also the name yendan jutsu (method of analysis), and the name kigen seihō already mentioned.

---

[1] So Euclid's Elements, II, devoted to arithmetic, includes a considerable part of algebra, such as the geometric proofs about $(a \pm b)^2$ and $(a + b)(a - b)$. The fact was recognized by Ramus (1569) when he stated that "Algebra est pars arithmeticae" (1586 ed., p. 322).

[2] Colebrooke (pp. 112, 325) transliterates this as cuttácára, cutta, cuttaca, cuttaca-vyavahára, and cuttacád'hyára, meaning the determination of a pulverizing multiplier p such that, if n_1, n_2, and n_3 are given numbers, then $pn_1 + n_2$ shall be divisible by n_3.

[3] Or Vīja Ganita, Beej Gunnit. The spelling as given in the first printed edition (Calcutta, 1846) is Beej Gunita. The name Vīja-kriyā, meaning seed analysis, is also used.

[4] Sir M. Monier-Williams, Indian Wisdom, 4th ed., p. 174 (London, 1893).

[5] Nesselmann, loc. cit., p. 44; Colebrooke translation, p. 129. Bhāskara also used sama-sodanam (transposition) to include the two terms which had been used by al-Khowârizmî.

[6] T'ien-yuen-shu, celestial-element method. The Japanese called it tengen jutsu.

[7] Mikami, China, p. 157.

Algebra at one time stood a fair chance of being called *Fakhrî*, since this was the name given to the work of al-Karkhî (*c.* 1020), one of the greatest of the Arab mathematicians. Had his work been translated into Latin, as al-Khowârizmî's was, the title might easily have caught the fancy of the European world. Al-Karkhî relates that he was long and sorely hindered in his attempts to complete his work, because of the tyranny and violence endured by the people, until "God, may his name be hallowed and exalted, sent to their aid our protector, the vizir, the illustrious lord, the perfect one in government, the vizir of vizirs, clothed with double authority, Abû Gâlib," whose familiar name was Fakhr al-Mulk. In honor of this patron the name Fakhr gave rise to the title of the book, *al-Fakhrî*.

The Name "Algebra." Our real interest in the name centers around the word *algebra*, a word appearing, as we have seen, in the title of one of the works by al-Khowârizmî (*c.* 825),— *al-jabr w'al-muqâbalah*. It also appears in the early Latin translations under such titles as *Ludus algebrae almucgrabalaeque* and *Gleba mutabilia*. In the 16th century it is found in English as *algiebar and almachabel*, and in various other forms, but was finally shortened to *algebra*.[1] The words mean restoration and opposition,[2] and one of the clearest explanations of their use is given by Behâ Eddîn (*c.* 1600) in his *Kholâsat al-Ḥisâb* (*Essence of Arithmetic*):

The member which is affected by a minus sign will be increased and the same added to the other member, this being *algebra*; the homogeneous and equal terms will then be canceled, this being *al-muqâbala*.

That is, given $bx + 2q = x^2 + bx - q$,
al-jabr gives $bx + 2q + q = x^2 + bx$,
and *al-muqâbalah* gives $3q = x^2$.

---

[1] An English translation of al-Khowârizmî's work by F. Rosen appeared in London in 1831. A Latin version was published by Libri in his *Histoire*, Vol. I (Paris, 1835), and by Karpinski (1915) from a Scheubel (Scheybl) MS. at Columbia University, as already stated.

[2] Or redintegration and equation. *Jabr* is from *jabara* (to reunite or consolidate), possibly allied to the Hebrew *gâbar* (make strong).

This statement was put into verse, as was usual in the East, and thus became generally known in the Arab schools. It may be crudely translated thus:

> Cancel minus terms and then
> Restore to make your *algebra*;
> Combine your homogeneous terms
> And this is called *muqâbalah*.[1]

In a general way we may say that *al-jabr* or *al-jebr* has as the fundamental idea the transposition of a negative quantity, and *muqâbalah* the transposition of a positive quantity and the simplification of each member.[2] Al-Khowârizmî's title was adopted by European scholars,[3] appearing both in the Arabic, with many curious variants, and in Latin. The Moors took the word *al-jabr* into Spain, an *algebrista* being a restorer, one who resets broken bones.[4] At one time it was not unusual to see over the entrance to a barber shop the words "Algebrista y Sangrador" (bonesetter and bloodletter), and both the striped pole which is used in America as a barber's sign and the metal basin used for the same purpose in Europe today are relics of the latter phase of the haircutter's work. From Spain the word passed over to Italy, where, in the 16th century, *algebra* was used to mean the art of bonesetting.[5] Thence it found its way into France as *algèbre*, and so on to England, where one writer (1541) speaks of "the helpes of Algebra & dislocations,"

---

[1] From a Persian algebra written probably after the 12th century. Nesselmann (p. 50) put it into German verse, and the above English quatrain, taken from his translation, gives only a general idea of the wording.

[2] Rollandus (*c.* 1424) has *De arte dolandi*, the art of chipping off or cutting with an ax, probably meaning the chipping off or subtracting of equals from both members.

[3] Thus Robert of Chester's translation (*c.* 1140) begins his *Liber Algebrae et Almucabola* thus: "In nomine Dei χρij et misericordis incipit liber Restaurationis et Oppositionis numeri . . . filius Mosi Algoŭrizim dixit Mahometh."

[4] So in *Don Quixote* (II, chap. 15), where mention is made of "*un algebrista* who attended to the luckless Samson."

[5] Libri, *Histoire*, 1838 ed., II, 80. The question of the connection of *al-jabr* with the Hebrew root *sh-b-r* (from which comes *tishboreth*, fracture), and with the Hindu word for pulverizer, is worthy of study.

and another (1561) says: "This Araby worde Algebra sygni-
fyeth as well fractures of the bones, etc. as sometyme the
restauration of the same."[1]

As already said, the name was much distorted by the Latin
translators. Thus Guglielmo de Lunis (c. 1250?) gives it as
gleba mutabilia, and Roger Bacon (c. 1250) speaks of the
science as *algebra . . . et almochabala*.[2] A 15th century manuscript testifies both to the mystery of the subject and to the
uncertainty of name when it speaks of the subtleties of *largibra*.[3]
In the early printed books it appeared in equally curious forms,
such as *Gebra vnd Almuthabola*.[4]

Some of the late Latin writers attributed the name to one
Geber,[5] an Arab philosopher, whom they supposed to be the
inventor of the science;[6] and certain Arab writers speak of a
Hindu named Argebahr or Arjabahr, a name which may have
influenced the Latin translators.[7] Even as good a scholar as
Schoner went far astray in his interpretation of the title.[8]

---

[1] See Oxford Dictionary under *algebra*.

[2] "Algebra quae est negotiatio, et almochabala quae est census." *Opus Majus*,
ed. Bridges, I, p. lvii.

[3] Anon. MS. in Boncompagni's library: " . . . di subtili . ℞ . di largibra."
Narducci *Catalogo* (2d ed., 1892), No. 397 (2).

[4] A. Helmreich, *Rechenbüch*, 1561 (1588 ed.).

[5] There was an Arab scholar, Jabir ibn Aflah, Abû Mohammed, of Seville
(c. 1145), whose astronomy was translated by Gherardo of Cremona, his contemporary, and was printed in 1534.

[6] The name appears as *Greber* in Heilbronner's *Hist. Math.*, p. 340 (1742).
Ghaligai (1521) spoke of it as "composta da uno home Arabo di grâde intelli-
gentia," adding that "alcuni dicono essere stato uno il qual nome era Geber."

[7] Libri, *Histoire* (I, 122), thinks that this writer was Āryabhaṭa.

[8] Thus in a note on Ramus (1586 ed., p. 322) he says: "Nomen Algebrae
Syriacum putatur significans artem & doctrinam hominis excellentis. Nam *geber*
Syris significat virum . . . ut apud nos Magister aut Doctor . . . & ab Indis
harum artium perstudiosis dicitur *Aliabra* item *Alboret*, tametsi proprium autoris
nomen ignoretur."

As an example of still more uncertain history, A. Helmreich (*Rechenbüch*,
1561; 1588 ed., fol. b 2, r., of the *Vorrede*) asserts that algebra was due to "Ylem/
der grosse Geometer in Egypten/zur zeit desz Alexandri Magni, der da war ein
Praeceptor oder vorfahrer Euclidis, desz Fürsten zu Megarien." We also find
such forms as *Agabar, Algebra muchabila, Reghola della raibre mochabilᵒ', regola
del acabremᵒ', ghabile, dellacibra e muchabile, lacibra umachabille*, all in MSS. of
the 15th century; and in the 16th century such forms as *arcibra*.

FIRST PAGE OF AN ALGEBRA MANUSCRIPT OF *c*. 1460

Possibly by Raffaele Canacci, a Florentine mathematician. On the second line
may be seen the name for algebra,— *Algebra amucabale*. In Mr. Plimpton's library

Other Names. Because the unknown quantity was called *res* by the late Latin writers, which was translated into Italian as *cosa*,[1] the early Italian writers called algebra the *Regola de la Cosa*, whence the German *Die Coss* and the English *cossike arte*.[2]

The Italians of the 15th and 16th centuries often called algebra the greater art, to distinguish it from commercial arithmetic, which was the lesser art, just as we speak of higher arithmetic and elementary arithmetic. This distinction may have been suggested by the seven *arti maggiori* and the fourteen *arti minori* recognized by the merchants of medieval Florence.[3] Thus we have such names as *Ars Magna*, used by Cardan (1545),[4] *l'Arte Maggiore*, used by various other Italian writers, and *l'arte mayor*, used by Juan Diez, whose book has been mentioned as having appeared in Mexico in 1556. The title as given in the Mexican book is as follows:

¶ Quiſtiones ꝟel arte mayoꝛ tocantes al algebꝛa.

Vieta (*c.* 1590) rejected the name "algebra" as having no significance in the European languages, and proposed to use the term "analysis," and it is probably to his influence that the popularity of this term in connection with higher algebra is due.

Of the other names for algebra the only one that we need consider is "logistic." Since the term had dropped out of use as a name for computation about 1500, it was employed to

---

[1] From the Latin *causa*; compare the French *chose*.

[2] Thus Pacioli (1494 ed., fol. 67, *r.*): "Per loperare de larte magiore: ditta dal vulgo la regola de la cosa ouer algebra e amucabala." It will be recalled that Rudolff's title for his book (1525) on algebra was *Die Coss.* Helmreich says: " . . . vnd wird bey den Welschen [the Italians] genent das buch *Delacosa*/welchs wir Deutschē die *Regulā Cos* oder *Algebra* nennen." *Rechenbüch*, 1561 (1588 ed.), fol. b 2.

[3] E. G. Gardner, *The Story of Florence*, p. 42. London, 1900.

[4] Also by Gosselin, *De arte magna, seu de occulta parte numerorum quae & Algebra & Almucabala vulgo dicitur, libri IV*, Paris, 1577. The name was used, however, for various purposes, as in Kircher's *Ars Magna Lucis et Umbrae*, Amsterdam, 1671, and numerous other works.

designate a higher branch, just as "calculus" was appropriated a century or so later. Thus we find it used by Buteo and others[1] to cover advanced arithmetic and algebra, although it never became popular.[2]

3. TECHNICAL TERMS

Coefficient. Of the terms commonly used in algebra, it is possible at this time to mention only a few typical ones.

The coefficient was called by Diophantus (c. 275) the *ple'thos*[3] (multitude), and by Brahmagupta (c. 628) the *anca*, or *prakriti*,[4] but most early writers used no special name. The term "coefficient" and the use of literal coefficients are late developments, the former being due to Vieta.

Unknown Quantity. The unknown quantity was called by Ahmes (c. 1550 B.C.) *ahe*[5] or *hau* ("mass," "quantity," or "heap"). Diophantus called it "an undefined number of units."[6] Brahmagupta called it the *yāvat-tāvat*,[7] and possibly this suggested to the Arabs the use of *shā* (*sei, chaï,* meaning "thing" or "anything"), whence the medieval use of *res* (thing) for this purpose.[8] The Chinese used *yuen* (element),[9] as already stated, but they also used a word meaning "thing."[10]

Powers. Before the invention of a satisfactory symbol like x^2 it became necessary to have a special name for the square of the unknown, and in the Greek geometric algebra it was called

---

[1] H. Vitalis, *Lexicon Mathematicvm*, p. 25 (Rome, 1690); Nesselmann, *loc. cit.*, p. 57.

[2] There are many other names used for this purpose. *E.g.*, 'Alî ibn Velî ibn Hamza, a western Arab, then (1590/91) living in Mecca, wrote a work entitled *Tuḥfet al-a'dad li-davî al-roshd ve'l sadâd* (*The Gift of Numbers for the Possessors of Reason and Correct Insight*), relating to elementary algebra. See E. L. W. M. Curtze, *Abhandlungen*, XIV, 184. The name reminds one of Recorde's *Whetstone of witte*, London, 1557.

[3] Πλῆθος.   [4] Colebrooke's translation, pp. 246, 348.

[5] By Egyptologists, *'ḥ'w*. See Peet, *Rhind Papyrus*, p. 61.

[6] Πλῆθος μονάδων ἄλογον (*ple'thos mona'don a'logon*).

[7] Also given as *jabut tabut*, literally the "so much as," "as far, so far," "as much, so much," or "however much." Compare Bombelli's use of *tanto* (so much).

[8] The Arabs also used *jidr* (*dyizr*, root), whence the Latin *radix*.

[9] Mikami, *China*, pp. 81, 91.   [10] Compare *res* and *cosa*, page 392.

a "tetragon[1] number" or a "power."[2] Diophantus called the third power a cube,[3] the fourth power a "power-power,"[4] the fifth power a "power-cube,"[5] and the sixth power a "cube-cube,"[6] using the additive instead of the multiplicative principle. The Arab writers called the square of the unknown a *mal*, a word meaning "wealth," whence the medieval Latin *census* (evaluation of wealth, tax) was used for the same purpose, appearing in the early Italian algebras as *censo*, sometimes incorrectly written as *zenso*. Therefore algebra was not uncommonly called *Ars rei et census* as well as *Ars rei*[7] and *Regola o l'arte della cosa*.

Equation. The word "equation," while generally used as at present ever since the medieval writers set the standard, has not always had this meaning. It is used by Ramus in his arithmetic and by his commentator Schoner to denote a continued proportion, although in speaking of algebra Ramus (1567)[8] and Gosselin (1577) follow the ordinary usage.

Absolute Term. In the equation

$$a_0 x^n + a_1 x^{n-1} + \cdots + a_{n-1} x + a_n = 0$$

we speak of a_n as the absolute term. There have been various names for it, Diophantus (*c.* 275) calling it monads.[9] The

---

[1] Τετράγωνος ἀριθμός (*tetra'gonos arithmos'*, four-angled number). So in Euclid VII, def. 18.

[2] Δύναμις (*dy'namis*), from the same root as "dynamo," "dynamic," and "dynamite." So in Plato (*Timæus*, 31); but he also uses the term (*Theætetus*, 147 D) to mean the square root of a non-square number. When Diophantus speaks of any particular square number, he uses τετράγωνος ἀριθμός, but otherwise δύναμις. Heath, *Diophantus*, 2d ed., p. 38.

[3] Κύβος (*ku'bos*).   [5] Δυναμόκυβος (*dynamo'kubos*).
[4] Δυναμοδύναμις (*dynamody'namis*).   [6] Κυβόκυβος (*kubo'kubos*).
[7] Nesselmann, *Alg. Griechen*, pp. 55, 56.
[8] Thus Ramus, in his arithmetic (1567), says: "AEquatio est quando continuatae rationes continuantur iterum," Schoner giving as an example

$$
\begin{array}{ccc}
10. & 15. & 12. \\
20. & 30. & 24.
\end{array}
$$

meaning 10 : 15 : 12 = 20 : 30 : 24 (1586 ed., p. 188).
[9] Μονάδες(*mona'des*), with the abbreviation μ°. Heath, *Diophantus*, 2d ed., p. 39.

Hindus called it *rupa* or *ru*,[1] and the Chinese gave it the name *tai*,[2] an abbreviation of *tai-kieh* (extreme limit).

Commutative and Distributive Laws. The use of the terms "commutative" and "distributive" in the usual algebraic sense is due to the French mathematician Servois (1814). The use of the term "associative" in this sense is due to Sir William Rowan Hamilton.[3]

4. SYMBOLS OF ALGEBRA

Symbols of Operation. The symbols of elementary arithmetic are almost wholly algebraic, most of them being transferred to the numerical field only in the 19th century,[4] partly to aid the printer in setting up a page and partly because of the educational fashion then dominant of demanding a written analysis for every problem. When we study the genesis and development of the algebraic symbols of operation, therefore, we include the study of the symbols used in arithmetic. Some idea of the status of the latter subject in this respect may be obtained by looking at almost any of the textbooks of the 17th and 18th centuries. Hodder,[5] for example, gives no symbols before page 201,[6] then remarking: "Note that a + thus, doth signifie Addition, and two lines thus ⊏ Equality, or Equation, but a × thus, Multiplication," no other symbols being used. Even Recorde, who invented the modern sign of equality, did not use it in his arithmetic, the *Ground of Artes* (c. 1542), but only in his algebra, the *Whetstone of witte* (1557).

Earliest Symbols. The earliest symbols of operation that have come down to us[7] are Egyptian. In the Ahmes Papyrus (c. 1550 B.C.) addition and subtraction are indicated by

---

[1] Colebrooke's *Vīja Gaṇita*, p. 186 n.; E. Strachey, *Bija Ganita*, p. 117 (London, n.d.), where it is given as *roop*.

[2] Mikami, *loc. cit.*, p. 81. The word is also transliterated *taë*.

[3] Cajori, *Hist. of Math.*, 2d ed., p. 273. New York, 1919.

[4] There are, of course, exceptions. The Greenwood arithmetic (1729), for example, used the algebraic symbols.

[5] His was the first English arithmetic to be reprinted in the American colonies (Boston, 1719). [6] 1672 ed.

[7] Excepting those connected with notation, as in the subtractive principle of the Babylonians, already mentioned.

special symbols, but these are simply hieratic forms from the hieroglyphics and are not symbols in the sense in which we use the term. The symbol $\mathcal{A}$ was used to designate addition. It appears in the Ahmes Papyrus as Λ.[1] The symbol Λ was used to designate subtraction. It appears in the Ahmes Papyrus as $\mathcal{A}$.[2]

Diophantus (c. 275) represented addition by simple juxtaposition, as in $K^{Y}\bar{a}\Delta^{Y}\overline{\iota\gamma}$ for $x^3 + 13\,x^2$. For subtraction he seems to have used the symbol $\wedge$, although we are not certain as to its precise form. Since we have no manuscript of his *Arithmetica* earlier than the Madrid copy of the 13th century, we are also uncertain as to the authenticity of the following passage:

"Minus multiplied by minus makes plus, and minus by plus makes minus. The sign of negation is ψ turned upside down, $\wedge$."[3] It is fully as probable that the symbol is a deformed Λ (*lambda*), the Greek letter L and the initial for a word indicating subtraction.[4]

The Hindus at one time used a cross placed beside a number to indicate a negative quantity, as in the Bakhshālī manuscript of possibly the 10th century. With this exception it was not until the 12th century that they made much use of symbols of operation.[5] In the manuscripts of Bhāskara (c. 1150) a small circle or a dot is placed above a subtrahend, as in $\overset{\circ}{6}$ or $\overset{.}{6}$ for -6,[6] or the subtrahend is inclosed in a circle, just as children, in scoring a game, indicate 6 less than zero[7] by the symbol ⑥

---

[1] Ahmes wrote, as usual, from right to left, but hieroglyphic sentences are generally printed from left to right. See the Brit. Mus. facsimile, Pl. IX, row 5.

[2] Peet, *Rhind Papyrus*, p. 64; Eisenlohr, *Ahmes Papyrus*, p. 47.

[3] Tannery, *Diophantus*, I, 13, and *Bibl. Math.*, V (3), 5; Heath, *Diophantus*, 2d ed., p. 130. On the relation of this symbol to the symbol $\cap$, which is used in the Ayer Papyrus (c. 200–400), see *Amer. Journ. of Philology*, XIX, 25.

[4] Λιπόντες (*lipon'tes*, diminished by), or Λείπειν (*lei'pein*, to be missing).

Of course a further exception is also to be made of the representing of sums by juxtaposition and of division by means of fractions. For the Bakhshālī MS. see Volume I, page 164.

[6] See Colebrooke's edition, p. 131; Taylor's edition, Introduction, p. 11.

[7] C. I. Gerhardt, *Études historiques*, p. 8. Berlin, 1856.

European Symbols for Plus and Minus. The early European symbol for plus, used in connection with the Rule of False Position, was p,[1] P,[2] or p̄, the last being the most common of the three.[3] The word *plus*, used in connection both with addition and with the Rule of False Position, was also employed; but, strange to say, it is much later than the word *minus* as indicating an operation. The latter is found in the works of Fibonacci (1202), while the use of *plus* to indicate addition is not known before the latter part of the 15th century.[4]

Since p̄ or p̃ was used for *plus*, m̄ or m̃ was naturally used for *minus*, and this usage is found in many works of the 15th and 16th centuries. As usual, the bar simply indicated an omission, as in *Sūma* for *Summa*, in the title of Pacioli's work.[5] In the 15th century the symbol 1̄9̄ was often used for *minus*,[6] but most writers preferred the m̄.

Racial Preferences. We now come to one of the many cases of racial habit in determining mathematical custom. In the 16th century the Latin races generally followed the Italian School, using p̄ and m̄ or their equivalents,[7] while the German School

---

[1] So in the Rollandus MS. (*c.* 1424), where the terms are so arranged as to require no minus sign. The Rule of False Position is explained on pages 437–441.

[2] Clavius (1583) uses P and M for *plus* and *minus* in his Rule of False Position, which he gives in his arithmetic; but in his algebra (1608) he uses the cross, saying: "Plerique auctores pro signo + ponunt literam P, . . . sed placet nobis uti nostris signis." It must be remembered that the symbols in the Rule of False Position are hardly symbols of operation in the ordinary sense of the term.

[3] *E.g.*, Chuquet (1484) and Pacioli (1494), the latter first using it in his Rule of False Position (fol. 106).

[4] *Bibl. Math.*, XIII (2), p. 105. Fibonacci used it, however, in the Rule of False Position.

[5] Compare the French *hôtel* for *hostel* or *hospital*, and the German *über* for *ueber*. See also Cajori, "Varieties of Minus Signs," *Math. Teacher*, XVI, 295.

[6] This is the case in Mr. Plimpton's MS. of al-Khowârizmî, written in 1456. See also the Regiomontanus-Bianchini correspondence (*c.* 1464) in the *Abhandlungen*, XII, 233, 279; and Curtze in the *Bibl. Math.*, I (3), 506.

[7] *E.g.*, Pacioli (1494) writes m̄ for *minus* in his algebra, but he follows the general custom of using *de* in cases like "7. de 9"; Cardan (1530) writes ℞ . V . 7 . p̄ ℞ . 4 for $\sqrt{7 + \sqrt{4}}$; Feliciano Lazesio (1526) writes 6. piu. ℞ . 16 for 6 + 16 *x*, and 10. m̄ ℞ . 4 for 10 − 4 *x*; Tartaglia (1556) and Cataneo (1546) write *piu* and *men* for *plus* and *minus* in the Rule of False Position; Santa-Cruz (1594) uses the equivalent Spanish words *mas* and *menos*; and Peletier

preferred the symbols + and −, neither of which is found for this purpose, however, before the 15th century.[1]

Origin of our Plus and Minus Signs. In a manuscript of 1456, written in Germany,[2] the word *et* is used for addition and is generally written so that it closely resembles the symbol +. The *et* is also found in many other manuscripts, as in "5 et 7" for 5 + 7, written in the same contracted form,[3] as when we write the ligature & rapidly. There seems, therefore, little doubt that this sign is merely a ligature for *et*.

The origin of the minus sign has been more of a subject of dispute. Some have thought that it is a survival of the bar in $\overline{19}$ or in $\overline{m}$, but it is more probable that it comes from the habit of early scribes of using it as the equivalent of *m*, as in Sūma for Summa. Indeed, it is quite probable that the use of $\overline{X}$ for 10 thousand (X *mille*) is an illustration of the same tendency, the bar (—) simply standing for *m* (*mille*). In the uncial writing we commonly find ÷ for *m*, and in the Visigothic we find ÷ for the same purpose. It is quite reasonable, therefore, to think of the dash (−) as a symbol for *m* (minus), just as the cross (+) is a symbol for *et*. It is also possible that its use in this sense may have come from the habit of merchants in indicating a missing number in a case like 2 yd. − 3 in., where the number of feet is missing. We have the same habit in writing certain words today, using either a dash or a series of dots.

(1549) uses the French *plus* and *moins*, while Gosselin (1577) uses P and M. There are, of course, exceptions, as when Trenchant (1566) uses + and − in his work in the Rule of False Position, and when Ramus (1569) writes: "At si falsa conjectura sit, notatur excessus cum signo plus sic †, vel defectus cum signo minus −" (*Schol. Math.*, 1569, p. 138). When he comes to algebra, the plus takes the form —+— (*ibid.*, p. 269). Vieta (*c.* 1590) wrote x^4*. px^2 for $x^4 + px^2$, frequently using the dot as a sign of addition. He also used = for subtraction in certain cases. The asterisk denoted an absence of some power of x.

[1] Libri's surmise that they are due to Leonardo da Vinci is not warranted. See *Bibl. Math.*, XIII (2), 52, and his MSS. as published by Boncompagni. Similarly, Treutlein's idea that they are due to Peurbach (*Abhandlungen*, II, 29) is not substantiated.   [2] The al-Khowârizmî MS. in Mr. Plimpton's library.

[3] *E.g.*, the Regiomontanus-Bianchini correspondence, *Abhandlungen*, XII, pp. 233, 279; *Bibl. Math.*, I (3), 506. See also I. Taylor, *The Alphabet*, I, 8 (London, 1883); J. W. L. Glaisher, *Messenger of Math.*, LI, 1–148.

The signs + and − first appeared in print in an arithmetic, but they were not employed as symbols of operation. In the latter sense they appear in algebra long before they do in arithmetic. Their first appearance in print is in Widman's arithmetic (1489), the author saying: "Was − ist / das ist minus . . . vnd das + das ist mer."[1] He then speaks of "4 centner + 5 pfund" and also of "4 centner − 17 pfund," thus showing the excess or deficiency in the weight of boxes or bales. He does not use the symbols to indicate operations, but writes, for example, "$\frac{2}{5}$ $\frac{4}{5}$ $\frac{3}{5}$ adir fa $\frac{9}{5}$ ist $1\frac{4}{5}$," as we write $1\frac{4}{5}$ instead of $1 + \frac{4}{5}$, juxtaposition signifying addition.

Manifestly the minus sign was more important as a warehouse mark than the plus sign, since mere juxtaposition serves to express excess.[2]

The first one to make use of the signs + and − in writing an algebraic expression was the Dutch mathematician Vander Hoecke (1514),[3] who gave ℞ $\frac{3}{4}$ − ℞ $\frac{3}{5}$ for $\sqrt{\frac{3}{4}} - \sqrt{\frac{3}{5}}$, and ℞ 3 + 5 for $\sqrt{3} + 5$. The next writer to employ them to any extent was Grammateus (1518). He first used them in the Rule of False Position, where, as already stated, they expressed excess and deficiency

92
```
4 +  5   Wilebn das wyſ.
4——17   ſen oder deßgley.
3 +  30  chen/So ſumier
4——19   die zenttner vnd
3 +  44  ℔ vnnd was auß
3 +  22  —iſt/das iſt mi.
zentner 3——11 ℔ nus dz ſetz Beſon.
3 +  50  der vnnd werden
4——16   4539℔ (So
3 +  44  du die zendtner
3 +  29  zů ℔ gemachett
3 —1 2   haſt vnnd das /
3 +  9   das iſt meer
```
darzů Addiereſt)vnd >5 minus. Nun
ſolt du für holtz abſchlahen allweeg für
ain legel 24 ℔. Vnd das iſt 13 mal 24.
vnd macht 3 12 ℔ darzů addier das —
das iſt >5 ℔ vnd werden 387. Dyeſuß.
trahier von 4539.Vnd Bleyben 4152
℔.Nun ſprich 100 ℔ das iſt ein zentner
pro 4 ℔½ wie kumen 4152℔ vnd kumē
171 ℔½ ß 4heller½ Vñ iſt recht gmacht

Pfeffer
℞

[1] 1508 ed., fol. 59; in the 1526 edition, "was auss − ist / das ist minus . . . vnnd das / + das ist meer."

[2] E.g., Albert (1534) writes: "Item/Wie komen 11 cent. 3 stein 18 pfund Zien," and "Item / 12 centner 4 stein —— 6 pfund Talg." He frequently uses the long bar to indicate deficiency, but never uses the plus sign.

[3] Such a statement is likely to be invalidated at any time, and it simply means that no case is known to the author that can be placed earlier than that in Vander Hoecke's work. See the facsimile on page 401. There is a copy of the 1514 edition in the British Museum. For the 1537 edition see *Rara Arithmetica*, p. 183.

instead of operations to be performed.[1] When he wrote upon
algebra, however, he used them in the modern sense.[2]

These symbols seem to have been employed for the first time
in arithmetic, to indicate operations, by Georg Wälckl (1536),[3]
who used $+\frac{1}{3}$ 230 to indicate the addition of $\frac{1}{3}$ of 230, and
$-\frac{1}{5}$ 460 to indicate the subtraction of $\frac{1}{5}$ of 460. The algebraist
who did the most to bring them into general use was Stifel,
to whom the credit for their invention was formerly given.[4]
For $3\,x + 2$ he wrote "3 sum: +2.,"[5] and similarly for polynomials involving the minus sign.[6] From this time on the two symbols were commonly used by both German and Dutch writers,
the particular forms of the signs themselves not being settled
until well into the 18th century. Thus, for example, the 1752
edition of Bartjens has

$$xx \equiv \div \!\!-\!\!- 2375\,x \; \maltese \; 1785000$$
for
$$x^2 = -2375\,x + 1,785,000.$$

[1] "Ist zu vil / setze + Ist aber zu wenig / setze $-$" (1535 ed., fol. E 3).
Riese (*Rechnung auff der Linien vnd Federn*, 1522) used the symbols $-\!\!+\!\!-$ and
$-\!\!\div\!\!-$, the latter being also used by various other writers to indicate subtraction.
Thierfelder (1587), for example, has "25 fl. $\div$ 232 gl." (pp. 110, 229). There are
numerous variants, such as $\div$ (Coutereels, 1599, 1690 edition of the *Cyffer-Boeck*)
and $-\!\!\div\!\!-$ (Wilkens, 1669). On the present use of $\div$ for $-$, see R. Just, *Kaufmännisches Rechnen*, Leipzig, 1901.

[2] "Vnd mañ brauchet solche zeichen als + ist mehr/vnd—/minder." He
illustrates by adding $6x + 6$ and $12x - 4$, thus:

$$
\begin{array}{l}
6\,\text{pri.} + 6\,\text{N} \\
\underline{12\,\text{pri.} - 4\,\text{N}} \\
18\,\text{pri.} + 2\,\text{N}
\end{array}
$$

In one sense $6\,x + 6$ means an excess of 6 over $6\,x$, and we evidently find here the
transition from the excess stage to the addition stage.

[3] *Die Wälsch practica*, Strasburg (Nürnberg?), 1536.

[4] Probably because of this expression: "Darumb so gedenck nur nicht, das
dise ding schwer seyen zu lernen, oder zubehalten, und ist doch die gantz sach
diser meiner zeichen hiemit gantz auszgericht unnd an tag gebracht." *Deutsche
Arithmetica*, 1545. For a facsimile from his work, see page 403.

[5] ". . . das zeichen + / welches ich setzē muss zwischen sie / als 2 zu 3 sum:
machen 3 sum: + 2. das machstu denn also lesen / 3 summen vnd 2." *Ibid.*,
fol. 21.

[6] "Denn wo du dises zeichen $-$ findest/magstu darfur lesen/Weniger oder
Minder."

CItem woldi afrecken R $\frac{12}{27}$ van R $\frac{27}{48}$ so ad

deert beyde de quadraten coët 1 $\frac{1}{144}$ daer na multi

pliceert dye een quadrate metten anderen coemt 1 die

treet van 1 $\frac{1}{144}$ reft $\frac{1}{144}$ no dê R daer af is $\frac{1}{12}$

CIrrationale.

CDie irrationale seät met — oft met + naer den

epsch vanden wercke/ als treckt R $\frac{3}{5}$ van R $\frac{3}{4}$

reft R $\frac{3}{4}$ — R $\frac{3}{5}$

CItem wildi afrecken R $\frac{1}{5}$ van R $\frac{4}{5}$ reft R $\frac{1}{5}$

CItem wildi afrecken R $\frac{3}{4}$ van R 3 — $\frac{3}{16}$ reste

R $\frac{3}{16}$

CItem + van + oft — van — subtraheert ende +

van — oft — van + addeert inde subtractie so ver-

re als fi addeerlic sijn oft subtraheerlic.

C fMultiplicacie inden R van rationalen.

Uildi multipliceren inden R soe weet dat ghp

moet stellen alle de nommers van eender natu

re als R te multipliceren met simpelen nommer soe

moet gla den nommer multipliceren nac de qualiteyt

Des R Als wildi multipliceren R 9 met 4 so set 4 in

finen R multipliceert 4 in haer seluen coemt R 16 ende

multipliceert 9 met 16 coët 144 hier wt treet R coët

12 soe veel is R 9 ghemultipliceert met 4/ want R 9

so 3 dit multipliceert met 4 coemt 12 als voren.

Wildi multipliceret R^{3a} 8 met 5 so multipliceert 5 cu

VANDER HOECKE'S USE OF THE PLUS AND MINUS SIGNS

Early use of these signs in Belgium and Holland in 1514. This facsimile is from
the 1537 edition

England adopts the Symbols. England early adopted the Teutonic forms, and Recorde (*c.* 1542) says "thys fygure +, whiche betokeneth to muche, as this lyne, —— plaine without a crosse lyne, betokeneth to lyttle."[1] Baker (1568) made a vain attempt to change the plus sign, saying: "This Figure ×, betokeneth more: and this plaine line ——, signifieth lesse."[2] All this was in connection with the Rule of False Position, and not in connection with arithmetic operations. As symbols of operation most of the English writers of this period reserved the + and — for algebra.[3]

Variants of the Symbols. The variants of the plus sign (+) were naturally many, partly because the early printers had to make up the sign by combining lines that they had in their fonts. Occasionally, however, the religious question enters, as in certain Hebrew works of the 19th century, in which the Christian symbol of the cross is changed[4] to ⊥.

The expression "plus or minus" is very old, having been in common use by the Romans to indicate simply "more or less." It is often found on Roman tombstones, where the age of the deceased was given in some such form as AN · LXXXXIIII · P · M; that is, "94 years more or less."

Symbols of Multiplication. Symbols of multiplication were more slow in their development than symbols of addition and subtraction, the reason being the need for the latter as warehouse marks and in the popular Rule of False Position. The absence of a sign as in £5 and 3 ft. led naturally in the 16th century to a similar usage in such algebraic forms as 6 Pri. for $6x$ and 7 ʒ (7 zenzo or 7 censo) for $7x^2$. The late medieval

---

[1] *Ground of Artes*, ed. 1558, fol. Z 6.

[2] Ed. 1580, fol. 184 (numbered 194). Thierfelder (1587) uses × twice through a mistake of the printer (pp. 194, 246), and Wilkens (1669) uses it purposely in connection with + (pp. 190, 191).

[3] Thus Digges (1572), in his treatment of algebra: "Then shall you ioyne them with this signe + Plus"; and Hylles (1600) says: "The badg or signe of addition is +," stating the sum of 3 and 4 as "3 more 4 are 7," and writing 10——3 for "10 lesse 3."

[4] This is found in several such works. Among the latest writers to use the symbol was G. J. Lichtenfeld, *Yedeeoth ha-Sheurim*, Warsaw, 1865.

Von disen zweyen zeichen/
+ vnd —. VII.

O ich von zeychen reden werde/soltu mich verstehn
von disen zeichen + vnd —/Deñ solliche verzeich
nis/ Sum:oder Sum: A. oder ſſ. ꝛc. Werde ich
nicht zeychen nennen/sondern/namen/oder benen=
nung der zalen.Wa ich nu rede von gleichen zeichē/
soltu es verstehn von + vnd -|- / oder von — vnd —. Also
auch/wa ich von vngleichen zeichen rede / so verstehe es/von +
vnd —.

So haben nu dise zwey zeichen +vnd— / ein sonderlichen
Algorithmum/welchē ich hie stellen will auff 4 Regeln. Denn er
gehöret zum Algorithmo der vngerechneten zalen wie du woll se=
hen wirst/vnd alles was vorhin gesagt ist von disen namen sum:
sum:A. ꝛc. das gehöret alles hie her / als vnter ein einigen Algo=
rithmum.

Die erst Regel von dem Addiren
vnd Subtrahiren. VIII.

Wey gleiche zeichen / machen eben das sel=
big zeichen/ im Addiren vñ Subtrahiren/
ohn allein so du im subtrahiren die zal / die
du soltest subtrahirē/nicht kanst subtrahirē.

Exempla vom Addiren.

8	Sum:	+	7.		8	Sum:	—	18.
12	Sum:	+	11.		3	Sum:	—	6.
20	Sum:	+	18.		11	Sum:	—	24.

Hie sihest nu vor augen/wie + vnd + mache im ersten ex=
emplo

STIFEL'S USE OF THE SIGNS + AND — IN ALGEBRA

From Stifel's *Deutsche Arithmetica. Inhaltend. . . . Die Deutsche Coss,*
Nürnberg, 1545

writers usually arranged their multiplication tables for commercial use in columns, as in the two cases which follow:

$$2 \cdot 43 \cdot 86$$
$$2 \cdot 44 \cdot 88$$

17	18	306
17	19	323

In this arrangement no symbols of operation or equality were used,[1] the dot serving for both purposes, being really nothing but a symbol of separation, like the ruled lines.

In the first printed books no such symbols appear, the Treviso arithmetic, for example, giving the multiplication table in the form

2 via 5 fa 10.

Development of the Symbol ×. The common symbol × was developed in England about 1600. In the second edition of Edward Wright's translation of Napier's *Mirifici logarithmorum canonis descriptio* (London, 1618)[2] is "An Appendix to the

Logarithmes," and this contains the statement (p. 4): "The note of Addition is (+) of subtracting (−) of multiplying (×)," a statement that is very likely due to Samuel Wright. The larger symbol (×) is probably due to Oughtred.[3] It was not a new mathematical sign, having long been used in cross

---

[1] The first of these examples is from a MS. of Benedetto da Firenze written c. 1460, and the second from one of Luca da Firenze (c. 1475), both in Mr. Plimpton's library.

[2] But not in the 1616 edition.

[3] See F. Cajori, in *Nature*, (December 3) 1914, p. 364; *William Oughtred*, p. 27 (Chicago, 1916); "A List of Oughtred's Mathematical Symbols," *University of California Publications in Mathematics*, I, 171. This monograph should be consulted on the entire question of symbols. It contains a careful study of various algebraic signs. Samuel Wright was the son of Edward Wright. He entered Caius College, Cambridge, in 1612 and died c. 1616.

multiplication, in the check of nines,[1] in connection with the multiplication of terms in the division[2] or addition[3] of fractions, for the purpose of indicating the corresponding products in proportion,[4] and in the "multiplica in croce" of algebra as well as in arithmetic.[5] It was probably because of this last use that the symbol was suggested for multiplication, but we have no positive evidence on the subject. It was not readily adopted by arithmeticians, however, being of no practical value to them. In the 18th century some use was made of it in numerical work, but it was not until the second half of the 19th century that it became popular in elementary arithmetic. On account of its resemblance to x it was not well adapted to use in algebra, and so the dot came to be employed, as in $2 \cdot 3 = 6$ (America) and $2.3 = 6$ (Europe). This device seems to have been suggested by the old Florentine multiplication tables; at any rate Vlacq, the Dutch computer (1628), used it in some of his work, thus:

Factores	Faci
7.17	119

although not as a real symbol of operation.[6] Clavius (1583) had an idea of the dot as a symbol of multiplication, for he writes $\frac{3}{5} \cdot \frac{4}{7}$ for $\frac{3}{5} \times \frac{4}{7}$;[7] and Harriot (posthumous work of 1631) actually used the symbol in a case like $2.aaa$ for $2a^3$. The first

---

[1] In this connection Hylles (1600) speaks of it as the "byas crosse."

[2] As in $\frac{2}{3} \times \frac{3}{5} \frac{9}{10}$, for $\frac{3}{5} \div \frac{2}{3} = \frac{9}{10}$. See page 226.

[3] As in $\frac{4}{5} \times \frac{5}{6}$, for $\frac{4}{5} + \frac{5}{6} = \frac{24 + 25}{30}$.

[4] As in the case of $2 : 3 = 4 : 6$, as shown at the right. See Buteo, $\frac{2}{3} \times \frac{4}{6}$ *De Qvadratvra circuli*, p. 67, *et passim* (1559).

[5] Thus Ghaligai (1521; 1552 ed., fol. 76) gives

$$7 \text{ piu } \mathbb{R} \text{ } 48$$
$$7 \text{ piu } \mathbb{R} \text{ } 48$$

to indicate $(7 + \sqrt{48})(7 + \sqrt{48})$.

[6] In his text he uses a rhetorical form, thus: "3041 per 10002 factus erit 30416082."

[7] " . . . minutia minutiae ita scribēda est $\frac{3}{5} \cdot \frac{4}{7}$ pronūciaturque sic. Tres quintae quatuor septimarū vnius integri" (*Epitome*, 1583).

II

writer of prominence to employ the dot in a general way for algebraic multiplication seems to have been Leibniz (who also used the symbol $\frown$)[1] or possibly his contemporary, Christian Wolf, and subsequent algebraists have commonly used it where the absence of a sign does not suffice.[2]

The Symbol $\div$. The Anglo-American symbol for division ($\div$), as already stated, has long been used on the continent of Europe to indicate subtraction. Like most elementary combinations of lines and points, the symbol is old,[3] and toward the close of the 15th century the Lombard merchants used it to indicate a half, as in $4 \div$, $4 \div$, and similar expressions.[4] There is even a possibility that it was used by some Italian algebraists to indicate division,[5] but it first appeared in print in the *Teutsche Algebra*, by Johann Heinrich Rahn[6] (1622–1676), which appeared in Zürich in 1659. John Pell had been Cromwell's political agent in Switzerland (1654–1658), and Aubrey[7] tells us that "Rhonius was Dr. Pell's pupil at Zürich." He further asserts that "Rhonius's Algebra, in High Dutch, was indeed Dr. Pell's." At any rate, Rahn used the symbol and Pell made it known in England through his translation (London, 1688) of the work.

Symbol (:) for Ratio. The symbol (:) to indicate ratio seems to have originated in England early in the 17th century. It appears in a text entitled *Johnsons Arithmetick; In two*

---

[1] Gerhardt's edition of his works, II, 239; VII, 54.

[2] Wolf (1713) makes frequent use of the dot in cases like $1.2.3.4$ and $m - 2 . m - 3$, for 4! and $(m - 2)(m - 3)$ respectively. See the second edition of his *Elementa Matheseos*, I, 322 (Halle, 1730); also the facsimile of Leibniz's letter, Volume I, page 420.

[3] It was used for *est* as early as the 10th century, as in $i \div$ for *id est* and $\overline{it} \div$ for *interest*. If used in a case like *divisa $\div$*, for *divisa est*, it might possibly have suggested its independent use as a symbol of division.

[4] A. Cappelli, *Dizionario di abbreviature latine ed italiane*, 2d ed., pp. 415, 425. Milan, 1912.

[5] In a MS. in Mr. Plimpton's library, the *Aritmetica et Prattica*, by Giacomo Filippo Biôdi (Biondi) dal Anciso, copied in 1684, the symbol $\underline{\text{Ч}}^-$ stands for division, so that various forms of this kind were probably used.

[6] Latin Rhonius; see Volume I, page 412.

[7] *Brief Lives*, Oxford edition of 1898, II, 121.

Bookes,[1] but to indicate a fraction, $\frac{3}{4}$ being written 3 : 4. To indicate a ratio it appears in an astronomical work, the *Harmonicon Coeleste* (London, 1651), by Vincent Wing and an unknown writer, "R. B." In this work the forms $A : B :: C : D$ and $A . B :: C . D$ appear frequently as equivalent in meaning.[2] It is possible that Leibniz, who used it as a general symbol of division in 1684,[3] took it from these writers. The hypothesis that it came from $\div$ by dropping the bar has no historical basis. Since it is more international than $\div$, it is probable that the latter symbol will gradually disappear.

Various other symbols have been used to indicate division, but they have no particular interest at the present time.

The Radical Sign. The ancient writers commonly wrote the word for root or side,[4] as they wrote other words of similar kind when mathematics was still in the rhetorical stage. The symbol most commonly used by late medieval Latin writers to indicate a root was R,[5] a contraction of *radix*, and this, with numerous variations, was continued in the printed books for more than a century.[6] The symbol was also used for other pur-

---

[1] Title as in F. Cajori, "Oughtred's Mathematical Symbols," *Univ. of Calif. Pub. in Math.*, I, 181. De Morgan (*Arith. Books*, p. 104) gives it as *Johnson's Arithmatick In 2 Bookes*, 2d ed., London, 1633.

[2] F. Cajori, "Oughtred's Math. Symbols," *loc. cit.*, p. 181. See also W. W. Beman, in *L'Intermédiaire des math.*, IX, 229; F. Cajori, *William Oughtred*, p. 75. In his *Clavis Mathematicae* (1631) Oughtred used a dot to indicate either division or ratio, but in his *Canones Sinuum* (1657) the colon (:) is used for ratio, possibly by some editor or assistant. It appears in the proportion 62496 : 34295 :: 1 : 0/54.9 —. Oughtred ordinarily used the dot for ratio, as in $A . B :: C . D$.

[3] Gerhardt, edition of his works, 3. Folge, V, 223 : "$x : y$ quod idem est ac x divis. per y seu $\frac{x}{y}$."

[4] As in Euclid, X, 96. Schoner used l for the square root : "Quadrati latus inexplicabile retextum significatur praenota litera l" (*De numeris figuratis liber*, 1569; 1586 ed., p. 263). On the Egyptian symbol see Peet, *Rhind Papyrus*, p. 20.

[5] Thus Chuquet (1484) used both R and R^2 for square root, R^3 for cube root, R^4 for fourth root, and so on. See Boncompagni's *Bullettino*, XIII, 655. Regiomontanus (*c.* 1464) has "$\frac{2,5}{2}$ i9 R de $\frac{6\frac{2}{4}1}{}$" for $\frac{2,5}{2} - \sqrt{6\frac{2}{4}1}$, as in the *Abhandlungen*, XII, 234.

[6] Thus Pacioli (1494) has "E cosi la .R. de .20$\frac{1}{4}$. e .4$\frac{1}{2}$" (fol. 45, *v.*). He also uses R. 2^a for square root and R. 3^a for cube root, as on fol. 46, *r*. E. de la Roche (1520) used R and R^2 for square root, R$\square$ and R^3 for cube root, and ⊢R and R^4

poses, including *response*,[1] *res*,[2] *ratio*,[3] *rex*,[4] and the familiar *recipe* in a physician's prescription.[5]

Meanwhile the Arab writers had used various symbols for expressing a root, among them ⟩, as in the case of

$$\frac{I}{\text{ן}} \quad \overset{?}{I} \quad \frac{I}{\text{ן}} \quad \overset{?}{\underset{\frown}{}}$$

for $\sqrt{4\frac{1}{2}} + \sqrt{1\frac{1}{2}}$,[6] but none of these signs seem to have influenced European writers.

European Symbols for Roots. The symbol $\sqrt{}$ first appeared in print in Rudolff's *Coss* (1525),[7] but without our modern indices. When Stifel edited this work,[8] in 1553, he varied this symbolism, using ν for $\sqrt{}$, ν for $\sqrt[3]{}$, ν for $\sqrt[4]{}$, and so on. It is frequently said that Rudolff used $\sqrt{}$ because it resembled a small *r*, for *radix*, but there is no direct evidence that this is true. The symbol may quite as well have been an arbitrary invention. It is a fact, however, that in and after the 14th century we find in manuscript such forms as r r r r r and r used for the letter *r*.[9]

It was a long time after these writers that a simple method was developed for indicating any root, and then only as a result of many experiments. For example, Vlacq[10] used $\sqrt{}$ for square root, $\sqrt{③}$ for cube root, $\sqrt{}\sqrt{}$ for fourth root, and so on; Rahn[11]

---

for fourth root. See the *Abhandlungen*, I, 63. Cardan (1539) and Tartaglia (1556) used ℞ for square root and ℞ cu. for cube root, while Ghaligai (1521) used ℞ □ and ℞ □□, and Bombelli (1572) used ℞ .q and ℞ .c. respectively for the same purposes. There were also the usual run of eccentricities, as illustrated by the use of Rad. 300 for $\sqrt{300}$ by an Italian arithmetician, Bonini, in 1517.

[1] Trenchant (1566).

[2] For the unknown quantity, as in the Rollandus MS. (*c.* 1424). As representing *res* in general, it is found as early as the 8th century.

[3] As early as the 8th century.　　　　[4] As early as the 14th century.

[5] Also as early as the 14th century.

[6] F. Woepcke, *Recherches*, p. 15. The Arabic forms are read from right to left.

[7] " . . . vermerkt von kürtz wegen radix quadrata mit solchem character $\sqrt{}$ . . . radix cubica würt bedeut durch solchen character C $\sqrt{}$ $\sqrt{}$."

[8] *Die Coss Christoph Rudolffs*, fols. 61, 62 (Königsberg i. Pr., 1553). The title-page bears the date 1553; the colophon, 1554.

[9] For these and other forms consult A. Cappelli, *Dizionario*, 2d ed., p. 318.

[10] *Arithmetica Logarithmica*, p. 4. Gouda, 1628.　　[11] *Teutsche Algebra*, 1659.

(1622–1676) used $\sqrt{}$, $\sqrt{}C$, $\sqrt{}\sqrt{}$, $\sqrt{}CC$, and $\sqrt{}\sqrt{}\sqrt{}$ for the square, cube, fourth, sixth, and eighth roots respectively, and various writers used $\sqrt{}. \mathfrak{Z}.$, $\sqrt{}.$ cc., $\sqrt{}\mathfrak{Z}.\mathfrak{Z}.$, $\sqrt{}S\mathfrak{Z}$, and $\sqrt{}.\mathfrak{Z}.$ cc. for the square, cube, fourth, fifth, and sixth roots respectively.[1]

French, English, and Italian writers of the 16th century were slow in accepting the German symbol, and indeed the German writers themselves were not wholly favorable to it. The letter l (for *latus*, side; that is, the side of a square)[2] was often used. Thus we find the Ramus-Schoner work of 1592 using $l4$ for $\sqrt{4}$, lc5 for $\sqrt[3]{5}$, lsq6 and ll6 for $\sqrt[4]{6}$, l ∫ 3 for $\sqrt[5]{3}$, and other similar forms, and using the related forms 1 l., 1 q., 1 c., 1 bq., and 1 qc. for a, a^2, a^3, a^4, and a^5 respectively. For the binomial $12 + \sqrt{32}$ the work has b 12 + l 32, and for the residual $12 - \sqrt{32}$ it has r 12 — l 32. In a somewhat similar way Gosselin, in his *De Arte Magna* (1577), uses L 9 for $\sqrt{9}$, LC 8 for $\sqrt[3]{8}$, LL 16 for $\sqrt[4]{16}$, and LV 24 PL 9 for $\sqrt{24 + \sqrt{9}}$ (the V standing for *universale* and the P for *plus*).

General Adoption of the Radical Sign. In the 17th century our common square-root sign was generally adopted, of course with many variants. Thus Stevin[3] has substantially the same symbols as those used by Rudolff, but with $\sqrt{}③$ for cube root, $W③$ for the fourth root of the cube root, and so on, with $\sqrt{}3)(2$ for $\sqrt{3} \cdot x^2$ and $\sqrt{}3(2)$ for $\sqrt{3\,x^2}$. Antonio Biondini, whose algebra appeared in Venice in 1689, has such symbols as V8 x for $\sqrt{8}\,x$ and $\underset{c}{V}24\,xx$ for $\sqrt[3]{24\,x^2}$. The different variants of the root sign are too numerous to mention in detail in this work, particularly as they have little significance. Such forms as

$$\frac{\sqrt{a} \times 100}{\lceil\ \ 100} \quad \text{for} \quad \frac{\sqrt{a} \times 100}{\sqrt{100}}$$

are not uncommon. Newton used $\sqrt[3]{8}$, $\sqrt[4]{16}$, ... for $\sqrt[3]{8}$, $\sqrt[4]{16}$, ...,[4] but he also used $\sqrt[3]{a}$.

---

[1] *E.g.*, Cardinael, *Arithmetica*, Bk. I. Amsterdam, 1659.
[2] See page 407, n. 4. [3] *Arithmetiqve* (1585), Girard edition of 1634, pp. 10, 19.
[4] *Arithmetica Universalis*, p. 37 (Cambridge, 1707). Among other statements he has "quod $\sqrt{6}$ valeat $\sqrt{2} \times 3$." Later, as on page 273, he has $\sqrt[3]{-4 \pm \sqrt{8}}$.

By the close of the 17th century the symbolism was, therefore, becoming fairly well standardized. We have, however, in Ozanam's *Dictionnaire Mathematique* (Paris, 1691) such forms as $\sqrt{C}.aab$ for $\sqrt[3]{a^2b}$ and $\sqrt{C}.a^3 + 3\,abb$ for $\sqrt[3]{a^3 + 3ab^2}$, so that there still remained some work to be done. The 18th century saw this accomplished, and it also saw the negative and fractional exponent come more generally into use. The early history of these forms is considered later.

Symbols of Relation. One of the earliest known symbols of algebra is a sign of equality. This may be said to have appeared in the Ahmes Papyrus ($c.$ 1550 B.C.), although Ahmes simply used a hieratic form for a hieroglyphic. He commonly wrote $\doteq$ for the hieroglyphic ⌣, *ṭemṭ,* meaning "together," the result of addition. In hieroglyphics, for example, we should have ∩ | ⌣ ∩| for $10 + 1 = 11$.[1] The Egyptians also used ◠, *er,* meaning "it makes," as in

$$ ⸫ \quad ⌣_{⁅⁆} \quad ⸙ \quad ⌣_{⸭} \quad ◠|, $$

meaning $\qquad \frac{2}{3} \quad \frac{1}{5} \quad \frac{1}{10} \quad \frac{1}{30} \quad er \quad$ I,

or $\qquad\qquad \frac{2}{3} + \frac{1}{5} + \frac{1}{10} + \frac{1}{30} = 1.$[2]

There is no evidence of the use of a generally recognized symbol for equality until the Greeks employed the initials ι^σ or $\check\iota\sigma$ for ἴσος ($i'sos$),[3] equal. This symbol is found in the *Arithmetica* of Diophantus ($c.$ 275).[4] The Arabs, contrary to the Greek custom, used for this purpose the final letter of their word for equality.

In general the classical and medieval writers used the full word.[5] In the Middle Ages a general shorthand was adopted

[1] Eisenlohr ed., p. 39. But on all this consult Peet, *Rhind Papyrus.*

[2] Eisenlohr, *loc. cit.*, p. 41. They also had other forms, for which consult both Peet and Eisenlohr. The symbols used in equations are given on page 422.

[3] As in *isosceles, isoperimetry, isogonal,* etc.

[4] He also used ἴσος ἐστι. For a discussion of the symbol see Heath, *Diophantus,* 2d ed., p. 47. The small Greek letters here shown are modern.

[5] *E.g.,* Fibonacci, in his *Flos* ($c.$ 1225), used such forms as *equantur* and *equabitur. Scritti,* II, 235.

by university students in copying their texts. Partly as a result of this movement there slowly developed a set of mathematical symbols, other contributing causes being a commercial shorthand and the advantage of expressing an equation in a form easily held by the eye. Thus we have such symbols as ÷ for *est*, p for *per*, ĉ for *cento*, and ∝ or ∞ for equality.[1] This symbol for equality, ∝ or ∞, was used by Descartes (1637) and is found in various manuscripts of his period. It has generally been thought to come from æ, for *aequalis, aequales, aequalia,* or *aequantur*. This may be the case, although it is by no means certain.[2]

Various other symbols were used for the same purpose. Thus Buteo (1559) used [; Xylander (1575), ‖; and Hérigone (1634), 2/2. Leibniz (*c.* 1680) used =, ⌐, and other symbols with nearly the same meaning.

Modern Symbol of Equality. As a printed symbol our sign (=) is due to Recorde,[3] who says: "I will sette as I doe often in woorke vse, a paire of paralleles, or Gemowe[4] lines of one lengthe, thus: ══, bicause noe .2. thynges, can be moare equalle." If he had used shorter lines (=), there might be some reason for thinking that the symbol was suggested by the medieval use of = for *esse*,[5] but Recorde's clear statement of its arbitrary invention in the form ══ is conclusive.

The symbol was not immediately popular. When Rahn (1622–1676) wrote his algebra, a century later, he felt obliged

---

[1] The bar, ——, indicating equality, as used in the correspondence of Regiomontanus, can hardly be considered a symbol in the ordinary sense. See *Bibl. Math.,* I (3), 506.

[2] *E.g.,* in the algebra of Clavius (Rome, 1608, p. 39 seq.) there are expressions like "æquatio inter ⅚ ℞+7, & 1℞," and "sit æquatio inter 4 ℞, & 72 −8℞," so that the ∞ may possibly have come from &. It is quite as reasonable to think that it was a purely arbitrary invention.

[3] *Whetstone of witte,* London, 1557. See the facsimile on page 412.

[4] From O. F. *gemeus,* twins, from Lat. *gemellus,* twin. Recorde uses *gemowe* in his *Pathewaie to knowledg* (1551) to mean parallel, speaking of "Paralleles, or Gemowe lynes." The various zodiacal signs for the gemini may have suggested all these forms.

[5] But not for *est,* where ÷ was commonly used. We find it also in compounds like =nt for *essent.* See A. Cappelli, *Dizionario,* 2d ed., p. 407.

to explain its meaning as not familiar to mathematicians,[1] and the use of ∞ continued until well along in the 18th century.

The Arte

as their woạkes doe ertende) to diſtinate it onely into ttwo partes. Whereof the firſte is, *when one nomber is equalle vnto one other.* And the ſeconde is, *when one nomber is compared as equalle vnto.* 2. *other nombers.*

Alwaies willyng you to remēber, that you reduce your nombers , to their leaſte denominations , and ſmalleſte foạmes,befoạe you pạocede any farther.

And again,if your *equation* be ſoche, that the greateſte denomination *Coſike,* be ioined to any parte of a compounde nomber , you ſhall tourne it ſo , that the nomber of the greateſte ſigne alone , maie ſtande as equalle to the reſte.

And this is all that neadeth to be taughte , concernyng this woọ₂ke.

Howbeit,foạ eaſie altératiō of *equations.* I will pạopounde a fewe cráples,bicauſe the ertraction of their rootes,maie the moạe aptly bee wạoughte. And to auoide the tedioufe repetition of theſe woọạdes : is equalle to : I will ſette as I doe often in woọ₂ke bſe,a paire of paralleles,oạ Gemowe lines of one lengthe, thus:=====,bicauſe noe.2. thynges,can be moate equalle. And now marke theſe nombers.

1. $14.\mathcal{z}.\,-\!\!\mid\!\!-.15.\mathcal{g}=====71.\mathcal{g}.$

2. $20.\mathcal{z}.\,-\!\!-\!\!-.18.\mathcal{g}====.102.\mathcal{g}.$

3. $26.\mathcal{z}\,-\!\!\mid\!\!-10\mathcal{z}====9.\mathcal{z}\,-\!\!-10\mathcal{z}\,-\!\!\mid\!\!-213.\mathcal{g}.$

4. $19.\mathcal{z}\,-\!\!\mid\!\!-192.\mathcal{g}===10\mathcal{z}\,-\!\!\mid\!\!-108\mathcal{g}\,-\!\!-19\mathcal{z}$

5. $18.\mathcal{z}\,-\!\!\mid\!\!-24.\mathcal{g}.====8.\mathcal{z}.\,-\!\!\mid\!\!-2.\mathcal{z}.$

6. $34\mathcal{z}\,-\!\!-\!\!-12\mathcal{z}====40\mathcal{z}\,-\!\!\mid\!\!-480\mathcal{g}\,-\!\!-9.\mathcal{z}.$

RECORDE'S SIGN OF EQUALITY

From Recorde's *Whetstone of witte* (1557)

---

[1] "Bey disem anlaasz hab ich das namhafte gleichzeichen = zum ersten gebraucht, bedeutet *ist gleich*" (*Teutsche Algebra*, 1659). It was probably suggested to him by Pell, who was familiar with Recorde's works.

Symbol of Proportion. The symbol for the equality of ratios
(: :), now giving way to the common sign of equality, was introduced by Oughtred (*c.* 1628),[1] and Dr. Pell gave it still
more standing when he issued Rahn's algebra in English
(1668). It seems to have been arbitrarily chosen.

The symbol $\because$ for continued proportion was used by English writers of the 17th and 18th centuries[2] and is still commonly seen in French textbooks.

Symbols of Inequality. The symbols $>$, $<$, for greater and
less, are due to Harriot[3] (1631). They were not immediately
accepted, for many writers preferred $\sqsubset$ and $\sqsupset$, symbols which
Oughtred (1631) had suggested.[4]
The symbols $\neq$, $\not<$, and $\not>$ are modern and are not international, but in the 1647 edition of Oughtred's *Clavis* the somewhat analogous symbols $\sqsubset$ and $\sqsupset$ appear for *non majus* and
non minus respectively. On the Continent the symbols $\geqq$ and
$\leqq$, or some of their variants, apparently invented by Pierre
Bouguer[5] (1734), are commonly used.

Symbol for Infinity. The symbol for infinity (∞) is first
found in print in the *Arithmetica Infinitorum* published by
Wallis in 1655,[6] and may have been suggested by the fact that
the Romans commonly used this symbol for a thousand, just as
we use "myriad" for any large number, although in the Greek
it meant ten thousand.

---

[1] In his *Elementi decimi Euclidis declaratio,* added to the 1648 edition of his
Clavis, he gives the symbol for "proportio, sive ratio aequalis : :." F. Cajori,
William Oughtred, p. 26; "Oughtred's Math. Symbols," *loc. cit.,* p. 181, n. 8. It
appears also in the 1631 edition of the *Clavis* itself.

[2] *E.g.,* Barrow (*Lectiones Mathematicae,* Lect. XXVII, London, 1683). J. Ward
(*c.* 1706) says: "The character made Use of to signify continued Proportionals
is $\because$" (*The Young Mathematician's Guide,* London, 12th ed., 1771, p. 77). It
also appears in the American Greenwood arithmetic (1729).

[3] *Artis Analyticae Praxis.* London, 1631 (posthumous).

[4] *E.g.,* Barrow: "A$\sqsubset$B. A major est quam B. A$\sqsupset$B A minor est quam
B" (*Lectiones Opticae & Geometricae* (London, 1674), preface; and English
edition (1735), p. 310).

[5] See biographical note on page 327.

[6] This is seen, for example, in such expressions as "jam numerus incrementorum est ∞" (*Opera,* I, 453 (1695)).

Integral Exponents. Our present integral exponents may be said to have begun with Descartes (1637), although Hérigone (1634) had nearly anticipated him. Since the early methods of indicating powers relate naturally to the writing of equations, these are more appropriately considered in connection with that topic (page 421). It may simply be said at this time that Harriot (who died in 1621), in the transition period from the use of forms like Aq to forms like x^2, used aa for a^2 and aaa for a^3. This symbolism was commonly employed until well into the 18th century, even in writing a polynomial involving a^4; that is, before c. 1750 it was common to find expressions like $a^5 + a^4 + aaa + aa + 1$, or even $a^5 + aaaa + aaa + aa + 1$.

In his *Cursus Mathematicus* (1634–1637) Hérigone used $a2, a3$, and $a4$ for a^2, a^3, a^4, no doubt influenced by the fact that Girard (1629) used forms like 5 (2) for $5x^2$; and some of his contemporaries, like Dechales (c. 1660) and Jacques de Billy (1602–1679), did the same. Descartes (1637), however, wrote the exponents in the present manner;[1] but even without this symbolism Stevin (1585) had already given a systematic discussion of integral exponents.[2]

General Exponents. The general exponent was known in theory long before it came into practical use. Oresme (c. 1360) wrote

$$\tfrac{1}{2}\,2^{\prime} \quad \text{for} \quad 2^{\frac{1}{2}}$$

and

$$\boxed{1^{\prime}\ \tfrac{1}{2}}\ 4 \quad \text{for} \quad 4^{1\frac{1}{2}},$$

and used other similar forms, as already stated. He also gave rules for fractional exponents.

Chuquet (1484) used[3] 12^0 for 12, 12^1 for 12 times a "nombre linear," 12^2 for $12x^2$, and so on. For $9x^{-3}$ he wrote[4] .9.^{3} m̃,

---

[1] "Et *aa*, ou *a*², pour multiplier *a* par soy-même; Et *a*³, pour le multiplier encore une fois par *a*, & ansi à l'infini" (1705 ed., p. 4).

[2] *L'Arithmetiqve*, Girard edition of 1634, p. 53. See also the general discussion by Hérigone, *loc. cit.*

[3] As he says, "côe nôbre simpleт̃t pris sans aulcune denomĩacion ou dont sa denoĩacĩo est .o." (Boncompagni's *Bullettino*, XIII, 737).

[4] *Ibid.*, p. 742. See also Ch. Lambo, "Une Algèbre Française de 1484. Nicolas Chuquet," *Revue des Questions Scientifiques*, October, 1902.

thus showing that he had an idea of negative exponents, but it was more than two centuries before the theory was understood. As to fractional exponents, certain evidences show that the idea was developing during the 16th century. This is seen especially in Stifel's *Arithmetica Integra*[1] (1544), where there is given what amounts to the relation

$$\left(\tfrac{27}{8}\right)^{1\,:\,\frac{3}{4}} = \left(\tfrac{27}{8}\right)^{\frac{4}{3}} = \tfrac{81}{16}.$$

Albert Girard[2] (1629) employed the fractional exponent, representing it by such forms as $\left(\tfrac{1}{6}\right)$ 2000 for $\sqrt[6]{2000}$, and $\left(\tfrac{3}{2}\right)$ 49 for $49^{\frac{3}{2}}$; and the study of logarithms from the standpoint of exponents, undertaken at about the same time, tended to bring these general forms into wider use.

Wallis on General Exponents. The first of the writers of this period to explain with any completeness the significance of negative and fractional exponents, however, was Wallis (1655). He showed that x^0 should signify 1, and established relations of the following nature:[3]

$$x^{-1} = \frac{1}{x} \qquad\qquad x^{-n} = \frac{1}{x^n}$$

$$x^{\frac{1}{2}} = \sqrt{x} \qquad\qquad x^{\frac{1}{q}} = \sqrt[q]{x}$$

$$x^{\frac{2}{3}} = \sqrt[3]{x^2} \qquad\qquad x^{\frac{p}{q}} = \sqrt[q]{x^p}$$

Newton supplemented the work of Wallis and in 1669 made use[4] of such forms as $x^{\frac{5}{3}}$ and x^{-3}, and after this time the symbolism became universally recognized.

[1] H. Wieleitner, "Gebrochene Exponenten bei Michael Stifel," *Unterrichtsblätter für Mathematik und Naturwissenschaften*, 1922, No. 5.

[2] *Invention nouvelle en l'algebre*, pp. 97–101. Amsterdam, 1629.

[3] Thus he speaks of $\frac{1}{8}$ "cujus index − 3," that is, $2^{-3} = \frac{1}{8}$; of $\frac{1}{\sqrt{2}}$ "cujus index − $\frac{1}{2}$," that is, $2^{-\frac{1}{2}} = \frac{1}{\sqrt{2}}$ ("Arithmetica Infinitorum," in the *Opera* (1695), I, 410, 459, and in the earlier edition).

[4] In the "De Analysi per aequationes numero terminorum infinitas" sent by Collins to Barrow, July 31, 1669. See the *Commercium Epistolicum*, p. 67 (London, 1725). For interesting comments on Newton's use of exponents see G. A. Lecchi, *Arithmetica Universalis Isaaci Newtoni*, Liber II, Pars III, p. 118 (Milan, 1752).

Symbols of Aggregation. Symbols of aggregation first developed to any considerable extent in the 16th century, and in connection with the study of radicals. Tartaglia (1556) writes "22 men (22 men ℞ 6" for $22 - (22 - \sqrt{6})$. Bombelli (1572) used L, for *legato*, as a kind of symbol of aggregation, as in the squaring of $2 + x + \sqrt{20 - 6x + x^2}$, which appears as

$$2. \text{ p. } \overset{1}{\text{ĭ}}. \text{ p. R. q. L } 20. \text{ m. } \overset{1}{6}. \text{ p. } \overset{2}{\text{ĭ}}. \text{ ⌐,}$$

the result being given as

$$\overset{2}{2}. \text{ p. } 24. \text{ m. } \overset{1}{2}. \text{ p. R. q. L } \overset{4}{4}. \text{ m. } \overset{8}{8}. \text{ p. } 2\overset{1}{2}4. \text{ p. } 320 \text{ ⌐,}$$

in which the L and the reversed L are clearly symbols of aggregation and may naturally have suggested our square parentheses, first used by Girard (1629) for this purpose.

Other Italian writers frequently employed the letter V, the initial of *universalis*, to indicate that a root sign applied to all the expression which followed. Thus Cardan, in his first printed solution of the cubic equation, has

$$\text{℞ V : cu. ℞ } 108 \text{ p : } 10$$

$$\text{m : ℞ V : cu. ℞ } 108 \text{ m : } 10$$

for $\qquad \sqrt{\sqrt[3]{108} + 10} - \sqrt{\sqrt[3]{108} - 10}.$

By the time Clavius published his algebra (1608) the parentheses had apparently become common, for he uses them freely without any explanation.

5. FUNDAMENTAL OPERATIONS

Number of Operations. While there were certain operations in arithmetic that were looked upon as fundamental, the number varying from time to time, this was not the case in the early printed algebras. It was only when textbooks, based upon the early arithmetics, came into use, that such operations as addition and subtraction were given as distinct topics. For example, Pacioli (1494) begins his work on algebra[1] by consid-

---

[1] *Sūma*, fol. 111, *v.*

ering a few definitions, then the laws of signs, and then the operations with monomials, taking up the operations with polynomials somewhat incidentally as they arise;[1] and the same may be said of the other Italian algebraists[2] of the 16th century. Clavius was one of the first to consider the subject somewhat as we do at present.[3] He early introduces a chapter *De additione et svbtractione numerorum Cossicorum,* this being followed by *De mvltiplicatione & diuisione numerorum Cossicorum.*

The reason for this early neglect was that algebra was looked upon as a study for mathematicians, not for boys and girls in their school years. For any mature mind that is interested in mathematics these operations are too simple to require any special attention.

Amount of Work. For this reason the amount of work assigned to topics of this kind by those algebraists who gave them any attention was very slight. For example, Pacioli gives no examples involving numerical cases like $(-2)(-3) = +6$, except a few that are completely worked out,[4] and similarly when he comes to surds.[5] An illustration of his problems is seen in the following:

$$\text{via. 4. } \bar{\text{p}}. \text{ R. 6.}$$
$$4. \bar{\text{m}}. \text{ R. 6.}$$
$$\overline{16. \bar{\text{m}}. 6.}$$
$$\text{Productum 10}$$

meaning that $(4 + \sqrt{6})(4 - \sqrt{6}) = 16 - 6 = 10.$[6] Similarly, Tartaglia[7] solves a few typical problems involving signs, but gives no exercises for original work.

Difficulties Due to Poor Symbolism. A good idea of the general difficulties which characterized this period of poor symbolism is seen in the algebra of Pedro Nunes, of which the second

[1] *E.g.,* fol. 127, *v.,* seq.
[2] Tartaglia, *General Trattato,* II, fol. 81 (1556); Bombelli, *Algebra, libro primo* (1572); Cardan, *Ars Magna,* cap. i (1545). [3] *Algebra,* p. 16 (Rome, 1608).
[4] Fol. 111, *v.,* seq. [5] Fol. 115, *v.,* seq. [6] Fol. 123, *r.*
[7] *General Trattato, La seconda parte,* fol. 81 (Venice, 1556).

edition appeared at Antwerp in 1567. His multiplication of $3x^2 + 2\frac{1}{4}x + 5\frac{1}{16}$ by $4x + 3$ appears as follows:

$$
\begin{array}{l}
3 \cdot \text{ce} \cdot \tilde{p} \cdot \quad 2 \cdot \text{co} \cdot \frac{1}{4} \cdot \tilde{p} \cdot 5\frac{1}{16} \\
\underline{4 \cdot \text{co} \cdot \tilde{p} \cdot 3.} \\
12 \cdot \text{cu} \cdot \tilde{p} \cdot \quad 9 \cdot \text{ce} \cdot \tilde{p} \cdot 20 \cdot \text{co} \cdot \frac{1}{4} \\
\underline{\qquad 9 \cdot \text{ce} \cdot \tilde{p} \cdot \quad 6 \cdot \text{co} \cdot \frac{3}{4} \cdot \tilde{p} \cdot 15\frac{3}{16}} \\
12 \cdot \text{cu} \cdot \tilde{p} \cdot 18 \cdot \text{ce} \cdot \tilde{p} \cdot 27 \cdot \text{co} \cdot \quad \tilde{p} \cdot 15\frac{3}{16}
\end{array}
$$

Bombelli (1572) sets forth the work more after the modern plan, but gives no cases to be solved independently. The following is a type:

$$
\begin{array}{l}
\underset{\sim}{1} \quad \text{I} \ p \ 2 \\
\underline{\underset{\sim}{1} \quad \text{I} \ p \ 2} \\
\underset{\sim}{2} \ \text{I} \ p \ 4 \ \underset{\sim}{1} \ p \ \ 4 \\
\underline{\underset{\sim}{2} \ \text{I} \ p \ 4 \ \underset{\sim}{1} \ p \ \ 4} \\
\underset{\sim}{4} \ \text{I} \ p \ \ 8 \ \ 3 \ p \ 24 \ \underset{\sim}{2} \ p \ 32 \ \underset{\sim}{1} \ p \ 16 \\
\underline{\underset{\sim}{1} \ \text{I} \ p \ \ 2} \\
\underset{\sim}{5} \ \text{I} \ p \ 10 \ \underset{\sim}{4} \ p \ 40 \ \ 3 \ p \ 80 \ \underset{\sim}{2} \ p \ 80 \ \underset{\sim}{1} \ p \ 32,
\end{array}
$$

meaning that $(x + 2)^2 = x^2 + 4x + 4,$

$$(x^2 + 4x + 4)^2 = x^4 + 8x^3 + 24x^2 + 32x + 16,$$

and this multiplied by $x + 2$ gives the fifth power of $x + 2$.[1]

6. CONTINUED FRACTIONS

Early Ideas. It is not necessary to speak of the history of simple algebraic fractions, since these forms were transferred from arithmetic. When Euclid found the greatest common measure of two lines,[2] or when the same principle was applied to the finding of the greatest common divisor of two numbers,[3]

---

[1] 1572 ed., p. 69. The I seems to have been the coefficient of the highest power.
[2] *Elements*, X, 3 and 4, for commensurable magnitudes in general.
[3] *Elements*, VII, 2 and 3.

a process was used that is similar to that of converting a fraction into a continued fraction, as is evident from the following:

$$12) \, 38 \, (3$$
$$\frac{36}{}$$
$$2) \, 12 \, (6$$
$$\frac{12}{}$$

$$\frac{12}{38} = \frac{6}{19} = \frac{1}{3 + \frac{1}{6}}.$$

This is the earliest important step in the theory of continued fractions.[1] Further traces of the general idea are found occasionally in the Greek and Arab writings.

Beginning of the Modern Theory. Although the Greek use of continued fractions in the case of greatest common measure was well known in the Middle Ages, the modern theory of the subject may be said to have begun with Bombelli (1572). In his chapter relating to square root[2] he considered the case of $\sqrt{13}$. Substituting our modern symbolism, he showed that this number is equal to

$$3 + \cfrac{4}{6 + \cfrac{4}{6 + \cdots}}.$$

In other words, he knew essentially that

$$\sqrt{a^2 + b} = a + \cfrac{b}{2a + \cfrac{b}{2a + \cdots}}.$$

The next writer to consider these fractions, and the first to write them in substantially the modern form, was Cataldi[3] (1613), and to him is commonly assigned the invention of the theory. His method was substantially the same as Bombelli's,

---

[1] On the history in general, see S. Günther, *Beiträge zur Erfindungsgeschichte der Kettenbrüche*, Prog., Weissenburg, 1872; Italian translation, Boncompagni's *Bullettino*, VII.

[2] "Modo di formare il rotto nella estrattione delle Radici quadrate," *Algebra*, p. 35.

[3] *Trattato del modo brevissimo di trovare la radice quadra delli numeri*, Bologna, 1613.

but he wrote the result of the square root of 18 in the following form:

$$4. \ \& \ \frac{2}{8.} \ \& \ \frac{2}{8.} \ \& \ \frac{2}{8}.$$

This he then modified, for convenience in printing,[1] into the form[2]

$$4 \ \& \ \frac{2}{8.} \ \& \ \frac{2}{8.} \ \& \ \frac{2}{8.}$$

The third writer to take up the theory was Daniel Schwenter (1618). In attempting to find approximate values for $\frac{177}{233}$ he found the greatest common divisor of 177 and 233, and from this he determined the convergents[3] as $\frac{79}{104}, \frac{19}{25}, \frac{3}{4}, \frac{1}{1}$, and $\frac{0}{1}$.

The next writer of prominence to use these forms was Lord Brouncker,[4] who transformed the product

$$\frac{4}{\pi} = \frac{3 \cdot 3 \cdot 5 \cdot 5 \cdot 7 \cdot 7 \cdots}{2 \cdot 4 \cdot 4 \cdot 6 \cdot 6 \cdot 8 \cdots},$$

which had been discovered by Wallis, into the fraction

$$\frac{4}{\pi} = 1 + \cfrac{1}{2 + \cfrac{9}{2 + \cfrac{25}{2 + \cfrac{49}{2 + \cdots}}}},$$

as already stated on page 311. He made no further use of these forms, Wallis then taking up the work and using the name "continued fraction."[5]

---

[1] "Notisi, che nõ si potendo cõmodamẽte nella stampa formare i rotti . . . ." See Tropfke, *Geschichte*, II (1), 362.

[2] ". . . facendo vn punti all' 8. denominatore de ciascun rotto, à significare, che il sequente rotto è rotto d' esso denominatore" (p. 70).

[3] For details see Tropfke, *Geschichte*, II (1), 363.

[4] J. Wallis, *Opera Mathematica*, I, 469 (Oxford, 1695). See also *Commercium Epistolicum*, London, 1725 ed., p. 215.

[5] ". . . quae denominatorem habeat continue fractum" (*Opera*, I, 469). His symbolism is

$$1\tfrac{1}{2} \ \tfrac{9}{2} \ \tfrac{25}{2} \ \tfrac{49}{2} \ \tfrac{81}{2} \ \&c.$$

See also Euler, *Introductio in Analysin Infinitorum*, ed. nova, I, 305 (Lyons, 1797).

The next advance was made by Huygens in his work on the description of a planetarium,[1] the ratio 2,640,858 : 77,708,431, for example, being written as a continued fraction.

In some manner, perhaps through the missionaries in China, the idea of the continued fraction found its way to Japan at about this time.[2] Takebe Hikojirō Kenkō (1722) used such forms for the value of π, stating that the plan was due to his brother, Takebe Kemmei. The first few convergents given by him are $\frac{3}{1}$, $\frac{22}{7}$, $\frac{333}{106}$, $\frac{355}{113}$, and $\frac{103993}{33102}$.

Euler founds the Modern Theory. The first great memoir on the subject was Euler's *De fractionibus continuis* (1737), and in this work the foundation for the modern theory was laid. Among other interesting cases Euler developed e as a continued fraction,[3] thus:

$$e = 2 + \cfrac{1}{1 + \cfrac{1}{2 + \cfrac{1}{1 + \cfrac{1}{4 + \cfrac{1}{1 + \cdots}}}}}$$

Of the later contributors to the theory, special mention should be made of Lagrange[4] (1767) and Galois.[5]

7. THE WRITING OF EQUATIONS

Equations in One Unknown. In speaking of the symbols for unknown quantities we are brought directly in touch with the symbols for integral exponents and with the writing of equations, and so it is convenient to treat of these topics in their relation to one another.

---

[1] *Descriptio automati planetarii*, The Hague, 1698 (posthumous).

[2] Smith-Mikami, p. 145.

[3] *Comm. Acad. Petrop.* for 1737, IX, 120 (Petrograd, 1744). See Tropfke, *Geschichte*, II (1), 337. See also Euler's *Introductio*, I, 293, "De fractionibus continuis," especially the forms on pages 117, 307; and his "De formatione fractionum continuarum," in the *Acta I'etrop.* for 1779, I, 3 (Petrograd, 1782), and other essays.

[4] See Serret's edition of his works, II, 539, and VII, 3 (Paris, 1868).

[5] Gergonne's *Annales de Math. Pures et Appliquées*, XIX, 294; posthumous (1828–1829). *Œuvres mathématiques d'Evariste Galois*, pp. 1–8 (Paris, 1897).

As already stated, the Egyptians called the unknown quantity *ahe* or *hau*, meaning "mass."[1] This word was represented in the hieroglyphic[2] as 𓏺 𓈖 𓏤.

For example, the equation $x(\frac{2}{3}+\frac{1}{2}+\frac{1}{7}+1)=37$ would appear in hieroglyphics as

and in the hieratic of the Ahmes Papyrus[3] (*c.* 1550 B.C.) as

It will be observed that, although Ahmes knew of symbols for plus, minus, and equality, they are not commonly used in his equations. They are found, however, in No. 28 of both the Peet and the Eisenlohr translation.

Symbolism of Diophantus. The first writer to make much effort toward developing a symbolism for the powers of algebraic expressions was Diophantus (*c.* 275). He used the following abbreviations for the various powers of the unknown:

MODERN	DIOPHANTUS[4]		LATE EDITIONS
x^0	$\overset{\circ}{\text{M}}$	μονάδες, units	$\mu\hat{o}$
x^1	ϛ	ἀριθμός, number	ς' or $\varsigma^{o'}$
x^2	Δ^{Υ}	δύναμις, power	$\delta^{\hat{v}}$
x^3	K^{Υ}	κύβος, cube	κ^{ι}
x^4	$\Delta^{\Upsilon}\Delta$	δυναμοδύναμις, power-power	$\delta\delta^{\hat{v}}$
x^5	ΔK^{Υ}	δυναμόκυβος, power-cube	$\delta\kappa^{\hat{v}}$
x^6	$K^{\Upsilon}K$	κυβόκυβος, cube-cube	$\kappa\kappa^{v}$

[1] On this term see page 393, n. 5.

[2] There are several variants. See Eisenlohr, *Ahmes Papyrus*, p. 42.

[3] Eisenlohr, *ibid.*, p. 54, No. 33; in the British Museum facsimile, Plate **X**, row 5.

[4] See Heath, *Diophantus*, 2d ed., p. 32. Much has been written about the symbol ϛ for *x*, and Heath gives a careful discussion of the various theories and a statement of the various forms for the symbol as they occur in different MSS. He concludes that the original symbol was a contraction of the initial letters αρ of ἀριθμός (*arithmos'*, number), instead of being the final sigma. Originally the capitals Δ^{Υ} were used for δ^{v}, and similarly K^{Υ} instead of κ^{v}, and so on.

Diophantus wrote his equations quite as we do, except for the symbols; thus the equation

$$(2x+1)^2 = 4x^2 + 4x + 1$$

appears, in modern Greek letters, as

$$\varsigma\ \bar{\beta}\ \overset{\circ}{\text{M}}\ \bar{a}\ \square^{os}\ \dot{\epsilon}\sigma\tau\acute{i}\ \Delta^{\Upsilon}\ \bar{\delta}\ \varsigma\ \bar{\delta}\ \overset{\circ}{\text{M}}\ \bar{a},$$

and the equation $8x^3 - 16x^2 = x^3$

appears[1] as $\text{K}^{\Upsilon}\ \bar{\eta}\ \bigwedge\ \Delta^{\Upsilon}\ \overline{\iota F}\ \acute{\iota}\sigma\ \text{K}^{\Upsilon}\ \bar{a}.$

DIOPHANTUS ON EQUATIONS

From a manuscript of the 14th century showing the symbolism then in use. It begins: "We call the square δύναμις, and it has for its symbol a *delta* (Δ) surmounted by a *upsilon* (Υ)." [This character is seen in the middle of line 2.] The symbol ΔΔᵛ (for x^4) appears in line 5, Δκᵛ (for x^5) in line 8, and κκᵛ (for x^6) in the last line. From Rodet, *Sur les Notations Numériques*, Paris, 1881

In speaking of Diophantus, however, it should again be stated that the most ancient manuscript of his *Arithmetica* now extant was written in the 13th century,—about a thousand years after the original one appeared. We are therefore quite

---

[1] In the first equation ἐστί (*esti'*) stands for "is equal to," and in the second case ἰσ stands for ἴσος (*i'sos*, equal). See Tannery's *Diophantus*, I, 230–231, 258–259.

uncertain as to the symbols used by Diophantus himself and as to the various interpolations that may have been made by the medieval copyists.

ALGEBRAIC SOLUTION ACCORDING TO DIOPHANTUS

From a manuscript of the 14th century. The problem[1] is to find two numbers such that their sum is equal to 20 and the difference of their squares to 80

Oriental Symbols. It was from this Greek method of expressing the equality of the two members that the Arabs seem to have derived theirs, as in the case of

for
$$38 = 19\,x + x^2,$$

the equation being written from right to left after the Semitic custom, which obtains in writing Arabic.[2] From this form, reversed in order of writing, came the one that we use.

The Chinese and Hindus, however, had methods of writing their equations that were very different from those which the

---

[1] The solution reads substantially as follows:

Let there be	$x + 10$	$10 - x$.
Squares,	$x^2 + 20x + 100$	$x^2 + 100 - 20x$.
Difference of squares,	$40x$	$= 80$.
Division,	x	$= 2$,
whence	$x + 10 = 12$	$10 - x = 8$.

From Rodet

[2] The above is from al-Qalasâdî. See L. Matthiessen, *Grundzüge der antiken und modernen Algebra*, 2d ed., p. 269 (Leipzig, 1896); hereafter referred to as Matthiessen, *Grundzüge.*

Arabs and Persians adapted from the Greek works. The Far East depended more upon position. The Chinese commonly represented the coefficients by sticks, their so-called "bamboo rods" which they used in calculating, and these they placed in squares on a ruled board. Ch'in Kiu-shao (c. 1250), for example, represented the equation $x^3 + 15 x^2 + 66 x = 360$ as here shown; and if he needed to write the equation, he did so in the same manner.[1] The positive terms were represented by red sticks or marks, and the negative terms either by black ones or (as in the illustration) by a stick placed diagonally across some part of the numeral. The system is simply one of detached coefficients, the place values of the coefficients being indicated sometimes by the squares running horizontally, but ordinarily as shown in the illustration. The native Japanese mathematicians used the same method, having imported it from China.[2]

yuen (*element*)

tai (*extreme*)

The Hindu method was better than the Chinese, and in one respect was the best that has ever been suggested. Bhāskara (c. 1150) represented the equation $18 x^2 = 16 x^2 + 9 x + 18$ as follows:

न्यासः याव १८ या ० रू ०
याव १६ या ९ रू १८

This may be transliterated as

ya v 18	*ya* 0	*ru* 0
ya v 16	*ya* 9	*ru* 18

which means

$18 x^2$	$0 x$	0
$16 x^2$	$9 x$	18

or $18 x^2 + 0 x + 0 = 16 x^2 + 9 x + 18,$

or $2 x^2 - 9 x = 18.$

[1] The Chinese word *tai* (from *tai-kieh* or *tai-chi*, extreme limit, or great extreme) means the absolute term, and *yuen* (element) means the first power of the unknown. See Mikami, *China*, pp. 81, 82, 91; L. Vanhée, "La notation algébrique en Chine au XIIIᵉ siècle," *Revue des Questions Scientifiques*, October, 1913.

[2] Smith-Mikami, p. 50.

The word for the first power of the unknown, the *yāvat-tāvat*, already explained, is abridged to *ya*; and the word for the second power, *yāvat-varga*, to *ya v*.[1]

PAGE FROM BHĀSKARA'S BIJA GANITA

From the first printed edition, showing the method of writing equations. Lines 8 and 9 give the equation shown in the text. The next equation is interesting for the use of the dot above the Sanskrit 9 to indicate subtraction; thus:

> *ya v* 2   *ya* 9̇   *ru*   0
> *ya v* 0   *ya* 0   *ru* 18

Such a plan shows at a glance the similar terms one above another, and permits of easy transposition.

When the Arabic algebras were translated into Latin, the rhetorical form was used. Thus Robert of Chester (*c.* 1140), in his translation of al-Khowârizmî, wrote "Substantia et 10 radices 39 coaequantur drachmis" for "A square and 10 roots are equal to 39 units"; that is,

$$x^2 + 10x = 39.[2]$$

Al-Khowârizmî himself wrote his equations in rhetorical form, thus: "A square, multiply its root by four of its roots, and the product will be three times the square, with a surplus of fifty dirhems."[3]

Medieval Manuscripts. In the manuscript period of the Middle Ages we find

---

[1] Colebrooke, *loc. cit.*, p. 140; E. Strachey, *Bija Ganita* (often bound with Colebrooke), p. 117.

[2] In the Scheubel MS., translated by Karpinski, pp. 70–73. Somewhat the same form is used in the transcription in Libri's *Histoire*, I, 255, although the exact wording is "Census et decem radices equantur triginta novem dragmis."

[3] Rosen translation, p. 56; Arabic text in the Rosen edition, p. 40.

letters coming into use to represent algebraic as well as geometric quantities. This is seen in the work of Jordanus Nemorarius (*c.* 1225), a contemporary of Fibonacci.[1] This was not common, however, most writers preferring to use some such symbols as R for *res* (thing, the unknown), *ce.* for *census* (the second power of the unknown), and *cu.* for *cubus* (the third power of the unknown), with other shorthand abbreviations. Such symbolism is seen in the manuscripts of Regiomontanus (*c.* 1463), one problem from which is reproduced in facsimile on page 429. In an Italian manuscript of about the same period[2] the quadratic equation $x^2 + 10x = 39$ appears in the rhetorical form as follows: "lo censo e·10·sue cose cioe·10·sue rā. sono igualj a·39· drãe."

Equations in Printed Form. The following examples will suffice to show the general development of the symbolism of the equation from the first printed work containing algebra to the time when our present symbolism was fairly well settled:

Pacioli (1494)[3]: "Trouame .1. n°. che gióto al suo q̄drat° facia .12." Modern form: $x + x^2 = 12$.

Vander Hoecke (1514)[4]: 4 Se. − 51 Pri. − 30 N. dit is ghelijc $45\frac{3}{5}$. Modern form: $4x^2 - 51x - 30 = 45\frac{3}{5}$.

Ghaligai (1521)[5]: 1□e 32 c° − 320 numeri. Modern form: $x^2 + 32x = 320$.

---

[1] For example, in the second problem in his *De Numeris Datis* he says: "Datus numerus sit .a. qui diuidatur in .b.c.d.e. . . ." *Abhandlungen*, II, 135. See also A. Favaro, Boncompagni's *Bullettino*, XII, 129; M. Curtze, *Abhandlungen*, XII; A. Witting and M. Gebhardt, *Beispiele zur Gesch. der Math.*, II, 26 (Berlin, 1913).

[2] In Mr. Plimpton's library. See *Rara Arithmetica*, p. 459. The equation is from the MS., fol. 279, *v.*

[3] Fol. 145, *r.* In his solutions, but not in his problems, he used .co. (*cosa*, thing, as already explained) for *x*; .ce. (*census* or *zensus*; in Italian, *censo*, evaluation of wealth, tax) for x^2; .cu. (*cubus*) for x^3; .ce.ce. (*census census*) for x^4; and .p°.r°. (*primo relato*) for x^5. The Latin *census* (a registering of citizens and property) was conducted by the censors, who gave *censura*, censure, to those who incurred their disfavor, *census* coming probably from *centere*, to number by the *centum*, hundred.

[4] He used Pri., Se., 3^a, 4^a, and 5^a for *x*, x^2, x^3, x^4, and x^5 respectively. The system failed because of the difficulty in writing coefficients. See his 1537 edition, fol. 64, *v.*   [5] Fol. 96, *v.*

Rudolff $(1525)^1$: Sit $1\ \delta$ aequatus $12\ \mathcal{R} - 36$. Modern form: $x^2 = 12\,x - 36$.

Cardan $(1545)^2$: cub^9 p: 6 reb^9 aeqlis 20. Modern form: $x^3 + 6x = 20$.

Scheubel $(1551)^3$: 4 sex. aequantur 108 ter. Modern form: $4x^7 = 108x^4$.

Tartaglia $(1556)^4$: "Trouame uno numero che azontoli la sua radice cuba uenghi ste, cioe .6." Modern form: $x + \sqrt[3]{x} = 6$.

Buteo $(1559)^5$: $1\lozenge$ P6ρP9C$1\lozenge$P3ρP24. Modern form: $x^2 + 6x + 9 = x^2 + 3x + 24$.

Bombelli $(1572)^6$: $\overset{6}{\underset{.}{1}}$. p. $\overset{3}{8}$. Eguale à 20. Modern form: $x^6 + 8x^3 = 20$. In the text proper he would write this equation $1\overset{6}{\cup}$p. $8\overset{3}{\cup}$eguale à 20. In the same way he would write $1\overset{3}{\cup}$ eguale à R. q. 108. p. 10. for $x^3 = \sqrt{108} + 10$, which may be compared with Cardan's form on page 463.

---

^{1}From Stifel's edition of 1553 (1554), fol. 243, v. The symbols for the first five powers of the unknown, beginning with the first, are $\mathcal{R}$ (contraction of *radix*?), δ (*zensus*), $\mathcal{e}$ (contraction of cs, for *cubus*), $\delta\delta$ (*zensus zensus*), β (*sursolidus*). The German writers in general used this system until well into the 17th century. Although the symbol for the unknown is usually taken as a contraction for *radix*, it is quite as probable that it is the common ligature of the Greek γ and ρ. This stood for *gram'ma* ($\gamma\rho\acute\alpha\mu\mu\alpha$), a letter, or for *gramme'* ($\gamma\rho\alpha\mu\mu\acute\eta$), a line. In the medieval works it was a common thing to represent the unknown by a line. This is seen in an algebra as early as al-Khowârizmî's (c. 825) and in one as late as Cardan's (1545). For evidence of the frequent use of the symbol for *gramma*, see Michael Neander, $\Sigma\Upsilon\Nu$OΨIΣ *mensvrarvm et Pondervm*, Basel, 1555. It may, indeed, have suggested to Descartes the use of x, this being the letter most nearly resembling it.

^{2}*Ars Magna*, 1545 ed., fol. 30, r. His names or abbreviations for x, x^2, x^3, x^4, and x^5 are *res*, $\bar qd$ (*quadratum*), *cu'* or *cub^9* (*cubus*), $\bar qd'\ \bar qd'$, $\bar qd\ \bar qd$, or $\bar qd^{tt}\ qd^m$ (*quadrati quadratum*), and *relatum primum*, with necessary variants of these forms. For a facsimile, see pages 462, 463.

^{3}Also 1 pri. + 12 N aequales 8 ra., for $x^2 + 12 = 8x$. See *Abhandlungen*, IX, 455.

^{4}*La Noua Scientia*, 1554 ed., fol. 114, r. His problems are in rhetorical form, and his symbolism is substantially that of Cardan and other Italian contemporaries. For Recorde's equations (1557) see the facsimile on page 412.

^{5}The next step is $3\rho[15]$, sometimes with both brackets, sometimes with only the first. Buteo, *Logistica, quae & Arithmetica vulgò dicitur*, Lyons, 1559.

^{6}*Algebra*, p. 273. He indicated x, x^2, x^3, ... by $\overset{1}{\cup}, \overset{2}{\cup}, \overset{3}{\cup}, \cdots$. It is in this work that there appears for the first time in Italian any important approach to the modern symbolism of the equation.

SYMBOLISM OF THE EQUATION AS USED BY REGIOMONTANUS,
c. 1463

From a letter written by Regiomontanus. The problem is to find a number x such that

$$\frac{100}{x} + \frac{100}{x+8} = 40 \quad \text{or} \quad x^2 + 3x = 20,$$

from which $x = \sqrt{\frac{89}{4}} - \frac{3}{2}$, as stated in the third line from the bottom. The conclusion is that the first divisor is $\sqrt{22\frac{1}{4}}$ minus $1\frac{1}{2}$. See page 427

Gosselin (1577)[1]: 12LM1QP48 aequalia 144M24LP2Q. Modern form: $12x - x^2 + 48 = 144 - 24x + 2x^2$.

Stevin (1585)[2]: 3 ② + 4 egales à 2 ① + 4. Modern form: $3x^2 + 4 = 2x + 4$.

Ramus and Schoner (1586)[3]: 1q —+— 81 aequatus sit 65. Modern form: $x^2 + 8x = 65$.

Vieta (c. 1590)[4]: 1 Q C — 15 Q Q + 85 C — 225 Q + 274 N, aequatur 120. Modern form:

$$x^6 - 15x^4 + 85x^3 - 225x^2 + 274x = 120.$$

Clavius (1608)[5]: "Sit aequatio inter 1.3⤶ℓ & 800ℓ−156751." Modern form: $x^6 = 800x^3 - 156{,}751$.

Girard (1629)[6]: 1 (4) + 35 (2) + 24 = 10 (3) + 50 (1), or with the several exponents inclosed in circles. Modern form: $x^4 + 35x^2 + 24 = 10x^3 + 50x$.

Oughtred (1631)[7]: $\frac{1}{2}$ Z ± √q: $\frac{1}{4}$ Zq − AE = A. Modern form: $\frac{1}{2}Z \pm \sqrt{\frac{1}{4}Z^2 - AE} = A$.

Harriot (1631)[8]: aaa − 3 · bba══ + 2 · ccc. Modern form: $x^3 - 3b^2x = 2c^3$.

---

[1] Guillaume Gosselin was a native of Caen, but we know almost nothing of his life. He published an algebra, De arte magna, seu de occulta parte numerorum, quae & Algebra, & Almucabala vulgo dicitur; Libri QVATVOR, Paris, 1577; and a French translation of Tartaglia's arithmetic, Paris, 1578. See H. Bosmans, Bibl. Math., VII (3), 44. In the above equation he uses L for latus (the side of the square), Q for quadratus (square), and P and M for plus and minus.

[2] L'Arithmetiqve, p. 272. See also his Œuvres, Girard ed., p. 69 (Leyden, 1634). He used ①, ②, ③, ··· for x, x^2, x^3, ···.

[3] Algebrae Liber Primus, 1586 ed., p. 349.

[4] For a discussion of the dates of his monographs, see Cantor, Geschichte, II (2), 582. He also used capital vowels for the unknown quantities and capital consonants for the known, thus being able to express several unknowns and several knowns. The successive powers of A were then indicated by A, Aq, Acu, Aqq, Aqcu, and so on, the additive principle of exponents being followed. The above example is from his Opera Mathematica, ed. Van Schooten, p. 158 (Leyden, 1646).

[5] Algebra, p. 62. For his symbols, see ibid., p. 11.

[6] Invention nouvelle en l'algèbre, p. 131 (Amsterdam, 1629), with rules for the symmetric functions of the roots.

[7] Clavis, p. 50; Cajori, William Oughtred, p. 29.

[8] See his Artis Analyticae Praxis, London, 1631. He represented the successive powers of the unknown by a, aa, aaa, . . . ; Tropfke, Geschichte, III (2), 143, a work which should be consulted (pp. 119-148) on this entire topic.

Hérigone (1634)[1]: 154a⌣71a2 —+—14a3⌢a4 2/2 120. Modern form: $154a - 71a^2 + 14a^3 - a^4 = 120$.

Descartes (1637)[2]: yy ∞ cy $-\dfrac{cx}{b}$ y+ay−ac. Modern form:

$$y^2 = cy - \frac{cx}{b} y + ay - ac.$$

Wallis (1693): $x^4 + bx^3 + cxx + dx + e = 0$, which is the modern form, with the exception of xx; this, as already stated, was commonly written for x^2 until the close of the 18th century.[3]

Equating to Zero. It is difficult to say who it was who first recognized the advantage of always equating to zero in the study of the general equation. It may very likely have been Napier, for he wrote his *De Arte Logistica* before 1594 (although it was first printed in Edinburgh in 1839), and in this there is evidence that he understood the advantage of this procedure.[4] Bürgi (*c.* 1619) also recognized the value of making the second member zero, Harriot (*c.* 1621) may have done the same, and the influence of Descartes (1637) was such that the usage became fairly general.[5]

Several Unknowns. The ancients made little use of equations with several unknown quantities. The first trace that we find of problems involving such equations is in Egypt. There are

[1] *Cursus Mathematicus*, 5 vols., Paris, 1634–1637; 2d ed., 6 vols. in 4 (Paris, 1644), Vol. II, chap. xiv. He represented the successive powers of the unknown by $a, a2, a3, a4, \ldots$.

[2] *La Geometrie*, 1637; 1705 ed., p. 36. It will be seen that this form does not differ much from our own. Descartes used the last letters of the alphabet for the unknown quantities and the first letters for the known, and this usage has persisted except in the case of those formulas in which the initial letter serves a better purpose. That it was not immediately accepted, however, is seen by the fact that Rahn (Rhonius) used final letters for unknowns and large letters for knowns, as in his algebra of 1659 (English translation, 1688).

[3] *E.g.*, in Euler's *Algebra*, French ed., Petrograd, 1798, where such forms as $xx + yy = n$ are found.

[4] For example, on page 156 he takes the equation $4 ℞ - 6 = 5 ℞ - 20$ and reduces it to $- ℞ + 14 = 0$, "quae aequatio ad nihil est." In general all his higher equations have zero for the second member. See also Eneström, *Bibl. Math.*, III (3), 145.

[5] For a discussion see Tropfke, *Geschichte*, III (2), 26; Kepler's *Opera*, ed. Frisch, V, 104. The credit is often claimed for Stifel (*c.* 1525), but he, like Harriot, made no general practice of equating to zero.

two papyri[1] of the Middle Kingdom ($c.$ 2160–1700 B.C.) which contain problems of this nature. One of these problems is to divide 100 square measures into two squares such that the side of one of the squares shall be three fourths the side of the other; that is,

$$x^2 + y^2 = 100,$$
$$y = \tfrac{3}{4}\,x.$$

The other problems also involve quadratics, one found in 1903 being substantially[2]

$$x^2 + y^2 = 400,$$
$$x : y = 2 : 1\tfrac{1}{2}.$$

Simultaneous Linear Equations. The earliest of the Greek contributions to the subject of simultaneous linear equations are, according to the testimony of Iamblichus ($c.$ 325) in his work on Nicomachus, due to Thymaridas of Paros ($c.$ 380 B.C.?). He is said to have given a rule called ἐπάνθημα (*epan'thema*, flower), which he seems to have used in solving n special types of equations, namely, $x_0 + x_1 + x_2 + \cdots + x_{n-1} = s$, $x_0 + x_1 = a_1$, $x_0 + x_2 = a_2$, $\cdots x_0 + x_{n-1} = a_{n-1}$, the method being the ordinary one of adding. Iamblichus applied the rule to other cases.[3]

Some use of simultaneous linear equations is also found in the work of Diophantus ($c.$ 275), who spoke of the unknowns as the first number, the second number, and so on,[4] a method that was too cumbersome to admit of any good results.

Chinese and Japanese Methods. The subject was greatly extended by the Chinese. Using the "bamboo rods" as calculating sticks, they placed these in different squares on the table so as to represent coefficients of different unknowns, and hence

---

[1] The Berlin Papyrus No. 6619, published by H. Schack-Schackenburg, *Zeitschrift für ägyptische Sprache*, XXXVIII (1900), 135, and XL (1902), 65; and the Kahun Papyrus, also studied by Schack-Schackenburg in 1903 and described by F. Ll. Griffith in 1897.

[2] M. Simon, *Geschichte der Mathematik im Altertum*, pp. 41, 42 (Berlin, 1909); hereafter referred to as Simon, *Geschichte*.

[3] Heath, in R. W. Livingstone's *The Legacy of Greece*, p. 110 (Oxford, 1922).

[4] That is, ὁ πρῶτος ἀριθμός, ὁ δεύτερος ἀριθμός, and so on. *E.g.*, Book II, Prop. 17; Book IV, Prop. 37.

they needed no special symbols.[1] Indeed, we are quite justified in saying that the first definite trace that we have of simultaneous linear equations is found in China. In the *Arithmetic in Nine Sections*[2] there are various problems that require the solution of equations of the type $y = ax - b$, $y = a'x + b'$. A rule is given for the solution which amounts substantially to the following:

Arrange coefficients,
$$\begin{matrix} a & a' \\ b & b' \end{matrix}$$

Multiply crosswise,
$$\begin{matrix} ab' & a'b \\ b & b' \end{matrix}$$

Add,
$$\begin{matrix} ab' + a'b \\ b + b' \end{matrix}$$

Result,
$$\frac{ab' + a'b}{a - a'},$$

and
$$\frac{b + b'}{a - a'}.$$

The method of reasoning is not stated, but the work was probably done by the aid of bamboo rods to represent the coefficients.[3]

The next step of which we have evidence was taken much later by Sun-tzï, the date being uncertain but probably in the 1st century. He solved what is equivalent to the system $2x + y = 96$, $2x + 3y = 144$, and his method of elimination was substantially first to multiply the members of each equation, when necessary, by the coefficient of x in the other equation.[4]

From this time on, the solution of simultaneous linear equations was well known in China. The only improvement made upon the early methods consisted in the arrangement of the bamboo rods in such a way as to allow for a treatment of the coefficients similar to that found in the simplification of determinants. This was finally carried over to Japan and was amplified by Seki Kōwa (1683) into what may justly be called the first noteworthy advance in the theory of these forms.[5]

[1] Mikami, *China*, p. 73.
[2] *K'iu-ch'ang Suan-shu*, of uncertain date, possibly as early as 1100 B.C., and certainly pre-Christian. See Volume I, page 31.
[3] Mikami, *China*, p. 16.   [4] *Ibid.*, p. 32.   [5] *Ibid.*, p. 191.

Hindu Symbolism. The Hindus represented the various unknowns by the names of colors, calling them "black,"[1] "blue,"[2] "yellow,"[3] "red," and so on. They wrote the coefficients at the right of the abridged words and represented a negative term by a dot placed above the coefficient. For example,[4]

$$\text{yā 5} \quad \text{kā 8} \quad \text{nī 7} \quad \text{rū 90}$$
$$\text{yā 7} \quad \text{kā 9} \quad \text{nī 6} \quad \text{rū 62}$$

means $5x + 8y + 7z + 90 = 7x + 9y + 6z + 62$

and gives rise to
$$\frac{k\bar{a}\ \dot{\imath}\ n\bar{\imath}\ 1\ r\bar{u}\ 28}{y\bar{a}\ 2},$$

as it appears in the Colebrooke version, which means

$$-y + z + 28 = 2x.$$

Problems involving several unknowns did not possess much interest for the Arab and Persian writers, as may be seen from the algebras of al-Khowârizmî[5] and Omar Khayyam.

Early European Symbolism. The algebraists of the 16th century gave relatively little attention to simultaneous linear equations. The use of x, y, and z for unknown quantities was not suggested until the 17th century, and so it was the custom of some writers to use ordinary capital letters. For example, we find Buteo (1559) using

$$1\ A, \tfrac{1}{2}\ B, \tfrac{1}{2}\ C\ [\ 17,$$
$$1\ B, \tfrac{1}{3}\ A, \tfrac{1}{3}\ C\ [\ 17,$$
$$1\ C, \tfrac{1}{4}\ A, \tfrac{1}{4}\ B\ [\ 17,$$

for what we should now write as $x + \tfrac{1}{2}y + \tfrac{1}{2}z = 17$, etc. By multiplication he reduces these to three equivalent equations in

---

[1] *Kālaca*, abridged to *ka* (卐Ｔ). See Colebrooke's translation of Bhāskara, pp. 184 n., 227. The known number was *rūpa*, abridged to *rū*.

[2] *Nīlaca*, abridged to *nī* (ℐ卐).

[3] *Pītaca*, abridged to *pī* (ℐ卐).   [4] Colebrooke's translation, p. 231.

[5] But see E. Wiedemann, *Sitzungsberichte der Physikalisch-medizinischen Sozietät zu Erlangen*, 50.–51. Bd. (1920), p. 264.

which, however, the original symbolism changes slightly, the period replacing the comma to indicate addition, thus:

$$2A \cdot 1B \cdot 1C [34,$$
$$1A \cdot 3B \cdot 1C [51,$$
$$1A \cdot 1B \cdot 4C [68.$$

He then eliminates in the usual manner.[1] Gosselin, in his *De arte magna* (Paris, 1577), uses a similar arrangement.[2]

Literal Equations. The equations considered by the ancient and medieval writers were numerical. Even the early Renaissance algebraists followed the same plan, their crude symbolism allowing no other. It was not until the close of the 16th century that the literal equation made its appearance, owing largely to the influence of the new symbolism invented by Vieta and his contemporaries. For example, Adriaen van Roomen published in 1598 a commentary on the algebra of al-Khowârizmî[3] in which he distinguished between two types of equation, the *numerosa* and the *figurata*. The former was applied to problems with numerical data, while the latter resulted in general formulas.[4] Van Roomen asserts that writers on algebra up to his time used the *numerosa* method only, whereas he was the first to use the *figurata* one, although as a matter of fact Vieta seems to have preceded him. The actual dates of invention, but not of publication, are, however, obscure.

8. THE SOLUTION OF EQUATIONS

Linear Equations. The earliest solutions of problems involving equations were doubtless by trial. In the time of Ahmes (*c.* 1550 B.C.), however, the methods of making the trials

---

[1] G. Wertheim, "Die Logistik des Johannes Buteo," *Bibl. Math.*, II (3), 213.

[2] H. Bosmans, "Le 'De arte magna' de Guillaume Gosselin," *Bibl. Math.*, VII (3), 44.

[3] H. Bosmans, "Le fragment du Commentaire d'Adrien Romain sur l'algèbre de Mahumed ben Musa El-chowârezmî," *Annales de la Société Scientifique de Bruxelles*, XXX (1906), second part, p. 266.

[4] "Differentia igitur inter has duas talis statui potest, quod figurata inveniat regulam solvendi problema propositum; numerosa vero duntaxat regulae illius exemplum."

were fairly well simplified. Thus, his equation[1] $x + \frac{1}{7}x = 19$ is solved substantially as follows: Assume 7 as the number. Then, to use the form of the text,

"Once	gives	7
$\frac{1}{7}$	gives	1
$1\frac{1}{7}$	gives	8

"As many times as 8 must be multiplied to make 19, so many times must 7 be multiplied to give the required result.

"Once	gives	8
Twice	gives	16
$\frac{1}{2}$	gives	4
$\frac{1}{4}$	gives	2
$\frac{1}{8}$	gives	1

"Together, 2, $\frac{1}{4}$, $\frac{1}{8}$ gives   19 [in which he selects the addends making 19].

"Multiply 2, $\frac{1}{4}$, $\frac{1}{8}$ by 7 and obtain the required result.

"Once	gives	2, $\frac{1}{4}$, $\frac{1}{8}$
Twice	gives	4, $\frac{1}{2}$, $\frac{1}{4}$
4 times	gives	9, $\frac{1}{2}$

"Together,   7   gives   16, $\frac{1}{4}$, $\frac{3}{8}$, the result."[2]

The Greek methods are discussed later in connection with the quadratic.

The chief contribution to the solution of linear equations made by the Arab writers was the definite recognition of the application of the axioms to the transposition of terms and the reduction of an implicit function of x to an explicit one, all of which is suggested by the name given to the science by al-Khowârizmî (c. 825).

---

[1] In this discussion all equations will be given in the modern form. On the general history of solutions two of the best works for the student to consult are A. Favaro, "Notizie storico-critiche sulla Costruzione delle Equazioni," *Atti della R. Accad. di Scienze, Lettere ed Arti in Modena*, Vol. XVIII, 206 pages with extensive bibliography; Matthiessen, *Grundzüge*.

[2] For the translation I am indebted to Dr. A. B. Chace, of Providence, Rhode Island. For a slightly different version see Peet, *Rhind Papyrus*, p. 61.

False Position. To the student of today, having a good symbolism at his disposal, it seems impossible that the world should ever have been troubled by an equation like $ax + b = 0$. Such, however, was the case, and in the solution of the problem the early writers, beginning with the Egyptians, resorted to a method known until recently as the Rule of False Position. The ordinary rule as used in the Middle Ages seems to have come from India,[1] but it was the Arabs who made it known to European scholars. It is found in the works of al-Khowârizmî (*c.* 825), the Christian Arab Qosṭâ ibn Lûqâ al-Ba'albekî (died *c.* 912/13), Abû Kâmil (*c.* 900), Sinân ibn al-Fatḥ (10th century), Albanna (*c.* 1300),[2] al-Haṣṣâr (*c.* 12th century),[3] and various others. The Arabs called the rule the *ḥisab al-Khataayn*[4] and so the medieval writers used such names as *elchataym.*[5] When Pacioli wrote his *Sūma* (1494) he used the term *el cataym,*[6] probably taking it from Fibonacci. Following Pacioli, the European writers of the 16th century used the same term, often with a translation into the Latin or the vernacular.[7]

---

[1] There is a medieval MS., published by Libri in his *Histoire*, I, 304, and possibly due to Rabbi ben Ezra. It refers to this rule, "quem Abraham compilavit et secundum librum qui Indorum dictus est composuit." See M. Steinschneider, *Abhandlungen*, III, 120; F. Woepcke, "Mémoire sur la propagation des chiffres indiens," *Journal Asiatique* (Paris, 1863), I (6), 34, 180; Matthiessen, *Grundzüge*, p. 275; C. Košt'ál, *Regula falsae positionis*, Prog., Braunau, 1886.

[2] See Volume I, page 211. For the original and a translation of his process see F. Woepcke, *Journal Asiatique*, I (6), 511.

[3] See Volume I, page 210. For a translation of his arithmetic see H. Suter, *Bibl. Math.*, II (3), 12; on this rule see page 30.

[4] Rule of Two Falses. There are various transliterations of the Arabic name.

[5] Leonardo Fibonacci, in the *Liber Abaci*, cap. XIII, under the title *De regulis elchatayn* says: "Elchataieym quidem arabice, latine duarum falsarum posicionum regula interpretatur. . . . Est enim alius modus elchataym; qui regula augmenti et diminucionis appelatur." See the Boncompagni edition, p. 318; M. Steinschneider, *Abhandlungen*, III, 122; G. Eneström, *Bibl. Math.*, IV (3), 205.

[6] He speaks of it as a "certa regola ditta El cataym. Quale (secondo alcuni) e vocabulo arabo." Fol. 98, *v.*

[7] Thus we have " . . . per il Cataino detto alcuni modo Arabo" (Cataneo, *Le Pratiche*, Venice, 1567 ed., fol. 58); "Delle Regole del Cattaino ouero false positioni" (Pagani, 1591, p. 164); "Regola Helcataym (vocabulo Arabo) che in nostra lingua vuol dire delle false Positioni" (Tartaglia, *General Trattato*, I, fol. 238, *v.* (Venice, 1556)); "La Reigle de Faux, que les Arabes appellent la Reigle Catain" (Peletier, 1549; 1607 ed., p. 253); "La regola del Cataino" (G. Ciacchi, *Regole generali d' abbaco*, p. 278 (Florence, 1675)).

II

This name was not, however, the common one in the European books, and in the course of the 16th century it nearly disappeared. In general the method went by such names as Rule of False,[1] Rule of Position,[2] and Rule of False Position.[3]

Rule of Double False explained. The explanation of this rule, as related to the equation $ax + b = 0$, is as follows:

Let g_1 and g_2 be two guesses as to the value of x, and let f_1 and f_2 be the failures, that is, the values of $ag_1 + b$ and $ag_2 + b$, which would be equal to 0 if the guesses were right. Then

$$ag_1 + b = f_1 \qquad (1)$$

and
$$ag_2 + b = f_2; \qquad (2)$$

whence
$$a(g_1 - g_2) = f_1 - f_2. \qquad (3)$$

From (1),
$$ag_1g_2 + bg_2 = f_1g_2,$$

and from (2),
$$ag_1g_2 + bg_1 = f_2g_1;$$

whence
$$b(g_2 - g_1) = f_1g_2 - f_2g_1. \qquad (4)$$

Dividing (4) by (3),
$$-\frac{b}{a} = \frac{f_1g_2 - f_2g_1}{f_1 - f_2}.$$

But, since
$$-\frac{b}{a} = x,$$

we have here a rule for finding the value of x.[4]

---

[1] "La Reigle de Faux" (Trenchant, 1566; 1578 ed., p. 213); "Falsy" (Van der Schuere, 1600, p. 185); "Regula Falsi" (Coutereels, 1690 edition of the *Cyffer-Boeck*, p. 541).

[2] "Auch Regula Positionum genant" (Suevus, 1593, p. 377); "Reigle de Faux, mesmes d'une Position" (Peletier, 1549; 1607 ed., p. 269).

[3] "Rule of falshoode, or false positions" (Baker, 1568; 1580 ed., fol. 181); "False Positie" and "Fausse Position" (Coutereels, Dutch-French ed., 1631, p. 329); "Valsche Positie" (Eversdyck's Coutereels, 1658 ed., p. 360); "Reghel der Valsches Positien" (Wilkens, 1669 ed., p. 353).

[4] The formula is more elegantly derived by taking the eliminant of

$$\begin{aligned} ax + b + 0 &= 0 \\ ag_1 + b - f_1 &= 0 \\ ag_2 + b - f_2 &= 0, \end{aligned}$$

which is
$$\begin{vmatrix} x & 1 & 0 \\ g_1 & 1 & -f_1 \\ g_2 & 1 & -f_2 \end{vmatrix} = 0,$$

by the expansion of which the result at once appears.

Suppose, for example, that

$$5x - 10 = 0.$$

Make two guesses as to the value of x, say $g_1 = 3$ and $g_2 = 1$.

Then
$$5 \cdot 3 - 10 = 5 = f_1,$$
and
$$5 \cdot 1 - 10 = -5 = f_2.$$

Then
$$x = \frac{f_1 g_2 - f_2 g_1}{f_1 - f_2} = \frac{5 \cdot 1 - (-5) \cdot 3}{5 - (-5)} = \frac{20}{10} = 2.$$

Awkward as this seems, the rule was used for many centuries, a witness to the need for and value of a good symbolism. We have here placed two false quantities in the problem, and from these we have been able to find the true result.

Recorde's Rule in Verse. From the above formula for x it will be possible to interpret the doggerel rule given by Robert Recorde in his *Ground of Artes* (*c.* 1542):

> Gesse at this woorke as happe doth leade.
> By chaunce to truthe you may procede.
> And firste woorke by the question,
> Although no truthe therein be don.
> Suche falsehode is so good a grounde,
> That truth by it will soone be founde.
> From many bate to many mo,
> From to fewe take to fewe also.
> With to much ioyne to fewe againe,
> To to fewe adde to manye plaine.
> In crossewaies multiplye contrary kinde,
> All truthe by falsehode for to fynde.[1]

Recorde thought highly of the rule, and it was appreciated by writers generally until the 19th century.[2]

---

[1] *Ground of Artes*, 1558 ed., fol. Z 4.

[2] Thus Thierfelder (1587, p. 226) says: "Für allen Regeln der gantzen Arithmetic (ohn allein die Regel Cosz auzgenommen) ist sie die Kunstreichste weytgegreifflichste vñ schönste"; and Peletier (1549; 1607 ed., p. 269) remarks: "Gemme Phrissien a inuenté l'artifice de soudre par la Reigle de Faux, mesmes d'une Position, grand' partie des exemples subjects à l'Algebre." Even as late as 1884, in the *Instruction für den Unterricht an den Gymnasien in Österreich* (Vienna, 1884, p. 315), the rule is recommended.

Method of the Scales. The Arabs modified the rule by what they called the Method of the Scales,[1] a name derived from the following figure, used in the solution:

Suppose, for example, that we wish to solve the equation $x + \frac{2}{3}x + 1 = 10$, a problem set by Behâ Eddîn (c. 1600). We may make as our guesses $g_1 = 9$, whence $f_1 = 6$; $g_2 = 6$, whence $f_2 = 1$. Then place the figures thus:

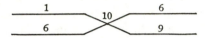

The lines now aid the eye to write the result according to the rule already set forth, as follows:[2]

$$x = \frac{6 \cdot 6 - 1 \cdot 9}{6 - 1} = \frac{27}{5} = 5\frac{2}{5}.$$

Rule of Single False. Thus far we have considered the Rule of Double False, where a double guess was made, but there was also a modification of the method known as the Rule of Single False.[3] Albanna (c. 1300) gives the latter in the form of a rule which, worked out in modern symbols, is as follows: Given that

$$ax + b = 0.$$

Make a guess, g, for the value of x, the failure being f; that is,

$$ag + b \doteq f.$$

---

[1] *'Alm bi'l kaffatain.* This name was translated into Latin as the *Regula lancium* or *Regula bilancis.*

[2] For variations of the method see Matthiessen, *Grundzüge*, p. 278; for the Arab proof by geometry see *ibid.*, p. 281.

[3] Tartaglia called it "Position Sempia," as distinct from "Position Doppia" (*General Trattato*, I, fol. 239, *v.*, and 266, *r.*); the Spanish had rules "De vna falsa posicion" and "De dos falsas posiciones" (Santa-Cruz, 1594; 1643 ed., fols. 210, 212); Clavius (1586 Italian ed., pp. 195, 203) has "Regola del falso di semplice positione" and "di doppia positione"; and Chuquet (1484, MS., fols. 32, 42) has "De la Rigle de vne posicion" and "de deux posicions."

Then to obtain the required rule we may proceed as follows:

$$a = \frac{f}{g - x},$$

whence

$$\frac{fx}{g - x} + b = 0$$

and

$$x = \frac{gb}{b - f} = \frac{g(ag) - gf}{ag} = \frac{g(f - b) - gf}{f - b},$$

this last indicating the rule used.

For example, in the equation $\frac{1}{5}x + \frac{1}{6}x = 20$, if we take $g = 30$, we have

$$\frac{1}{5} \cdot 30 + \frac{1}{6} \cdot 30 = 11,$$

which is 9 too small, whence $f = -9$. Then

$$x = \frac{30(-9 + 20) - 30(-9)}{-9 + 20} = 54\frac{6}{11}.$$

Apologies for the Name of Rule of False. The name "Rule of False" was thought to demand an apology in a science whose function it is to find the truth, and various writers made an effort to give it. Thus Humphrey Baker (1568) says:

The Rule of falsehoode is so named not for that it teacheth anye deceyte or falsehoode, but that by fayned numbers taken at all aduentures, it teacheth to finde out the true number that is demaunded, and this of all the vulgar Rules which are in practise) is ỹ most excellence.[1]

Besides the "Rule of False" the method was also called the "Rule of Increase and Diminution,"[2] from the fact that the error is sometimes positive and sometimes negative. Indeed, as already stated, in the 16th century the symbols + and − were much more frequently used in this connection than as symbols of operation.

---

[1] 1580 ed., fol. 181. Similar excuse is offered by Thierfelder (1587, p. 225): "Darum̄ nicht dasz sie falsch oder vnrecht sey"; by Apianus (1527): "Vnd heisst nit darum falsi dass sie falsch vnd vnrecht wehr, sunder, dass sie auss zweyen falschen vnd vnwahrhaftigen zalen, vnd zweyen lügen die wahrhaftige vnd begehrte zal finden lernt." Like explanations are given by many other writers.

[2] "Regula augmenti et decrementi" or "diminutionis."

Regula Infusa. Rabbi ben Ezra (c. 1140) tells us of a substitution method due to another Hebrew writer, Job ben Salomon, of unknown date, which was called in Latin translation the *Regula infusa*.[1] This may be illustrated as follows:

Given	$m(ax + b) + c = 0$,
let	$ax + b = y$,
and then	$my + c = 0$,
whence	$y = -c/m$,
and so	$ax + b = -c/m$,

which can now be solved. Rabbi ben Ezra illustrates this by taking

$$x - \tfrac{1}{3}x - 4 - \tfrac{1}{4}(x - \tfrac{1}{3}x - 4) = 20$$

and letting	$x - \tfrac{1}{3}x - 4 = y$;
whence	$y - \tfrac{1}{4}y = 20$
and	$y = 26\tfrac{2}{3}$,
and so	$x - \tfrac{1}{3}x - 4 = 26\tfrac{2}{3}$,

which can now be solved.[2] Although the method is very artificial, it is occasionally found in the algebras of today, especially in connection with radical equations.

Classification of Equations. Our present method of classifying equations according to their degree is a modern one. The first noteworthy attempt at a systematic classification is found in the algebra of Omar Khayyam (c. 1100), but the classification there given is not our present one. Omar considers equations of the first three degrees as either simple or compound. The simple equations are of the type $r = x$, $r = x^2$, $r = x^3$, $ax = x^2$, $ax = x^3$, $ax^2 = x^3$. Compound equations are first classified as trinomials, and these include the following twelve forms: (1) $x^2 + bx = c$, $x^2 + c = bx$, $bx + c = x^2$; (2) $x^3 + bx^2 = cx$,

---

[1] ". . . secundum regulam que vocatur infusa. Et ipsa est regula Job, filii Salomonis." Libri, *Histoire*, 1838 ed., I, 312. There is some doubt, however, as to whether Libri was right in referring this work to Rabbi ben Ezra.

[2] Matthiessen, *Grundzüge*, p. 272.

$x^3 + cx = bx^2, cx + bx^2 = x^3, x^3 + cx = d, x^3 + d = cx, cx + d = x^3$; $x^3 + bx^2 = d, x^3 + d = bx^2, bx^2 + d = x^3$. They are then classified as quadrinomials, as follows: (3) $x^3 + bx^2 + cx = d$, $x^3 + bx^2 + d = cx$; (4) $x^3 + bx^2 = cx + d$, $x^3 + cx = bx^2 + d$, $x^3 + d = bx^2 + cx$. Of this early plan of classifying equations by the number of terms we still have a trace in our chapter on binomial equations.

Classification according to Degree. Such was the general method of classifying equations, naturally with variations in details, until after books began to be printed. Pacioli (1494), for example, has a similar system.[1] It was not until about the beginning of the 17th century that the classification according to degree, with a recognition that a literal coefficient might be either positive or negative, was generally employed, and this was due in a large measure to the influence of such writers as Stevin (1585), Vieta (c. 1590), Girard (1629), Harriot (1631, posthumous), Oughtred (1631), and Descartes (1637). In particular, Descartes set forth in his *Géométrie* the idea of the degree of an equation, or, as he says, of the dimensions of an equation,[2] reserving the word "degree" for use with respect to lines.[3]

Quadratic Equations. The first known solution of a quadratic equation is the one given in the Berlin Papyrus mentioned on page 432. The problem reduces to solving the equations

$$x^2 + y^2 = 100, \qquad y = \tfrac{3}{4}\,x,$$

and the solution is substantially as follows:

Make a square whose side is 1 and another whose side is $\tfrac{3}{4}$. Square $\tfrac{3}{4}$, giving $\tfrac{9}{16}$. Add the squares, giving $\tfrac{25}{16}$, the square root of which is $\tfrac{5}{4}$. The square root of 100 is 10. Divide 10 by $\tfrac{5}{4}$, giving 8, and $\tfrac{3}{4}$ of 8 is 6. Then

$$8^2 + 6^2 = 100 \quad \text{and} \quad 6 = \tfrac{3}{4} \text{ of } 8,$$

---

[1] *Sūma*, 1494 ed., fol. 145 seq.

[2] "Sçachez donc qu'en chaque Equation, autant que la quantité inconnuë a de dimensions, autant peut-il y avoir de diverses racines, c'est à dire de valeurs de cette quantité" (1705 ed., p. 106). He then speaks of "$x^3 - 9xx + 26x - 24 \infty 0$, . . . en laquelle x ayant trois dimensions a aussi trois valeurs qui sont 2, 3, & 4."

[3] ". . . distingué divers degrez entre ces lignes" (*ibid.*, p. 27).

so that the roots of the two implied equations are 6 and 8. The solution is therefore a simple case of false position.[1]

The Greeks were able to solve the quadratic equation by geometric methods. As already stated, Euclid (*c.* 300 B.C.) has in his *Data* three problems involving quadratics. Of these the first (Prob. 84) is as follows:

If two straight lines include a given area in a given angle and the excess of the greater over the less is given, then each of them is given.

Expressed in algebraic form with reference to the rectangle, if $xy = k^2$ and $x - y = a$, then x and y can be found. Euclid solves the problem geometrically.[2] He also gives in the *Elements* such geometric problems as the following:

To cut a given straight line so that the rectangle contained by the whole and one of the segments shall be equal to the square on the remaining segment.[3]

This may be represented algebraically by the equation $a(a - x) = x^2$ or by $x^2 + ax = a^2$.

Quadratics among the Hindus. It is possible that the altar constructon of the Hindus involved the solution of the equation $ax^2 + bx = c$, and this may date from the *Śulvasūtra* period (roughly speaking, say 500 B.C.); but whether or not this is the case, we have no record of the method of solution.[4]

When we come to the time of Āryabhaṭa (*c.* 510), we find a rule, relating to the sum of a geometric series, which shows that the solution of the equation $ax^2 + bx + c = 0$ was known, but we have no rule for the solution of the equation itself.[5]

It should be repeated, however, that up to the 17th century an equation of the type $x^2 + px = q$, for example, was looked

---

[1] Schack-Schackenburg, *Zeitschrift für ägyptische Sprache*, XXXVIII, 135; XL, 65. See also Cantor, *Geschichte*, I (3), 95, and Simon, *Geschichte*, p. 41.

[2] The other two have already been given on page 381.

[3] *Elements*, II, 11. See also VI, 28, 29.

[4] G. Milhaud, "La Géométrie d'Apastamba," in the *Revue générale des sciences*, XXI, 512–520.

[5] The rule for the summation is No. XX in Rodet's *Leçons de Calcul d'Āryabhaṭa*, pp. 13, 33 (Paris, 1879). In all such cases the possibility of the younger Āryabhaṭa must be considered.

upon as distinct from one of the type $x^2 - px = q$; the idea that p might be either positive or negative did not occur to algebraists until some time after the invention of a fairly good symbolism. This accounts for the special rules for different types that are found in the Middle Ages and the early Renaissance.

Brahmagupta's Rule. Brahmagupta ($c.$ 628) gave a definite rule for the quadratic. For example, he gave the equation

$$ya\ v\ 1\ ya\ 1\overset{.}{0}$$
$$ru\ \overset{.}{9},$$

that is, $x^2 - 10x = -9$,[1] with the solution substantially as follows:

Here absolute number ($\overset{.}{9}$) multiplied by (1) the [coefficient of the] square ($\overset{.}{9}$), and added to the square of half the [coefficient of the] middle term, namely, 25, makes 16; of which the square root 4, less half the [coefficient of the] unknown ($\overset{.}{5}$), is 9; and divided by the [coefficient of the] square (1) yields the value of the unknown 9.

Expressed in modern symbols,

$$x = \frac{\sqrt{-9 \cdot 1 + (-5)^2} - (-5)}{1} = 9.$$

Mahāvīra's Rule. Mahāvīra ($c.$ 850) gave no rule for the quadratic, but he proposed a problem involving the equation

$$c + \frac{c}{3} + 3 \cdot \frac{c^2}{36} = m,$$

adding the following statement:

In relation to the combined sum [of the three quantities] as multiplied by 12, the quantity thrown in so as to be added is 64. Of this [second] sum the square root diminished by the square root of the quantity thrown in gives rise to the measure . . .

Expressed in modern symbols, this means that

$$c = \sqrt{12\,m + 64} - \sqrt{64},$$

which shows that Mahāvīra had substantially the modern rule for finding the positive root of a quadratic.[2]

---

[1] Colebrooke, p. 347, his transliteration being followed.
[2] See his work, p. 192.

The Hindu Rule. Srīdhara (c. 1025) was the first, so far as known, to give the so-called Hindu Rule for quadratics. He is quoted by Bhāskara (c. 1150) as saying:

Multiply both sides of the equation by a number equal to four times the [coefficient of the] square, and add to them a number equal to the square of the original [coefficient of the] unknown quantity. [Then extract the root.][1]

This rule, although stated by Bhāskara, is not the first one given by him. He begins by saying:

[Its re-solution consists in] the elimination of the middle term, as the teachers of the science denominate it. . . . On this subject the following rule is delivered. . . . When a square and other [term] of the unknown is involved in the remainder; then after multiplying both sides of the equation by an assumed quantity, something is to be added to them, so as the side may give a square-root. Let the root of the absolute number again be made equal to the root of the unknown; the value of the unknown is found from that equation.

It will be observed that this is simply a more general form of Srīdhara's rule. The method has been the subject of much discussion by the various commentators on Bhāskara.[2]

Al-Khowârizmî's Rules. Al-Khowârizmî (c. 825) used two general methods in solving the quadratic of the form

$x^2 + px = q$, both based upon Greek models. Given $x^2 + 10x = 39$, he constructed a square as here shown. Then the unshaded part is $x^2 + px$, and is therefore equal to q. In order to make it a square we must add the four shaded squares, each of which is $(\frac{1}{4}p)^2$ and the sum of

---

[1] *Vija-Ganita*, p. 209. That is, given $ax^2 + bx = c$, we have first $4a^2x^2 + 4abx = 4ac$. Then $4a^2x^2 + 4abx + b^2 = b^2 + 4ac$, whence

$$2ax + b = \sqrt{b^2 + 4ac},$$

the negative root being neglected. The purpose of the multiplication by $4a$ was to avoid fractions.

[2] See the *Vija-Ganita*, pp. 207–209.

which is $\frac{1}{4}p^2$, which in this case is 25. Since $25 + 39 = 64$, we have

$$x + \tfrac{1}{2}p = 8 \; ;$$

whence

$$x + 5 = 8$$

and

$$x = 3.$$

His statement as to the second method is as follows:

You halve the number of the roots, which in the present instance yields five. This you multiply by itself; the product is twenty-five. Add this to thirty-nine; the sum is sixty-four. Now take the root of this, which is eight, and subtract from it half the number of the roots, which is five; the remainder is three. This is the root of the square which you sought for ; the square itself is nine.[1]

The negative root was neglected, as was regularly the case until modern times.

His second method was similar to our common one. In the figure the unshaded part is $x^2 + px$, and he adds the square of $\frac{1}{2}p$.

He then has

$$x^2 + px + \tfrac{1}{4}p^2 = \tfrac{1}{4}p^2 + q,$$

whence

$$x = \sqrt{\tfrac{1}{4}p^2 + q} - \tfrac{1}{2}p,$$

of which he takes only the positive root.[2]

Al-Khowârizmî also considers other forms, his solution[3] of the type $x^2 + q = px$ being based upon the identity

$$(\tfrac{1}{2}p)^2 - (\tfrac{1}{2}p - x)^2 = x(p - x) = px - x^2 = q,$$

from which it follows that

$$x = \tfrac{1}{2}p + \sqrt{\tfrac{1}{4}p^2 - q}.$$

Omar Khayyam's Rule. Omar Khayyam's rule ($c.$ 1100) for solving the quadratic $x^2 + px = q$ is as follows:

Multiply half of the root by itself; add the product to the number and from the square root of this sum subtract half the root. The remainder is the root of the square.[4]

[1] Rosen ed., p. 8.

[2] For a discussion of his methods see Matthiessen, *Grundzüge*, p. 299.

[3] For discussion and for the geometric proof see Rosen's edition, p. 16; Matthiessen, *Grundzüge*, p. 304; Libri, *Histoire*, I, 236.

[4] That is, $x = \sqrt{\tfrac{1}{4}p^2 + q} - \tfrac{1}{2}p$. By "half the root" is meant $\tfrac{1}{2}p$, and by "the number" is meant q. He used the equation $x^2 + 10x = 39$, which was the one

He also gave rules for other types, that for $x^2 + q = px$ being based upon the identity

$$x(p - x) + (x - \tfrac{1}{2}p)^2 = (\tfrac{1}{2}p)^2,$$

and that for $px + q = x^2$ upon the identity[1]

$$x(x - p) + (\tfrac{1}{2}p)^2 = (x - \tfrac{1}{2}p)^2.$$

Chinese Work in Quadratics. The Chinese gave some attention to quadratic equations in the Middle Ages, including those of the form

$$x^2 - sx - \left(\frac{s^2 - H^2}{2}\right) = 0,$$

but how far they were original in their work has not yet been scientifically determined.[2]

With respect to the quadratic equation the medieval algebraists added nothing of importance to the work of the Arabic writers from whom they derived their inspiration, and the Renaissance algebraists did little except in their improvement of the symbolism. It was not until the close of the 16th century that the next noteworthy contribution was made.

Harriot treats of Equations by Factoring. The first important treatment of the solution of quadratic and other equations by factoring is found in Harriot's *Artis Analyticae Praxis* (1631). He takes as his first case the equation

$$aa - ba + ca = + bc$$

and writes it in the form

$$
\begin{array}{l|l}
a - b & = aa - ba \\
a + c & \quad + ca - bc,
\end{array}
$$

also used by al-Khowârizmî and was apparently a favorite problem of the schools. He also considered the arithmetically impossible solutions. For a discussion of his methods and proofs see Woepcke's translation, p. 17; Matthiessen, *Grundzüge*, p. 301.

[1] See Woepcke's translation, pp. 20, 23; Matthiessen, *Grundzüge*, pp. 305, 309.
[2] L. Vanhée, in *T'oung-pao*, XIII, 291; XII, 559; XV, 111.

where the first member stands for $(a - b)(a + c)$ and the equation becomes $(a - b)(a + c) = 0$. From this fact he finds that $b = a$.[1]

In a similar way the equation

$$aaa + baa - cda = + bcd$$

is factored into $(a + b)(aa - cd) = 0$, and the solution is given that $aa = cd$.[2]

Vieta advances the Theory. In the work of Vieta the analytic methods replaced the geometric, and his solutions of the quadratic equation were therefore a distinct advance upon those of his predecessors. For example, to solve the equation[3] $x^2 + ax + b = 0$ he placed $u + z$ for x. He then had

$$u^2 + (2z + a)u + (z^2 + az + b) = 0.$$

He now let $2z + a = 0$, whence $z = -\tfrac{1}{2}a$, and this gave

$$u^2 - \tfrac{1}{4}(a^2 - 4b) = 0,$$
$$u = \pm\tfrac{1}{2}\sqrt{a^2 - 4b},$$

and
$$x = u + z = -\tfrac{1}{2}a \pm \tfrac{1}{2}\sqrt{a^2 - 4b}.$$

---

[1] "Nam si ponatúr $a === b$ erit $a - b === 0$," p. 16. The relation $a + c = 0$ is neglected.

[2] P. 19. The relation $a = -b$ is neglected.

[3] The symbolism used here is, of course, modern. Vieta's own solution is as follows:

"Si A quad. + B 2 in A, aequatur Z plano. A + B esto E. Igitur E quad. aequabitur Z plano + B quad.

"Consectarium. Itaque $\sqrt{Z \text{ plani} + B \text{ quad.}} - B$ fit A, de qua primum quaerebatur.

"Sit B 1. Z planum 20. A 1 N. 1 Q + 2 N, aequatur 20. et fit 1 N $\sqrt{21} - 1$."

That is, if $A^2 + 2BA = Z$, we may represent this in modern form as $x^2 + 2ax = b$, where $A = x$, $B = a$, $Z = b$.

Let $A + B = E$, that is, let $x + a = u$.

It follows that $u^2 = x^2 + 2ax + a^2 = b + a^2$, and so $x = \sqrt{b + a^2} - a$.

In particular, he says, let $B = 1$ and $Z = 20$. The equation is then $x^2 + 2x = 20$, whence $x = \sqrt{21} - 1$.

He has similar solutions for the following:

"Si A quad. − B in A 2, aequatur Z plano,"

and   "Si D 2 in A − A quad., aequatur Z plano,"

showing that at this time in his work he had not grasped the idea of such a general quadratic equation as $x^2 + a_1x + a_2 = 0$. In his *De numerosa potestatum*

Modern Methods. Of the modern methods[1] for obtaining the formula for the solution of the quadratic, interesting chiefly from the standpoint of theory, a single one may be mentioned. This method uses determinants and is due to Euler and Bézout, but was improved by Sylvester (1840) and Hesse (1844).

Given $$x^2 + px + q = 0,$$

let $$x = u + z\,;$$

whence $$x^2 = (u + z)x.$$

Then $$x^3 \qquad + px^2 \qquad + qx = 0,$$
$$x^2 - (u + z)x = 0,$$

and $$x^3 - (u + z)x^2 \qquad\qquad = 0\,;$$

whence

$$\begin{vmatrix} 1 & p & q \\ 0 & 1 & -(u+z) \\ 1 & -(u+z) & 0 \end{vmatrix} = 0.$$

Expanding, $$-p(u+z) - (u+z)^2 - q = 0,$$

and hence $$u^2 + (2z + p)u + (z^2 + pz + q) = 0.$$

Letting $$2z + p = 0,$$

we find that $$u = \pm \tfrac{1}{2}\sqrt{p^2 - 4q},$$

and $$x = -\frac{p}{2} \pm \tfrac{1}{2}\sqrt{p^2 - 4q}.$$

Simultaneous Quadratic Equations. Problems involving the combination of a linear and a quadratic equation were, as we have seen, familiar to the Egyptians, and the Greeks were fully able to apply their geometry to such cases. The algebraic treatment of two quadratics was not seriously considered, however, until it was taken up by Diophantus ($c.$ 275) for indeterminate forms. He speaks of equations like

$$y^2 = ax^2 + bx + c$$

and $$z^2 = a'x^2 + b'x + c'$$

. . . *resolutione tractatus* (Paris, 1600), however, he uses the terms "affected" and "pure" with respect to quadratic equations. See Volume I, page 311. See also his *De aequationum recognitione et emendatione libri duo*, Tract. II, cap. i (Paris, 1615); Matthiessen, *Grundzüge*, p. 311.

[1] For a list of modern methods consult Matthiessen, *Grundzüge*, p. 315 seq.

as "double equations."[1] Among his more difficult equations of this type is the pair

$$x^2 + x - 1 = u^2,$$
$$x^2 - 1 = w^2,$$

which Diophantus expresses as follows:

To find three numbers such that their solid content minus any one gives a square.[2]

The subject never interested the medieval writers particularly, and not until the 17th century do we find much attention paid to it. By that time the symbolism was such that the only question involved was that of stating the cases in which a solution is possible.

Indeterminate Quadratic Equations. The study of indeterminate quadratic equations begins with such cases as $x^2 + y^2 = z^2$. The finding of formulas for the sides of a Pythagorean Triangle occupied the attention of various Greek writers. Proclus (c. 460) tells us that Pythagoras (c. 540 B.C.) himself gave a rule, and tradition says that it was, as expressed in modern symbols,

$$n^2 + \left(\frac{n^2 - 1}{2}\right)^2 = \left(\frac{n^2 + 1}{2}\right)^2$$

where n is an odd number. Plato (c. 380 B.C.) gave the rule

$$(2n)^2 + (n^2 - 1)^2 = (n^2 + 1)^2,$$

which, like the one attributed to Pythagoras, is connected with Euclid's proposition[3] to the effect that

$$\left(\frac{a + b}{2}\right)^2 - \left(\frac{a - b}{2}\right)^2 = ab,$$

a relation that forms the basis of the theory of quarter squares

---

[1] Διπλοϊσότης, διπλῆ ἰσότης, διπλῆ ἰσωσις. See Heath, *Diophantus*, 2d ed., p. 73

[2] Book IV, 23. That is, the first number is x, the second is 1, and the third is $x + 1$, the "solid content" being $x \cdot 1 \cdot (x + 1)$. The results are $\frac{17}{8}$, 1, and $\frac{25}{8}$. For further explanation see Heath, *loc. cit.*, p. 184.

[3] Heath, *Diophantus*, 2d ed., pp. 116, 242 n.

Diophantus on Indeterminate Equations. It was Diophantus, however, who may properly be called the father of the study of indeterminate equations, which were generally limited in his *Arithmetica* to quadratic types. With these equations the object was to obtain rational results, while with indeterminate equations of the first degree the object was usually to obtain integral results. The problem proposed by Diophantus is that of solving either one or two equations of the form

$$Ax^2 + Bx + C = y^2.$$

His simpler types may be represented by the following:

To add the same [required] number to two given numbers so as to make each of them a square.[1]

One of the more difficult problems is as follows:

To find three numbers such that their sum is a square and the sum of any pair is a square.[2]

Pell Equation. One of the most famous indeterminate quadratic equations is of the form

$$Ay^2 + 1 = x^2.$$

This form is commonly attributed to John Pell (1668) but is really due to Fermat (c. 1640) and Lord Brouncker[3] (1657). The problem itself is apparently much older than this, however, for it seems involved in various ancient approximations to the square roots of numbers. Thus the Greek approximation $\frac{7}{5}$ for the ratio of the diagonal to the side of a square goes back to Plato's time at least, and 7 and 5 are the roots of the equation

$$x^2 - 2y^2 = -1.$$

---

[1] III, 11. That is, if the given numbers are 2 and 3, then $x + 2$ and $x + 3$ must both be squares. He finds that $x = \frac{97}{64}$.

[2] III, 6. His results are 80, 320, 41. For solution see Heath, *Diophantus*, 2d ed., pp. 68, 158.

[3] E. E. Whitford, *The Pell Equation*, New York, 1912; H. Konen, *Geschichte der Gleichung* $t^2 - Du^2 = 1$, Leipzig, 1901; G. Wertheim, "Ueber den Ursprung des Ausdruckes 'Pellsche Gleichung,'" *Bibl. Math.*, II (3), 360; Heath, *Diophantus*, 2d ed., p. 286.

Theon of Smyrna ($c.$ 125) considered a relation that would now be written as the equation

$$x^2 - 2y^2 = \pm 1,$$

carrying his computations as far as the case of

$$17^2 - 2 \cdot 12^2 = 1$$

and stating a rule for finding the solutions.

The special case of the Cattle Problem, doubtfully attributed to Archimedes, requires the number of bulls of each of four colors, white (W), blue (B), yellow (Y), and piebald (P), and the number of cows of the same colors (w, b, y, p) such that B = $(\frac{1}{4} + \frac{1}{5})$(Y + P),   P = $(\frac{1}{6} + \frac{1}{7})$(W + Y),   w = $(\frac{1}{3} + \frac{1}{4})$(B + b), b = $(\frac{1}{4} + \frac{1}{5})$(P + p), p = $(\frac{1}{5} + \frac{1}{6})$(Y + y), and y = $(\frac{1}{6} + \frac{1}{7})$(W + w). Reduced to a single equation, the problem involves the solution of the indeterminate quadratic equation

$$x^2 - 4,729,494y^2 = 1,$$

and the number of yellow bulls, for example, has 68,848 periods of three figures each.[1]

The general problem may have been discussed in the lost books of Diophantus,[2] perhaps in the form

$$x^2 - Ay^2 = 1,$$

and its equivalent is clearly stated in the works of Brahmagupta[3] ($c.$ 628).

Fermat ($c.$ 1640) was the first to state that the equation $x^2 - Ay^2 = 1$, where A is a non-square integer, has an unlimited number of integral solutions,[4] and from that time on the problem attracted the attention of various scholars, among the most prominent being Euler (1730), who stated that the solu-

[1] Heath, *History*, II, 97; Whitford, *loc. cit.*, p. 20, with bibliography.

[2] P. Tannery, "L'arithmétique des Grecs dans Pappus," in the *Mémoires de la Soc. des sci. de Bordeaux*, III (2), 370.

[3] Colebrooke ed., p. 363.

[4] *Œuvres*, ed. Tannery and Henry, II, 334 (Paris, 1894); Whitford, *loc. cit.*, p. 46.

tion of the equation $ax^2 + bx + c = y^2$ requires the solution of the equation $x^2 - Ay^2 = 1$.[1] It was he who, through an error, gave to the general type the name of the Pell Equation.

Cubic Equation. The oldest known cubic equation of the form $x^3 = k$ is possibly due to Menæchmus ($c.$ 350 B.C.), although tables of cubes had been worked out by the Babylonians two thousand years earlier. It had been recognized since the time of Hippocrates ($c.$ 460 B.C.) that the solution of the problem of the duplication of the cube depended on the finding of two mean proportionals between two given lines. Algebraically this means the finding of x and y in the equations

$$\frac{a}{x} = \frac{x}{y} = \frac{y}{b}.$$

From these relations it is evident that

$$y^2 = bx \text{ (a parabola)}$$

and $\qquad\qquad xy = ab \text{ (an equilateral hyperbola)};$

whence $\qquad\qquad y^3 = ab^2 \text{ (a cubic equation)}.$

Menæchmus is said to have solved the cubic by finding the intersection of the two conics. If $b = 2a$, then $y^3 = 2a^3$, and the problem becomes the well-known one of the duplication of the cube, which interested so many Greek writers.[2]

The next reference to the cubic among the Greeks is in a certain problem of Archimedes, to cut a sphere by a plane so that the two segments shall have a given ratio.[3] This reduces to the proportion

$$\frac{c - x}{b} = \frac{c^2}{x^2}$$

and to the equation $\qquad x^3 + c^2 b = cx^2.$

---

[1] P. H. von Fuss, *Correspondance mathématique et physique de quelques célèbres géomètres du XVIIIième siècle*, I, 37 (Petrograd, 1843).

[2] For a partial list of these writers see Woepcke, translation of Omar Khayyam, p. xiij. The reference to Menæchmus is not certain.

[3] *De sphaera et cylindro*, Lib. II. See also Heath's *Archimedes*, chap. vi.

Eutocius (*c.* 560) tells us that Archimedes solved the problem by finding the intersection of two conics, namely,

$$x^2 = \frac{a^2}{c} y \quad \text{(a parabola)}$$

and $\qquad y(c - x) = bc$ (a hyperbola).

Diophantus solved a single cubic equation, $x^3 + x = 4x^2 + 4$. This equation arises in connection with the following problem:

To find a right-angled triangle such that the area added to the hypotenuse gives a square, while the perimeter is a cube.[1]

His method is not given, the statement, expressed in modern language, being that x "is found to be" 4. Possibly Diophantus saw that $x(x^2 + 1) = 4(x^2 + 1)$; whence $x = 4$.

The Cubic among the Arabs and Persians. Nothing more is known of the cubic equation among the Greeks, but the problem of Archimedes was taken up by the Arabs and Persians in the 9th century. In a commentary on Archimedes, Almâhânî (*c.* 860) considered the question, but so far as known he contributed nothing new. He brought the problem into such prominence, however, that the equation $x^3 + a^2b = cx^2$ was known among the Arab and Persian writers as Almâhânî's equation.[2]

One of his contemporaries, Ṭâbit ibn Qorra (*c.* 870), considered special cases of cubic equations, as in the duplication of the cube. These equations he solved by geometric methods, but he was unable to contribute to the general algebraic theory.

A little later Abû Ja'far al-Khâzin (*c.* 960), a native of Khorâsân, considered the problem and, as Omar Khayyam tells us, "solved the equation by the aid of conic sections."[3]

The last of the Arabs to give any particular attention to the solution was Alhazen[4] (*c.* 1000). Omar Khayyam[5] refers to

---

[1] Bk. VI, prob. 17. See also Heath's *Diophantus*, 2d ed., p. 69.
[2] Matthiessen, *Grundzüge*, p. 367; Cantor, *Geschichte*, I, chap. xxxv.
[3] Woepcke translation, p. 3.
[4] Al-Ḥasan ibn al-Ḥasan ibn al-Ḥaitam.
[5] Woepcke's translation, p. 73, with discussion; Matthiessen, *Grundzüge*, p. 367.

his method. Alhazen solved the equation by finding the intersection of

$$x^2 = ay \text{ (a parabola)}$$

and

$$y(c - x) = ab \text{ (a hyperbola)},$$

a method not unlike the one attributed to Archimedes.

The last of the Persian writers to consider the cubic equation with any noteworthy success was Omar Khayyam[1] ($c.$ 1100). In his list of equations he specified thirteen forms of the cubic that had positive roots, this being a decided advance in the general theory. He solved equations of the type $x^3 + b^2x = b^2c$ by finding the intersection of the conics $x^2 = by$ and $y^2 = x(c-x)$; of the type $x^3 + ax^2 = c^3$ by finding the intersection of $xy = c^2$ and $y^2 = c(x + a)$; and of the type $x^3 \pm ax^2 + b^2x = b^2c$ by finding the intersection of $y^2 = (x \pm a)(c - x)$ and $x(b \pm y) = bc$.

It is said, but without proof from the sources, that Omar Khayyam stated that it was impossible to solve in positive integers the equation $x^3 + y^3 = z^3$, the simplest of the family of equations of the type $x^n + y^n = z^n$ with which Fermat's name is connected.

In general it may be said that the Arab writers believed that the cubic equation was impossible of solution.[2]

Chinese and Hindu Interest in the Cubic. The Chinese algebraists did nothing worthy of note with the general cubic equation. Their interests lay in applied problems, and these all led to numerical equations. The numerical cubic first appears in a work by Wang Hs'iao-t'ung, about 625.[3]

He gave the following problem:

There is a right-angled triangle the product of the sides of which is $706\frac{1}{4}$ and the hypotenuse of which is greater than one side by $36\frac{9}{10}$. Find the lengths of the three sides.

Wang used a numerical equation of the form $x^3 + ax^2 - b = 0$ and stated the answer incorrectly as $14\frac{7}{20}$, $49\frac{1}{5}$, and $51\frac{1}{4}$, although there is doubt as to the validity of the copy.

---

[1] Ball, *Hist. of Math.*, 6th ed., p. 159; Woepcke's translation, p. 25 seq.

[2] Cantor, *Geschichte*, I (2), 736.

[3] In the *Ch'i-ku Suan-king*. See Mikami, *China*, p. 54.

Various later Chinese algebraists treated of numerical equations, but it was not until the 18th century, when European influences were powerful, that any attempt was made by them to classify equations of the third degree. In a work prepared under the direction of Emperor Kanghy, who ruled China from 1662 to 1722, nine types are given:

$$x^3 \pm bx = c \qquad\qquad x^3 \pm ax^2 = c$$
$$x^3 \pm ax^2 \pm bx = c \qquad -x^3 + ax^2 = c,$$

but in every case the solution is numerical and only a single positive root is given.[1]

The Hindus paid little attention to cubic equations except as they entered into relatively simple numerical problems relating to mensuration. Bhāskara (c. 1150) gave one example,

$$x^3 + 12x = 6x^2 + 35,$$

the root being 5,[2] but such a result is easily found by trial, the equation being made for this purpose.

Medieval Interest in the Cubic. In the Middle Ages various sporadic attempts were made by European scholars to solve the cubic equation. Fibonacci, for example, attacked the problem in his *Flos* of c. 1225. He states that one Magister Johannes, a scholar from Palermo, proposed to him the problem of finding a cube which, with two squares and ten roots, should be equal to 20.[3] That is, the problem is to solve the equation

$$x^3 + 2x^2 + 10x = 20,$$

a numerical equation discussed later (p. 472). Another attempt was made by an anonymous writer of the 13th century whose work has been described by Libri.[4] He took two cubics, one of the type $ax^3 = cx + k$ and the other of the type $ax^3 = bx^2 + k$.

---

[1] The work was the *Lü-li Yüan-yüan*. See Mikami, *China*, pp. 117–119.

[2] Colebrooke, *loc. cit.*, p. 214.

[3] "Altera uero questio á predicto magistro Iohanne proposita fuit, vt inueniretur quidam cubus numerus, qui cum suis duobus quadratis et decem radicibus in unum collectis essent uiginti" (*Flos*, Boncompagni ed., p. 228).

[4] *Histoire*, 1838 ed., II, 213, 214. The MS. is probably Florentine.

In each case he displayed great ignorance, possibly because he was unable, on account of his unfamiliarity with radicals to check his results. It is also possible that he sought only approximate results, although this is not stated. His method in the first case was as follows:

Given
$$ax^3 = cx + k,$$

we have
$$x^3 = \frac{c}{a}x + \frac{k}{a};$$

whence he assumed that
$$x = \frac{c}{2a} + \sqrt{\left(\frac{c}{2a}\right)^2 + \frac{k}{a}},$$

which is the root of $ax^2 = cx + k$ but not of the given equation.

His method in the second case was equally fallacious.

Slight attempts at numerical cubics were also made by Regiomontanus,[1] who gave, for example, the equation

$$\tfrac{1}{324} x^3 + \tfrac{540}{29} = 3\,x;$$

but he contributed nothing of value to the theory.

The Cubic in Printed Books. Pacioli (1494) asserted substantially that the general solution is impossible.[2]

Of the early German writers only one made any noteworthy attempt at the solution, and this was a failure. Rudolff (1525) suggested three numerical equations, each with one integral root and each being easily solved by factoring.[3] His method in connection with one of these equations is interesting. In modern symbols it is substantially as follows:

Given
$$x^3 = 10\,x^2 + 20\,x + 48,$$

we have
$$x^3 + 8 = 10\,x^2 + 20\,x + 56;$$

whence
$$x^2 - 2\,x + 4 = 10\,x + \frac{56}{x+2},$$

---

[1] Cantor, *Geschichte*, II, chap. lv.

[2] Fol. 149, *r*., has the following:

> Impossibile. Censo de censo: e cĕso equale. a cosa.
> Impossibile. Censo de censo e cosa. equale. a censo.

That is, the solution of equations like $ax^4 + cx^2 = dx$ and $ax^4 + dx = cx^2$ is impossible.   [3] *Die Coss*, 1553 ed., fol. 477, *r*.

all of which is correct. He now assumes that he can, in general, split the two members and say that

$$x^2 - 2x = 10x$$

and
$$4 = \frac{56}{x+2}.$$

Both of these equations are satisfied if $x = 12$, but the method is not otherwise general.

Similar solutions of special cases are found in various works of the 16th century, notably in a work by Nicolas Petri of Deventer,[1] published at Amsterdam in 1567. This writer was highly esteemed by his contemporaries.[2]

A few special cases, such as

$$x^3 = 38\tfrac{1}{2}x - 90,$$

he solves by factoring, and he then proceeds to a more elaborate discussion of certain cases that are mentioned later.

The Italian Algebraists and the Cubic. The real interest in the cubic lies, however, in the work of the Italian algebraists of the 16th century, and notably in the testimony of Cardan and Tartaglia. Cardan (1545) says that Scipio del Ferro discovered the solution of the type $x^3 + bx = c$ thirty years earlier ($c.$ 1515), revealing the secret to his pupil Antonio Maria Fior (Florido).[3] The source of the solution is unknown. Ferro may have re-

---

[1] *Arithmetica. Practicque omne cortelycken te leré chijpheré* . . . *Door my Nicolaum petri F. Dauentriensem*, Amsterdam, 1567. The name also appears as Nicolas Peetersen or Pietersz (Pieterszoon), Petri F. meaning Petri Filius (son of Peter).

[2] H. Bosmans, "La 'Practiqve om te leeren cypheren' de Nicolas Petri de Deventer," *Annales de la Société scientifique de Bruxelles*, XXXII, 2e Partie, Reprint, 1908.

[3] "Verùm temporibus nostris, Scipio Ferreus Bononiensis, capitulum cubi & rerum numero aequalium inuenit, rem sanè pulchram & admirabilem. . . . Huius æmulatiõe Nicolaus Tartalea Brixellensis, amicus noster, cũ in certamẽ cũ illius discipulo Antonio Maria Florido uenisset, capitulum idem, ne uinceretur, inuenit, qui mihi ipsum multis precibus exoratus tradidit" (*Ars Magna*, fol. 3, r.). On the general work of the Italians with respect to the development of algebra see E. Bortolotti, "Italiani scopritori e promotori di teorie algebriche," in the *Annuario della R. Università di Modena*, Anno 1918–1919.

ceived it from some Arab writer, or he may have discovered it himself in spite of his apparent lack of mathematical ability. Tartaglia agrees with Cardan's statement except as to time, placing it somewhat earlier (in 1506),[1] a matter of little consequence. Cardan further says that Florido had a contest with Tartaglia which resulted in the latter's discovery of the method for solving this particular type, and that Tartaglia, at Cardan's request, revealed it to him.

Tartaglia states his side of the case rather differently and more explicitly. He says that Zuanne de Tonini da Coi[2] (see Volume I, page 295) sent him, in 1530, two problems, namely,

$$x^3 + 3x^2 = 5$$

and
$$x^3 + 6x^2 + 8x = 1000,$$

neither of which he could solve; but that in 1535[3] he found the method of solving any equation of the type $x^3 + ax^2 = c$. Tartaglia further states that he had a contest with Florido in 1535 and knew that he had only to set problems of this type to defeat his opponent, provided he could first find the latter's method of solving problems of the type $x^3 + bx = c$. He therefore exerted himself and succeeded in discovering it just before the contest,[4] thus being able to solve anything that Florido could set, and being able to propose problems that the latter could not master.

Tartaglia and Cardan. Da Coi now importuned Tartaglia to publish his method, but the latter declined to do so. In 1539 Cardan wrote to Tartaglia, and a meeting was arranged at which,

---

[1] ". . . se auantaua che gia trenta anni tal secreto gli era stato mostrato da un gran mathematico." From Qvesito XXV, dated December 10, 1536. See the 1554 edition of the *Qvesiti*, fol. 106, *v*.

[2] Also known as Giovanni Colle and Joannes Colla.

[3] In his statement of December 10, 1536 (Qvesito XXV), he says: " . . . & questo fu l'anno passato, cioe del .1535. adi .12. di. Febraro (uero è in Venetia ueneua à esser del .1534.) . . ." See also A. Oliva, *Sulla soluzione dell' equazione cubica di Tartaglia*, Milan, 1909.

[4] "Per mia bona sorte, solamente .8. giorni auanti al termine . . . Io haueua ritrouata la regola generale." *Qvesiti*, libro nono, Qvesito XXV; 1554 ed., fol. 106, *v*.

Tartaglia says, having pledged Cardan to secrecy, he revealed the method in cryptic verse[1] and later with a full explanation.[2]

Cardan admits that he received the solution from Tartaglia, but says that it was given to him without any explanation.[3] At any rate, the two cubics $x^3 + ax^2 = c$ and $x^3 + bx = c$ could now be solved. The reduction of the general cubic $x^3 + ax^2 + bx = c$ to the second of these forms does not seem to have been considered by Tartaglia at the time of the controversy. When Cardan published his *Ars Magna* (1545), however, he transformed the types $x^3 = ax^2 + c$ and $x^3 + ax^2 = c$ by the substitutions $x = y + \frac{1}{3} a$ and $x = y - \frac{1}{3} a$ respectively, and transformed the type $x^3 + c = ax^2$ by the substitution $x = \sqrt[3]{c^2}/y$, thus freeing the equations of the term in x^2. This completed the general solution, and he applied the method to the complete cubic in his later problems.

Cardan's Originality. Cardan's originality in the matter seems to have been shown chiefly in four respects. First, he reduced the general equation to the type $x^3 + bx = c$; second, in a letter written August 4, 1539, he discussed the question of the irreducible case; third, he had the idea of the number of roots to be expected in the cubic; and, fourth, he made a beginning in the theory of symmetric functions.[4]

[1] Quando chel cubo con le cose appresso
    Se aggualia à qualche numero discreto $x^3 + bx = c$
    Trouan dui altri differenti in esso. $u - v = c$
Dapoi terrai questo per consueto
    Che 'l lor produtto sempre sia eguale $uv = \left(\dfrac{b}{3}\right)^3$
    Al terzo cubo delle cose neto,
El residuo poi suo generale
    Delli lor lati cubi ben sostratti
    Varra la tua cosa principale. $x = \sqrt[3]{u} - \sqrt[3]{v}$

Qvesiti, 1554 ed., fol. 120, v. There are sixteen lines more. See also Gherardi in Grunert's *Archiv*, LII, 143 seq. and 188.

[2] Substantially this: If $x^3 + bx = c$, let $u - v = c$ and $uv = \left(\dfrac{b}{3}\right)^3$. Then $x = \sqrt[3]{u} - \sqrt[3]{v}$, for $\left(\sqrt[3]{u} - \sqrt[3]{v}\right)^3 + b\left(\sqrt[3]{u} - \sqrt[3]{v}\right) = u - v$. See the second part of his Qvesito XXXV, *Qvesiti*, 1554 ed., fol. 121, v.

[3] ". . . ut Nicolaus inuenerit & ipse, qui cum nobis rogantibus tradidisset, suppressa demonstratione . . ." *Ars Magna*, 1545 ed., fol. 29, v., shown in facsimile on pages 462, 463.

[4] See also Eneström's summary in *Bibl. Math.*, VII (3), 293.

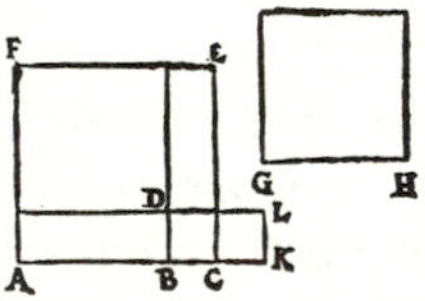

in quadratum A C ter,eſt m:& reliquum quod ei æquatur eſt p:igitur
triplum C B in q̄dratum A B,& triplum A C in q̄dratū C B, & ſexcuplū
A B nihil faciunt. Tanta igitur eſt differentia,ex cōmuni animi ſentẽ-
tia,ipſius cubi A C,à cubo B C,quantum eſt quod cōflatur ex cubo A C,
& triplo A C in quadratum C B,& triplo C B in quadratum A C m:& cu
bo B C m:& ſexcuplo A B,hoc igitur eſt 20,quia differentia cubi A C,à
cubo C B,fuit 20,quare per ſecundum ſuppoſitum 6' capituli , poſita
B C m:cubus A B æquabitur cubo A C , & triplo A C in quadratum B C,
& cubo B C m:& triplo B C in quadratum A C m:cubus igitur A B,cum
ſexcuplo A B,per communem animi ſententiam , cum æquetur cubo
A C & triplo A C in quadratum C B, & triplo C B in quadratum A B m:
& cubo C B m:& ſexcuplo A B , quæ iam æquatur 20 , ut probatum
eſt,æquabuntur etiam 20,cum igitur cubus A B & ſexcuplum A B æ-
quentur 20,& cubus G H,cum ſexcuplo G H æquentur 20,erit ex com
muni animi ſententia,& ex dictis,in 35² p'& 31² undecimi elementorum,G H æqualis A B,igitur G H eſt differentia A C & C B , ſunt autem
A C & C B,uel A C & C K,numeri ſeu liniæ continentes ſuperficiem , æ-
qualem tertiæ parti numeri rerum,quarum cubi differunt in numero
æquationis,quare habebimus regulam.

Regvla.

Deducito tertiam partem numeri rerum ad cubum , cui addes
quadratum dimidij numeri æquationis,& totius accipe radicem , ſcili
cet quadratam,quam ſeminabis,uniq̃ dimidium numeri quod iam
in ſe duxeras,adijcies,ab altera dimidium idem minues,habebisq̃ Bi
nomium cum ſua Apotome, inde detracta ꝶ cubica Apotomæ ex ꝶ
cubica ſui Binomij,reſiduū quod ex hoc relinquitur,eſt rei eſtimatio.

Exemplum.cubus & 6 poſitiones, æquantur 20,ducito 2 , tertiam partem 6 , ad cubum,fit 8,duc 10 dimidium numeri in ſe,
fit 100,iunge 100 & 8,fit 108,accipe radicem quæ eſt ꝶ 108, & eam geminabis,alte
ri addes 10,dimidium numeri,ab altero mi
nues tantundem,habebis Binomiū ꝶ 108
p:10,& Apotomen ꝶ 108 m:10 , horum
accipe ꝶⁱᶜ cubⁱⁱ & minue illam quę eſt Apo
tomæ,ab ea quæ eſt Binomij, habebis rei æſtimationem, ꝶ v: cub: ꝶ
108 p:10 m:ꝶ v: cubica ꝶ 108 m:10.

cub⁹ p:6 reb⁹ æq̄lis 20		
2		20
8	———	10
	108	
ꝶ 108 p:10		
ꝶ 108 m:10		
ꝶ v: cu.ꝶ 108 p:10		
m:ꝶ v:cu.ꝶ 108 m:10		

Aliud,cubus p: 3 rebus æquetur 10,duc 1 ,tertiam partem 3 , ad
cubum,fit 1,duc 5,dimidium 10,ad quadratum,fit 25,iunge 25 & 1,

CARDAN'S SOLUTION OF THE CUBIC
Continuation of the solution as given on page 462. For the meaning of the
symbols see page 428

With respect to the irreducible case, his solution of the type $x^3 + bx = c$ is

$$x = \sqrt[3]{\sqrt{\tfrac{1}{4}c^2 + \tfrac{1}{27}b^3} + \tfrac{1}{2}c} - \sqrt[3]{\sqrt{\tfrac{1}{4}c^2 + \tfrac{1}{27}b^3} - \tfrac{1}{2}c},$$

and if b is negative and is such that $\tfrac{1}{4}c^2 + \tfrac{1}{27}b^3$ is also negative, then we have the cube root of a complex number, thus reaching an expression that is irreducible even though all three values of x turn out to be real.[1]

With respect to the number of roots to be expected in the cubic, he gave[2] the equations $x^3 + 10x = 6x^2 + 4$ with roots $2, 2 \pm \sqrt{2}$; $x^3 + 21x = 9x^2 + 5$ with roots $5, 2 \pm \sqrt{3}$; and $x^3 + 26x = 12x^2 + 12$ with roots $2, 5 \pm \sqrt{19}$; but before this time only two roots were ever found,[3] negative roots being generally rejected.

As to the question of symmetric functions, he stated that the sum of the roots is minus the coefficient of x^2.[4]

Cardan's solution, with part of his explanation, is shown in facsimile on pages 462 and 463. In the solution he states that the root of the equation $x^3 + 6x = 20$ is

$$x = \sqrt[3]{\sqrt{108} + 10} - \sqrt[3]{\sqrt{108} - 10}.$$

He also gave thirteen forms of the cubic which have positive roots, these having already been given by Omar Khayyam.

---

[1] The reality of the roots for this case was shown by Kästner (1745) and A. C. Clairaut (1746). As an example of the irreducible case, in the equation $x^3 - 63x - 162 = 0$ the rule gives

$$x = \sqrt[3]{81 + 30\sqrt{-3}} + \sqrt[3]{81 - 30\sqrt{-3}},$$

which we cannot reduce, although as a matter of fact the solution is

$$x = (-3 + 2\sqrt{-3}) + (-3 - 2\sqrt{-3}) = -6.$$

[2] Cap. XVIII, Exemplum quintum: "Cubus & 10 res, aequatur 6 quadratis p : 4 (1545 ed., fol. 39, r.). The roots are "2 p : ℞ 2, uel 2 m: ℞ 2, potest etiam esse 2." The folio is incorrectly numbered 36.

[3] It was Euler (1732) who gave the first noteworthy modern discussion of the cubic, insisting on the recognition of all three roots and stating how these roots were found. "De formis radicum aequationum cuiusque ordinis conjectatio," in *Comment. Petropol. ad annos 1732–1733*, printed in 1738; VI, 217.

[4] ". . . uelut in quinto exemplo, 2 p: ℞ 2, & 2, & 2 m: ℞ 2, componunt 6, numerum quadratorum," and so for other cases. That is, in the case of $x^3 + 10x = 6x^2 + 4$ the sum of $2 + \sqrt{2}$, 2, and $2 - \sqrt{2}$ is 6 (fol. 39, v.).

Nicolas Petri and the Cubic. In his work of 1567 Nicolas Petri of Deventer, as already mentioned, gave some attention to the cubic equation. This is found in a subdivision on *Cubicq Coss*,[1] in which he gives eight cubic equations such as

$$x^3 = 9x + 28,$$

$$23x^3 + 32x = 905\tfrac{5}{9}, \quad \text{and} \quad x^3 = 3x^2 + 5x + 16,$$

all of which he solves by Cardan's method.

In the same year that Petri's work appeared, Pedro Nuñez (to take the form of his name used in the treatise here mentioned) published his *Libro de algebra en arithmetica y geometria* at Antwerp.[2] In this work he considers such equations as

$$x^3 + 3x = 36 \quad \text{and} \quad x^3 + 9x = 54,$$

and seeks to show that Tartaglia's rule is not practical where one root is easily found by factoring. He shows a familiarity with the works of both Tartaglia and Cardan.

Vieta generalizes the Work. Although Cardan reduced his particular equations to forms lacking a term in x^2, it was Vieta[3] who began with the general form

$$x^3 + px^2 + qx + r = 0$$

and made the substitution $x = y - \tfrac{1}{3}p$, thus reducing the equation to the form

$$y^3 + 3by = 2c.$$

He then made the substitution

$$z^3 + yz = b, \quad \text{or} \quad y = \frac{b - z^2}{z},$$

which led to the form

$$z^6 + 2cz^3 = b^3,$$

a sextic which he solved as a quadratic.

---

[1] "Volgen sommighe exempelen ghesolueert deur die Cubicq Coss."

[2] H. Bosmans, "Sur le 'Libro de algebra' de Pedro Nuñez," *Bibl. Math.*, VIII (3), 154. The original name is Nunes. See Volume I, page 348.

[3] *Opera mathematica. IV. De aequationum recognitione et emendatione libri duo*, Tract. II, cap. vii (Paris, 1615). His equation is stated thus: "Proponatur A cubus + B plano 3 in A, aequari Z solido 2"; that is, $A^3 + 3 \, BA = 2 \, Z$, or, in our symbols, $y^3 + 3by = 2c$. The problem as worked out by Vieta is given in Matthiessen, *Grundzüge*, p. 371.

He also gave two or three other solutions, but the one here shown is particularly clear and simple. In his work in equations he was greatly aided by his new symbolism (p. 430).

Hudde's Contribution. Although Descartes contributed to the solution of the cubic equation by his convenient symbolism and by his work on equations in general, he made no specific contribution of importance. The next writer to materially simplify the work of Vieta was Hudde (c. 1658). Taking advantage of Descartes's symbolism, he brought the theory of the cubic equation to substantially its present status. He is also the first algebraist who unquestionably recognized that a letter might stand for either a positive or a negative number.[1]

His method of solving the cubic equation is to begin with

$$x^3 = qx + r$$

and let

$$x = y + z,$$

so that $\quad y^3 + 3y^2z + 3yz^2 + z^3 = qx + r.$

He then lets $\quad y^3 + z^3 = r$

and $\quad 3zy^2 + 3z^2y = qx,$

which gives $\quad y = \tfrac{1}{3}q/z.$

Hence $\quad y^3 = r - z^3 = \tfrac{1}{27}q^3/z^3,$

and so $\quad z^3 = \tfrac{1}{2}r \pm \sqrt{\tfrac{1}{4}r^2 - \tfrac{1}{27}q^3} = A$

and $\quad y^3 = \tfrac{1}{2}r \mp \sqrt{\tfrac{1}{4}r^2 - \tfrac{1}{27}q^3} = B.$

Hence $\quad x = \sqrt[3]{A} + \sqrt[3]{B},$

which satisfies both his assumptions.[2]

Equation of the Fourth Degree. After the cubic equation had occupied the attention of Arab scholars, with not very significant results, the biquadratic equation was taken up. Abû'l-Faradsh[3] completed the *Fihrist c.* 987, and in this he refers to

---

[1] Eneström, in *Bibl. Math.*, IV (3), pp. 208, 216.

[2] The problem, as worked out by Hudde, is given in Matthiessen, *Grundzüge*, p. 374.

[3] Abû'l-Faradsh (Faraj) Mohammed ibn Isḥâq, known as Ibn Abî Ya'qûb al-Nadîm. The title is *Kitâb al-Fihrist (Book of Lists)*. See the *Abhandlungen*, VI, 1.

the following problem by Abû'l-Wefâ (c. 980): "On the method of finding the root of a cube and of a fourth power and of expressions composed of these two powers."[1] The last means that we are to solve the equation $x^4 + px^3 = q$. The equation could have been solved by the intersection of the hyperbola $y^2 + axy + b = 0$ and the parabola $x^2 - y = 0$, but the work in which Abû'l-Wefâ's problem appeared is lost and we do not know what he did in the way of a solution.

Woepcke, a German-French orientalist (c. 1855), has called attention to an anonymous MS. of an Arab or Persian algebraist in which there is given the biquadratic equation

$$(100 - x^2)(10 - x)^2 = 8100.$$

This is solved by taking the intersection of $(10 - x)y = 90$ and $x^2 + y^2 = 100$, but there is no evidence that the author was concerned with the algebraic theory.[2]

It may therefore be said that the Arabs were interested in the biquadratic equation only as they were in the cubic, that is, from the standpoint of the intersection of two conics.

The Italian Algebraists and the Biquadratic. The problem of the biquadratic equation was laid prominently before Italian mathematicians by Zuanne de Tonini da Coi, who in 1540 proposed the problem, "Divide 10 into three parts such that they shall be in continued proportion and that the product of the first two shall be 6." He gave this to Cardan[3] with the statement that it could not be solved, but Cardan denied the assertion, although himself unable to solve it. He gave it to Ferrari, his

---

[1] *Abhandlungen*, VI, 73, note 253. See also Matthiessen, *Grundzüge*, p. 543; F. Woepcke, *Recherches . . . Constructions géom. par Aboûl Wafâ*, p. 36, 8, n. 2 (Paris, 1855).

[2] Woepcke's translation of Omar Khayyam, Addition D, p. 115. The problem was to construct an isosceles trapezium (trapezoid) $ABCD$ such that $AB = AD = BC = 10$, and the area is 90.

[3] Cardan states it thus: "Exemplum. Fac ex 10 tres partes proportionales, ex quarum ductu primæ in secundam, producantur 6. Hanc proponebat Ioannes Colla, & dicebat solui non posse, ego uero dicebam, eam posse solui, modum tamē ignorabam, donec Ferrarius eum inuenit." *Ars Magna*, cap. xxxix, qvæstio v; 1545 ed., fol. 73, v.

pupil (Vol. I, p. 300), and the latter, although then a mere youth, succeeded[1] where the master had failed.

Ferrari's method[2] may be summarized in its modern form as follows: Reduce the complete equation

$$x^4 + a_1 x^3 + a_2 x^2 + a_3 x + a_4 = 0$$

to the form

$$x^4 + p x^2 + q x + r = 0$$

and thence to

$$x^4 + 2 p x^2 + p^2 = p x^2 - q x - r + p^2,$$

or

$$(x^2 + p)^2 = p x^2 - q x + p^2 - r.$$

Write this as

$$(x^2 + p + y)^2 = (p + 2 y) x^2 - q x + (p^2 - r + 2 p y + y^2).$$

Now determine y so that the second member shall be a square. This is the case when

$$4 (p + 2 y)(p^2 - r + 2 p y + y^2) = q^2,$$

which requires the solution of a cubic in y, which is possible. The solution then reduces to the mere finding of square roots.

This method soon became known to algebraists through Cardan's *Ars Magna*, and in 1567 we find it used by Nicolas Petri in the work already mentioned. Petri solves four equations, the first being

$$x^4 + 6 x^3 = 6 x^2 + 30 x + 11.$$

Of this he gives only the root $1 + \sqrt{2}$, neglecting the roots $1 - \sqrt{2}$, $-4 \pm \sqrt{5}$ because they are negative.

---

[1] The proportion is $\dfrac{6}{x} : x = x : \dfrac{1}{6} x^3$, and the other condition is that

$$\frac{6}{x} + x + \frac{1}{6} x^3 = 10,$$

the two conditions reducing to $x^4 + 6 x^2 + 36 = 60 x$. Ferrari's method makes this depend upon the solution of the equation $y^3 + 15 y^2 + 36 y = 450$, or, as Cardan (*Ars Magna*, fol. 74, *r.*) states the problem, "1 cubum p: 15 quadratis p: 36 positionibus æquantur 450."

[2] Cardan, *Ars Magna*, 1545, cap. xxxix, qvæstio v, fol. 73, *v.*; Bombelli, *Algebra*, 1572, p. 353; Matthiessen, *Grundzüge*, p. 540. Bombelli's first special case is "1 ᶜ p. 20 ᵖ eguale à 21"; that is, $x^4 + 20 x = 21$.

Vieta and Descartes. Vieta (*c.* 1590) was the first algebraist after Ferrari to make any noteworthy advance in the solution of the biquadratic.[1] He began with the type $x^4 + 2gx^2 + bx = c$, wrote it as $x^4 + 2gx^2 = c - bx$, added $g^2 + \frac{1}{4}y^2 + yx^2 + gy$ to both sides, and then made the right side a square after the manner of Ferrari. This method also requires the solution of a cubic resolvent.

Descartes[2] (1637) next took up the question and succeeded in effecting a simple solution of problems of the type

$$x^4 + px^2 + qx + r = 0,[3]$$

a method considerably improved (1649) by his commentator Van Schooten.[4] The method was brought into its modern form by Simpson (1745).[5]

Equation of the Fifth Degree. Having found a method differing from that of Ferrari for reducing the solution of the general biquadratic equation to that of a cubic equation, Euler had the idea that he could reduce the problem of the quintic equation to that of solving a biquadratic, and Lagrange made the same attempt. The failures of such able mathematicians led to the belief that such a reduction might be impossible. The first noteworthy attempt to prove that an equation of the fifth degree could not be solved by algebraic methods is due to Ruffini (1803, 1805),[6] although it had already been considered by Gauss.

The modern theory of equations in general is commonly said to date from Abel and Galois. The latter's posthumous (1846) memoir on the subject established the theory in a satisfactory manner. To him is due the discovery that to each equation there

---

[1] *De aequationum recognitione et emendatione libri duo*, Tract. II, cap. vi, prob. iii (Paris, 1615). For solution see Matthiessen, *Grundzüge*, p. 547.

[2] *La Geometrie*, Lib. III; 1649 ed., p. 79; 1683 ed., p. 71; 1705 ed., p. 109.

[3] For examples see Matthiessen, *Grundzüge*, p. 549.

[4] 1649 ed., p. 244.

[5] For the various improvements see Matthiessen, *Grundzüge*, p. 545 seq.

[6] "Della insolubilità delle equazioni algebraiche generali di grado superiore al quarto," *Mem. Soc. Ital.*, X (1803), XII (1805).

corresponds a group of substitutions (the "group of the equation") in which are reflected its essential characteristics. Galois's early death left without sufficient demonstration several important propositions, a gap which has since been filled.

Abel[1] showed that the roots of a general quintic equation cannot be expressed in terms of its coefficients by means of radicals.

Lagrange had already shown that the solution of such an equation depends upon the solution of a sextic, "Lagrange's resolvent sextic," and Malfatti and Vandermonde had investigated the construction of resolvents.

The transformation of the general quintic into the trinomial form $x^5 + ax + b = 0$ by the extraction of square and cube roots only was first shown to be possible by Bring (1786) and independently by Jerrard[2] (1834). Hermite (1858) actually effected this reduction by means of a theorem due to Tschirnhausen, the work being done in connection with the solution by elliptic functions.[3]

Symmetric Functions. The first formulas for the computation of the symmetric functions of the roots of an equation seem to have been worked out by Newton, although Girard (1629) had given, without proof, a formula for a power of the sum, and Cardan (1545) had made a slight beginning in the theory. In the 18th century Lagrange (1768) and Waring (1770, 1782) made several valuable contributions to the subject, but the first tables, reaching to the tenth degree, appeared in 1809 in the Meyer-Hirsch *Aufgabensammlung*. In Cauchy's celebrated memoir on determinants (1812) the subject began to assume new prominence, and both he and Gauss (1816) made numerous and important additions to the theory. It is, however, since the discoveries by Galois that the subject has become one of great

---

[1] *Mémoire sur les équations algébriques*, Christiania, 1824, and Crelle's *Journal*, 1826.

[2] R. Harley, "A contribution to the history . . . of the general equation of the fifth degree . . .," *Quarterly Journal of Mathematics*, VI, 38.

[3] For a bibliography of much value in the study of the history of equations see G. Loria, in *Bibl. Math.*, V (2), 107.

significance. Cayley (1857) gave a number of simple rules for the degree and weight of symmetric functions, and he and Brioschi simplified the computation of tables.

Harriot's Law of Signs. The law which asserts that the equation $X = 0$, complete or incomplete, can have no more real positive roots than it has changes of sign, and no more real negative roots than it has permanences of sign, was apparently known to Cardan[1]; but a satisfactory statement is possibly due to Harriot (died 1621)[2] and certainly to Descartes.[3]

Numerical Higher Equations. The solution of the numerical higher equation for approximate values of the roots begins, so far as we know, in China. Indeed, this is China's particular contribution to mathematics, and in this respect her scholars were preëminent in the 13th and 14th centuries.[4] In the *Nine Sections*, written apparently long before the Christian era, there is found the "celestial element method."[5] This was a method of solving numerical higher equations; it is found in various early Chinese works, reaching its highest degree of perfection in the works of Ch'in Kiu-shao (1247). Here it appears, as already stated, in a form substantially equivalent to Horner's Method (1819).[6]

Fibonacci on Numerical Equations. The first noteworthy work upon numerical higher equations done in Europe is due to Fibonacci (1225), and relates to the case of the cubic equation

---

[1] Cantor, *Geschichte*, II (2), 539; Eneström, *Bibl. Math.*, VII (3), 293.

[2] *Artis analyticae praxis. Ad aequationes Algebraicas . . . resolvendas*, London, 1631 (posthumous); Matthiessen, *Grundzüge*, 2d ed., pp. 18, 268.

[3] *La Geometrie*, 1637; 1649 ed., p. 78; 1705 ed., p. 108, with the statement: "On connoît aussi de ceci combien il peut y avoir de vrayes racines, & combien de fausses en chaque Equation; à sçavoir, il y en peut avoir autant de vrayes que les signes + & — s'y trouvent de fois être changez, & autant de fausses qu'il s'y trouve de fois deux signes + , ou deux signes — qui s'entresuivent." The law usually bears the name of Descartes.

[4] Y. Mikami, *China*, 25, 53, 76, *et passim*; L. Matthiessen, "Zur Algebra der Chinesen," in *Zeitschrift für Math. und Phys.*, XIX, Hl. Abt., 270. For doubts as to the originality of this work and as to the authenticity of the text of the *Nine Sections* see G. Loria, "Che cosa debbono le matematiche ai Cinesi," *Bollettino della Mathesis*, XII (1920), 63.

[5] *T'ien-yuen-shu*, the Japanese *tengen jutsu*.

[6] See Volume I, page 270. For a detailed solution see Mikami, *China*, p. 76 seq.

$x^3 + 2x^2 + 10x = 20$, already mentioned. His method of attack was substantially as follows:

Since $$x^3 + 2x^2 + 10x = 20,$$

we have $$10(x + \tfrac{1}{10}x^3 + \tfrac{1}{5}x^2) = 20,$$

or $$x + \tfrac{1}{10}x^3 + \tfrac{1}{5}x^2 = 2,$$

so that $$x < 2.$$

But $$1 + 2 + 10 = 13 < 20,$$

and so $$x > 1.$$

But x is not fractional; for if $x = a/b$, then

$$\frac{a}{b} + \frac{a^3}{10\,b^3} + \frac{a^5}{5\,b^5}$$

cannot be integral, and so x must be irrational.

Further, x cannot be the square root of an integer; for, from the given equation,

$$x = \frac{20 - 2x^2}{10 + x^2},$$

and if x were equal to $\sqrt{a}$ we should have

$$\sqrt{a} = \frac{20 - 2a}{10 + a},$$

which is impossible.

Fibonacci here closes his analysis and simply makes a statement which we may express in modern symbols as

$$x = 1° \ 22' \ 7'' \ 42''' \ 33^{iv} \ 4^v \ 40^{vi},$$

a result correct to $1\tfrac{1}{2}^{vi}$; that is, the value is only $\frac{1}{31104000000}$ too large. How this result was obtained no one knows, but the fact that numerical equations of this kind were being solved in China at this time, and that intercourse with the East was possible, leads to the belief that Fibonacci had learned of the solution in his travels, had contributed what he could to the theory, and had then given the result as it had come to him.

Vieta and Newton contribute to the Theory. About the year 1600 Vieta suggested that a particular root of a numerical equation could be found by a process similar to that of obtaining a root of a number. By substituting in $f(x)$ a known approximate root of $f(x) = n$ he was able to find the next figure by division.[1]

Newton (1669) simplified this method of Vieta's, and the plan of procedure may be seen in his solution of the equation $y^3 - 2y - 5 = 0$. He first found by inspection that $2 < y < 3$. He then let $2 + p = y$;

whence $\qquad y^3 - 2y - 5 = -1 + 10p + 6p^2 + p^3 = 0$,

and $p = 0.1$, approximately.

Letting $0.1 + q = p$, we have

$$0.061 + 11.23\,q + 6.3\,q^2 + q^3 = 0;$$

whence $q = -0.0054$, approximately.

Letting $-0.0054 + r = q$, we have

$$0.000541708 + 11.16196\,r + 6.3\,r^2 = 0;$$

whence $r = -0.00004854$, approximately.

Similarly, we could let $-0.00004854 + s = r$, and proceed as before. We could then reverse the process and find p.

In this way he finds[2] the approximate value

$$y = 2.09455147.$$

As already stated (page 471), in 1819 William George Horner carried this simplification still farther, the root being developed figure by figure. The process terminates if the root is commensurable, and it may be carried to any required number of decimal places if it is incommensurable.[3]

Fundamental Theorem. The Italian algebraists of the 16th century tacitly assumed that every rational integral equation has a root. The later ones of that century were also aware that a quadratic equation has two roots, a cubic equation three roots,

[1] Burnside and Panton, *Theory of Equations*, 4th ed., I, 275. Dublin, 1899.

[2] "De analysi per aequationes numero terminorum infinitas," extract of 1669 in the *Commercium Epistolicum*, p. 76 (London, 1725). Wallis also gave an approximation method in 1685.

[3] For a simple presentation see Burnside and Panton, *loc. cit.*, I, 227, and consult that work (I, 275) for further information on the subject.

and a biquadratic equation four roots. The first writer to assert positively that every such equation of the nth degree has n roots and no more seems to have been Peter Roth, a Nürnberg Rechenmeister, in his *Arithmetica philosophica* (Nürnberg, 1608).[1] The law was next set forth by a more prominent algebraist, Albert Girard, in 1629.[2] It was, however, more clearly expressed by Descartes (1637), who not only stated the law but distinguished between real and imaginary roots and between positive and negative real roots in making the total number.[3] Rahn (Rhonius), also, gave a clear statement of the law in his *Teutschen Algebra* (1659).[4]

After these early steps the statement was repeated in one form or another by various later writers, including Newton (*c.* 1685) and Maclaurin (posthumous publication, 1748). D'Alembert attempted a proof of the theorem in 1746, and on this account the proposition is often called d'Alembert's Theorem. Other attempts were made to prove the statement, notably by Euler (1749) and Lagrange, but the first rigorous demonstration is due to Gauss (1799, with a simple treatment in 1849).

Trigonometric Solutions. In the 16th century Vieta[5] suggested (1591) the treatment of the numerical cubic equation by trigonometry, and Van Schooten later elaborated the plan. Girard[6] (1629) was one of the first, however, to attack the problem scientifically. He solved the equation $1 \; ③ \propto 13 \; ① + 12$, that is, $x^3 = 13x + 12$, by the help of the identity

$$\cos 3\,\phi = 4\cos^3\phi - 3\cos\phi.$$

---

[1] In modern works the name also appears as Rothe. See Tropfke, *Geschichte*, III (2), 95, with a quotation from the original work. Roth died at Nürnberg in 1617. See Volume I, page 421.

[2] "Toutes les équations d'algèbre reçoivent autant de solutions, que la dénomination de la plus haute quantité le demonstre." *Invention nouvelle en l'algèbre*, Amsterdam, 1629; quoted in Tropfke, *Geschichte*, III (2), 95, to which refer for further details.

[3] "Au reste tant les vrayes racines que les fausses ne sont pas toujours réelles, mais quelquefois seulement imaginaires." *La Geometrie* (1705 ed.), p. 117.

[4] English translation, London, 1668. See Volume I, page 412.

[5] See Van Schooten's edition of his *Opera*, p. 362 (Leyden, 1646).

[6] *Invention nouvelle en l'algèbre*, Amsterdam, 1629. On the primitive Arab method see Matthiessen, *Grundzüge*, p. 894; on Girard, see *ibid.*, p. 896.

A single solution of a quadratic equation by trigonometric methods will show the later development of the subject.[1]

Fischer's Solution. Let

$$x^2 - px + q = 0. \qquad\qquad p^2 \gtreqless 4q$$

Then let $\qquad\qquad x_1 = p \cos^2\phi$

and $\qquad\qquad x_2 = p \sin^2\phi.$

Then $\qquad\qquad x_1 + x_2 = p(\cos^2\phi + \sin^2\phi) = p$

and $\qquad\qquad x_1 x_2 = p^2(\cos\phi \sin\phi)^2$

$$= \tfrac{1}{4} p^2 \sin^2 2\phi.$$

The angle ϕ can now be found from the relation

$$\sin 2\phi = 2\sqrt{q}/p.$$

For example, given the equation

$$x^2 - 93.7062\,x + 1984.74 = 0,$$

we find $\qquad\qquad 2\phi = 71° 57' 44.6'',$

whence $\qquad\qquad \phi = 35° 58' 52.3'';$

and hence $\qquad\qquad x_1 = 61.3607^-$

and $\qquad\qquad x_2 = 32.3454.$

Such methods have been extensively used with the cubic and biquadratic equations.[2]

9. Determinants

Among the Chinese. The Chinese method of representing the coefficients of the unknowns of several linear equations by means of rods on a calculating board naturally led to the discovery of simple methods of elimination. The arrangement of the rods was precisely that of the numbers in a determinant. The Chinese, therefore, early developed the idea of subtracting columns and rows as in the simplification of a determinant.[3]

---

[1] This is due to Fischer, *Die Auflösung der quadratischen und kubischen Gleichungen durch Anwendung der goniometrischen Functionen*, Elberfeld, 1856. See Matthiessen, *Grundzüge*, p. 885, and consult this work for a detailed history of the subject. [2] For a list of writers see Matthiessen, *Grundzüge*, p. 888 seq. [3] Mikami, *China*, pp. 30, 93.

Among the Japanese. It was not until Chinese science had secured a firm footing in Japan, and Japanese scholars had begun to show their powers, that the idea of determinants began to assume definite form. Seki Kōwa, the greatest of the Japanese mathematicians of the 17th century, is known to have written a work called the *Kai Fukudai no Hō* in 1683. In this he showed that he had the idea of determinants and of their expansion. It is strange, however, that he used the device only in eliminating a quantity from two equations and not directly in the solution of a set of simultaneous linear equations.[1]

Determinants in Europe. So far as Western civilization is concerned, the theory of determinants may be said to have begun with Leibniz[2] (1693), who considered these forms solely with reference to simultaneous equations, as the Chinese had already done.

It was Vandermonde (1771) who first recognized determinants as independent functions. To him is due the first connected exposition of the theory, and he may be called its formal founder. Laplace (1772) gave the general method of expanding a determinant in terms of its complementary minors, although Vandermonde had already considered a special case. Immediately following the publication by Laplace, Lagrange (1773) treated of determinants of the second and third orders and used them for other purposes than the solution of equations.

The next considerable step in advance was made by Gauss (1801). He used determinants in his theory of numbers, introduced the word "determinant"[3] (though not in the present signification,[4] but rather as applied to the discriminant of a quantic), suggested the notion of reciprocal determinants, and came very near the multiplication theorem.

---

[1] T. Hayashi, "The Fukudai and Determinants in Japanese Mathematics," in the *Proc. of the Tokyō Math. Soc.*, V (2), 257; Mikami, *Isis*, II, 9.

[2] Sir Thomas Muir, *Theory of Determinants in the Historical Order of Development* (4 vols., London, 1890, 1911, 1919; 2d ed., 1906, 1911, 1920, 1923), which consult on the whole question; M. Lecat, *Histoire de la théorie des Déterminants à plusieurs dimensions*, Ghent, 1911. [3] Laplace had used "resultant."

[4] "Numerum bb − ac, cuius indole proprietates formae (a, b, c) imprimis pendere in sequentibus docebimus, determinantem huius uocabimus."

The next great contributor was Jacques-Philippe-Marie Binet,[1] who formally stated (1812) the theorem relating to the product of two matrices of m columns and n rows, which for the special case of $m = n$ reduces to the multiplication theorem.

On the same day (November 30, 1812) that he presented his paper to the Académie, Cauchy presented one on the same subject. In this paper he used the word "determinant" in its present sense, summarized and simplified what was then known on the subject, improved the notation, and gave the multiplication theorem with a proof more satisfactory than Binet's. He may be said to have begun the theory of determinants as a distinct branch of mathematics.

Aside from Cauchy, the greatest contributor to the theory was Carl Gustav Jacob Jacobi.[2] With him the word "determinant" received its final acceptance. He early used the functional determinant which Sylvester has called the Jacobian, and in his famous memoirs in Crelle's *Journal* for 1841 he considered these forms as well as that class of alternating functions which Sylvester has called alternants.

About the time of Jacobi's closing memoirs Sylvester (1839) and Cayley began their great work in this field. It is impossible to summarize this work briefly, but it introduced the most important phase of the recent development of the theory.

10. RATIO, PROPORTION, AND THE RULE OF THREE

Nature of the Topics. It is rather profitless to speculate as to the domain in which the concept of ratio first appeared. The idea that one tribe is twice as large as another and the idea that one leather strap is only half as long as another both involve the notion of ratio; both are such as would develop early in the history of the race, and yet one has to do with ratio of numbers and the other with ratio of geometric magnitudes. Indeed, when we come to the Greek writers we find Nicomachus including ratio in his arithmetic, Eudoxus in his geometry, and

---

[1] Born at Rennes, February 2, 1786; died in Paris, May 12, 1856.
[2] See Volume I, page 506.

Theon of Smyrna in his chapter on music.[1] Still later, Oriental merchants found that they could easily secure results to certain numerical problems by a device which, in the course of time, became known as the Rule of Three, and so this topic found place in commercial arithmetics, although fundamentally it is an application of proportion. Since ratio, proportion, and variation are now considered as topics of algebra, however, it is appropriate to treat of these subjects, as well as the Rule of Three, in the present chapter.[2]

Technical Terms. The word "ratio" as commonly used in school, while sanctioned by ancient usage,[3] has never been a favorite outside the mathematical classroom. It is a Latin word[4] and was commonly used in the arithmetic of the Middle Ages to mean computation. To represent the idea which we express by the symbols $a:b$ the medieval Latin writers generally used the word *proportio*, not the word *ratio*; while for the idea of an equality of ratio, which we express by the symbols $a:b=c:d$, they used the word *proportionalitas*.[5] That these terms were thoroughly grounded in the vernacular is seen today in the common use of such expressions as "divide this in the proportion of 2 to 3," and "your proportion of the expense," and in

[1] P. Tannery, "Du rôle de la musique grecque dans le développement de la mathématique pure," *Bibl. Math.*, III (3), 161.

[2] It may be said that medieval writers looked upon ratio and proportion as a branch of mathematics quite distinct from geometry and arithmetic. See also S. Günther, *Geschichte der Mathematik*, p. 180 (Leipzig, 1908) ; hereafter referred to as Günther, *Geschichte*.

[3] For a discussion of the terms λόγος (*lo'gos*), *ratio*, and *proportio*, see Heath, *Euclid*, Vol. II, pp. 116–129. See also Boethius, ed. Friedlein, p. 3 *et passim*.

[4] From the verb *reri*, to think or estimate; past participle, *ratus*. Hence *ratio* meant reckoning, calculation, relation, reason.

[5] Thus Boethius (*c.*510): "Proportionalitas est duarum vel plurium proportionum similis habitudo," ed. Friedlein, p. 137; Jordanus Nemorarius (*c.* 1225): "Proportionalitas est si'litudo ₚportionū" (1496 ed., Lib. 2).
In the Biblioteca Laurenziana at Florence is a MS. (Codex S. Marco Florent. 184) of Campanus (*c.*1260) *De proportione et proportionalitate* with the inscription *Tractatus Campani de proporcione et proporcionabilitate*. Included with it is a MS. of al-Miṣrî (*c.*900), *Epistola Ameti filii Joseph de proporcione et proportionalitate*. See *Bibl. Math.*, IV (3), 241; II (2), 7. The title of Pacioli's work affords another example: *Sūma de Arithmetica Geometria Proportioni & Proportionalita*, Venice, 1494.

the occasional use of an expression like "the proportionality of the cost is the same as that of the amount."

That the word "proportion" was commonly used in medieval and Renaissance times to mean ratio is seen in most mathematical works of those periods.[1] It was so used by the American Greenwood (1729)[2] and has by no means died out in our language.[3]

The use of *proportio* for *ratio* was not universal in the early days of printing, however, for various writers used both terms as we use them today.[4]

General Types of Ratio. From the time of the Greeks to the 17th century the writers on theoretical arithmetic employed a set of terms and ideas in connection with ratio that seem to mathematicians of the present time unnecessarily complicated. A few of these have survived in our algebra, most of them have disappeared, and all of them had, under ancient conditions, good reasons for being. Of those which are still found in some of our textbooks there may be mentioned three general types of ratio of integers: namely, a ratio of equality,[5] like $a:a$; a ratio

---

[1] Thus Campanus (*c.*1260): "Proportio est duarum quantitatum eiusdem generis ad inuicem habitudo" (Codex S. Marco Florent. 184); Jordanus Nemorarius (*c.*1225): "Proportio est duąɟ quātitatum eiusdem generis vnius ad alteram certa in quātitate relatio" (1496 ed., Lib. 2); Leonardo of Cremona: "la proporcion del diametro a la circonferentia" (original MS. in Mr. Plimpton's library; see *Rara Arithmetica*, p. 474); Chuquet (1484): "Proporcion cest labitude qui est entre deux nõbres quant est compare (lung) a laultre" (the Marre MS. in the author's possession, used by Boncompagni, *Bullettino*, XIII, 621); Rudolff (1526): "Die proporcion oder schicklgkeit der ersten gegē der andern" (1534 ed., fol. E viij). Barrow (1670) used the expression in his lectures on geometry, and most other writers of the period did the same.

[2] ". . . the Proportion that each Figure bears to its neighbouring Figure" (p. 50).

[3] *E.g.*, Alison and Clark, *Arithmetic*, chap. xxi (Edinburgh, 1903).

[4] Thus Fine (Finæus): "Ratio igitur . . . est duarum quantitatum eiusdem speciei adinuicem comparatarū habitudo. . . . Proportio est, contingens inter comparatas adinuicem quantitates rationum similitudo" (*Protomathesis*, 1530–1532; 1555 ed., fols. 38 and 57). See also L. L. Jackson, *The Educational Significance of Sixteenth Century Arithmetic*, p. 119 (New York, 1906).

[5] The "aequalitatis proportio" of the Latin writers; *e.g.*, Scheubel (1545). In a numerical ratio like $a:b$, both a and b were generally considered integral unless the contrary was stated, but the incommensurable ratio of lines was recognized by the Pythagoreans and by all subsequent geometers.

of greater inequality, like $a:b$ when $a>b$; and a ratio of lesser inequality,[1] like $a:b$ when $a<b$. Of the last two there were recognized various subspecies, such as multiple ratio,[2] like $ma:a$, where m is integral; superparticular ratio, like $(m+1):m$ (which had several types, such as sesquialteran, as in the case of $3:2$, sesquitertian, as in the case of $4:3$, and so on); superpartient, like $(m+n):n$, where $m>n>1$, as in the cases of $5:3$ and $7:3$; multiple superparticular, like $(mn+1):m$, as in the cases of $7:3$ and $15:7$; and multiple superpartient, like $(mn+k):m$, where $m>k>1$, as in the cases of $14:3$ and $19:5$. These terms were capable of a large number of combinations and were essentially, from our present point of view, the result of an effort to develop a science of general fractions at a time when the world had no good symbolism for the purpose. With the introduction of our common notation and the invention of a good algebraic symbolism such terms disappeared.[3] This disappearance was hastened by such writers as Stifel (1546), who spoke out plainly against their further use, although his own acts were not always consistent with this statement.[4] When $\frac{9}{76}$ had to be called "suboctupla subsuperquadripartiens nonas" by a writer as late as 1600,[5] it was evident that the ancient usage must give way, and that ratios must be considered with respect to the modern fractional notation instead of depending upon the ancient Roman method.

[1] Boethius, ed. Friedlein, p. 238.

[2] Boethius (ed. Friedlein, p. 46) speaks of such relations: "Maioris vero inaequalitatis .V. sunt partes. Est enim una, quae vocatur multiplex, alia superparticularis, . . . " It should be observed that the Greeks did not consider ratios as numbers in the way that we do; that is, they did not consider $6:3$ as identical with 2 but as a relation of 6 to 3, this relation being a multiple ratio.

[3] For a full treatment of the subject see Pacioli, *Sūma*, 1494 ed., fol. 72.

[4] In his *Rechenbuch* (p. 35) he says: "Von den Proportzen. Zvm ersten | des Boetius | Stapulensis | Apianus | Christoff Rudolff | vnd andere gelerte Leuth | die proportiones leren | mit solche worten Multiplex | Duplex | Tripla | Superparticularis | Sesquialtera . . . vnnd der gleichen wort ohn zal | ist wol recht vnd nutzlich gelert Aber das man ein Teutschen Leser | dem die Lateinisch sprach ist vnbekant | will man solchen worten beladen | das ist ohn not | vnnd ohn nutz." In his *Coss Christoffs Rudolffs* (chap. 12), however, he gives "die fünferley proportionirte zalen," the multiple, superparticular, superpartient, multiplexsuperparticular, and the multiplexsuperpartient.

[5] Van der Schuere, 1624 ed., fol. 193.

Other Greek Ideas of Ratio. Certain other Greek ideas have come down to us and still find a place in our algebras. For example, we speak of $a^2 : b^2$ as the duplicate ratio[1] of a to b, although to double $a : b$ would give $2a : b$. To the Greek, however, the ratio $a_1 : a_n$ was considered as compounded or composed[2] of the ratios $a_1 : a_2, a_2 : a_3 \cdots, a_{n-1} : a_n$; and since $a^2 : b^2$ is similarly compounded of $a^2 : ab$ and $ab : b^2$ or of $a : b$ and $a : b$, it was called[3] the duplicate of $a : b$.

In like manner we have from the Greeks the idea of ratios compounded by addition when as a matter of fact they have been, according to our conception, multiplied.[4]

In the Middle Ages the distinction between ratios and fractions, or ratio and division, became less marked, and in the Renaissance period it almost disappeared except in cases of incommensurability.[5] An illustration of this fact is seen in the way in which Leibniz speaks of "ratios or fractions."[6]

Proportion as Series. The early writers often used *proportio* to designate a series,[7] and this usage is found as late as the 18th century.[8] The most common use of the word, however, limited it to four terms. Thus the early writers spoke of an arithmetic proportion, meaning $b - a = d - c$, as in 2, 3, 4, 5; and of a geometric proportion, meaning $a : b = c : d$, as in 2, 4, 5, 10. To

---

[1] Euclid's διπλασίων (*diplasi'on*), but commonly given by other Greek writers as διπλάσιος λόγος (*dipla'sios lo'gos*). See Heath, *Euclid*, Vol. II, p. 133.

[2] Heath, *Euclid*, Vol. II, p. 133.

[3] Euclid, *Elements*, VI, def. 5, apparently an interpolation. See Heath, *Euclid*, Vol. II, p. 189.

[4] See Heath, *Euclid*, Vol. II, p. 168. Similarly, Scheubel (1545, Tract. II) speaks "de proportionum Additione . . . siue ut alij Compositione," saying that the ratios $9 : 4$ and $5 : 3$ "componunt" $45 : 12$.

[5] A modern Arab arithmetic, published at Beirut in 1859, remarks: "This division is called by the Magrebiner [West Arabs] 'the denomination,' but the Persians call it *al-nisbe* [the ratio]." H. Suter, *Bibl. Math.*, II (3), 17.

[6] ". . . aut in rationibus vel Fractionibus." Letter to Oldenburg, 1673.

[7] Thus Pacioli (1494): "che tu prendi li numeri . . . 1. 2. 3. 4. 5. 6. 7. 8. 9. 10 . . . hauerai la prima specie de la proportion" (*Suma*, fol. 72, *r.*).

[8] When Vitalis (Geronimo Vitale) published his *Lexicon Mathematicvm* (Rome, 1690), the usage was apparently unsettled. He says: "Igitur Proportio Arithmetica est cum tres, vel plures numeri per eandem differentiam progrediuntur; vt 4. 7. 10. 13. 16. 19. 22. & sic procedendo in infinitum" (p. 681); but he also uses *proportio* in the modern sense (p. 732).

these proportions the Greeks[1] added the harmonic proportion

$$\frac{1}{b} - \frac{1}{a} = \frac{1}{d} - \frac{1}{c},$$

as where $a=\frac{1}{2}$, $b=\frac{1}{3}$, $c=\frac{1}{4}$, and $d=\frac{1}{6}$. These three names are now applied to series. To them the Greeks added seven others, all of which go back at least to Eudoxus ($c.$ 370 B.C.).[2] The Renaissance writers began to exclude several of these,[3] and at the present time we have only the geometric proportion left, and so the adjective has been dropped and we speak of proportion alone.

Types of Proportion. The fact that geometric proportion has survived, in algebra at least, is largely due to Euclid's influence, since algebraically a proportion is nothing more than a fractional equation and might be treated as such. Especially is this true of such expressions as "by alternation," "by inversion," "by composition," "by division," and "by composition and division," three of which are now misnomers in the modern use of the words. They come to us directly from the Arabs,[4] who received them from Greek sources.[5] There were also various other types of geometric proportion besides the one commonly seen in textbooks,[6] but most of these types are now forgotten,

[1]On the theory in Euclid, see Heath, *Euclid*, Vol. I, p. 137; Vol. II, pp. 113, 119, 292.

[2]Boethius and certain of his predecessors gave all ten forms, those besides the three above mentioned being as follows:

$$a:c = b - c:a - c, \qquad a:c = a - c:b - c,$$
$$b:c = b - c:a - b, \qquad a:c = a - c:a - b,$$
$$a:b = b - c:a - b, \qquad b:c = a - c:b - c,$$
$$b:c = a - c:a - b.$$

See his *Arithmetica*, ed. Friedlein, p. 137 seq.; Günther, *Math. Unterrichts*, p. 85; Cantor, *Geschichte*, I, chap. xi, for the earlier knowledge of these forms.

[3]Thus Ramus: "Genera autē proportionis duo tantum instituimus, quia haec sola simplicia & mathematica sunt. Nicomachus fecit decem. Jordanus addidit undecimam." *Scholarvm Mathematicarvm, Libri vnvs et triginta*, p. 134 (Basel, 1569).

[4]*E.g.*, see al-Karkhî ($c.$ 1020), the *Kâfî fîl Hisâb*, ed. Hochheim, II, 15.

[5]Heath, *Euclid*, Vol. I, p. 137; Vol. II, pp. 113, 119, 133, 168, 189, 292.

[6]*E.g.*, Scheubel (1545): "Sex sunt species proportionalitatis, permuta, conuersa, coniuncta, disiuncta, euersa, & aequa" (Tractatus II).

although "continued proportion," but with a change in the older meaning,[1] has survived both in algebra and in geometry.

Terms Used in Proportion. The terms "means," "antecedent," and "consequent" are due to the Latin translators of Euclid.[2] There have been attempts at changing them, as when the antecedent was called a leader and the consequent a comrade,[3] but without success. It would be quite as simple to speak of them as the first, second, third, and fourth terms.[4]

Rule of Three. The mercantile Rule of Three seems to have originated among the Hindus. It was called by this name by Brahmagupta (*c.* 628) and Bhāskara[5] (*c.* 1150), and the name is also found among the Arab and medieval Latin writers.

Brahmagupta and Mahāvīra state the Rule. Brahmagupta stated the rule as follows: "In the Rule of Three, Argument, Fruit, and Requisition are the names of the terms. The first and last terms must be similar. Requisition multiplied by Fruit, and divided by Argument, is the Produce."[6] Mahāvīra (*c.* 850) gave it in substantially the same form, thus: "*Phala* multiplied by *Icchā* and divided by *Pramāṇa* becomes the answer, when the *Icchā* and *Pramāṇa* are similar."[7]

For example: "A lame man walks over $\frac{1}{8}$ of a *krośa* [32,000 feet] together with $\frac{1}{5}$ [thereof] in $7\frac{1}{2}$ days. Say what [distance] he [goes over] in $3\frac{1}{5}$ years [at this rate],"—a very good illustration of the absurdity of the Oriental problem.

---

[1] *E.g.*, Fine (Finæus, 1530) defines a *proportio continua* as one like "8/4/2/1: ut enim 8 ad 4, sic 4 ad 2, atq3 2 ad 1" (*De Arithmetica Practica*, 1555 ed., fol. 59), and a *proportio disjuncta* as (to use his symbolism) one like 8/4, 6/3.

[2] Euclid used μεσότητες (*mesot'etes*, means), ἡγούμενα (*hegou'mena*, leading [terms], antecedents), and ἑπόμενα (*hepom'ena*, following [terms], consequents), but he had no need for "extremes." See *Elements*, VII, 19.

[3] Thus Scheubel (1545): " . . . alter antecedens uel dux, alter consequens uel comes appellatur." The use of *dux* (duke, leader) comes from Euclid's term.

[4] *E.g.*, see Clavius (1583), *Epitome Arithmeticae Practicae*, chap. xvii.

[5] One of the scholiasts of Bhāskara called it the *Trairāśica*, the "three rule." See Colebrooke's translation, pp. 33, 283. On the general subject see Taylor's translation, p. 41. [6] Colebrooke's translation, p. 283.

[7] *Gaṇita-Sāra-Saṇgraha*, p. 86. The *phala* is the given quantity corresponding to what is to be found; the *pramāna* was a measure of length, but in proportion it is the term corresponding to the *icchā*; the *icchā* is the third term in the rule.

Bhāskara (c. 1150) gave the rule in much the same form as that used by Brahmagupta, thus: "The first and last terms, which are the argument and requisition, must be of like denomination; the fruit, which is of a different species, stands between them; and that, being multiplied by the demand [that is, the requisition] and divided by the first term, gives the fruit of the demand [that is, the Produce]."[1]

As an example, Bhāskara gave the following: "Two palas and a half of saffron are purchased for three sevenths of a niska: How many will be purchased for nine niskas?"

His work appears[2] as
$$3 \quad 5 \quad 9$$
$$7 \quad 2 \quad 1$$

In our symbolism it might be represented as $\frac{3}{7} N. \ 2\frac{1}{2} P. \ 9 N.$ It is thus seen that the idea of equal ratios is not present, as would be the case if we should write $x : 2\frac{1}{2} = 9 : \frac{3}{7}$. Proportion was thus concealed in the form of an arbitrary rule, and the fundamental connection between the two did not attract much notice until, in the Renaissance period, mathematicians began to give some attention to commercial arithmetic. One of the first to appreciate this connection was Widman[3] (1489), and in this he was followed by such writers as Tonstall[4] (1522), Gemma Frisius[5] (1540), and Trenchant[6] (1566).

Names for the Rule of Three. Recorde (c. 1542) calls the Rule of Three "the rule of Proportions, whiche for his excellency is called the Golden rule,"[7] although his later editors called it by the more common name.[8] Its relation to algebra was first strongly emphasized by Stifel[9] (1553-1554).

---

[1] Colebrooke's translation, p. 33.   [2] See Taylor's translation, p. 41.

[3] "Sy ist auch recht genāt regula proportionū/wā in d' regel werdē erkāt vn erfundē alle ꝑportiones" (1508 ed., fol. 50).   [4] *De Arte Supputandi*, Lib. III.

[5] He calls his chapter "De Regvla Proportionvm, siue Trium Numerorum" (1575 ed., fol. C6).

[6] "La regle de troys, qui est la regle des proportions ou proportionaux" (1578 ed., p. 120).   [7] 1558 ed., fol. M4.   [8] *E.g.*, John Mellis, 1594 ed., p. 449.

[9] "Gar wunderbarlich wickeln vnd verknüpffen sich zusammen die Detri vnd die Coss also dass die Coss im grund auch wol möchte genennt werden die Detri. . . . Vn steckt also die gantz Coss in der Regel Detri/widervmb steckt die Gantz Detri in der Coss." See the *Abhandlungen*, I, 86.

RULE OF THREE IN THE SIXTEENTH CENTURY

From an Italian MS. of 1545. Notice the arrangement of terms; also the early per cent sign as given at the end of the fifth line. From a manuscript in Mr. Plimpton's library

When the rule appeared in the West, it bore the common Oriental name,[1] although the Hindu names for the special terms were discarded. So highly prized was it among merchants, however, that it was often called the Golden Rule,[2] a name apparently in special favor with the better mathematical writers.[3] Hodder, the popular English arithmetician of the 17th century, justifies this by saying: "The Rule of Three is commonly called, *The Golden Rule*; and indeed it might be so termed; for as Gold transcends all other Mettals, so doth this Rule all others in Arithmetick."[4] The term continued in use in England until the end of the 18th century at least,[5] perhaps being abandoned because of its use in the Church.

[1] Thus Pacioli (1494) calls it the "regula trium rerum la regola ditta dl .3. ouer de le .3. cose," and "la regola del .3."; the Treviso arithmetic (1478), "La regula de le tre cose"; Pellos (1492), the "Regula de tres causas." Chuquet (1484) remarks, "La rigle de troys est de grant recōmandacion. . . . La rigle de troys est ansi appellee pource quelle Requiert tousiours troys nombres"; Grammateus (1518) speaks of the "Regula de tre in gantzē" and "in prüchen," and Rudolff (1526) of "Die Regel de Tri," a term often abridged by German writers into "Regeldetri," as in the work of Licht (1500). Klos, the Polish writer of 1538, also calls it the "Regula detri."

[2] Thus Petzensteiner (1483): "Vns habñ die meyster der freyn kunst vō d' zal ein regel gefundē die heist gulden regel Dauo das sie so kospar vnd nucz ist. . . . Sie wirdet auch geneñet regula d' tre nach welsischer [*i.e.*, Italian] zungen. . . . Sie hat auch vile āder namē"; and Köbel (Augsburg edition of 1518) speaks of "Die Gulden Regel (die von den Walen de Try genant wirt)." In Latin it often appears as *regula aurea*. The Swedish savant, Peder Månsson, writing in Rome *c.* 1515, speaks of the rule "quam nonnulli regulam auream dixere: Itali vero regulam de tri" (see *Bibl. Math.*, II (2), 17).

The French writers used the same expression. Thus Peletier (1549): "La Reigle de Trois . . . vulgairement ansi dite. . . . Les ancients l'ont appellee la Reigle d'or: parce que l'invention en est tres ingenieuse, & l'usage d'icelle infini" (1607 ed., p. 68).

[3] Thus Ciacchi (1675): "La Regola del Tre così chiamata da' Practici vulgari, e da' Mattematici regola d' oro, o pure delle quattro proporzioni è principalissima, ed apporta vn' inestimabile benefizio, ed vna gran comodità a' Mercanti." *Regole generali d' abbaco*, p. 121 (Florence, 1675).

[4] See the tenth edition, 1672, p. 87. This simile was a common one with writers; thus Petzensteiner (1483): "als golt vbertrifft alle ander metall."

Vitalis (Geronimo Vitale), in his *Lexicon Mathematicvm*, p. 748 (Rome, 1690), says: "Quare meritò *Aurea* appellata est; namque plus auro valet: & non Arithmeticis modo, Geometris . . . necessaria est; sed & vniuerso hominum generi, in commercijs ineundis, . . ."

[5] *E.g.*, the 1771 edition of Ward's *Young Mathematician's Guide* (p. 85) speaks "of Proportion Disjunct; commonly called the Golden Rule."

Regula de Tri
oder gulden Regul.

Egula de Tri MERCATORum/
Genannt aurea proportionum
Darumb das sie gar bequentlich
Jm Kauff/begert drei dmg/namlich
Den Kauff/das Werth/die Frag zum dritten/
Das Werth soll stet stehen in der mitten/
Der Kauff vornen/die Frag dahinden/
Wilt du die Frag vnd Facit finden/
Hinden/vornen gleich Namen richt/
Die klein die groß allzeit zerbricht/
Mult plicier die hinder Zahl
Mit der mitleren allemahl/
Theil sProduckt mit dem vordern ab/
So kombt dir dein Frag vnd Auffgab.
Verkehrd Regul wilt dein Prob finden/
Was erstlich gstanden ist dahinden
Muß vornen/sforder hinden stehen/
S Facit muß in die mitten gehen/
Die nuler Zahl rauß kommen soll/
Kombt sie/so hasts getroffen wol.
rc.

RULE OF THREE, OR THE GOLDEN RULE, IN VERSE

From Lautenschlager's arithmetic (1598)

The Merchants' Rule. Its commercial uses also gave to the Rule of Three the name of the Merchants' Key or Merchants' Rule,[1] and no rule in arithmetic received such elaborate praise as this one which is now practically discarded as a business aid.[2]

A Rule without Reason. The rule was usually stated with no explanation; thus Digges (1572) merely remarked, "Worke by the Rule ensueing. . . . Multiplie the last number by the seconde, and diuide the Product by the first number," and similar statements were made by most other early arithmeticians, occasionally in verse.[3]

The arrangement of the terms was the same as in the early Hindu works, the first and third being alike. As Digges expressed it, "In the placing of the three numbers this must be observed, that the first and third be of one Denomination."[4] This custom shows how completely these writers failed to recognize the relation between the Rule of Three and proportion.

---

[1] Thus Licht (1500, fol. 9) says: "Regula Mercatorum. [Q]uam detri. quia de trib⁹. per apocopam appellamus. Regula Aurea docte ac perite ab omib⁹ appellari videt deberi"; Clavius (1583) calls it the "clavis mercatorum"; Peletier (1549) says, "Mesmes, aucuns l'ont nommee la Clef des Marchands" (1607 ed., p. 68); Wentsel (1599) speaks of the "Regvla avrea mercatoria/ regvla de tri, Regvla van dryen/ regle a trois, &c."; and Lautenschlager, in his arithmetic in rime (1598), has, as shown in the facsimile on page 487,

> REGULA DE TRI | ODER GULDEN REGUL.
> REgnla de Tri MERCATORum |
> Genannt aurea proportionum.

Of the various other names, *Schlussrechnung* has continued among the Germans. See R. Just, *Kaufmännisches Rechnen*, I. Teil, p. 75 (Leipzig, 1901).

[2] A few of the hundreds of eulogies given to the rule are as follows: Gemma Frisius (1540), "Res breuis est & facilis, vsus immensus, cum in vsu communi, tum in Geometria ac reliquis artibus Mathematicis" (1563 ed., fol. 18); Adam Riese (1522), "Ist die furnameste vnder allē Regeln" (1550 ed. of *Rechenung*, fols. 13, 59); Clavius (1583; *Opera*, 1611, II, 35), "Primo autem loco sese offert regula illa nunquam satis laudata, quae ob immensam vtilitatem, Aurea dici solet, vel regula Proportionum, propterea quod in quatuor numeris proportionalibus, quorum priores tres noti sunt . . . vnde & regula trium apud vulgus appellata est," showing that he, like Stifel, recognized the relation to proportion; Van der Schuere (1600), "Den Regel van Drien, die van vele ten rechte den Gulden Regel genaemd word/ overmits zyne weerdige behulpzaemheyd in alle andere Regelen" (1634 ed., fol. 12).

[3] *E.g.*, Lautenschlager (1598) and Sfortunati (1534; 1545 ed., fol. 33).

[4] 1579 ed., p. 29.

Once set down, it was the custom to connect the terms by curved lines, as in the following cases:[1]

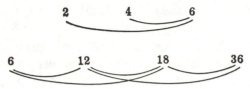

The Arabs, however, used such forms as

I	5
3	15

and

2	4
3	6

to indicate a proportion, paying no attention to the labels on the numbers.[2]

Arrangement of Terms. The rule being purely arbitrary, it became necessary to have this arrangement in the proper order, and the early printed books gave much attention to it. Borghi (1484) gave a whole chapter to this point,[3] and Glareanus (1538) arranged[4] an elaborate scheme to help the student.[5] Later writers, however, recognized that if the rule were to be considered as a case of proportion, it would be necessary to re-arrange the terms so that the first two should be alike. Thus in place of a form like

12 yards————20s.————6 yards,

[1] The first of these is from a 17th century MS. in the author's library; the second is from Werner's *Rechenbuch*, 1561, fol. B 2.

[2] E. Wiedemann, "Über die Wage des Wechselns von Châzinî und über die Lehre von den Proportionen nach al-Bîrûnî," *Sitzungsberichte der Physik.-med. Societät zu Erlangen*, 48. u. 49. Bd., p. 4.

[3] "¶Como le tre cose contenute in detta regola sono ordinate, e quale debbi esser prima, e qual seconda, e qual terza" (1540 ed., fol. 36).

[4] 1543 ed., fol. 20.

[5] He arranged his rule thus:

Sinistra	Medius locus	Dextera
Res empta	Numer⁹ pretij	Numer⁹ qstionis
Diuisor	Multiplicãdus	Multiplicator
2	36	7

as given in Hodder's 17th century work,[1] we find Blassière
(1769)[2] and others of the 18th century using such forms as

Ellen	Ellen	Guld.	Guld.
3	:	36	=	4	:	x

In the old Rule of Three the result was naturally written at
the right, and for this reason the unknown quantity came to be
placed at the right in the commercial problems in proportion.[3]

Inverse Proportion. Of the various special forms of the Rule
of Three the one known as inverse proportion is the simplest. It
results when the ratio of two quantities is equal to the reciprocal
of the ratio of two quantities which seem to correspond to them.
Bhāskara (c. 1150) gives an illustration: "Bullocks which have
plowed four seasons cost four niskas: what will bullocks which
have plowed twelve seasons cost?"[4]

This rule went by such names as the inverse, converse, or
everse Rule of Three.[5] Recorde (c. 1542) used a name that
became quite common in England, remarking: "But there is
a contrarye ordre as thys: That the greater the thyrde summe is
aboue the fyrste, the lesser the fourthe summe is beneth the
second. and this rule you maye call the Backer rule."[6]

---

[1] *Arithmetick*, 10th ed. (1672), p. 89.   [2] 1790 ed., p. 149.
[3] Thus Rabbi ben Ezra (c. 1140) wrote

47 63
7 0

for 47:7 = 63:x, the 0 standing for the unknown. In the translation of certain
Arab works the unknown is placed first, as in x:84 = 12:7, because the Arabs
wrote from right to left.

[4] Taylor's *Lilawati*, p. 42, with spelling as given by Taylor. The result is
written "niska 1, and fraction $\frac{1}{3}$."

[5] Thus Köbel (1514), "Die Regel de Tri verkert/ im Latein Regula conuersa
genant" (1549 ed., fol. 68); Gemma Frisius (1540), "Regvla Trivm Euersa";
Albert (1534), "Regula Detri Conuersa"; Thierfelder (1587), "Regula Conversa/
oder vmbekehrte Regel de Try"; Van der Schuere (1600), "Verkeerden Reghel
van Drien"; Digges (1572), "The Rule of Proportion Inuersed."

[6] 1558 ed., fol. M 6. In some later editions it is called "the Backer or Reverse
Rule" (1646 ed., p. 180). See also Baker (1568; 1580 ed., fol. 46). In French it
appeared as "La regle de troys rebourse" (Trenchant, 1566; 1578 ed., p. 155),
"La Reigle de Trois Reuerse ou Rebourse" (Peletier, 1549; 1607 ed., p. 74), and
"La Regle Arebourse" (Coutereels, Dutch and French work, 1631 ed., p. 204).
It was occasionally called the bastard Rule of Three. Thus Santa-Cruz (1594):
"Exemplo de la regla de tres bastarda."

In this rule it was the custom to leave the terms as in simple proportion but to change the directions for solving. Hylles (1592) gives the rule as follows:

> The Golden rule backward or conuerst,
> Placeth the termes as dooth the rule direct:
> But then it foldes[1] the first two termes rehearst,
> Diuiding the product got by that effect.
> Not by the first, but onely by the third,
> So is the product the fourth at a word.[2]

Compound Proportion. What has been called, for a century or two, by the name of compound proportion originally went by such names as the Rule of Five[3] when five quantities were involved, the Rule of Seven if seven quantities were used, and so on. Bhāskara, for example, gives rules of five, seven, nine, and eleven.[4] Peletier (1549) speaks of Ptolemy as the inventor of the Rule of Six, referring, however, to the proposition in geometry relating to a transversal of three sides of a triangle.[5] The names beyond that for five were rarely used;[6] indeed, all beyond that for three were more commonly called by the general name of Double Rule of Three,[7] Compound Rule of

---

[1] An interesting translation of *plicare* as found in *multiplicare* (to manifold).

[2] *Arithmeticke*, 1600 ed., fol. 135.

[3] It appears in Bhāskara (*c.* 1150) as *pancha-rásica* (five-rule). See Taylor's translation, p. 43; but the spellings in Colebrooke's translation, p. 37, are here followed. In Europe we find such names as "Regvle de cinqve parte" (Ortega, 1512; 1515 ed.); "Regula Quinque/oder zwyfache Regel de Try" (Thierfelder, 1587); "Regel von fünffen" (Rudolff, 1526); "Die Regel von fünff zalen" (Köbel, 1514; 1531 ed.); "*Regvla dvplex* Auch *Regula Quinque* genant . . . die zwyfache Regel . . . von fünff Zalen" (Suevus, 1593); "den Zaamengestelden Reegel van Drien. Anders Genaamd Den Reegel van Vyven" (Blassière, 1769); and "Den Regel van Vyven, of anders genaamt den Dobbelen Regel van Drien" (Bartjens, 1792 ed.).

[4] *Pancha-rásica, sapta-rásica, nava-rásica,* and *écádasa-rásica,* as spelled by Colebrooke.

[5] "La Reigle de six Quantités a esté inuentee par Ptolemee" (1607 ed., p. 220). On this see also Cardan's *Practica* (1539) with its "Caput 46. de regula 6. quātitatum."

[6] Thus Ciacchi (1675) says: "Non è molto vsitata da' practici Arrimetici la regola del sette."

[7] As by Recorde (*c.* 1542). The name also appears under such forms as "Regula duplex" (Gemma Frisius, 1540) and "la regle double"(Trenchant, 1566).

Three,[1] Conjoint Rule,[2] Plural Proportion,[3] and, finally, Compound Proportion, a term which became quite general in the 18th century.

Artificial Nature of the Problems. The artificial nature of the problems in compound proportion has been evident from the beginning. Thus Mahāvīra (c. 850) gives this case: "He who obtains 20 gems in return for 100 gold pieces of 16 *varṇas* — what [will he obtain] in return for 288 gold pieces of 10 *varṇas*?"[4] And Bhāskara (c. 1150) has this type: "If eight best, variegated silk scarfs measuring three cubits in breadth and eight in length cost a hundred [nishcas]; say quickly, merchant, if thou understand trade, what a like scarf three and a half cubits long and half a cubit wide will cost."[5]

Practice. There appears in certain English arithmetics of the present day a chapter on Practice, a kind of modification of the Rule of Three. In the manuscripts of the later Middle Ages and in the early printed books of Italy the word is used to mean simply commercial arithmetic in general, whence possibly the origin of our phrase "commercial practice" today.[6] When northern writers of the 16th century spoke of Italian practice they usually referred merely to Italian commercial arithmetic in general.[7] In the 17th century the Dutch writers generally used the term *Practica* (*Practijcke*) to mean that part of arith-

---

[1] "Regula Trivm Composita" (Clavius, 1583); "The Golden Rule compound" (Recorde, c. 1542; 1646 ed., p. 195); "Den menichvuldigen Regel" (Coutereels, 1631 ed., p. 213); "Gecomposeerden dubbelden Regel" (Houck, 1676, with a distinction between this and the mere "Dubbelden Regel" and "Regel van Conjuncte").

[2] "Den Versamelden Reghel" or "La Regle Conjoincte" (Coutereels, 1631 ed., p. 219).

[3] "The Double Rule of Three . . . Under this Rule is comprehended divers Rules of Plural Proportion" (Hodder, *Arithmetick*, 10th ed., 1672, p. 131).

[4] English translation, p. 91.

[5] Colebrooke's translation, p. 37.

[6] So Tartaglia (1556) has a chapter "Della Prattica Fiorentina" for Florentine commercial arithmetic, and Riese (1522) has "Rechenung . . . die practica genandt."

[7] So Stifel (1544): "Praxis Italica Praxis illa quam ab Italis ad nos devolutam esse arbitramur." Even as late as 1714 there was a chapter in Starcken's arithmetic on "Die Italianische Practiça/ oder Kurtze Handels-Rechnung."

metic relating to financial problems,[1] and they also used the expression "Italian Practice," as in the work by Wentsel (1599) of which part of the title is here shown in facsimile. The term "Welsh practice" had a similar meaning, the word "Welsh" ("Welsch") signifying foreign.[2] This expression is often found in the German arithmetics of the 16th century.

T.' FONDAMENT
Uan Arithmetica : mette Jta=
liaenfche Pzactijck / midtfgaders d'aller
nootwendichste stucken van den Reghel van
Jnterest.
Beydes in Nederdupts ende in Franchois/
met redelicke ouereenstemminghe ofte
Concozdantien.
Alles
Doo₂ MARTINVM VVENCESLAVM,
AQVISGRANENSEM.

WELSH OR ITALIAN PRACTICE

Part of the title-page of the arithmetic of Wentsel (Wenceslaus) (1599)

In England practice came to mean that part of commercial arithmetic in which short processes were used. Baker (1568) mentions it in these words:

Some there be, whiche doe call these rules of practise, breefe rules: for that by them many questions may bee done with quicker expedition, then by the Rule of three. There be others which call them the small multiplication, for because that the product is always lesse in quantity, than the number whiche is to be multiplyed.

---

[1] Thus Eversdyck's edition of Coutereels (1658 ed., p. 91), Stockmans (1676 ed., p. 173), Van der Schuere (1600, fol. 49), and Mots (1640, fol. G 5).

[2] Modern German *wälsch*, foreign; particularly Gallic, Roman, Italian. So we find Rudolff (1526) speaking of "Practica oder Wellisch Rechnung" and Helmreich (1561) having a chapter on "De Welsche Practica oder Rechnung." Dutch writers commonly used "foreign" instead of "Welsch," and so we often find chapters on "Buyten-lantsche Rekeninghe."

This rather indefinite statement gave place to clearer definitions as time went on, and Greenwood (1729) speaks of practice as follows:

THIS *Rule* is a contraction or rather an Improvement of the *Rule of Three*; and performs all those Cases, where *Unity* is the *First Term*; with such Expedition, and Ease, that it is, in an extraordinary manner, fitted to the *Practice* of Trade, and Merchandise; and from thence receives its *Name*.

A single example from Tartaglia (1556) will show the resemblance of the Italian solution to that which was called by American arithmeticians the "unitary method," especially in the 19th century:

If 1 pound of silk costs 9 *lire* 18 *soldi*, how much will 8 ounces cost?[1]

The solution is substantially as follows:

> 1 lb. costs 9 *lire* 18 *soldi*,
> 4 oz. cost $\frac{1}{3}$ of this, or 3 *lire* 6 *soldi*,
> 8 oz. cost twice this, or 6 *lire* 12 *soldi*.

11. SERIES

Kinds of Series. Since the number of ways in which we may have a sequence of terms developing according to some kind of law is limitless, like the number of laws which may be chosen, there may be as many kinds of series or progressions[2] as we wish. The number to which any serious attention has been paid in the development of mathematics, however, is small. The arithmetic and geometric series first attracted attention, after which the Greeks brought into prominence the harmonic series. These three were the ones chiefly studied by the ancients. Boethius (*c.* 510) tells us that the early Greek writers knew these three, but that later arithmeticians had suggested three others which had no specific names.[3]

---

[1] 1 lb. = 12 oz., 1 *lira* = 20 *soldi*.

[2] For our purposes we shall not distinguish at present between these terms.

[3] "Vocantur autē quarta: quinta: vel sexta" (*Arithmetica*, 1st ed., 1488, II, cap. 41; ed. Friedlein, p. 139).

Occasionally some special kind of series is mentioned, as when Stifel speaks of the "astronomical progression" $1, \frac{1}{60}, \frac{1}{3600},$ $\cdots,$[1] one of the few instances of a decreasing series in the early European books.

Most of the Hindu writers used only two elementary series, but Brahmagupta (c. 628), Mahāvīra (c. 850),[2] and Bhāskara (c. 1150) all considered the cases of the sums of squares and cubes.[3] The Arab[4] and Jewish[5] writers also gave some attention to these several types.

Medieval Treatment of Series. In the medieval works a series was generally considered as ascending, although descending series had been used by Ahmes, Archimedes, and certain Chinese writers long before the time in which these works were written.[6] The same custom was followed by the early Renaissance writers.[7]

Somewhat better known than the classification of series as arithmetic, geometric, and harmonic, at least before the 17th century, was the classification into natural, nonnatural, continuous, and discontinuous, these terms being used rather loosely

---

[1] *Arithmetica Integra,* 1544, fol. 64, the name being *astronomica progressio.* It is simply the natural series of "astronomical fractions."

[2] English translation, p. 170. His rule for the sum of the squares is substantially

$$s = n\left\{\left[\frac{(2n-1)d^2}{6} + ad\right](n-1) + a^2\right\}.$$

[3] Colebrooke's translation, p. 52. Bhāskara remarks: "Former authors have stated that the sum of the cubes of the terms one, &c. is equal to the square of the summation"; that is, $\Sigma n^3 = (\Sigma n)^2$. (Taylor's translation, p. 60.)

[4] *E.g.,* al-Ḥaṣṣâr (c. 1175?); see *Bibl. Math.,* II (3), 32. They are also given by al-Qalaṣadî (c. 1475); see Boncompagni's *Bullettino,* XIII, 277.

[5] *E.g.,* Rabbi ben Ezra (c. 1140), in the *Sefer ha-Mispar,* Silberberg translation, p. 120.

[6] So Fibonacci (1202) says: ". . . colligere numeros quotcumque ascendentes ab ipso dato numero equaliter, ut per ascensionem unitatis, uel binarii, uel ternarii . . ." (*Liber Abaci,* p. 166 (fol. 70, *r.*)).

[7] Thus Stifel, in his edition of Rudolff's *Coss* (1553): "Es ist aber Progressio (eygentlich zu reden nach der Arithmetica) ein ordnung vieler zalen so nach einander auffsteygen oder absteygen nach eyner rechten richtigen Regel" (fol. 7, *v.*). So Trenchant (1566) states definitely that the terms must increase, and Chuquet (1484) says: "Progression est certaine ordonnance de nombre par laquelle le premier est surmonte du second dautant que le second est surmonte du tiers et ꝑ sequēment les ault's se plus en ya" (fol. 20, *r.*).

by early writers. For example, the series 1, 2, 3, . . . was called a natural series,[1] from which we have the expression "natural series of numbers."[2] A discontinuous, or intercised,[3] progression was one in which the difference was not unity.[4]

Name for Series. The Greek name for a series, as used first by the early Pythagoreans, was ἔκθεσις (ek'thesis),[5] literally a selling out, and the name for a term of the series was ὅρος (hor'os),[6] literally a boundary. Boethius (c. 510), like the other Latin writers, used the word progressio,[7] and this was generally the custom until modern times.

The Teutonic writers followed their usual plan of avoiding international names based upon the Latin, and so we find various terms used by the Dutch[8] and German[9] mathematicians.

[1] Thus Chuquet (1484): "Et doit on sauoir que progression se fait en plusieurs et diuerses manieres. Car aulcunesfoiz elle cōmance a .1. et progredyst par .1. cōme .1. 2. 3. 4. ꝯc. telle est appellee par les anciens progression naturelle ou continue pgression" (fol. 20, r.). Similarly, Pellos (1492) speaks of " pgression natural" (1, 2, 3, . . .), " pgression nō natural" (1, 3, 5, . . .), and " pgression ni part natural ꝯ ni part nō natural" (8, 9, 10, . . .). Van der Schuere (1600), however, calls any series like 1, 2, 3, . . . or 1, 3, 5, . . . a "natuerlikke overtredinghe /oft Aritmetische Progressio," speaking of a geometric series as "onnatuerlikke overtredinghe."

[2] This is found in Stifel (1544): "naturalis numerorum Progressio, est Progressio Arithmetica progrediens ab unitate per binarium ad reliquos numeros secundum differentiam unitatis. ut

$$1. \overset{1.}{\cdot} 2. \overset{1.}{\cdot} 3. \overset{1.}{\cdot} 4. \overset{1.}{\cdot} 5. \ . \ . \ ."$$

He also speaks "de Progressione naturali numerorum imparium," viz., 1, 3, 5, 7, . . . (Arithmetica Integra, fols. 20, 21).

[3] Intercissa (Huswirt, 1501), int'cise (Chuquet, 1484), vnderschnitten (Köbel, 1514).

[4] "Alcune cōmāce a .1. mais el progredist par aultre nombre que .1. cōe 1. 3. 5. ꝯc. ou .1. 4. 7.ꝯc. et est ceste appellee Int'cise progression ou prog'ssion discontinuee" (Chuquet (1484), fol. 20, r.; Boncompagni's Bullettino, XIII, 617). Santa-Cruz (1594) says that any progression "començando de la vnidad dicha continua." [5] A word also meaning exhibition or exposition.

[6] A word also meaning a limit, marking stone, rule, standard, or boundary between two objects. [7] Ed. Friedlein, I, pp. 9, 10, et passim.

[8] Dutch writers of the 16th century used progressio together with such terms as overtredinghe (stepping over) and opklimminge (ascending; literally, upclimbing).

[9] Although modern writers use Reihe, Köbel (1514), for example, says : "Die acht species ist Progressio zu Latein/vnnd ist Fürzelen geteutscht," in later German Fürzählung. The terms Aufsteigung, Fortgehung, Reihe, and Progression are also used.

The change to the name "series" seems to have been due to writers of the 17th century. James Gregory, for example, writing in 1671, speaks of "infinite serieses," and it was in connection with infinite sequences that it was at first used by the British algebraists. Even as late as the 1693 edition of his algebra, however, Wallis used the expression "infinite progressions" for infinite series.

Extent of Treatment. Although series was commonly looked upon as one of the fundamental operations,[1] it was rarely accorded much attention in the early printed books. Tzwivel (1505), for example, gives only 32 lines to both arithmetic and geometric progressions, including all definitions and rules; while Huswirt (1501) allows only one page and Digges (1572) only two pages to the subject.

Nearly all the early writers limited the work to finding the sum of the series,[2] although a few gave a rule for finding the last term of an arithmetic or a geometric series. With these writers there was no attempt to justify the rule, the mere statement sufficing. It was only through the influence of a better algebraic symbolism in the 17th century that the various cases could easily be discussed and the development of rules for all these cases made simple.

Relation to Proportion. The ancient writers commonly connected progression with proportion, or rather with proportionality, to use a name which, as already stated, was at one time popular; and they applied the names "arithmetic," "geometric," and "harmonic" to each. Some of the early printed books call attention to this relation, saying that a proportion is merely a progression of four terms.[3]

---

[1] Thus Pacioli (1494): "la sexta e penultīa specie dilla pratica d' arithcᵃ. laq̇le e chiamata pgressiōe" (*Sūma*, fol. 37).

[2] Johann Albert (1534) distinctly states that this is the sole purpose of the work: "Progredirn leret/wie man viel zaln (welche nach natŭrlicher ordnung oder durch gleiche mittel/nach einander folgen) in eine Summa/auffs kŭrtzest vnd behendest bringen sol" (1561 ed., fol. E 4). See also Treutlein, *Abhandlungen*, I, 60.

[3] Thus Trenchant (1566): ". . . car Progression n'est qu'vne continuation des termes d'vne proportion" (1578 ed., p. 274).

Arithmetic Series. The first definite trace that we have of an arithmetic series as such is in the Ahmes Papyrus (*c.* 1550 B.C.), where two problems are given involving such a sequence. The first[1] of the problems is as follows: "Divide 100 loaves among five persons in such a way that the number of loaves which the first two receive shall be equal to one seventh of the number that the last three receive."

The solution shows that an arithmetic progression is understood, in which $n = 5$, $s_5 = 100$, and

$$\frac{(a+4\,d)+(a+3\,d)+(a+2\,d)}{7} = (a+d)+a.$$

Then, by modern methods, $\qquad 2\,d = 11\,a$.

Therefore $\qquad 100 = s = \dfrac{2\,a+4\,d}{2} \cdot 5 = 60\,a$,

whence $\qquad a = 1\frac{2}{3} \qquad$ and $\qquad d = 9\frac{1}{6}$.

Therefore the series is $1\frac{2}{3}$, $10\frac{5}{6}$, 20, $29\frac{1}{6}$, $38\frac{1}{3}$, although the method here given is not the one followed by Ahmes.

The second problem,[2] with its solution as given by Ahmes, reads as follows:

Rule of distributing the difference. If it is said to thee, corn measure 10, among 10 persons, the difference of each person in corn measure is $\frac{1}{8}$. Take the mean of the measures, namely 1. Take 1 from 10, remains 9. Make one half of the difference, namely, $\frac{1}{16}$. Take this 9 times. This gives to thee $\frac{1}{2}\frac{1}{16}$. Add to it the portion of the mean. Then subtract the difference $\frac{1}{8}$ from each portion, [this is in order] to reach the conclusion. Make as shown:

1 $\frac{1}{2}$ $\frac{1}{16}$,   1 $\frac{1}{4}$ $\frac{1}{8}$ $\frac{1}{16}$,   1 $\frac{1}{4}$ $\frac{1}{16}$,   1 $\frac{1}{8}$ $\frac{1}{16}$,   1 $\frac{1}{16}$,

$\frac{1}{2}$ $\frac{1}{4}$ $\frac{1}{8}$ $\frac{1}{16}$, $\frac{1}{2}$ $\frac{1}{4}$ $\frac{1}{16}$, $\frac{1}{2}$ $\frac{1}{8}$ $\frac{1}{16}$, $\frac{1}{2}$ $\frac{1}{16}$, $\frac{1}{4}$ $\frac{1}{8}$ $\frac{1}{16}$.

This may be stated in modern form as follows: Required to divide 10 measures among 10 persons so that each person shall have $\frac{1}{8}$ less than the preceding one.

---

[1] Problem 40 in the Eisenlohr translation, p. 72; Peet, *Rhind Papyrus*, p. 78.

[2] No. 64 of the Eisenlohr translation. The version here given is furnished by Dr. A. B. Chace. For another translation see Peet, *loc. cit.*, p. 107.

That is, $n = 10$, $s_n = 10$, $d = -\frac{1}{8}$, so that

$$s_n = 10 = \frac{2a + (n-1)d}{2} \cdot n = (2a - \tfrac{9}{8}) \cdot 5,$$

whence $a = 1\frac{9}{16}$, and the series is the descending progression

$$1\tfrac{9}{16}, \quad 1\tfrac{7}{16}, \quad 1\tfrac{5}{16} \cdots, \quad \tfrac{9}{16}, \quad \tfrac{7}{16}.$$

Connection with Polygonal Numbers. The Greeks knew the theory of arithmetic series, but they usually treated it in connection with polygonal numbers. For example, the following are the first four triangular numbers:

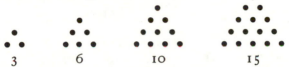

| 3 | 6 | 10 | 15 |

It is evident that each triangular number is the sum of the series $\sum_1^n n$, and the Greeks were well aware of the rule for this summation.[1]

Chinese Work in Series. Nowhere in the very early Chinese works do we find any attempt to sum either an arithmetic or a geometric series.[2] In the *Wu-ts'ao Suan-king*, written about the beginning of the Christian era, or possibly earlier, we find the following problem:

There is a woman who weaves 5 feet the first day, her weaving diminishing day after day until, on the last day, she weaves 1 foot. If she has worked 30 days, how much has she woven in all?

The unknown author then gives this rule:

Add the amounts woven on the first and last days, take half the sum, then multiply by the number of days.

It is interesting to see that this earliest Chinese problem that we have yet found on the subject is, like the second case in the Ahmes Papyrus, one involving a descending series.[3]

---

[1] Heath, *Diophantus*, 2d ed., 247; Gow, *loc. cit.*, p. 103; Nesselmann, *Alg. Griechen*, chap. xi.   [2] Mikami, *China*, p. 18.   [3] Mikami, *China*, p. 41.

In Europe the rule for the sum was naturally the same as in the East, allowing for the difference in language,[1] and was occasionally put in verse for easy memorizing.[2]

The rule for finding any specified term is given by Cardan in his *Practica* (1539) and by Clavius in his *Epitome* (1583).

Geometric Series. The first examples of a geometric series yet found are due to the Babylonians, *c.* 2000 B.C., and tablets containing such examples are still extant.[3] In Egyptian mathematics the first problem on this subject thus far found is in the Ahmes Papyrus (*c.* 1550 B.C.)[4] and reads as follows:

The *one* scale.		Household	7
Once gives	2801	Cats	49
Twice gives	5602	Mice	343
Four times gives	11204	Barley [spelt]	2301 [*sic*]
Together	19607	Hekt measures	16807
		Together	19607

The left-hand column seems to be intended as a deduction of a rule for summing a geometric progression. Probably Ahmes saw that if the ratio is equal to the first term, $s_n = (s_{n-1} + 1)r$. Thus he found the sum of four terms to be 2800, and to this he added 1 and multiplied the result by 7 in order to obtain the sum of five terms. Possibly this is the significance of the expression "The *one* scale." Similarly, in the right-hand column

---

[1] Thus in an old MS. at Munich: "Addir albeg zesam daz erst vnd das leczt, vnd daz selb multiplicir mit dem halben der zal des posicionum" (Curtze, *Bibl. Math.*, IX (2), 113).

[2] Thus Huswirt (1501):

Si primus numerus cum postremo faciat par
Eius per mediū loca singula multiplicabis
Ast impar medium vult multiplicari locorum.

That is, $s = \frac{1}{2}n\,(a + l)$ if $(a + l)$ is odd, but $s = n \cdot \frac{1}{2}(a + l)$ if $(a + l)$ is even. The rule for the two cases goes back at least to Fibonacci (1202). See the Boncompagni edition, I, 166. By the time of Stifel (1544) a single rule answered for both cases.　　　　　　　[3] Hilprecht, *Tablets*, p. 17.

[4] Eisenlohr translation, p. 184, No. 79. The author has used a MS. translation from the hieratic, by Dr. A. B. Chace. On this section consult Tropfke, *Geschichte*, II (1), 315; Peet, *Rhind Papyrus*, p. 121. As in all such cases, reference to Ahmes means to the original from which he copied.

it is quite possible that Ahmes added four terms, then added 1, making the 2801 of the left-hand column, and finally multiplied by 7; but all this is merely conjectural.

The problem suggests the familiar one of the seven cats, although here stated quite differently. There is some doubt as to the word "household," the original word *pir* (*pr*) possibly having a different meaning. The *hekt* (*hekat*) was a measure of capacity. Essentially, therefore, Ahmes uses a rule based upon the formula $s = a \ (r^n - 1)/(r - 1)$. It is interesting to observe that a similar problem is given by Fibonacci (1202) and is solved in much the same way.[1]

The Greeks had rules for summing such a series,[2] and Euclid gave one that may be expressed as follows:

$$\frac{a_{n+1} - a_1}{a_n + a_{n-1} + \cdots + a_1} = \frac{a_2 - a_1}{a_1},$$

which amounts to saying that

$$\frac{ar^n - a}{s_n} = \frac{ar - a}{a},$$

whence would come our common formula

$$s_n = \frac{ar^n - a}{r - 1}.$$

The Hindus showed their interest in geometric series chiefly in the summation problems. The following typical problem is taken from Bhāskara (*c.* 1150):

A person gave a mendicant a couple of cowry shells first; and promised a twofold increase of the alms daily. How many *nishcas* does he give in a month?[3]

---

[1] *Scritti*, I, 311; Tropfke, *Geschichte*, VI (2), 15.

[2] Nesselmann, *Alg. Griechen*, p. 160; Euclid, *Elements*, IX, 35, 36; Heath, *Euclid*, Vol. II, p. 420.

[3] The wording and spelling is that of Colebrooke, *Bhāskara*, §128, p. 55. A *niska* (to take the better spelling) is 16 × 16 × 4 × 20 cowry shells. The cowry shell was then used as a small unit of value. The answer given in the translation is 2,147,483,646 cowry shells = 104,857 *nishcas*, 9 *drammas*, 9 *panas*, 2 *cácinis*, and 6 shells. See also *ibid.*, p. 291.

II

The Arabs apparently obtained the rule for summation from the Greeks, and it appears in an interesting form in the chessboard problem in the works of Albêrûnî (*c.* 1000).

Medieval European Rule. The medieval writers apparently obtained the rule from the Arabs, for it appears in the *Liber Abaci* of Fibonacci (1202).[1] The first modern treatment of the case is found in the *Algorithmus de Integris* (1410) of Prosdocimo de' Beldamandi.[2] Prosdocimo's treatment is as follows:

$$a + ar + ar^2 + \cdots + ar^{n-1} = ar^{n-1} + \frac{ar^{n-1} - a}{r - 1},$$

which is but little more complicated than our ordinary formula.[3] The same rule is given by Peurbach[4] (*c.* 1460). It is given by Chuquet (1484) in the form

$$s = \frac{rar^{n-1} - a}{r - 1},$$

and this is the plan used by Simon Jacob (1560), Clavius (1583), and others. Stifel (1544) gave the rule in the awkward form of

$$s = \frac{(rar^{n-1} - a)a}{ar - a},$$

a method used by Tartaglia[5] (1556), although he ordinarily preferred the one given by Prosdocimo de' Beldamandi.[6]

The ordinary type of puzzle problem in series, running through all the literature of the subject from the time of the Hindus to the 19th century, may be illustrated by the following from Baker (1568): "A Marchante hath solde 15 yeardes of Satten, the firste yarde for 1 s̄, the second 2 s̄, the thyrd 4 s̄, the

---

[1] Boncompagni ed., I, 309, under "de duplicatione scacherii."
[2] First printed at Padua in 1483.
[3] As in all such cases it is to be understood that the rule is stated rhetorically in the original work, the modern algebraic notation being then unknown.
[4] It appears in his *Elementa Arithmetices Algorithmvs de numeris integris,* Vienna, 1492.   [5] *General Trattato,* II, fol. 6, *r.*
[6] *Ibid.*; see the last problem on the same page.

fourth 8s̄, and so increasing by double progression Geometri-
call . . . ," the total cost being then required.[1]

Other problems relate to the buying of orchards in which the
value of the trees increases in geometric series, or to buying a
number of castles on the same plan. Problems of this kind are
mentioned later.

The rule for the sum of n terms is given by Clavius[2] (1583)
and was undoubtedly known to various earlier writers. If we
designate the elements by a, r, n, l, and s, and if any three of
these elements are known, then the others can be found. This
general problem was first stated by Wallis[3] (1657) and was
solved for all cases not requiring logarithms. His formula[4] for
S, one of the earliest stated in a form analogous to the one
used at present, is

$$\frac{VR - A}{R - 1} = S.$$

The first infinite geometric series known to have been
summed is the one given by Archimedes ($c.$ 225 B.C.) in his
quadrature of the parabola.[5] The series summed is

$$1 + \tfrac{1}{4} + (\tfrac{1}{4})^2 + \cdots + (\tfrac{1}{4})^n + \cdots.$$

The general formula for summing the infinite series a, ar,
$ar^2, \cdots, ar^n, \cdots$, where $r < 1$, was given by Vieta ($c.$ 1590).

Harmonic Series. Pythagoras and his school gave much attention to the cultivation of music, not only as a means of
exciting or subduing the passions but as an abstract science.
This led to, or at any rate was connected with, the important

---

[1] 1580 ed., fol. 40. Substantially the same problem is given in Trenchant (1566;
1578 ed., p. 292).

[2] "Detrahatur primus terminus ab vltimo, & reliquus numerus per numerum,
qui vna unitate minor sit, quam denominator, diuidatur. Si enim Quotienti
vltimus terminus, siue maius extremum adiiciatur, componetur summa omnium
terminorum" (*Opera*, 1611, II, 68, of the *Epitome Arithmeticae Practicae*); that
is, $s = (l - a)/(r - 1) + l$.

[3] *Opera*, I, cap. xxxi, p. 158 seq.

[4] " . . . si terminus primus seu minimus dicatur A, maximus V, communis
rationis Exponens R, & progressionis summa S " (p. 158).

[5] Heath, *Archimedes*, chap. vii; Kliem translation, p. 137.

discovery of the relation of the tone to the length of the vibrating string, and hence to the introduction of harmonic proportion,[1] which later writers developed into harmonic series.

Higher Series. The first instances of the use of arithmetic series of higher order were confined to special cases. The series of squares was the earliest to attract attention. Archimedes[2] used geometry to show that

$$3\,[a^2 + (2\,a)^2 + (3\,a)^2 + \cdots + (na)^2]$$
$$= (n+1)\,(na)^2 + a\,(a + 2\,a + 3\,a + \cdots + na).$$

For $a = 1$ this reduces to

$$1^2 + 2^2 + 3^2 + \cdots + n^2 = \tfrac{1}{6}\,n\,(n+1)\,(2\,n+1),$$

which appears substantially in the *Codex Arcerianus* (6th century). It is also found in the Hindu literature as shown by the works of Mahāvīra (*c.* 850).[3]

The sum of the cubes appears in the *Codex Arcerianus* in the form

$$1^3 + 2^3 + 3^3 + \cdots + 10^3 = (\tfrac{1}{2} \cdot 10 \cdot 11)^2.$$

The Hindus had rules for finding this sum, and they appear in the works of Brahmagupta (*c.* 628),[4] Mahāvīra (*c.* 850),[5] and Bhāskara (*c.* 1150).[6]

Among the Arabs similar rules are found, as in the works of al-Karkhî (*c.* 1020),[7] where

$$\sum_1^{10} n^2 = (1 + 10) \cdot 10 \cdot (\tfrac{10}{3} + \tfrac{1}{6}) = 385$$

and

$$\sum_1^{10} n^3 = \left(\sum_1^{10} n\right)^2.$$

[1] T. Gomperz, *Les penseurs de la Grèce*, p. 112 (Lausanne, 1904); H. Hankel, *Geschichte*, p. 105; Gow, *Greek Math.*, p. 68.
[2] See Tropfke, *Geschichte*, II (1), 318, on the entire topic. On this point see the Heiberg edition of *Archimedes*, II (1), 34.
[3] English translation, p. 170.   [4] Colebrooke's translation, p. 293.
[5] P. 171.   [6] Colebrooke's translation, p. 53.
[7] See Woepcke's translation of the *Fakhrî*, pp. 60, 61.

Fibonacci[1] (1220) and various other medieval scholars gave the same treatment of the subject. In the *Liber Quadratorum*[2] (1225) Fibonacci also gave the related forms

$$12(1^2 + 3^2 + 5^2 + \cdots + n^2) = n(n+2)(2n+2) \text{ when } n \text{ is odd}$$

and

$$12(2^2 + 4^2 + 6^2 + \cdots + n^2) = n(n+2)(2n+2) \text{ when } n \text{ is even.}$$

That the sum of the cubes may be found by adding the odd numbers is apparent from the following relations:

$$1^3 = 1,$$
$$2^3 = 3 + 5,$$
$$3^3 = 7 + 9 + 11,$$

and so on. This method of finding $\sum_1^n n^3$ was known to Nicomachus (*c.* 100). The general formula

$$1^3 + 2^3 + 3^3 + \cdots + n^3 = [\tfrac{1}{2} n(n+1)]^2$$

appears in substance in Pacioli's *Sūma*[3] (1494), but was already known.

A rule for summing the fourth powers, which may be expressed by

$$\sum_1^n n^4 = \left(\frac{\sum n - 1}{5} + \sum n\right) \sum n^2,$$

appears in the *Key of Computation* of al-Kashî (*c.* 1430).[4]

Bernoulli Numbers. The case of $\sum n^m$ attracted attention in the 17th century, but the rule is first found in the *Ars Conjectandi*[5] (1713) of Jacques Bernoulli and involves what Euler[6]

---

[1] *Scritti*, I, 167 (fol. 70, *v.*).   [2] *Scritti*, II, 263, 264.   [3] Fol. 44, *r.*, l. 29.
[4] The *Miftâh al-ḥisâb* of Jemshîd ibn Mes'ûd ibn Maḥmud Giyât̲ ed-dîn al-Kashî (died *c.* 1436).
[5] II, cap. 3, p. 97; Tropfke, *Geschichte*, VI (2), 24.
[6] *Institutiones calculi differentialis*, II, § 122 (Petrograd, 1755). Euler's words are "ab inventore Jacobo Bernoulli vocari solent Bernoulliani."

designated as the "Bernoulli Numbers." These numbers (A, B, C, D) appear in the following summation of powers as given by Bernoulli:[1]

$$\int n^c \infty \frac{1}{c+1} n^{c+1} + \frac{1}{2} n^c + \frac{c}{2} A n^{c-1} + \frac{c \cdot c - 1 \cdot c - 2}{2 \cdot 3 \cdot 4} B n^{c-3}$$

$$+ \frac{c \cdot c - 1 \cdot c - 2 \cdot c - 3 \cdot c - 4}{2 \cdot 3 \cdot 4 \cdot 5 \cdot 6} C n^{c-5}$$

$$+ \frac{c \cdot c - 1 \cdot c - 2 \cdot c - 3 \cdot c - 4 \cdot c - 5 \cdot c - 6}{2 \cdot 3 \cdot 4 \cdot 5 \cdot 6 \cdot 7 \cdot 8} D n^{c-7},$$

where $\quad A \infty \frac{1}{6}, \quad B \infty -\frac{1}{30}, \quad C \infty \frac{1}{42}, \quad D \infty -\frac{1}{30},$

and where ∞ expresses equality; and the method of deriving these values is also given.

Revival of Infinite Series. The interest in the infinitesimal as an element in analysis, which manifested itself about the beginning of the 17th century, carried with it the notion of an infinite number of elements. Partly, no doubt, on this account the study of series with an infinite number of terms, already known to the Greeks, was revived, and the idea of products with an infinite number of factors was suggested.

The first of these products of any special interest has already[2] been mentioned as due to Vieta (1593). It may be expressed in modern form[3] as

$$\frac{2}{\pi} = \sqrt{\frac{1}{2}} \cdot \sqrt{\frac{1}{2} + \frac{1}{2}\sqrt{\frac{1}{2}}} \cdot \sqrt{\frac{1}{2} + \frac{1}{2}\sqrt{\frac{1}{2} + \frac{1}{2}\sqrt{\frac{1}{2}}}} \cdots,$$

and this, with others of the same nature, has already been considered in this work.[4]

There are three general periods in the later development of infinite series:[5] (1) the period of Newton and Leibniz,—that

---

[1] *Ars Conjectandi*, p. 97.　　　　　　　　　　　[2] Vol. I, p. '312.

[3] See the Van Schooten edition of Vieta's works, p. 300.

[4] For logarithmic series, see page 513 and Volume I, page 434.

[5] R. Reiff, *Geschichte der unendlichen Reihen*, Tübingen, 1889. See also, for comparison, H. Wieleitner, "Zur Geschichte der unendlichen Reihen im christlichen Mittelalter," *Bibl. Math.*, XIV (3), 150; Tropfke, *Geschichte*, VI (2), 54.

of its introduction; (2) the period of Euler,—the formal stage; (3) the modern period,—that of the scientific investigation of the validity of infinite series. This third period, which may be designated as the critical one, began in 1812 with the publication of Gauss's celebrated memoir on the series

$$1 + \frac{\alpha \cdot \beta}{1 \cdot \gamma} x + \frac{\alpha \cdot (\alpha + 1) \cdot \beta \cdot (\beta + 1)}{1 \cdot 2 \cdot \gamma \cdot (\gamma + 1)} x^2 + \cdots.$$

Euler had already considered this series, but Gauss was the first to master it, and under the name of "hypergeometric series," due to Pfaff (1765–1825), it has since occupied the attention of a large number of mathematicians. The particular series is not so important as the standard of criticism which Gauss set up, embodying the simpler criteria of convergence and the questions of remainders and the range of convergence.

Cauchy (1821) took up the study of infinite series and elaborated the theory of convergence which James Gregory (1668) had already begun and to which Maclaurin, Euler, and Gauss had made noteworthy contributions.[1] The term "convergent series" is due to Gregory (1668) and the term "divergent series" to Nicolas (I) Bernoulli (1713).[2]

Abel (1826) gave careful study to the series

$$1 + \frac{m}{1} x + \frac{m(m-1)}{2!} x^2 + \cdots,$$

correcting certain of Cauchy's conclusions and giving a scientific summation of the series for complex values of m and x.

Binomial Theorem. The development of $(a + b)^n$ for any integral value of n, or at least a device for finding the coefficients, was known in the East long before it appeared in Europe. The case of $n = 2$ was also known to Euclid (c. 300 B.C.),[3] but any evidence of the generalization of the law for other values

---

[1] On the history of criteria of convergence see F. Cajori, in the *Bulletin of the New York Math. Soc.*, II, 1; see also III, 186.

[2] F. Cajori, *Bulletin of the Amer. Math. Soc.*, XXIX, 55.

[3] *Elements*, II. For a summary of his work on algebraic identities see Nesselmann, *Alg. Griechen*, p. 154.

of *n* first appears, so far as we know, in the algebra of Omar Khayyam (*c.* 1100). This writer did not give the law, but he asserted that he could find the fourth, fifth, sixth, and higher roots of numbers by a law that he had discovered and which did not depend upon geometric figures.[1] He states that this law was set forth by him in another work, but of this work there seems to be no copy extant.

Pascal Triangle. In one of the works of Chu Shï-kié (1303), the greatest of the Chinese algebraists of his time, the triangular arrangement of the coefficients is given in the following form,

$$
\begin{array}{ccccccc}
 & & & 1 & & & \\
 & & 1 & & 1 & & \\
 & & 1 & 2 & 1 & & \\
 & 1 & 3 & & 3 & 1 & \\
 1 & 4 & & 6 & & 4 & 1 \\
1 & 5 & 10 & & 10 & 5 & 1 \\
 & & \cdot & \cdot & \cdot & \cdot & \\
\end{array}
$$

a form now commonly known as the Pascal Triangle.[2]

This triangular array first appeared in print on the title-page of the arithmetic of Apianus (1527), as shown in the illustration on page 509.[3] In the form

$$
\begin{array}{ccccc}
1 & 2 & 1 & & \\
1 & 3 & 3 & 1 & \\
1 & 4 & 6 & 4 & 1 \\
1 & 5 & 10 & 10 & 5 & 1 \\
\end{array}
$$

[1] "J'ai composé un ouvrage sur la démonstration de l'exactitude de ces méthodes. . . . J'en ai, en outre, augmenté les espèces, c'est-à-dire que j'ai enseigné à trouver les côtés du carré-carré, du quadrato-cube, du cubo-cube, etc., à une étendue quelconque, ce qu'on n'avait pas fait précédemment. Les démonstrations que j'ai données à cette occasion ne sont que des démonstrations arithmétiques." Translated by F. Woepcke, *L'Algèbre d'Omar Alkhâyyamî,* p. 13 (Paris, 1851).

[2] Mikami, *China,* p. 106.

[3] On the general subject see H. Bosmans, "Note historique sur le Triangle arithmétique, dit de Pascal," *Annales de la Société scientifique de Bruxelles,* XXXI, October, 1906; Tropfke, *Geschichte,* VI (2), 37.

PASCAL TRIANGLE AS FIRST PRINTED, 1527

Title-page of the arithmetic of Petrus Apianus, Ingolstadt, 1527, more than a century before Pascal investigated the properties of the triangle

it is first found in Stifel's *Arithmetica Integra* (1544), appearing a year later in the *De Nvmeris et Diversis Rationibvs* of Scheubel (1545). It also appears in the various editions of Peletier's arithmetic (Poitiers, 1549 and later). Tartaglia (1556) gave it as his own invention,[1] and soon after his time it became common property. Bombelli (1572), for example, gave the coefficients for all powers of $a + b$ up to the seventh, using them in finding corresponding roots,[2] and Oughtred (1631) gave them up to the tenth power.[3] The triangular array was investigated by Pascal (1654) under a new form, substantially as follows:[4]

	1	2	3	4	5	6	7	8	9	10
1	1	1	1	1	1	1	1	1	1	1
2	1	2	3	4	5	6	7	8	9	
3	1	3	6	10	15	21	28	36		Rangs Paralleles
4	1	4	10	20	35	56	84			
5	1	5	15	35	70	126				
6	1	6	21	56	126					
7	1	7	28	84						
8	1	8	36							
9	1	9								
10	1									

Rangs Perpendiculaires

He made numerous discoveries relating to this array and set them forth in his *Traité du triangle arithmétique*,[5] published

---

[1] *General Trattato*, II, fols. 69, *v*.; 71, *v*. (Venice, 1556).   [2] *Algebra*, p. 64.
[3] F. Cajori, *William Oughtred*, p. 29 (Chicago, 1916).
[4] This is from the plate in Pascal's *Œuvres*, Vol. V (Paris, 1819). The description is given on pages 1–56. In the original there are diagonals in the above figure. See also Tropfke, *Geschichte*, VI (2), 37.
[5] "Le nombre de chaque cellule est égal à celui de la cellule qui la précède dans son rang perpendiculaire, plus à celui de la cellule qui la précède dans son rang parallèle" (*Œuvres*, V, 3) (Paris, 1819).

posthumously in 1665, and among these was essentially our present Binomial Theorem for positive integral exponents. After this time the triangular array was common in the East as well as in the West.

Generalization of the Binomial Theorem. The generalization of the binomial theorem for negative and fractional values of n is due to Newton, who set it forth in letters which he wrote to Oldenburg on June 13, 1676, and October 24, 1676.[1]
The proof of the Binomial Theorem was slowly developed by later writers. Among those who contributed to a satisfactory demonstration were Maclaurin[2] for rational values of n, Giovanni Francesco M. M. Salvemini (de Castillon)[3] and Kästner (1745)[4] for integral values,

PASCAL TRIANGLE IN JAPAN

From Murai Chūzen's *Sampō Dōshi-mon* (1781), showing also the sangi forms of the numerals

Euler[5] (1774) for fractional exponents, and Abel[6] (c. 1825) for general values of n, taking n as a complex number.

[1] See *Commercium Epistolicum*, London, 1712; 1725 ed., pp. 131, 142. In his letter of October 24 he proceeds from $(1 - x^2)^{\frac{0}{2}}$, $(1 - x^2)^{\frac{2}{2}}$, $(1 - x^2)^{\frac{4}{2}}$, $\cdots$ to $(1 - x^2)^{\frac{1}{2}}$ and $(1 - x^2)^{\frac{3}{2}}$, "vel generaliter $\overline{1 - xx}\,^m$," and finds, for example, that "$\overline{1 - xx}\,|^{\frac{1}{2}}$ valeret $1 - \frac{1}{2}x^2 - \frac{1}{8}x^4 - \frac{1}{16}x^6$ &c." On the doubtful assertion that Pascal may have anticipated this discovery, see G. Eneström, *Bibl. Math.*, V (3), 72.

[2] *Treatise of Fluxions*, p. 607 (1742).

[3] Born at Castiglione, 1708; died 1791. See *Phil. Trans.*, XLII (1742), 91. He used the theory of combinations. See page 326.   [4] Cantor, *Geschichte*, III, 660.

[5] *Novi comment. Petrop.*, XIX, 103; see Tropfke, *Geschichte*, II (1), 331. See also the English translation of Euler's *Algebra*, I, 172, 177 (London, 2d ed., 1810).

[6] The article appeared posthumously in Crelle's *Journal*, I (1826), 311. See also Abel's *Œuvres*, I, 219 (Christiania, 1881).

The generalization of the Binomial Theorem into the Polynomial Theorem was due chiefly to Leibniz (1695), Jacques Bernoulli, and De Moivre.[1]

Finite Differences. The treatment of series by the method of finite differences appeared in the 17th century. In 1673 Leibniz wrote to Oldenburg concerning the following scheme of treating the series of cubes:

```
        o     o     o
     6     6     6     6
  6    12    18    24    30
 1    7    19    37    61    91
o    1    8    27    64   125   216
```

He said that John Pell attributed the discovery to Gabriel Mouton, of Lyons.[2]

Taylor's Formula and Maclaurin's Formula. In 1715 Brook Taylor published the formula which bears his name, and which we now express as follows:

$$f(x + h) = f(x) + hf'(x) + \frac{h^2}{2!}f''(x) + \cdots.[3]$$

It was not until 1742 that Colin Maclaurin published the corresponding formula

$$f(x) = f(o) + xf'(o) + \frac{x^2}{2!}f''(o) + \cdots,$$

a relation that is easily derived from the preceding one.[4]

Trigonometric Series. The development of trigonometric functions in series first attracted the attention of mathematicians

---

[1] De Moivre's articles appeared in the *Phil. Trans.*, XIX (1697), 619; XX, 190.

[2] *Commercium Epistolicum*, p. 109 (London, 1712; 1725 ed.). Gabriel Mouton, born at Lyons, 1618; died at Lyons, September 28, 1694. He suggested (1670) a system of measures not unlike the metric system.

[3] *Methodus Incrementorum directa et inversa*, prop. 7 (London, 1715). The series had already been announced by him in 1712.

A Complete System of Fluxions, Edinburgh, 1742.

in the 17th century. To James Gregory (1671) are due the following:[1]

$$x = \tan x - \tfrac{1}{3}\tan^3 x + \tfrac{1}{5}\tan^5 x - \tfrac{1}{7}\tan^7 x + \cdots,$$

$$\tan x = x + \tfrac{1}{3}x^3 + \tfrac{2}{15}x^5 + \tfrac{17}{315}x^7 + \cdots,$$

$$\sec x = 1 + \tfrac{1}{2}x^2 + \tfrac{5}{24}x^4 + \tfrac{61}{720}x^6 + \cdots.$$

He also gave the important series

$$\text{arc } \tan x = x - \tfrac{1}{3}x^3 + \tfrac{1}{5}x^5 - \cdots,$$

but this is easily deduced from the one given above for tan x. Newton[2] gave (c. 1669) the antitrigonometric series for arc sin x, essentially as follows:

$$\text{arc } \sin x = \sin^{-1} x = x + \tfrac{1}{6}x^3 + \tfrac{3}{40}x^5 + \tfrac{5}{112}x^7 + \cdots.$$

Logarithmic Series. The idea of expressing a logarithm by means of a series seems to have originated with Gregory and to have been elaborated by Nicolaus Mercator[3] (1667), who discovered, for a special case at least, the relation

$$\log(1 + a) = a - \tfrac{1}{2}a^2 + \tfrac{1}{3}a^3 - \tfrac{1}{4}a^4 + \cdots,$$

where

$$1 \geqq a > -1.$$

The value of Mercator's and Gregory's contributions was recognized by Wallis in reviews which he wrote of their works.[4]

12. Logarithms

Technical Terms. The word "logarithm"[5] means "ratio number" and was an afterthought with Napier. He first used the expression "artificial number," but before he announced his discovery he adopted the name by which it is now known.[6]

---

[1] These were communicated to Collins in a letter from Gregory. See the *Commercium Epistolicum*, London, 1712; 1725 ed., pp. 98, 210 n.

[2] *Commercium Epistolicum*, pp. 97, 126; Tropfke, *Geschichte*, VI (2), 46.

[3] *Logarithmotechnia sive methodus construendi logarithmos nova*, London, 1668. The theory was worked out the year before.

[4] *Phil. Trans.*, 1668, pp. 640, 753.

[5] From the Greek λόγος (log'os), ratio, + ἀριθμός (arithmos'), number.

[6] This fact is evident from his *Descriptio*, 1619 ed.

Briggs introduced (1624) the word "mantissa." It is a late Latin term of Etruscan origin, originally meaning an addition, a makeweight, or something of minor value, and was written *mantisa*. In the 16th century it came to be written *mantissa* and to mean "appendix,"[1] and in this sense it was probably considered by Briggs. The name also appears in connection with decimals in Wallis's *Algebra* (1685), but it was not commonly used until Euler adopted it in his *Introductio in analysin infinitorum* (1748). Gauss suggested using it for the fractional part of all decimals.[2]

The term "characteristic" was suggested by Briggs (1624) and is used in the 1628 edition of Vlacq.[3]

The characteristic was printed in the early tables, and it was not until well into the 18th century that the custom of printing only the mantissas became generally established.

Napier's Invention. So far as Napier's invention is concerned, Lord Moulton expressed the fact very clearly when he said:[4]

The invention of logarithms came on the world as a bolt from the blue. No previous work had led up to it, foreshadowed it or heralded its arrival. It stands isolated, breaking in upon human thought abruptly without borrowing from the work of other intellects or following known lines of mathematical thought.

Napier worked at least twenty years upon the theory. His idea was to simplify multiplications involving sines, and it was a later thought that included other operations, applying logarithms to numbers in general. He may have been led to his discovery by the relation

$$\sin A \sin B = \tfrac{1}{2}(\cos \overline{A-B} - \cos \overline{A+B}),$$

[1] With this meaning it appeared as late as 1701 in J. C. Sturm, *Mathesis juvenalis*.

[2] "Si fractio communis in decimalem convertitur, seriem figurarum decimalium . . . fractionis *mantissam* vocamus . . ." See E. Hoppe, "Notiz zur Geschichte der Logarithmentafeln," *Mittheilungen der math. Gesellsch. in Hamburg*, IV, 52.

[3] ". . . prima nota versus sinistram, quam Characteristicam appellare poterimus . . ." It again appeared in Mercator's *Logarithmotechnia* (1668).

[4] "Inaugural Address: The Invention of Logarithms," *Napier Tercentenary Memorial Volume*, p. 1 (London, 1915).

for, as Lord Moulton says, in no other way can we "conceive that the man to whom so bold an idea occurred should have so needlessly and so aimlessly restricted himself to sines in his work, instead of regarding it as applicable to numbers generally."

Napier published his *Descriptio*[1] of the table of logarithms in 1614. This was at once translated into English by Edward Wright,[2] but with the logarithms contracted by one figure.

In Napier's time sin ϕ was a line, not a ratio. The radius was called the *sinus totus*, and when this was equal to unity the length of the sine was simply stated as sin ϕ. If r was not unity, the length was r sin ϕ. With this statement we may consider Napier's definition of a logarithm:

The Logarithme therefore of any sine is a number very neerely expressing the line, which increased equally in the meane time, whiles the line of the whole sine decreased proportionally into that sine, both motions being equal-timed, and the beginning equally swift.[3]

From this it follows that the logarithm of the *sinus totus* is zero. Napier saw later that it was better to take log 1 = 0.[4]

Napier then lays down certain laws relating to proportions, which may be stated symbolically as follows:

1. If $a:b = c:d$, then $\log b - \log a = \log d - \log c$.
2. If $a:b = b:c$, then $\log c = 2 \log b - \log a$.
3. If $a:b = b:c$, then $2 \log b = \log a + \log c$.
4. If $a:b = c:d$, then $\log d = \log b + \log c - \log a$.
5. If $a:b = c:d$, then $\log b + \log c = \log a + \log d$.
6. If $a:b = b:c = c:d$, then $3 \log b = 2 \log a + \log d$ and $3 \log c = \log a + 2 \log d$.

---

[1] *Mirifici Logarithmorum Canonis Descriptio*, Edinburgh, 1614.

[2] *A Description of the Admirable Table of Logarithmes*, London, 1616, published after Wright's death.

[3] Wright's translation of the *Descriptio*, pp. 4, 5.

[4] As to the priority of this idea, see G. A. Gibson, "Napier's logarithms and the change to Briggs's logarithms," in the *Napier Tercentenary Memorial Volume*, p. 114 (London, 1915); this volume should be consulted on all details of this kind. See also Dr. Glaisher's article on logarithms in the eleventh edition of the *Encyclopædia Britannica*.

The system was, therefore, designed primarily for trigonometry, but would also have been valuable for purposes of ordinary computation had not a better plan been suggested.

Napier also wrote a work on the construction of a table,[1] which was published posthumously as part of the 1619 edition of the *Descriptio*.

Napier's logarithms are not those of the so-called Napierian, or hyperbolic, system, but are connected with this system by the relation $\log_n a = 10^7 \cdot \log_e 10^7 - 10^7 \cdot \log_e a$. The relation between the sine and its logarithm in Napier's system is

$$\sin \phi = 10^7 \cdot e^{\frac{-\log_n \sin \phi}{10^7}}$$

so that the sine increases as its logarithm decreases.

Briggs's System. Henry Briggs, professor of geometry at Gresham College, London, and afterwards Savilian professor of geometry at Oxford, was one of the first to appreciate the work of Napier. Upon reading the *Descriptio* (1614) he wrote:

Naper, lord of Markinston, hath set my head and hands at work with his new and admirable logarithms. I hope to see him this summer, if it please God; for I never saw a book which pleased me better, and made me more wonder.

He visited Merchiston in 1615 and suggested another base, of which, however, Napier had already been thinking. In Briggs's *Arithmetica Logarithmica* the preface, written by Vlacq, contains the following statement[2] by the author of the work itself:

That these logarithms differ from those which that illustrious man, the Baron of Merchiston published in his *Canon Mirificus* must not surprise you. For I myself, when expounding their doctrine publicly in London to my auditors in Gresham College, remarked that it would be much more convenient that o should be kept for the logarithm of

---

[1] *Mirifici ipsius canonis constructio.*

[2] *Arithmetica Logarithmica sive Logarithmorum Chiliades Triginta* (London, 1624), preface. The original is in Latin; the translation of the statement is from the *Napier Tercentenary Memorial Volume*.

the whole sine (as in the *Canon Mirificus*). . . . And concerning that matter I wrote immediately to the author himself; and as soon as the season of the year and the vacation of my public duties of instruction permitted I journeyed to Edinburgh, where, being most hospitably received by him, I lingered for a whole month. But as we talked over the change in logarithms he said that he had for some time been of the same opinion and had wished to accomplish it. . . . He was of the opinion that . . . o should be the logarithm of unity.

The real value of the proposition made by Briggs at this time was that he considered the values of log $10^n a$ for all values of n. The relation between the two systems as they first stood may be indicated as follows:

Napier, $\log y = r(\log_e r - \log_e y)$, where $r = 10^7$;

Briggs, $\log y = 10^{10}(10 \cdot - \log_{10} y)$;

Napier (later suggestion), $\log y = 10^9 \log_{10} y$.

The first table of logarithms of trigonometric functions to the base 10 was made by Gunter, a colleague of Briggs at Gresham College, and was published in London in 1620.[1]

The Base e. In the 1618 edition of Edward Wright's translation of the *Descriptio* there is printed an appendix, probably written by Oughtred, in which there is the equivalent of the statement that $\log_e 10 = 2.302584$, thus recognizing the base e.

Two years later (1620) John Speidell[2] published his *New Logarithmes*, also using this base. He stated substantially that

$$\log n = 10^{-1} (\text{nap log } 1 - \text{nap log } n),$$

or $$\log n = 10^5 (10 + \log_e 10^{-5} x).$$

Continental Recognition. The same year (1624) that Briggs published his *Arithmetica Logarithmica* Kepler's first table appeared. A year later Wingate's *Arithmetiqve Logarithmiqve* (Paris, 1625) gave the logarithms of numbers from 1 to 1000, together with Gunter's logarithmic sines and tangents.

---

[1] *Canon Triangulorum, sive Tabulae Sinuum et Tangentium.*

[2] See *Napier Tercentenary Memorial Volume*, pp. 132, 221; F. Cajori, *History of Elem. Math.*, p. 164, rev. ed. (N.Y., 1917), and *History of Math.*, p. 153, rev. ed. (N.Y., 1919).

II

Holland was the third Continental country to recognize the work of Napier and Briggs. In 1626 there was published a work[1] by Adriaen Vlacq,[2] assisted by Ezechiel de Decker. In 1628 Vlacq republished Briggs's tables,[3] filling the gap from 20,000 to 90,000. The tables in this work were reprinted in London by George Miller in 1631.[4] It is interesting to note that the next complete edition of Vlacq's tables appeared in China.[5]

In Germany the theory was first made known by Johann Faulhaber[6] (1630).

Logarithms in Arithmetic. By the middle of the seventeenth century, logarithms found their way into elementary arithmetics, as is seen in Hartwell's (1646) edition of Recorde's *Ground of Artes*,[7] where it is said that "for the extraction of all sorts of roots, the table of Logarithmes set forth by M. Briggs are most excellent, and ready." Thereafter they were occasionally found in textbooks of this kind, both in Great Britain and on the Continent.

Forerunners of Bürgi. Napier approached logarithms from the standpoint of geometry, whereas at the present time we approach the subject from the relation $a^m a^n = a^{m+n}$. This relation was known to Archimedes[8] and to various later writers.

More generally, if we take the two series

	o	1	2	3	4	5	6	7
and	1	2	4	8	16	32	64	128,

[1] *Eerste Deel van de Nieuwe Telkonst*, Gouda, 1626. See D. Bierens de Haan in Boncompagni's *Bullettino*, VI, 203, 222; J. W. L. Glaisher, "Notice respecting some new facts in the early history of logarithmic tables," *Philosoph. Mag.*, October, 1872.

[2] Born at Gouda, *c.* 1600; died after 1655. The common Dutch spelling is now Vlack.

[3] *Arithmetica Logarithmica*, Gouda, 1628. It was also published with a French title-page.   [4] *Logarithmicall Arithmetike*, London, 1631.

[5] *Magnus Canon Logarithmorum* . . . *Typis Sinensibus in Aula Pekinensi* . . ., 1721.

[6] *Ingineurs-Schul, erster Theil, darinen durch den Canonem Logarithmicum* . . ., Frankfort, 1630. Faulhaber was born at Ulm, May 5, 1580; died at Ulm, 1635.   [7] Also the editions of 1662 and 1668.

[8] *Opera omnia*, ed. Heiberg, 2d ed., II, 243; Heath, *Archimedes*, p. 230.

the first one being arithmetic and the second one being geometric, we see that the latter may be written as follows:

$$2^0 \quad 2^1 \quad 2^2 \quad 2^3 \quad 2^4 \quad 2^5 \quad 2^6 \quad 2^7.$$

From this it is evident that

$$2^3 \cdot 2^4 = 2^7, \qquad\qquad (2^2)^3 = 2^6,$$
$$2^7 : 2^3 = 2^4, \qquad\qquad (2^4)^{\frac{1}{2}} = 2^2,$$

which are the fundamental laws of logarithms.
Most writers[1] refer to Stifel as the first to set forth these
basal laws, and we shall see that he did set them forth very
clearly; but he was by no means the first to do so, nor did they
first appear even in his century. Probably the best of the statements concerning them which appeared in the 15th century
were those of Chuquet in *Le Triparty en la Science des Nombres*, written in 1484, from which Estienne de la Roche copied
so freely in his *Larismethique* of 1520. Chuquet expressed
very clearly the relations

$$a^m a^n = a^{m+n}$$
and $$(a^m)^n = a^{mn}$$

in connection with the double series to which reference[2] has
been made, calling special attention to the latter law as "a
secret" of proportional numbers.[3]

---

[1] Among them is Kästner, *Geschichte der Mathematik*, I, 119, who has been
generally followed in this matter. See also Th. Müller, *Der Esslinger Mathematiker
Michael Stifel*, Prog., p. 16 (Esslingen, 1897), where the author states: "'Dies ist
das älteste Buch,' sagt Strobel, 'in welchem die Vergleichung des arithmetischen
Reihe mit der geometrischen als der Grund der Logarithmen vorkommt.'"
Much of the work on this topic appeared in the author's paper published in
the *Napier Tercentenary Memorial Volume*, p. 81 (London, 1915).

[2] "Il convient poser plus3s nōbres ,pporcional3 cōmancans a 1. constituez en
ordonnance continuee cōme 1.2.4.8.16.32. &c. ou .1.3.9.27. &c. ⊄ Maintenant con-
uient scauoir que .1. represente et est ou lieu des nombres dōt le̥ denōīa⁰ⁿ est .o./2.
represente et est ou lieu des premiers dont leur denomīacion est .1./4. tient le lieu
des second3 dont leur denomīacion est .2. Et .8. est ou lieu des tiers .16. tient la
place des quartz" (fol. 86, v., of the *Triparty*). This is taken from the copy made
by A. Marre from the original manuscript. Boncompagni published it in the
Bullettino, XIII, 593 seq., fol. 86, v., being on page 740.

[3] "⊄ Semblement qui multiplie .4. qui est nombre second par .8. qui est nombre
tiers montent .32. qui est nombre quint . . . ⊄ En ceste consideration est maĭfeste

It is difficult to say when a plan of this kind first appears in print, because it is usually hinted at before it is stated definitely. Perhaps it is safe, however, to assign it to Rudolff's *Kunstliche rechnung* of 1526, where the double series is given and the multiplication principle is clearly set forth;[1] and inasmuch as this work had great influence on Stifel, who in turn influenced Jacob, Clavius, and Bürgi, it was somewhat epochmaking.

The next writer to refer to the matter was probably Apianus (1527), who followed Rudolff so closely as to be entitled to little credit for what he did.

Following Apianus, the first arithmetician of any standing who seems to have had a vision of the importance of this relation was Gemma Frisius (1540), who gave the law with relation to the double array

3	9	27	81	243	729
0	1	2	3	4	5,

saying that the product of two numbers occupies a place indicated by the sum of their places (3×9 occupying the place indicated by $1 + 2$, or 3), and that the square of a number in the fifth place occupies the 2×5th place.[2]

The first arithmetician to take a long step in advance of Rudolff was Stifel (1544), the commentator (1553) on *Die*

---

vng secret qui est es nombres pporcionalz. Cest que qui multiplie vng nombre pporcional en soy Il en viēt le nombre du double de sa denomīacion come qui m̄ltiplie .8. qui est tiers en soy Il en vient .64. qui est sixᵉ. Et .16. qui est quart multiplie en soy. Il en doit venir 256. qui est huytᵉ. Et qui multiplie .128. qui est le .7ᵉ. pporcional par .512. qui est le 9ᵉ. Il en doit venir 65536. qui est le 16ᵉ " (*ibid.*, p. 741).

[1] "Nun merck wenn du zwo zalen mit einander multiplicirst / wiltu wissen die stat des quocients / addir die zalen der natürlichen ordnung so ob den zweyen mit einander gemultiplicirten zalen gefunden / dʒ collect bericht dich. Als wen ich 8 multiplicir mit 16. muss komen 128. darumb das 3 vnd 4 so vber dem 8 vnnd 16 geschriben zusamen geaddirt 7 machen." He gives several examples, but goes no farther with the law.

[2] "Si enim duos quoscunque ex his numeris inuicem multiplicaueris, productúmque per primum diuiseris, producetur numerus eo loco ponendus, quē duo facta indicabunt . . ." (ed. 1553, fol. 17, r., and note by Peletier (Peletarius), fol. 78, v.). The relation is not so clear as in some of the other texts, on account of the arrangement of the series.

Coss. It is not, however, in this work that the theory is set forth, but in the *Arithmetica Integra* of 1544. Stifel here refers several times to the laws of exponents. At first he uses the series

0	1	2	3	4	5	6	7	8
1	2	4	8	16	32	64	128	256,

distinctly calling the upper numbers *exponents*, and saying that the exponents of the factors are added to produce the exponent of the product and subtracted to produce the exponent of the quotient.[1] Moreover, he expressly lays down four laws, namely, that addition in arithmetic progression corresponds to multiplication in geometric progression, that subtraction corresponds to division, multiplication to the finding of powers, and division to the extracting of roots. Furthermore, Stifel not only set forth the laws for positive exponents but also saw the great importance of considering the negative exponents of the base which he selected, using the series

-3	-2	-1	0	1	2	3	4	5	6
$\frac{1}{8}$	$\frac{1}{4}$	$\frac{1}{2}$	1	2	4	8	16	32	64

and making the significant remark: "I might write a whole book concerning the marvellous things relating to numbers, but I must refrain and leave these things with eyes closed."[2] What these mysteries were we can only conjecture.

---

[1] "Qualicunq3 facit Arithmetica progressio additione, & subtractione, talis facit progressio Geometrica multiplicatione, & diuisione. ut plene ostendi lib. 1. capite de geomet. progres. Vide ergo,

0.	1.	2.	3.	4.	5.	6.	7.	8.
1.	2.	4.	8.	16.	32.	64.	128.	256.

Sicut ex additione (in superiore ordine) 3 ad 5 fiunt 8, sic (in inferiore ordine) ex multiplicatione 8 in 32 fiunt 256. Est autem 3 exponens ipsius octonarij, & 5 est exponens 32 & 8 est exponens numeri 256. Item sicut in ordine superiori, ex subtractione 3 de 7, remanent 4, ita in inferiori ordine ex diuisione 128 per 8, fiunt 16" (fols. 236, 237).

It will be noticed that he speaks of 8 as "exponens numeri 256," and not as the exponent of 2, but this has no significance with respect to the theory.

[2] "Posset hic fere nouus liber integer scribi de mirabilibus numerorum, sed oportet ut me hic subducā, & clausis oculis abeā."

A number of French writers of this period were also aware of the law, and Peletier[1] (1549) stated it clearly for the case of multiplication. Five years later Claude de Boissière elaborated this treatment and spoke of the "marvellous operations" which can be performed by means of the related series. Two years after Boissière's work was published the theory was again given by Forcadel (1565), with a statement that the idea was due to Archimedes, that it was to be found in Euclid, and that Gemma Frisius had written upon it. Ramus recognized its value but added nothing to it or to its possible applications. When, however, Schoner came to write his commentary on the work of Ramus, in 1586, a decided advance was made, for not only did he give the usual series for positive exponents, but, like Stifel, he used the geometric progressions with fractions as well, although, as stated above, not with negative exponents. Further, he used the word "index" where Stifel had used "exponent," and, like this noteworthy writer, gave evidence of an appreciation of the importance of the law. In general the French writers already named (and in the list should also be included the name of Chauvet) paid no attention to any of the laws except that of multiplication, while the German writers, following the lead of Stifel, took the broader view of the theory. This was not always the case, for Sigismund Suevus, a German arithmetician who wrote as late as 1593, did not go beyond the limits set by most of the French arithmeticians; but in general the German writers were in the lead. This is particularly true of Simon Jacob (1565), who followed Stifel closely, recognizing all four laws, and, as is well known, influencing Jobst Bürgi. These writers did not use the general exponents essential to logarithms, but the recognition of the four laws is significant.

[1] The extract here given is from the 1607 edition of *L'Arithmetiqve*, p. 67. In speaking of the series

3	6	12	24	48	96
0	1	2	3	4	5

he says: "Ie sçauoir qui est le nôbre qui eschet au neufieme lieu en ceste Progression Double Ie diuise 48, qui est sur 4, par le premier nombre de la Progression, 3: prouiennent 16: lesquels je multiplie par 96, qui est sur 5: (car 4 & 5 font 9) prouiendront 1536, qui sera le nombre à mettre au neufieme lieu."

Bürgi and the Progress Tabulen. In 1620 Jobst Bürgi published his Progress Tabulen, a work conceived some years earlier. As stated above, it is well known that he was influenced by Simon Jacob's work. The tables were printed at Prag and are simply lists of antilogarithms with base 1.0001. The logarithm is printed in red in the top line and the left-hand column, and the antilogarithms are in black, and hence Bürgi calls the logarithm *Die Rothe Zahl.* The first part of his table is as follows:

	0	500	1000	1500	2000
0	100000000	100501227	101004966	101511230	102020032
10	10000	11277	15067	21381	30234
20	20001	21328	25168	31534	40437
30	30003	31380	35271	41687	50641

The manuscripts of Bürgi are at the Observatory at Pulkowa, but none seem to be of a date later than 1610, so that he probably developed his theory independently of Napier. It is evident that he approached the subject algebraically, as Napier approached it geometrically.[1]

The only extensive table of antilogarithms is due to James Dodson (London, 1742).

Logarithms in the Orient. Logarithms found their way into China through the influence of the Jesuits. The first treatise upon the subject published in that country was a work by one Sié Fong-tsu, a pupil of the Polish Jesuit John Nicolas Smogolenski (1611–1656). This treatise was published about 1650, although Smogolenski had already mentioned the theory in one of his works.[2] Vlacq's tables (1628) were reprinted in Peking, as already stated, in 1713.

---

[1] Thus Kepler says: " . . . qui etiam apices logistici Justo Byrgio multis annis ante editionem Neperianam, viam praeiverunt, ad hos ipsissimos Logarithmos. Etsi homo cunctator et secretorum suorum custos, foetum in partu destituit, non ad usus publicos educavit" (*Opera Omnia*, VII, 298) (Frankfort a. M., 1868).

[2] The *T'ien-pu Chên-yüan*, as stated in Volume I, page 436.

13. PERMUTATIONS, COMBINATIONS, PROBABILITY

Permutations and Combinations. The subject of permutations may be said to have had a feeble beginning in China in the *I-king* (*Book of Changes*), the arrangements of the mystic trigrams, as in ☰☷, furnishing the earliest known example.[1] It is not improbable that it was the *I-king* that suggested to a certain Japanese daimyō of the 12th century that he write a book, now lost, upon permutations.[2]

Greek Interest in the Subject. The subject received some slight attention at the hands of certain Greek writers. Plutarch[3] (1st century) tells us[4] that Xenocrates (*c.* 350 B.C.), the philosopher, computed the number of possible syllables as 1,002,000,000,000, but it does not seem probable that this represents an actual case in combinations.[5] Plutarch also states that Chrysippus (*c.* 280–*c.* 207 B.C.), a Stoic philosopher, found the number of combinations of ten axioms to be more than 1,000,000, and that Hipparchus (*c.* 140 B.C.) gave the number as 101,049[6] if admitted and 310,925 if denied; but we have no evidence of any theory of combinations among the Greeks.[7]

Interest of Latin Writers in the Subject. The Latin writers, having little interest in any phase of mathematics except the practical, paid almost no attention to the theory of combinations. The leading exception was Boethius (*c.* 510). He gives a rule for finding the combinations of n things taken two at a time which we should express as $\frac{1}{2}n(n-1)$.[8]

---

[1] See Volume I, page 25.

[2] The theory is referred to as *Keishizan* in Volume I, page 274.

[3] *Quaestiones Conviv.*, Lib. VIII, 9, iii, 12; ed. Dübner, II, 893 (Paris, 1877).

[4] Tropfke, *Geschichte*, II (1), 351.

[5] Gow, *Greek Math.*, pp. 71 n., 86; Tropfke, *Geschichte*, II (1), 351.

[6] "Centena millia atque insuper mille et quadraginta novem." The number in Tropfke is incorrect.

[7] With respect to the single possible case in Pappus, see ed. Hultsch, II, 646–649: "Nam ex tribus dissimilibus generis triades diversae inordinatae existunt numero decem."

There is a slight trace of interest in the subject in the works of Plato and Aristotle, but not enough to be worthy of discussion in this chapter. See J. L. Heiberg, *Philologus*, XLIII, 475, with references.   [8] J. L. Heiberg, *ibid.*

Hindu Interest in the Subject. The Hindus seem to have given the matter no attention until Bhāskara (*c.* 1150) took it up in his *Lilāvati.* In this work he considered the subject twice. He asserted that an idea of permutations "serves in prosody . . . to find the variations of metre; in the arts [as in architecture] to compute the changes upon apertures [of a building] ; and [in music] the scheme of musical permutations; in medicine, the combinations of different savours."[1] He gave the rules for the permutations of *n* things taken *r* at a time, with and without repetition, and the number of combinations of *n* things taken *r* at a time without repetition.[2]

Early European Interest in the Subject. Early in the Christian Era there developed a close relation between mathematics and the mystic science of the Hebrews known as the cabala. This led to the belief in the mysticism of arrangements and hence to a study of permutations and combinations. The movement seems to have begun in the anonymous *Sefer Jezira* (*Book of Creation*), and shows itself now and then in later works.

It seems to have attracted the attention of the Arabic and Hebrew writers of the Middle Ages in connection with astronomy. Rabbi ben Ezra (*c.* 1140), for example, considered it with respect to the conjunctions of planets, seeking to find the number of ways in which Saturn could be combined with each of the other planets in particular, and, in general, the number of combinations of the known planets taken two at a time, three at a time, and so on. He knew that the number of combinations of seven things taken two at a time was equal to the number taken five at a time, and similarly for three and four and for six and one. He states no general law, but he seems to have been aware of the rule for finding the combinations of *n* things taken *r* at a time.[3]

---

[1] Colebrooke translation, p. 49. [2] *Ibid.*, p. 123.

[3] D. Herzog, *Zophnath Paneach* (in Hebrew), Cracow, 1911. It is an edition of Josef ben Eliezer's supercommentary (that is, a commentary on a commentary by Rabbi ben Ezra) on the Bible. The passage occurs in an extract from Rabbi ben Ezra's astrological manuscript *ha-Olam*, now in Berlin. The title of the book means "the revealer of secrets." See also J. Ginsburg, "Rabbi ben Ezra on Permutations and Combinations," *Mathematics Teacher*, XV, 347.

Levi ben Gerson, in his *Maassei Choscheb* (*Work of the Computer*), written in 1321, carried the subject considerably farther. He gave rules for the permutation of n things taken all together and also taken r at a time, and for the combination of n things taken r at a time.[1]

A few years later Nicole Oresme (*c.* 1360) wrote a work[2] in which he gave the sum of the numbers representing the combinations of six things taken 1, 2, 3, 4, and 5 at a time. He also gave these combinations in detail, as that $_2C_6 = 15$, $_3C_6 = 20$, and so on, of course in the rhetorical form, and seems to have known the general law involved, although he did not state it.

First Evidence of Permutations in Print. The first evidence of an interest in the subject to be found in the printed books is given in Pacioli's *Sūma* (1494), where he showed how to find the number of permutations of any number of persons sitting at a table.[3] In England the subject was touched upon by W. Buckley (*c.* 1540), who gave special cases of the combinations of n things taken r at a time. Tartaglia (1523) seems first to have applied the theory to the throwing of dice.[4]

In the 16th century the learned Rabbi Moses Cordovero[5] wrote the *Pardes Rimmonim* (*Orchard of Pomegranates*),[6] in which he gave an interesting treatment of permutations and combinations and showed some knowledge of the general laws.[7]

---

[1] Eneström, *Bibl. Math.*, XIV (3), 261; G. Lange, German translation of the treatise, published at Frankfort a. M., 1909; Tropfke, *Geschichte*, VI (2), 64.

[2] *Tractatus de figuratione potentiarum et mensurarum difformitatum.* See H. Wieleitner, "Ueber den Funktionsbegriff und die graphische Darstellung bei Oresme," *Bibl. Math.*, XIV (3), 193.

[3] Fol. 43, *v.* He gives the results for $n = 1, 2, \ldots, 11$, and adds "Et sic in infinitum."

[4] In the *General Trattato*, II, fol. 17, *r.*, he states that he discovered the rule: "Regola generale del presente auttore ritrouata il primo giorno di quarasima l'anno 1523. in Verona, di sapere trouare in quanti modi puo variar il getto di che quantita di dati si voglia nel tirar quelli." See also *L'Enseignement Mathématique*, XVI (1914), 92.

[5] Born at Safed, Palestine, in 1522; died at Safed, June 25, 1570.

[6] Salonika, 1552, with later editions.

[7] M. Turetsky, "Permutations in the 16th century Cabala," *Mathematics Teacher*, XVI, 29.

At about the same time Buteo not only discussed the question of the number of possible throws with four dice[1] but took up the problem of a combination lock with several movable cylinders like those shown in the illustration of the lock below.

EARLY COMBINATION LOCK

From Buteo's *Logistica*, 1560 ed., p. 313

As would naturally be expected, special cases of combinations of various kinds occur in the works of the 17th century. An illustration is found in the *Artis Analyticae Praxis* (p. 13) of Harriot, where the following symbolism is used for the product of binomials:

$$
\begin{aligned}
\left.\begin{array}{l} a - b \\ a - c \\ a - d \\ a - f \end{array}\right| = {} & aaaa - baaa + bcaa \\
& \quad - caaa + bdaa \\
& \quad - daaa + cdaa - bcda \\
& \quad - faaa + bfaa - bcfa \\
& \quad + cfaa - bdfa \\
& \quad + dfaa - cdfa + bcdf
\end{aligned}
$$

The first writer to give the general rule that

$$
_nC_r = \frac{n(n-1)(n-2)\cdots(n-r+1)}{r!}
$$

was Hérigone[2] (1634).

[1] "Ludens aleator tesseris quatuor, quaero quibus & quot modis inter se diuersis iacere possit?" *Logistica*, Lyons, 1559; 1560 ed., p. 305.

[2] *Cursus mathematicus*, II, 102. Paris, 1634.

In his work on the arithmetic triangle[1] Pascal showed the relation between the formation of the binomial coefficients and the theory of combinations, a subject also treated of by Fermat and others. Among the early writers upon the theory were Huygens, Leibniz,[2] Frénicle,[3] and Wallis,[4] and there is a brief tract on the subject which is thought to be due to Spinoza (1632–1677).[5]

The first work of any extent that is devoted to the subject was Jacques Bernoulli's *Ars Conjectandi.*[6] This work contains the essential part of the theory of combinations as known to-day. In it appears in print for the first time, with the present meaning, the word "permutation."[7] For this concept Leibniz had used *variationes* and Wallis had adopted *alternationes.* The word "combination" was used in the present sense by both Pascal and Wallis.[8] Leibniz used *complexiones* for the general term, reserving *combinationes* for groups of two elements and *conternationes* for groups of three,—words which he generalized by writing *con2natio, con3natio,* and so on.

Probability.[9] The theory of probability was mentioned in connection with the throwing of dice by Benvenuto d' Imola, a commentator on Dante's *Divina Commedia*, printed in the

---

[1] Written *c.* 1654 but printed posthumously in 1665. Beginning at this point, the reader may profitably consult the *Encyklopädie der math. Wissensch.*, I, 29.

[2] *Ars combinatoria,* 1666.

[3] "Abrégé des combinaisons" (1676), published in the *Mém. de l'acad. royale des sciences,* Paris, V (1729), 167.

[4] *De combinationibus, alternationibus, et partibus aliqotis, tractatus* (1685), in his *Opera,* II, 483 (Oxford, 1693).

[5] D. Bierens de Haan, "Twe zeldzame Werken van Benedictus Spinoza," *Nieuw Archief voor Wiskunde,* Amsterdam, XI (1884), 49. The title of the tract is *Reeckening van Kanssen,* and the work appeared in 1687, ten years after Spinoza's death.

[6] Posthumously printed at Basel in 1713. There is an English edition of 1795 under the title: *Permutations and Combinations: Being an Essential and Fundamental Part of the Doctrine of Chances.*

[7] "*De Permutationibus. Permutationes* rerum voco variationes. . . ."

[8] In the latter's *De Combinationibus,* English ed., 1685; *Opera* (1693), II, 483. His definition of combinations is on page 489 of that work.

[9] I. Todhunter, *History of the Mathematical Theory of Probability,* Cambridge, 1865; C. Gouraud, *Histoire du Calcul des Probabilités,* Paris, 1848.

Venice edition of 1477.[1] The gambling question first appears in a mathematical work, however, in Pacioli's *Sūma*[2] (1494). Here two gamblers are playing for a stake which is to go to the one who first wins n points, but the play is interrupted when the first has made p points and the second q points. It is required to know how to divide the stakes. The general problem also appears in the works of Cardan[3] (1539) and, as already stated, of Tartaglia[4] (1556). It first attracted wide attention in connection with the question proposed to Pascal (*c.* 1654) and by him sent to Fermat. The statement was substantially the one given in Pacioli to the effect that two players of equal skill left the table before completing the game. The stakes, the necessary score, and the score of each person being known, required to divide the stakes. Pascal and Fermat agreed upon the result, but used different methods in solving. As a result of the discussion so much interest was aroused in the theory that the doctrine of probability is generally stated to have been founded by Pascal and Fermat.

The first printed work on the subject was probably a tract of Huygens that appeared in 1657.[5] There also appeared an essay upon the subject by Pierre Rémond de Montmort in 1708.[6] The first book devoted entirely to the theory of probability was the *Ars Conjectandi* (1713) of Jacques Bernoulli, already mentioned. The second book upon the subject was De Moivre's *Doctrine of Chances: or, A Method of Calculating the Probability of Events in Play* (1718); and the third,

---

[1] Cantor, *Geschichte*, II (2), 327; Tropfke, *Geschichte*, II (1), 356. This is the fifth or sixth printed edition, Hain 5942; Copinger, I, 185, No. 5942.

[2] "Una brigata gioca apalla a .60. el gioco e .10 p caccia. e fãno posta duc .10. acade p certi accidēti che nõ possano fornire e luna $p$te a .50. e laltra .20. se dimanda che tocca p $p$te de la posta." Fol. 197, *r.*

[3] *Practica*, Milan, 1539, "Caput 61. De extraordinariis & ludis," No. 17 of the chapter.

[4] *General Trattato* (1556), I, fol. 265, *r.*, where he quotes Pacioli under the title "Error di fra Luca dal Borgo." On a trace of the theory in a writing by Giovanni Francesco Peverone (*c.* 1550), see L. Carlini, *Il Pitagora*, VII, 65.

[5] "De ratiociniis in ludo aleae," in Van Schooten's *Exercitationum mathematicarum libri quinque*, Leyden, 1657. See also the *pars prima* of Bernoulli's *Ars Conjectandi*.

[6] *Essai d'analyse sur les jeux d'hasard*, Paris, 1708; 2d ed., *ibid.*, 1714.

Thomas Simpson's *Laws of Chance* (1740). One of the best-known works on the theory is Laplace's *Théorie analytique des probabilités*, which appeared in 1812. In this is given his proof of the method of least squares.

The application of the theory to mortality tables in any large way may be said to have started with John Graunt, whose *Natural and Political Observations* (London, 1662) gave a set of results based upon records of deaths in London from 1592. The first tables of great importance, however, were those of Edmund Halley, contained in his memoir on *Degrees of Mortality of Mankind*,[1] in which he made a careful study of annuities. It should be said, however, that Cardan seems to have been the first to consider the problem in a printed work, although his treatment is very fanciful. He gives a brief table in his proposition "Spatium vitae naturalis per spatium vitae fortuitum declarare," this appearing in the *De Proportionibvs Libri V*,[2] p. 204.

Although a life-insurance policy is known to have been underwritten by a small group of men in London in 1583, it was not until 1699 that a well-organized company was established for this purpose.

Besides the early work of Graunt and Halley there should be mentioned the *Essai sur les probabilités de la vie humaine* (Paris, 1746; supplementary part, 1760) by Antoine Deparcieux the elder (1703–1768). The early tables were superseded in the 18th century by the Northampton Table. Somewhat later the Carlisle Table was constructed by Joshua Milne (1776–1853). In 1825 the Equitable Life Assurance Society of London began the construction of a more improved table, since which time other contributions in the same field have been made by the Institute of Actuaries of Great Britain in coöperation with similar organizations, by Sheppard Homans (*c.* 1860) of New York,—the so-called American Experience Table,—and by Emory McClintock (1840–1916), also of New York.

[1] *Phil. Trans.*, London, 1693.
[2] Basel, 1570. For a sketch of the later tables see the articles on Life Insurance in the encyclopedias.

TOPICS FOR DISCUSSION

1. Leading steps in the development of algebra.
2. General racial characteristics shown in the early development of algebra.
3. The early printed classics on algebra.
4. Various names for algebra, with their origin and significance.
5. Development of algebraic symbolism relating to the four fundamental operations and to aggregations.
6. Development of symbolism relating to powers and roots.
7. Development of symbolism relating to the equality and to the inequality of algebraic expressions.
8. Methods of expressing equations, with a discussion of their relative merits.
9. Methods of solving linear equations.
10. Methods of solving quadratic equations.
11. History of the discovery of the method of solving cubic and biquadratic equations.
12. History of continued fractions and of their uses.
13. General steps in the development of the numerical higher equation.
14. History of the Rule of False Position, with the reasons for the great popularity of the rule.
15. Development of the idea of classifying equations according to degree instead, for example, according to the number of terms.
* 16. Development of the indeterminate equation.
17. General steps in the application of trigonometry to the solution of the quadratic and cubic equations.
18. General steps in the early development of determinants.
19. History of the Rule of Three and of its relation to proportion.
20. General nature of series in the early works on mathematics.
21. History of infinite products in the 17th century.
22. The historical development of the Binomial Theorem.
23. History and applications of Taylor's and Maclaurin's formulas.
24. History of the Pascal Triangle and of its applications.
25. The invention of logarithms and the history of their various applications.
26. History of permutations, combinations, and the theory of probability. .

CHAPTER VII

ELEMENTARY PROBLEMS

I. Mathematical Recreations

Purpose of the Study. In this chapter we shall consider a few of the most familiar types of problems that have come down to us. Some of these types relate to arithmetic, while others have of late taken advantage of algebraic symbolism, although at one time they were solved without the modern aids that algebra supplies.

Mathematical Recreations. Ever since problems began to be set, the mathematical puzzle has been in evidence. Without defining the limits that mark the recreation problem it may be said that the Egyptians and Orientals proposed various questions that had no applications to daily life, the chief purpose being to provide intellectual pleasure. The Greeks were even more given to this type of problem, and their geometry was developed partly for this very reason. In the later period of their intellectual activity they made much of indeterminate problems, and thereafter this type ranked among the favorite ones.

In the Middle Ages there developed a new form of puzzle problem, one suggested by the later Greek writers and modified by Oriental influences. This form has lasted until the present time and will probably continue to have a place in the schools.

Problems of Metrodorus. So far as the Greeks were concerned, the source book for this material is the *Greek Anthology*.[1] This contains the arithmetical puzzles supposed to be due to

[1] The first noteworthy edition was that of Friedrich Jacobs, Leipzig, 1813–1817. There is an English translation by W. R. Paton, London, 1918, being Volume V of the Loeb Classical Library. In this translation the arithmetic problems begin on page 25, and from these the selections given here have been made.

Metrodorus about the year 500(?). A few of these problems will serve to show the general nature of the collection.

Polycrates Speaks: "Blessed Pythagoras, Heliconian scion of the Muses, answer my question: How many in thy house are engaged in the contest for wisdom performing excellently?"

Pythagoras Answers: "I will tell thee, then, Polycrates. Half of them are occupied with belles lettres; a quarter apply themselves to studying immortal nature; a seventh are all intent on silence and the eternal discourse of their hearts. There are also three women, and above the rest is Theano. That is the number of interpreters of the Muses I gather round me."

The following problem relates to a statue of Pallas:

"I, Pallas, am of beaten gold, but the gold is the gift of lusty poets. Christians gave half the gold,[1] Thespis one eighth, Solon one tenth, and Themison one twentieth, but the remaining nine talents and the workmanship are the gift of Aristodicus."

The following relates to the finding of the hour indicated on a sundial and still appears in many algebras, modified to refer to modern clocks:

"Best of clocks,[2] how much of the day is past?"
"There remain twice two thirds of what is gone."

The next problem involves arithmetic series, as follows:

Crœsus the king dedicated six bowls weighing six minæ,[3] each [being] one drachma heavier than the other.[4]

A type that has long been familiar in its general nature is seen in the following:

A. "Where are thy apples gone, my child?"
B. "Ino has two sixths, and Semele one eighth, and Autonoe went off with one fourth, while Agave snatched from my bosom and carried

---

[1] It should be recalled that this was written probably about the time of such Christian scholars as Capella and Cassiodorus.

[2] Literally, hour indicator.

[3] A mina contained 100 drachmas.

[4] That is, than the one next smaller. Find the weight of each.

away a fifth. For thee ten apples are left, but I, yes I swear it by dear Cypris, have only this one."[1]

The following problem has more of an Oriental atmosphere:

"After staining the holy chaplet of fair-eyed Justice that I might see thee, all-subduing gold, grow so much, I have nothing; for I gave forty talents under evil auspices to my friends in vain, while, O ye varied mischances of men, I see my enemy in possession of the half, the third, and the eighth of my fortune."[2]

One of the remote ancestors of a type frequently found in our algebras appears in the following form:

"Brick-maker, I am in a great hurry to erect this house. Today is cloudless, and I do not require many more bricks, but I have all I want but three hundred. Thou alone in one day couldst make as many, but thy son left off working when he had finished two hundred, and thy son-in-law when he had made two hundred and fifty. Working all together, in how many days can you make these?"

This collection of puzzles, now attributed entirely to Metrodorus, contains numerous enigmas, one of which is numerical enough to deserve mention:

If you put one hundred in the middle of a burning fire, you will find the son and slayer of a virgin.[3]

Comparison with Oriental Problems. Such problems seem more Oriental than Greek in their general form, but if we could ascertain the facts we should probably find that every people cultivated the somewhat poetic style in the recreations of mathematics. It happens, however, that we have more evidence of it in India and China than we have in the Mediterranean countries, and hence we are led to believe it was more frequently found among the higher class of mathematicians in the East than among those of the West.

---

[1] There were 120, for $120 = 40 + 15 + 30 + 24 + 10 + 1$.

[2] $480 + 320 + 120 + 40 = 960$.

[3] The answer is Pyrrhus, son of Deidameia and slayer of Polyxena; for if ρ, the Greek symbol for 100, is inserted in the middle of the genitive form πυρός (fire), it becomes πυρρός (Pyrros, Pyrrhus). This is the mythological Pyrrhus (Neoptolemus), son of Achilles and Deidameia.

Medieval Collections. The first noteworthy collection of recreations, after the one in the *Greek Anthology*, is the *Propositiones ad acuendos iuvenes*, of which there is extant no manuscript written before the year 1000. This collection is attributed to Alcuin of York (*c.* 775), who is known to have sent a list of such recreations to Charlemagne.[1] It contains many stock problems such as those of the hare and hound, and the cistern pipes. Rabbi ben Ezra (*c.* 1140), Fibonacci (1202), Jordanus Nemorarius (*c.* 1225), and many other medieval writers made use of these standard types.

Printed Books. The first noteworthy collection of recreative problems to appear in print was that of Claude-Gaspar Bachet (1612).[2] While not so popular as various later works, and containing much that is trivial, it was a pioneer and is much better than some of those that went through many more editions.

From the bibliographical standpoint the most interesting of the printed collections is that of a Jesuit scholar, Jean Leurechon (1624).[3] He published his work under the name of H[endrik] van Etten at Pont-à-Mousson in 1624. It was a poor collection of trivialities,[4] but it struck the popular fancy and went through at least thirty-four editions before 1700, some of these being published under other names.

The next writer of note was Jacques Ozanam (1640–1717), a man who was self-taught and who had a gift for teaching others. He had faith in the educational value of recreations, and this fact, together with his familiarity with the subject and his success as a teacher, enabled him to write one of the most popular works on the subject that has ever appeared.

---

[1] "Misi aliquas figuras Arithmeticae subtilitatis laetitiae causa" (Cantor, *Geschichte*, I (2), 784).

[2] *Problemes plaisans et delectables, qui se font par les nombres. Partie recueillis de diuers autheurs, & inuentez de nouueau auec leur demonstration*, Lyons, 1612. There is a copy of this edition in the Harvard Library. Later editions: Lyons, 1624; Paris, 1874, 1879, 1884.

[3] Born at Bar-le-Duc, *c.* 1591; died at Pont-à-Mousson, January 17, 1670. He wrote on astronomy.

[4] Montucla, in his revision of Ozanam, speaks of it as "une pitoyable rapsodie."

The work was first published in 1692 or 1694[1] and since then there have been at least twenty different editions.

There have been many other works on the subject,[2] but none of them has had the popular success of those of Leurechon and Ozanam.

Japanese Geometric Problems. The Japanese inherited from the Chinese a large number of curious geometric problems, and by their own ingenuity and perseverance elaborated these tests of skill until they far surpassed their original teachers. Some of these problems were mentioned in Volume I, and the circle problem will be referred to in Chapter X of this volume; but in this connection it is proper to refer to one type of interesting problems frequently found in the early Japanese works. These problems refer to the inscribing and measuring of circles inscribed in various figures such as semicircles, fans, and ellipses.

A FAN PROBLEM FROM JAPAN

From Takeda Shingen's *Sampō Benran*, 1824

2. TYPICAL PROBLEMS

Pipes filling the Cistern. Few problems have had so extended a history as the familiar one relating to the pipes filling a cistern,[3] and the traveler who is familiar with the Mediterranean

---

[1] The date 1692 is on the testimony of Montucla, in his 1790 edition of Ozanam. It is probable that he was in error on this point. See *L'Intermédiare des Mathématiciens*, VI, 112, and various histories of mathematics.

[2] Bibliographies that are fairly complete may be found in E. Lucas, *Récréations Mathématiques*, 4 vols., I, 237 (Paris, 1882–1894); W. Ahrens, *Mathematische Unterhaltungen und Spiele*, p. 403 (Leipzig, 1901; 2d ed., 1918). These are the leading modern contributors to the subject, the works of Lucas being probably the best that have as yet appeared.

[3] See the author's article in the *Amer. Math. Month.*, XXIV, 64, from which extracts are here made.

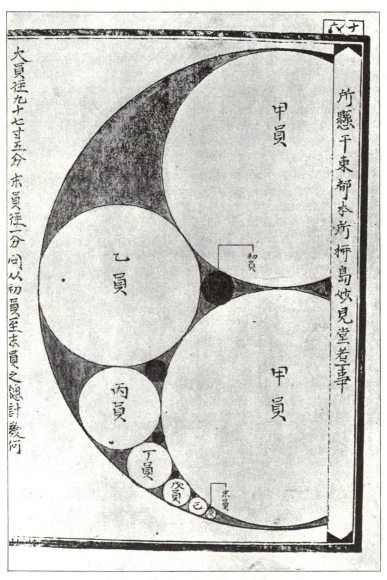

PROBLEM OF TANGENT CIRCLES

From a manuscript by Iwasaki Toshihisa (*c.* 1775)

lands cannot fail to recognize that here is its probable origin. Not a town of any size that bears the stamp of the Roman power is without its public fountain into which or from which several conduits lead. In the domain of physics, therefore, this would naturally be the most real of all the problems that came within the purview of every man, woman, or child of that civilization. Furthermore, the elementary clepsydra[1] may also have suggested this line of problems, the principle involved being the same.

The problem in definite form first appears in the Μετρήσεις (metre'seis) of Heron (c. 50?), and although there is some question as to the authorship and date of the work, there is none as to the fact that this style of problem would appeal to such a writer as he. It next appears in the writings of Diophantus (c. 275)[2] and among the Greek epigrams of Metrodorus (c. 500?), and soon after this it became common property in the East as well as the West. It is found in the list attributed to Alcuin (c. 775); in the Lilāvati of Bhāskara[3] (c. 1150); in the best-known of all the Arab works on arithmetic, the Kholâsat al-Ḥisâb of Behâ Eddîn (c. 1600); and in numerous medieval manuscripts. When books began to be printed it was looked upon as one of the standard problems of the schools, and many of the early writers gave it a prominent position, among them being men like Petzensteiner (1483), Tonstall (1522), Gemma Frisius (1540), and Robert Recorde (c. 1542).[4]

---

[1] Attributed to Plato (c. 380 B.C.) but improved by Ctesibius of Alexandria in the second century B.C. On the subject of clepsydræ see Chapter IX of this volume.

[2] In Bachet's edition (the Fermat edition of 1670, p. 271) appears this metrical translation:

> Totum implere lacum tubulis è quatuor, uno
> Est potis iste die, binis hic & tribus ille,
> Quatuor at quartus.
> Dic quo spatio simul omnes.

[3] See Taylor's translation, p. 50; Colebrooke translation, p. 42.

[4] In Recorde it appears for the first time in English: "Ther is a cestern with iiij. cocks, conteinyng 72 barrels of water, And if the greatest cocke be opened, the water will auoyde cleane in vj howers," etc. (Ground of Artes, 1558 ed., fol. A 7, v.).

Variants of the Problem. Such, then, was the origin of what was once a cleverly stated problem of daily life. This problem, like dozens of others, went through many metamorphoses, of which only a few will here be mentioned.

In the 15th century, and probably much earlier, there appeared the variant of a lion, a dog, and a wolf, or other animals, eating a sheep,[1] and this form was even more common in the 16th century.[2]

In the 16th century we also find in several books the variant of the case of men building a wall or a house, and this form has survived to the present time. It appeared in Tonstall's *De Arte Supputandi* (1522)[3] and in Cataneo's work (1546),[4] and in due time became modified to the form beginning, "If *A* can do a piece of work in 4 days, *B* in 3 days," and so on.

The influence of the wine-drinking countries shows itself in the variant given by Gemma Frisius (1540),[5] who states that a man can drink a cask of wine in 20 days, but if his wife drinks with him it will take only 14 days, from which it is required to find the time it would take his wife alone.

The influence of a rapidly growing commerce led one of the German writers of 1540 to consider the case of a ship with

---

[1] Johann Widman (1489) under the chapter title "Eyn fasz mit 3 zapffen." His form is: "Lew Wolff Hunt Itm̄ des gleichen 1 lew vnd 1 hunt vn̄ 1 wolff diese essen mit einander 1 schaff. Vnd der lew esz das schaff allein in einer stund. Vnd d' wolf in 4 stunden. Vnd der hunt in 6 stunden. Nun ist die frag wan sy dass schaff all 3 mit einåder essen/ in wie låger zeit sy das essen" (1509 ed., fol. 92; 1519 ed., fol. 112).

[2] Thus Cataneo, *Le Pratiche*, 1546; Venice edition of 1567, fol. 59, *v*.: "Se un Leone mangia in 2. hore una pecora, & l' Orso la mangia in 3. hore, & il Leopardo la mangia in 4. hore, dimandasi cominciando a mangiare una pecora tutti e 3. a un tratto in quanto tempo la finirebbono."

This form is also found in J. Albert's work of 1534 (1561 ed., fol. N viii), in Coutereels (1631 ed., p. 352), and in the works of numerous other writers.

In this chapter a few authors of textbooks will be mentioned whose names are not of sufficient importance to entitle them to further attention. The dates will serve to show their relative chronological position. For names of major importance consult Volume I.

[3] With the statement that it is similar to the one about the cistern pipes: "Questio hæc similis est illi de cisterna tres habēte fistulas : et simili modo soluenda" (fol. f 1). [4] See fol. 60, *v*., of the Venice edition of 1567.

[5] 1563 edition of his arithmetic, fol. 38.

3 sails, by the aid of the largest of which a voyage could be made in 2 weeks, with the next in size in 3 weeks, and with the smallest in 4 weeks, it being required to find the time if all

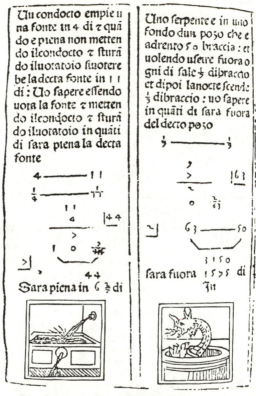

FROM CALANDRI'S WORK OF 1491

The problems of the pipe filling the cistern and of the serpent crawling out of the well. Calandri's was the first arithmetic printed with illustrations

three were used. Unfortunately several factors were ignored, such as that of one sail blanketing the others and the fact that the speed is not proportional to the power.[1]

[1] "Item/ 1 ein Schiff mit 3 Siegeln gehet vom Sund gen Riga/ Mit dem grösten allein/ in 2 wochen/ Mit dem andern/ in 3 wochen/ Vnnd mit dem kleinsten/in 4 wochen," etc. (J. Albert (1540; 1561 ed., fol. N vii)).

The agricultural interests changed the problem to that of a mill with four "Gewercken,"[1] and other interests continued to modify it further until, as is usually the case, the style of problem has tended to fall from its own absurdity. Its varied history may be closed by referring to a writer of the early 19th century,[2] moved by a bigotry which would hardly be countenanced today, who proposed to substitute a problem relating to priests praying for souls in purgatory.

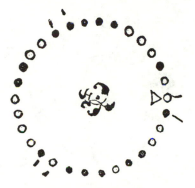

Turks and Christians. There is a well-known problem which relates that fifteen Turks and fifteen Christians were on a ship which was in danger, and that half had to be sacrificed.

THE TURKS AND CHRISTIANS

From Buteo's *Logistica*, Lyons, 1559 (1560 ed., p. 304). The problem begins: "In naui vectores quindecim Christiani totidēq; Iudei, suborta tēpestate magna"

It being necessary to choose the victims by lot, the question arose as to how they could be arranged in a circle so that, in counting round, every fifteenth one should be a Turk.

It is probable that the problem goes back to the custom of *decimatio* in the old Roman armies,[3] the selection by lot of every tenth man when a company had been guilty of cowardice, mutiny, or loss of standards in action. Both Livy (II, 59) and Dionysius (IX, 50) speak of it in the case of the mutinous army of the consul Appius Claudius (471 B.C.), and Dionysius further speaks of it as a general custom. Polybius (VI, 38) says that it was a usual punishment when troops had given way to

---

[1] "Ein Mülmeister hat ein Müle mit vier Gewercken/ Mit dem ersten mehlt er in 23 stūden 35 Scheffel/Mit dem andern 39 Scheffel/ Mit dem dritten 46 Scheffel/Vnnd mit dem vierten 52 Scheffel," etc. The question then is, How long it will take them together to grind 19 Wispel (1 Wispel = 24 Scheffel) (*ibid.*).

[2] R. Hay, *The Beauties of Arithmetic*, p. 218 (1816).

[3] E. Lucas, *Arithmétique Amusante*, p. 17 (Paris, 1895).

panic. The custom seems to have died out for a time, for when Crassus resorted to decimation in the war of Spartacus he is described by Plutarch (*Crassus*, 10) as having revived an ancient punishment. It was extensively used in the civil wars and was retained under the Empire, sometimes as *vicesimatio* (every twentieth man being taken), and sometimes as *centesimatio* (every hundredth man).

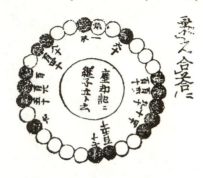

THE JOSEPHUS PROBLEM IN JAPAN

From Muramatsu Kudayū Mosei's *Mantoku Jinkō-ri* (1665)

Now it is very improbable that those in charge of the selection would fail to have certain favorites, and hence it is natural that there may have grown up a scheme of selection that would save the latter from death. Such customs may depart, but their influence remains.

In its semimathematical form the problem is first referred to in the work of an unknown author, possibly Ambrose of Milan (*c.* 370), who wrote, under the nom de plume of Hegesippus, a work *De bello iudaico*.[1] In this work he refers to the fact that Josephus was saved on the occasion of a choice of this kind.[2] Indeed, Josephus himself refers to the matter of his being saved by lucky chance or by the act of God.[3]

---

[1] Edited by C. F. Weber and J. Caesar, Marburg, 1864. See W. Ahrens, *Math. Unterhaltungen und Spiele*, p. 286 (Leipzig, 1901; 2d ed., 1918).

[2] "Itaque accidit ut interemtis reliquis Iosephus cum altero superesset neci" (quoted from Ahrens, *loc. cit.*).

[3] Καταλείπεται δὲ οὗτος, εἴτε ὑπὸ τύχης χρὴ λέγειν εἴτε ὑπὸ Θεοῦ προνοίας, σὺν ἑτέρῳ.

The oldest European trace of the problem, aside from that of Hegesippus, is found in a manuscript of the beginning of the 10th century. It is also referred to in a manuscript of the 11th century and in one of the 12th century. It is given in

THE JOSEPHUS PROBLEM IN JAPAN

From Miyake Kenryū's *Shojutsu Sangaku Zuye* (1795 ed.), showing the problem of the stepmother, referred to on page 544

the *Ta'hbula* of Rabbi ben Ezra (*c.* 1140), and indeed it is to this writer that Elias Levita, who seems first to have given it in printed form (1518), attributes its authorship.

The problem, as it came to be stated, related that Josephus, at the time of the sack of the city of Jotapata by Vespasian, hid himself with forty other Jews in a cellar. It becoming necessary to sacrifice most of the number, a method analogous to the old Roman method of *decimatio* was adopted, but in such a way as to preserve himself and a special friend. It is on this account that German writers still call the ancient puzzle by the name of Josephsspiel.

Chuquet (1484) mentions the problem, as does at least one other writer of the 15th century.[1] When, however, printed works on algebra and higher arithmetic began to appear, it became well known. The fact that such writers as Cardan[2] and Ramus[3] gave it prominence was enough to assure its coming to the attention of scholars.[4]

Like so many curious problems, this one found its way to the Far East, appearing in the Japanese books as relating to a stepmother's selection of the children to be disinherited. With characteristic Japanese humor, however, the woman was described as making an error in her calculations, so that her own children were disinherited and her stepchildren received the estate.

Testament Problem. There is a well-known problem which relates that a man about to die made a will bequeathing $\frac{1}{3}$ of his estate to his widow in case an expected child was a son, the son to have $\frac{2}{3}$; and $\frac{2}{3}$ to the widow if the child was a daughter, the daughter to have $\frac{1}{3}$. The issue was twins, one a boy and the other a girl, and the question arose as to the division of the estate.

The problem in itself is of no particular interest, being legal rather than mathematical, but it is worthy of mention because it is a type and has an extended history. Under both the Roman and the Oriental influence these inheritance problems played a very important rôle in such parts of analysis as the ancients had developed. In the year 40 B.C. the *lex Falcidia* required at least $\frac{1}{4}$ of an estate to go to the legal heir. If more than $\frac{3}{4}$ was otherwise disposed of, this had to be reduced by the rules of partnership. Problems involving this "Falcidian

---

[1] Anonymous MS. in Munich. See *Bibl. Math.*, VII (2), 32; Curtze, *ibid.*, VIII (2), 116; IX (2), 34; *Abhandlungen*, III, 123.

[2] In his *Practica* of 1539.

[3] See his edition of 1569, p. 125.

[4] It is also in Thierfelder's arithmetic (1587, p. 354), in Wynant van Westen's *Mathemat. Vermaecklyckh* (1644 ed., I, 16), in Wilkens's arithmetic of 1669 (p. 395), and in many other early works.

fourth" were therefore common under the Roman law, just as problems involving the widow's dower right were and are common under the English law.

The problem as stated above appears in the writings of Juventius Celsus (c. 75), a celebrated jurist who wrote on testamentary law; in those of Salvianus Julianus, a jurist in the reigns of Hadrian (117–138) and Antoninus Pius (138–161); and in those of Cæcilius Africanus (c. 100), a writer who was celebrated for his knotty legal puzzles.[1]

In the Middle Ages it was a favorite conundrum, and in the early printed arithmetics[2] it is often found in a chapter on inheritances which reminds one of the Hindu mathematical collections. It went through the same later development that characterizes most problems, and finally fell on account of its very absurdity. That is, Widman (1489) takes the case of triplets, one boy and two girls,[3] and in this he is followed by Albert (1534) and Rudolff (1526). Cardan (1539) complicates it by supposing 4 parts to go to the son and 1 part to the mother, or 1 part to the daughter and 2 parts to the mother, and in some way decides on an 8, 7, 1 division.[4] Texeda (1546) supposes 7 parts to go to the son and 5 to the mother, or 5 to the daughter and 6 to the mother, while other writers of the 16th century complicate the problem even more.[5] The final complications of the "swanghere Huysvrouwe" or "donna grauida" are found in some of the Dutch books, and these and

---

[1] Coutereels (Eversdyck edition of 1658, p. 382) traces the problem back to lib. 28, title 2, law 13, of the *Digest* of Julianus. He gives the usual 4, 2, 1 division as followed by Tartaglia, Rudolff, Ramus, Trenchant, Van der Schuere, and others. Coutereels, however, argues for the 4, 3, 2 division, and in this he has the support of various writers. Peletier gives 2, 2, 1, and others give 9, 6, 4. Brief historical notes appear in other books, as in the Schoner edition of Ramus (1586 ed., p. 186).

[2] Thus we have "Ein Testament" (Widman), "Erbteilung vnd vormundschaft" (Riese), "Erf-Deelinghe" (Van der Schuere), and "Erbtheilugs-Rechnung" (Starcken).

[3] Edition of 1558, fol. 97. He then divides the property in the proportion 4, 2, 1, 1.   [4] *Practica*, cap. 66, ex. 87.

[5] Ghaligai (1521), Köbel (1514), Riese (*Rechnung nach lenge*, 1550 ed.), Trenchant (1566), Van der Schuere (1600), Peletier (1607 ed., p. 244), Coutereels (1631 ed., p. 358), Starcken (1714 ed., p. 444), Tartaglia (*Tvtte l'opere d'aritmetica*, 1592 ed., II, 136).

the change in ideas of propriety account for the banishment of the problem from books of our day.[1] The most sensible remark about the problem to be found in any of the early books is given in the words of the "Scholer" in Robert Recorde's *Ground of Artes* (*c.* 1542): "If some cunning lawyers had this matter in scanning, they would determine this testament to be quite voyde, and so the man to die vntestate, because the testament was made vnsufficient."[2]

Problems of Pursuit. Problems of pursuit are among the most interesting elementary ones that have had any extended history. It would be difficult to conceive of problems that seem more real, since we commonly overtake a friend in walking, or are in turn overtaken. It would therefore seem certain that this problem is among the ancient ones in what was once looked upon as higher analysis. We have a striking proof that this must be the case in the famous paradox of Achilles and the Tortoise.[3]

It is a curious fact, however, that the simplest case, that of one person overtaking another, is not found in the Greek collections, although it appears in China[4] long before it does in the West. It is given, perhaps for the first time in Europe, among the *Propositiones ad acuendos juvenes* attributed to Alcuin, in the form of the hound pursuing the hare.[5] Thereafter it was looked upon as one of the necessary questions of European mathematics, appearing in various later medieval manuscripts. It is given in Petzensteiner's work of 1483, Calandri[6] used it in 1491, Pacioli has it in his *Sūma*[7] (1494),

---

[1] "Soo ontfangt sy ter tijdt haerder baringhe eenen Sone met een Dochter/ eñ een Hermaphroditus, dat is/half Man/half Vrouwe." Van der Schuere, 1600, fol. 98. In this case he divides 3175 guldens thus: d. 254, m. 508, s. 1524, h. 889. The same problem appears in Clausberg, *Demonstrative Rechen-Kunst*, 1772.

[2] 1558 ed., fol. X 8.

[3] For a study of this problem see F. Cajori, *Amer. Math. Month.*, XXII, 1 seq.

[4] For example, in the *Nine Sections* (*c.* 1105 B.C.?) and in Liu Hui's commentary (*c.* 263) on this classic. See also Volume I, page 32.

[5] "De cursu canis ac fuga leporis."

[6] "Una lepre e inanzi aun chane 3000 passi et ogni 5 passi delcane sono ϸ 8 diquegli della lepre uosapere inquanti passi elcane ara giūto lalepre."

[7] "Vna lepre e dinançe a vn cane passa .60. e per ogni passa .5. che fa el cane la lepre ne fa .7. e finalmente el cane lagiongni [la giongi in the edition of 1523,

and most of the prominent writers on algebra or higher arithmetic inserted it in their books from that time on.[1]

In those centuries in which commercial communication was chiefly by means of couriers who traveled regularly from city to city (a custom still determining the name of *correo* for a postman in certain parts of the world) the problem of the hare

PROBLEM OF THE HARE AND HOUND

From a MS. of Benedetto da Firenze, *c.* 1460. It begins, "Vna lepre e înanzi a .1°. cane"

and hound naturally took on the form of, or perhaps paralleled, the one of the couriers. This problem was not, however, always one of pursuit, since the couriers might be traveling either in

from la giugnere, to overtake her] dimando in quanti passa el cane giôgera la lepre." Fol. 42, *v.* He says that the problem is not clear, because we do not know whether the "passa .60." are leaps of the dog or of the hare, showing that he felt bound to take the problem as it stood, without improving upon the phraseology.

[1] Thus Rudolff (*Kunstliche rechnung*, 1526; 1534 ed., fol. N vj); Köbel (*Rechenbuch*, 1514; 1549 ed., fol. 88, under the title "Von Wandern über Landt," with a picture in which the hare is quite as large as the hound); Cardan (*Practica*, 1539, cap. 66); Wentsel (1599, p. 51); Ciacchi (*Regole generali d'Abbaco*, p. 130, Florence, 1675); Coutereels (*Cyffer-Boeck*, 1690 ed., p. 584), and many others.

the same direction or in opposite directions.[1] This variant of the problem is Italian, for even the early German writers gave it with reference to Italian towns.[2] As a matter of course, it was also varied by substituting ships for couriers.[3]

It was natural to expect that the problem should have a further variant, namely, the one in which the couriers should not start simultaneously. In this form it first appeared in print in Germany in 1483,[4] in Italy in 1484,[5] and in England in 1522,[6] although doubtless known much earlier.

The invention of clocks with minute hands as well as hour hands gave the next variant, as to when both hands would be together,—a relatively modern form of the question, as is also the astronomical problem of the occurrence of the new moon. One of the latest forms has to do with the practical question of a railway time-table, but here graphic methods naturally take the place of analysis, so that of all the variants those of the couriers and the clock hands seem to be the only ones that will survive. Neither is valuable per se, but each is interesting, each is real within the range of easy imagination, and each involves a valuable mathematical principle,—a fairly refined idea of function.

[1] See Pacioli's *Sūma*, 1494, fol. 39, for various types.

[2] Thus Petzensteiner (1483, fol. 53), in his chapter "Von wandern," makes the couriers go to "rum" (Rome), thus: "Es sein zween gesellen die gand gen rum. Eyner get alle tag 6 meyl der ander geth an dem ersten tage 1 meyl an dem andern zwue etc. unde alle tag eyner meyl mer dan vor. Nu wildu wissen in wievil tagen eyner als vil hat gangen als der ander." Günther, *Math. Unterrichts*, p. 304.

[3] Thus Calandri (1491) says: "Una naue ua da Pisa a Genoua in 5 di: unaltra naue uiene da genoua a pisa in 3 di. uo sapere partendosi in nun medesimo tempo quella da Pisa per andare a Genoua et quella da Genoua ꝑ andare a pisa in quanti di siniscon terrano insieme."

[4] Petzensteiner's arithmetic, printed at Bamberg.

[5] Borghi's arithmetic.

[6] Tonstall's *De Arte Supputandi*, fol. 4, "Cvrsor ab Eboraco Londinvm proficiscens," etc. See also Cardan (*Practica*, 1539, cap. 66, with various types); Ghaligai (1521; 1552 ed., fol. 64); Albert (1540; 1561 ed., fol. P i); Baker (1568; 1580 ed., fol. 36); Coutereels (1631 ed., p. 371, and Eversdyck edition of 1658, p. 403); Trenchant (1566; 1578 ed., p. 280); Wentsel (1599, p. 51); Peletier (1549; 1607 ed., p. 290); Van der Schuere (1600, fol. 179); Schoner (notes on Ramus, 1586 ed., p. 174), and many others.

The Chessboard Problem. One of the best-known problems of the Middle Ages is that relating to the number of grains of wheat that can, theoretically speaking, be placed upon a chessboard, one grain being put on the first square, two on the second, four on the third, and so on in geometric progression, the total number being $2^{64} - 1$, or 18,446,744,073,709,551,615. The problem is Oriental. A chessboard problem of a different character appeared in the writings of one I Hang, a Chinese Buddhist of the T'ang Dynasty (620–907),[1] so that games on a checkered board had already begun to attract the attention of mathematicians in the East.

Ibn Khallikân,[2] one of the best known of the Arab biographers (1256),

yon Wandern.

⟨ yon wandern vber Lande.

Ween Bürger auß Oppenheym/ Einer Son Heynrich/der ander Contz vő Treber genant/ wolten mit einander gen Rom geben/ Vnd Heynrich was alt / vnd mocht einen tag nicht mehr denn zehen meilen gehen/ Aber Contz von Treber was jung vnd starck/ der mocht einen tag 13. meilen gehen/ Deßhalben gieng Son Heynrich neun tag ehe auß Oppenheym denn Contz von Treber / Also war Son Heynrich Contzen 90. meilen fürgangen/ ehe Contz angehabt hat außzugehen.

Nun ist die frag / in wie vil tagen Contz von Treber Son Heynrichen vbergangen/ vnd die zwen zusamen kommen seind?

PROBLEM OF THE COURIERS

From Köbel's *Rechenbuch* (1514), the edition of 1564

relates[3] that when Sissah ibn Dahir invented the game of Chess, the king, Shihrâm, was filled with joy and commanded that chessboards should be placed in the temples. Furthermore, he commanded Sissah to ask for any reward he pleased. Thereupon Sissah asked for one grain of wheat for the first square, two for the next, and so on in geometric progression.

[1] G. Vacca, *Note Cinesi*, p. 135 (Rome, 1913). This problem is rather one of permutations.

[2] Or Challikân. Born September 22, 1211; died October 29, 1282.

[3] In his *Biographical Dictionary* (translation from the Arabic by Mac Guckin de Slane, 4 vols., Paris and London, 1843–1871), III, 69.

The result of the request is not recorded, but as an old German manuscript remarks, "Daz mecht kain kayser bezalen."[1]

The problem goes back at least as far as Mas'ûdî's *Meadows of Gold*[2] of the 10th century. It also appeared in the works of various other Arab writers,[3] and thence found its way into Europe through the *Liber Abaci*[4] (1202) of Fibonacci. It is found in numerous manuscripts of the 13th, 14th, and 15th centuries and in various early printed books.[5] The problem was much extended by later writers.[6] It found a variant in the problem of the horseshoe nails which appears in several manuscripts of the 15th and 16th centuries.

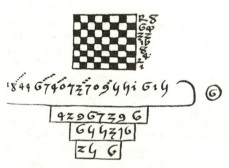

CHESSBOARD PROBLEM, *c.* 1400

From an Italian manuscript of *c.* 1400, now in the Columbia University Library

A Dutch arithmetician, Wilkens,[7] takes the ratio in the chessboard problem as three instead of two, and considers not only the number of grains but also the number of ships necessary to carry the total amount, the value of the cargoes, and the impossibility that all the countries of the world[8] should produce such an amount of wheat.[9]

[1] "No emperor could pay all that." Curtze, *Bibl. Math.*, IX (2), 113.

[2] Mas'ûdî died at Cairo in 956. A French translation in nine volumes, with Arabic text, appeared in Paris, 1861–1877. See also Boncompagni's *Bullettino*, XIII, 274.

[3] H. Suter, *Bibl. Math.*, II (3), 34.

[4] Boncompagni ed., I, 309.

[5] *E. g.*, Pacioli's *Sūma* (1494), fol. 43; Cardan's *Practica* (1539), cap. 66.

[6] As by Clavius, *Epitome* (1585), p. 297.

[7] 1669 ed., p. 112.

[8] "Al de Provintien van de gheheele werelt."

[9] For further historical notes see J. C. Heilbronner, *Historia Matheseos Universæ*, p. 440 (Leipzig, 1742).

Un cavallo che ha quatro ferri soto li piedi et per ogni
ferro ha sei chiodi che sono chiodi 24. Io dimando a un
quatrino l'uno quanti quatrini valera quelli chiodi 24
cominciando dal primo sino al ultimo sempre radopiando
de mano in mano che segaita li numeri sino al 24 come
qui soto ci vede descritta

```
 1 — 1
 2 — 2
 4 — 3
 8 — 4
 16 — 5
 32 — 6
 64 — 7
128 — 8
256 — 9
512 — 10
1024 — 11
2048 — 12
4096 — 13
8192 — 14
16384 — 15
32768 — 16
65536 — 17
131072 — 18
262144 — 19
524288 — 20
1048576 — 21
2097152 — 22
4194304 — 23
8388608 — 24
16777215
```

Si che ci vede che 24 chiodi, sempre
radopiando li numeri, come al incontro
ci vede l'esempio, fanno quatrini cento sesa-
nta sete, milioni, e setanta sete milia, e
due cento, e quindeci dico 16777215
Che fanno gazete, a quatrini sei per
gazeta dico 2796202: e quatrini 3
Che sono berlingoti: 2796 20: et gazete
due, e quatrini 3
Che fanno scudi trenta nove milia
nove cento, e quaranta cinque et
berlingoti cinque, e gazete due, e
quatrini tre dico 39945:5:2:3
dico a quatrini sei per gazeta
et a berlingoti 7 per scudo
Ma volendo che la gazeta valia
quatrini 7 per gazeta faranno
se non scudi dico, trenta quatro
milia due cento e trenta nove et
un berlingoto dico 34239:i:—
ci che da li sei quatrini alli sete
per gazeta vi è di diferentia dico
scudi cinque milia sete cento e sei
berlingoti quatro, gazete due e quatrini 3
dico ————— 5706:4:2:3

THE PROBLEM OF THE HORSESHOE NAILS

From an anonymous MS. written in Italy c. 1535. As the problem is usually
stated, the blacksmith receives one penny for the first nail, two pence for the
second, four for the third, and so on, and there are twenty-four nails. This manuscript is in Mr. Plimpton's library

The Mule and the Ass. Among the recreational problems that have come down to us there is one which appears in the form of an epigram with the name of Euclid attached. Rendered in English verse it is as follows:

A mule and an ass once went on their way with burdens of wine-skins;
Oppressed by the weight of her load, the ass was bitterly groaning.
The mule, observing her grievous complaints, addressed her this
   question:
"Mother, why do you murmur, with tears, for a maiden more fitting?
For give me one measure of wine, and twice your burden I carry;
But take one measure from me, and still you will keep our loads equal."
Tell me the measure they bore, good sir, geometry's master.[1]

3. COMMERCIAL PROBLEMS

Economic Problems. For the student of economics there is an interesting field in the problems of the 15th and 16th centuries, as may be seen from a few illustrations. The manuscripts and early printed books on arithmetic tell us that Venice was then the center of the silk trade, although Bologna, Genoa, and Florence were prominent. Florence was the chief Italian city engaged in the dyeing of cloth. "Nostra magnifica Città di Venetia," as Tartaglia so affectionately and appropriately called her, carried on her chief trade with Lyons, London, Antwerp, Paris, Bruges, Barcelona, Montpellier, and the Hansa towns, besides the cities of Italy. Chiarino (Florence, 1481) indicates the following as the most important cities with which Florence had extensive trade, his spelling being here preserved: Alessandria degypto, Marsilia, Mompolieri, Lisbona, Parigi, Bruggia, Barzalona, Londra, Gostatinopoli, and Dommasco, with the countries of Tunizi, Cypri, and Candia. Tartaglia gives Barcelona, Paris, and Bruges as the leading cities connected with Genoa in trade a half century later.

---

[1] *Euclidis Opera*, ed. Heiberg and Menge, VIII, 286 (Leipzig, 1916). The translation is by Professor Robbins, University of Michigan. See *The Classical Journal*, XV, 184.

[2] See the author's article in the *Amer. Math. Month.*, XXIV, 221, from which extracts are here made.

We also know from Chiarino the most important commodities of Florentine trade in the decade before America was discovered. These were rame (brass), stoppa (tow), zolphi (sulphur), smeriglio (emery), lana (wool), ghalla (gall), trementina (turpentine), sapone (soap), risi (rice), zucchari (sugar), cannella (cinnamon), piombo (lead), lini (flax), pece (pitch), acciai (thread), canapa (hemp), incenso (incense), indachi (indigo), mace (mace), cubeba (cubebs), borage (borax), and the ever-present saffron, the "king of plants," then everywhere used as a *sine qua non* in daily life, and now almost forgotten.

℄ Ein ander Exempel.

Ein Fraw oder Haußmutter gehet auff den marckt / kaufft vberhaupt ein Körblin mit Rebnerbyrn / darumb gibt sie achtzehen pfenning / so sie heim kompt / findet sie im körblin hundert vnd achtzig byrn / Ist die frag / wie vil byren sie vmb ein pfenning habe? Thu / als obgelert / so kompt dir zehen / Also vil byren hat sie vmb einen pfenning / Vnd ist wolfeyl drumb.

THE PROBLEM OF THE MARKET WOMAN

From Köbel's *Rechenbüchlein* of 1514 (1564 edition)

The problems also tell us the cost of the luxuries and the necessities of life. Spanish linen was worth, for example, from 94 to 120 ducats per hundredweight, while Italian linen ran as high as 355 ducats and Saloniki linen as high as 380 ducats. French linen was much cheaper than the latter, selling for 140 ducats. The arithmetics tell us that the linen was baled and sent from Venice to towns like Brescia on muleback.

The problems "delle pigione" tell us that the houses of the bourgeoisie rented in Siena, in 1540, at about 25 to 30 lire per year, while a century later they rented in Florence for from 120 to 300 lire. We also have the prices of sugar, ginger, pepper, and other commodities, showing that these three, for example, were only within the reach of the wealthy.

Hotel life in a grand establishment is also revealed in various problems, of which this one, printed in 1561, is a fair type:

Item/Wenn in einem Gasthause weren 8 Kamern/in jglicher Kamer stünden 12 Bette/in jglichem Bette legen 3 Geste/vnd ein jglicher Gast gebe dem Hausgesinde 6 ₰ trinckgelt/Wie viel thuts in einer Summa?

That these conditions of 12 beds in a room and 3 guests in a bed are not exaggerated, many travelers in remote parts of the world today can testify.

Partnership. There are three historic stages in the conduct of mercantile business: (1) that of individual enterprise, (2) that of partnership,[1] and (3) that of corporations.[2] The first of these has always existed, but in extensive business affairs it early gave way to partnerships in which the profits were divided according to the money invested, the time that it was employed in the business, or both. As business operations became still more extensive the partnership generally gave place to the corporation. Although the corporation has only recently come into great prominence, there were *societates publicanorum*[3] in Rome, each directed by the *magister societatis* and made up of members who received shares of the profits in proportion to their investments. These societies were not formed for the conduct of general business, however, but only for collecting taxes for the censors.[4] The division of profits according to amounts invested goes back to the Babylonian merchants and is frequently mentioned in ancient records.[5]

Partnerships and Usury. Aside from the necessity of joining capital in large business enterprises there was another reason

---

[1] "Partner" is from the Latin *partionarius*, from *partitio*, a share or part. It comes through the Old French *parsonnier* and Middle English *parcener*.

[2] Latin *corporatio*, from *corpus*, a body. Compare "corporeal," "corps," "corse," and "corpse."

[3] That is, societies of the farmers-general of the revenues.

[4] From *censere*, to value or tax, whence our "census."

[5] A. H. Sayce, *Social Life among the Assyrians and Babylonians*, p. 63 (London, n. d.).

why partnerships flourished so extensively in the Middle Ages. The laws and the popular prejudice in Christendom against taking interest on money placed the "pope's merchants"[1] at a disadvantage with respect to the Jews. Merchants in need of money were generally helped by their guilds, ordinary borrowing being resorted to only in cases of emergency, as in the *Merchant of Venice*.[2] Hence, if a man had money lying idle for a time, it was natural that he should join with others in some temporary venture and take his share of the profit. He thus secured interest on his capital without incurring popular odium. A man might even be taken into partnership for a limited time only, or he might be compelled by his partners to withdraw;[3] in these cases it became necessary to divide the profits according to the amount invested and the time.

Various Names for Partnership. There is hardly a medieval writer on business arithmetic who does not give this subject an important place,[4] and nearly every printed commercial book for a period of four hundred years devoted a chapter to the topic. The Latin arithmeticians called it the *Regula de societate*[5] or *Regula consortii*,[6] while the Italian writers commonly used the plural term *compagnie*.[7] When the services of a

[1] W. Cunningham, *The Growth of English Industry and Commerce during the Early and Middle Ages*, pp. 329, 364 (London, 1896).

[2] The Christian laws had forced the business of money-lending into the hands of the Jews, as in the case of Shylock.

[3] An interesting case is told in the records of the famous business house of Kress, in Nürnberg: "Und do die rechnung geschah, do zalt man Paulus Forchtel sein gelt und wolten sein nit länger in unser gesellschaft haben." The records also relate: "Item wir haben gantze rechnung gemacht an sant barbara obent do man zelt von gotes gepurt 1395 yar und es westund [belonged to] yeden Iᶜ XXXI gld. zu gewinn." G. von Kress, *Beiträge zur Nürnberger Handelsgeschichte aus den Jahren 1370 bis 1430*. See Günther, *Math. Unterrichts*, p. 291 n.

[4] *E.g.*, Fibonacci (1202), *Liber Abaci*, I, 139; Johannes Hispalensis (*c.*1140), *Liber algorismi* (No. II of Boncompagni's *Trattati*), p. 111; and many others.

[5] Thus Huswirt (1501): "Regula de societate mercatorum et lucro" and "De societate et intercessione tpis" (temporis). Cardan (1539, cap. 52) has "De societatibus."

[6] Thus Gemma Frisius (1540): "Regula consortij, siue, vt dicunt, Societatis."

[7] Thus Feliciano (1526) has a chapter *Dele compagnie* (1545 ed., fol. 30). In Spanish the word appears as *compania* and in French as *compagnie*, but the word *société* was also used, as by Peletier (1549).

partner were considered instead of any money contribution that he might have made, they used the term *soccite*.[1]

English writers, following the Italian practice, often used the word "company,"[2] although in general the word "fellowship"[3] was preferred.

538 Two marchants made a companie, A put in 300 pound for 2 monethes, and then putteth yet in 100 pound, and 6 monethes after that taketh out 200 poūd, and with the reſt remaineth vntill the yeares end.B put in 100 pound for one moneth, and then putteth yet in 700 pound, and 6 monethes after that taketh out a certaine ſumme of money, and with the reſt remaineth vntill the yeares end. and then finde to haue gained together 400 poūd, whereof B muſt haue 80 pound more then A, the queſtion is how much money B tooke out of the companie, without reckoning intereſt vpon intereſt.

```
300..2.. 600          400
100                     80
─────────            ──────
400..6..2400          320
200                   160
─────────
200..4.. 800
160 .. 3800..240
      19
     ────
     5700
100..1.. 100
700
─────
800.. 6 ..4800   Facit B tooke out 640 pound.
     4900
      800
      160
```

If

PROBLEM IN PARTNERSHIP

From Masterson's *Arithmetike* (1592)

[1] Thus Cataneo (1546) follows his chapter *Delle Compagnie* by one *Delle Soccite*, saying: "lequali son simili alle compagnie," but that the latter consider "il capitale e non la persona & l' altro mette solo la persona senza altro capitale." Practically, he says, these problems have to do with the case in which some "gentil' huomo" puts in his cattle and some "uillano o soccio minore" puts in his time. The 16th century books also use the form *soccide*. The modern form is *soccio*, *soccita*.

[2] "Two men Company, and make a Stock of 700 l," in Hodder's *Arithmetick*, 10th ed., p. 152 (1672); but he calls the subject "The Rule of Fellowship."

[3] Thus Recorde (c. 1542) speaks of "the rule of Fellowshyppe . . . or Company" (1558 ed., fol. N 1), and Baker (1568) gives "the rule of Felowship." The term is used by the American Greenwood (1729) and in Pike's well-known arithmetic.

The Germans ordinarily preferred the term *Gesellschaft*[1] and the Dutch writers followed their lead.[2]

Pasturage Problems. Akin to partnership problems, and often classified with them, are pasturage problems.[3] These may have begun with the custom of the Roman *publicani* of renting to stock owners sections of the estates which the government had farmed out to them, payment being made in proportion to the number of cattle.[4] It is probable that the early use of commons by the shepherds was regulated according to the principles inherited from the early Roman conquerors.

The importance attached to the subject in the 16th and 17th centuries may be inferred from the fact that Clavius (1583) devotes thirty-two pages to it and Coutereels (1599) allows forty pages.[5]

Profit and Loss. The expression "profit and loss," still found in our arithmetics, although not always used in commercial parlance in quite the same sense, is an old Italian one. The books written in the vernacular used the term *guadagni e perdite*,[6] while those written in Latin called their chapter on the subject *De lucris & damnis*.[7] The term passed over into German as *Gewin und Verlust*,[8] into Dutch as *Winst ende*

---

[1] Thus Köbel (1514) has *Gesellschaft der Kaufleit*, and Albert (1534) has *Gesellschaft / oder der Kauffleut Regel / von eigelegtem Gelde*. Suevus (1593) gives the Latin form also, *Regvla societatis. Regel der Gesellschaft.*

[2] Thus Van der Schuere (1600) has *Reghel van Gheselschap.*

[3] Cardan (1539) speaks of them under the head *De societatibus bestiarum*, and Ortega (1512, the Rome edition of 1515) speaks of *compagnia pecoraria*. In the Dutch books of the 17th century the subject commonly went under the name *Vee-Weydinghe.*

[4] Ramsay and Lanciani, *Manual of Roman Antiquities*, 17th ed., p. 548 (London, 1901).

[5] Similarly, Pagani (1591), twenty-four pages; Werner (1561), twenty-six pages; Van der Schuere (1600), twenty-six pages; and Cardinael (1674 ed.), twenty-five pages.

[6] "Gains and losses," as in Sfortunati (1534), Cataneo (1546), and Pagani (1591).

[7] As in Cardan, 1539, cap. 59.

[8] Thus Rudolff (1526) gives an "Exēpel von gewin vñ verlust," Riese (1522) has "Vom gewin vnd vorlust," and Köbel (1514) has "Regel vnd frag / von gewiñ der Kauffleut angekaufftet wahr / Regula Lucri."

verlies,[1] and into French as *gain & perte*.[2] The English writers used "loss and gain," Recorde (*c.* 1542) saying that "the fourth Chapter treateth of Losse and Gaine, in the trade of Marchandise."

The early American texts followed the English phraseology, speaking of "loss and gain." Thus Greenwood (1729) remarks that "the Intention of this *Rule* is, to discover what is *Lost*, or *Gained per Cent.* in the *Sale* or *Purchase* of any Quantity of *Goods*: in Order to raise, or fall the *Price* thereof accordingly."

The popularity of the subject in the 16th century may be inferred from the fact that Werner's *Rechenbuch* (1561) devotes forty-seven pages to it, and that other commercial arithmetics were similarly generous.

Commission and Brokerage. Although the subject of commission and brokerage is not new, these terms are relatively modern. The early printed books use such terms as "factorage"[3] and "factorie,"[4] from "factor,"[5] a middleman in the purchase and sale of products. The term "factor" was used in this sense in the Middle Ages, when the father of Fibonacci was (*c.* 1175) a factor in Bougie, and in the Renaissance period.[6] It is still used, although less commonly, in America and Great Britain, and warehouses for goods to be exported are still called "factories" in various parts of the world.

---

[1] As in Van der Schuere (1600). It also appears as *Winningh en verlies*, as in Bartjens (1633).

[2] So the Dutch-French work of Coutereels (1631) has "Comptes de gain & perte."

[3] Even as late as the 19th century Pike's arithmetic (8th ed., p. 204 (New York, 1816)) has the definition: "Factorage, Is an allowance of so much per cent. to a Factor or Correspondent, for buying and selling goods." It defines a broker as a merchant's assistant in buying or selling.

[4] Thus Rudolff (1526) and Werner (1561) have *Factorey*. Of the Dutch writers, Bartjens (1633) has *Factorie*; Raets (1580), *Rekeninghen van Facteurijen*; and Van der Schuere (1600), *Facteur-Rekeninghe*.

[5] That is, operator, from the Latin *facere*, to act or do. Compare the factor of a number.

[6] Thus Werner (1561): "Item ein Kauffman macht seinem Factor ein geding"; Trenchant (1566): "Aux compagnies d'entre marchans & facteurs"; Recorde (*c.* 1542): "A Merchant doth put in 800 pound into the hands of his Factor" (1646 ed., p. 519).

The word "broker" is not so common as "factor" in problems before the 19th century, although it appears in Middle English[1] to designate one who does business for another or acts as his agent.[2]

The term "commission," as now used to indicate a percentage, is relatively modern.

Equation of Payments. The absence of banking facilities to the extent now known in America, the difficulties in transmitting money, and the scarcity of currency before the great improvements in gold-mining in the 19th century rendered necessary until very recently an extensive credit system. Importing houses bought on credit and exported goods on credit to those from whom they bought, balancing their accounts from time to time. The process of finding the balance due, so that neither party should lose any interest, was the problem of the equation of payments. The subject is found in many manuscripts of the 14th and 15th centuries, and when textbooks began to appear in print it was looked upon as of great importance. Thus Recorde (c. 1542) says:

Rules of Payment, which is a right necessarie Rule, and one of the chiefest handmaydes that attendeth vpon buying and selling.[3]

The subject went by various names,[4] but the later English and American writers generally used the expression "Equation of Payments."

Interest. The taking of interest is a very old custom, going back long before the invention of coins, to the period in which values were expressed by the weight of metal or by the quantity of produce. The custom of paying interest was well known

---

[1] *Brocour* or *broker*. It probably came from the Anglo-Saxon *brúcan*, to use or employ. The root is found in the Scandinavian and Teutonic languages, referring to business in general.

[2] A word coined in the 16th century from the Latin *agere*, to act or to do.

[3] Mellis ed., p. 478 (1594).

[4] Thus Hodder (1672 ed., p. 163) calls the chapter "Of Equation," and the Dutch works have such names as "Den Regel van Paeyement of Betalinghe" (Eversdyck's Coutereels, 1658 ed., p. 181) or simply "Reghel van Payementen" (Stockmans, 1609 ed., fol. Q 4; Houck, 1676 ed., p. 108).

in ancient Babylon. In Sumerian tablets of the period before 2000 B.C. the rate is often given as varying from the equivalent of 20 per cent to that of 30 per cent, according to whether it was paid on money, that is, on precious metals, or on produce. In general, in the later Babylonian records, the rate ran from $5\frac{1}{2}$ per cent to 20 per cent on money and from 20 per cent to $33\frac{1}{3}$ per cent on produce, although not expressed in per cents.[1] Even princes engaged in trade and insisted upon their interest, for one of the tablets relates the following:

Twenty manehs of silver, the price of wool, the property of Belshazzar, the son of the king. . . . All the property of Nadin-Merodach in town and country shall be the security of Belshazzar, the son of the king, until Belshazzar shall receive in full the money as well as the interest upon it.[2]

Tablets of Nineveh as old as the 7th century B.C. have the following records:

The interest [may be computed] by the year.
The interest may be computed by the month.

The interest on ten drachmas is two drachmas.

Four manehs of silver . . . produce five drachmas of silver per month.[3]

Interest in Ancient India. The custom was also known in ancient India, appearing in the early legal writings of the *Sūtra* period, some centuries before the beginning of our era.[4] In the

---

[1] M. Jastrow, Jr., *The Civilization of Babylonia and Assyria*, pp. 323, 326, 338 (Philadelphia, 1915); A. H. Sayce, *Zeitschrift für Assyriologie*, V (1890), 276; T. G. Pinches, *ibid.*, I, 198, 202; A. H. Sayce, *Social Life among the Assyrians and Babylonians*, chap. v, p. 67 (London, n. d.) (hereafter referred to as Sayce, *Social Life*); G. Billeter, *Geschichte des Zinsfusses im griechisch-röm. Altertum* (Leipzig, 1898),—the leading authority. [2] Sayce, *Social Life*, p. 65.

[3] Since four manehs was about $180, and five drachmas was about $2, the interest on $180 was $24 a year, the rate being 13 per cent. See J. Menant, *La Bibliothèque du palais de Ninive*, p. 71 (Paris, 1880).

[4] Thus the *Dharma-sāstras* state that "5 *Māshās* for every 20 [*Kārshāpanas*] may be taken every month." Since 20 *māshās* were probably equal to a *kārshāpana*, the rate was $1\frac{1}{4}$ per cent per month, or 15 per cent annually. See R. C. Dutt, *A History of Civilization in Ancient India*, I, 174, 237 (London, 1893).

medieval period there are many evidences of the taking of interest. For example, Mahāvīra (c. 850) has various problems of the following type:

O friend, mention, after calculating the time, by what time 28 will obtain as interest on 80, lent out at the rate of $3\frac{1}{2}$ per cent [per month].

Bhāskara (c. 1150) also paid much attention to the subject, giving such problems as the following:

If the interest of a hundred[1] for a month be five, say what is the interest of sixteen for a year.

If the interest of a hundred for a month and one third be five and one fifth, say what is the interest of sixty-two and a half for three months and one fifth.

If the principal sum, with interest at the rate of five on the hundred by the month, amount in a year to one thousand, tell the principal and interest respectively.[2]

Interest Customs in Greece. The rate in Greece seems not to have been restricted by law and to have varied from 12 per cent to 18 per cent. In the time of Demosthenes 12 per cent was thought to be low. There were two general plans for computing interest: (1) at so much per month per mina, and (2) at such a part of the principal per year. Interest was usually paid at the end of each month.[3]

Interest in Rome. In Rome the rate of interest was at first unrestricted.[4] The Twelve Tables[5] limited the interest charged

---

[1] That is, the rate per cent.

[2] These extracts show that the rate of interest in India in the 12th century was about 60 per cent, and that interest was computed on a percentage basis. See Colebrooke's translation of the *Lilāvati*, pp. 36, 39.

[3] F. B. Jevons, *A Manual of Greek Antiquities*, p. 397 (London, 1895); Harper's *Dict. of Class. Lit. and Antiq.*, p. 665.

Interest was called *faenus*, or *fenus*, a later term being *usura* (from *uti*, to use), commonly expressed in the plural, *usurae*. So Cicero has "pecuniam pro usuris auferre." From this came the French *usure* and our "usury." Capital was *caput* (head, originally a head of cattle) or occasionally *sors* (lot or chance).

[5] *Duodecim Tabulae*, the first code of Roman law, 451–449 B.C., and the foundation of that law up to the time of the *Corpus Iuris* of Justinian, c. 530.

to Romans to one twelfth ($8\frac{1}{3}$ per cent) of the capital, and
later (*c.* 100 B.C.) this limitation was extended to aliens as well.
The *Lex Genucia* (342 B.C.) prohibited the taking of interest
altogether,[1] but like the medieval canon law this seems not to
have been enforced.

In later Roman times the Eastern custom of monthly interest came into use, the ordinary rate being 1 per cent per
month, payable in advance, or 12 per cent per year. In Cicero's
time 48 per cent per year was allowed, and under the first
emperors 25 per cent was common. A little later 12 per cent
per year was made the maximum; Justinian reduced this rate
to $\frac{1}{2}$ per cent per month, which gave rise to the common rate
of 6 per cent. In classical Latin works the rates of interest are
usually mentioned either as *fenus unciarum*[2] or as *usurae
centesimae*.[3]

Interest in the Middle Ages. In medieval Europe the canon
law forbade the taking of usury, that is, the payment in advance for the use of money. The time had not come for borrowing money for such remunerative purposes as extensive
manufacturing or as building railways and steamships, and so
the principal was often consumed by usury instead of being
increased. Usury would therefore have speedily resulted in
the enslavement of the peasants, who were without money or
financial ability. Hence the Church came to recognize a distinction between loans for production, which might reasonably
have carried some remuneration, and those for consumption,
which were contrary to public policy.[4]

---

[1] M. Cantor, *Politische Arithmetik*, p. 2 (Leipzig, 1898).

[2] Uncial interest, that is, interest by twelfths, $\frac{1}{12}$ being the common rate. This
was $\frac{1}{120}$ per month when the ancient year consisted of ten months. When the
year was later divided into twelve months the rate was still $\frac{1}{120}$ per month or
$\frac{12}{120}$ per year. Since interest was paid by the month, this made the former rate
$8\frac{1}{3}$ per cent and the latter 10 per cent per year. See Ramsay and Lanciani,
Manual of Roman Antiquities, 17th ed., p. 472 (London, 1901).

[3] Hundredth interest, or 1 per cent a month. This was the ancient "per cent."
If the security was poor, this was raised to *binae centesimae* (2 per cent per
month) or even to *quaternae centesimae* (4 per cent per month).

[4] See the decree of the fifth Lateran Council (1512-1517) in Janet, *Le capital,
la spéculation, et la finance, au XIXe siècle*, p. 81 (Paris, 1892).

There was, however, another reason which was not so openly stated, namely, the desire of the Church and of the ruling classes to prevent the dangerous rivalry to authority which would have resulted from the accumulation of too large fortunes; in other words, to avoid the dangers of capitalism.

Origin of the Term "Interest." To overcome this restriction there accordingly developed a new economic custom. The borrower paid nothing for the use of the money if it was repaid at the time specified. If, however, he failed so to pay the principal, he was held to compensate his creditor by a sum which represented the difference, or "that which is between" ("id quod *interest*") the latter's position because of the delay and what his position would have been had he been paid promptly. *Id quod interest* was recognized by the Roman law, but as a certain per cent agreed to in advance it first appears in the 13th century, possibly suggested from the East.[1] Speaking of this method, Matthew Paris (1253) tells us that in his time 10 per cent was exacted every two months, and adds that in this way unscrupulous men "circumvented the needy in their necessities, cloking their usury under the show of trade."[2]

Among the economic movements of the Renaissance period was a serious questioning of the validity of the canon law against usury and a determination to recognize a new type of interest, namely, usury paid at the end of the term of borrowing.[3] As a result of this feeling the subject of interest found place in many of the early printed books, particularly in Italy,[4]

[1] Compare Bhāskara's *Lilāvati* (*c.* 1150), Colebrooke's translation, p. 39. Fibonacci (1202) gives problems involving 20 per cent interest, but the Hindu works give rates as high as 60 per cent.

[2] *Chronica Majora*, III, 329, published in the *Rolls Series*. See also W. Cunningham, *The Growth of English Industry and Commerce* . . . , p. 329 (London, 1896). One of the best historical sketches to be found in the early arithmetics is given by Sfortunati, *Nvovo Lvme*, 1534 (1544-1545 ed., fol. 60).

[3] See such works as F. de Platea, *Opus restitutionum usurarum et excommunicationum*, Venice, *c.* 1472 (de Platea lived *c.* 1300); J. Nider, *Tractatus de contractibus mercatorum* (s. l. a., but Cologne, with at least seven editions before 1501); and many other similar works of the period.

[4] So Calandri (1491), who uses thirty days to the month, sometimes using per cent ("per 3 anni a 10 per cento lanno") and sometimes stating the rate as the equivalent of so many pence in the pound.

although sometimes against the protest of the author.[1] This spirit of protest showed itself in the people's literature of England, as in Francis Thynne's (16th century) epigram:

> Stukelie the vsurer is dead, and bid vs all farwell,
> who hath a Iourney for to ride vnto the court of hell.

A similar testimony is found in Lauder (1568):

> Credit and frist [delay] is quyte away,
> No thing is let but for Usure;
> For euerie penny thay wyll haue tway:
> How long, Lord, will this warld indure?[2]

In spite of these protests the English parliament in 1545 sanctioned the taking of interest,[3] fixing the maximum rate at 10 per cent. The protest was such that the law was repealed in 1552, but it was reënacted in 1571,[4] and since that time all works on commercial mathematics have included the topic.

In Germany the opposition to interest was also very strong, and Martin Luther published a sermon on the subject in 1519.[5]

Compound Interest. The compounding of interest was known to the Romans and was not forbidden until rather late.[6] The late medieval and the Renaissance Italians, from whom we derive so much of our modern business arithmetic, used the word *merito*[7] for interest in general, and where it was computed "simply by the year"[8] it was called simple interest.[9]

---

[1] So Cataneo (*Le Pratiche*, 1546), under the title *De semplici meriti vsvreschi*, speaks of the practice as often "diabolical," and Pagani (1591) calls it "Cosa in vero molto biasmeuole, & diabolica."

[2] W. Lauder, *The Lamentationn of the Pure, twiching the Miserabill Estait of this present World*, published by the Early Eng. Text Soc., p. 28 (London, 1870).   [3] 37 Hen. VIII, c. 9.

[4] E. P. Cheyney, *Industrial and Social History of England*, p. 172 (New York, 1901).   [5] *Eyn Sermon von den Wucher*, Wittenberg, 1519.

[6] Harper's *Dict. of Class. Lit. and Antiq.*, p. 665.

[7] It passed into the French as *merite*, although the word *interest* was also used. Thus Trenchant (1566): "A calculer les merites ou interestz" (1578 ed., p. 299). The Italians also used the term *usura*. The 1515 Italian edition of Ortega (1512) has *Regula de lucro*.

[8] "Simplicemente all' anno," as Tartaglia (1556) says (1592 ed., II, fol. 95). See also Cardan's *Practica*, 1539, capp. 57 and 58.

[9] Ciacchi (1675, pp. 80, 228), a later writer, speaks *De' meriti semplici.*

Compound interest among the early Italians was computed from the beginning of each year[1] or period[2] and was called by the English writers of the 17th century "interest upon interest."[3] The taking of such interest was frequently charged against the Jews,[4] although unjustly so,[5] and is even characterized by their name.[6]

In the 15th and 16th centuries interest was usually computed either on a percentage basis or at so many pence to the pound.[7] The rate varied from the 60 per cent mentioned by Bhāskara (c. 1150) and his European contemporaries to smaller limits.[8]

The difficulty in computing interest gave rise in the 16th century to the use of tables. These were extended in the 17th century, a table of compound interest appearing in Richard Witt's *Arithmeticall questions* (London, 1613).

Discount. The computing of discount[9] for the payment of money due at a future time is relatively modern. It is found

[1] "À capo d' anno," as Tartaglia (1556) describes it (1592 ed., II, fol. 95).

[2] "À capo d' alcun tempo," as Cataneo (1546) describes it (1567 ed., fol. 53). Similarly Tartaglia (1592 ed., II, fol. 119): "Del meritar à capo d' anno, ò altro termine che d' alcuni è detto vsura." The expression passed over into French as "merite à chef de terme" (Trenchant, 1566; 1578 ed., p. 299).

[3] So in Hodder, 10th ed., 1672, p. 139. The Dutch commonly called it "interest op interest" or "Wins-ghewin (VVinsts-Gewin)."

[4] Thus Pagani (1591, p. 147): "e questo modo di meritare e communemente vsitato da gl' hebrei ne suoi Banchi."

[5] Günther, *Math. Unterrichts*, p. 290. Pagani also says that the Christians were equally to blame.

[6] "Ma d' altra sorte è la ragion dell' usura, che chiamano Guidaica" (*sic*, for "Giudaica"), as in the Italian edition (1567, fol. 32) of Gemma Frisius, but not in the Latin edition. Similarly Van der Schuere (1600, fol. 127) speaks of "een Ioodtsch profijt."

[7] "Meriter est bailler ses deniers pour profiter à raison d'vn tant pour ₺. ou pour 100: par an" (Trenchant, 1566; 1578 ed., p. 298).

[8] For example, Sfortunati (1534) gives the rate "a denari .2. la libra il mese," which is 10 per cent a year, and goes even as high as 4 pence per pound per month, or 20 per cent a year; Trenchant (1566) gives one problem at 12 per cent; Tartaglia (1556) gives 10 per cent, 16 per cent, 20 per cent, and other rates; one of the Dutch writers, Raets (1580) gives from 8 per cent to 14 per cent, 10 per cent being stated thus: "Soo 100 winnen in een iaer 10"; and Cardinael (1674 ed.) gives rates ranging from 10 per cent to 20 per cent.

[9] Formerly "discompt," from the Old French *descompter*, to reckon off, from *des-*, away, *compter*, to count.

II

in some of the 16th century arithmetics[1] but is more common in the century following, appearing under various names.[2]

Assize of Bread. One of the standard problems of the 16th century books related to the variation in size of a loaf of bread as the wheat varied in value. For example, if a 10-cent loaf weighs 14 oz. when wheat is worth $1.80 a bushel, how much should it weigh when wheat is worth $2.20 a bushel?[3] The problem had its genesis in real conditions. Loaves were formerly of two kinds: (1) "assized bread," always sold at the same price but varying in weight according to the price of wheat, and (2) "prized bread," always of the same weight but varying in price.[4] The legal regulation for the assized bread goes back at least to 794, being found at that time in a Frankfort capitulary, and is probably of Roman origin. London regulations are found as early as the 12th century, and in the "assize of bread" of Henry II (1154–1189) these are worked out by inverse proportion.[5] As a result of these regulations, tables of the assize of bread were prepared and their use was

---

[1] Thus Cataneo (1546) has a chapter "Del semplice sconto" and one (corresponding to compound interest) "Dello sconto a capo d' alcvn tempo." Trenchant (1566) discounts an amount due in four years "à raison de 12 pour 100 par an."

[2] Thus in Coutereels's *Cyffer-Boeck* (1690 ed., p. 289) it appears as "Rabatteeren, Disconteeren, of af-korten," and in Hodder's *Arithmetick* (1672 ed., p. 175) as "The Rule of Rebate, or Discount."

[3] Thus Ortega (1512; 1515 ed., fol. 59): "℄ Si de vno misura de grano che costa 10 carlini mi dano 4 vnze de pane per vno dinaro si voi sapere se de vna altra misura che costera 20 carlini quante vnze ne darano per uno medesimo dinaro." See also Gemma Frisius (1540; 1555 ed., fol. 66), Rudolff (*Kunstliche rechnung*, 1526; 1534 ed., fol. K 4), Albert (1534; 1561 ed., fol. N 1, under *Regula Detri Conuersa*), Suevus (1593, p. 320, with two pages "Vom Brodgewichte in thewren vnd wolfeihlen Jaren"), and many other writers of the 16th century and later.

[4] J. Nasmith, *An examination of statutes . . . the assize of bread*, Wisbech, 1800; S. Baker, *Artachthos Or a New Booke declaring the Assise or Weight of Bread*, London, 1621.

[5] "Quando *quartierium* frumenti se vendit *pro* sex sol.; tunc debet panis *esse bonus* et albus et ponderare sexdecim sol. de xx[ti] lores [*i.e.*, 20 d. to 1 oz.]. . . . Quando *pro quatuor* solidis tunc debet ponderare *triginta* sex sol. et *alius* quadraginta sex sol. . . . ," and so on for different weights. W. Cunningham, *The Growth of English Industry and Commerce during the Early and Middle Ages*, p. 568 (London, 1896).

made obligatory.[1] This problem of the size of lòaves was a common one in the early printed books and is often found as late as the second half of the 19th century.[2] The following, from the 1837 edition of Daboll's well-known American arithmetic, illustrates the type: "If when wheat is 7s. 6d. the bushel, the penny loaf will weigh 9 oz. what ought it to weigh when wheat is 6s. per bushel?"

Tare and Tret. Until the middle of the 19th century the subject of "tare and tret" was found in most of the English and American commercial arithmetics. "Tare" meant an allowance of a certain weight or quantity from the weight or quantity of a commodity sold in a box, cask, bag, or the like. The word came from the Arabic *ṭarḥa*[3] (what is thrown away) through the Spanish *tara* and the French *tare*, and shows the commercial influence exerted by the Arabs in Spain.

"Tret" meant about the same thing, but the word shows the Italian influence, meaning originally an allowance on things transported.[4] In England it was an allowance of 4 lb. in every 104 lb.

There was also a third term that was related to "tare" and "tret" and is commonly found in the English books of the 16th century. This term is "cloff," meaning an allowance of 2 lb. made on every 3 cwt. of certain goods in order that the weight might hold out in retailing. Thus Recorde (*c.* 1542) has problems of this type:

"Item at 3ŝ 4d the pound weight, what shal 254½ be worth, in giuing 4 l weight vpon euery 100 for treate."

"Item if 100 l be worth 36ŝ 8d, what shall 800 l be worth in rebating 4 pound upon euery 100 for tare and cloffe."[5]

---

[1] Such a table, from the Record Book of the city of Hull, is reproduced in facsimile in E. P. Cheyney, *Industrial and Social History of England*, p. 67 (New York, 1901).

[2] A rare and interesting tract on the subject is that of J. Powel, *Assize of Bread*, London, 1615, a guide for those who had to interpret the old law.

[3] From *ṭaraḥa*, he threw down.

[4] The word is from the Latin *trahere*, to draw or pull, whence *tractus*, Italian *tratto*, and French *trait*. From the same root we have "tract" and "traction."

[5] *Ground of Artes*, 1594 ed., p. 487. The origin of the term is uncertain.

In Baker's arithmetic (1568) "the eyght chapter treateth of Tares and allowances of Marchandise solde by weight," and other arithmetics of the period also presented the subject at considerable length.[1]

Cutting of Cloth. Problems relating to the cutting of cloth correlated so closely with the needs of merchants that the commercial schools seem generally to have included them in the 16th century. Thus Grammateus (1518) has problems on the cutting of cloth by tailors,[2] and Tartaglia (1556) also devoted considerable attention to them.[3] No attention was paid to the pattern, and the problems show that drapers had flexible consciences with respect to advising as to the amount needed for a garment.[4] Baker (1568) says that "the 5 Chapter treateth of lengthes and bredthes of Tapistrie, & other clothes,"[5] and John Mellis has a similar chapter in his addition (1582) to Recorde's *Ground of Artes*.

The custom of carpeting rooms, which reached its highest point in the 19th century, led to the inclusion of problems relating to this subject. The return to rugs in the 20th century is leading to a gradual elimination of the topic in America.

Barter. Of the applications of arithmetic none has had a more interesting history than barter,[6] a subject now very nearly obsolete in textbooks, although temporarily revived among nations as a result of the World War of 1914–1918, owing to conditions of exchange. There are three fairly well defined periods in the exchange of products. The first is that of pure barter, seen today in the exchange of guns and ammunition for a tusk of ivory in remote parts of Africa,—a period lasting throughout the era of savage life. This is also seen in the ancient method of paying taxes "in kind," so many fowls out

---

[1] Thus Ortega (1512; 1515 ed., fol. 53), "Regvla de tre de tara"; Stockmans (1589), "Reghel van Tara"; Ciacchi (1675), "Delle tare a vn tanto per cento"; Coutereels (1690 ed.), "Tara-Rekeningh."

[2] "℀ Schneider regel" (1535 ed., fol. C6).

[3] 1592 ed., II, fol. 79.   [4] Tartaglia, *loc. cit.*, fol. 81.   [5] 1580 ed., fol. 126.

[6] Possibly from the Old French *barat, barate, barete*, whence *bareter*, to cheat or beguile. It appears in Italian as *baratto* (Ortega, 1512; 1515 Italian ed., fol. 78) or *baratti* (Feliciano da Lazesio, 1536).

of a dozen, or one cow out of a given number.[1] The second is that wherein a fixed value was assigned to certain products, such as grain or dates,[2] these products acting as media of exchange or as bases for determining values in bartering other products,—a period lasting until money was invented and indeed until currency became common. The third period is that of the adoption of money as a medium of exchange, this medium taking such forms as wampum, shells, coins, ingots, and government certificates.

Two influences perpetuated barter long after the first of these periods and indeed down to the present time, namely, the scarcity of currency[3] and the international fairs. In these fairs[4] the merchants found that barter was a necessity on account of the scarcity and diversity of money.[5]

Various Names for Barter. Barter also went by the name of "exchange," quite as we use the word "trade" at present. Thus an English writer of 1440 has the expression "Bartyrn or changyn or chafare[6] oone thynge for a othere, *cambio*,"[7] and

[1] For example, in Egypt. See H. Maspero, *Les Finances de l'Égypte sous les Lagides*, p. 29 (Paris, 1905).

[2] A. H. Sayce, *Social Life among the Assyrians and Babylonians*, chap. v (London, n. d.) ; W. Cunningham, *The Growth of English Industry and Commerce*, p. 114 (London, 1896).

[3] Ciacchi mentions this effect in his *Regole generali d'abbaco*, p. 114 (Florence 1675). It should be observed that the output of gold from 1850 to 1900 was greater than that of the preceding three hundred and fifty years, which accounts in part for the greater amount of currency now available.

[4] Compare the fair of Nijni Novgorod and the smaller fairs of Leipzig, Munich, and Lyons, all of which still continue, and the various international expositions which are modern relics of the ancient gatherings of merchants. In a MS. on arithmetic, written in Italy in 1684, nine pages are given to a list of great fairs, mostly European, which Italian merchants of that time were in the habit of attending.

[5] Thus Cataneo (1546), speaking "De baratti," says: "E Necessario al buon mercante non uolendo receuer danno esser molto experto nel barattere" (1567 ed., fol. 49).

[6] Middle English *chaffare*, *chepefare*, from the Anglo-Saxon *cēapian* (to buy) + *fare* (to go). There is the same root in the word "cheap," originally a bargain, and in "Cheapside," the well-known London street.

[7] Italian for "exchange." So Ghaligai (1521) speaks of "*Barattare*, ouer cambiare una Mercantia a un' altra"; and Pellos (1492) has "lo .xiij. capitol qui ensenha cābiar aut baratar vna causa per lautra."

Baker (1568) gives fifteen pages to his chapter which "treateth of the Rules of Barter: that is to say to change ware for ware." The early German writers had a similar usage.[1]

The French writers of the 16th century often used the interesting word *troquer*,[2] a word meaning to barter, the chapter being called *Des Troques*. From this word we have the English "truck," the material bartered, a word which came to mean the most common objects of exchange, such as garden truck, and the cart (truck) in which the dealer (truckman) carried it, and finally to mean worthless material in general.

Since merchandise was often bartered for future delivery, as in the case of goods from Damascus or China,[3] the question of interest, or its equivalent, often had to be considered. This gave rise to the distinction between barter without time and barter with time.[4]

Barter in America. In American colonial life the subject of barter played an important part. A diary of 1704, kept by one Madam Knight of Boston, gives an idea of the arithmetic involved, as the following extract will show:

They give the title of merchant to every trader; who Rate their Goods according to the time and spetia they pay in: viz. Pay, mony, Pay as mony, and trusting. Pay is Grain, Pork, Beef, &c. at the prices sett by the General Court that Year; mony is pieces of Eight, Ryalls, or Boston or Bay shillings (as they call them,) or Good hard money, as sometimes silver coin is termed by them; also Wampom, vizt. Indian beads wch serves for change. Pay as mony is provisions, as aforesd one Third cheaper than as the Assembly or Genel Court sets it; and Trust as they and the mercht agree for time.

---

[1] Using their word *Stich*, meaning exchange. Thus Petzensteiner (1483): "Nu merck hübsch rechnung von stich." Rudolff (1526) tells how "zwen stechen mit einander," a phrase now used with respect to dice, and Albert (1534) relates that "Zween wöllen miteinander stechen."

[2] Thus Trenchant (1566): "Deux marchans veulent troquer leurs marchandises." Compare the Dutch *Mangelinge* (*Manghelinge*, *Mangelingh*) as a synonym for *troques* and *change* in the 16th and 17th century arithmetics.

[3] These two cases are mentioned in a MS. of 1684, written at Ancona, now in the library of Mr. Plimpton.

[4] Thus Ortega (1512; 1515 ed., fol. 78): " . . . baratto . . . p tempo como senza tēpo."

Now, when the buyer comes to ask for a commodity, sometimes before the merchant answers that he has it, he sais, is Your pay redy? Perhaps the Chap Reply's Yes: what do You pay in? say's the merchant. The buyer having answered, then the price is set; as suppose he wants a sixpenny knife, in pay it is 12d—in pay as money eight pence, and hard money its own price; viz. 6d. It seems a very Intricate way of trade and what Lex Mercatoria had not thought of.

Another diary, kept by one Jeremiah Atwater, a New Haven (Connecticut) merchant, about 1800, had various entries of a similar nature, among which is the following:

To 5 yds Calico at 2s 6d per yard.
To be paid in turnips at 1s 6d and remainder in shoes. As far as the turnips pay, the calico is to be 2s 6d and the remainder toward shoes at 2s 8d.

Taxes. Of all the applications of arithmetic, taxation is one of the oldest. The tax collector is mentioned in the ancient papyri of Egypt,[1] in the records of Babylon,[2] in the Bible,[3] and, indeed, in the histories of all peoples. His methods reached the extreme of cruelty among the Saracens, and a decided trace of this cruelty is still seen among some of their descendants.

In Greece the tax[4] was levied directly on property or indirectly by tolls or customs. Resident aliens paid a poll[5] tax, and an indirect tax[6] of 2 per cent was levied at the customhouses.

Rome had an elaborate system of taxation,[7] and this was the source of our present systems. It included the tariff,[8] the

---

[1] See Volume I, page 45.   [2] Sayce, *Social Life*, p. 68.
[3] *E.g.*, 2 Kings, xxiii, 35; Luke, ii, 1.   [4] Τέλος (*tel'os*).
[5] Head, from the Danish *bol*, a ball, bowl, bulb, or head. Hence the "polls," where the heads of the electors are counted. The Greek term was μετοίκιον (*metoi'kion*).
[6] The πεντηκοστή (*pentecoste'*), fiftieth. This is the same word as our "pentecost," which refers to the fiftieth day after the Passover in the Jewish calendar. In Greek taxation the word referred to the tax of $\frac{1}{50}$, or 2 per cent, on exports and imports.
[7] Latin *taxatio*, from *taxare*, to estimate or evaluate.
[8] Spanish *tarifa*, a price list or book of rates; from the Arabic *ta'rif*, giving information, from the root *'arf*, knowing. The word shows the Arab influence, through the Spanish, upon modern business. The Spanish town Tariffa was

ground tax,[1] the poll tax,[2] the tithes (still familiar in certain parts of the world),[3] and, in later times, the tax on traders.[4]

It is a curious and interesting fact that the subject of taxation commanded but little attention in the early textbooks. It is possible that authors hesitated to touch upon such a sensitive spot because of the necessity for receiving an *imprimatur* from the taxing powers. Although it is occasionally found in the 16th century,[5] it was not common until textbook writers were more free in their offering.

Banking and Exchange. In the days when Europe was made up of a large number of small principalities, each with its own system of coinage, the subject of exchange was much more familiar to the average business man than it is today. How recently this was the case may be seen in a remark of Metternich's in 1845, that Italy "represents simply a group of independent states united under the same geographic term." The European traveler gets some idea of the early situation today, for at the railway stations on the borders he finds the exchange office,[6] where he may exchange the money of the country he is leaving for that of the country he is entering.

So important was the subject considered that the 1594 edition of Recorde's *Ground of Artes* devotes twenty-one pages to it, saying that it is of great value to the merchants dealing with Lyons, inasmuch as "there are 4 faires in a yeere, at which they do commonly exchange."[7]

---

named from the fact that it was the leading customhouse at one time. The Latin term was *portorium,* from *portare,* to carry, as in "import," "export," and "transport."

[1] *Tributum soli,* tribute of the land, *tributum* coming from *tribuere,* to bestow or pay.   [2] *Tributum capitis.*   [3] *Decimae.*

[4] *Collatio lustralis.* This was especially prominent in the 5th century, when the great social upheaval led to the aggrandizement of the aristocracy. See S. Dill, *Roman Society in the Last Century under the Western Empire,* p. 204 (London, 1898).

[5] As in Savonne (1563; 1571 ed., fol. 41), with the name "Reigles des impositions."

[6] *Bureau des changes, Wechselbureau, Wissel Bureau, Cambio.*

[7] The editor, Mellis, gives a long list of the leading fairs which an English merchant might attend.

Chain Rule. In the days when the value of coins varied greatly from city to city as well as from country to country, money changers employed a rule, probably of Eastern origin, which was known by various names[1] but was most commonly

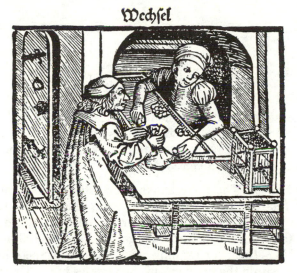

Wechsel

RENAISSANCE DEALER IN EXCHANGE
From the 1500 edition of Widman's arithmetic (1489)

called the chain rule[2] or continued proportion. The following problem is adapted from Widman (1489) and illustrates the type and the solution:

A man went to a money changer in Vienna with 30 Nürnberg pence and asked that they be exchanged for Viennese money. Since the money changer was ignorant of their value, he proceeded thus: 7 pence of Vienna are worth 9 pence of Linz, and 8 of Linz are worth 11 of Passau, and 12 of Passau are worth

---

[1] For example, *Regula conjuncta, Regel conjoinct, Te Zamengevoegden Regel, Regel van Vergelykinge,* and *De Gemenghde Regel,* these terms being taken from various Dutch and Dutch-French books of the 17th and 18th centuries.

[2] *Den Kettingh-Regel, Den Ketting Reegel,* in the early Dutch books. "Gleichsam wie die Glieder einer 'Kette,'" as R. Just, a modern German writer, has it in his *Kaufmännisches Rechnen,* I, 81 (Leipzig, 1901).

13 of Wilsshof, and 15 of Wilsshof are worth 10 of Regensburg, and 8 of Regensburg are worth 18 of Neumarkt, and 5 of Neumarkt are worth 4 of Nürnberg. Then

$$
\begin{array}{ccccccc}
7 & 9 & 12 & 13 & 8 & 18 & 30 \\
& 8 & 11 & 15 & 10 & 5 & 4
\end{array}
$$

so that
$$
\frac{7 \cdot 8 \cdot 12 \cdot 15 \cdot 8 \cdot 5 \cdot 30}{9 \cdot 11 \cdot 13 \cdot 10 \cdot 18 \cdot 4} = 13\frac{23}{429},
$$

the value of the Nürnberg money in pence of Vienna.[1]

Like many Eastern problems it is found in the works of Fibonacci (1202),[2] and thereafter it was common until the latter part of the 19th century.

Early Banks. The early banks were established in places of greatest relative safety, and these were usually the temples. All kinds of valuables were thus protected from the depredations of thieves, both private and governmental, civil and military. This is seen, for example, in the great business interests carried on within the precincts of the temples in the Ur Dynasty (c. 2450–2330 B.C.) of Babylon.[3] On the tablets of this period may be found the records of loans, receipts, promissory notes, leases, mortgages, taxes, and other commercial activities. A little later, in the first millennium B.C., drafts appear[4] in quite the form used even today.

For the reason above stated, the priests in the Greek temples were frequently money lenders.[5] It was also on this account

---

[1] The rule closes: "Vñ multiplicir in kreucz durchauss auff 2 teyl vñ dividir" (fol. 152).

[2] "De baractis monetarum cum plures monete inter similes" (*Liber Abaci*, p. 126).

[3] M. Jastrow, Jr., *The Civilization of Babylonia and Assyria*, p. 318 (Philadelphia, 1915).

[4] A. T. Clay, *Babylonian Records in the Library of J. Pierpont Morgan*, Part I (New York, 1912).

[5] J. P. Mahaffy, *Old Greek Life*, p. 38 (London, 1885); hereafter referred to as Mahaffy, *Greek Life*.

that coins were struck[1] at the temple of Juno Moneta[2] in Rome. The later Greek bankers and money changers were called τραπεζῖται,[3] a word derived from τράπεζα,[4] a table, just as "bank" comes from "bench."[5] Their tables were placed in the agoræ (public places) and the finding of "the tables of the money changers"[6] in the Temple at Jerusalem was not at all unusual. Indeed, one may see similar sights in the temples of Southern India today, or in the entrance to the great pagoda at Rangoon, Burma.

The business of the *trapezitai* included buying foreign money at a discount and selling it at a premium, paying interest on deposits, acting as pawnbrokers, and performing the duties of modern notaries.[7]

Banks in Rome. Bankers are mentioned by Livy (IX, 40, 16) as carrying on business as early as the 4th century B.C. At first a private banker was called an *argentarius* (silver dealer), an officer connected with the mint being a *nummularius*.[8] Somewhat later these terms were used along with *mensarius*[9] and *collectarius*[10] to represent any kind of banker.

---

[1] In early days they were stamped by the stroke of a hammer, and the word has remained in use.

[2] Juno the Admonisher (Adviser, Instructor). On this account the Romans used the word *moneta* to mean money, whence also our word "mint."

[3] *Trapezi'tai*, literally, "tablers." The Hebrew usage is the same.

[4] *Tra'peza*, whence our "trapezium," a figure representing a table (originally with two parallel sides), and "trapezoid," a figure shaped like a trapezium (originally with no parallel sides). The bankers were also called ἀργυραμοιβοί, argyramoiboi', money changers. See Harper's *Dict. Class. Lit.*, p. 1597; Mahaffy, *Greek Life*, p. 38; F. B. Jevons, *Manual of Greek Antiq.*, p. 395 (London, 1895).

[5] Late Latin *bancus*, a bench; French *banc*, a long seat or table. So we have a bench of judges and the bank of a stream. "Banquet," simply a diminutive form, came to mean a feast instead of the table. In Italy *banca* came to mean a tradesman's stall, a counter, and a money changer's table, as well as a bank.

[6] As late as 1567 we find English writers telling of how "Christ overthrew the Exchaungers bankes."

[7] M. S. Koutorga, *Essai historique sur les trapézites ou banquiers d'Athènes*, Paris, 1859.

[8] Coin-man. Originally an officer of the mint who tested the silver before it was coined.

[9] Or *mensularius*, from *mensa*, table, influenced by the Greek name for banker.

[10] Late usage, found in Justinian's *Institutes*.

The exchange bureau was called the *permutatio*.[1] In the banking department the funds of the creditor were called the *depositum*,[2] whence our "deposit." This was subject to a *perscriptio*,[3] a check, quite as it is at present. The *depositum* drew no interest,[4] being like our common open accounts; but there developed also a kind of savings-bank department in which a deposit known as a *creditum*[5] drew interest.[6]

Letters of Credit. The ancient bankers issued letters of credit quite like those issued at present,[7] and also made drafts on one another.[8] The idea is therefore without foundation that it was the Jews who, driven from France to Lombardy in the 7th century, first made use of foreign drafts.[9]

Stockholders. Among the "seven greater arts" recognized in medieval Florence was that of the money changers.[10] As early as 1344 the city government, finding itself unable to pay some $300,000 that it owed, formed a bank[11] and issued shares of stock[12] which were transferable as in modern corporations.[13]

---

[1] Compare our "permutation." J. Marquardt, *La vie privée des Romains*, French translation, p. 15 (Paris, 1893).

[2] *De* + *ponere*, to place. A *depositum* is that which is placed down.

[3] *Per* + *scribere*, to write; a written order.

[4] It was *vacua pecunia*, unproductive money.

[5] From *credere*, to trust, to have confidence in, to believe, whence our "creed"; in banking business a sum held in trust.

[6] Because of this interest feature, the claims of the *depositarii* were legally preferred to those of the *creditores* in case of the failure of a bank. See the *Digest*, XVI, 3, 7, 2.

[7] Mahaffy, *Greek Life*, p. 38. They were called by the Greeks συστατικαί ἐπιστολαί (*systatikai' epistolai'*), letters of introduction.

[8] Cicero uses "permutare Athenas" to mean "to draw on Athens"; and "ab Egnatio solvat" to mean "to pay by draft on Egnatius."

[9] Unger, *Die Methodik*, p. 90.

[10] *Arte del cambio* or *dei cambiatori*. The other six *arti maggiori* were those of (1) dressers of foreign cloth, (2) dealers in wool, (3) judges and notaries, (4) physicians and apothecaries, (5) dealers in silk, (6) furriers. E. G. Gardner, *The Story of Florence*, p. 28 (London, 1900).

[11] *Monte*, mount, bank, money; compare the French *mont-de-piété*.

[12] Originally the word meant a thing that was stuck or fixed, and hence a post, the Anglo-Saxon *stocc*, as in "stockade." The same root is found in "etiquette" (Old French *estiquet*), a little note "stuck up" on the gate of a court. From the same source we have "stack" and "ticket." See also page 194.

[13] C. A. Conant, *History of Modern Banks of Issue*, p. 21 (New York, 1896).

The idea of issuing notes payable in coin but only partly covered by a reserve was a development of the 17th century, beginning in Amsterdam (1609) and developing into more modern form in Stockholm (1661).[1]

Bills of Exchange. Probably the first bill of exchange to appear in a printed work on mathematics is the one given by Pacioli (1494), the form being substantially the same as the one now in use.[2]

Four Kinds of Exchange. In the early printed arithmetics there were four kinds of exchange, of which we have preserved two. The four types were as follows:

1. Common exchange, the mere interchange of coins, the work of the "money changers."[3]

2. Real exchange, by means of drafts.[4]

3. Dry exchange,[5] a method of evading usury laws by means of fictitious bills[6] of exchange,—drafts that bore no fruit.[7]

[1] *Ibid.*, p. 24.

[2] 1494 adi 9 agosto I v[a].

Pagate per questa prima nostra a Lodouico de francesco da fabriano e compagni once cento doro napolitane insu la proxima fiera de fuligni per la valuta daltretanti receuuti qui dal Magnifico homo miser Donato da legge quondā miser Priamo. E ponete p noi. Idio da mal ve guardi.

> vostro Paganino de paganini da Brescia ss. . . .
> Domino Alphano de Alphanis e cōpagni in peroscia. . . .
> *Sūma*, 1494 ed., fol. 167, *v*.

For later examples, see Cardan's *Practica*, 1539, cap. 56; Trenchant's arithmetic of 1566 (1578 ed., p. 350), and other commercial works.

[3] The Italians called it "cambio menuto, ouer commune," as in Tartaglia's work (1592 ed., II, fol. 174). In Spanish it appeared as "cambio por menudo" (Saravia, 1544), and in French as "change menu ou commun" (Trenchant, 1566). It is this form that is referred to in 1335, in the English act of 9 Edw. III, stat. 2, c. 7: "Et que table deschange soit a Dovorri & aillours, ou & q[a]nt il semblera a nos & notre consail per faire eschange,"—our "bureau des changes."

[4] Saravia (1544) makes two divisions of this type. See his Italian translation of 1561, fols. 108, 110.

[5] Italian, *cambio secco*; French, *change sec*; German, *trockener Wechsel*, as the terms appear in the 16th century.

[6] From the Latin *bulla*, a bubble, a leaden seal that looked like a bubble, and hence the sealed document, like a papal bull. From the same root we have "bullet," "bulletin," "bowl," and "bullion" (a mass of sealed or stamped metal).

[7] So called, as Saravia (Italian translation, 1561) says, from their resemblance to an "albero secco, il quale non ha humore, ne foglie ne frutto."

Such exchange was placed under the ban by an English statute of 1485/6: "eny bargayne . . . by the name of drye exchaunge . . . be utterly voide." English writers sometimes spoke of it as "sick" exchange,[1] confusing the French *sec* (dry) with an English word of different meaning.

FROM THE MARGARITA PHYLOSOPHICA

(1503)

Showing geometry as largely concerned with gaging and similar practical work

4. Fictitious exchange,[2] the plan of collecting a debt by drawing on the debtor.

As coinage came to be better settled a definite par of exchange was recognized, and so, beginning in the arithmetics of the latter part of the 16th century, we find various rules relating to this subject. Thus Recorde (*c.* 1542) says that "as touching the exchange, it is necessary to vnderstand or know the Pair, which the Italians call Pari."[3]

Days of Grace. The Italian cities had fixed rules as to the number of days after sight or after date at which drafts should be paid. Drafts between Venice and Rome were payable ten days after sight; between Venice and London, three months after date; of Venice on Lyons, at the next succeeding quarterly fair, and so on, thus giving the payer time to obtain money. In these customs is to be found the origin of the

[1] So T. Wilson, writing on usury in 1584, speaks of "sicke and drie exchange."
[2] The *cambio fittitio* of the Italian writers of the 16th century, and the *change fict* of the French.
[3] 1594 ed., p. 557. We also have such Dutch terms as *Rekeninghe vander Pary* (Raets, 1580) and *Den Reghel Parij* (Stockmans, 1676 ed.), and such French terms as *le per* (Savonne, 1563).

"days of grace," formerly allowed in England and America, but generally, owing to improved banking facilities, abandoned in the latter country about the opening of the 20th century.

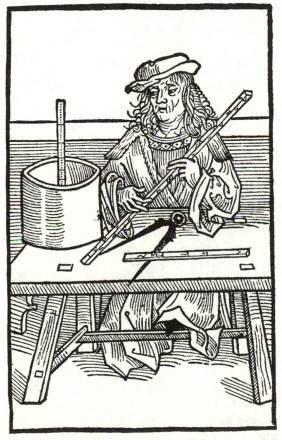

THE GAGER (GAUGER)

From Köbel's *Vysirbůch*, 1515, showing the tools of the art

Meaning of a Check. At present the check (in England, cheque) is extensively used instead of a bank draft. The word has an interesting history, coming from the Persian *shah*, a king. In the game of chess the player called out "*shah*"

when the king was in danger, and *"shah-mat"* ("the king is dead") when the king could no longer move. From this we have our "check" and "checkmate" in chess, "check"

THE GAGER AT WORK

From Johann Frey's *Ein new Visier büchlein*, Nürnberg [1543]

being thought to mean simply "stop." Hence we have the Middle English *chek*, French *échec* (a check or defeat), Italian *scacco* (a chess board), "checkers," and "check" (a stop in one's account at a bank).

Gaging. Before the size of casks was standardized as the result of manufacturing in quantities or of general laws affecting large territories, the subject of gaging[1] (gauging) played

---

[1] The Middle English was *gagen* or *gawgen*, to gage. The *u* came in through the Old French *gauger*. In medieval Latin *gaugia* meant a standard wine cask, but the origin of the word is uncertain.

an important rôle in applied mathematics. The word relates to the finding of the capacity of casks and barrels. In Germany a gager was called a Visierer,[1] and in the 15th century there appeared numerous manuals with such titles as *Vysirbůch*, *Vysierbuch*, and *Visyrbuechleynn*.[2] The custom was carried over to England, and even in our early American arithmetics there were chapters on gaging. The first notable German book on the subject was Köbel's work (see page 579) of 1515, although the *Margarita phylosophica* of 1503 pictured geometry (see page 578) as chiefly concerned with such measurements.

Other Applications. It would be difficult to give a satisfactory list of all the applications of elementary mathematics to the manifold interests of man that have developed in the centuries past. These applications include, besides those already mentioned, such topics as the adulteration of goods,[3] accountancy and exchange,[4] small commerce from town to town,[5] the leather trade,[6] grazing, and baking,—a list that might be extended to include many other topics and that illustrates the way in which arithmetic has met human needs.

[1] From *visieren*, to visé, to indorse a standard, to show that it has been seen (subjected to vision) and approved.

[2] *E.g.*, Grammateus, 1523.

[3] Thus Köbel: "Die Regel Fusci mit jhrer Erklårung/folget hernach. . . . Das wort Fusci/bedeut nicht anders/dann ein zerbrochen gût gemülb/oder andere vnreynigkeit/so in der Specerei funden wirt/als under den Negelin/Imber/Saffran/&c. Auch Silber vnderm Golt/Kupffer vnderm Silber." *Rechenbüchlein*, 1514; 1549 ed., fol. 77. See also other German works of the period, such as Albert (1534; 1561 ed., fol. D viii) and Thierfelder (1587, p. 116). The word also appears as *Fusti*.

[4] Frequently found in the early Dutch books as "Rekeninghe voor cassiers" (Van der Schuere, 1600, fol. 65), "Reductio, ofte Cassiers Rekeninge" (Cock, 1696 ed., p. 96), "Cassiers Rekeninghe" (Mots, 1640, fol. 1), "Den Regel vander Muntē Oft den Reghel gheheeten Regula Cassiers" (Stockmans, 1676 ed., p. 205).

[5] A common topic in the French, Dutch, and German books of the 16th and 17th centuries under such titles as "Rechnung vberland" (Rudolff, *Kunstliche rechnung* (1526; 1534 ed., fol. k 1)); "Overlantsche Rekeninghe" and "Comptes de Voyage" (Coutereels, Dutch and French arithmetic, 1631, p. 283); "Uytlandtsche Rekeningen" and "Voyages" (Eversdyck's edition of Coutereels, 1658, p. 299); and "Voyagien" (Houck, 1676 ed., p. 148).

[6] Particularly in such centers as Leipzig, where we find arithmetics of about 1700 with such chapters as "Leder und Rauchwahren-Rechnung."

4. APPLICATIONS OF ALGEBRA

General Nature. The first applications of algebra were in the nature of number puzzles. Such was the first algebraic problem[1] of Ahmes (*c.* 1550 B.C.), already mentioned,—"Mass, its whole, its seventh, it makes 19." Such are, in general, the problems of Diophantus (*c.* 275), and the problems which still form the large majority of those given in current textbooks. When a pupil is called upon today to solve the equation $x + \frac{1}{7}x = 19$, he is really solving the first problem of Ahmes, and all our abstract work in equations is a development of this type.

The second general application is geometric, and this characterizes the works of the Greek writers, with the exception of Diophantus.

The third general application is to fanciful problems relating to human affairs, and this is essentially Oriental in spirit.

The fourth type is characterized by the attempt to relate algebra actually to the affairs of life. The first steps in this direction were taken when algebra was more or less a part of arithmetic, problems often being given that were essentially algebraic but were solved without any further symbolism than that afforded by the medieval algorism. It is with this type that we are working at present and are making some advance.

We shall now consider the first three of these stages in detail.

The Number Puzzle. It was with the number puzzle that algebra seems to have taken its start. The desire of the early philosophers to unravel some simple numerical enigma was similar to the child's desire to find the answer to some question in the puzzle column of a newspaper. A few types will be selected involving linear equations, the quadratic being considered later.

Ahmes (*c.* 1550 B.C.) gave numerous problems like "Mass, its $\frac{2}{3}$, its $\frac{1}{2}$, its $\frac{1}{7}$, its whole, it makes 33."[2]

---

[1] No. 24 in Peet, *Rhind Papyrus*, p. 61, with slightly different translation.
[2] *Ibid.*, No. 31. In modern symbols,

$$x + \tfrac{2}{3}x + \tfrac{1}{2}x + \tfrac{1}{7}x = 33.$$

The problems of Diophantus (*c.* 275) were often of this type, as in the first one given in Book I:[1] "To divide a given number into two having a given difference," with the particular case of $2x + 40 = 100$.

In the Middle Ages in Europe the standard algebraic problem was of the same general nature. This is seen, for example, in the *De Numeris Datis* of Jordanus Nemorarius (*c.* 1225), where all the problems are abstract, as in the following case:

"If there should be four numbers in proportion and three of them should be given, the fourth would be given."[2]

From early times to the present the number puzzle has played a leading part in algebra, and under current conditions, when algebra is required as a school subject, this is not altogether fortunate.

Simultaneous Linear Equations. Problems involving simultaneous linear equations were more numerous in the Orient in early times than they were in Europe. Thus we find in India a considerable number of such problems, together with rules that amount to directions for solving various types of simultaneous equations. For example, Mahāvīra (*c.* 850) has the following problem:

The mixed price of 9 citrons and 7 fragrant wood-apples is 107; again, the mixed price of 7 citrons and 9 fragrant wood-apples is 101. O you arithmetician, tell me quickly the price of a citron and of a wood-apple here, having distinctly separated those prices well.

His rule for the solution is similar to the one used today for eliminating one unknown.[3]

Another of Mahāvīra's problems, evidently suggested by one that appeared in the Greek epigrams and was there attributed to Euclid, is as follows:

Three merchants saw [dropped] on the way a purse [containing money]. One [of them] said [to the others], "If I secure this purse,

---

[1] Heath, *Diophantus*, 2d ed., p. 131.

[2] Problem 30: "Si fuerit [*sic*] IIII numeri proporcionales et tres eorum dati fuerint et quartus datus erit" (*Abhandlungen*, II, 143).

[3] See his work, p. 130.

I shall become twice as rich as both of you with your moneys on hand." Then the second [of them] said, "I shall become three times as rich." Then the other [, the third,] said, "I shall become five times as rich." What is the value of the money in the purse, as also the money on hand [with each of the three merchants]?[1]

Indeterminate Problems. The indeterminate problem, leading to what is now called the indeterminate equation, is very old. It is probable that it formed a type of recreation long before the time of Archimedes, since in the problems assigned to him there appears one of very great difficulty. This problem has already been discussed on page 453.[2]

Although Diophantus (c. 275) proposed many indeterminate equations, they were not in the form of applied problems. Cases of the latter kind seem to have come chiefly from the Orient, at least in early times.

In the *Greek Anthology* (c. 500?) there are two problems involving indeterminate linear equations. The first (XIV, 48) is as follows:

The three Graces were carrying baskets of apples and in each was the same number. The nine Muses met them and asked them for apples, and they gave the same number to each Muse, and the nine and three had each of them the same number. Tell me how many they gave and how they all had the same number.[3]

The second (XIV, 144) is a dialogue between two statues:

A. How heavy is the base on which I stand, together with myself!
B. My base together with myself weighs the same number of talents.
A. I alone weigh twice as much as your base.
B. I alone weigh three times the weight of yours.[4]

---

[1] See his work, p. 155.

[2] The problem was discovered by the German dramatist G. E. Lessing, in 1773, while he was serving as court librarian at Wolfenbüttel. On the history of the problem see B. Lefebvre, *Notes d'Histoire des Mathématiques*, p. 33 n. (Louvain, 1920). On a few indeterminate problems due to Heron, see Heath, *History*, II, 444.

[3] The equations reduce to $x = 4y$. There were $12n$ apples in all.

[4] The equations are $x + y = u + v$, $x = 2v$, $u = 3y$.

With their usual desire to give a fanciful but realistic touch to algebra, the Chinese applied the indeterminate equation to a problem commonly known as that of the Hundred Fowls. This problem goes back at least to the 6th century[1] and differs so greatly in its nature from those of the Greeks that it seems to have originated in the East. The problem is as follows:

If a cock is worth 5 *sapeks*; a hen, 3 *sapeks*; and 3 chickens together, 1 *sapek*, how many cocks, hens, and chickens, 100 in all, will together be worth 100 *sapeks*?

From China the problem apparently found its way to India, for it appears in Mahāvīra's work (c. 850) in the following form:

Pigeons are sold at the rate of 5 for 3 [*paṇas*], *sārasa* birds at the rate of 7 for 5 [*paṇas*], and peacocks at the rate of 3 for 9 [*paṇas*]. A certain man was told to bring at these rates 100 birds for 100 *paṇas* for the amusement of the king's son, and was sent to do so. What [amount] does he give for each [of the various kinds of birds that he buys]?

Mahāvīra gave a method for solving such problems that was sufficient to satisfy those who were interested in puzzles, but which had little merit otherwise.[2]

This fanciful type of problem was probably made known in Europe at the time of the general penetration of Oriental ideas, and here it developed into a form somewhat like this:

20 persons, men, women, and girls, have drunk 20 pence worth of wine; each man pays 3 pence, each woman 2 pence, and each girl ½ penny; required the number of each.

Algebraically, $m + w + g = 20$

and $3m + 2w + \frac{1}{2}g = 20,$

---

[1] Kaye, *Indian Math.*, p. 40; L. Vanhée, "Les cents volailles ou l'analyse indéterminée en Chine," *T'oung-pao*, XIV, and reprint; L. Matthiessen, "Vergleichung der indischen Cuttaca- und der chinesischen Tayen-Regel, unbestimmte Gleichungen und Congruenzen ersten Grades aufzulösen," *Sitzungsberichte der math.-naturwiss. Section in der Verhandl. der Philol. Vers. zu Rostock*, 1875.

[2] See his work, pp. 133–135.

which would be indeterminate if it were not for the fact that
there were some of each and that the result must be in positive
integers, thus admitting of only one set of answers, namely,
1, 5, and 14. In general, all problems of this type are of the
form

$$x + y + z = m,$$
$$ax + by + cz = n,$$

reducing to $\qquad px + qy = r,$

thus being indeterminate unless the physical conditions are
such as to exclude all but one set of values.

The Regula Coecis. Such problems are known in European
works as early as the 9th century, and thereafter they become
common.[1] In the 15th century they begin to mention the persons as being at a *cecha*,[2] and hence the rule for solving such
problems became known to 16th century writers as the *Regula
Coecis*. There has been much dispute as to the origin of the
term *coecis*,[3] but from such historic evidence as we now have it
seems to relate to the fact above stated, namely, that the persons were at a *cecha*.[4] The problem seems to relate to drinking
where each paid his own share,[5] very likely from the fact that
zeche meant originally the money[6] paid for the drinks.

[1] Thus a 14th century MS. in Munich (cod. lat. Monac. 14684): "Sint hic
milites, pedites et puelle, et sint in universo 12, et habeant 12 panes, parciendos,
et quilibet miles accipiat duos, quilibet pedes quartam partem panis, quilibet
puella medietatem panis: queritur, quot erunt milites, pedites et puelle." *Bibl.
Math.*, IX (2), 79.

[2] Curtze, for example, found, in a MS. of 1460, a problem beginning: "ponam
casus, quod sint 20 persone in una cecha" (*Abhandlungen*, VII, 35).

[3] Which also appears in such forms as *zekis, zeches, cekis, ceci, coeci, caeci,* and
caecis. Bibl. Math., XIII (2), 54, and VI (3), 112.

[4] We find similar expressions in the 16th century. Thus Rudolff (*Kunstliche
rechnung*, 1526) has a topic "Von mancherley person an einer zech," and says
further: "Es sitzen 20 person an einer zech/man/frawen/vnd jungfrawen . . .,"
adding that this is the "regel/welche sie nennen Cecis oder Virginū/" (1534 ed.,
fol. Nvij).

[5] L. Diefenbach, in his *Novum Glossarium* (Frankfort a.M., 1867, p. 339),
says: "symbolum . . . vulgo zecha quo quisque suam portionem confert." In
American slang, a *zecha* was a "Dutch treat."

[6] Italian *zecca*; compare *zecchiere*, a mint master, as in Ciacchi, *Regole generali d' Abbaco* (Florence, 1675, p. 247).

There have been various other speculations as to the word. One of these is that it comes from an Arabic expression signifying not content with one but demanding many, referring to the many possible solutions.[1] It has also been thought to come from the Latin *caecus* (blind), with such fanciful explanations as that a problem of this kind was solved by the blind Homer, or that the solver went blindly to work to find the solution. Indeed, it was often called the Blind Rule in the 16th and 17th centuries.[2]

Another name for the rule was *Regula potatorum* (rule of the drinkers), and this gives added reason for the interpretation of *zecha* as a drinking bout.[3] A still more common name was *Regula virginum*, rule of the girls, usually explained by the fact that the solutions show more girls than men or women.[4]

In spite of the fact that this style of problem is interesting, the arithmeticians often discouraged its use because of its indeterminateness,[5] although there were found others who increased the difficulty by using more than three unknowns.

Alligation. Beginning apparently in the Renaissance period as an application of indeterminate equations, the *Regula Alli-*

---

[1] "Cintu Sekîs, hoc est adulteram indigetarunt: propterea, ut opinor, quod uno ac legitimo quaestionis enodatu non contenta, plures plerumque admittat solutiones." J. W. Lauremberg, *Arithmetica* (Sorö, 1643), quoted by Zeuthen in *L'Intermédiaire des Mathématiciens*, p. 152 (Paris, 1896) (hereafter referred to as *L'Intermédiaire*); *Bibl. Math.*, X (2), 96. Carra de Vaux (*Bibl. Math.*, XI (2), 32) says that Lauremberg's expression should have the word *sikkîr*, which means toper, and this is more reasonable.

[2] Thus Thierfelder asks: "Warumb wirdt dise Regel Cecis genannt?" and explains that the problem is indeterminate and that "ein Ungeübter nicht bald finden kan/darumb ist es jm ein blinde Regel" (1587, p. 211). Cardinael speaks of it as "Den Blinden-Reghel" (1674 ed., p. 88).

[3] The Dutch writer, Bartjens, also speaks of it as "Bachus-rekeninge" (1752 ed., p. 213). See also Unger, *Die Methodik*, p. 101.

[4] This origin is stated by Jacobus Micyllus (1555): "regula quam ab eo, ut videtur, appellarunt, quod virginum personae ac nomen inter exempla illius subinde repetuntur." So we find "Coecis oft Virginum" (Van der Schuere, 1600, fol. 174), "Rekeninge Coecis, ofte Virginum" (Bartjens, 1633), and similar forms, expecially among the early Dutch writers.

[5] Thus Stockmans: "Desen regel niet soo seker en is in zijn werckinge als de ander voorgaēde" (1676 ed., p. 380); and Coutereels remarks that it is "meer vermakelijkheyd als sekerheyd" (1690 edition of the *Cyffer-Boeck.* p. 559).

gationis,[1] or Rule of Alligation, attracted considerable attention for nearly three hundred years. Problems in alligation were sometimes indeterminate and sometimes not. The following, for example, is indeterminate:

How many Raisins of the Sun, at *7d. per lb.* and Malaga Raisins at *4d. per lb.* may be mixed together for *6d. per lb.*?[2]

Such a problem becomes determinate by the addition of some further appropriate condition, as in the following case:

A tobacconist mixed 36 lb. of tobacco, at 1s. 6d. per lb. 12 lb. at 2s. a pound, with 12 lb. at 1s. 10d. per lb.; what is the price of a pound of this mixture?[3]

In general, such problems were simply ingenious efforts to make algebra seem real, but they were usually solved without the aid of algebraic symbols, and hence they found place in higher arithmetics until the close of the 19th century, when they generally disappeared except as a few remained in the form of mixture problems[4] in the elementary algebras. Indeed, Recorde (*c.* 1542) remarked that the rule of alligation "might be well called the rule of Myxture."[5] He was the first English writer to suggest other applications than those referring to alloys, saying: "it hath great vse in composition of medicines,

---

[1] From *ad*, to, + *ligare*, to bind. From the same roots come the French *alliage*, and our words *alloy* and *ally*. So we have in French "La regle des aliages" (Trenchant, 1566; 1578 ed., p. 191) and "La Reigle d'Alligation" (Peletier, 1549; 1607 ed., p. 247).

[2] T. Dilworth, *The Schoolmasters Assistant*, new ed., p. 97 (London, 1793). This was one of the most celebrated English arithmetics of the period; it had great influence on American textbooks.

[3] N. Daboll, *Schoolmaster's Assistant*, 1837 ed., p. 177. This was one of the most celebrated of the early American arithmetics.

[4] The Dutch writers called them problems solved by "Den Reghel van Menginghe" (Cardinael, 1674 ed., p. 66, with 23 pages to the subject). So we find "Allegatio, Menginghe" (Van der Schuere, 1600), "Rekeninghen van Mengelingen" (Raets, 1580 ed., fol. K3), "Alligationis, ofte Menginghe" (Bartjens, 1676 ed., p. 165), and "Alligatio, Alliage, of Mengingh" (Coutereels, *Cyffer-Boeck*, 1690 ed., p. 484). The Italian writers sometimes had a chapter *De' mescoli* (on mixtures), and this may have suggested the German *Mischungsrechnung*. For a recent use of the topic in Germany, see R. Just, *Kaufmännisches Rechnen*, I, p. 86 (Leipzig, 1901). [5] *Ground of Artes*, 1558 ed., fol. Y3.

and also in myxtures of metalles, and some vse it hath in myxtures of wines, but I wshe it were lesse vsed therin than it is now a daies." These practices rendered the subject so popular that Baker (1568) gave forty-eight pages to it in his 1580 edition.

Mint Problems. One of the leading applications of alligation was found in the general need for the mixing of chemicals and metals by the alchemists of the Renaissance,[1] by bell founders, and by mint masters. As to the coining of money, it should be remembered that this was not in general a government monopoly in the Middle Ages, so that it was looked upon as something unheard of that Ferdinand and Isabella should assert this right in 1496.[2] The privilege belonged rather to cities or districts and even in a single small country was often claimed by several people,[3] often as an inherited right.[4] Add to this fact the great awakening in the mining industry of Germany in the latter part of the 15th century and the extensive importations of gold and silver from the Americas in the first half of the 16th century, and it will be seen that the subject of alligation naturally had at that time a new and popular field in the mixing of alloys for purposes of coinage.[5] This explains the interest in the subject of coinage in the 16th century.[6]

[1] "Il consolare oro, ed argento non è altro, che vn' allegazione di que' due metalli, per li quali la maggior parte degli Alchimisti son diuenuti miseri, e mendichi, per volere inuestigare la congelazione del Mercurio in vera, ed ottima Luna, o Sole, la quale senza il diuino aiuto in vano da gli Alchimisti vien tentata." Ciacchi, *Regole generali d' Abbaco*, p. 244 (Florence, 1675).

[2] H. B. Clarke, in the *Cambridge Mod. Hist.*, I, chap. xi.

[3] W. Cunningham, in the *Cambridge Mod. Hist.*, I, chap. xv.

[4] E. P. Cheyney, *Documents Illustrative of Feudalism*, p. 34 (Philadelphia, 1898).

[5] Of course alloys for this purpose had been known and used to some extent ever since the early coinage from natural electron in Asia Minor and from bronze in Rome. On the latter, see F. Gnecchi, *Monete Romane*, 2d ed., p. 86 (Milan, 1900).

[6] So Rudolff (1526; 1534 ed., fol. M 6) and Grammateus (1518; 1535 ed., fols. C 8, D 4, etc.) have chapters on *Muntzschlag*, and various writers speak of the problems of the *müntzmeister* (Rudolff), *Munt-meester* (Van der Schuere, 1600), and *mint-master* (Hodder, 1672 ed.). The Germans also had chapters on *Silber Rechnung*, *Goldt Rechnung*, and *Kupfer Rechnung* (Riese, 1522; 1550 ed., fols. 37-40); the Dutch, on *Rekeninghe van Goudt end Silber*, with *Comptes d'or & d'argent* (Coutereels's Dutch and French editions of 1631, p. 298, and other dates); and the Italians, on *Del consolare dell' oro et dell' argento* (Cataneo, 1546; Tartaglia, 1556).

Problem of Hiero's Crown. Closely related to this subject is the problem of Hiero's crown, which Robert Recorde (*c.* 1542) states in quaint language as follows:

Hiero kynge of the Syracusans in Sicilia hadde caused to bee made a croune of golde of a wonderfull weight, to be offered for his good successe in warres: in makynge wherof, the goldsmyth frauduletly toke out a certayne portion of gold, and put in syluer for it. [Recorde then relates the usual story of Archimedes and the bath, telling how the idea of specific gravity came] as he chaunced to entre into a bayne full of water to washe hym, [and] reioycing excedingly more then if he had gotten the crown it self, forgat that he was naked, and so ranne home, crying as he ranne εὕρηκα, εὕρηκα,[1] I haue foūd, I have found.

Cardan[2] asserts that the story is due to Vitruvius, who simply transmitted the legend. It appears in various books of the 16th century and is still found in collections of algebraic problems.

First Problems in the New World. To those especially who live in the New World some local interest attaches to the problems that appear in the first mathematical work (1556) to be printed there. Among these problems are the following:[3]

I bought 10 varas of velvet at 20 pesos less than cost, for 34 pesos plus a vara of velvet. How much did it cost a vara? Add 20 pesos to 34 pesos, making 54 pesos, which will be your dividend. Subtract one from 10 varas, leaving 9. Divide this into 54, giving 6, the price per vara.

I bought 12 varas of velvet at 30 pesos less than cost, for 98 pesos minus 4 varas. How much was the cost per vara? The following is a short method: add the 30 pesos and the 98 pesos, making 128; add the number of varas, 12 and 4, making 16; divide 16 into 128, giving 8, the price per vara.

---

[1] *Eu'reka, eu'reka*; more precisely, *heu'reka*.

[2] *Practica*, 1539, cap. 66, ex. 45. He relates that Archimedes "nudus e balneo exultās domū reuertebatur," adding, "nescio an ob amorē veritatis potius laudandus quā ob importunā & impudicā nuditatē vituperādus."

[3] See also pages 385, 392, and Volume I, pages 353–356.

I bought 9 varas of velvet for as much more than 40 pesos as 13 varas at the same price is less than 70 pesos. How much did a vara cost? Add the number of pesos, 40 and 70, making 110. Add the number of varas, 9 and 13, making 22. Dividing 110 by 22, the quotient is 5, the price of each vara.

A man traveling on a road asks another how many leagues it is to a certain place. The other replies: "There are so many leagues that, squaring the number and dividing the product by 5, the quotient will be 80." Required to know the number of leagues.

A man is selling goats. The number is unknown except that it is given that a merchant asked how many there were and the seller replied: "There are so many that, the number being squared and the product quadrupled, the result will be 90,000." Required to know how many goats he had.

The work contains other and more difficult problems in algebra and the theory of numbers, but the above are types of the ordinary puzzles which the author places before his readers.

5. MAGIC SQUARES

Oriental Origin. The magic square seems unquestionably to be of Chinese origin. The first definite trace that we have of it is in the *I-king*, where it appears as one of the two mystic arrangements of numbers of remote times. This particular one, the *lo-shu*, is commonly said to have come down to us from the time of the great emperor Yu, *c.* 2200 B.C. The tradition is that when this ruler was standing by the Yellow River a divine tortoise appeared, and on its back were two mystic

4	9	2
3	5	7
8	1	6

symbols, one being the *lo-shu* already described in Volume I, Chapter II. As may be seen from the illustration there given, it is merely the magic square here shown.

This particular square is found in many recent Chinese works, and every fortune teller of the East makes use of it in his trade. Little by little the general knowledge of magic squares seems to have been extended, and when Ch'êng Tai-wei

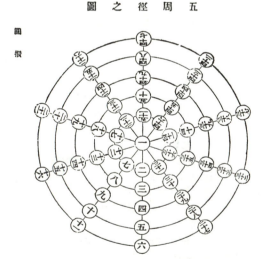

MAGIC SQUARES IN JAPAN

Half of a magic square as given in Hoshino Sanenobu's *Kō-ko-gen Shō*, 1673

MAGIC CIRCLE FROM SEKI'S WORKS

From the reprint of the works of the great Japanese mathematician, Seki Kōwa
(*c.* 1661)

wrote his *Systematized Treatise on Arithmetic*[1] in 1593, he included not only a discussion of magic squares but also one of magic circles.

Japan. The Japanese became particularly interested in the subject in the 17th century, as the illustrations from the works

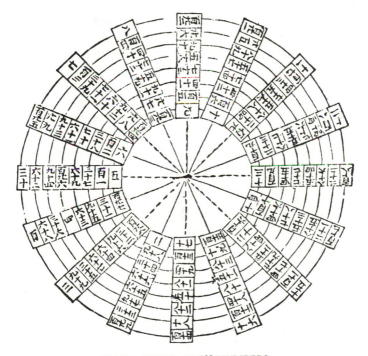

MAGIC CIRCLE OF 129 NUMBERS

From Muramatsu Kudayū Mosei's *Mantoku Jinkō-ki*, 1665. The numbers in each radius add to 524, or 525 with the center 1

of some of their leading writers show.[2] Among the prominent scholars who gave attention to these forms were Muramatsu Kudayū Mosei, who wrote several works on arithmetic and

---

[1] *Suan-fa Tong-tsung.* See Volume I, page 352.
[2] For their treatment of the subject see Smith-Mikami, pp. 57, 69, 71, 73, 79, 116, 120, 177.

geometry, beginning in 1663; Hoshino Sanenobu, whose *Triangular Extract*[1] appeared in 1673; Isomura Kittoku, whose *Ketsugi-shō* appeared in 1660; and the great Seki Kōwa, who devoted one of his *Seven Books*[2] to the theory of magic squares and magic circles.

7	12	1	14
2	13	8	11
16	3	10	5
9	6	15	4

India. From China the magic square seems to have found its way into India and the adjacent southern countries, but whether this was direct or through the Arab influence we can only conjecture. It appears in a Jaina inscription in the ancient town of Khajurāho, India, where various ruins bear records of the Chandel dynasty (870–1200), and is probably not older than the 11th or 12th century. This Indian square, shown above, displays a somewhat advanced knowledge of the subject, for not only has it an even number of squares on a side, but each of the four minor squares has a relation to the others, as may be seen by the illustration at the right.

This is perhaps the earliest trace of such fantastic elaborations of the magic square,[3] although no careful study of the history has yet been made.[4] It is probable that the astrologers carried such ideas to the West, where their influence upon the medieval mathematics of Europe is apparent.

9	19 15	25	9	15 19	25
25	19 15	9	25	15 19	9

Today the magic square is used as a charm all through India, being found in fortune bowls, in medicine cups, and in amulets. In Thibet it is particularly in evidence, being found in the

[1] *Kō-ko-gen Shō.*

[2] Or *Shichibusho*. The particular book is the *Hōjin Yensan*, revised in MS. in 1683.

[3] F. Schilling, *Jahresbericht der deutschen Math. Verein.*, XIII, 383.

[4] But see such works as W. S. Andrews, *Magic Squares* (Chicago, 1907), and subsequent articles in *The Open Court*. For Roman and Egyptian claims see E. Falkener, *Games, Ancient and Oriental*, p. 277 (London, 1892).

"wheel of life"[1] and worn as an amulet to ward off evil. It is also seen in Sumatra, in the Malay Peninsula, and in the other countries which have had close relations with India and China.

The squares are not always of the pure type, however; that is, the sums of the rows, columns, and diagonals are not always constant. In some of them, for example, the columns add successively to 300, 200, 100, and the like,[2] and in many of them the numbers are repeated whenever it was necessary to make the sums come as desired.

Connection with Alchemy. It seems that the numbers must often have been connected with the old alchemistic idea of the planets and the metals,—an idea that

THIBETAN TALISMAN

With signs of the zodiac, the ancient *pa-kua* or trigrams, and the *lo-shu* in Thibetan numerals

permeated the doctrines of many of the medieval mystics. Of the three triads made up of the nine digits, the first had the following relations:

1 = gold = the sun, ⊙

2 = silver = the moon, ☽

3 = tin = Jupiter, ♃ (the hand grasping the thunderbolt)

[1] See the illustration in Volume I, page 27.
[2] S. S. Stitt, "Notes on some Maldivian Talismans," *Journ. Royal Asiatic Soc.,* pp. 121, 130 (London, 1906).

The second triad was as follows:

4 = gold again = the sun, ☉
5 = mercury = Mercury, ☿
6 = copper = Venus, ♀

The third triad was as follows:

7 = silver again = the moon, ☽
8 = lead = Saturn, ♄
9 = iron = Mars, ♂

With such a relationship it is possible to understand such talismans as one found in the Maldivian Islands, in the Indian Ocean, where the charm to protect a virgin sums to 18, whose digits sum to 9, the number of Mars, the protector. In these numbers there also enters the idea of congruence, particularly to the modulus 9. Thus a talisman to keep out Satan has its rows 80 ≡ 8, 1600 ≡ 7, 180 ≡ 9, · · ·, which have for their sum 69 ≡ 15 ≡ 6 = Venus,—a vagary that can be explained only by conjecture.[1]

Hebrews. The magic square played an important part in the cabalistic writings of the Hebrews. Rabbi ben Ezra (*c.* 1140) mentions it, although it had been used by Hebrew writers long before his time.[2] The Eastern Jews early found in the ancient Chinese *lo-shu*, with its constant sum 15, a religious symbol, inasmuch as 15 in Hebrew is naturally יה (10 + 5), which is made up of the first two letters of Jahveh (Jehovah), that is, יהוה, although, lest they should be guilty of profanity, they always wrote 9 + 6 for 15. If the corner numbers of the common form of square are suppressed, the even (feminine) elements are eliminated and there remain only the odd numbers, this cruciform arrangement serving as a charm among various Oriental peoples.[3]

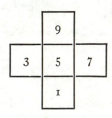

---

[1] Stitt, *loc. cit.*, p. 144.   [2] M. Steinschneider, *Abhandlungen*, III, 98.
[3] D. Martines, *Origine e progressi dell' Aritmetica*, p. 39 (Messina, 1865), with considerable information upon this subject.

Arabs. Whether the magic square reached the Arabs from India, from China, or from Persia, we do not know. It might readily have come from any one of these countries through either of the others, but at any rate it was well known to various late Arab writers, appearing, for example, in the works of the philosopher Gazzâlî[1] about 1100.

Christians. In general the magic square found no recognition as a Christian symbol, although the occult writers naturally made use of it. When it appears in medieval mathematical works it is usually in the form of a problem whose solution requires the arrangement of the ordinary 9-celled square shown below.[2]

The textbook writers of the 16th century paid considerable attention to the subject. Cardan, for example, gives seven different squares bearing respectively the names of the sun, the moon, and the five planets then known, and gives some directions for the formation of such squares.[3]

In art the first instance of the use of these forms is probably the one in the well-known *Melancholia* by Dürer.

One of the most elaborate examples found in architectural decoration is cut in the wall in the Villa Albani at Rome.[4] It contains eighty-one cells and is dated 1766.

---

[1] Mohammed ibn Mohammed ibn Mohammed, Abû Ḥâmid, al-Gazzâlî, born at Ṭûs, 1058/59; died at Ṭûs, 1111. The Arabic name of the magic square is *shakal turâbi*. See E. Rehatsek, "Explanations and Facsimiles of eight Arabic Talismanic Medicine-cups," *Journ. of the Bombay Branch of the Royal Asiatic Soc.*, X, 150.

[2] For example, Günther reports a 13th century MS. with this problem: "Tres erant fratres in Colonia, habentes 9 vasa vini. Primum vas continet 1 amam, secundum 2, tertium 3, · · · nonum 9. Divide vinum illud aequaliter, inter illos tres, vassis inconfractis."

The division is $1 + 5 + 9$, $3 + 4 + 8$, $2 + 6 + 7$, but that this is indeterminate is seen by the magic square here represented. [3] *Practica*, 1539, capp. 42, 66.

2	7	6
9	5	1
4	3	8

[4] The inscription is "Caetanus Gilardonus Romanus philotechnos inventor. A.D. MDCCLXVI." On the recent mathematical investigations the reader will do well to consult such works on mathematical recreations as those of Lucas, Ball, Schubert, and Ahrens; the work by W. S. Andrews, mentioned on page 594; and the references in G. A. Miller, *Historical Introduction to Mathematical Literature*, p. 20 (New York, 1916).

II

DÜRER'S "MELANCHOLIA," WITH MAGIC SQUARE

One of the first magic squares to appear in print. Chiefly interesting because it is an even-celled square. Dürer (1471–1528) also wrote on higher plane curves in connection with art

TOPICS FOR DISCUSSION

1. The general progress of mathematical recreations, with a consideration of the leading works upon the subject.

2. The influence of mathematical recreations upon the development of mathematics.

3. Traces of the ancient puzzle problems in the elementary mathematical literature of the present time, with a consideration of the value of these problems.

4. Types of puzzle problems dependent upon indeterminate equations, with a study of their history.

5. Mathematics as an aid to the study of the history of economics and commerce.

6. The historical development of the corporation, as traced through the problems of arithmetic.

7. Racial influences upon business customs, as seen in commercial problems.

8. The standard applications of arithmetic from remote times to the present, with a consideration of their important changes.

9. Commercial problems at one time important and now nearly obsolete, with a consideration of the causes of their rise and fall.

10. Certain commercial problems and customs, at one time important but now nearly or quite obsolete, left their impress upon our language. Consider a few such cases.

11. Ancient problems of the bank compared with those of the present time.

12. The need for barter in ancient times and the reasons for its retention, even in highly civilized countries, until recently.

13. The history of various types of taxation in ancient, medieval, and modern times.

14. The reason why gaging was considered as an important branch of mathematics in the early printed arithmetics and geometries, and why it lost its standing in the 19th century.

15. Reasons for the study of magic squares in ancient times, in the Middle Ages, and at the present time, with typical illustrations.

16. Instances of the use of the magic square in the East as a talisman or amulet.

17. The relation of magic squares to alchemy, and the relation of alchemy to modern science.

CHAPTER VIII

TRIGONOMETRY

1. GENERAL DEVELOPMENT OF TRIGONOMETRY

Meaning of the Term. If we take trigonometry to mean the analytic science now studied under this name, we might properly place its origin in the 17th century, after the development of a satisfactory algebraic symbolism. If we take it to mean the geometric adjunct to astronomy in which certain functions of an angle are used, we might look for its real origin in the works of Hipparchus (*c.* 140 B.C.), although there are earlier traces of its use. If we take it to mean literally "triangle measurement,"[1] the origin would naturally be placed much earlier, say in the second or third millennium B.C. Since this third phase is considered under geometry, we may properly confine our work to the development of the idea of the functions of an angle, giving first a brief sketch of the rise of the science and then the history of certain of its details.

Egypt. In the Ahmes Papyrus (*c.* 1550 B.C.) there are five problems[2] relating to the mensuration of pyramids, and four of these make mention of the *seqt*[3] of an angle. Ahmes is not at all clear in expressing the meaning of this word, but from the context it is thought that the *seqt* of

---

[1] Τρίγωνον (*tri'gonon*, triangle) + μέτρον (*met'ron*, measure).

[2] Nos. 56–60 in the list. See Peet, *Rhind Papyrus*, p. 97; Eisenlohr, *Ahmes Papyrus*, pp. 134–148. Throughout this chapter much use has been made of A. Braunmühl, *Geschichte der Trigonometrie*, 2 vols., Leipzig, 1900, 1903; hereafter referred to as Braunmühl, *Geschichte*.

[3] Eisenlohr takes the word to mean ratio number. It is also transliterated *śḳd* and *seqet*. It may be significant that the Hebrew *sgd* means "bowing."

the regular pyramid shown on page 600 is probably equivalent to cot ∠*OMV*.[1] The Egyptian pyramids were generally constructed so that ∠*OMV* was approximately constant (about 52°) and ∠*OAV* was about 42°. At present we are without means of knowing what use was made of this function.

Babylon. The relation between the mathematical knowledge of the Egyptians and that of the Babylonians in the third millennium B.C., as seen in the unit fraction (p. 210), leads us to suppose that the latter people may have known of the primitive Egyptian trigonometry. We have, however, no direct knowledge that this was the case. There are evidences of angle measure at a very early date, as witness fragments of circles which seem to have been used for this purpose and which have come down to us. These fragments seem to have been parts of primitive astrolabes, as stated on page 348.

There is also extant an astrological calendar of King Sargon, of the 28th century B.C., and a table of lunar eclipses beginning 747 B.C., so that evidence of an interest in astronomy is not lacking throughout a long period of Babylonian history. All this involved a certain amount of angle measure, but there is no direct evidence of any progress in what we commonly understand as trigonometry.

The Gnomon. Herodotus (*c.* 450 B.C.) tells us that the Greeks obtained their sundial from Babylon. This is very likely true, for we know that the Egyptians used a sun clock as early as 1500 B.C., and the Babylonians could hardly have been behind them in the knowledge of such a device. The relation of the sundial to trigonometry is seen in the fact that it is an instrument for a form of astronomical

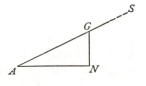

observation. A staff *GN*, called by the Greeks a *gnomon* (p. 16), is erected and the shadow *AN* observed. It is longest at noon when *S*, the sun, is farthest south, this being at the winter

---

[1] For discussion see Braunmühl, *Geschichte*, I, 2: Eisenlohr, *Ahmes Papyrus*, p. 137. Peet, *Rhind Papyrus*, p. 98, gives cot ∠*OMV*.

solstice, and shortest when it is farthest north, at the summer solstice; and hence an examination of its limits enables the observer to measure the length of the year. The daily lateral motion of the point A allows for the measure of diurnal time, quite as the motion of noon along AN allows for the measure of annual time. The gnomon being constant, the length of AN at noon varies with $\angle A$, and to us this means a recognition that AN, or $AN: GN$, is a function of $\angle A$, namely, the cotangent. We have no trace, however, of any name (except the *seqt*) for such a relation in the period of which we are speaking.

China. In the *Chóu-peï Suan-king* (*c.* 1105 B.C.)[1] the right-angled triangle is frequently used in the measure of distances, heights, and depths, and it is quite probable that the ratios of the sides were recognized. One passage reads, "The knowledge comes from the shadow, and the shadow comes from the gnomon," so that possibly a primitive plane trigonometry was known in China in the second millennium B.C. Aside from this there is no evidence that the early Chinese had names for any functions of an angle. The early astronomical interests of the Chinese, however, like those of other ancient peoples, necessitated some kind of angle measure.

Greece. When Thales measured the height of a pyramid by means of its shadow, he used what was already known, probably in various parts of the world, as "shadow reckoning."[2] In his "Banquet of the Seven Wise Men" Plutarch speaks of Nilax, one of the guests, as saying to Thales:

Whereas he[3] honors you, he particularly admires you for divers great accomplishments and particularly for the invention whereby, with little effort and by the aid of no mathematical instruments, you found so accurately the height of the pyramids. For, having fixed your staff erect at the point of the shadow cast by the pyramid, two triangles

---

[1] But see Volume I, page 30; also Mikami, *China*, p. 4.

[2] We have this term, substantially at least, in the *Chóu-peï Suan-king*; in the works of Brahmagupta (*c.* 628), under "Measure by Shadow," p. 317; in Mahāvīra (*c.* 850), under "Calculations relating to Shadows," p. 275; and in Bhāskara, under "Ch'háyá-vyavahára" (determination of shadow), p. 106.

[3] The king of Egypt, called by the Greeks Amasis, *c.* 570 B.C.

were formed by the tangent rays of the sun, and from this you showed that the ratio of one shadow to the other was equal to the ratio of the [height of the] pyramid to the staff.

Essentially the measure of heights by means of shadows involves the knowledge that, in this figure, $BC : AB = B'C' : AB'$. To us it seems as if $\tan A$ would be suggested by such a relation, but we have no evidence that this was the case in the time of Thales. We only know that, centuries later, AB was called the *umbra recta* (right shadow), showing that the relation of AB to BC entered trigonometry through shadow reckoning.

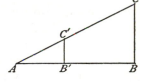

It is said that Anaximander (*c.* 575 B.C.) erected near Sparta the first gnomon in Greece. It was probably in the form of an obelisk, a mere post placed perpendicular to the apparent plane of the earth's surface, and not the triangular form later in use. It could have been used for determining the meridian line, and tradition says that this was done; but besides this it served, as it probably did in Egypt and Babylon, to measure the year, the seasons, and the time of day.

Relation to Astronomy. In this early work of Anaximander, as in similar cases among the Babylonians and Egyptians, it is evident that the real purpose in view was the study of astronomy, the unraveling of the mysteries of the universe. This led to the study of the celestial sphere, the triangles being, therefore, spherical figures. This accounts for the fact that the study of spherical triangles kept pace with that of plane triangles in the Greek trigonometry. We find, however, no tangible evidence of the definition or even of the idea of a spherical triangle before the appearance of the work of Menelaus on spherics (*c.* 100).[1]

Early Works on Spherics. The oldest extant works on spherics, and indeed the oldest Greek mathematical texts that have come down to us, are two astronomical treatises by Autol'ycus[2]

---

[1] Heath, *History*, II, 262.   [2] Αὐτόλυκος. See Heath, *History*, I, 348.

of Pitane (*c.* 330 B.C.). The first is on a moving sphere[1] and consists of twelve elementary propositions relating to the principal circles. The second work was on the risings and settings of the fixed stars,[2] in two books. Neither of these works shows, however, any knowledge of spherical trigonometry.

Aristarchus. The next important step in the development of trigonometry was taken by the astronomer Aristarchus of Samos (*c.* 260 B.C.).[3] He attempted to find the distances from the earth to the sun and the moon, and also the diameters of these bodies. His geometric reasoning was accurate, but his instruments were so crude that he could come no nearer the ratio of the distance of the moon to that of the sun than to say that it was between $\frac{1}{18}$ and $\frac{1}{20}$. In his proof he makes use of ratios which are suggestive of the tangent of an angle.[4]

Hipparchus. In his commentary on the *Almagest*,[5] Theon of Alexandria (*c.* 390) asserts that Hipparchus (*c.* 140 B.C.), the greatest of the Greek astronomers, wrote twelve books on the computation of chords[6] of angles, but of these books we have no further trace. Hipparchus himself, in the fragment of his work that has come down to us, leads us to believe that he was engaged in such computations and in the graphic solution of spherical triangles.[7] It therefore seems reasonable to assert,

[1] Περὶ κινουμένης σφαίρας.

[2] Περὶ ἐπιτολῶν καὶ δύσεων. The two works were edited by Hultsch in 1885.

[3] P. Tannery thinks that this step, which is usually attributed to Aristarchus, was taken by Eudoxus (*c.*370 B.C.). See his "Aristarque de Samos," *Mém. de la Soc. des sciences de Bordeaux*, V (2), 241; *Mémoires Scientifiques*, I, 371; Heath, *History*, II, 1.

[4] This proof is given in Braunmühl, *Geschichte*, I, 8, and by Tannery, *Mémoires Scientifiques*, I, 376. See also the Commandino edition of Aristarchus, 1572; R. Wolf, *Geschichte der Astronomie*, p. 172 (Munich, 1877). The work of Aristarchus, Περὶ μεγεθῶν καὶ ἀποστημάτων ἡλίου καὶ σελήνης, was translated by A. Nokk and published, with a commentary, in a Programm, Freiburg, 1854. See also Heath, *History*, II, 4.

[5] Halma's French translation, p. 110 (Paris, 1821).

[6] The Greeks called the chord εὐθεῖα (*euthei'a*), the Latin *chorda* being from the Greek χορδή (*chorde'*, intestine), whence it meant a string made of dried intestine used in a lyre, and hence a straight chord of a bow (arc).

[7] Braunmühl, *Geschichte*, I, 10; Heath, *History*, II, 257.

from the evidence that we have, that the science of trigonometry begins with Hipparchus.[1] It has been asserted, but the proof is unsatisfactory, that the formulas for sin $(A \pm B)$ and cos $(A \pm B)$, and for the radius of the circumscribed circle $\left(R = \dfrac{abc}{4\Delta}\right)$, were essentially known to him.[2] In order to solve a triangle Hipparchus and other early writers always supposed it inscribed in a circle. The sides were then considered as chords, and these were computed as functions of the radius. In this way the table of chords was of special value. Triangles on a sphere were always decomposed into right-angled triangles, and these were solved separately. Although not mentioning the subject of spherical triangles in any of his works now extant, Hipparchus solves a certain problem in which he must have used the equivalent of the formula tan $b = $ cos A tan c, where $C = 90°$, and both he and Ptolemy ($c.$ 150) knew the relation which we express by the equation $\sin^2 A + \cos^2 A = 1$.[3]

The treatise of Theodosius of Tripoli ($c.$ 100) on the sphere[4] may be passed with mere mention, since it contains no work on trigonometry.

Heron of Alexandria. Although Heron ($c.$ 50?)[5] showed much ingenuity in his mensuration of the triangle, and was thoroughly conversant with the art of surveying as practiced in Egypt, it cannot be said that he gave any evidence of appreciating the significance of trigonometry. He made use of certain rules which we should express in formulas for finding the area of regular polygons, giving in each case the product of the square of a side by a certain number, and these rules afford some evidence of a kind of prognosis of trigonometric functions. That

---

[1] On his astronomical work, see J. B. J. Delambre, *Histoire de l'astronomie ancienne*, I, 106 (Paris, 1817); P. Tannery, *Recherches sur l'histoire de l'astronomie ancienne*, Paris, 1893.

[2] That they are essentially involved in Euclid's *Elements*, VI, 16, is shown in the Simson additions, Props. C and D. See Heath's *Euclid*, Vol. II, pp. 224, 225.

[3] Heath, *History*, II, 259.

[4] Latin ed., Paris, 1529; Greek ed., Beauvais, 1558. See Volume I, page 125.

[5] Or possibly as late as $c.$ 200. See Volume I, page 125.

is, taking A_n as the area of a regular n-gon and s_n as the side, he stated that the following relations exist:

$$A_8 = \tfrac{13}{30}\, s_3^2, \qquad\qquad A_8 = \tfrac{29}{6}\, s_8^2,$$

$$A_5 = \tfrac{5}{3}\, s_5^2, \quad \text{or} \quad \tfrac{12}{7}\, s_5^2, \qquad A_9 = \tfrac{51}{8}\, s_9^2, \quad \text{or} \quad \tfrac{88}{6}\, s_9^2,$$

$$A_6 = \tfrac{13}{5}\, s_6^2, \qquad\qquad A_{10} = \tfrac{15}{2}\, s_{10}^2,$$

$$A_7 = \tfrac{43}{12}\, s_7^2, \qquad\qquad A_{12} = \tfrac{45}{4}\, s_{12}^2.$$

Since $A_n = \tfrac{1}{4}\, n s_n^2 \cot \dfrac{180°}{n}$, it might be inferred that Heron had some knowledge of $\cot \dfrac{180°}{n}$, but there is nothing in the coefficients to indicate this knowledge.[1]

Menelaus. About 100 A.D. the astronomer Menelaus of Alexandria, then living in Rome, took up the study of spherical triangles, a subject which, as we have seen, may have occupied the attention of Hipparchus. He wrote a work in six

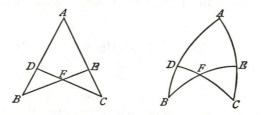

books on chords, and although this is lost we have his treatise on spherics,[2] which not only forms the oldest known work on spherical trigonometry but reveals a remarkable knowledge of geometry and trigonometry in general.

[1] P. Tannery, "Arithmétique des Grecs dans l'Héron d'Alexandrie," *Mémoires de la Soc. des sciences de Bordeaux*, IV; *Mémoires Scientifiques*, I, 189; Heath, *History*, II, 326.

[2] *Menelai Sphaericorum Libri III*, translated by Maurolycus from Arabic and Hebrew sources and published at Messina in 1558. Mersenne (c. 1630) published an edition in 1644, and Halley's edition appeared posthumously at Oxford in 1758. For other editions see A. A. Björnbo, "Studien über Menelaos' Sphärik," *Abhandlungen*, XIV, 1, especially p. 17. See Heath, *History*, II, 261.

In the plane and spherical triangles shown on page 606 he proved the following relations:

Plane Triangle

$$\frac{CE}{AE} = \frac{CF}{DF} \cdot \frac{DB}{AB}$$

$$\frac{CA}{AE} = \frac{CD}{DF} \cdot \frac{FB}{BE}$$

Spherical Triangle

$$\frac{cd\,2\,CE}{cd\,2\,AE} = \frac{cd\,2\,CF}{cd\,2\,DF} \cdot \frac{cd\,2\,DB}{cd\,2\,AB}$$

$$\frac{cd\,2\,CA}{cd\,2\,AE} = \frac{cd\,2\,CD}{cd\,2\,DF} \cdot \frac{cd\,2\,FB}{cd\,2\,BE}$$

where $cd\,2\,CE$ stands for the chord of twice the arc CE, that is, for what we call $2 \sin CE$. Since six quantities are involved in each equation, this was known in the Middle Ages as the *regula sex quantitatum* and was looked upon as the fundamental theorem of the Greek trigonometry. Whether it is due to Menelaus, to Hipparchus, or possibly to Euclid is a matter of dispute, but it is found first in definite form in the *Spherics* of Menelaus,[1] the proposition on the plane triangle being a lemma for the other one.

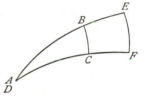

Menelaus also gave a *regula quatuor quantitatum*, as follows:[2] If the two triangles ABC and DEF have $\angle A = \angle D$ and $\angle C = \angle F$ then

$$\frac{cd\,2\,AB}{cd\,2\,BC} = \frac{cd\,2\,DE}{cd\,2\,EF}.$$

Ptolemy. The original contributions of Ptolemy ($c.$ 150) to trigonometry are few, if any; but we are greatly indebted to him for his summary, in the *Almagest*, of the theorems known to Hipparchus.[3] Like other Greek writers, he used chords of angles instead of sines, but the idea of the sine seems to have been in his mind.[4] He extended the table of chords begun by Hipparchus, and it is quite probable that this is the source of the table of sines used by the early Hindu writers.

[1] On this point consult M. Chasles, *Aperçu historique*, 2d ed., 291; Delambre, *Histoire de l'astronomie ancienne*, I, 245 (Paris, 1817); A. A. Björnbo, *Abhandlungen*, XIV, 96, 99; Heath, *History*, II, 266.

[2] For other features see Braunmühl, *Geschichte*, I, 17; A. A. Björnbo, *loc. cit.*, p. 124. [3] Heath, *History*, II, 276. [4] *Ibid.*, II, 283.

Hindu Trigonometry. Although the Hindus had already produced the *Sūrya Siddhānta* (*c.* 400), and although this work treated of the ancient astronomy, gave a table of half chords apparently based, as stated above, upon Ptolemy's work, and showed some knowledge of trigonometric relations,[1] it was not until Āryabhaṭa wrote his *Āryabhaṭīyam* (*c.* 510) that we had in Oriental literature a purely mathematical treatise containing definite traces of the functions of an angle. In this work he speaks of the half chord, as the *Sūrya Siddhānta* had done before him.[2]

The subsequent work of the Hindus was concerned chiefly with the construction of tables, and this will be mentioned later.[3]

Arab and Persian Trigonometry. The chief interest that the Arab and Persian writers had in trigonometry lay, as with their predecessors, in its application to astronomy. On this account we find a growing appreciation of the science, beginning with the founding of the Bagdad School and extending to the close of the Mohammedan supremacy in scientific matters.

The chief Arab writer on astronomy was Albategnius[4] (*c.* 920), who ranked as the Ptolemy of Bagdad. Like the Hindus, he used half chords instead of chords. He also gave the rule for finding the altitude of the sun, which we express by the formula

$$x = \frac{l\sin(90° - \phi)}{\sin\phi},$$

which is simply equivalent to saying that

$$x = l\cot\phi,$$

but there is no evidence to show that he had any real knowledge of spherical trigonometry.

---

[1] See Volume I, pages 34, 145. There is a translation by Burgess in the *Journal of the American Oriental Society*, VI.

[2] L. Rodet, *Leçons de Calcul d'Āryabhata*, pp. 11, 24 (Paris, 1879).

[3] Colebrooke, *Aryabhaṭa*, pp. 90 n., 309 n.

[4] Al-Battânî. His work on the movements of the stars was translated by Plato of Tivoli (*c.* 1120) under the title *De motu stellarum*. It is known to us through the writings of Regiomontanus, and was published at Nürnberg in 1537. There was also a Bologna edition of 1645.

Abû'l-Wefâ (*c.* 980) did much to make the *Almagest* known, computed tables with greater care than his predecessors, and began a systematic arrangement of the theorems and proofs of trigonometry. With him the subject took on the character of an independent science.

It was, however, Naṣîr ed-dîn al-Ṭûsî (*c.* 1250), a Persian astronomer, who wrote the first work[1] in which plane trigonometry appears as a science by itself.

Ulugh Beg (*c.* 1435) of Samarkand was better known as an astronomer than as a writer on trigonometry, but the tables of sines and tangents computed under his direction helped to advance the science.

Arab Influence in Europe. With the decline of Bagdad the study of trigonometry assumed greater importance in Spain, particularly as related to those spherical triangles needed in the work in astronomy. The most important writers were the astronomers Ibn al-Zarqâla[2] (*c.* 1050), who constructed a set of tables, and Jabir ibn Aflah[3] (*c.* 1145). In the 13th century Alfonso X (*c.* 1250) directed certain scholars at Toledo to compute a new set of tables, chiefly for astronomical purposes; these Alfonsine Tables were completed *c.* 1254[4] and were long held in high esteem by later astronomers.

Fibonacci (1220) was acquainted with the trigonometry of the Arabs and, in his *Practica Geometriae*, applied the subject to surveying.

Peurbach and Regiomontanus. By the 14th century England knew the Arab trigonometry, and in the 15th century, thanks largely to Peurbach (*c.* 1460), who computed a new table of sines, and to his pupil Regiomontanus (*c.* 1464), European scholars in general became well acquainted with it. The work

---

[1] *Shakl al-qaṭṭâʿ* (*Theory of Transversals*). There was a French translation published at Constantinople in 1891.

[2] Ibrâhîm ibn Yaḥyâ al-Naqqâsh, Abû Isḥâq, known as Ibn al-Zarqâla, or, in the translations, as Arzachel. He lived in Cordova.

[3] Or Jeber (Geber) ibn Aphla, of Seville. The German transliteration is Dschabir ibn Aflah. His astronomical work was published at Nürnberg in 1543.

[4] See page 232 and also Volume I, page 228.

of Regiomontanus[1] had great influence in establishing the science as independent of astronomy. He computed new tables and may be said to have laid the foundation for the later works on plane and spherical trigonometry. In this general period there were also various minor writers, like Leonardo of Cremona (c. 1425), but they contributed little of value.[2] Copernicus (c. 1520) completed some of the work left unfinished by Regiomontanus and embodied it in a chapter *De Lateribus et Angulis Triangulorum*, later (1542) published separately by his pupil Rhæticus.

Influence of Printed Books. The first printed work on the subject may be said to be the *Tabula directionum* of Regiomontanus, published at Nürnberg before 1485.[3]

The first book in which the six trigonometric functions were defined as functions of an angle instead of an arc, and substantially as ratios, was the *Canon doctrinae triangulorum* of Rhæticus (Leipzig, 1551), although it gives no names for sin ϕ, cos ϕ, and csc ϕ except *perpendiculum, basis*, and *hypotenusa*.[4] Rhæticus was the first to adopt the semiquadrantal arrangement of the tables, giving the functions to 45° and then using the cofunctions. He found sin $n\phi$ in terms of sin ϕ, sin $(n-1)\phi$, and cos $(n-2)\phi$, a subject elaborated by Jacques Bernoulli[5] (1702).

Vieta and his Contemporaries. Vieta (c. 1580) added materially to the analytic treatment of trigonometry. He also computed sin 1′ to thirteen figures and made this the basis for the rest of the table. With him begins the first systematic development of the calculation of plane and spherical triangles by the

---

[1] *De triangulis omnimodis Libri V*, written c. 1464, first printed at Nürnberg in 1533. He also edited Ptolemy's *Almagest*, the first edition appearing at Venice in 1496. On his indebtedness to Naṣîr ed-dîn and others see A. von Braunmühl, "Nassîr Eddîn Tûsi und Regiomontan," *Abh. der Kaiserl. Leop.-Carol. Deutschen Akad. der Naturforscher*, LXXI, p. 33 (Halle, 1897).

[2] See J. D. Bond, *Isis*, IV, 295.

[3] Second ed., Venice, 1485; 3d ed., Augsburg, 1490. See Hain, 13,799.

[4] On his double use of this term for secant and cosecant, see Braunmühl, *Geschichte*, I, 147.

[5] See page 629; Tropfke, *Geschichte*, II (1), 229.

aid of all six functions. In one of his tracts there appears the important formula

$$\frac{a+b}{a-b} = \frac{\tan \frac{1}{2}(A+B)}{\tan \frac{1}{2}(A-B)},$$

which had already been discovered by Fincke, as mentioned below.

Albert Girard published at The Hague in 1626 a small but noteworthy work on trigonometry, and in this he made use of the spherical excess in finding the area of a spherical triangle. This was also given in his algebra of 1629. It also appeared at about the same time in Cavalieri's *Directorium generale* (Bologna, 1632) and, a little later, in his *Trigonometria plana et spherica*[1] (Bologna, 1643).

Thomas Fincke,[2] a Danish mathematician, published an important work, the *Geometria Rotundi*, in Basel in 1583 (2d ed., 1591). He gave the law relating to $a+b : a-b$, expressing it as

$$\frac{\frac{1}{2}(a+b)}{\frac{1}{2}(a+b)-b} = \frac{\tan \frac{1}{2}(180° - C)}{\tan \left[\frac{1}{2}(180° - C) - B\right]}.$$

The equivalent of our present form is due to Vieta, as already stated.

Pitiscus (1595) published an important trigonometry in which he corrected the tables of Rhæticus and modernized the treatment of the subject. In this work the word "trigonometry" appears for the first time as the title of a book on the subject.

British Writers. Besides his invention of logarithms, which has already been considered, Napier replaced the rules for spherical triangles by one clearly stated rule, the Napier Analogies,[3] published posthumously in his *Constructio* (Edinburgh, 1619).

Oughtred's trigonometry appeared in 1657. In this work he attempted to found a symbolic trigonometry; and although

---

[1] "In omnibus verò triangulis sphaericis tres eorum anguli simul sumpti superant duos rectos. Et excessus eorum est . . ." (p.29). He adds: "quod ego probaui in meo Directorio P. 3, Cap. 8."

[2] See N. Nielsen, *Matematiken i Danmark, 1528–1800* (Copenhagen, 1912).

[3] For a discussion of this subject see Braunmühl, *Geschichte*, II, 16.

algebraic symbolism was now so advanced as to make this pos‧ sible, the idea was not generally accepted until Euler's influence was exerted in this direction in the 18th century.

John Newton (1622-1678) published in 1658 a treatise on trigonometry[1] which, while based largely on the works of Gellibrand and other writers, was the most complete book of the kind that had appeared up to that time. Newton and Gellibrand even went so far as to anticipate our present tendency by giving tables with centesimal divisions of the angle.

The greatest contribution to trigonometry made by John Wallis (1616–1703) was probably his encouragement of the statement of formulas by equations instead of by proportions, and his work on infinite series. The former advanced the analytic feature and the latter made possible the calculation of functions by better methods.

Sir Isaac Newton[2] (1642–1727) made many improvements in trigonometry, as in all other branches of mathematics. He expanded $\sin^{-1}x$, or arc sin x, in series, and by reversion he then deduced a series for sin x. He also communicated to Leibniz the general formulas for sin nx and cos nx.

The first to derive general formulas for tan nx and sec nx directly from the right-angled triangle was a French writer, Thomas-Fantet de Lagny (c. 1710). He was also the first to set forth in any clear form the periodicity of the functions. The word "goniometry" was first used by him (1724), although more in the etymological sense of mere angle measure than is now the case.

The Imaginary recognized in Trigonometry. The use of the imaginary in trigonometry is due to several writers of the first half of the 18th century. Jean Bernoulli discovered (1702) the relation between the arc functions[3] and the logarithm of an imaginary number. In his posthumous work of 1722 Cotes showed that

$$\phi i = \log (\cos \phi + i \sin \phi),$$

---

[1] *Trigonometria Britannica, or the doctrine of triangles in two books*, London, 1658.

[2] Braunmühl, *Geschichte*, II, chap. iii.   [3] Such as arc sin x, or $\sin^{-1} x$.

although no writers at that time used this particular symbolism. As early as 1707 De Moivre knew the relation

$$\cos \phi = \tfrac{1}{2}(\cos n\phi + i \sin n\phi)^{\frac{1}{n}} + \tfrac{1}{2}(\cos n\phi - i \sin n\phi)^{\frac{1}{n}},$$

which is obviously related to the theorem

$$(\cos \phi + i \sin \phi)^n = \cos n\phi + i \sin n\phi,$$

published in 1722 and usually called by his name.[1] Euler gave (1748) the equivalent of the formula

$$e^{\phi i} = \cos \phi + i \sin \phi,$$

but this was no longer new. His use of i for $\sqrt{-1}$ (1777) was, however, a welcome contribution. Lambert (1728–1777) extended this phase of trigonometry and developed the theory of hyperbolic functions which Vincenzo Riccati[2] had already (*c.* 1757) suggested and which Wallace[2] elaborated later.

Functions as Pure Number. The first writer to define the functions expressly as pure number was Kästner[3] (1759), although they had already been used as such by various writers.[4]

Trigonometry becomes Analytic. As already stated, through the improvements in algebraic symbolism European trigonometry became, in the 17th century, largely an analytic science, and as such it entered the field of higher mathematics.

In the Orient, however, the science continued in its primitive form, largely that of shadow reckoning, until the Jesuits carried European methods to China, beginning about the year 1600. From that time on the Western influence generally prevailed, not merely in such centers as Nanking and Peking but also, somewhat later, in Japan.

---

[1] Braunmühl, *Geschichte*, II, 76.   [2] See Volume I, page 458.

[3] A. G. Kästner, *Anfangsgründe der Arithmetik Geometrie ebenen und sphärischen Trigonometrie und Perspectiv*, Göttingen, 1st ed., 1759; 2d ed., 1764; 3d ed., 1774. He remarks : "Bedeutet also nun *x* den Winkel in Graden ausgedruckt, so sind die Ausdrückungen sin *x*; cos *x*; tang *x* u. s. w. Zahlen, die für jeden Winkel gehören" (3d ed., p. 380).

[4] Thus Regiomontanus (c. 1463) speaks of the tangents as *numeri*. This occurs in his *tabula foecunda*, prepared for astronomical purposes, and so called "quod multifariam ac mirandam utilitatem instar foecundae arboris parare soleat."

With this brief summary of the development of the science we may proceed to a consideration of a few of the special features which the teacher will meet in elementary trigonometry.

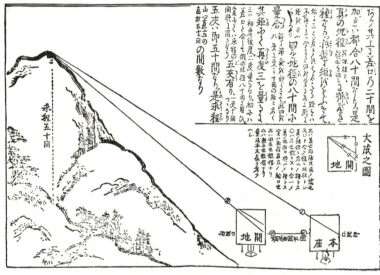

JAPANESE TRIGONOMETRY OF *c.* 1700

From Murai Masahiro's *Riochi Shinan*, early in the 18th century, showing European influence

2. Trigonometric Functions

Sine. The most natural function for the early astronomer to consider was the chord of an arc of a circle having some arbitrary radius. Without any good notation for fractions it was not convenient to take a radius which would give difficult fractional values for the approximate lengths of the chords. A convenient radius, such as 60, being taken, the chord of the arc, being considered purely as a line, was the function first studied by the astronomers.

The first table of chords of which we have any record was computed by Hipparchus (*c.* 140 B.C.), but this table is lost and we have no knowledge as to its extent or its degree of accuracy. The next table of chords of which we have good evidence was that of Menelaus (*c.* 100), but this is also lost, although his work on spherics shows his use of the function. The third important table of chords is that of Ptolemy (*c.* 150). He divided the circle into 360° and the diameter into 120 equal parts,[1] a relation doubtless suggested both by the numerous factors of 120 and, since 3 × 120 = 360, by the ancient use of 3 for π. Influenced like Hipparchus by Babylonian precedents, he used sexagesimal fractions, the radius consisting of 60 *moirai*, each *moira* of 60 minutes, and so on.[2]

Origin of the Sine. A special name for the function which we call the sine is first found in the works of Āryabhaṭa (*c.* 510). Although he speaks of the half chord,[3] he also calls it the chord half[4] and then abbreviates the term by simply using the name *jyā* or *jīva* (chord). He follows Ptolemy in dividing the circle into 360°, and gives a table of sines, of which a portion is shown on page 626.

It is further probable, from the efforts made to develop simple tables, that the Hindus were acquainted with the principles which we represent by the formulas

$$\sin^2 \phi + \cos^2 \phi = 1,$$

$$\sin^2 \phi + \text{versin}^2 \phi = 4 \sin^2 \frac{\phi}{2},$$

and

$$\sin \frac{\phi}{2} = \sqrt{\frac{1 - \cos \phi}{2}},$$

the last two of these appearing in the *Pañca Siddhāntikā* of Varāhamihira (*c.* 505).

The table of sines given by Āryabhaṭa was reproduced by Brahmagupta (*c.* 628), but he did nothing further with trigonometry. Bhāskara (*c.* 1150), however, in his *Siddhānta Śiromāni*, gave a method of constructing a table of sines for every degree.[5]

---

[1] Τρήματα (*tre'mata*, literally "holes," and hence the holes, or pips, of dice). These parts were also called μοῖραι (*moi'rai*, parts), usually translated as degrees.

[2] The minutes were ἑξηκοστὰ πρῶτα (*hexekosta' pro'ta*), first sixtieths; the seconds were ἑξηκοστὰ δεύτερα (*hexekosta' deu'tera*), second sixtieths.

[3] *Ardhā-jyā, ardhajyā,* or *ardhā-djyā.*

[4] *Jyā-ardhā.*   [5] Tropfke, *Geschichte,* II (1), 102.

Name for Sine. The *jyā*[1] of Āryabhaṭa found its way into the works of Brahmagupta as *kramajyā*, that is, straight sine, or *sinus rectus*, as distinguished from the *sinus versus*, the versed sine. This was changed to *karaja* when it went over into Arabic, and as such appears in the Bagdad School of the 9th century. In particular, al-Khowârizmî used it in the extracts which he made from the *Brahmasiddhāntā* of Brahmagupta, probably the work known as the *Sindhind*. It is also found, with natural variants in form, in the writings of the Spanish Arab Ibn al-Zarqâla[2] (*c.* 1050). The sine also appears in the *Pañca Siddhāntikā* of Varāhamihira (*c.* 505), where a table is computed with the Greek diameter of 120. Indeed, the probability of Greek influence upon the methods used by the Hindus is very strong.

The Arabs used the meaningless word *jîba*, phonetically derived from the Hindu *jyā*. The consonants of the word permitted the reading *jaib*, which means bosom, and so this was adopted by later Arabic writers.

Sine in Latin Works. When Gherardo of Cremona (*c.* 1150) made his translations from the Arabic[3] he used *sinus* for *jaib*, each word meaning a fold,[4] and this usage, possibly begun even earlier, was followed by other European scholars. The word "chord" was also used for the same purpose.[5]

---

[1] It has such forms as *djyā, dschyā* (German transliteration), *jîva* and *jîba*.

[2] Or al-Zarkâla, the Latin Arzachel. In the Latin translation there is a chapter "De inventione sinus et declinationis per Kardagas." On his use of *kardagas* see Braunmühl, *Geschichte*, I, 78; on such variants as *gardaga* and *cada*, see *ibid.*, page 102; and on such special uses as *cardaga* for arc 15°, see *ibid.*, pages 110, 120.

[3] *E.g.*, the *Canones sive regulae super tabulas Toletanas* of al-Zarqâla : "Sinus cuius libet portionis circuli est dimidium corde duplicis portionis illius." See also the *Astronomia Gebri filii Affla Hispalensis*, which Apianus edited and published at Nürnberg in 1533.

[4] *Jaib* means bosom, breast, bay ; and *sinus* means bosom, bay, a curve, the fold of the toga about the breast, the land about a gulf, a fold in land.

[5] Thus Plato of Tivoli: ". . . sive mentionẽ cordarũ de medietatis cordis opportere intelligi, nisi aliquo proprio nomine signauerimus, quod & cordā integram appellabimus." On this subject, which has caused much controversy, see Braunmühl, *Geschichte*, I, 49, with bibliography. For the absurd suggestion that sinus = s. ins. = semissis inscriptae [chordae], with bibliography, see Tropfke, *Geschichte*, II (1), 212.

Abû'l-Wefâ (*c.* 980) defined clearly the chord, sine, and versed sine (sinus versus). He showed that

$$\sin \phi = \tfrac{1}{2} \operatorname{cd} 2\,\phi,$$

$$\frac{2\,r - \operatorname{cd}(180° - \phi)}{\operatorname{cd}\dfrac{\phi}{2}} = \frac{\operatorname{cd}\dfrac{\phi}{2}}{r}, \quad \text{our} \quad 2\sin^2\frac{\phi}{2} = 1 - \cos\phi,$$

$$\frac{\operatorname{cd}\phi}{\operatorname{cd}\dfrac{\phi}{2}} = \frac{\operatorname{cd}\left(180° - \dfrac{\phi}{2}\right)}{r}, \quad \text{our} \quad \sin\phi = 2\sin\frac{\phi}{2}\cos\frac{\phi}{2}$$

and $\sin(\phi \pm \phi') = \sqrt{\sin^2\phi - \sin^2\phi\,\sin^2\phi'} \pm \sqrt{\sin^2\phi' - \sin^2\phi\,\sin^2\phi'}.$

He also constructed a table of sines for every $15'$.

Ibn al-Zarqâla, mentioned on page 616, computed a table of sines and versed sines, using as his arbitrary radius $150'$ and also, following Ptolemy, 60^u, where μ stands for μοῖραι (*moi̓-rai*)[2] and is here used so as not to confuse units of line measure with degrees of angle measure. It is probable that such tables were known to Rabbi ben Ezra (*c.* 1140).

In his *Practica Geometriae* (1220) Fibonacci defines the *sinus rectus arcus* and *sinus versus arcus,*[3] and from that time on the terms were generally recognized in the Middle Ages. Tables of sines were given in various works thereafter,[4] so that their use became common.

Other Names for Sine. The term "sine" was not, however, universally recognized, for Rhæticus (*c.* 1560) preferred *perpendiculum*. Of the special terms which appeared from time to time there may be mentioned the *sinus totus* and *sinus perfectus*, both of which were used for sin 90°.[5]

[1] See page 616, n. 2.   [2] See page 232, n. 3.

[3] ". . . .be. uocatur sinus rectus utriusque arcus .ab. et .bc.; et recta .ae. uocatur sinus uersus arcus .ab." (*Scritti*, II, 94).

[4] *E.g.*, Johannes de Lineriis (*c.* 1340). Ulugh Beg's tables (*c.* 1435) were computed for every minute of arc.

[5] *E.g.*, by Johann von Gmünden (*c.* 1430). Regiomontanus (1463) used *sinus totus rectus*. Rhæticus (*c.* 1550) used *sinus totus*, as did most other writers of the time.

Abbreviations for Sine. The first writer to make any general use of a satisfactory abbreviation for sine was Girard (1626). He designated the sine of A by A, and the cosine of A by a.[1] As early as 1624 the contraction *sin* appears on a drawing representing Gunter's scale, but it does not appear in Gunter's work published in that year.[2] In a trigonometry published by Richard Norwood (London, 1631) the author states that "in these examples *s* stands for *sine*: *t* for *tangent*: *sc* for *sine complement*: *tc* for *tangent complement*: *sec* for *secant*." The first writer to use the symbol *sin* for *sine* in a book seems to have been the French mathematician Hérigone (1634). Cavalieri (1643) suggested *Si*, and in the 1647 edition of Oughtred the symbol *S* is used. In 1654, Seth Ward, Savilian professor of astronomy at Oxford, himself a pupil of Oughtred's, used *s*, taking *S'* for the *sinus complementi*. Oughtred's symbol was adopted by various English writers of the 17th century.[3] The symbols $\sin^{-1}x$, $\cos^{-1}x$, $\cdots$, for arc sin x, arc cos x, $\cdots$, were suggested by the astronomer Sir John F. W. Herschel (1813).

Versed Sine. The next function to interest the astronomer was neither the cosine nor the tangent, but, strange as it may seem to us, the versed sine. This function, already occasionally mentioned in speaking of the sine, is first found in the *Sūrya Siddhānta* (*c.* 400) and, immediately following that work, in the writings of Āryabhaṭa, who computed a table of these functions. A sine was called the *jyā*; when it was turned through 90° and was still limited by the arc, it became the turned (versed) sine, *utkramajyā*,[4] so that the versin $\phi = 1 - \cos \phi$.

From India it passed over to the Arab writers, and Albategnius (al-Battânî, *c.* 920), for example, expressly states that he uses the expression "turned chord"[5] for the versed sine.

Since the early writers were given to fanciful resemblances and spoke of the bow (ACB) and string (AB), or the *arcus*

---

[1] Cantor, *Geschichte*, II (2), 709; Tropfke, *Geschichte*, II (1), 217.

[2] F. Cajori, "Oughtred's Mathematical Symbols," *Univ. of Calif. Pub. in Math.*, I, 185. Consult this article also for the rest of this topic.

[3] *E.g.*, Sir Charles Scarburgh (1616–*c.* 1696), a name also given as Scarborough.

[4] Or *utramadjyā*. [5] In some of the Latin translations, *chorda versa*.

and *chorda*, it was natural for them to speak of the versed sine as the arrow. So the Arabs spoke of the *saḥem*, or arrow, and the word passed over into Latin as *sagitta*, a term used by Fibonacci (1220)[1] to mean versed sine and commonly found in the works of other medieval writers.[2] Among the Renaissance writers there was little uniformity. Maurolico (1558) used *sinus versus major* of φ to designate versin (180° − φ), but others preferred the briefer term *sagitta*.

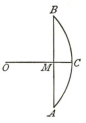

Cosine. Since the Greeks used the chord of an arc as their function, they had no special use for the chord of the complement. When, however, the right-angled triangle was taken as the basis of the science, it became convenient to speak of the sine of a complement angle. Thus there came into use the *koṭijyā* of Āryabhaṭa (*c.* 510),[3] although the sine of 90° − φ commonly served the purpose then as it did later with the Arabs.[4] Even when a special name became necessary it was developed slowly. Plato of Tivoli (*c.* 1120) used *chorda residui* or spoke of the complement angle.[5] Regiomontanus (*c.* 1463) used *sinus rectus complementi*. Rhæticus (1551) preferred *basis*, Vieta (1579) used *sinus residuae*, Magini (1609) used *sinus secundus*, while Edmund Gunter (1620) suggested *co.sinus*, a term soon modified by John Newton (1658) into *cosinus*, a word which was thereafter received with general favor. Cavalieri (1643) used the abbreviation *Si.2*; Oughtred, *s co arc*; Scarburgh, *c.s.*; Wallis,[6] Σ; William Jones (1706), *ś*; and Jonas Moore[7] (1674), *Cos.*, the symbol generally adopted by later writers.

[1] *Scritti*, II, 94, ll. 10, 16, 18.
[2] *E.g.*, Levi ben Gerson (*c.* 1330), in his *De sinibus, chordis, et arcubus*, where his translator also uses *sinus versus*; and Johann von Gmünden (*c.* 1430).
[3] The possible *seqt, śkd*, or *seqet* of the Egyptians (p. 600). Also transliterated *koṭidjyā*.
[4] As with Albategnius (al-Battânî) and others.
[5] "Quod ad perficiendum 90 deficit."
[6] "Σ, co-sinus, seu sinus complementi" (*Opera*, 1603. II, 591).
[7] *Bibl. Math.*, I (3), 69.

Tangent and Cotangent. While the astronomers found the chord and sine the functions most useful in their early work, and so developed them first, the more practical measurements of heights and distances first required the tangent and cotangent,—the gnomon and shadow respectively. It is possible that Ahmes (*c.* 1550 B.C.) knew the tangent, but in any case we know that shadow reckoning was an early device for finding heights, and that it was related to the sundial which Anaximander (*c.* 575 B.C.) introduced into Greece. Unlike the sine and cosine, the tangent and cotangent developed side by side, the reason being that the gnomon and shadow were equally important, the complementary feature playing no part at first. The Greeks, however, made no use of these functions of an angle, so far as we know, except as Thales measured the heights of pyramids by means of shadows and similar triangles.

The Umbra Recta and Umbra Versa. The *Sūrya Siddhānta* (*c.* 400) and other Hindu works speak of the shadow, particularly in connection with astronomical rules, but it was the Arabs who first made any real use of it as a function.

It was Aḥmed ibn ʿAbdallâh,[1] commonly known as Ḥabash al-Ḥâsib, "the computer" (*c.* 860), who constructed the first

U mbra recta

table of tangents and cotangents,[2] but it exists only in manuscript. The Arab writers distinguished the straight shadow, translated by the later medieval Latin writers as *umbra*, *umbra recta*, or *umbra extensa*, and the turned shadow, the *umbra versa* or *umbra stans*, the terms varying according as the gnomon was perpendicular to a horizontal plane, as in ordinary dials, or to a vertical wall, as in sundials on a building. They were occasionally called the horizontal and vertical shadows.[3] The shadow names were also used by most of the later Latin authors and by writers in general until relatively modern times, being frequently found as late as the 18th century.

---

[1] Or al-Mervazî. See Volume I, page 174.

[2] These are given in a MS. of his astronomical tables preserved at Berlin. See Suter, *Abhandlungen*, X, 209.

[3] *E.g.*, by Abû'l-Ḥaṣan Ali ibn ʿOmar al-Marrâkoshî, of Morocco (*c.* 1260).

These functions do not seem to have interested the western Arab writers, no trace of either *umbra* being found in the works of Jabir ibn Aflah (*c.* 1145).

The terms *umbra recta* and *umbra versa* were not used by Gerbert, but Robertus Anglicus (*c.* 1231) speaks of the *umbra*, so that by his time it had come to be somewhat recognized. Thereafter the names *umbra recta* and *umbra versa* were in fairly common use.

Table of Shadows. The first writer whose table of shadows is generally known is Albategnius (al-Battânî, *c.* 920), the table giving the cotangents for each degree of the quadrant.[1]

Abû'l-Wefâ (*c.* 980) constructed a table of tangents for every 15′, the first table of tangents that is known to us; and about this time there was computed a table of cotangents for every 10′. Under the direction of Ulugh Beg (*c.* 1435) there was prepared a table of tangents for every 1′ from 0° to 45° and of every 5′ from 45° to 90°, but his table of cotangents was constructed only for every 1°.[2]

Names and Symbols. Although Rhæticus (1551) did not use these common names for tangent and cotangent, he defined each as a ratio and gave the most complete table that had appeared up to that time.

Vieta (*c.* 1593) called the tangent the *sinus foecundarum* (abridged to *foecundus*[3]) and also the *amsinus* and *prosinus*. It was not until Thomas Fincke wrote his *Geometria Rotundi* (1583) that the term "tangent" appeared as the equivalent of *umbra versa*.[4] The name was adopted by Pitiscus (1595), and the reputation of this great writer gave it permanent standing. Magini (1609) used *tangens secunda* for cotangent. The term *cotangens* was first used for this function by Edmund Gunter (1620).

---

[1] That is, he gave the value of $u = l \cdot \dfrac{\cos \phi}{\sin \phi}$ for $\phi = 1°, 2°, \cdots$, l being the length of the gnomon.

[2] Braunmühl, *Geschichte*, I, 75.

[3] On the origin of the term see page 613 and Braunmühl, *Geschichte*, I, 161 n.; Tropfke, *Geschichte*, II (1), 210.

[4] "Recta sinibus connexa est tangens peripheriae, aut eam secans" (*Geometria Rotundi*, p. 73 (Basel, 1583)).

As abbreviations for tangent and cotangent, Cavalieri (1643) used Ta and $Ta.2$; Oughtred (1657), t arc and t co arc; Sir Charles Scarburgh, $t.$ and $ct.$; and Wallis (1693), T and τ. The abbreviation tan, as in $\overset{tan}{A}$, was first used by Girard (1626), and $Cot.$ was suggested by Jonas Moore (1674), but even yet we have no generally accepted universal symbols for tangent and cotangent.

Secant and Cosecant. Since neither the astronomer nor the surveyor of early times had any need for the secant and cosecant, except as the hypotenuse of a right-angled triangle, these functions were developed much later than the others. The secant seems first to have been considered by al-Mervazî (Ḥabash, $c.$ 860), although the two functions first appear in definite form in the works of Abû'l Wefâ ($c.$ 980), but without special names. Little was done with them by the Arabs, however, and it was not until tables for navigators were prepared in the 15th century that secants and cosecants appeared in this form.[1] Although Copernicus (1542) knew the secant, speaking of it as the *hypotenusa* and computing a set of values of these functions, it was his pupil Rhæticus who first included secants in a printed table. The secant and cosecant appear with the other four functions in his *Canon doctrinae triangulorum* (Leipzig, 1551), although Rhæticus speaks of each in that work as a hypotenuse. The name "cosecant" seems to have appeared first in his posthumous *Opus Palatinum* (1596). Maurolico (1558) included in his tables[2] the secants from 0° to 45°.

Names for Secant and Cosecant. The name "secant" was first used by Fincke (1583) and, although Vieta (1593) called this function the *transsinuosa*, the more convenient and suggestive name soon came into general use. The cosecant was called the *secans secunda* by Magini (1592) and Cavalieri (1643). Pitiscus (1613) gave the secants and cosecants in his tables, and since then they have been commonly found in similar publications.

---

[1] Braunmühl, *Geschichte*, I, 114; 115 n.

[2] *Tabula benefica*, in his work on spherics.

By way of abbreviations for secant and cosecant, Cavalieri (1643) used *Se* and *Se.2*; Oughtred (1657), *se arc* and *sec co arc*; Wallis (1693), *s* and *σ*; but the more convenient symbol *sec*, suggested by Girard (1626) in the form $\overset{sec}{A}$ soon came into general use. There is as yet no international symbol for cosecant, *cosec* and *csc* both being used.

Relation between Functions. Although the functions themselves were not specifically named, various early writers make statements which involve in substance many of the relations that we now recognize. Thus the formula

$$\sin \phi = \sqrt{1 - \cos^2 \phi}$$

or $\qquad \sin^2 \phi + \cos^2 \phi = 1$

is essentially the Pythagorean Theorem and as such was known to the Greeks.

Abu'l-Wefâ (*c.* 980) knew substantially the formulas

$$\tan \phi : 1 = \sin \phi : \cos \phi,$$
$$\cot \phi : 1 = \cos \phi : \sin \phi,$$
$$\sec \phi = \sqrt{1 + \tan^2 \phi},$$

and $\qquad \csc \phi = \sqrt{1 + \cot^2 \phi}.$

Rhæticus (1551) knew the relations

$$\sec \phi : 1 = 1 : \cos \phi$$

and $\qquad \csc \phi : 1 = 1 : \sin \phi.$

Vieta (1579) gave the following proportions:[1]

$$1 : \sec \phi = \cos \phi : 1 = \sin \phi : \tan \phi,$$
$$\csc \phi : \sec \phi = \cot \phi : 1 = 1 : \tan \phi,$$

and $\qquad 1 : \csc \phi = \cos \phi : \cot \phi = \sin \phi : 1.$

3. Trigonometric Tables

Early Methods of Computing. The more important of the earliest trigonometric tables have been mentioned in connection with the several functions. A brief statement will now be

---

[1] Tropfke, *Geschichte*, II (1), 225.

made as to the general methods of computing these tables and as to the early printed tables themselves.

The first methods of which we have definite knowledge are those of Ptolemy (*c.* 150).[1] His computation of chords depends on four principles:

I. From the sides of the regular inscribed polygons of 3, 4, 5, 6, and 10 sides he obtained the following:

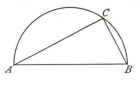

$$\text{cd } 36° = 37^\mu\ 4'\ 55'',$$
$$\text{cd } 72° = 70^\mu\ 32'\ 3'',$$
$$\text{cd } 60° = 60^\mu,$$
$$\text{cd } 90° = 84^\mu\ 51'\ 10'',$$
and $$\text{cd } 120° = 103^\mu\ 55'\ 23''.$$

In a semicircle as here shown, $\overline{BC}^2 + \overline{CA}^2 = \overline{AB}^2$, and so

$$\text{cd}(180° - 36°) = \text{cd } 144° = \sqrt{120^2 - \text{cd}^2\ 36°} = 114^\mu\ 7'\ 37''.$$

II. In an inscribed quadrilateral the sum of the rectangles of the two pairs of opposite sides is equal to the rectangle of the two diagonals. This is known as Ptolemy's Theorem and is found in the *Almagest*.[2]

III. The chord of a half arc can be found from the chord of the arc; that is, from cd 12° it is possible to find cd 6°, and then cd 3°, and so on.[3]

IV. By a scheme of interpolation it is possible to approximate the chord of $\frac{1}{3}\phi$, when cd ϕ is known.[4]

With the help of these principles Ptolemy was able to find the chords of all angles to a fair degree of approximation. Thus he found that
$$\text{cd } 1°\ 0' = 1^\mu\ 2'\ 50'',$$
which would make
$$\sin 30' = \tfrac{1}{2}\text{ cd } 1°\ 0' = 0^\mu\ 31'\ 25'' = 0.0087268,$$

---

[1] The tables are given at the end of Lib. I, cap. ix, of the *Almagest*.

[2] Halma ed., I, 29; Heiberg ed., p. 36; Braunmühl, *Geschichte*, I, 19.

[3] Essentially he has $\sin\frac{\phi}{2} = \sqrt{\dfrac{1 - \cos\phi}{2}}$. For the mathematical discussion, see Braunmühl, *Geschichte*, I, 20.

[4] For mathematical discussion, see Braunmühl, *Geschichte*, I, 21; Tropfke, *Geschichte*, II (1), 206.

whereas our seven-place tables give it as 0.0087265. Abû'l-
Wefâ (c. 980) computed this result as $0^\mu\ 31'\ 24''\ 55'''\ 54^{iv}\ 55^v$,
—a value which is correct as far as the tenth decimal place.

द्वितीयाऽध्याय ३३

घटाने पर शेष २२४ होगा।इस २२४ को प्रथम ज्याद्धं २२५ के साथ जोड़ देने
से योगफल ४४९ होगा । यही द्वितीय ज्याद्धं है ।।१५।। उक्त द्वितीय ज्याद्धं ४४९
को प्रथम ज्याद्धं से भाग करके भागफल २ठेकर यह २ इसके साथ पूर्व द्वितीय
ज्याद्धं निष्कासन भाग फल से जो १ मिला है, जोड़ने से ३ होगा । इस ३
को उक्त भाजक २२५ से घटाने पर २२२ बचेंगा,इसी २२२ को द्वितीय ज्याद्धं ४४९
के साथ जोड़ने से ६७१ हुंगा, यही तृतीय ज्याद्धं है । इसी प्रकार क्रमशः
२४ ज्याद्धं गणना करनी होगी ।।१६।। किसी वृत्त के चतुर्थांश जिस का व्यासाद्धं
३४३८ उस के ३४ अंश की ज्याद्धं निम्नलिखित होंगी ।।

	अंश वा कला		ज्या		अंश वा कला		ज्या
प्रथम कोण	३$\frac{3}{4}$	२२५	२२५	१३ वां कोण	४८।।।	२४९५	२२६७
द्वितीय ,,	७$\frac{1}{2}$	४५०	४४९	१४ वां ,,	५२$\frac{1}{2}$	२७००	२४३१
तृतीय ,,	११$\frac{1}{4}$	६७५	६७१	१५ वां ,,	५६$\frac{1}{4}$	२८७५	२५८२
चतुर्थ ,,	१५	८००	८८०	१६ वां ,,	६०	३१५०	२७२८
पञ्चम ,,	१८$\frac{3}{4}$	११२५	११०५	१७ वां ,,	६३।।।	३३२५	२८५९
छठा ,,	२२$\frac{1}{2}$	१३५०	१३१५	१८ वां ,,	६७$\frac{1}{2}$	३६००	२९७८
सप्तम ,,	२६$\frac{1}{4}$	१५७५	१५२०	१९ वां ,,	७१$\frac{1}{4}$	३६२४	३०८४
अष्टम ,,	३०	१८००	१७१९	२० वां ,,	७५	४०५०	३१७७
नवम ,,	३३$\frac{3}{4}$	२०२५	१९१५	२१ वां ,,	७८।।।	४२२४	३३७२
दशम ,,	३७$\frac{1}{2}$	२२५०	२०९३	२२ वां ,,	८२$\frac{1}{2}$	४४५०	३४०९
एकादश ,,	४१$\frac{1}{4}$	२४७५	२२६७	२३ वां ,,	८६$\frac{1}{4}$	५१७४	३४३१
द्वादश ,,	४५	२७००	२४३१	२४ वां ,,	९०	५४००	३४३८

पूर्वोक्त ज्याद्धं परिमाण सब को उलटे प्रकार से ३४३८ व्यासाद्धं से पृथक्
पृथक् घटाने पर जो अङ्क घटाने से बचेंगे उनको उत्क्रमज्या कहते हैं । प्रति
३४ अंश में इस प्रकार उत्क्रमज्या हो जाती हैं । १६-२२ श्लोक तक ।।

मुनयोरन्ध्रयमला रसपट्टकामुनीश्वराः । द्व्यष्टैकारूप-
षड्दस्राः सागरार्थहुताशनाः ॥२५॥ खर्तुवेदा नवाद्रघर्यो
दिङ्नगास्त्र्यर्थकुङ्कराः। नगाम्बरवियच्चन्द्रारूपभूधरश-

FROM THE FIRST PRINTED EDITION OF THE SŪRYA SIDDHĀNTA

Printed at Meerut, India, c. 1867. This is the oldest Hindu work on astronomy.
It shows the table of which a portion is given on page 626

A table of sines is given in the *Sūrya Siddhānta* (c. 400),
and Āryabhaṭa (c. 510) gives a table of sines and versed sines.
The following portion of the table of sines, substantially as in

some of the manuscripts of the *Sūrya Siddhānta* and as given also in Āryabhaṭa's work, will serve to show the degree of accuracy:

ARC	SINE	MODERN VALUE
$3° 45'$	$225'$	$224.84'$
$7° 30'$	$449'$	$448.72'$
$11° 15'$	$671'$	$670.67'$
$15°$	$890'$	$889.76'$

Āryabhaṭa's method of working out his table was to take sin $3° 45'$[1] as equal to arc $3° 45'$, and from this to find the sines of multiples of this angle by the rule already given in the *Sūrya Siddhānta*,[2]

$$\sin (n + 1) \phi = \sin n\phi + \sin n\phi - \sin (n - 1) \phi - \frac{\sin n\phi}{\sin \phi},$$

which is correct except for the last term.

Arab Methods. The early Arabs used the Hindu results, but later scholars developed original methods of attack. Of these, one of the best known is given by Mîram Chelebî (*c.* 1520)[3] in his commentary on Ulugh Beg (*c.* 1435).[4] He gives two methods, the first being somewhat similar to the one used by Ptolemy. The second method is interesting because it involves the approximate solution of a cubic equation of the form

$$ax - b = x^3.$$

European Computers. Of the later computers of the Middle Ages and early Renaissance, Regiomontanus (1546) stands at the head, but his methods were not new. Indeed, there was no particular originality shown in the computations from the time of Ptolemy to the invention of the modern methods based on series.

---

[1] Known by the special name *kramajyā*.　　[2] Braunmühl, *Geschichte*, I, 35.

[3] Mûsâ ibn Mohammed ibn Maḥmûd ibn Qâḍizâdeh al-Rûmî, a teacher in Gallipoli, Adrianople, and Brusa; died 1524/25.

[4] M. Woepcke, "Discussion de deux méthodes arabes pour déterminer une valeur approchée de sin 1°," *Journal de math. pures et appliquées*, XIX (1854), 153; A. Sédillot, *Prolégomènes des Tables astronomiques d'Ouloug-Beg*, Paris, 1853; Braunmühl, *Geschichte*, I, 72, with incorrect date.

In general, all ancient tables were constructed with Ptolemy's radius of 60; that is, the *sinus totus*, or sin 90°, was 60. This was due to the necessity of avoiding fractions in the period before the invention of decimals. The first to adopt the simpler form, sin 90° = 1, was Jobst Bürgi (*c.* 1600), but his tables computed on this basis are not extant. Although the invention of decimal fractions had now made the use of unity possible for the *sinus totus*, this idea was not fully appreciated until a memoir by de Lagny was written in 1719.[1] It was nearly thirty years later that the plan received its first great support at the hands of Euler.[2]

Early Printed Tables. Of the early printed tables there may be mentioned as among the more important the table of sines with the radius divided decimally, published by Apianus in 1533; the table of all six functions based on a semiquadrantal arrangement, published by Rhæticus in 1551, calculated to every 10′ and to seven places; Vieta's extension of the tables of Rhæticus to every minute (1579, but the printing began in 1571); the table of tangents by Reinhold (1511–1553) to every minute, printed in 1554; the table of all six functions, published in England by Blundeville in 1594; the *Opus Palatinum*, with the functions for every 10″ to ten decimal places, with tables of differences, compiled by Rhæticus and published by Valentin Otto (or Otho) in 1596. Dr. Glaisher, referring to the work of Rhæticus, speaks of him as "by far the greatest computer of pure trigonometrical tables" and as one "whose work has never been superseded." The *Opus Palatinum* was so named in honor of the elector palatine, Friedrich IV, who paid for its publication.[3]

The serious use of tables based upon the centesimal division of the angle was a result of the movement that led to the metric system. An elaborate set of such tables was prepared in Paris at about the close of the 18th century, and little by little the plan found favor. Such a set of tables appeared even in Japan

---

[1] *Histoire et Mémoires de l'Acad. d. sci.*, Paris, 1721, p. 144; 1726, p. 292; 1727, p. 284; 1729, p. 121.

[2] *Introductio in analysin infinitorum*, I, § 127. Lausanne, 1748.

[3] For a summary of such tables see the *Encyc. Brit.*, 11th ed., XXVI, 325.

as early as 1815, but it was not until the close of the 19th century that the idea took any firm hold upon the mathematics of Europe, and then with the French schools still in the lead.

切　餘	弦　餘
⊃ ⊃ ⊃ ⊃ ⊃ ⊃ ⊃ ⊙ ⊙ ⊍	⊍ ⊍ ⊍ ⊍ ⊍ ⊍ ⊍ ⊙ ―
七八三五六六九二七五	五八九九九九九
五一五九一八四六八二	九三九九九九九
六八六八二八九〇九一一	三六八九九九九
七二〇三六三二三四一	六五七九九九九
六三一一九八五四一一	九一六九九九九
五三三五九八四五九	二五四九九九九
二九四〇七四八一八	四五二九九九九
六七六九四一六一七	五二〇九九九九
三九〇八六五六三六	六六七八九九九
切　正	弦　正

CENTESIMAL TABLES OF JAPAN

From a manuscript of a work on trigonometry, by Miju Rakusai, written in 1815, showing a table of natural functions on the decimal division of the angle. This page shows the cotangents and cosines, beginning at the top with 0°

4. TYPICAL THEOREMS

Addition Theorem of Sines. It is impossible, in the space allowed, to mention more than a few of the important theorems of trigonometry, and these will now be considered.

The Greeks knew essentially that

$$\sin(\phi \pm \phi') = \sin\phi \cos\phi' \pm \cos\phi \sin\phi'.$$

Stated as a proposition involving chords, it is probable that Hipparchus (*c.* 140 B.C.) knew it. It was certainly known to Ptolemy (*c.* 150), and it often bears his name. Bhāskara (*c.* 1150) also gives the theorem. As already stated on page 617, Abû'l-Wefâ (*c.* 980) gave it essentially under the form

$$\sin(\phi \pm \phi') = \sqrt{\sin^2\phi - \sin^2\phi \sin^2\phi'} \pm \sqrt{\sin^2\phi' - \sin^2\phi \sin^2\phi'}.$$

Functions of Multiple Angles. The formula

$$\sin 2\phi = 2 \sin\phi \cos\phi$$

is a corollary of the general case of $\sin(\phi + \phi')$. It is first expressly given as a rule by Abû'l-Wefâ, the form being, as already stated, chord ϕ : chord $\frac{1}{2}\phi$ = chord $(180° - \frac{1}{2}\phi)$: r.

Vieta[1] (1591) first gave the formulas

$$\sin 3\phi = 3 \cos^2\phi \sin\phi - \sin^3\phi,$$
$$\cos 3\phi = \cos^3\phi - 3 \sin^2\phi \cos\phi,$$

and connected $\sin n\phi$ with $\sin\phi$ and $\cos\phi$.

Rhæticus (1569) found the relation

$$\cos n\phi = \cos(n-2)\phi - 2 \sin\phi \sin(n-1)\phi.$$

Newton (1676) gave the well-known relation

$$\sin n\phi = n \sin\phi + \frac{(1-n^2)n}{3!} \sin^3\phi + \cdots,$$

and Jacques Bernoulli (1702) showed that

$$\sin n\phi = \cos^n\phi - \frac{n(n-1)}{2!} \cos^{n-2}\phi \sin^2\phi + \cdots,$$
$$\cos n\phi = \frac{n}{1} \cos^{n-1}\phi \sin\phi - \frac{n(n-1)(n-2)}{3!} \cos^{n-3}\phi \sin^3\phi + \cdots.$$

Functions of Half an Angle. Ptolemy[2] (*c.* 150) knew substantially the sine of half an angle, expressed as half a chord, and it is probable that Hipparchus (*c.* 140 B.C.) and certain that Varāhamihira (*c.* 505) knew the relation which we express as

$$\sin\frac{\phi}{2} = \sqrt{\frac{1 - \cos\phi}{2}}.$$

[1] Tropfke, *Geschichte*, II (2), 57-61, with bibliography.
[2] See the Heiberg edition of Ptolemy, p. 39.

II

After the development of analytic trigonometry in the 17th century, these relations were greatly extended. Four others may be mentioned as typical, the first two, due to Euler (1748), being

$$\tan 2\phi = \frac{2\tan\phi}{1-\tan^2\phi},$$

$$\cot 2\phi = \frac{\cot\phi-\tan\phi}{2};$$

and the others, due to Lambert (1765), being

$$\sin 2\phi = \frac{2\tan\phi}{1+\tan^2\phi},$$

$$\cos 2\phi = \frac{1-\tan^2\phi}{1+\tan^2\phi}.$$

Theorem of Sines. The important relation now expressed as

$$\frac{a}{\sin A} = \frac{b}{\sin B} = \frac{c}{\sin C}$$

was known to Ptolemy (c. 150) in substance, although he expressed it by means of chords.[1]

While recognized by Albêrûnî and other Oriental writers, it was Naṣîr ed-dîn (c. 1250) who first set it forth with any clearness. A little later Levi ben Gerson (c. 1330) stated the law in his work *De sinibus, chordis, et arcubus*;[2] but the first of the Renaissance writers to express it with precision was Regiomontanus[3] (writing c. 1464).

---

[1] Thus, if C is a right angle in triangle ABC, then

$$a = \frac{c \text{ chord } 2A}{120^\mu},$$

where 120 μοῖραι is the diameter of the circumcircle.

[2] "... omnium triangulorum rectilineorum talem proportionem una linea habet ad aliam, qualem proportionem unus sinus angulorum, quibus dictae lineae sunt subtensae, habet ad alium." See Braunmühl, *Geschichte*, I, 106.

[3] "In omni triangulo rectilineo proportio lateris ad latus est, tamquam sinus recti anguli alterum eorum respicientis, ad sinum recti anguli reliquum latus respicientis" (Lib. II, prop. 1). See Tropfke, *Geschichte*, V (2), 74.

Theorem of Cosines. The fact that

$$c^2 = a^2 + b^2 - 2\,ab\cos C$$

is essentially a geometric theorem of Euclid.[1] In that form it was known to all medieval mathematicians. In the early printed books it appears in various forms, Vieta (1593) giving it substantially as

$$2\,ab/(a^2 + b^2 - c^2) = 1/\sin(90° - C),$$

and W. Snell (1627) as

$$2\,ab/[c^2 - (a - b)^2] = 1/(1 - \cos C).$$

Theorem of Tangents. The essential principle of the law of tangents, which was given by Vieta[2] (p. 611) and improved by Fincke (1583), was known to Ptolemy (c. 150). Regiomontanus (c. 1464) expressed it by a rule which we should state as

$$\frac{\sin A + \sin B}{\sin A - \sin B} = \frac{\tan\frac{1}{2}(A + B)}{\tan\frac{1}{2}(A - B)}.$$

Areas. The first evidence of the rule which resulted in the formula for the area of a triangle, which we know as

$$\Delta = \tfrac{1}{2}\,ab\sin C,$$

is found in the trigonometry of Regiomontanus (c. 1464),[3] but the theorem is not explicitly stated by him. Snell (1627) gave it in the form

$$1 : \sin A = bc : 2\,\Delta.$$

Right-angled Spherical Triangle. The Greek mathematicians made use of the right-angled spherical triangle in their computations, but nowhere do we find a systematic treatment of the subject. Taking the hypotenuse as c, we have the following six cases:

1. $\cos c = \cos a \cos b.$
2. $\cos c = \cot A \cot B.$
3. $\cos A = \cos a \sin B.$
4. $\cos A = \tan b \cot c.$
5. $\sin b = \sin c \sin B.$
6. $\sin b = \tan a \cot A.$

---

[1] *Elements*, II, 12, 13.

[2] "Ut aggregatum crurum ad differentiam eorundem, ita prosinus dimidiae angulorum ad basin ad prosinum dimidiae differentiae" (*Opera*, Schooten ed., p. 402). See Tropfke, *Geschichte*, V (2), 80.

[3] *De triangulis omnimodis*, Nürnberg, 1533.

In his astronomical problems Ptolemy (*c.* 150) makes use essentially of the first, fourth, fifth, and sixth of these cases,[1] although without the functions he could not give the rules.

The third case is essentially given by Jabir ibn Aflah (*c.* 1145), and so it was commonly known as Jabir's Theorem.[2] The first writer to set forth essentially all six cases was Naṣîr ed-dîn (*c.* 1250). Napier's Rules for the right-angled spherical triangle appeared in his tables of 1614.[3]

Oblique-angled Spherical Triangle. The oblique-angled spherical triangle was not seriously studied by itself until the Arabs began to consider it in the 10th century.

The Theorem of Sines,

$$\frac{\sin a}{\sin A} = \frac{\sin b}{\sin B} = \frac{\sin c}{\sin C},$$

and the Theorem of Cosines of Sides,

$$\cos a = \cos b \cos c + \sin b \sin c \cos A,$$

may have been known to them, but they are first found in print in the *De triangulis* written by Regiomontanus *c.* 1464.[4]

The Theorem of Cosines of Angles,

$$\cos A = - \cos B \cos C + \sin B \sin C \cos a,$$

was given in substance by Vieta in 1593, although he had used it before this date.[5] It was first proved by Pitiscus in 1595.

The Theorem of Cotangents,

$$\frac{1}{\sin b \csc A} = \frac{\cot c + \cos A \cot b}{\cot C},$$

was also given in substance by Vieta, but was modified by Adriaen van Roomen (1609) and proved by Snell (1627).[6]

---

[1] Tropfke, *Geschichte*, V (2), 131, with references. Consult this work and Braunmühl's *Geschichte* for further details.

[2] It was possibly known to Tâbit ibn Qorra (*c.* 870).

[3] *Mirifici logarithmorum canonis descriptio*, 1614, Lib. II, cap. iv.

[4] On the Theorem of Sines see Tropfke, V (2), 133; on the Theorem of Cosines see *ibid.*, p. 139.

[5] For the priority question see Tropfke, *Geschichte*, V (2), 139.

[6] For the general literature on this subject see Braunmühl, *Geschichte*, I (1), 25; Tropfke, *Geschichte*, V (2), 137 seq., especially p. 143.

TOPICS FOR DISCUSSION

1. The etymology of the words "trigonometry," "geometry," "mensuration," "agrimensor," "survey," "geodesy," and other terms having a related meaning.

2. Primitive needs that would naturally tend to the development of trigonometry.

3. The relation of shadow reckoning to plane trigonometry in various countries and at various times.

4. The influence of astronomy upon the development of the science of trigonometry.

5. The Greek astronomers who contributed most to the study of trigonometry, the function which they developed, and the reason why this function was selected.

6. The contributions of Menelaus to the study of trigonometry.

7. The Hindu contributions to the science.

8. The assertion that the chief contribution to mathematics made by the Arab scholars was to the science of trigonometry, and that this contribution was important.

9. The Arab and Persian writers on trigonometry, and the important features of their work.

10. Influence of Peurbach and Regiomontanus.

11. The change of trigonometry from being essentially geometric to being largely analytic, and the influence of this change upon the later development of the science.

12. Development of the concept of the sine of an angle, and the origin of the name.

13. Development of the concept and name of the cosine.

14. Development of the concepts of the versed sine and the coversed sine, with reasons for their gradual disappearance.

15. The favorite functions in astronomy and those used in practical mensuration.

16. Development of the tangent and cotangent.

17. Development of the secant and cosecant, and the causes leading to their gradual disuse in modern times.

18. The relation between trigonometric functions.

19. Development of the leading methods of computing trigonometric tables.

20. The history of typical and important theorems of trigonometry.

CHAPTER IX

MEASURES

1. Weight

Measures in General. The subject of metrology is so extensive that it is impossible, in a work like this, to do more than give a few notes relating to the measures in common use. The purpose of this chapter, therefore, is simply to lay before the student some of the points of interest in the history of the most familiar of our several units of measure, to suggest the significance of the names of these units, and to indicate some of the works on the subject to which he may go for further information.[1]

EGYPTIAN WEIGHT

A porphyry weight found near the North Pyramid at Lisht and now in the Metropolitan Museum. The inscription reads, "Senusert, giving life eternally, 70 gold debens." It was used for ʰing gold

Egypt. The use of the balance for purposes of weighing is doubtless prehistoric, for weights are found in remains of the first dynasty of Egypt (*c.* 3400 B.C.).[2] The first inscribed weight that has been found is of the 4th dynasty, the time of the Great

[1] An excellent summary of the history is given under "Weights and Measures" in the *Encyc. Brit.*, 11th ed. The measure of angles and arcs has been already considered in Chapter VIII.

[2] See W. M. Flinders Petrie, *Proceedings of the Soc. of Biblical Archæol.*, London, XXIII, 385. See also *Bulletin of the Metrop. Mus. of Art*, New York, XII, 85.

Pyramid.[1] The earliest Egyptian scales were simple balances, either held in the hand or supported on a standard. These are frequently illustrated in the temple wall pictures. The steelyard with its sliding weight and fixed fulcrum was used as early as 1350 B.C.

The Egyptian weights of which the names and values are known with certainty were the *deben* (*dbn*, formerly read *uten*, about 13.6 grams, but commonly taken as 15 grams)and the *kidet* (*kdt*, *kite*, 0.1 of a *deben*).[2]

Babylonia. The Babylonians used a cubic foot of rain water to establish their unit of weight, the standard talent.[3] The chief subdivision of the talent was called a *maneh* and was $\frac{1}{60}$ of a talent. They also had a unit of weight known as the *she*, about 45 mg. Our knowledge of the Babylonian measures in general is derived from a number of inscribed tablets such as the one here shown.

TABLE OF BABYLONIAN MEASURES OF CAPACITY AND WEIGHT

A fragment of a clay tablet found at Nippur and dating from *c.* 2200 B.C. The reverse side contains tables of weight, length, and area. Courtesy of the University of Pennsylvania

Hebrews. The Hebrew standards were kept in the temple, as was also the case in other nations. Thus we read[4] of "the shekel of the Sanctuary," that is, the standard shekel, about a quarter of an ounce in early times, or a half ounce after the

[1] *Proceedings of the Soc. of Biblical Archæol.*, XIV, 442.
[2] Peet, *Rhind Papyrus*, p. 26; *Encyc. Brit.*, 11th ed., XXVIII, 480.
[3] Mahaffy, *Greek Life*, p. 67. See also J. Brandis, *Das Münz- Mass- und Gewichtswesen in Vorderasien*, pp. 1, 41 (Berlin, 1866); G. A. Barton, *Haverford Library Collection of Cuneiform Tablets*, Philadelphia, 1905, 1909.
[4] Exodus, xxx, 13.

time of the Maccabees (1st century B.C.). This shekel of the Hebrews was the *sicilicus* of the Romans.[1] The Hebrew *maneh* was 100 shekels, or $\frac{1}{30}$ of a talent. The shekel was also used as a unit of capacity, and with the Babylonians it was equivalent to 0.07 liter.[2]

Greece. The Greek unit of weight in Homer's time was the talent,[3] a standard that varied from country to country. The ancient Greek talent weighed about 57 pounds, but the Hebrews used the term for a unit of about $93\frac{3}{4}$ pounds. It was also used as a unit of value, generally the value of a talent of silver, this being about $1180 in Greece and from about $1650 to about $1900 among the Hebrews, according to present standards.

For a smaller weight the Greeks used the drachma,[4] originally "a handful" but used by the ancients to designate both a weight and, as in the case of the talent, a unit of value. In modern Greece it is a coin identical in value with the franc at the normal rate of exchange.

The later Greek weights may be thus summarized: 1 *talanton* = 60 *mnai* = 6000 *drachmai* = 36,000 *oboloi* = 288,000 *chalkoi*.[5]

Rome. The Roman unit of weight was the pound. This was divided into twelfths (*unciae*).[6] The usual Roman weights may be summarized as follows:

1 *libra* = 12 *unciae* = 48 *sicilici* = 288 *scripula* = 576 *oboli* = 1728 *siliquae*.

The ounce was about 1.09 oz. avoirdupois, or 412 grains.

---

[1] Or *siclus*; Greek σίγλος. [2] Barton, *loc. cit.*, II, 18.

[3] Τάλαντον (*tal'anton*). It was originally smaller than the later talent. See F. Hultsch, *Griechische und römische Metrologie*, p. 104 (Berlin, 1882); Harper's *Dict. Class. Lit.*; Pauly-Wissowa; A. Böckh, *Metrologische Untersuchungen*, Berlin, 1838.

[4] Δραχμή (*drachme'*). The Lydian drachma of the 7th century B.C. was $\frac{1}{4}$ of a shekel.

[5] In Greek, τάλαντον; μνᾶ, μναῖ; δραχμή, δραχμαί; ὀβολός, ὀβολοί; χαλκοῦς, χαλκοῖ. The *chalkous* was about 0.091 g. As a measure of value it was a copper coin worth $\frac{1}{8}$ of an obol, somewhat less than $\frac{1}{2}$ of an American cent. The *talanton* was about 26,196 g.

[6] Hultsch, *loc. cit.*, p. 144; Ramsay and Lanciani, *Manual of Roman Antiqs.*, 17th ed., p. 461 (London, 1901). There are many works on the subject. Among

The Romans had a table known as the *mensa ponderaria*, in the stone top of which were cavities like washbasins, with a plug in the bottom of each cavity. These were standards of capacity, or of capacity with respect to weight.[1]

Far East. In India and other parts of the Far East the weights and currencies were commonly based upon the weights of certain seeds. The favorites were the *abrus precatorius*, a creeper having a small, bright-red seed with a black spot on it,[2] and the *adenanthera pavonina*, a large pod-bearing tree with a bright-red seed which is conventionally taken as weighing twice an abrus seed.[3]

England. In England the grain was originally the weight of a barleycorn, a barley grain. The Latin *granum* has the same root (*gar*) as our word "corn."[4]

the earlier ones are A. Alciatus, *Libellvs. De Ponderibvs et mensuris*, Copenhagen, 1530; L. Portius, *De sestertio pecvniis ponderibvs et mensvris antiqvis libri duo*, s. l. a. (Venice?, *c.* 1500), with editions at Florence (1514?) and Basel (1520 and 1530); G. Budaeus, *De asse, et partibvs eivs, libri V*, Paris, 1514 (title as in Lyons edition, 1551); G. Agricola, *Libri quinque de Mensuris & Ponderibus*, Paris, 1533; Venice, 1533 and 1535; Basel, 1549 and 1550; an epitome, Lyons, 1552; H. Uranius, *De re nvmaria, mensvris et ponderibus Epitome ex Budaeo, Portio*, . . ., Solingen, 1540; M. Neander, ΣΥΝΟΨΙΣ *mensvrarvm et pondervm*, . . . Basel, 1555. These show the interest taken in the subject in the 16th century. Of the 18th century works one of the best is J. Arbuthnot, *Tables of Antient Coins, Weights, and Measures*, 2d. ed., London, 1754.

[1] A Naples specimen is illustrated in Mau's *Pompeii*, Kelsey's second edition, New York, 1902.

[2] Often seen for sale in European and American shops. The name *precatorius* (from *precator*, one who prays) comes from the fact that certain Buddhists use these as beads for their rosaries.

[3] R. C. Temple, "Notes on the Development of Currency in the Far East," *Indian Antiquary*, 1899, p. 102. Other seeds were used, as is shown by H. T. Colebrooke, "On Indian Weights and Measures," *Asiatic Researches*, V (1799), 91, with tables. See also his *Lilāvati*, p. 1, §2.

[4] Whence also "garner," to gather grain; "pomegranate," from the French *pomme* (apple) and *grenate* (seeded); "granite," a grained or spotted stone; "garnet"; "grange"; and the Spanish *granada*. The Scotch statute required that the inch be "iii bear cornys gud and chosyn but tayllis" (tailless). The Latin statute of England read: "Tria grana ordei sicca et rotunda faciunt pollicem." On the history of British measures in general see F. W. Maitland, *Domesday Book and Beyond*, Cambridge, 1897, p. 368; J. H. Ramsay, *The Foundations of England*, London, 1898, I, 533; F. Seebohm, *The English Village Community*, 4th ed., p. 383 (London, 1896); R. Potts, *Elementary Arithmetic*, London, 1886.

The word "pound" comes from the Latin *pondo* (by weight),[1] and the ounce, as already stated, from the Latin *uncia*, a twelfth of the Roman pound.[2]

England had developed a system of weights before the Troy weight was introduced from the French town of Troyes, one of the many places in which fairs were held in the Middle Ages. This introduction seems to have taken place as early as the second half of the 13th century, for Grafton's *Chronicles*[3] has this to say of the matter:

About this tyme[4] was made the statute of weightes and measures, that is to say, that a sterlyng penny should waye .xxxij. graynes of wheate drie and round, and taken in the middes of the eare,[5] and .xx. of those pence shoulde make an ounce, and .xij. ounces make a pound Troy: And .viij. pound Troy weight make a gallon of Wine, and .viij. wyne galons to make a London bushell, which is the .viij. part of a quarter. Also three barly Cornes dry and round should make an ynch, & .xij. ynches a foote, and thre foote a yard, a fiue yards, halfe a perch, or poll, & .xl. pol in length & thre in bredth an acre of land.

And these standardes of weight and measures were confirmed in the .xv. yere of king Edward the thirde, and also in the tyme of Henry the sixt and of Edward the fourth, and lastly confirmed in the last yere of Henry the seuenth. But in the time of king Henry the sixt it was ordeyned that the same ounce should be deuided into .xxx. pence, and in the tyme of king Edward the fourth, into .xl. pence, and in the tyme of king Henry the eight into .xliiij. pence: But the weight of the ounce Troy, and the measure of the foote continued always one.

In due time the Troy weight was replaced by the avoirdupois for general purposes and was thenceforth limited chiefly to

---

[1] From *pendere*, to weigh. From the same root we have such words as "depend," "spend," and "pendulum," and the French *poids* and our "poise."

[2] Lack of space precludes any discussion of the relation of the apothecaries' weight to the ancient Greek and Roman systems and symbols. There is an extensive literature on the subject.

[3] 1569 ed.; 1809 reprint, p. 277.   [4] "The LIJ Yere of Henry III," *i.e.*, 1268.

[5] So Recorde (*c.* 1542) says: "Graine, meaninge a grayn of corne or wheat drye, and gathered out of the myddle of the eare." *Ground of Artes*, 1558 ed., fol L 4.

the use of goldsmiths.[1] These goldsmiths also used in this connection the carat, a weight consisting of 12 grains.[2] The word had a variety of meanings, being commonly used to express the purity of gold, "22 carats fine" meaning an alloy that is $\frac{22}{24}$ "fine gold." It appears in various forms,[3] and its meaning in this sense comes from the fact that a gold mark was 24 carats, so that a mark that had only 18 carats of gold was only $\frac{18}{24}$ pure. So Recorde (c. 1542) says: "The proofe of gold is made by Caracts, whereof 24 maketh a Marke of fine gold: the Caract is 24 graines."[4]

Avoirdupois Weight. The word "avoirdupois" is more properly spelled "averdepois," and it so appears in some of the early books. It comes from the Middle English *aver de poiz,*[5] meaning "goods of weight." In the 16th century it was commonly called "Haberdepoise," as in most of the editions of Recorde's (c. 1542) *Ground of Artes.* Thus in the Mellis edition of 1594 we have:

At London & so all England through are vsed two kinds of waights and measures, as the Troy waight & the Haberdepoise.

---

[1] So the Dutch arithmetics of the 17th century speak of it as *Assay-gewicht.* E.g., Coutereels's *Cyffer-Boeck,* 1690 ed., p. 16. The Dutch writers also called it *Trois gewicht,* as in Bartjens's arithmetic, 1676 ed., p. 155.

[2] So the Dutch arithmetics of Petrus (1567), Van der Schuere (1600), and others give 12 grains = 1 karat, 24 karats = 1 marck (for gold), and 20 angels = 1 ounce, 8 ounces = 1 marck (for silver). Trenchant (1566) says: "Per ansi le marc d'or sans tare est à 24 kar. de fin aloy." In this sense it comes from the Arabic *qīrāt,* a weight of 4 barleycorns; but the Arabs derived it from the Greek κεράτιον (*kera'tion*), the fruit of the locust tree, L. Latin *cerates.* Perhaps the Arabic use is responsible for the carat weight's being 4 diamond grains, now taken as 200 milligrams.

[3] Italian *carato,* French *carat,* and Spanish *quilate.* So Texada (1546): "24. quilates son de puro oro"; Sfortunati (1534): "Io mi trouo oro di .24. charatti"; Trenchant (1566): "18 karats de fin"; and Rudolff (1526): "fein 18 karat."

[4] Compare "4 marcx d'Or a 14 Carats de fin," in Coutereels's Dutch-French arithmetic, 1631 ed., p. 309.

[5] *Aver de pois, peis,* etc. The English *aver,* from the Old French *aveir* or *avoir,* meant goods, and *poiz* was the French *peis* or *pois,* Latin *pensum,* from *pendere,* to weigh. About the year 1500 the old Norman *peis* was superseded by the modern *pois.* The incorrect *du,* for *de,* came in about 1650. Even as late as 1729 the American Greenwood used "averdupois."

The system was introduced into England from Bayonne
c. 1300, but is essentially Spanish. The name is limited to the
English-speaking countries, the pound of 16 ounces being called
on the Continent by various names, such as the pound merchant.[1] Troy weight was the more popular until the 16th
century, when, as Digges (1572) tells us, "Haberdepoyse"
became the "more vsuall weight." Even a century later, however, the Troy weight was given first and was used for weighing such commodities as figs and tobacco and even lead
and iron.[2] There was also the Tower pound of 11.25 Troy
ounces, but this was abandoned about the year 1500. In the
latter part of the 18th century a popular writer[3] thus refers
to the matter:

When *Averdupois Weight* became first in Use, or by what *Law* it
was at first settled, I cannot find out in Statute Books; but on the
contrary, I find that there should be but one Weight (and one
Measure) used throughout this Realm, viz. that of Troy, (Vide 14
Ed. III, and 17 Ed. III). So that it seems (to me) to be first introduced by Chance, and settled by Custom, viz. from giving good or
large weight to those Commodities usually weighed by it, which are
such as are either very Coarse and Drossy, or very subject to waste;
as all kinds of Grocery Wares.

2. LENGTH

Babylonia and Egypt. The Babylonian measures, like those
of most early peoples, were derived to a considerable extent
from the human body. For example, one of the world's primitive measures was the cubit,[4] the length of the *ulna*, or forearm,
whence the English *ell* and French *aune*, but applied to various
lengths. This standard is found among the Babylonians, the

[1] Thus Trenchant (1566) says, "La liure marchande vaut 16 onces." The Dutch
writers sometimes called it "Holland weight," as in Coutereels's *Cyffer-Boeck*,
1690 ed., p. 17, where 1 pound = "2 Marck" or "16 once" or "32 loot."
[2] Hodder's arithmetic, 1672 ed., pp. 15, 66, 68.
[3] J. Ward, *The Young Mathematician's Guide*, 12th ed., p. 32. London, 1771.
[4] Latin *cubitum*, elbow. Sir Charles Warren, *The Ancient Cubit*, London,
1903; a scholarly and extended treatment of the subject

length varying from 525 mm. to 530 mm.[1] It was known some-what earlier in Egypt and numerous specimens are still extant.[2]

Greece and Rome. The method of fixing standards by meas-urement of the human body naturally led to many variations. Thus the Attic foot[3] averaged 295.7 mm.; the Olympic, 320.5 mm.; and the Æginetan, 330 mm. A similar variation is found in Western Europe, the Italian foot being 275 mm.; the Ro-man, 296 mm. (substantially the same as the Attic); and the *pes Drusianus*, 333 mm.[4] The foot was not a common measure until *c*. 280 B.C., when it was adopted as a standard in Pergamum.

The fingerbreadth[5] was used by both Greeks and Romans, as was also the palm[6] of four digits. The cubit was six palms, or twenty-four digits, the Roman foot was $13\frac{1}{2}$ digits, and the fathom[7] was the length of the extended arms. The mile[8] was, as the name indicates, a thousand units, the unit being a double step.

In general, the most common Roman measures of length may be summarized as follows: the *pes* (foot) was 0.296 m. long, and 5 *pedes* (feet) made 1 *passus*; 125 *passus* made 1 *stadium*, about 185 m.; and 8 *stadia* made 1 mile, about 1480 m.[9]

---

[1] J. Brandis, *Das Münz- Mass- und Gewichtswesen in Vorderasien* (Berlin, 1866), p. 21; Hilprecht, *Tablets*, p. 35. The Babylonian name was *ammatu*, and this unit was divided into 30 *ubânu* (*ubânê*). To use our common measures, we may say that the average Roman cubit was 17.4 inches; the Egyptian, 20.64 inches; and the Babylonian, 20.6 inches. See Peet, *Rhind Papyrus*, p. 24.

[2] As the *mahi*, three of which made the *xylon*, the usual length of a walking staff, about 61.5 inches, and 40 of which made the *khet*. Other measures are also known, such as the foot, which was equivalent to about 12.4 of our inches.

[3] Ποῦς (*pous*). The general average as given by Hultsch (*loc. cit.*, p. 697) is 308 mm. [4] K. R. Lepsius, *Längenmasse der Alten*, Berlin, 1884.

[5] Δάκτυλος (*dak'tylos*); Latin, *digitus*.

[6] Δοχμή (*dochme'*); Latin, *palmus*. This is our "hand," used in measuring the height of a horse's shoulders. Homer speaks of handbreadths (δῶρον, *do'ron*) and cubits (πυγών, *pygon'*).

[7] Anglo-Saxon *fœthm*, embrace. The Greek word was ὀργυια (*or'guia*), the length of the outstretched arms; Latin, *tensum*, stretched.

[8] *Mille passuum* (colloquially *passum*). The pace was a double step, and hence a little over 5 Anglo-American feet.

[9] The Greek stadium (στάδιον, *sta'dion*) varied considerably in different cities. The Athenian stadium was about 603–610 Anglo-American feet.

Far East. The finger appears in India as "eight breadths of a *yava*" (barleycorn), four times six fingers making a cubit,[1] as in Greece. The other Oriental units have less immediate interest.

England. In England there was little uniformity in standards before the Norman Conquest. The smaller units were determined roughly by the thumb,[2] span,[3] cubit, ell,[4] foot, and pace.

A relic of this primitive method is seen in the way in which a woman measures cloth, taking eight fingers to the yard, or the distance from the mouth to the end of the outstretched arm.

For longer distances and for farm areas it was the custom to use time-labor units, as in a day's journey or a morning's plowing, such terms being still in use in various parts of the world. The furlong (40 rods, or an eighth of a mile) probably came from the Anglo-Saxon *furlang,* meaning "furrow long."

The word "yard" is from the Middle English *yerd* and the Anglo-Saxon *gyrd,* meaning a stick or a rod, whence also a yardarm on a ship's mast. That the standard was fixed in England by taking the length of the arm of Henry I (1068–1135) is not improbable. Thus an old chronicle relates: "That there might be no Abuse in Measures, he ordained a Measure made by the Length of his own Arm, which is called a *Yard.*"

The words "rod" and "rood" may have had a common origin. The rod was used for linear measure and the rood came to be used for a fourth of an acre.[5]

---

[1] *Hasta, cara,* the forearm. Colebrooke's *Lilāvati,* p. 2, § 35. The word *cubit* appears in India and Siam as *covid,* in Arabia as *covido,* and in Portugal as *covado.*

[2] Latin *pollex,* whence the French *pouce,* an inch. The word "inch," like "ounce," is (as already stated) from the Latin *uncia,* the twelfth of a foot or the twelfth of a pound. Originally the word meant a small weight and is allied to the Greek ὄγκος (*on'kos*), bulk, weight. The old Scotch inch was averaged from the thumbs of three men, "þat is to say, a mekill man and a man of messurabill statur and of a lytill man." See Maitland, *loc. cit.,* p. 369.

[3] The distance spanned by the open hand, from thumb to little finger; finally taken as 9 inches.

[4] The ell has varied greatly. In England it is 45 inches, that is, 1¼ yards. The old Scotch ell was 37.2 inches, and the Flemish ell was 27 inches.

[5] For a bibliography of the subject of measures of length consult the encyclopedias. Among the most ingenious studies of the subject is W. M. F. Petrie, *Inductive Metrology,* London, 1877.

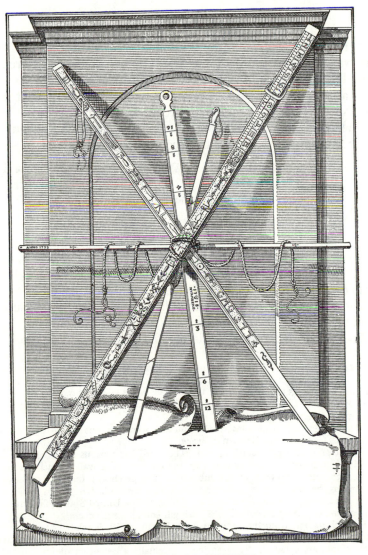

MEASURES OF LENGTH

European of the 18th century, showing the general appearance of the common
measuring sticks of the period. The three shortest pieces are ells

3. AREAS

Acre. The common unit of land measure known as the acre[1] has varied greatly in different countries and at different periods.[2] It was commonly taken to mean a morning's plowing,[3] a strip of land 4 rods wide and one furrow long, that is, 4 rods by 1 furlong, 4 rods by 40 rods, or 160 square rods.

The rood was a fourth of an acre and was also called a perch.[4] It is thus described by Recorde (*c.* 1542):

5 Yardes and a halfe make a Perche . . . [and] 1 Perche in bredth & 40 in length, do make a Rodde of land, which some cal a Rood, some a Yarde lande, and some a Forthendale.[5]

4. CAPACITY

Modern Times. The subject of measures of capacity is so extensive as to make it impossible to mention more than a few facts concerning our modern British and American units.

The gill was the Old French *gelle*, a sort of wine measure, from the Middle Latin *gillo* or *gellus*, a wine vessel.[6]

The quart[7] is, of course, simply the quarter of a gallon.[8]

---

[1] Anglo-Saxon *æcer*. The ancient units of area have no particular significance at the present time. It is, therefore, sufficient, merely to mention the Greek *plethron*, about 0.235 of an acre, and the Roman *jugerum*, about 0.623 of an acre.

[2] For some of these variations see F. W. Maitland, *Domesday Book and Beyond*, p. 374 (Cambridge, 1897).

[3] The cattle used in plowing in the morning were put out to pasture in the afternoon. Compare the German *Morgen*.

[4] Latin *pertica*, a pole, staff, or rod. The word has various other uses, as in the case of a perch of stone or masonry, the contents of a wall 18 inches thick, 1 foot high, and 1 rod long, or 24¾ cubic feet. The perch as a unit of length was the same as the rod.   [5] 1558 ed., fol. L 6.

[6] It may come from the same root as "gallon." The United States gill contains 7.217 cu. in., or 118.35 cu. cm. The British gill contains 142 cu. cm.

[7] French *quarte*, from the Latin *quartus*, fourth, which is related to *quattuor*, four, and to such words as "quadrilateral" (four-sided), "quarry" (a place where stones are squared), "quarantine" (originally a detention of forty days), "quarto," "quire" (Low Latin *quaternum*, a collection of four leaves), "square" (probably Low Latin *ex* (intensive) + *quadrare*, to square), "squad," and "squadron."

[8] The United States gallon contains 231 cu. in., like the old English wine gallon. The imperial (British) gallon contains 277.274 cu. in.

The pint may possibly receive its name from the Spanish *pinta*,[1] a mark, referring to a marked part of a larger vessel.[2] The word "bushel"[3] means a small box, but the origin of the word "peck," as applied to a measure, is obscure.

5. VALUE

Early Units. In the measure of value it became necessary at an early period to develop media of exchange of one kind or

another. The primitive pastoral people naturally used cattle of some kind, whence the Latin noun *pecunia*[4] (money) and the English adjective "pecuniary."

For media of exchange the Greeks often used copper utensils, and ingots of silver and gold. The Babylonians and Egyptians also made use of ingots and rings of the precious

EARLY ROMAN MONEY (PECUNIA)

Showing how a coin was stamped to represent the value of an ox, 4th century B.C. From Breasted's *Survey of the Ancient World*

metals, selling these by weight, whence came the *aes infectum*[5] of the Latins. From this relation of money to value came the double use of such measures as the talent and the pound. Even

---

[1] Latin *picta*, marked or painted, whence "picture." The Middle English form is *pynte*. The origin is, however, uncertain.

[2] Of the British measures whose names are still heard in the colonies and in America, kilderkin was the Dutch *kindeken* (German *Kinderchen*), a babekin, that is, a mere baby in bulk as compared with a tun or vat. "Tun" and "ton" are the same word (Middle English *tonne*, Low Latin *tunna*), meaning a large barrel and hence also a great weight.

[3] Middle English *buschel* or *boischel*; Low Latin *bussellus*, or *bustellus*, diminutive related to Old French *boiste*, a box. Compare "pyx," Greek πυξίς (*pyxis'*), a box, particularly one made of πύξος (*pyx'os*, Lat. *buxus*), boxwood. The imperial (British) bushel contains 2218.192 cu. in., and the Winchester bushel (which became the legal standard in the United States) contains 2150.42 cu. in.

[4] From Latin *pecus*, sheep, cattle. For discussion, see T. Gomperz, *Les penseurs de la Grèce*, French translation, p. 8 (Lausanne, 1904); F. Hultsch, *Griechische und römische Metrologie*, p. 162 (Berlin, 1882).

[5] *Aes*, bronze, copper, money; *infectum, in + facere*, to make; that is. uncoined money. See F. Gnecchi, *Monete Romane*, 2d ed., p. 86 (Milan, 1900).

II

at present, in certain mining districts, the ounce of gold is commonly spoken of as a unit of value. The pound[1] became the *libra*[2] in most Latin countries.

From the *aes infectum* as pieces of metal came the large and heavy metal disks of the early Romans. When these were stamped they became the *aes signatum*.[3]

Coins. The earliest stamped coins found in the Mediterranean countries were probably struck in Lydia in the 7th century B.C.,[4] or possibly in Ægina in the 8th century. They seem to have appeared in China at about the same time.

The first silver money coined in Rome (268 B.C.) was based upon the relation of 10 *asses* to the *denarius*, but the number was changed at a later date. Pliny tells us that the first gold money coined in Rome appeared in 217 B.C.[5] The *aureus*, or gold *denarius*, was first coined under Augustus (31 B.C.–14 A.D.) as $\frac{1}{40}$ of a pound, but it underwent gradual changes until, under Constantine (306–337), it became $\frac{1}{72}$ of a pound, then taking the name *solidus*.[6]

Great Britain. When Cæsar went to Britain (*c.* 55 B.C.) he found the natives using certain weights of metal as media of exchange. Coinage was introduced soon thereafter, based on Roman values. The figure of Britannia, which is still seen

---

[1] Latin, *pondo libra*, "a pound by weight." From the same root as *pondus*, a weight, we also have such words as "ponder," and such units as the Spanish *peso*; see also page 638, note 1.

[2] Latin *libra*, a balance, a pound weight, from the Greek λίτρα (*li'tra*), a pound, whence *litre*, liter. The constellation Libra has for its symbol ♎, the scale beam. From the same root we have such words as "deliberate," to weigh our thoughts. The *libra* appears in French as *livre* and in Italian as *lira*, the old pound in weight as well as in value. The French *livre* was also called a *franc*, as in Trenchant (1566): "la liure autremēt appellee *frāc*."

[3] Gnecchi, *loc. cit.*, p. 89.

[4] Herodotus, I, 94. Judging from the museum pieces the early coins seem to have been both of gold and of silver, and both circular and oblong. See an illustration in Volume I, page 56.

[5] Gnecchi, *loc. cit.*, chap. xiv and p. 145.

[6] *I.e.*, a solid piece of money. From this we have the word "soldier," a man who fought for money, and such words as the Italian *soldo* and the French *sol* and *sou*. The English symbol for shilling (/) comes from the old form of *s* (ſ) and was the initial for *solidus*, just as £ is the initial of *libra* (pound) and as *d* (for penny) is the initial of *denarius*.

on British coins as mistress of the seas, is not at all modern. It appears on one of the pieces of Hadrian's time (*c.* 130).[1]

The most ancient coin of the Saxon period (*c.* 600) is the *sceat*, a silver coin weighing about 1 gram. The word *scilling*, for shilling, appears at this time, the word possibly meaning a little scar.[2] The origin is doubtful, and the word may mean the "clinking coin."[3]

The word "penny" may have come from the Latin *pannus*, a cloth, and hence is the value of a certain piece; but, as in many similar cases, the origin is obscure. Since a piece of cloth was a convenient pledge for money borrowed, the word "pawn" may have come from the same source.[4]

The word "farthing," the Anglo-Saxon *feorthling*, is simply the "little fourth" of a penny.[5]

The Saxon coins were regulated by the pound weight. This pound was commonly known as the Cologne pound, having been brought from that city, and was probably the same as the weight known after the Conquest as the Tower pound, so called because the mint of London was in the Tower. A pound sterling was this weight of silver coins.

United States. The word "dollar" comes from the Low German *Daler*, German *Thaler*, from *Joachimsthaler*, since these large silver pieces first appeared in the Thal[6] of St. Joachim,

---

[1] For a brief résumé of the history of British coins, see R. Potts, *Elementary Arithmetic with brief notices of its history*, § II (London, 1886).

[2] *Skil* means to divide and comes from *skal* or *skar*, to cut; so that *scar-ling* means a little cut on a tally stick, to distinguish the mark from the larger scar (score) which indicated 20 shillings, or a pound. *Skar* is the root of Greek κείρειν (*kei'rein*), to cut close, and is allied to the Anglo-Saxon *scær* and the German *scheren*, whence our "shear." The English "jeer" may be from the same source through the Dutch phrase *den gheck sheeren*, "to shear the fool." A "share" of stock is from the same root.

[3] Compare the German *schellen*, to sound or tinkle. See Greenough and Kittredge, *Words and their Ways*, p. 140 (New York, 1901).

[4] Similarly "panel," a piece of anything marked off. The Middle English of "penny" is *peni*, plural *penies* and *pens*. Compare the Anglo-Saxon *pening* and the German *Pfennig*.

[5] The word is substantially the same as "firkin," from the Dutch *vier* (four) + *kin* (diminutive, as in "lambkin"), once a fourth of a barrel.

[6] English "dale" and "dell." Coined there because of the silver mines in the valley.

Bohemia, in ᴌne 16th century. "Dime" is from the French *disme* and Latin *decem* (ten), "cent" is from *centum* (hundred), and "mill" is from *mille* (thousand). It took over fifty years to replace the English system by the "Federal" in the United States. The origin of the symbol $ is uncertain. It seems to have first appeared in print in Chauncey Lee's *The American Accomptant* (Lansingburgh, 1797) (although in a form very different from the one familiar at the present time), but it was used in manuscripts before that date. The Italian and British merchants had long used ℔ for pounds, writing it quite like our dollar sign (see Volume I, page 233), and it is possible that our merchants in the closing years of the 18th century simply adopted this symbol, just as we have adopted the English word "penny" to mean a cent, which is only a halfpenny.[1]

6. METRIC SYSTEM

Need for the System. The ancient systems of measures were open to two serious objections: (1) they were planned on a varying scale instead of the scale of ten by which the civilized world always counts, and (2) they were not uniform even in any single country. Before the metric system was adopted there were, in northern France alone, eighteen different *aunes*,[2] and in the entire country there were nearly four hundred ways of expressing the area of land.

This condition was not unique in France; it was found in all European countries. Before the days of good roads and easy communication from place to place the difference in standards was not very troublesome, but by the end of the 18th century it became evident that some uniformity was essential.

Early Attempts at Reform. As early as 650 there was an effort made at uniformity in France, a standard of measure being kept in the king's palace. Under Charlemagne (*c.* 800) there

---

[1] See *Rara Arithmetica*, p. 470. There are various hypotheses as to the origin of the symbol $, most of them obviously fanciful.

[2] The cloth measure, the old English ell, as already given. On the metric system as a whole, see the excellent historical work of G. Bigourdan, *Le système métrique,* Paris, 1901.

was nominal uniformity throughout the kingdom, the standards of the royal court being reproduced for use in all leading cities. After Charlemagne's death, however, the numerous feudal lords adopted scales to suit their own interests. Attempts were made at various other times, as in 864, 1307, and 1558, to unify the systems in France, but none of these were successful.

Rise of the Metric System. In 1670, Gabriel Mouton, vicar of the church of St. Paul, at Lyons, proposed a system which should use the scale of 10, and which took for its basal length an arc 1′ long on a great circle of the earth. This unit he called a *milliare* or *mille*, 0.001 of a *mille* being called a *virga* and 0.1 of a *virga* being called a *virgula*. It will be seen that this was, in general plan, not unlike the metric system.

In England, Sir Christopher Wren (*c.* 1670) proposed as the linear unit the length of a pendulum beating half seconds. In France, Picard suggested (1671) the length of a pendulum beating seconds, and Huygens (1673) approved of this unit.

In order to avoid the difficulty involved in the varying length of the second pendulum in different latitudes, La Condamine suggested (1747) the use of a pendulum beating seconds at the equator, a proposal which would, if adopted, have given a standard approximating the present meter. In 1775 Messier determined with great care the length of the second pendulum for 45° of latitude, and an effort was made to adopt this as the linear unit, but it met with no success.

France works out the Metric System. In 1789 the French Académie des Sciences appointed a committee to work out a plan for a new system of measures, and the following year Sir John Miller proposed in the House of Commons a uniform system for Great Britain. About the same time Thomas Jefferson proposed to adopt a new system in the United States, taking for a basal unit the length of the second pendulum at 38° of latitude, this being the mean for this country. In 1790 the French National Assembly took part in the movement, and as a result of the widespread agitation it was decided to proceed at once with the project of unification. The second pendulum was given up and an arc of one ten-millionth of a quarter of a meridian was selected as the basal unit. A careful survey was made of the length of the meridian from Barcelona to Dunkirk, but troubles with the revolutionists (1793) delayed the work. The committees which began and carried on the enterprise were changed from time to time, but they included some of the greatest scientists of France, such as Borda, Lagrange, Lavoisier, Tillet, Condorcet, Laplace, Monge, Cassini, Meusnier, Coulomb, Haüy, Brisson, Vandermonde, Legendre, Delambre, Berthollet, and Méchain. Owing to a slight error in finding the latitude of Barcelona, the original idea of the unit was not carried out, but a standard meter was fixed, and from this copies were made for use in all civilized countries.

The system was merely permissive in France until 1840, when it was made the only legal one. The expositions held in London (1851) and Paris (1855, 1867) aided greatly in making the system known outside of France. In 1919 the system was the only legal one in thirty-four countries, with a population of about 450,000,000; was optional in eleven countries, with a population of about 730,000,000; and was recognized by twenty-six countries, with a population of about 690,000,000, this recognition taking the form of assisting in the support of the International Bureau of Weights and Measures at Paris.

The modern plan of determining such standards adopts as the unit the length of a light wave of a defined type.[1]

---

[1] Owing to the importance of the subject and the necessity for condensing the treatment in this chapter, the following bibliographical references are added: W. S. B. Woolhouse, *Measures, Weights, and Moneys of all Nations*, 6th ed., London, 1881; F. Hultsch, *Griechische und römische Metrologie*, Berlin, 1882; A. Böckh, *Metrologische Untersuchungen*, Berlin, 1838; F. W. Clarke, *Weights, Measures, and Money of all Nations*, New York, 1875; William Harkness, "The Progress of Science as exemplified in the Art of Weighing and Measuring," *Bulletin of the Philosophical Society of Washington*, X, p. xxxix; E. Noel, *Science of Metrology*, London, 1889; C. E. Guillaume, *Les récents progrès du système métrique*, Paris, 1913. See also the various *Procès-verbaux des séances* of the *Comité international des poids et mesures*, Paris, 1876 to date; W. Cunningham, *The Growth of English Industry and Commerce*, p. 118 (London, 1896); A. De Morgan, *Arithmetical Books*, p. 5 (London, 1847); Albêrûnî's *India*, translated by E. C. Sachau, 2 vols., London, 1910, for the metrology of India *c.* 1000 (see particularly Volume I, chapter xv).

7. TIME

Problem Stated. Before the time of printed calendars, when astronomical instruments were crude affairs and the astronomer was merely a court astrologer, and when the celebration on a movable feast day of the anniversary of a fixed religious event did not seem to arouse adverse criticism, even among the enemies of the various faiths, the regulation of the calendar[1] naturally ranked as one of the chief problems of mathematics.

The Computus. Accordingly there arose in all ecclesiastical schools of any standing in the Christian church the necessity for instructing some member or group of the priestly order in the process of computing the dates of Easter and the other holy days. For this purpose there were prepared short treatises on the subject. A book of this kind was generally known as a *Computus Paschalis*, *Computus Ecclesiasticus*, or, more commonly, a *Computus* or *Compotus*.[2]

General Nature of the Computi. Briefly stated, the modern form of the ancient *computus* begins with the assertion that Easter day, on which the other movable feasts of the Church depend, is the first Sunday after the full moon which happens upon or next after March 21. If the full moon happens upon a Sunday, Easter day is the Sunday following. The full moon

---

[1] From the Latin *kalendarium*, a list of interest payments due upon the first days of the months, the *kalendae*. The word was not used in its present sense by the Romans. They used *fasti* to indicate a list of days in which the holidays were designated.

Of the many works on the calendar, one of the latest and most extensive is F. K. Ginzel, *Handbuch der math. und techn. Chronologie. Das Zeitrechnungswesen der Völker*, 3 vols., Leipzig, 1906–1914. See also J. T. Shotwell, "The Discovery of Time," *Journal of Philosophy, Psychology, and Scientific Methods*, XII, Nos. 8, 10, 12, and *Records of Civilization, Sources and Studies*, chapter iv (New York, 1922). For a popular essay see M. B. Cotsworth, *The Evolution of Calendars*, Washington, 1922.

[2] Hieronymus Vitalis (Girolamo Vitali), *Lexicon Mathematicvm*, Paris, 1688 (Rome edition of 1690, p. 173), thus defines *computus*: "Significatio pressiùs accepta est, atque antonomasticè haesit annorum ratiocinio, & temporum distributioni, quod proprie Chronologos, & Astronomos spectat." The spelling *compotus* was at one time the more common,—possibly a kind of pun upon the convivial habits of the computers, a *compotatio*, the Greek *symposium* (συμπόσιον, *sympos'ion*), meaning a "drinking together."

is taken as the 14th day of a lunar month, "reckoned according to an ancient ecclesiastical computation and not the real or astronomical full moon."

In order to use this ancient computation it becomes necessary to be able to find the Golden Number of the year. This is done by adding 1 to the number of the year and dividing by 19, the remainder being the number sought. If the remainder is 0, the Golden Number is 19. Thus the Golden Number of 1930 is found by taking the remainder of $(1930 + 1) \div 19$, which is 12.

Taking the seven letters A, B, C, D, E, F, G, the letter A belongs to January 1, B to January 2, and so on to G, which belongs to January 7, after which A belongs to January 8, and so on. If January 2 is Sunday, the Dominical letter of the year is B. By means of the Dominical letter it is possible to find the day of the week of any given date. The finding of this letter depends upon a few simple calculations connected with tables given in the *computi*.[1]

Universality of the Problem. The problem was not confined to the Christian church. Since most early religions were connected with sun worship or with astrology, work somewhat similar to that of preparing the *computus* was needed in all religious organizations. Hence we find a problem analogous to the Christian one in the routine work of the Hebrew, Mohammedan, Brahman, and ancient Roman priests, all of whom acted as guardians of the calendar.[2] We shall now consider some of the astronomical difficulties in the way of making a scientific calendar.

---

[1] For a full discussion see A. De Morgan, "On the Ecclesiastical Calendar," *Companion to the Almanac for 1845*, p. 1 (London, n.d.), and "On the Earliest Printed Almanacs," *Companion to the Almanac for 1846*, p. 1.

One of the best studies of the *computus* is C. Wordsworth, *The Ancient Kalendar of the University of Oxford*, Oxford, 1904. This represents the calendar as it stood *c.* 1340.

[2] There is extant an Egyptian papyrus of about the beginning of the Christian Era that evidently was intended to serve the same purpose as the later *computi*. See W. M. Flinders Petrie, "The geographical papyrus (an almanack)," in *Two Hieroglyphic Papyri from Tanis*, published by the Egypt Exploration Fund, London, 1889.

The Day. Of the chief divisions of time the most obvious one is the day. This was, therefore, the primitive unit in the measure of time and the one which for many generations must have been looked upon as unvarying. As the race developed, however, various kinds of day were distinguished. First from the standpoint of invariability is the sidereal day, the length of time of a revolution of the earth as shown by observations on the fixed stars, namely, 23 hours 56 minutes 4.09 seconds[1] of our common time. First from the standpoint of the casual observer, however, is the true solar day, the length of time between one passage of the sun's center across the meridian and the next passage. This varies with the season, the difference between the longest and shortest days being 51 seconds; but for common purposes the solar day sufficed for thousands of years, the sundial being the means by which it was most frequently measured. As clocks became perfected a third kind of day came into use, the artificial mean solar day, the average of the variable solar days of the year, equal to 24 hours 3 minutes 56.56 seconds of sidereal time. In addition to these general and obvious kinds of day, writers on chronology distinguish others which do not concern the present discussion.[2]

The day began with the Babylonians at sunrise; with the Athenians, Jews, and various other ancient peoples, and with certain Christian sects, at sunset; with the Umbrians, at noon; and with the Roman and Egyptian priests, at midnight.[3]

The Month. The next obvious division of time was the month, originally the length of time from one new moon to the next, and one that served as the greater unit for many thousands of years. As science developed, however, it became apparent, as in the case of the day, that there are several kinds of months. There is the sidereal month, the time required for a passage of the moon about the earth as observed with reference to the

---

[1] All such figures are approximations, varying slightly with different authorities.

[2] See, for example, A. Drechsler, *Kalenderbüchlein*, p. 19 (Leipzig, 1881).

[3] Thus Pliny: " Ipsum diem alii aliter observavere. Babylonii inter duos solis exortus; Athenienses inter duos occasus; Umbri a meridie ad meridiem; . . . Sacerdotes Romani, et qui diem diffiniere civilem, item Aegyptii et Hipparchus, a media nocte in mediam " (*Hist. Nat.*, II, cap. 79).

fixed stars, namely, 27 days 7 hours 43 minutes 11.5 seconds. There is also the synodical month, from one conjunction of the sun and moon to the next one, averaging 29 days 12 hours 44 minutes 3 seconds, or 2 days 5 hours 0 minutes 51.5 seconds more than the sidereal month. This is the month of those who use a lunar calendar, and is the basis of the artificial month, twelve of which make our common year.[1]

The Year. Less obvious than the day or the month was the year, a period observable only about one three hundred sixty-fifth as often as the day and about one twelfth as often as the month. It took the world a long time to fix the length of the year with any degree of accuracy, and the attempt to harmonize time-reckoning by days, months, and years has given rise to as many different calendars as there have been leading races. First of all there has to be considered what constitutes a year. The sidereal year is the period of revolution of the earth about the sun, namely, 365 days 6 hours 9 minutes 9.5 seconds (365.256358 days). The tropical year is the period of apparent revolution of the sun about the earth from the instant of one vernal equinox to the next, and would be the same as the sidereal year if it were not for the slight precession of the equinox, amounting to about 50″ a year. This precession makes the length of the tropical year 365 days 5 hours 48 minutes 46.43 seconds (365.242204 days).[2] There is also the anomalistic year of 365 days 6 hours 14 minutes 23 seconds, measured from the time when the earth is nearest the sun to the next time[3] that they are in the same relative position,—a year that is slightly longer than the sidereal. There is also the lunar year of twelve synodical months, probably the first one recognized by the primitive observers of nature, and in addition to this there are various other periods which have gone by the same general name.[4]

---

[1] For other types of month see Drechsler, *loc. cit.*, p. 24.
[2] This was the length in the year 1800. It varies about 0.59 seconds a century. B. Peter, *Kalenderkunde*, 2d ed., p. 20 (Leipzig, 1901).
[3] From perihelion to perihelion.
[4] For the list, see Drechsler, *loc. cit.*, p. 26. On the cycle, see *ibid.*, p. 30; on the era, p. 44.

The Week. The week was less obvious than the day, the month, or even the year, having no astronomical events by which to mark its limits. It seems very likely that it arose from the need for a longer period than the day and a shorter period than the month. Hence we have the half month, known as the fortnight (fourteen nights), and the quarter month, known as the week.

Early Attempts in making a Calendar. Of the various attempts to perfect a calendar only a few will be mentioned, and in general those that had some bearing upon the Christian system.[1]

The Babylonians, whose relation to the invading Chaldeans was such as to make their later calendars substantially identical, seem to have been the first of the world's noteworthy astronomers. Aristotle relates that before 2200 B.C. they attempted scientific observations of the heavens, and Porphyrius (*c.* 275) tells us that Callisthenes (*c.* 330 B.C.) took to the Greek sage the results of a series of Chaldean observations extending over 1903 years. The Chaldeans knew the length of the year as 365 days 6 hours 11 minutes, but used both the lunar month and lunar year for civil purposes. They divided both the natural day and the natural night into twelve hours each, and in quite early times the sundial and water clock were known, the latter for use at night. For astronomical purposes the day was divided into twenty-four equal hours. They probably very early used a fourth of a month as a convenient division of time, or rather the half of the half, as was the customary way of thinking in the ancient world.

The Chinese Calendar. We are ignorant of the nature of the primitive Chinese calendars.[2] Certainly earlier than 2000 B.C.

---

[1] A good treatment of the subject is given by L. Ideler, *Handbuch der math. und techn. Chronologie*, 2 vols., 1825; 2d ed., Breslau, 1883, a facsimile of the first edition. Good résumés are given by Drechsler, *loc. cit.*, p. 56; Peter, *loc. cit.*, p. 5 (chiefly on the Christian calendar). From all these sources information has freely been taken. On the general question of chronology see also J. B. Biot, "Résumé de Chronologie Astronomique," *Mémoires de l'Académie des Sciences*, XXII, 209–476 (Paris). See especially Ginzel's work mentioned on page 651 n.

[2] Mikami, *China*, pp. 5, 45.

the subject occupied the attention of the astrologers. It is, however, very difficult to unravel a system which changed with each emperor, and only a few general principles can be set forth. Under the emperor Yau (*c.* 2357–*c.* 2258 B.C.) an effort was made to establish a scientific calendar for the whole country, and possibly this was done even earlier, under the emperor Huang-ti (*c.* 2700 B.C.).[1] In accordance with a decree of Wu-wang (1122 B.C.)[2] the day seems to have begun with midnight, although before this time, under the Shang dynasty (1766–1122 B.C.), it began at noon. The civil day has twelve hours, and the middle of the first hour is midnight.[3] Each hour is divided into eight parts (*khe*), each being our quarter, and each of these into fifteen *fen*, each therefore being our minute. In modern times the *fen* has been divided into sixty *miao* (seconds) under European influence. At present also the American clock is becoming common in China, so that the ancient system seems destined soon to pass away. Nevertheless the connection between the old Chinese calendar and that of Europe seems apparent. The Chinese days were named in such a way as to give seven-day periods corresponding to our weeks, and the month began, as was so often the case in early times, with the new moon. The twelve lunar months were supplemented in such a way as to harmonize the lunar and solar years, the Chang Cycle[4] being used by the Chinese before the Greeks knew of the same system under the name of the Metonic Cycle.

The Egyptian Calendar. In the ancient Egyptian calendar, which influenced all the Mediterranean countries beginning with Crete and the mainland of Greece, the business day included the night, the natural day and night being each divided into twelve hours, these hours varying in length with the

---

[1] Or Hoang-ti. His reign seems to have begun in the year 2704 B.C. See Volume I, page 24; Mikami, *China*, p. 2.

[2] Founder of the Chau (Cheu, Choi) dynasty, 1122–225 B.C.

[3] These hours (*shi*) are each 120 European minutes in length.

[4] For details of the complex system, and for the influence of China on Japan, see Drechsler, *loc. cit.*, pp. 71, 88.

season. The civil day seems to have commonly begun at sunset, although the priests are said by Pliny to have begun theirs at midnight. In later times, for astronomical purposes, the day began at noon[1] and was divided into twenty-four equal hours. The Romans considered the planets as ruling one hour of each day, in the following order, beginning with the first hour of Saturday: Saturn, Jupiter, Mars, Sun, Venus, Mercury, Moon, the sun and moon being placed among the "wanderers."[2] From these planets they named the days by the following plan: Taking Saturn for the first hour of Saturday and counting the hours forward, it will be seen that the second hour is ruled by Jupiter, and so on to the twenty-fourth, which is ruled by Mars. Then the next hour, the first of Sunday, is ruled by the Sun, the first hour of the next day by the Moon, and so on. Thus the days of the week were named by the ruling planets of their first hours, and we have Saturn's day, Sun's day, Moon's day, Mars's day,[3] Mercury's day,[4] Jupiter's day,[5] Venus's day,[6] a system that has come down to our time and seems destined to continue indefinitely, in spite of the fact that we are using names of heathen gods in the calendars of various religions.

Each month except the last (*Mesori*) in the native Egyptian calendar contained thirty days, five days being added to *Mesori* so as to make the year one of three hundred sixty-five days. Since this gave an error of about one fourth of a day, the year was a changing one, coming back to its original position with respect to the heavenly bodies once in 4 × 365 common years, or 1460 years (1461 Egyptian years). The year began with the first day of the month of Thoth, the god who, according to Plato's *Phædrus*, introduced the calendar and numbers into Egypt. As early as the 14th century B.C. the Egyptians recognized the value of a fixed year, but the changing one was too strongly implanted in the religious canons of the people to be given up. The fixed year was used to the extent of a division

---

[1] So with Ptolemy the astronomer, c. 150.
[2] "Planet" is from the Greek πλανήτης (*plane'tes*), originally a wanderer.
[3] French, *Mardi*.   [4] French, *Mercredi*.
[5] In the Northern lands, Thor's day.
[6] In the Northern lands, Frigg's day, Frigg being the goddess of marriage.

into three seasons, regulated by the river,—the Water Season, the Garden Season, and the Fruit Season,[1]—these being easily determined by the temple observers. From the temple, too, came the announcements of the turn in the rise or fall of the river, the nilometers being under the observation of the priests. These early nilometers may be seen in the temples today; they were concealed from the observation of the common people, the water being admitted by subterranean channels.

The Alexandrian Calendar. After Egypt became a Roman province (*c.* 30 B.C.) the Alexandrian calendar, including the fixed year, was introduced, although the varying year of the ancients remained in popular use until the 4th century. The Alexandrian system was used until the first half of the 7th century, when the country yielded (638) to the Mohammedan conqueror, with an attendant change of the calendar except in Upper Egypt, where the Coptic, Abyssinian, or Ethiopian churches maintained their supremacy. Since 1798, when the French obtained brief control of the country, the European system has been used side by side with the Mohammedan.

The Athenian Calendar. The Athenian calendar followed the Egyptian in beginning the new day at sunset and in dividing both day and night into twelve hours. The seven-day week was not used, however, the lunar month being divided into three parts. Of these the first consisted of ten days, numbered in order, the "5th day of the beginning of the month" being the fifth. Then followed nine days, numbered as before, but with the designation "over ten." From there to the end of the month the numbers were 20, 1 over 20, and so on, these days also being numbered backwards from the end of the month. In the popular calendar the month began with the new moon, and twelve of these months made three hundred fifty-four days, requiring the insertion of a new month[2] every three years.[3] Meton (432 B.C.) constructed a nineteen-year cycle

---

[1] Namely, June 21 to October 20; October 21 to February 20; February 21 to June 20.　　　[2] A second month of Poseideon, known as Poseideon II.

[3] The detailed variations of this plan need not be considered here. See any work on the calendar.

in which the third, fifth, eighth, eleventh, thirteenth, sixteenth, and nineteenth years should contain the extra month,[1] a plan which Callippus, a century later (325 B.C.), modified to include four nineteen-year cycles.[2] Still later Hipparchus (150 B.C.) suggested the use of four of the cycles of Callippus, less a day, or 110,036 days in all, but neither of the last two calendars came into popular use.

Roman Calendar. The oldest of the Roman calendars seems to have been the one attributed to Romulus. The year probably consisted of ten months of varying length, or of 304 days, beginning with March. Numa Pompilius (715–672 B.C.) is said to have added two other months, January and February, and his year was probably lunar. The Decemvirs (5th century B.C.) decreed a solar year, the regulation of which was left in the hands of the priests. The calendar was so mismanaged, however, that by the time of Julius Cæsar each day was eighty days out of its astronomical place, and radical measures were necessary for its reform. Cæsar therefore decreed that the year 46 B.C. should have four hundred forty-five days[3] and that thereafter the year should consist of three hundred sixty-five days, with a leap year every fourth year.[4]

Names of the Months. Following the older custom as to the beginning of the year in March, Caesar at first used the following plan for the calendar:

1. Martius,	31 days.	7. Septembris,	30 days.
2. Aprilis,	30 days.	8. Octobris,	31 days.
3. Maius,	31 days.	9. Novembris,	30 days.
4. Junius,	30 days.	10. Decembris,	31 days.
5. Quintilis,	31 days.	11. Januarius,	31 days.
6. Sextilis,	31 days.	12. Februarius,	28 days.

---

[1] 19 years = 235 months = 6939¾ days. The months as arranged, however, contained 6940⅔ days.

[2] 4 × 19 years = 76 years = 940 months. The months were 29 or 30 days and totaled 27,759 days. [3] Hence called *annus confusionis*.

[4] A calendar of *c.* 100 B.C. was recently found at Anzio, in the Campagna, based upon a lunar year of 353 days with an intercalary month of 27 days on alternate years.

This accounts for our names "September" (7th month), "October," "November," and "December." On his original plan every alternate month, beginning with March, had thirty-one days, the others having thirty days, except that February received its thirtieth day only once in four years. Cæsar later decreed that the year should begin with January, and finally, but during his life, the name of Quintilis, the month in which he was born, was changed to Julius. He also changed the number of days in certain months, and the result appears in our present calendar. After his death, in the second year of his calendar, a further confusion arose, apparently through a misunderstanding on the part of the priests as to the proper date for leap year. This was corrected by Augustus, and in his honor the name of Sextilis was changed to bear his name. From that time on the Julian calendar remained in use until its reformation under Gregory XIII in 1582, and it was used by the Greek Catholics, including the Russians, until the World War of 1914–1918, the dates until that time differing by thirteen days from those of the calendar of Western Europe.

Christian Calendar. The indebtedness of the present European calendar to those already described is apparent, and it is also evident that our calendar has had an extensive history. The beginning of the year, for example, has not been uniform from time to time and in different countries. In the early centuries the year usually began with April in the East[1] and with March in the West, although sometimes with the Feast of the Conception, Christmas day,[2] Easter, or Ascension day, or at other times according to the fancy of the popes. Finally Innocent XII again decreed that the year should begin on

---

[1] Although the Byzantine calendar began with September 1.

[2] As in Spain until the 16th century and in Germany from the 11th century. March 1 and March 25 (the Annunciation) were favorite dates, although Advent Sunday (the fourth Sunday before Christmas) has generally been recognized as the beginning of the ecclesiastical year. March 1 was used generally in medieval France, in Oriental Christendom, and (until 1797) in Venice. March 25 was used by the medieval Pisans and Florentines. Most of the Italian states adopted January 1 in 1750. For further details consult Ginzel (see page 651 n.) or such works as A. Drechsler, *Kalenderbüchlein*, p. 77 (Leipzig, 1881).

January 1, beginning with 1691, as Philip II had done for the Netherlands in 1575,[1] and as Julius Cæsar had done before the Christian Era.

Numbering the Years. Following the Roman custom, the years in the early centuries of Christianity were dated from the accession of the emperor or consul. We have a relic of this in the dating of acts of parliament in England and of presidential proclamations in America.[2] It was not until the abbot Dionysius Exiguus[3] (533) arranged the Christian calendar that the supposed[4] date of the birth of Christ was generally taken for the beginning of our era, Christmas day being therefore appropriately selected as the first day of the year 1. This calendar was adopted in Rome in the 6th century,[5] in various other Christian countries in the 7th century, and generally throughout Europe in the 8th century.[6]

Changes in Easter. Not only the beginning of the year but the determination of Easter has been the subject of much change. We now consider Easter as the first Sunday after the first full moon following the vernal equinox,[7] as decreed in 325 by the Council of Nice (Nicæa). Formerly it fell on the date

---

[1] England adopted January 1 in 1752.

[2] As "in the 150th year of our independence."

[3] Dionysius the Little. He went to Rome c. 500 and died there in 540.

[4] He considered the birth of Christ as taking place in the year 754 of the founding of Rome, although early Christians placed it in the year 750.

[5] There are, however, no extant inscriptions of the 6th century which bear dates in the Christian Era. See M. Armellini, *Archeologia Cristiana*, p. 479 (Rome, 1898). Sporadic efforts had been made before the 6th century to use a Christian calendar. The oldest known specimen of such a calendar dates from 354. See B. Peter, *Kalenderkunde*, 2d ed., p. 4 (Leipzig, 1901).

[6] The exceptions were the Spanish peninsula and Southern France. Charlemagne was the first great ruler to use (783) the Dionysian calendar.

[7] In Rome. It is possible to have a difference of a week between this Sunday in Rome and in (say) Honolulu, the full moon occurring on Sunday in Rome when it is still Saturday in some places to the west. This has occasionally occurred as an astronomical fact although not as an ecclesiastical one. It should be understood that, for Church purposes, March 21 is taken as the date of the vernal equinox, and that the full moon is not determined by modern astronomy but by certain rules as laid down, say, in the Book of Common Prayer. Easter, therefore, now varies from March 22 to April 25. For a good résumé of the Easter problem see Peter, *loc. cit.*, p. 58.

of the Jewish Passover, but in order to avoid this coincidence the Church readjusted its calendar. Justinian, with this in view, decreed (547) that Easter should be 21 days (instead of 14 days) after the first new moon after March 7. In this way, in general, the Passover and Easter do not come together, although occasionally they synchronize.[1]

The Gregorian Calendar. The present calendar of Western Europe and the Americas, the so-called Gregorian calendar, was necessitated by the fact that the year is not $365\frac{1}{4}$ days long, as recognized by Cæsar, but is about 11 minutes 14 seconds shorter than this. Therefore once in 128 years the Julian calendar receded one day from the astronomical norm, and by the close of the 14th century the departure of Easter day from its traditional position became so noticeable that it was the subject of much comment. It was not until Gregory XIII, however, consulting with such scientists as Aloysius Lilius[2] and Christopher Clavius ($c.$ 1575), determined on a reform, that anything was really accomplished. He decreed that October 4, 1582, should be called October 15, and that from the total number of leap years there should be dropped three in every four centuries. In particular he decreed that only such centennial years as are divisible by 400 (1600, 2000, 2400, etc.) should be leap years. This requires no further adjusting of the calendar for over 3000 years. Italy, Spain, Portugal, Poland, France, and a part of the Netherlands adopted this calendar in 1582. In 1583 it was recognized in part of Germany, the old style being also used until 1700. Part of Switzerland adopted it in 1584, and the other part, together with Denmark and the rest of the Netherlands, in 1700. It was also adopted in Hungary in 1587, in Prussia in 1610, in England in 1752, and in Sweden in 1753. So fixed had the Julian calendar become in the minds of the people, however, that even as late as the opening of the 19th century O. S. (old style) and N. S. (new style)

---

[1] As in 1805, 1825, 1828, 1832, and early in the present century, on April 12, 1903. Among the many suggestions for Easter is that of Jean Bernoulli that it should be the first Sunday after March 21, without reference to the moon.

[2] Ludovico Lilio, Luigi Lilio Ghiraldi (1510–1576).

were used in dating letters in America, while writers on arithmetic felt it necessary to include a description of the Julian and Gregorian calendars as late as the close of that century. The changes brought about by the World War of 1914–1918 led to a more general adoption of the Gregorian calendar in the few countries which had continued to use the Julian or other types.

The Calendar in the French Revolution. In the early days of the French Revolution an attempt was made to impose a new calendar upon the country, partly as a protest against the Christian church. It was hoped that this reform, like that which resulted in the metric system, would receive international recognition. The new era was to begin with the autumnal equinox which occurred on September 22, 1792. There were twelve months of thirty days each, and these months were divided into decades in which the days were named numerically,—Primidi, Duodi, and so on. The extra five or six days of the year were grouped at the end as holidays. The months were named according to natural conditions, thus: In autumn, Vendémiaire (vintage), September 22–October 21; Brumaire (fog), October 22–November 20; Frimaire (sleet), November 21–December 20. In winter, Nivôse (snow), December 21–January 19; Pluviôse (rain), January 20–February 18; Ventôse (wind), February 19–March 20. In spring, Germinal (seed), March 21–April 19; Floréal (blossom), April 20–May 19; Prairial (pasture), May 20–June 18. In summer, Messidor (harvest), June 19–July 18; Fervidor or Thermidor (heat), July 19–August 17; Fructidor (fruit), August 18–September 16. As might have been known, the scheme failed, and on August 30, 1805, a decree was signed reëstablishing the Gregorian calendar, beginning January 1, 1806.[1]

Other Calendars. The other calendars are of no special interest in the history of mathematics. The Hindus began their year with the day of the first new moon after the vernal equinox. The Jews begin their day at sunset, their week on Saturday night (*i.e.*, when their holy day ends and Sunday be-

---

[1] This decree may still be seen in the Musée des Archives Nationales, in Paris.

gins), and their year with *Tishri* 1.[1] Their calendar, more lunar than ours, is quite complicated.

The Maya civilization[2] had a curious system, the year beginning with the winter solstice and being divided into eighteen months, entirely independent of astronomical considerations. Scholars have recently asserted that their calendar goes back as far as the 34th century B.C.

The Mohammedans begin their day with sunset, and, like many other Eastern peoples, divide both daytime and nighttime into twelve hours, the length of the hour varying with the season. The week begins on Sunday, and Friday is the day of rest. Their month begins with the new moon, and the year is purely lunar, of 354 or 355 days. The year 1343 A.H. began on August 2, 1924, of the Christian calendar. The era began with the Hejira, the flight of Mohammed from Mecca on July 15 or 16, 622. On account of frequent references to the Mohammedan calendar in literature, it may be added that a simple rule, accurate enough for practical purposes, for translating a year of the Hejira into a Christian year is as follows: To 97 per cent of the number of the year add 622; the result is the Christian year. Thus 1326 A.H. = 97 per cent × 1326 + 622 = 1908 A.D.

Early Christian Computi. The first noteworthy Christian work[3] on the calendar was that of Victorius of Aquitania (457). About a century later a second *Computus Paschalis* appeared, probably written by Cassiodorus (562). In the next century the question of Easter had become so complicated as to cause (664) a dispute between the church in England and the authorities in Rome. The best of the early works on the

---

[1] *I.e.*, the first new moon after the autumnal equinox. Their year formerly began with *Nisan*, their seventh month, thus using, like that of the Hindus, the vernal equinox. In 1908 *Tishri* 1 was September 26 of the Christian calendar. See S. B. Burnaby, *Elements of the Jewish and Muhammedan Calendars*, London, 1901.

[2] S. G. Morley, *An Introduction to the Study of the Maya Hieroglyphs*, Washington, 1915; C. P. Bowditch, *The Numeration, Calendar Systems, and Astronomical Knowledge of the Mayas*, Cambridge, Massachusetts, 1910.

[3] On this work and the works of later scholars on the same subject see B. Lefebvre, *Notes d'Histoire des Mathématiques*. p. 39 (Louvain, 1920).

computus is the one written by Bede[1] in the 8th century. This contains a precise statement as to the method of finding the date of Easter in any year.

In the 9th century both Hrabanus Maurus (*c.* 820) and Alcuin (*c.* 775) wrote upon the problem, and Charlemagne thought the subject so important that he urged that it be considered in every monastery.[2]

Medieval Works. Lectures were held upon the subject in the 13th and 14th centuries in the various European universities. Sacrobosco (*c.* 1250) wrote a work on it,[3] such a practical calculator as Paolo Dagomari (*c.* 1340) did not hesitate to do the same, and even the Jewish scholars contributed treatises on the Christian calendar as well as their own.[4] It is to a commentary by Andalò di Negro (*c.* 1300) on a work by Jacob ben Machir (d. 1307) that we owe the first prominent use in Europe of the Arabic word *almanac*,[5] later brought into general use by such writers as Peurbach (*c.* 1460) and Regiomontanus (*c.* 1470).

Printed Works. The first printed *computus* was that of Anianus.[6] In this work there appears the original of the familiar rime beginning "Thirty days hath September."[7]

---

[1] *De temporum ratione.*

[2] D. C. Munro, *Selections from the Laws of Charles the Great*, p. 15 (Philadelphia, 1900), "Admonitio generalis," 789. See also T. Ziegler, *Geschichte der Pädagogik*, p. 28 (Munich, 1895); Günther, *Math. Unterrichts*, p. 66.

[3] *Libellus de anni ratione, seu ut vocatur vulgo computus ecclesiasticus.*

[4] There is a MS. now in Petrograd, written by Jechiel ben Josef (1302), under the title *Injan Sod ha-Ibbur*, with a chapter on the Christian *computus*. See M. Steinschneider, "Die Mathematik bei den Juden," *Bibl. Math.*, XI (2), 16, 38, 74; XII (2), 5, 33.

[5] Heb. אלמנך, from the Arab. *al-manākh*. The word is not pure Arabic, however, and the real origin is unknown. See Boncompagni's *Bullettino*, IX, 595; Günther, *Math. Unterrichts*, p. 190 n.

[6] *Cōputus manualis magri aniani. metricus cū꞉mēto* (Strasburg, 1488). There is said to have been an edition printed at Rome in 1486. For bibliography of Anianus see C. Wordsworth, *The Ancient Kalendar of the University of Oxford*, p. 113 (Oxford, 1904). See also the facsimile on page 668.

[7] The Latin form as given by Anianus is as follows:

> Junius aprils september et ipse nouember
> Dant triginta dies reliquis suⱶpadditur vnus,
> De quorum numero februarius excipiatur.—Fol. B 8

See *Rara Arithmetica*, p. 33.

This was not original with Anianus, however, for it is found in various medieval manuscripts.[1] It first appeared in English verse in 1590.

A COMPUTUS OF 1393

In this MS. there appears, in Italian, the verse "Thirty days hath September"

[1] *E.g.*, in the above anonymous Italian MS. of 1393, beginning:
"Trenta di a nouembre apile giugno & settembre."
The MS. is in Mr. Plimpton's library. See *Rara Arithmetica*, p. 443.

Sunt aries thaurus gemini cancer leo virgo libra qz scorpius arcintenens caper amphora pisic

FROM A COMPUTUS OF 1476

This page shows the usual verses beginning "Sunt aries, thaurus, gemini, cancer."
The work also contains the verses, in Latin, beginning "Thirty days hath September," which are found as early as the 13th century. From Mr. Plimpton's library

Liber qui Compotus inscribitur: vna cum figuris et ma/
nibus necessariis tam in suis locis qꝫ in fine libri positis.
Incipit feliciter

Lux orta est iusto Psalmista Ista ver
ba possunt dupliciter considerari. primo pnt
dici de deo qꝫ est lux vera. ideo dicebat dauid.
lux orta est iusto. Et de ista luce dicit Joba.i.
Erat lux vera qꝫ illuminat omnē hominē veniē/
tē in hunc mundū. Secūdo ponit de scientia.
Et dicit lux quasi scientē reddētē lucidū. quia facit hominē sciē/
rem esse lucidū. Jn quibus verbis ad cōmendationē scie duo bre
uiter tangūtur. primo eꝰn tangit sciētie altitudo preciosa p hoc
quod dicit lux. Secundo largitudo gloriosa per hoc qꝫ dicit or/
ta est. primū probatur auctoritate et ratioe. auctoritate Jsidori
sic dicentis. Scientia est fons indeficiens. bonitatis via. sui sal/
uatoris cognitio. Ratioe sic. Illud est validū et preciosum quod
de inualido ꞇ imperfecto facit validum ꞇ pfectum. sciētia ē baiuꝰ
modi. ergo ꞇꞇ. maior est manifesta. minor declaratur p phm ter/
cio de aīa sic dicentē: Aīa in pncipio sue creationis ē tanqꝫ tabu
la rasa in qua nihil depictū est. depingibilis tn sciētiis ꞇ virtutí/
bus. primum probat auctoritate boetii ꞇ ratoe in prologo arís
metrice. Scientia ē cor̄ que vera sunt ꞇ impmutabilis essentie no
straqꝫ cōphensione veritatis. Ratioe sic: illud est tanqꝫ sūmū
bonū quod habet largitionē gloriosam scia ē hmōi. ergo ꞇꞇ. ma
ior est vera. minor pbatur p diffinitionē sciētie qꝫ talis ē. Scia est
qꝫdā habitus aīe rōnalis nō innatus sed acqꝰsitus oīm hūanarꝰ re
rum ỹndagatrix ꞇ totius humane vite gubernatrix. qꝫ scia sit ha/
bitus ptꝫ. qr scia est aliquod existens in aīa. sed omne illud quod
est in aīa aut ē hitus aut potētia aut passio. Et hoc testat Aristo.
in secūdo ethicor̄. qꝫ scia nō sit passio ptꝫ qr passiones sunt in vo
luntate scia nō ē hmōi. ergo ꞇꞇ. qꝫ nō sit potētia ptꝫ. qr qꝫlꝫbꝫ po/
ten ia sit a natura. sicut irascibilis. ꞇ cōcupiscibilis. ꞇ sic relinqui
tur qꝫ nō sit potentia. qꝫ sit hitus aīe rōnalis ptꝫ p pdicta. qꝫ aūt

a ii

THE COMPUTUS OF ANIANUS (1488)

First page of the edition of c. 1495. The work of Anianus was the first one
printed on the *computus*. See page 665

Among the prominent *computi* printed in the 16th century is that of Arnaldo de Villa Nova (*c.* 1275).[1] Many of the early arithmetics also gave a brief treatment of the *computus*.[2]

The *computus* finally found a place in various liturgical works, and at present can be conveniently studied in pages prefixed to most editions of the Book of Common Prayer.

Early Timepieces. One of the general problems in connection with the calendar has to do with the finding of the hours of the day. For this purpose the shadow cast by some obstruction to the sun's rays was probably used by all primitive peoples. At first it is probable that a prominent tree, a rock, or a hill was selected, but in due time an artificial gnomon[3] was erected and lines were drawn on the earth to mark off the shadows. Since the hour shadow is longer when the sun is near the horizon, either concave surfaces or curve lines on a plane were placed at the foot of the gnomon.

Hours. The ancients usually had twelve hours in the day and the same number in the night. There have been various speculations as to why twelve was selected for this purpose, among them being one which referred the custom to the Babylonian knowledge of the inscribed hexagon.[4] It is probable, however, that twelve was used in measuring time for the same reason that it was used for measuring length and weight,—because the common fractional parts (halves, thirds, and fourths) were easily obtained. The day hours were longer than the night hours in the summer and shorter in the winter, a fact referred to by several ancient writers.[5]

---

[1] *Computus Ecclesiasticus & Astronomicus*, Venice, 1501. There was another work by this name printed at Venice in 1519.

[2] *E.g.*, the Treviso arithmetic (1478), fol. 57. Köbel's *Rechenbüchlin* (1531 ed.) devotes ten pages to the subject.

[3] Herodotus uses this term γνώμων (*gno'mon*), and it is common in Greek literature. Later writers sometimes called it a horologe (ὡρολόγιον, *horolog'ion*) when used for the sundial specifically. In still later times it was called the pole (πόλος, *pol'os*).

[4] G. H. Martini, *Abhandlung von den Sonnenuhren der Alten*, p. 18 (Leipzig, 1777).

[5] So Vitruvius: "Brumalis horae brevitates"; St. Augustine: "Hora brumalis aestivae horae comparata, minor est."

Although Herodotus (II, 109) speaks of the "12 parts" of the day among the Babylonians and the Greeks, the word "hour"[1] was not used either by him or by Plato or Aristotle. It was apparently a later idea to give these divisions a special name.

Early Dials. The sundial seems to have been used first in Egypt, but it is found also at an early date in Babylonia. Herodotus (II, 109) says that it was introduced from Babylon into Greece, and tradition says that this was done by Anaximander (c. 575 B.C.), the gnomon being placed at the center of three concentric circles. The early Egyptian dial has already been mentioned in Volume I, page 50. The first concave dial to be used in Greece is said to have been erected on the island of Cos by Berosus.[2] Several such dials have been found in the Roman remains,[3] and the early ones have no numerals on the hour lines, these lines being easily distinguished without such aids.[4]

Besides the plane dials and the concave spherical dials there were both concave and convex cylindric forms. Vitruvius (c. 20 B.C.) tells us that one Dionysodorus[5] invented the cylindric

---

[1] Ὥρα (ho'ra). There may be some relation between the word and the name of the Egyptian Horus, god of the rising sun, and the Hebrew *or* (light).

[2] A priest of Belus at Babylon. The name was probably Bar (Ber) Oseas or Barosus, that is, son of Oseas. Fl. c. 250 B.C. Vitruvius (IX, 4; X, 7, 9) says that Berosus went to Cos in his later years, founded a school of astrology, and invented what seems to have been a hemispherical sundial. For a general description of early dials see G. H. Martini, *loc. cit.*, pp. 24, 70. On Anaximander's connection with the gnomon, as recorded by Diogenes Laertius, Favorinus, and Herodotus, see W. A. Heidel, "Anaximander's Book, the earliest known geographical treatise," *Proceedings of the American Academy of Arts and Sciences*, LVI, 239.

[3] The first one was discovered (1741) in a Tusculan villa. For early descriptions see G. L. Zuzzeri, *D'una antica villa scoperta sul dosso del Tusculo, e d'un antico orologio a sole*, Venice, 1746. Boscovich also described it in the *Giornale de' Letterati per . . . 1746*, art. 14. The second one was found (1751) at Castelnuovo, near Rome; the third, also near Rome; and the fourth (1764), at Pompeii, although apparently made in Egypt.

[4] Thus Persius (*Satires*, III, 4) : ". . . quinta dum linea tangitur umbra," the shadow resting on the fifth line of the day, an hour before noon.

[5] Also spelled Dionysiodorus. He is said by Pliny to have found the radius of the earth to be c. 5000 miles, but nothing is known of his life. He lived c. 50 B.C.

form of dial, but we do not know whether it was convex or concave. We also know from Vitruvius that there were various other forms in use by the Romans.[1]

Difficulties with the Gnomon. One difficulty that was experienced with the large sundials of the ancients was that the gnomon did not cast a distinct shadow. The size of the sun is such as to have the shadow terminate in a penumbra which rendered the determination of the solstice, for example, a difficult matter. This is one reason why it is not probable that the Egyptian obelisks were used by scientific observers as gnomons. To overcome the difficulty the Greeks often used a column with a sphere on top, the center of the sphere corresponding to the center of the shadow and the center of the sun. Such gnomons are found on medals of the time of Philip of Macedon, and it is possible that this is the explanation of the column on the coin of Pythagoras shown on page 70 of Volume I. Dials of this type were introduced in Rome by Menelaus (*c.* 100), or at least were improved by him.[2]

As would naturally be expected, there were many special forms of dials. The "dial of Ahaz" (Isaiah, xxxviii), for example, was probably a flight of stairs, very likely curved, upon which a ray of sunlight fell. This dates from about the 8th century B.C.

It is impossible to do more at this time than to refer thus briefly to the use of dials among the ancients. The literature of the subject is very extensive.

Hourglasses and Clepsydræ. The need was early felt for some kind of device to tell the hours at night as well as during the day, and in cloudy weather as well as when the sun's direct rays gave their aid. Various methods were employed, such as burning tapers, hourglasses, and water clocks. The hourglass was known probably as early as 250 B.C. Plato (*c.* 380 B.C.) gave much thought to the matter, and his conclusions may have

---

[1] "Aliaque genera et qui supra scripti sunt, et alii plures inventa reliquerunt."
[2] On this type see Bigourdan, *L'Astronomie*, p. 91.

suggested to Ctesib'ius (*c.* 150 B.C.)[1] the idea of a water clock, the clepsydra,[2] which the second P. Cornelius Scipio Nasica ("Scipio with the pointed nose") is said to have introduced into Rome (*c.* 159 B.C.). In the early forms of these clocks the water trickled from one receptacle to another in a given time, much as the sand flows in an hourglass, but the later forms were more complicated. It was such an instrument that Harun al-Rashid sent to Charlemagne in 807. The clepsydra was known in Egypt as early as 1400 B.C., and numerous specimens are still to be seen, although not in use, in the Orient.

Influence on Later Timepieces. Since the priesthood, which composed the learned class, kept account of the official time in the early days of civilization, the Church continued to undertake this task until the modern period. The priest tolled the hour as determined by the dial or hourglass. The dial was put in a conspicuous place in the town, on the church tower, and hence in modern times the clock is often seen in the church tower and the hours are struck on the church bell. Because of this fact we have our name "clock," a word probably derived from the Celtic and meaning bell, whence the French *cloche*, a bell.

When the hour lines were marked on the dial, Roman numerals were used, always with IIII instead of IV for four, and hence we see the same markings upon modern clocks. The ancient gnomon was under the care of the priests, and brass plates are still to be seen in the floors of some of the churches in the Mediterranean countries, the sun shining through a certain window and telling the seasons as marked upon them.[3]

---

[1] Κτησίβιος, a native of Alexandria. None of his works are extant, but he is said to have invented not only a water clock but also a hydraulic organ and other machines.

[2] Κλεψύδρα (*klepsy'dra*). See Volume I, page 69 n.; Vitruvius, *De Architectura*, IX, cap. 9; Pliny, *Hist. Nat.*, VII.

[3] Those who wish to obtain further information upon the subject should consult such encyclopedias as the *Britannica* and such works as the following: A. Fraenkel, "Die Berechnung des Osterfestes," Crelle's *Journal*, CXXXVIII, 133; E. M. Plunket, *Ancient Calendars and Constellations*, London, 1903; C. P. Bowditch, "Memoranda on the Maya Calendars," *American Anthropologist*, III (N.S.), 129.

Clocks. It should not be thought that clocks of the general form known at the present time date only from Galileo's discovery of the isochronal property of the pendulum. As a matter of fact, wheel clocks go back to Roman times, and Boethius is said to have invented one (*c.* 510). Such clocks are known to have been used in churches as early as 612. The invention of those driven by weights is ascribed to Pacificus, archdeacon of Verona, in the 9th century, although a similar claim is made on behalf of various others. Clocks involving an assemblage of wheels are medieval in origin, and one was set up in St. Paul's, London, as early as 1286. Small portable clocks were in use in the 15th century, as witness a letter of 1469, written by Sir John Paston and containing the following admonition:

I praye you speke wt Harcourt off the Abbeye ffor a lytell clokke whyche I sent him by James Gressham to amend and yt ye woll get it off him an it be redy.

The oldest mechanical clock of which we have any complete description was made by a German named Heinrich De Vick and was set up in the tower of the palace of Charles V of France in 1379.[1] The principle employed was that of a weight suspended by a cord which was wound about a cylinder. This cylinder communicated power to a train of geared wheels which, in turn, transformed by means of a "scape wheel" the rotary motion to a backward-and-forward motion controlling the hands. The tendency of the weight to descend too rapidly was checked by a device for regulating the action of the wheels.

The pendulum clock was introduced about 1657 and seems chiefly due to Huygens. The principle of the pendulum, properly attributed to Galileo, had been observed as early as the 12th century by Ibn Yûnis (*c.* 1200), and had been employed

---

[1] On the general topic see W. I. Milham, *Time and Timekeepers*, New York, 1923, with full bibliography; E. von Bassermann-Jordan, *Die Geschichte der Zeitmessung und der Uhren* (Berlin, 1920–   ); R. T. Gunther, *Early Science in Oxford*, Vol. II (Oxford, 1923); H. T. Wade, "Clocks," *New International Cyclopædia*, V, 470 (2d ed., New York, 1914).

by astronomers to estimate intervals of time elapsing during an observation, but it had not been applied to a clock. It was made known in England through Ahasuerus Fromanteel, a Dutch clockmaker, about 1662.

John Harrison's[1] great construction of a ship's chronometer with a high degree of precision was made in the second half of the 18th century and finally secured for him the prize of £20,000 offered by the British government in 1714 for a method of ascertaining within specified limits the longitude of a ship at sea.[2] At the present time the noon of Greenwich mean time (G. M. T.) is communicated to ships by wireless, and so, in the case of the larger vessels, the finding of longitude no longer depends upon the chronometer alone.

Harrison's contribution to practical navigation was so important as to warrant a brief statement about the nature of his work. Although he was by trade a carpenter, his mechanical tastes led him to experiment with clocks. Having observed the need for a pendulum of constant length, he devised (1726) the "gridiron pendulum," in which the downward expansion of the steel rods compensated for the upward expansion of the brass ones. After the British government (1714)[3] had offered the prize already mentioned, Harrison gave his attention to the perfection of a watch that should serve to assure Greenwich time to a ship at sea. By 1761 he had constructed one that, after a voyage of several months, had lost only 1 min. 54½ sec. and assured the longitude within 18 miles. The government paid him £10,000 in 1765, and a like sum in 1767,—a modest reward for an invention of such great value to the world, even though the degree of accuracy would now be considered very unsatisfactory.

---

[1] Born at Foulby, parish of Wragby, Yorkshire, early in 1693; died in London, March 24, 1776.

[2] For a list of such prizes see Bigourdan, *L'Astronomie*, p. 166.

[3] The original act reads, "At the Parliament to be Held at *Westminster*, the Twelfth Day of *November*, *Anno Dom.* 1713," but was printed (and doubtless enacted) in 1714. It is entitled "An Act for Providing a Publick Reward for such Person or Persons as shall Discover the Longitude at Sea," there being nothing "so much wanted and desired at Sea, as the Discovery of the Longitude." On the entire topic see R. T. Gould, *The Marine Chronometer, its History and Development*, London, 1923.

TOPICS FOR DISCUSSION

1. Additional information concerning weight, length, area, and volume as found in the various encyclopedias.

2. The cubit, shekel, talent, and various other measures referred to in Biblical literature.

3. The development of measures, including weights, in accord with human needs.

4. An etymological study of such words as *metric, groschen, doubloon, measure, watch, day, month,* and *year.*

5. Influence of the Roman system of measures, including weights and values, upon other European systems.

6. The universality of certain primitive units of measure such as the cubit and inch.

7. Primitive customs as related to units of measure, such as "a day's journey," "a watch in the night," and a "Morgen."

8. Primitive measures still found in various parts of the country, having been transmitted from generation to generation like the folklore of the people.

9. Supplementary information on the calendar, as found in the various encyclopedias.

10. Meaning and use of the "golden number" and "dominical letter" as set forth in various encyclopedias or in works on the Church calendar.

11. Methods of finding the date of Easter as given in the books referred to in the preceding topic.

12. The reform of the calendar under Gregory XIII, including a study of earlier attempts at reform.

13. Significance of such technical terms as *computus Paschalis, calendar, a red-letter day,* and *almanac.*

14. History of the discovery of the approximate length of the year, and the method of ascertaining it.

15. Relation among calendars used by various peoples of ancient and medieval times.

16. Influence of Rome upon the calendars of European countries, and the effort of France, during the Revolution, to break away from tradition with respect to the divisions of the year and the names of the days and months.

17. The mathematics of the sundial.

CHAPTER X

THE CALCULUS

1. Greek Ideas of a Calculus

General Steps Described. There have been four general steps in the development of what we commonly call the calculus, and these will be mentioned briefly in this chapter. The first is found among the Greeks.[1] In passing from commensurable to incommensurable magnitudes their mathematicians had recourse to the method of exhaustion, whereby, for example, they "exhausted" the area between a circle and an inscribed regular polygon, as in the work of Antiphon (*c.* 430 B.C.).

The second general step in the development, taken two thousand years later, may be briefly called the method of infinitesimals. This method began to attract attention in the first half of the 17th century, particularly in the works of Kepler (1616) and Cavalieri (1635), and was used to some extent by Newton and Leibniz.

The third method is that of fluxions and is the one due chiefly to Newton (*c.* 1665). It is this form of the calculus that is usually understood when the invention of the science is referred to him.

The fourth method, that of limits, is also due to Newton, and is the one now generally followed.

Contributions of the Greeks. As stated above, the Greeks developed the method of exhaustion about the 5th century B.C. The chief names connected with this method have already been

---

[1] Sir T. L. Heath, "Greek Geometry with special reference to infinitesimals," *Mathematical Gazette*, March, 1923; D. E. Smith, *Mathematics*, in the series "Our Debt to Greece and Rome," Boston, 1923; G. H. Graves, "Development of the Fundamental Ideas of the Differential Calculus," *The Mathematics Teacher*, III, 82.

mentioned, but a few details of their work and that of their contemporaries will now be given.

Zeno of Elea (*c.* 450 B.C.) was one of the first to introduce problems that led to a consideration of infinitesimal magnitudes. He argued that motion was impossible, for this reason:

Before a moving body can arrive at its destination it must have arrived at the middle of its path; before getting there it must have accomplished the half of that distance, and so on *ad infinitum*: in short, every body, in order to move from one place to another, must pass through an infinite number of spaces, which is impossible.[1]

Leucippus (*c.* 440 B.C.) may possibly have been a pupil of Zeno's. Very little is known of his life and we are not at all certain of the time in which he lived, but Diogenes Laertius (2d century) speaks of him as the teacher of Democritus (*c.* 400 B.C.). He and Democritus are generally considered as the founders of the atomistic school, which taught that magnitudes are composed of indivisible elements[2] in finite numbers. It was this philosophy that led Aristotle (*c.* 340 B.C.) to write a book on indivisible lines.[3]

Democritus is said to have written on incommensurable lines and solids, but his works are lost, except for fragments, and we are ignorant of his method of using the atomic theory.

Method of Exhaustion. Antiphon (*c.* 430) is one of the earliest writers whose use of the method of exhaustion is fairly well known to us. In a fragment of Eudemus (*c.* 335 B.C.), conjecturally restored by Dr. Allman,[4] we have the following description:

Antiphon, having drawn a circle, inscribed in it one of those polygons[5] that can be inscribed: let it be a square. Then he bisected each side of this square, and through the points of section drew straight lines at right angles to them, producing them to meet the cir-

---

[1] Allman, *Greek Geom.*, p. 55.
[2] Ἄτομοι (*a'tomoi*). Allman, *Greek Geom.*, p. 56.
[3] Περὶ ἀτόμων γραμμῶν (first edition, Paris, 1557). The work is also attributed to Theophrastus.   [4] Allman, *Greek Geom.*, p. 65.
[5] That is, according to the usage of the time, regular polygons.

II

cumference; these lines evidently bisect the corresponding segments of the circle. He then joined the new points of section to the ends of the sides of the square, so that four triangles were formed, and the whole inscribed figure became an octagon. And again, in the same way, he bisected each of the sides of the octagon, and drew from the points of bisection perpendiculars; he then joined the points where these perpendiculars met the circumference with the extremities of the octagon, and thus formed an inscribed figure of sixteen sides. Again, in the same manner, bisecting the sides of the inscribed polygon of sixteen sides, and drawing straight lines, he formed a polygon of twice as many sides; and doing the same again and again, until he had exhausted the surface, he concluded that in this manner a polygon would be inscribed in the circle, the sides of which, on account of their minuteness, would coincide with the circumference of the circle.

We have in this method a crude approach to the integration of the 17th century.

Bryson (c. 450 B.C.), who seems to have lived just before Antiphon's period of greatest activity, was at one time thought to have used a method that had the merit of circumscribing as well as inscribing regular polygons and exhausting the area between them. This was probably not the case (Vol. I, p. 84), although the method was used by some of his successors. There is also no reliable evidence to prove the assertion that Bryson assumed that the area of the circle is the arithmetic mean between the areas of two similar polygons, one circumscribed and the other inscribed.

The Contribution of Eudoxus. Eudoxus of Cnidus (c. 370 B.C.) is probably the one who placed the theory of exhaustion on a scientific basis. It is uncertain just how much reliance is to be placed upon the tradition which asserts that Book V of Euclid's *Elements* (the book on proportion) is due to him, but it is thought that the fundamental principles there laid down are his. The fourth definition in Book V is: "Magnitudes are said to have a ratio to one another which are capable, when multiplied, of exceeding one another," and this excludes the relation of a finite magnitude to a magnitude of the same kind which is either infinitely great or infinitely small.[1] It is in this definition and the related axiom that Dr. Allman finds a basis for the scientific method of exhaustion and discerns the probable influence of Eudoxus. According to Archimedes, this method had already been applied by Democritus (*c.* 400 B.C.) to the mensuration of both the cone and the cylinder.

It is known that Hippocrates of Chios (*c.* 460 B.C.) proved that circles are to one another as the squares on their diameters, and it seems probable that he also used the method of exhaustion,—a subject which was evidently much discussed about that time. Archimedes tells us that the "earlier geometers" had proved that spheres have to one another the triplicate ratio of their diameters, so that the method was probably used by others as well.

Archimedes and Integration. It is to Archimedes himself (*c.* 225 B.C.) that we owe the nearest approach to actual integration to be found among the Greeks.[2] His first noteworthy advance in this direction was concerned with his proof that the area of a parabolic segment is four thirds of the triangle with the same base and vertex, or two thirds of the circumscribed parallelogram. This was shown by continually inscribing in each segment between the parabola and the inscribed figure a triangle with the same base and the same height as the segment. If A is the area of the original inscribed triangle, the process adopted by him leads to the summation of the series

$$A + \tfrac{1}{4} A + (\tfrac{1}{4})^2 A + \cdots,$$

or to finding the value of

$$A[1 + \tfrac{1}{4} + (\tfrac{1}{4})^2 + (\tfrac{1}{4})^3 + \cdots],$$

so that he really finds the area by integration and recognizes, but does not assert, that

$$(\tfrac{1}{4})^n \to 0 \text{ as } n \to \infty,$$

this being the earliest example that has come down to us of the summation of an infinite series.

---

[1] Heath, *Euclid*, Vol. II, p. 120; see also his *Archimedes*, p. xlvii.
[2] Heath, *Archimedes*, p. cxlii.

Area of the Parabola. In his proof relating to the quadrature of the parabola Archimedes first proves two propositions numbered 14 and 15 in his treatise on this curve. These assert that, with respect to the figure here shown,[1]

$$\triangle EqQ < 3(FO_1 + F_1O_2 + \cdots + F_{n-1}O_n + \triangle E_nO_nQ,$$
and $\quad \triangle EqQ > 3(R_1O_2 + R_2O_3 + \cdots + R_{n-1}O_n + \triangle R_nO_nQ.$

He then states (Prop. 16) that the area of the segment of the parabola is equal to $\frac{1}{3}\triangle EqQ$. The proof is by a *reductio ad absurdum* and is given by Heath substantially as follows:

I. Suppose that the area of the segment is greater than $\frac{1}{3}\triangle EqQ$. Then the excess can, if continually added to itself, be made to exceed

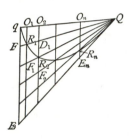

$\triangle EqQ$. And it is possible to find a submultiple of the triangle EqQ less than the said excess of the segment over $\frac{1}{3}\triangle EqQ$.

Let the triangle FqQ be the submultiple of the triangle EqQ. Divide Eq into equal parts each equal to qF, and let all the points of division including F be joined to Q meeting the parabola in $R_1, R_2, \cdots, R_n$ respectively. Through $R_1, R_2, \cdots, R_n$ draw diameters of the parabola meeting qQ in $O_1, O_2, \cdots, O_n$ respectively.

Let O_1R_1 meet QR_2 in F_1.
Let O_2R_2 meet QR_1 in D_1 and QR_3 in F_2.
Let O_3R_3 meet QR_2 in D_2 and QR_4 in F_3, and so on.
We have, by hypothesis,

$$\triangle FqQ < \text{area of segment} - \tfrac{1}{3}\triangle EqQ,$$
or $\qquad$ area of segment $-\triangle FqQ > \tfrac{1}{3}\triangle EqQ.$ $\qquad$ (1)

Now, since all the parts of qE, such as qF and the rest, are equal, we have $O_1R_1 = R_1F_1$, $O_2D_1 = D_1R_2 = R_2F_2$, and so on; therefore

$$\triangle FqQ = FO_1 + R_1O_2 + D_1O_3 + \cdots$$
$$= FO_1 + F_1D_1 + F_2D_2 + \cdots + F_{n-1}D_{n-1} + \triangle E_nR_nQ. \qquad (2)$$

But area of segment $< FO_1 + F_1O_2 + \cdots + F_{n-1}O_n + \triangle E_nO_nQ.$ (3)

---

[1] For the proof, see Heath, *Archimedes*, p. 241, preferably F. Kliem's German translation, pp. 361–365 (Berlin, 1914). The proof of the next proposition is taken from the same work, p. 244.

Subtracting the equation (2) from the inequality (3), we have

area of segment $-\triangle FqQ < R_1O_2 + R_2O_3 + \cdots + R_{n-1}O_n + \triangle R_nO_nQ$;

whence, *a fortiori*, by (1),

$$\tfrac{1}{3}\triangle EqQ < R_1O_2 + R_2O_3 + \cdots + R_{n-1}O_n + \triangle R_nO_nQ.$$

But this is impossible, since [Props. 14, 15]

$$\tfrac{1}{3}\triangle EqQ > R_1O_2 + R_2O_3 + \cdots + R_{n-1}O_n + \triangle R_nO_nQ.$$

Therefore $\qquad$ area of segment $\not> \tfrac{1}{3}\triangle EqQ$.

II. If possible, suppose the area of the segment less than $\tfrac{1}{3}\triangle EqQ$.

Take a submultiple of the triangle EqQ (as the triangle FqQ), less than the excess of $\tfrac{1}{3}\triangle EqQ$ over the area of the segment, and make the same construction as before.

Since $\triangle FqQ < \tfrac{1}{3}\triangle EqQ$ — area of segment, it follows that

$$\triangle FqQ + \text{area of segment} < \tfrac{1}{3}\triangle EqQ$$
$$< FO_1 + F_1O_2 + \cdots + F_{n-1}O_n + \triangle E_nO_nQ.$$
$$[\text{Props. } 14, 15]$$

Subtracting from each side the area of the segment, we have

$$\triangle FqQ < \text{sum of spaces } qFR_1, R_1F_1R_2, \cdots, E_nR_nQ,$$
$$< FO_1 + F_1D_1 + \cdots + F_{n-1}D_{n-1} + \triangle E_nR_nQ, \text{ a fortiori};$$

which is impossible, because, by (2) above,

$$\triangle FqQ = FO_1 + F_1D_1 + \cdots + F_{n-1}D_{n-1} + \triangle E_nR_nQ.$$

Hence area of segment $\not< \tfrac{1}{3}\triangle EqQ$.

Since, then, the area of the segment is neither greater than nor less than $\tfrac{1}{3}\triangle EqQ$, it is equal to it.

The Method of Archimedes. As to the working of the mind of Archimedes in arriving at the conclusion in regard to the area of the parabola (a conclusion which led to the above proof) we have some interesting evidence. In a manuscript discovered in Constantinople in 1906 by Professor Heiberg, the editor of the works of Archimedes, the latter's method of approach to certain propositions is set forth. In particular the first proposition relates to the steps taken in arriving at the conclusion with respect to the quadrature of the parabola.[1]

---

[1] The Heiberg edition was translated by Lydia G. Robinson, Chicago, 1909, and by Sir Thomas L. Heath, Cambridge, 1912.

The following is the translation as given in Heath's edition:

Let ABC be a segment of a parabola bounded by the straight line AC and the parabola ABC, and let D be the middle point of AC. Draw the straight line DBE parallel to the axis of the parabola and join AB, BC.

Then shall the segment ABC be $\frac{4}{3}$ of the triangle ABC.

From A draw AKF parallel to DE, and let the tangent to the parabola at C meet DBE in E and AKF in F. Produce CB to meet AF in K, and again produce CK to H, making KH equal to CK.

Consider CH as the bar of a balance, K being its middle point.

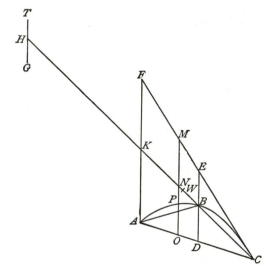

Let MO be any straight line parallel to ED, and let it meet CF, CK, AC in M, N, O, and the curve in P.

Now, since CE is a tangent to the parabola and CD the semiordinate,

$$EB = BD;$$

for this is proved in the *Elements* [*of Conics*].[1]

Since FA, MO are parallel to ED, it follows that

$$FK = KA,$$

and
$$MN = NO.$$

[1] *I.e.*, the works on conics by Aristæus and Euclid. See the similar expression in *On Conoids and Spheroids*, Prop. 3, and *Quadrature of Parabola*, Prop. 3.

Now, by the property of the parabola, which is proved in a lemma,

$$MO : OP = CA : AO \qquad [\textit{Quadrature of Parabola}, \text{ Prop. } 5]$$
$$= CK : KN = HK : KN. \qquad [\text{Eucl. VI. } 2]$$

Take a straight line TG equal to OP, and place it with its center of gravity at H, so that $TH = HG$; then, since N is the center of gravity of the straight line MO, and $MO : TG = HK : KN$, it follows that TG at H and MO at N will be in equilibrium about K.

[*On the Equilibrium of Planes*, I, 6, 7]

Similarly, for all other straight lines parallel to DE and meeting the arc of the parabola, (1) the portion intercepted between FC, AC with its middle point on KC and (2) a length equal to the intercept between the curve and AC placed with the center of gravity at H will be in equilibrium about K.

Therefore K is the center of gravity of the whole system consisting (1) of all the straight lines as MO intercepted between FC, AC and placed as they actually are in the figure and (2) of all the straight lines placed at H equal to the straight lines as PO intercepted between the curve and AC.

And, since the triangle CFA is made up of all the parallel lines like MO, and the segment CBA is made up of all the straight lines like PO within the curve, it follows that the triangle, placed where it is in the figure, is in equilibrium about K with the segment CBA placed with its center of gravity at H.

Divide KC at W so that $CK = 3\,KW$; then W is the center of gravity of the triangle ACF: for this is proved in the books on equilibrium.

[*On the Equilibrium of Planes*, I, 15]

Therefore $\triangle ACF$: segment $ABC = HK : KW$

$$= 3 : 1.$$

Therefore segment $ABC = \tfrac{1}{3} \triangle ACF$.

But $\triangle ACF = 4 \triangle ABC$.

Therefore segment $ABC = \tfrac{4}{3} \triangle ABC$.

Now the fact here stated is not actually demonstrated by the argument used, but that argument has given a sort of indication that the conclusion is true. Seeing, then, that the theorem is not demonstrated, but at the same time suspecting that the conclusion is true, we shall have recourse to the geometrical demonstration which I myself discovered and have already published.

Archimedes anticipates Modern Formulas. In his treatment of solids bounded by curved surfaces he arrives at conclusions which we should now describe by the following formulas: [1]

Surface of a sphere, $4\,\pi a^2 \cdot \frac{1}{2} \int_0^\pi \sin\theta\,d\theta = 4\,\pi a^2.$

Surface of a spherical segment,

$$\pi a^2 \int_0^a 2\sin\theta\,d\theta = 2\,\pi a^2 (1 - \cos\alpha).$$

Volume of a segment of a hyperboloid of revolution,

$$\int_0^b (ax + x^2)\,dx = b^2(\tfrac{1}{2}\,a + \tfrac{1}{3}\,b).$$

Volume of a segment of a spheroid,

$$\int_0^b x^2 dx = \tfrac{1}{3}\,b^3.$$

Area of a spiral, $\dfrac{\pi}{a} \int_0^a x^2 dx = \tfrac{1}{3}\,\pi a^2.$

Area of a parabolic segment,

$$\frac{1}{A^2} \int_0^A \Delta^2 d\Delta = \tfrac{1}{3}\,A.$$

2. MEDIEVAL IDEAS OF THE CALCULUS

Relation to Mensuration. The only traces that we have of an approach to the calculus in the Middle Ages are those relating to mensuration and to graphs. The idea of breaking up a plane surface into infinitesimal rectangles was probably present in the minds of many mathematicians at that time in the West as well as in the East, but it was never elaborated into a theory that seemed worth considering. For example, a Jewish writer, Jehudah Barzilai, living in Barcelona in the 13th century,[2] as-

---

[1] Heath, *Archimedes*, p. cxlvi seq.; G. Loria, *Le scienze esatte nell' antica Grecia*, 2d ed., p. 108 (Milan, 1914); Heath, *Method of Archimedes*, p. 8 (1912).
[2] *Sefer Jezira*, commentary by Judah ben Barzilai, p. 255 (Berlin, 1884).

serts that "it has been said that there is no form in the world except the rectangle, for every triangle or rectangle is composed of rectangles too small to be perceived by the senses."

The next important step in the preparation for the calculus taken in the Middle Ages is the one already described in connection with the geometric work of Oresme (c. 1360). His method of latitudes and longitudes gave rise to what we should now call a distribution curve or graph,—a step that is fundamental to the modern method of finding the area included between a curve and certain straight lines.

3. Modern Forerunners of the Calculus

Early Writers. As is usual in such cases, it is impossible to determine with certainty to whom credit belongs, in modern times, for first making any noteworthy move in the calculus, but it is safe to say that Stevin is entitled to serious consideration. His contribution is seen particularly in his treatment of the subject of the center of gravity of various geometric figures, anticipating as it did the work of several later writers.[1] Other writers, even in medieval times, had solved various problems in mensuration by methods which showed the influence of the Greek theory of exhaustion and which anticipated in some slight degree the process of integration. Among them may be mentioned the name of Ṭâbit ibn Qorra (c. 870), who found the volume of a paraboloid. Soon after Stevin wrote, Luca Valerio[2] published his *De quadratura parabolae* (Rome, 1606), using a method of attack that was essentially Greek in its spirit.

Kepler. Among the more noteworthy attempts at integration in modern times were those of Kepler (1609). In his notable work on planetary motion he asserted that a planet describes equal focal sectors of ellipses in equal times. This naturally demands some method for finding the areas of such sectors,

---

[1] *De Beghinselen der Weeghconst*, Leyden, 1686. For a summary see H. Bosmans, "Le calcul infinitésimal chez Simon Stevin," *Mathesis*, XXXVII (1923).

[2] Born *c.* 1552; died in 1618. He was professor of mathematics and physics at Rome.

and the one invented by Kepler was called by him the method of the "sum of the radii," a rude kind of integration. He also became interested in the problem of gaging, and published a work on this subject and on general mensuration as set forth by Archimedes.[1] Far from being an elementary treatment of gaging, this was a scientific study of the measurement of solids in general. Kepler considers solids as composed "as it were" (*veluti*) of infinitely many infinitely small cones or infinitely thin disks, the summation of which becomes the problem of the later integration.

Cavalieri. It was Kepler's attempts at integration that led Cavalieri to develop his method of indivisibles,[2] a subject which may also have been suggested to him by Aristotle's tract *De lineis insecabilibus*, to take the common Latin title.[3] It may also have been suggested by one of the fragments of Xenocrates (*c.* 350 B.C.), an Athenian, who wrote upon indivisible lines.[4]

Cavalieri's Lack of Clearness. Cavalieri was not always clear in his statements respecting the nature of an indivisible magnitude. In general, however, he seems to have looked upon a solid as made up practically of superposed surfaces, a surface as made up of lines, and a line as made up of points, these component parts being the ultimate possible elements in the decom-

---

[1] *Ausszug auss der vralten Messe-Kunst Archimedes, . . . Erklärung vnd Bestättigung der Oesterreichischen Weinvisier-Ruthen*, Linz, 1616; ed. Frisch, V, 497, 614 (Frankfort a. M., 1864). Kepler's letters (*ibid.*, p. 626) show that he was working on the subject as early as 1605. On this entire period see C. R. Wallner, *Bibl. Math.*, V (3), 113.

[2] *Geometria indivisibilibus continuorum nova quadam ratione promota*, Bologna, 1635; 2d ed., *ibid.*, 1653; *Exercitationes geometricae sex*, Bologna, 1647.

[3] H. Vogt referred to this tract in the *Bibl. Math.*, X (3), 146, and F. Cajori called attention to it more prominently in *Science* (U.S.), XLVIII (N.S.), 577. See also Heath, *History*, I, 346.

[4] " . . . in infinitum vero dividi non posse, sed in atomos quasdam desinere: has porro atomos non esse partium expertes et minimas, sed pro quantitate et materia dividi posse et partes habere : caeteroqui specie atomos et prima naturae statuens esse primas quasdam lineas insecabiles et ex his facta plana et solida prima."
This Latin translation by Simplicius is given, with the original Greek, in the "Xenocratis Fragmenta," F. W. A. Mullach, *Fragmenta Philosophorum Graecorum*, III, 118, § 21 (Paris, 1881).

position of the magnitude. He then proceeded to find lengths, areas, and volumes by the summation of these "indivisibles," that is, by the summation of an infinite number of infinitesimals. Such a conception of magnitude cannot be satisfactory to any scientific mind, but it formed a kind of intuitive step in the development of the method of integration and undoubtedly stimulated men like Leibniz to exert their powers to place the theory upon a scientific foundation.[1]

Illustration of Cavalieri's Method. Some idea of Cavalieri's method may be obtained by considering his comparison of a triangle with a parallelogram having the same base and the same altitude. Calling the smallest element of the triangle 1, the next will be 2, the next 3, and so on to n, the base. The area is therefore $1 + 2 + 3 + \cdots + n$, or $\frac{1}{2}n(n+1)$. But each element of the parallelogram is n, and there are n of them, as in the triangle, and so the area is n^2. Then the ratio of the area of the triangle to the area of the parallelogram is

$$\frac{1}{2}n(n+1) : n^2 \quad \text{or} \quad \frac{1}{2}\left(1 + \frac{1}{n}\right).$$

But $\qquad \dfrac{1}{2}\left(1 + \dfrac{1}{n}\right) \to \dfrac{1}{2} \quad$ as $\quad n \to \infty,$

and so the triangle is half the parallelogram.

By means of his method Cavalieri was able to solve various elementary problems in the mensuration of lengths, areas, and volumes, and also to give a fairly satisfactory proof of the theorem of Pappus with respect to the volume generated by the revolution of a plane figure about an axis.[2]

---

[1] For a discussion of Cavalieri's work and its relation to the calculus, see H. Bosmans, "Sur une contradiction reprochée à la théorie des 'indivisibles' chez Cavalieri," *Annales de la Société scientifique de Bruxelles*, XLII (1922), 82. For his work on the center of gravity see E. Bortolotti, "Le prime applicazioni del calcolo integrale alla determinazione del centro di gravità di figure geometriche," *Rendiconto . . . della R. Accad. delle Scienze*, Bologna, 1922, reprint.

[2] For a translation of Cavalieri's Theorem relating to the volumes of solids, see G. W. Evans, in *Amer. Math. Month.*, XXIV, 447. On the method in general see also H. Bosmans, "Un chapitre de l'œuvre de Cavalieri," *Mathesis*, XXXVI, 365.

Roberval. At the same time that Cavalieri was working on the problem of indivisibles Roberval[1] was proceeding upon a similar hypothesis. He considered the area between a curve and a straight line as made up of an infinite number of infinitely narrow rectangular strips, the sum of which gave him the required area. In the same way he attacked the problems of rectification and of cubature. He also found the approximate value of $\int_0^1 x^m dx$, m being a positive integer, by finding the value of

$$\frac{0^m + 1^m + 2^m + \cdots + (n-1)^m}{n^{m+1}},$$

asserting that this approaches $1/(m+1)$ as $n \to \infty$.

Fermat. Fermat (1636) reached the same conclusion, basing his treatment upon a method set forth by Archimedes, and also extended the proof to include substantially the cases in which m is fractional (1644) or negative (1659), although not using either the fractional or the negative exponent in his work. He also attacked (1636) the problem of maxima and minima, that is, of finding the points on a curve at which the tangent is parallel to the x-axis. It was probably because of this step that Lagrange expressed himself as follows:[2]

One may regard Fermat as the first inventor of the new calculus. In his method *De maximis et minimis* he equates the quantity of which one seeks the maximum or the minimum to the expression of the same quantity in which the unknown is increased by the indeterminate quantity. In this equation . . . he divides . . . by the indeterminate quantity which occurs in them as a factor; then he takes this quantity as zero and he has an equation which serves to determine the unknown sought. . . . His method of tangents depends upon the same principle.[3]

[1] *Traité des indivisibles*, mémoire, Paris, 1634. See A. E. H. Love, "Infinitesimal Calculus," *Encyc. Brit.*, 11th ed.

[2] *Œuvres de Lagrange*, ed. Serret, X, 294. See Cajori in *Amer. Math. Month.*, XXVI, 16.

[3] For confirmation of this opinion by Laplace and Tannery, see Cajori, *loc. cit.*, p. 17. On the work *De maximis et minimis* consult the *Supplément* to Volumes I–IV of the *Œuvres de Fermat*, edited by C. de Waard, Paris, 1922, and the review by H. Bosmans, *Revue des Questions Scientifiques*, Brussels, April, 1923.

With his name should be joined that of a later writer, Antonio di Monforte[1] (1644–1717), a Neapolitan mathematician who worked along similar lines.

Problem of Tangents. The problem of tangents, the basic principle of the theory of maxima and minima, may be said to go back to Pappus (c. 300).[2] It appears indirectly in the Middle Ages, for Oresme (c. 1360) knew that the point of maximum or minimum ordinate of a curve is the point at which the ordinate is changing most slowly. It was Fermat, however, who first stated substantially the law as we recognize it today, communicating (1638)[3] to Descartes a method which is essentially the same as the one used at present, that of equating $f'(y)$ to zero. Similar methods were suggested by René de Sluze[4] (1652) for tangents, and by Hudde[5] (1658) for maxima and minima.

Other Writers. From then until Newton finally brought the work to a climax various efforts were made in the same direction by such writers as Huygens, Torricelli, Pascal,[6] and Mersenne. The fact that the area of the hyperbola $xy = 1$, found by Grégoire de Saint-Vincent[7] (1647), is related to logarithms was recognized by Fermat, and Nicolaus Mercator[8] made use of the principle in his calculation of these functions.

Wallis. The first British publication of great significance bearing upon the calculus is that of John Wallis, issued in 1655.

[1] F. Amodeo, "La Regola di Fermat-Monforte per la ricerca dei massimi e minimi," *Periodico di Matematica*, XXIV, fasc. VI.

[2] There is a good summary of the history of tangents as related to the calculus in a work by Anibal Scipião Gomes de Carvalho, *A Teoria das Tangentes antes da Invenção do Cálculo Diferencial*, Coimbra, 1919.

[3] *Opera varia*, Toulouse, 1679.

[4] See also "A short and easy method of drawing tangents to all geometrical curves," *Phil. Trans.*, 1672.

[5] *De reductione aequationum et de maximis et minimis*, in a letter published in 1713.

[6] H. Bosmans, *Archivio di Storia della Scienza*, IV, 369.

[7] *Opus geometricum quadraturae circuli et sectionum coni*, 2 vols., Antwerp 1647.

[8] *Logarithmotechnia*, London, 1668 and 1674.

It is entitled *Arithmetica Infinitorum, sive Nova Methodus In-quirendi in Curvilineorum Quadraturam, aliaque difficiliora Matheseos Problemata,* and is dedicated to Oughtred. By a method similar to that of Cavalieri the author effects the quadrature of certain surfaces, the cubature of certain solids, and the rectification of certain curves. He speaks of a triangle, for example, "as if" (*quasi*) made up of an infinite number of parallel lines in arithmetic proportion, of a paraboloid "as if" made up of an infinite number of parallel planes, and of a spiral as an aggregate of an infinite number of arcs of similar sectors, applying to each the theory of the summation of an infinite series. In all this he expresses his indebtedness to such writers as Torricelli and Cavalieri. He speaks of the work of such British contemporaries as Seth Ward and Christopher Wren, who were interested in this relatively new method, and, indeed, his dedication to Oughtred is the best contemporary specimen that we have of the history of the movement just before Newton's period of activity.[1] All this, however, was still in the field of integration, the first steps dating, as we have seen, from the time of the Greeks.

Barrow. What is considered by us as the process of differentiating was known to quite an extent to Barrow (1663). In

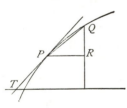

his *Lectiones opticae et geometricae*[2] he gave a method of tangents in which, in the annexed figure, Q approaches P as in our present theory, the result being an indefinitely small (*indefinite parvum*) arc. The triangle PRQ was long known as "Barrow's differential triangle,"[3] a name which, however, was not due to him. It is evident that this method, and the figure as well, must have had a notable influence upon the mathematics of his time.

[1] The work also appears in his *Opera Mathematica*, I, 255–470 (Oxford, 1695).

[2] London, 1669. The work seems to have been written in 1663 and 1664. Love, *loc. cit.*; J. M. Child, *Geometrical Lectures of Isaac Barrow*, Chicago, 1916.

[3] On this close approach to the later calculus see Whewell's edition of Barrow's *Mathematical Works*, p. xii (Cambridge, 1860), and Child, *loc. cit.*

It is quite probable that Barrow had advised Newton of his work on this figure as early as 1664.[1] Pascal had already published a figure of somewhat the same shape,[2] so that the study of triangles of the general nature illustrated above was being undertaken and discussed at this time in both England and France. The triangles given by both Barrow and Pascal were apparently known to Leibniz, and they assisted him in developing his own theory.[3]

Barrow also recognized the fact that integration is the inverse of differentiation, but he did not use this relation to aid him in solving the quadrature problem.

Period of the Invention of the Newtonian Method. We now approach the period which is popularly thought to be the one in which the calculus was invented. It is evident, however, that a crude integral calculus was already in use and that some approach had been made to the process of differentiation. It is also evident that the lines of approach to the calculus in general have been two in number, one representing the static phase as seen in the measurement of fixed lengths, areas, or volumes, and in the making use of such ideas as those of infinitesimals and indivisibles; the other representing the dynamic phase as seen in the motion of a point. To the former belong such names as Kepler and Cavalieri and, in general, Archimedes; to the latter belong the great leaders in the mathematics of the time of Newton and Leibniz.[4]

[1] J. M. Child, *The Early Mathematical Manuscripts of Leibniz*, p. 11 (Chicago, 1920), a work which students of the history of the calculus should consult, not merely for its translations but for its notes. On the figures used by Barrow, Pascal, and Leibniz, see *ibid.*, p. 15. This work is hereafter referred to as Child, *Leibniz Manuscripts*.

[2] In his *Lettres de A. Dettonville* (Paris, 1659), the part relating to the triangle having been written in 1658.

[3] Child, *Leibniz Manuscripts*, p. 16. See Leibniz's admission as to Barrow in his letter to Jacques Bernoulli (1703), *ibid.*, p. 20.

[4] On the general history of the development of the calculus in the 17th century the following works may be consulted with profit:

W. W. R. Ball, *History of Mathematics*, 10th ed., London, 1922, the treatment of the calculus being particularly complete; A. De Morgan, *On a point connected with the dispute between Keill and Leibniz about the Invention of Fluxions*, London, 1846; also the *Companion to the British Almanack*, 1852, and *Philo-*

4. NEWTON AND LEIBNIZ

Newton. Newton's great contribution to the theory consists in part in his extension of the method to include the other functions then in common use, in his recognition of the fact that the inverse problem of differentiation could be used in solving the problem of quadrature, in his introduction of a suitable notation, and in his wide range of applications of the subject. Starting with the knowledge already acquired by Barrow, he developed, beginning in 1665, his method of "fluxions." This he afterward set forth in three tracts,[1] which, in accordance with his unfortunate plan of avoiding publicity in his discoveries, were not printed until many years later.

Newton's Three Types. Newton recognized three types of the calculus. In his *Principia* (1687) he made some use of

sophical Magazine, June and November, 1852; G. J. Gerhardt, *Die Entdeckung der Differenzialrechnung durch Leibniz*, Halle, 1848; J. Raphson, *The History of Fluxions*, London, 1715; Latin edition the same year; H. Sloman, *Leibnizens Ansprüche auf die Erfindung der Differenzialrechnung*, Leipzig, 1857; English translation with additions, Cambridge, 1860; M. Cantor, *Geschichte der Mathematik*, III, chap. 97; W. T. Sedgwick and H. W. Tyler, *A Short History of Science*, chap. xiii (New York, 1917); H. Weissenborn, *Die Principien der höheren Analysis, als historisch-kritischer Beitrag zur Geschichte der Mathematik*, Halle, 1856; D. Brewster, *Memoirs of the Life, Writings, and Discoveries of Sir Isaac Newton*, 2 vols., Edinburgh, 1885; 2d ed., 1860; J. Collins, *Commercium Epistolicum de Varia Re Mathematica*, London, 1712; 2d ed., London, 1722 (two editions); 3d ed., London, 1725; French ed., Paris, 1856; G. Vivanti, *Il concetto d' infinitesimo e la sua applicazione alla matematica*, Mantua, 1894; 2d ed., Naples, 1901; J. M. Child, *The Early Mathematical Manuscripts of Leibniz*, Chicago, 1920.

On the history of fluxions in Great Britain, beginning with Newton, the best work is F. Cajori, *A History of the Conceptions of Limits and Fluxions in Great Britain from Newton to Woodhouse*, Chicago, 1919. On the general history of the later development of the calculus, see J. A. Serret and G. Scheffers, *Lehrbuch der Differential- und Integralrechnung*, II, 581–626 (5th ed., Leipzig, 1911), and III, 694–720 (Leipzig, 1914); Sedgwick and Tyler, *loc. cit.*, chap. xv; E. W. Brown, "Mathematics," in *The Development of the Sciences*, New Haven, 1923.

[1] (1) *De Analysi per Equationes numero terminorum infinitas*, written in 1666 and sent to Barrow, who made it known to John Collins, who allowed Lord Brouncker to copy it; it was not published, however, until (London) 1711. (2) *Method of Fluxions and Infinite Series*, written in 1671 but not printed until (London) 1736, and then in John Colson's translation with this title. (3) *Tractatus de Quadratura Curvarum*, apparently written in 1676, but not published until (London) 1704 (appendix to his *Opticks*).

infinitely small quantities,[1] but he apparently recognized that this was not scientific, for it is not the basis of his work in this field.

Method of Fluxions. His second method was that of fluxions. For example, he considered a curve as described by a flowing point, calling the infinitely short path traced in an infinitely short time the moment of the flowing quantity, and designated the ratio of the moment to the corresponding time as the "fluxion" of the variable, that is, as the velocity. This fluxion of x he denoted by the symbol $\dot{x}$. In his *Method of Fluxions*[2] he states that "the moments of flowing quantities are as the velocities of their flowing or increasing,"—a statement which may be expressed in the Leibnizian symbolism as

$$\frac{dy}{dx} = \frac{dy}{dt} : \frac{dx}{dt}.$$

His treatment of fluxions may be illustrated by the following extract from his work:[3]

If the moment of x be represented by the product of its celerity $\dot{x}$ into an indefinitely small quantity o (that is $\dot{x}o$), the moment of y will be $\dot{y}o$, since $\dot{x}o$ and $\dot{y}o$ are to each other as $\dot{x}$ and $\dot{y}$. Now since the moments as $\dot{x}o$ and $\dot{y}o$ are the indefinitely little accessions of the flowing quantities, x and y, by which these quantities are increased through the several indefinitely little intervals of time, it follows that these quantities, x and y, after any indefinitely small interval of time, become $x + \dot{x}o$ and $y + \dot{y}o$. And therefore the equation which at all times indifferently expresses the relation of the flowing quantities will as well express the relation between $x + \dot{x}o$ and $y + \dot{y}o$ as between x and y; so that $x + \dot{x}o$ and $y + \dot{y}o$ may be substituted in the same equation for those quantities instead of x and y.

Therefore let any equation

$$x^3 - ax^2 + axy - y^3 = 0$$

[1] F. Cajori, in *Amer. Math. Month.*, XXIV, 145; *ibid.*, XXVI, 15.
[2] Colson translation, p. 24 (London, 1736).
[3] Pp. 24, 25. See also G. H. Graves, *loc. cit.*, III, 82.

be given, and substitute $x + \dot{x}o$ for x and $y + \dot{y}o$ for y, and there will arise

$$x^3 + 3\,x^2\dot{x}o + 3\,x\dot{x}\dot{x}oo + \dot{x}^3o^3$$
$$- ax^2 - 2\,ax\dot{x}o - a\dot{x}^2oo$$
$$+ axy + ax\dot{y}o + a\dot{x}oy + a\dot{x}o\dot{y}o$$
$$- y^3 - 3\,\dot{y}oy^2 - 3\,\dot{y}^2ooy - \dot{y}^3o^3 = o.$$

Now, by supposition,

$$x^3 - ax^2 + axy - y^3 = o,$$

which therefore being expunged and the remaining terms being divided by o, there will remain

$$3\,\dot{x}x^2 + 3\,\dot{x}^2ox + \dot{x}^3oo - 2\,a\dot{x}x - a\dot{x}^2o + a\dot{y}x + a\dot{x}y$$
$$+ a\dot{x}\dot{y}o - 3\,\dot{y}y^2 - 3\,\dot{y}^2oy - \dot{y}^3oo = o.$$

But whereas o is supposed to be infinitely little that it may represent the moments of quantities, the terms which are multiplied by it will be nothing in respect to the rest. Therefore I reject them and there remains:

$$3\,\dot{x}x^2 - 2\,a\dot{x}x + a\dot{y}x + a\dot{x}y - 3\,\dot{y}y^2 = o.$$

Method of Limits. Newton's third method, that of limits, appears in his *Tractatus de Quadratura Curvarum* (1704). In the introduction he says:

Let a quantity x flow uniformly and let it be required to find the fluxion of x^n. In the time in which x by flowing becomes $x + o$, the quantity x^n becomes $\overline{x + o}\,]^n$; *i.e.*, by the method of infinite series,

$$x^n + nox^{n-1} + \frac{nn - n}{2}\,oox^{n-2} +, \text{ etc.,}$$

and the increment o and

$$nox^{n-1} + \frac{nn - n}{2}\,oox^{n-2} +, \text{ etc.,}$$

are to each other as 1 and

$$nx^{n-1} + \frac{nn - n}{2}\,ox^{n-2} +, \text{ etc.}$$

Now let the increment vanish and their last ratio will be 1 to nx^{n-1}.

He also gives the interpretation of these ratios as the slopes of a secant through two points on a curve and of the tangent which is the limiting position of this secant.[1] He adds:

If the points are distant from each other by an interval however small, the secant will be distant from the tangent by a small interval. That it may coincide with the tangent and the last ratio be found, the two points must unite and coincide altogether. In mathematics, errors, however small, must not be neglected.

In the *Principia* (Section I) Newton set forth his idea of these ultimate ratios as follows:

Ultimate ratios in which quantities vanish, are not, strictly speaking, ratios of ultimate quantities, but limits to which the ratio of these quantities, decreasing without limit, approach, and which, though they can come nearer than any given difference whatever, they can neither pass over nor attain before the quantities have diminished indefinitely.

In the fluxional notation Newton represented the fluent of x by $\dot{x}o$, or simply by $\dot{x}$. The fluent of $\dot{x}$ he represented by $\ddot{x}$, and so on,[2] a notation first published in the *Algebra* of John Wallis (1693).

Summary of Newton's Method. Ball has clearly summarized Newton's general method of treatment as follows:

There are two kinds of problems. The object of the first is to find the fluxion of a given quantity, or, more generally, "the relation of the fluents being given, to find the relations of their fluxions." This is equivalent to differentiation. The object of the second, or inverse, method of fluxions is, from the fluxion or some relations involving it, to determine the fluent; or, more generally, "an equation being proposed exhibiting the relation of the fluxions of quantities, to find the relations of those quantities, or fluents, to one another." This is

[1] Graves, *loc. cit.*

[2] "Sint v, x, y, z fluentes quantitates, & earum fluxiones his notis $\dot{v}, \dot{x}, \dot{y}, \dot{z}$, designabuntur respective. . . . Qua ratione $\dot{v}$ est fluxio quantitatis v, & $\ddot{v}$ fluxio ipsius $\dot{v}$, & $\dddot{v}$ fluxio ipsius $\ddot{v}$ " (*Opera*, II, 392).

equivalent either to integration, which Newton termed the method of quadrature, or to the solution of a differential equation which was called by Newton the inverse method of tangents.[1]

Leibniz. Leibniz (1684) was well aware of the work of men like Barrow, Huygens, Grégoire de Saint-Vincent, Pascal, and Cavalieri. He was in London in 1673, and there he probably met with scholars who were perfectly familiar with the discoveries of Barrow and Newton, and with Barrow himself he had extended correspondence. After leaving England he set to work upon the problems of tangents and quadratures and invented a notation which was original and at the same time was generally more usable than that of Newton,—the "differential notation." He proposed to represent the sum of Cavalieri's indivisibles by the symbol $\int$, the old form of s, the initial of *summa*, using this together with Cavalieri's *omn.* (for *omnia*), and to represent the inverse operation by d. By 1675 he had settled this notation,[2] writing $\int y\,dy = \frac{1}{2}y^2$ as it is written at present.

Leibniz published his method in 1684[3] and 1686,[4] speaking of the integral calculus as the *calculus summatorius*, a name connected with the *summa* ($\int$) sign. In 1696 he adopted the term *calculus integralis*, already suggested by Jacques Bernoulli in 1690.

His Conception of the Differential. Some idea of his conception of the differential may be obtained from a statement in a letter written by him to Wallis on March 30, 1699:

It is useful to consider quantities infinitely small such that when their ratio is sought, they may not be considered zero, but which are rejected as often as they occur with quantities incomparably greater.

[1] W. W. R. Ball, *Hist. of Math.*, 6th ed., p. 344 (London, 1915), to which the reader is referred for further details, Mr. Ball having given special attention to the work of Newton. See also A. von Braunmühl, *Bibl. Math.*, V (3), 355.

[2] But published in 1686. Love, *loc. cit.*

[3] "Nova methodus pro maximis et minimis, itemque tangentibus . . . ," in the *Acta Eruditorum.*

[4] "De geometria recondita et analysi indivisibilium atque infinitorum," also in the *Acta Eruditorum.* On an early case of integration (1599) before the symbolism appeared, see F. Cajori, in *Bibl. Math.*, XIV (3), 312.

Thus if we have $x + dx$, dx is rejected. But it is different if we seek the difference between $x + dx$ and x, for then the finite quantities disappear. Similarly we cannot have xdx and $dxdx$ standing together. Hence if we are to differentiate xy we write:

$$(x + dx)(y + dy) - xy = xdy + ydx + dxdy.$$

But here $dxdy$ is to be rejected as incomparably less than $xdy + ydx$. Thus in any particular case the error is less than any finite quantity.[1]

As to the approximate period at which he began to arrive at his laws for the differentiation of algebraic functions, we have a manuscript of his which was written in November, 1676, and in which he gives the following statements:[2]

$$\overline{dx} = 1, \ \overline{dx^2} = 2\,x, \ \overline{dx^3} = 3\,x^2, \text{ etc.};$$

$$\overline{d\,\frac{1}{x}} = -\frac{1}{x^2}, \ \overline{d\,\frac{1}{x^2}} = -\frac{2}{x^2}, \ \overline{d\,\frac{1}{x^3}} = \frac{3}{x^2}, \text{ etc.};$$

$$\overline{d\,\sqrt{x}} = \frac{1}{\sqrt{x}}, \text{ etc.};$$

$$dx^e = ex^{e-1}, \text{ and conversely } \int \overline{x^e} = \frac{x^{e+1}}{e+1}.$$

Hence $\overline{d\,\frac{1}{x^2}} = \overline{dx^{-2}}$ will be $-2\,x^{-3}$ or $-\frac{2}{x^3}$, and $d\,\sqrt{x}$ or $dx^{\frac{1}{2}}$ will be $-\frac{1}{2}x^{-\frac{1}{2}}$ or $-\frac{1}{2}\sqrt{\frac{1}{x}}$.

Some of these results are incorrect, probably because of careless writing, and some appear in his earlier manuscripts, but they all serve to show how the mind of Leibniz was working in this period. By the end of the year 1676 he had developed the rule for differentiating a product, and by July, 1677, he had the differentiation of algebraic functions well in hand.[3]

[1] *Leibnitzens Mathematische Schriften*, Gerhardt ed., IV, 63 (Series III, in *Leibnitzens Gesammelte Werke*, Pertz ed., Halle, 1859) (this portion translated by Mr. Graves).

[2] J. M. Child, *The Early Mathematical Manuscripts of Leibniz*, p. 124 (Chicago, 1920), the results being as there stated, including errors.

[3] Child, *loc. cit.*, p. 116.

His notation for differentiation was used in England by John Craig as early as 1693,[1] and the same writer used his sign for integration ten years later.[2] Both symbols were somewhat familiar to English mathematicians throughout the 18th century, although it was not until the 19th century that their use in Great Britain became general.

Priority Dispute. The dispute between the friends of Newton and those of Leibniz as to the priority of discovery was bitter and rather profitless. It was the subject of many articles[3] and of a report by a special committee of the Royal Society.[4]

English readers of the 18th century were so filled with the arguments respecting the controversy as set forth in the *Commercium Epistolicum* (1712) and Raphson's *History of Fluxions* (1715), that they gave Leibniz little credit for his work. It was not until De Morgan (1846) reviewed the case that they began generally to recognize that they had not shown their usual spirit of fairness. On the other hand, Leibniz was so stung by the accusations of his English critics that he too showed a spirit that cannot always be commended.

Leibniz states his Case. It is interesting to read the words of Leibniz in his own defense, as presented in his *Historia et Origo Calculi Differentialis*:[5]

Since therefore his[6] opponents, neither from the *Commercium Epistolicum* that they have published, nor from any other source, brought forward the slightest bit of evidence whereby it might be established that his rival used the differential calculus before it was published by our friend;[7] therefore all the accusations that were brought against him by these persons may be treated with contempt

[1] *Methodus Figurarum* (London, 1693). [2] *Tractatus Mathematicus.*

[3] Beginning with a publication by a Swiss scholar, Nicolas Fatio de Duillier (1664–1753), whose *Lineae brevissimi descensus investigatio geometrica duplex* appeared in London in 1699. See Child, *Leibniz Manuscripts*, pp. 22, 23.

[4] The report appeared in 1712. See Collins, *Commercium Epistolicum*. It was also edited by Biot and Lefort and published at Paris in 1856.

[5] Found in MS. by Dr. C. I. Gerhardt in the Royal Library at Hannover and published in Latin in 1846; English translation by Child, *Leibniz Manuscripts*, pp. 22, 57.

[6] *I.e.*, Leibniz's, the work being written in the third person.

[7] *I.e.*, himself.

as beside the question. They have used the dodge of the pettifogging advocate to divert the attention of the judges from the matter on trial to other things, namely to infinite series. But even in these they could bring forward nothing that could impugn the honesty of our friend, for he plainly acknowledged the manner in which he made progress in them; and in truth in these also, he finally attained to something higher and more general.

Brief Summary of the Dispute. The facts are that Leibniz knew of Barrow's work on the "differential triangle" before he began his own investigations, or could have known of it, and that he was also in a position to know something of Newton's work. The evidence is also clear that Newton's discovery was made before Leibniz entered the field; that Leibniz saw some of Newton's papers on the subject as early as 1677; that he proceeded on different lines from Newton and invented an original symbolism; and that he published his results before Newton's appeared in print. With these facts before us, it should be possible to award to each his approximate share in the development of the theory.[1]

Successors of Newton and Leibniz. Most of the British writers of the period 1693-1734, failing to comprehend Newton's position, considered a fluxion as an infinitely small quantity.[2] The first noteworthy improvement in England is due to Bishop Berkeley, who, in his *Analyst* (1734), showed the fallacy of this method of approach and attempted to prove that even Newton was at fault in his logic. Berkeley provoked great discussion in England, and the result was salutary, not that it affected Newton's standing, but that it put an end to much of the lax reasoning of his followers.[3]

---

[1] On the general controversy see the summary given in Ball, *Hist. of Math.*, 6th ed., pp. 356-362; H. Sloman, *The Claim of Leibniz to the Invention of the Differential Calculus*, English translation, London, 1860.

[2] F. Cajori, *Amer. Math. Month.*, XXIV, 145; XXVI, 15; to these articles the reader is referred for valuable details relating to this period.

[3] On the gradual improvement of the Leibniz theory through the laying of a scientific foundation for the doctrine of limits, see F. Cajori, "Grafting of the theory of limits on the calculus of Leibniz," *Amer. Math. Month.*, XXX, 223, with excellent bibliography.

Cauchy's Contribution. Perhaps the one to whom the greatest credit is due for placing the fundamental principle of the calculus on a satisfactory foundation is Cauchy.[1] He makes the transition from

$$\frac{dy}{dx} = \frac{f(x+i)-f(x)}{i} = f'(x)$$

to
$$dy = f'(x)\,dx$$

as follows:

Let $y = f(x)$ be a function of the independent variable x; i, an infinitesimal, and h, a finite quantity. If we put $i = \alpha h$, α will be an infinitesimal and we shall have the identity

$$\frac{f(x+i)-f(x)}{i} = \frac{f(x+\alpha h)-f(x)}{\alpha h},$$

whence we derive

(1) $$\frac{f(x+\alpha h)-f(x)}{\alpha} = \frac{f(x+i)-f(x)}{i}\,h.$$

The limit toward which the first member of this equation converges when the variable α approaches zero, h remaining constant, is what we call the "differential" of the function $y = f(x)$. We indicate this differential by the characteristic, d, as follows:

$$dy \text{ or } df(x).$$

It is easy to obtain its value when we know that of the derived function, y' or $f'(x)$. In fact, taking the limits of both members of equation (1), we have in general:

(2) $$df(x) = hf'(x).$$

In the particular case where $f(x) = x$, equation (2) reduces to

$$dx = h.$$

Thus the differential of the independent variable, x, is simply the finite constant, h. Substituting, equation (2) will become

$$df(x) = f'(x)\,dx,$$

or, what amounts to the same thing,

$$dy = y'dx.$$

[1] *Résumé des Leçons sur le Calcul Infinitésimal, Quatrième Leçon*, Paris, 1823; *Œuvres Complètes, Sér. II, Tome IV*, Paris, 1899.

5. JAPAN

The Yenri. There developed in Japan in the 17th century a native calculus which may have been the invention of the great Seki Kōwa (1642–1708), as tradition asserts, although we have no positive knowledge that he ever wrote upon the subject. This form of the calculus is known as the *yenri*, a word meaning "circle principle" or "theory of the circle" and possibly

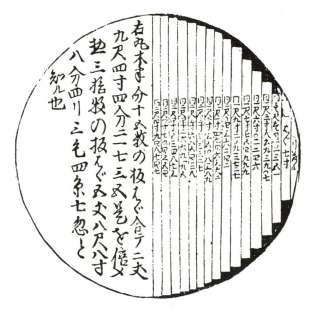

EARLY STEPS IN THE CALCULUS IN JAPAN

Crude integration, from Sawaguchi Kazuyuki's *Kokon Sampō-ki*, 1670. Sawaguchi was a pupil of Seki Kōwa, the Newton of Japan

suggested by an earlier Chinese title or by the fact that the method was primarily used in the measurement of this figure.

The mensuration of the circle by crude forms of integration is found in various works of the 18th century, such as the one illustrated above and the one shown on page 702, published by Mochinaga and Ōhashi in 1687. A similar use of the theory is found in connection with the mensuration of the sphere in

Isomura's work of 1684, and thereafter it appeared in numerous works in the closing years of the 17th century and the early part of the century following.

In a general way it may be said that the *yenri* was an application of series to the ancient method of exhaustion. For example, Takebe Kenkō (1722) found the approximate value of π by inscribing regular polygons up to 1024 sides, and probably more, giving the value to upwards of forty decimal places. In this work Takebe states that his method of approximation

EARLY STEPS IN THE CALCULUS IN JAPAN

From the *Kaisan-ki Kōmoku*, by Mochinaga and Ōhashi, representatives of the Seki School. The work was published in 1687. The method is essentially that of Sawaguchi

was not the one used by Seki Kōwa. In fact we know that the latter found an approximate value of π by computing successive perimeters, whereas Takebe based his work upon the squares of the perimeters, π^2 being taken as the square of the perimeter of a regular polygon of 512 sides. The value of π is expressed as a continued fraction, a plan which he states was due to his brother, Takebe Kemmei. Some of the formulas and series used by Takebe were very ingenious.[1]

[1] Smith-Mikami, p. 143.

TOPICS FOR DISCUSSION

1. General steps in the development of the calculus from the time of the Greeks to the present.

2. Zeno's paradoxes, their purpose, their fallacies, and their relation to the calculus.

3. The study of indivisible elements among the Greek philosophers, and its influence upon mathematics.

4. The atomistic philosophy of the Greeks, its founder and advocates; its bearing upon ancient mathematics and its relation to the modern calculus.

5. The relation of the method of exhaustion, especially as developed by Archimedes, to the integral calculus.

6. The various Greek writers on the method of exhaustion, together with a consideration of its results.

7. The contributions of Archimedes to the making of the calculus. His methods of proof.

8. The method employed by Archimedes in discovering his geometric propositions.

9. Formulas of the modern calculus anticipated by Archimedes and any other Greek writers.

10. The contributions of the Greeks to the subject of mechanics, and especially those of Aristotle and Archimedes.

11. Influence of Oresme with respect to the calculus.

12. Causes leading to Kepler's study of the problem of the calculus, together with a statement of the results of his work.

13. General nature of Cavalieri's contribution; the problems studied; the weakness of his method; the special results that he accomplished; and his influence upon Leibniz.

14. Fermat's contributions to the calculus compared with those of Cavalieri, Barrow, and Roberval.

15. The contributions of Roberval, Barrow, and other immediate predecessors of Newton.

16. Newton's discoveries in the calculus, with particular reference to the fundamental principles employed by him.

17. Leibniz's discoveries and the question of priority.

18. General nature of the developments in the calculus after Newton and Leibniz.

19. General nature of the early Japanese calculus.

INDEX

Since certain proper names are mentioned many times in this volume, only such page references have been given as are likely to be of considerable value to the reader, the first reference being to the biographical note in case one is given. In general, the biographies and bibliographies are to be found in Volume I. As a rule, the bibliographical references give only the page on which some important book or reference is first mentioned. Except for special reasons (such as a quotation, a discovery, or a contribution to which a reader may be likely to refer), no references are given to elementary textbooks or to the names of authors which are already given in Volume I and are mentioned only incidentally in Volume II. Obsolete terms are usually indexed only under modern forms. For further information consult the index to Volume I.

INDEX

INDEX

INDEX

INDEX

INDEX

INDEX

INDEX

INDEX

A CATALOG OF SELECTED
DOVER BOOKS
IN ALL FIELDS OF INTEREST

A CATALOG OF SELECTED DOVER
BOOKS IN ALL FIELDS OF INTEREST

CONCERNING THE SPIRITUAL IN ART, Wassily Kandinsky. Pioneering work by father of abstract art. Thoughts on color theory, nature of art. Analysis of earlier masters. 12 illustrations. 80pp. of text. 5⅜ × 8½. 23411-8 Pa. $3.95

ANIMALS: 1,419 Copyright-Free Illustrations of Mammals, Birds, Fish, Insects, etc., Jim Harter (ed.). Clear wood engravings present, in extremely lifelike poses, over 1,000 species of animals. One of the most extensive pictorial sourcebooks of its kind. Captions. Index. 284pp. 9 × 12. 23766-4 Pa. $11.95

CELTIC ART: The Methods of Construction, George Bain. Simple geometric techniques for making Celtic interlacements, spirals, Kells-type initials, animals, humans, etc. Over 500 illustrations. 160pp. 9 × 12. (USO) 22923-8 Pa. $9.95

AN ATLAS OF ANATOMY FOR ARTISTS, Fritz Schider. Most thorough reference work on art anatomy in the world. Hundreds of illustrations, including selections from works by Vesalius, Leonardo, Goya, Ingres, Michelangelo, others. 593 illustrations. 192pp. 7⅛ × 10¼. 20241-0 Pa. $8.95

CELTIC HAND STROKE-BY-STROKE (Irish Half-Uncial from "The Book of Kells"): An Arthur Baker Calligraphy Manual, Arthur Baker. Complete guide to creating each letter of the alphabet in distinctive Celtic manner. Covers hand position, strokes, pens, inks, paper, more. Illustrated. 48pp. 8¼ × 11.
24336-2 Pa. $3.95

EASY ORIGAMI, John Montroll. Charming collection of 32 projects (hat, cup, pelican, piano, swan, many more) specially designed for the novice origami hobbyist. Clearly illustrated easy-to-follow instructions insure that even beginning papercrafters will achieve successful results. 48pp. 8¼ × 11. 27298-2 Pa. $2.95

THE COMPLETE BOOK OF BIRDHOUSE CONSTRUCTION FOR WOOD-WORKERS, Scott D. Campbell. Detailed instructions, illustrations, tables. Also data on bird habitat and instinct patterns. Bibliography. 3 tables. 63 illustrations in 15 figures. 48pp. 5¼ × 8½. 24407-5 Pa. $1.95

BLOOMINGDALE'S ILLUSTRATED 1886 CATALOG: Fashions, Dry Goods and Housewares, Bloomingdale Brothers. Famed merchants' extremely rare catalog depicting about 1,700 products: clothing, housewares, firearms, dry goods, jewelry, more. Invaluable for dating, identifying vintage items. Also, copyright-free graphics for artists, designers. Co-published with Henry Ford Museum & Greenfield Village. 160pp. 8¼ × 11. 25780-0 Pa. $9.95

HISTORIC COSTUME IN PICTURES, Braun & Schneider. Over 1,450 costumed figures in clearly detailed engravings—from dawn of civilization to end of 19th century. Captions. Many folk costumes. 256pp. 8⅜ × 11¼. 23150-X Pa. $11.95

STICKLEY CRAFTSMAN FURNITURE CATALOGS, Gustav Stickley and L. & J. G. Stickley. Beautiful, functional furniture in two authentic catalogs from 1910. 594 illustrations, including 277 photos, show settles, rockers, armchairs, reclining chairs, bookcases, desks, tables. 183pp. 6½ × 9¼. 23838-5 Pa. $8.95

AMERICAN LOCOMOTIVES IN HISTORIC PHOTOGRAPHS: 1858 to 1949, Ron Ziel (ed.). A rare collection of 126 meticulously detailed official photographs, called "builder portraits," of American locomotives that majestically chronicle the rise of steam locomotive power in America. Introduction. Detailed captions. xi + 129pp. 9 × 12. 27393-8 Pa. $12.95

AMERICA'S LIGHTHOUSES: An Illustrated History, Francis Ross Holland, Jr. Delightfully written, profusely illustrated fact-filled survey of over 200 American lighthouses since 1716. History, anecdotes, technological advances, more. 240pp. 8 × 10¾. 25576-X Pa. $11.95

TOWARDS A NEW ARCHITECTURE, Le Corbusier. Pioneering manifesto by founder of "International School." Technical and aesthetic theories, views of industry, economics, relation of form to function, "mass-production split" and much more. Profusely illustrated. 320pp. 6⅛ × 9¼. (USO) 25023-7 Pa. $8.95

HOW THE OTHER HALF LIVES, Jacob Riis. Famous journalistic record, exposing poverty and degradation of New York slums around 1900, by major social reformer. 100 striking and influential photographs. 233pp. 10 × 7⅞. 22012-5 Pa $10.95

FRUIT KEY AND TWIG KEY TO TREES AND SHRUBS, William M. Harlow. One of the handiest and most widely used identification aids. Fruit key covers 120 deciduous and evergreen species; twig key 160 deciduous species. Easily used. Over 300 photographs. 126pp. 5⅜ × 8½. 20511-8 Pa. $3.95

COMMON BIRD SONGS, Dr. Donald J. Borror. Songs of 60 most common U.S. birds: robins, sparrows, cardinals, bluejays, finches, more—arranged in order of increasing complexity. Up to 9 variations of songs of each species.
Cassette and manual 99911-4 $8.95

ORCHIDS AS HOUSE PLANTS, Rebecca Tyson Northen. Grow cattleyas and many other kinds of orchids—in a window, in a case, or under artificial light. 63 illustrations. 148pp. 5⅜ × 8½. 23261-1 Pa. $3.95

MONSTER MAZES, Dave Phillips. Masterful mazes at four levels of difficulty. Avoid deadly perils and evil creatures to find magical treasures. Solutions for all 32 exciting illustrated puzzles. 48pp. 8¼ × 11. 26005-4 Pa. $2.95

MOZART'S DON GIOVANNI (DOVER OPERA LIBRETTO SERIES), Wolfgang Amadeus Mozart. Introduced and translated by Ellen H. Bleiler. Standard Italian libretto, with complete English translation. Convenient and thoroughly portable—an ideal companion for reading along with a recording or the performance itself. Introduction. List of characters. Plot summary. 121pp. 5¼ × 8½. 24944-1 Pa. $2.95

TECHNICAL MANUAL AND DICTIONARY OF CLASSICAL BALLET, Gail Grant. Defines, explains, comments on steps, movements, poses and concepts. 15-page pictorial section. Basic book for student, viewer. 127pp. 5⅜ × 8½. 21843-0 Pa. $3.95

BRASS INSTRUMENTS: Their History and Development, Anthony Baines. Authoritative, updated survey of the evolution of trumpets, trombones, bugles, cornets, French horns, tubas and other brass wind instruments. Over 140 illustrations and 48 music examples. Corrected and updated by author. New preface. Bibliography. 320pp. 5⅜ × 8½. 27574-4 Pa. $9.95

HOLLYWOOD GLAMOR PORTRAITS, John Kobal (ed.). 145 photos from 1926–49. Harlow, Gable, Bogart, Bacall; 94 stars in all. Full background on photographers, technical aspects. 160pp. 8⅜ × 11¼. 23352-9 Pa. $11.95

MAX AND MORITZ, Wilhelm Busch. Great humor classic in both German and English. Also 10 other works: "Cat and Mouse," "Plisch and Plumm," etc. 216pp. 5⅜ × 8½. 20181-3 Pa. $5.95

THE RAVEN AND OTHER FAVORITE POEMS, Edgar Allan Poe. Over 40 of the author's most memorable poems: "The Bells," "Ulalume," "Israfel," "To Helen," "The Conqueror Worm," "Eldorado," "Annabel Lee," many more. Alphabetic lists of titles and first lines. 64pp. 5³⁄₁₆ × 8¼. 26685-0 Pa. $1.00

SEVEN SCIENCE FICTION NOVELS, H. G. Wells. The standard collection of the great novels. Complete, unabridged. First Men in the Moon, Island of Dr. Moreau, War of the Worlds, Food of the Gods, Invisible Man, Time Machine, In the Days of the Comet. Total of 1,015pp. 5⅜ × 8½. (USO) 20264-X Clothbd. $29.95

AMULETS AND SUPERSTITIONS, E. A. Wallis Budge. Comprehensive discourse on origin, powers of amulets in many ancient cultures: Arab, Persian, Babylonian, Assyrian, Egyptian, Gnostic, Hebrew, Phoenician, Syriac, etc. Covers cross, swastika, crucifix, seals, rings, stones, etc. 584pp. 5⅜ × 8½. 23573-4 Pa. $12.95

RUSSIAN STORIES/PYCCKNE PACCKA3bl: A Dual-Language Book, edited by Gleb Struve. Twelve tales by such masters as Chekhov, Tolstoy, Dostoevsky, Pushkin, others. Excellent word-for-word English translations on facing pages, plus teaching and study aids, Russian/English vocabulary, biographical/critical introductions, more. 416pp. 5⅜ × 8½. 26244-8 Pa. $8.95

PHILADELPHIA THEN AND NOW: 60 Sites Photographed in the Past and Present, Kenneth Finkel and Susan Oyama. Rare photographs of City Hall, Logan Square, Independence Hall, Betsy Ross House, other landmarks juxtaposed with contemporary views. Captures changing face of historic city. Introduction. Captions. 128pp. 8¼ × 11. 25790-8 Pa. $9.95

AIA ARCHITECTURAL GUIDE TO NASSAU AND SUFFOLK COUNTIES, LONG ISLAND, The American Institute of Architects, Long Island Chapter, and the Society for the Preservation of Long Island Antiquities. Comprehensive, well-researched and generously illustrated volume brings to life over three centuries of Long Island's great architectural heritage. More than 240 photographs with authoritative, extensively detailed captions. 176pp. 8¼ × 11. 26946-9 Pa. $14.95

NORTH AMERICAN INDIAN LIFE: Customs and Traditions of 23 Tribes, Elsie Clews Parsons (ed.). 27 fictionalized essays by noted anthropologists examine religion, customs, government, additional facets of life among the Winnebago, Crow, Zuni, Eskimo, other tribes. 480pp. 6⅛ × 9¼. 27377-6 Pa. $10.95

FRANK LLOYD WRIGHT'S HOLLYHOCK HOUSE, Donald Hoffmann. Lavishly illustrated, carefully documented study of one of Wright's most controversial residential designs. Over 120 photographs, floor plans, elevations, etc. Detailed perceptive text by noted Wright scholar. Index. 128pp. 9¼ × 10¾.
27133-1 Pa. $11.95

THE MALE AND FEMALE FIGURE IN MOTION: 60 Classic Photographic Sequences, Eadweard Muybridge. 60 true-action photographs of men and women walking, running, climbing, bending, turning, etc., reproduced from rare 19th-century masterpiece. vi + 121pp. 9 × 12.
24745-7 Pa. $10.95

1001 QUESTIONS ANSWERED ABOUT THE SEASHORE, N. J. Berrill and Jacquelyn Berrill. Queries answered about dolphins, sea snails, sponges, starfish, fishes, shore birds, many others. Covers appearance, breeding, growth, feeding, much more. 305pp. 5¼ × 8¼.
23366-9 Pa. $7.95

GUIDE TO OWL WATCHING IN NORTH AMERICA, Donald S. Heintzelman. Superb guide offers complete data and descriptions of 19 species: barn owl, screech owl, snowy owl, many more. Expert coverage of owl-watching equipment, conservation, migrations and invasions, etc. Guide to observing sites. 84 illustrations. xiii + 193pp. 5⅜ × 8½.
27344-X Pa. $7.95

MEDICINAL AND OTHER USES OF NORTH AMERICAN PLANTS: A Historical Survey with Special Reference to the Eastern Indian Tribes, Charlotte Erichsen-Brown. Chronological historical citations document 500 years of usage of plants, trees, shrubs native to eastern Canada, northeastern U.S. Also complete identifying information. 343 illustrations. 544pp. 6½ × 9¼.
25951-X Pa. $12.95

STORYBOOK MAZES, Dave Phillips. 23 stories and mazes on two-page spreads: Wizard of Oz, Treasure Island, Robin Hood, etc. Solutions. 64pp. 8¼ × 11.
23628-5 Pa. $2.95

NEGRO FOLK MUSIC, U.S.A., Harold Courlander. Noted folklorist's scholarly yet readable analysis of rich and varied musical tradition. Includes authentic versions of over 40 folk songs. Valuable bibliography and discography. xi + 324pp. 5⅜ × 8½.
27350-4 Pa. $7.95

MOVIE-STAR PORTRAITS OF THE FORTIES, John Kobal (ed.). 163 glamor, studio photos of 106 stars of the 1940s: Rita Hayworth, Ava Gardner, Marlon Brando, Clark Gable, many more. 176pp. 8⅜ × 11¼.
23546-7 Pa. $10.95

BENCHLEY LOST AND FOUND, Robert Benchley. Finest humor from early 30s, about pet peeves, child psychologists, post office and others. Mostly unavailable elsewhere. 73 illustrations by Peter Arno and others. 183pp. 5⅜ × 8½.
22410-4 Pa. $5.95

YEKL and THE IMPORTED BRIDEGROOM AND OTHER STORIES OF YIDDISH NEW YORK, Abraham Cahan. Film Hester Street based on Yekl (1896). Novel, other stories among first about Jewish immigrants on N.Y.'s East Side. 240pp. 5⅜ × 8½.
22427-9 Pa. $6.95

SELECTED POEMS, Walt Whitman. Generous sampling from *Leaves of Grass.* Twenty-four poems include "I Hear America Singing," "Song of the Open Road," "I Sing the Body Electric," "When Lilacs Last in the Dooryard Bloom'd," "O Captain! My Captain!"—all reprinted from an authoritative edition. Lists of titles and first lines. 128pp. 5³⁄₁₆ × 8¼.
26878-0 Pa. $1.00

THE BEST TALES OF HOFFMANN, E. T. A. Hoffmann. 10 of Hoffmann's most important stories: "Nutcracker and the King of Mice," "The Golden Flowerpot," etc. 458pp. 5⅜ × 8½. 21793-0 Pa. $8.95

FROM FETISH TO GOD IN ANCIENT EGYPT, E. A. Wallis Budge. Rich detailed survey of Egyptian conception of "God" and gods, magic, cult of animals, Osiris, more. Also, superb English translations of hymns and legends. 240 illustrations. 545pp. 5⅜ × 8½. 25803-3 Pa. $11.95

FRENCH STORIES/CONTES FRANÇAIS: A Dual-Language Book, Wallace Fowlie. Ten stories by French masters, Voltaire to Camus: "Micromegas" by Voltaire; "The Atheist's Mass" by Balzac; "Minuet" by de Maupassant; "The Guest" by Camus, six more. Excellent English translations on facing pages. Also French-English vocabulary list, exercises, more. 352pp. 5⅜ × 8½. 26443-2 Pa. $8.95

CHICAGO AT THE TURN OF THE CENTURY IN PHOTOGRAPHS: 122 Historic Views from the Collections of the Chicago Historical Society, Larry A. Viskochil. Rare large-format prints offer detailed views of City Hall, State Street, the Loop, Hull House, Union Station, many other landmarks, circa 1904–1913. Introduction. Captions. Maps. 144pp. 9⅜ × 12¼. 24656-6 Pa. $12.95

OLD BROOKLYN IN EARLY PHOTOGRAPHS, 1865–1929, William Lee Younger. Luna Park, Gravesend race track, construction of Grand Army Plaza, moving of Hotel Brighton, etc. 157 previously unpublished photographs. 165pp. 8⅞ × 11¾. 23587-4 Pa. $13.95

THE MYTHS OF THE NORTH AMERICAN INDIANS, Lewis Spence. Rich anthology of the myths and legends of the Algonquins, Iroquois, Pawnees and Sioux, prefaced by an extensive historical and ethnological commentary. 36 illustrations. 480pp. 5⅜ × 8½. 25967-6 Pa. $8.95

AN ENCYCLOPEDIA OF BATTLES: Accounts of Over 1,560 Battles from 1479 B.C. to the Present, David Eggenberger. Essential details of every major battle in recorded history from the first battle of Megiddo in 1479 B.C. to Grenada in 1984. List of Battle Maps. New Appendix covering the years 1967–1984. Index. 99 illustrations. 544pp. 6½ × 9¼. 24913-1 Pa. $14.95

SAILING ALONE AROUND THE WORLD, Captain Joshua Slocum. First man to sail around the world, alone, in small boat. One of great feats of seamanship told in delightful manner. 67 illustrations. 294pp. 5⅜ × 8½. 20326-3 Pa. $5.95

ANARCHISM AND OTHER ESSAYS, Emma Goldman. Powerful, penetrating, prophetic essays on direct action, role of minorities, prison reform, puritan hypocrisy, violence, etc. 271pp. 5⅜ × 8½. 22484-8 Pa. $5.95

MYTHS OF THE HINDUS AND BUDDHISTS, Ananda K. Coomaraswamy and Sister Nivedita. Great stories of the epics; deeds of Krishna, Shiva, taken from puranas, Vedas, folk tales; etc. 32 illustrations. 400pp. 5⅜ × 8½. 21759-0 Pa. $9.95

BEYOND PSYCHOLOGY, Otto Rank. Fear of death, desire of immortality, nature of sexuality, social organization, creativity, according to Rankian system. 291pp. 5⅜ × 8½. 20485-5 Pa. $7.95

A THEOLOGICO-POLITICAL TREATISE, Benedict Spinoza. Also contains unfinished Political Treatise. Great classic on religious liberty, theory of government on common consent. R. Elwes translation. Total of 421pp. 5⅜ × 8½. 20249-6 Pa. $8.95

MY BONDAGE AND MY FREEDOM, Frederick Douglass. Born a slave, Douglass became outspoken force in antislavery movement. The best of Douglass' autobiographies. Graphic description of slave life. 464pp. 5⅜ × 8½.   22457-0 Pa. $8.95

FOLLOWING THE EQUATOR: A Journey Around the World, Mark Twain. Fascinating humorous account of 1897 voyage to Hawaii, Australia, India, New Zealand, etc. Ironic, bemused reports on peoples, customs, climate, flora and fauna, politics, much more. 197 illustrations. 720pp. 5⅜ × 8½.   26113-1 Pa. $15.95

THE PEOPLE CALLED SHAKERS, Edward D. Andrews. Definitive study of Shakers: origins, beliefs, practices, dances, social organization, furniture and crafts, etc. 33 illustrations. 351pp. 5⅜ × 8½.   21081-2 Pa. $8.95

THE MYTHS OF GREECE AND ROME, H. A. Guerber. A classic of mythology, generously illustrated, long prized for its simple, graphic, accurate retelling of the principal myths of Greece and Rome, and for its commentary on their origins and significance. With 64 illustrations by Michelangelo, Raphael, Titian, Rubens, Canova, Bernini and others. 480pp. 5⅜ × 8½.   27584-1 Pa. $9.95

PSYCHOLOGY OF MUSIC, Carl E. Seashore. Classic work discusses music as a medium from psychological viewpoint. Clear treatment of physical acoustics, auditory apparatus, sound perception, development of musical skills, nature of musical feeling, host of other topics. 88 figures. 408pp. 5⅜ × 8½. 21851-1 Pa. $9.95

THE PHILOSOPHY OF HISTORY, Georg W. Hegel. Great classic of Western thought develops concept that history is not chance but rational process, the evolution of freedom. 457pp. 5⅜ × 8½.   20112-0 Pa. $9.95

THE BOOK OF TEA, Kakuzo Okakura. Minor classic of the Orient: entertaining, charming explanation, interpretation of traditional Japanese culture in terms of tea ceremony. 94pp. 5⅜ × 8½.   20070-1 Pa. $2.95

LIFE IN ANCIENT EGYPT, Adolf Erman. Fullest, most thorough, detailed older account with much not in more recent books, domestic life, religion, magic, medicine, commerce, much more. Many illustrations reproduce tomb paintings, carvings, hieroglyphs, etc. 597pp. 5⅜ × 8½.   22632-8 Pa. $10.95

SUNDIALS, Their Theory and Construction, Albert Waugh. Far and away the best, most thorough coverage of ideas, mathematics concerned, types, construction, adjusting anywhere. Simple, nontechnical treatment allows even children to build several of these dials. Over 100 illustrations. 230pp. 5⅜ × 8½.   22947-5 Pa. $7.95

DYNAMICS OF FLUIDS IN POROUS MEDIA, Jacob Bear. For advanced students of ground water hydrology, soil mechanics and physics, drainage and irrigation engineering, and more. 335 illustrations. Exercises, with answers. 784pp. 6⅛ × 9¼.   65675-6 Pa. $19.95

SONGS OF EXPERIENCE: Facsimile Reproduction with 26 Plates in Full Color, William Blake. 26 full-color plates from a rare 1826 edition. Includes "The Tyger," "London," "Holy Thursday," and other poems. Printed text of poems. 48pp. 5¼ × 7.   24636-1 Pa. $4.95

OLD-TIME VIGNETTES IN FULL COLOR, Carol Belanger Grafton (ed.). Over 390 charming, often sentimental illustrations, selected from archives of Victorian graphics—pretty women posing, children playing, food, flowers, kittens and puppies, smiling cherubs, birds and butterflies, much more. All copyright-free. 48pp. 9¼ × 12¼.   27269-9 Pa. $5.95

PERSPECTIVE FOR ARTISTS, Rex Vicat Cole. Depth, perspective of sky and sea, shadows, much more, not usually covered. 391 diagrams, 81 reproductions of drawings and paintings. 279pp. 5⅜ × 8½. 22487-2 Pa. $6.95

DRAWING THE LIVING FIGURE, Joseph Sheppard. Innovative approach to artistic anatomy focuses on specifics of surface anatomy, rather than muscles and bones. Over 170 drawings of live models in front, back and side views, and in widely varying poses. Accompanying diagrams. 177 illustrations. Introduction. Index. 144pp. 8⅜ × 11¼. 26723-7 Pa. $7.95

GOTHIC AND OLD ENGLISH ALPHABETS: 100 Complete Fonts, Dan X. Solo. Add power, elegance to posters, signs, other graphics with 100 stunning copyright-free alphabets: Blackstone, Dolbey, Germania, 97 more—including many lower-case, numerals, punctuation marks. 104pp. 8⅛ × 11. 24695-7 Pa. $7.95

HOW TO DO BEADWORK, Mary White. Fundamental book on craft from simple projects to five-bead chains and woven works. 106 illustrations. 142pp. 5⅜ × 8. 20697-1 Pa. $4.95

THE BOOK OF WOOD CARVING, Charles Marshall Sayers. Finest book for beginners discusses fundamentals and offers 34 designs. "Absolutely first rate . . . well thought out and well executed."—E. J. Tangerman. 118pp. 7¾ × 10⅝. 23654-4 Pa. $5.95

ILLUSTRATED CATALOG OF CIVIL WAR MILITARY GOODS: Union Army Weapons, Insignia, Uniform Accessories, and Other Equipment, Schuyler, Hartley, and Graham. Rare, profusely illustrated 1846 catalog includes Union Army uniform and dress regulations, arms and ammunition, coats, insignia, flags, swords, rifles, etc. 226 illustrations. 160pp. 9 × 12. 24939-5 Pa. $10.95

WOMEN'S FASHIONS OF THE EARLY 1900s: An Unabridged Republication of "New York Fashions, 1909," National Cloak & Suit Co. Rare catalog of mail-order fashions documents women's and children's clothing styles shortly after the turn of the century. Captions offer full descriptions, prices. Invaluable resource for fashion, costume historians. Approximately 725 illustrations. 128pp. 8⅜ × 11¼. 27276-1 Pa. $11.95

THE 1912 AND 1915 GUSTAV STICKLEY FURNITURE CATALOGS, Gustav Stickley. With over 200 detailed illustrations and descriptions, these two catalogs are essential reading and reference materials and identification guides for Stickley furniture. Captions cite materials, dimensions and prices. 112pp. 6½ × 9¼. 26676-1 Pa. $9.95

EARLY AMERICAN LOCOMOTIVES, John H. White, Jr. Finest locomotive engravings from early 19th century: historical (1804–74), main-line (after 1870), special, foreign, etc. 147 plates. 142pp. 11⅜ × 8¼. 22772-3 Pa. $8.95

THE TALL SHIPS OF TODAY IN PHOTOGRAPHS, Frank O. Braynard. Lavishly illustrated tribute to nearly 100 majestic contemporary sailing vessels: Amerigo Vespucci, Clearwater, Constitution, Eagle, Mayflower, Sea Cloud, Victory, many more. Authoritative captions provide statistics, background on each ship. 190 black-and-white photographs and illustrations. Introduction. 128pp. 8⅞ × 11¾. 27163-3 Pa. $13.95

EARLY NINETEENTH-CENTURY CRAFTS AND TRADES, Peter Stockham (ed.). Extremely rare 1807 volume describes to youngsters the crafts and trades of the day: brickmaker, weaver, dressmaker, bookbinder, ropemaker, saddler, many more. Quaint prose, charming illustrations for each craft. 20 black-and-white line illustrations. 192pp. 4⅜ × 6.   27293-1 Pa. $4.95

VICTORIAN FASHIONS AND COSTUMES FROM HARPER'S BAZAR, 1867–1898, Stella Blum (ed.). Day costumes, evening wear, sports clothes, shoes, hats, other accessories in over 1,000 detailed engravings. 320pp. 9⅜ × 12¼.
22990-4 Pa. $13.95

GUSTAV STICKLEY, THE CRAFTSMAN, Mary Ann Smith. Superb study surveys broad scope of Stickley's achievement, especially in architecture. Design philosophy, rise and fall of the Craftsman empire, descriptions and floor plans for many Craftsman houses, more. 86 black-and-white halftones. 31 line illustrations. Introduction. 208pp. 6½ × 9¼.   27210-9 Pa. $9.95

THE LONG ISLAND RAIL ROAD IN EARLY PHOTOGRAPHS, Ron Ziel. Over 220 rare photos, informative text document origin (1844) and development of rail service on Long Island. Vintage views of early trains, locomotives, stations, passengers, crews, much more. Captions. 8⅞ × 11¾.   26301-0 Pa. $13.95

THE BOOK OF OLD SHIPS: From Egyptian Galleys to Clipper Ships, Henry B. Culver. Superb, authoritative history of sailing vessels, with 80 magnificent line illustrations. Galley, bark, caravel, longship, whaler, many more. Detailed, informative text on each vessel by noted naval historian. Introduction. 256pp. 5⅜ × 8½.   27332-6 Pa. $6.95

TEN BOOKS ON ARCHITECTURE, Vitruvius. The most important book ever written on architecture. Early Roman aesthetics, technology, classical orders, site selection, all other aspects. Morgan translation. 331pp. 5⅜ × 8½. 20645-9 Pa. $8.95

THE HUMAN FIGURE IN MOTION, Eadweard Muybridge. More than 4,500 stopped-action photos, in action series, showing undraped men, women, children jumping, lying down, throwing, sitting, wrestling, carrying, etc. 390pp. 7⅞ × 10⅝.
20204-6 Clothbd. $24.95

TREES OF THE EASTERN AND CENTRAL UNITED STATES AND CANADA, William M. Harlow. Best one-volume guide to 140 trees. Full descriptions, woodlore, range, etc. Over 600 illustrations. Handy size. 288pp. 4½ × 6⅜.
20395-6 Pa. $5.95

SONGS OF WESTERN BIRDS, Dr. Donald J. Borror. Complete song and call repertoire of 60 western species, including flycatchers, juncoes, cactus wrens, many more—includes fully illustrated booklet.   Cassette and manual 99913-0 $8.95

GROWING AND USING HERBS AND SPICES, Milo Miloradovich. Versatile handbook provides all the information needed for cultivation and use of all the herbs and spices available in North America. 4 illustrations. Index. Glossary. 236pp. 5⅜ × 8½.   25058-X Pa. $5.95

BIG BOOK OF MAZES AND LABYRINTHS, Walter Shepherd. 50 mazes and labyrinths in all—classical, solid, ripple, and more—in one great volume. Perfect inexpensive puzzler for clever youngsters. Full solutions. 112pp. 8⅛ × 11.
22951-3 Pa. $3.95

PIANO TUNING, J. Cree Fischer. Clearest, best book for beginner, amateur. Simple repairs, raising dropped notes, tuning by easy method of flattened fifths. No previous skills needed. 4 illustrations. 201pp. 5⅜ × 8½. 23267-0 Pa. $5.95

A SOURCE BOOK IN THEATRICAL HISTORY, A. M. Nagler. Contemporary observers on acting, directing, make-up, costuming, stage props, machinery, scene design, from Ancient Greece to Chekhov. 611pp. 5⅜ × 8½. 20515-0 Pa. $11.95

THE COMPLETE NONSENSE OF EDWARD LEAR, Edward Lear. All nonsense limericks, zany alphabets, Owl and Pussycat, songs, nonsense botany, etc., illustrated by Lear. Total of 320pp. 5⅜ × 8½. (USO) 20167-8 Pa. $6.95

VICTORIAN PARLOUR POETRY: An Annotated Anthology, Michael R. Turner. 117 gems by Longfellow, Tennyson, Browning, many lesser-known poets. "The Village Blacksmith," "Curfew Must Not Ring Tonight," "Only a Baby Small," dozens more, often difficult to find elsewhere. Index of poets, titles, first lines. xxiii + 325pp. 5⅜ × 8¼. 27044-0 Pa. $8.95

DUBLINERS, James Joyce. Fifteen stories offer vivid, tightly focused observations of the lives of Dublin's poorer classes. At least one, "The Dead," is considered a masterpiece. Reprinted complete and unabridged from standard edition. 160pp. 5³⁄₁₆ × 8¼. 26870-5 Pa. $1.00

THE HAUNTED MONASTERY and THE CHINESE MAZE MURDERS, Robert van Gulik. Two full novels by van Gulik, set in 7th-century China, continue adventures of Judge Dee and his companions. An evil Taoist monastery, seemingly supernatural events; overgrown topiary maze hides strange crimes. 27 illustrations. 328pp. 5⅜ × 8½. 23502-5 Pa. $7.95

THE BOOK OF THE SACRED MAGIC OF ABRAMELIN THE MAGE, translated by S. MacGregor Mathers. Medieval manuscript of ceremonial magic. Basic document in Aleister Crowley, Golden Dawn groups. 268pp. 5⅜ × 8½. 23211-5 Pa. $8.95

NEW RUSSIAN-ENGLISH AND ENGLISH-RUSSIAN DICTIONARY, M. A. O'Brien. This is a remarkably handy Russian dictionary, containing a surprising amount of information, including over 70,000 entries. 366pp. 4½ × 6⅛. 20208-9 Pa. $9.95

HISTORIC HOMES OF THE AMERICAN PRESIDENTS, Second, Revised Edition, Irvin Haas. A traveler's guide to American Presidential homes, most open to the public, depicting and describing homes occupied by every American President from George Washington to George Bush. With visiting hours, admission charges, travel routes. 175 photographs. Index. 160pp. 8¼ × 11. 26751-2 Pa. $10.95

NEW YORK IN THE FORTIES, Andreas Feininger. 162 brilliant photographs by the well-known photographer, formerly with *Life* magazine. Commuters, shoppers, Times Square at night, much else from city at its peak. Captions by John von Hartz. 181pp. 9¼ × 10¾. 23585-8 Pa. $12.95

INDIAN SIGN LANGUAGE, William Tomkins. Over 525 signs developed by Sioux and other tribes. Written instructions and diagrams. Also 290 pictographs. 111pp. 6⅛ × 9¼. 22029-X Pa. $3.50

ANATOMY: A Complete Guide for Artists, Joseph Sheppard. A master of figure drawing shows artists how to render human anatomy convincingly. Over 460 illustrations. 224pp. 8⅜ × 11¼. 27279-6 Pa. $9.95

MEDIEVAL CALLIGRAPHY: Its History and Technique, Marc Drogin. Spirited history, comprehensive instruction manual covers 13 styles (ca. 4th century thru 15th). Excellent photographs; directions for duplicating medieval techniques with modern tools. 224pp. 8⅜ × 11¼. 26142-5 Pa. $11.95

DRIED FLOWERS: How to Prepare Them, Sarah Whitlock and Martha Rankin. Complete instructions on how to use silica gel, meal and borax, perlite aggregate, sand and borax, glycerine and water to create attractive permanent flower arrangements. 12 illustrations. 32pp. 5⅜ × 8½. 21802-3 Pa. $1.00

EASY-TO-MAKE BIRD FEEDERS FOR WOODWORKERS, Scott D. Campbell. Detailed, simple-to-use guide for designing, constructing, caring for and using feeders. Text, illustrations for 12 classic and contemporary designs. 96pp. 5⅜ × 8½. 25847-5 Pa. $2.95

OLD-TIME CRAFTS AND TRADES, Peter Stockham. An 1807 book created to teach children about crafts and trades open to them as future careers. It describes in detailed, nontechnical terms 24 different occupations, among them coachmaker, gardener, hairdresser, lacemaker, shoemaker, wheelwright, copper-plate printer, milliner, trunkmaker, merchant and brewer. Finely detailed engravings illustrate each occupation. 192pp. 4⅝ × 6. 27398-9 Pa. $4.95

THE HISTORY OF UNDERCLOTHES, C. Willett Cunnington and Phyllis Cunnington. Fascinating, well-documented survey covering six centuries of English undergarments, enhanced with over 100 illustrations: 12th-century laced-up bodice, footed long drawers (1795), 19th-century bustles, 19th-century corsets for men, Victorian "bust improvers," much more. 272pp. 5⅜ × 8¼. 27124-2 Pa. $9.95

ARTS AND CRAFTS FURNITURE: The Complete Brooks Catalog of 1912, Brooks Manufacturing Co. Photos and detailed descriptions of more than 150 now very collectible furniture designs from the Arts and Crafts movement depict davenports, settees, buffets, desks, tables, chairs, bedsteads, dressers and more, all built of solid, quarter-sawed oak. Invaluable for students and enthusiasts of antiques, Americana and the decorative arts. 80pp. 6½ × 9¼. 27471-3 Pa. $7.95

HOW WE INVENTED THE AIRPLANE: An Illustrated History, Orville Wright. Fascinating firsthand account covers early experiments, construction of planes and motors, first flights, much more. Introduction and commentary by Fred C. Kelly. 76 photographs. 96pp. 8¼ × 11. 25662-6 Pa. $8.95

THE ARTS OF THE SAILOR: Knotting, Splicing and Ropework, Hervey Garrett Smith. Indispensable shipboard reference covers tools, basic knots and useful hitches; handsewing and canvas work, more. Over 100 illustrations. Delightful reading for sea lovers. 256pp. 5⅜ × 8½. 26440-8 Pa. $7.95

FRANK LLOYD WRIGHT'S FALLINGWATER: The House and Its History, Second, Revised Edition, Donald Hoffmann. A total revision—both in text and illustrations—of the standard document on Fallingwater, the boldest, most personal architectural statement of Wright's mature years, updated with valuable new material from the recently opened Frank Lloyd Wright Archives. "Fascinating"—*The New York Times*. 116 illustrations. 128pp. 9¼ × 10¾. 27430-6 Pa. $10.95

CATALOG OF DOVER BOOKS

PHOTOGRAPHIC SKETCHBOOK OF THE CIVIL WAR, Alexander Gardner. 100 photos taken on field during the Civil War. Famous shots of Manassas, Harper's Ferry, Lincoln, Richmond, slave pens, etc. 244pp. 10⅝ × 8¼.
22731-6 Pa. $9.95

FIVE ACRES AND INDEPENDENCE, Maurice G. Kains. Great back-to-the-land classic explains basics of self-sufficient farming. The one book to get. 95 illustrations. 397pp. 5⅜ × 8½. 20974-1 Pa. $7.95

SONGS OF EASTERN BIRDS, Dr. Donald J. Borror. Songs and calls of 60 species most common to eastern U.S.: warblers, woodpeckers, flycatchers, thrushes, larks, many more in high-quality recording. Cassette and manual 99912-2 $8.95

A MODERN HERBAL, Margaret Grieve. Much the fullest, most exact, most useful compilation of herbal material. Gigantic alphabetical encyclopedia, from aconite to zedoary, gives botanical information, medical properties, folklore, economic uses, much else. Indispensable to serious reader. 161 illustrations. 888pp. 6½ × 9¼. 2-vol. set. (USO) Vol. I: 22798-7 Pa. $9.95
Vol. II: 22799-5 Pa. $9.95

HIDDEN TREASURE MAZE BOOK, Dave Phillips. Solve 34 challenging mazes accompanied by heroic tales of adventure. Evil dragons, people-eating plants, bloodthirsty giants, many more dangerous adversaries lurk at every twist and turn. 34 mazes, stories, solutions. 48pp. 8¼ × 11. 24566-7 Pa. $2.95

LETTERS OF W. A. MOZART, Wolfgang A. Mozart. Remarkable letters show bawdy wit, humor, imagination, musical insights, contemporary musical world; includes some letters from Leopold Mozart. 276pp. 5⅜ × 8½. 22859-2 Pa. $6.95

BASIC PRINCIPLES OF CLASSICAL BALLET, Agrippina Vaganova. Great Russian theoretician, teacher explains methods for teaching classical ballet. 118 illustrations. 175pp. 5⅜ × 8½. 22036-2 Pa. $4.95

THE JUMPING FROG, Mark Twain. Revenge edition. The original story of The Celebrated Jumping Frog of Calaveras County, a hapless French translation, and Twain's hilarious "retranslation" from the French. 12 illustrations. 66pp. 5⅜ × 8½.
22686-7 Pa. $3.95

BEST REMEMBERED POEMS, Martin Gardner (ed.). The 126 poems in this superb collection of 19th- and 20th-century British and American verse range from Shelley's "To a Skylark" to the impassioned "Renascence" of Edna St. Vincent Millay and to Edward Lear's whimsical "The Owl and the Pussycat." 224pp. 5⅜ × 8½.
27165-X Pa. $4.95

COMPLETE SONNETS, William Shakespeare. Over 150 exquisite poems deal with love, friendship, the tyranny of time, beauty's evanescence, death and other themes in language of remarkable power, precision and beauty. Glossary of archaic terms. 80pp. 5³⁄₁₆ × 8¼. 26686-9 Pa. $1.00

BODIES IN A BOOKSHOP, R. T. Campbell. Challenging mystery of blackmail and murder with ingenious plot and superbly drawn characters. In the best tradition of British suspense fiction. 192pp. 5⅜ × 8½. 24720-1 Pa. $5.95

THE WIT AND HUMOR OF OSCAR WILDE, Alvin Redman (ed.). More than 1,000 ripostes, paradoxes, wisecracks: Work is the curse of the drinking classes; I can resist everything except temptation; etc. 258pp. 5⅜ × 8½. 20602-5 Pa. $5.95

SHAKESPEARE LEXICON AND QUOTATION DICTIONARY, Alexander Schmidt. Full definitions, locations, shades of meaning in every word in plays and poems. More than 50,000 exact quotations. 1,485pp. 6½ × 9¼. 2-vol. set.
Vol. 1: 22726-X Pa. $15.95
Vol. 2: 22727-8 Pa. $15.95

SELECTED POEMS, Emily Dickinson. Over 100 best-known, best-loved poems by one of America's foremost poets, reprinted from authoritative early editions. No comparable edition at this price. Index of first lines. 64pp. 5³⁄₁₆ × 8¼.
26466-1 Pa. $1.00

CELEBRATED CASES OF JUDGE DEE (DEE GOONG AN), translated by Robert van Gulik. Authentic 18th-century Chinese detective novel; Dee and associates solve three interlocked cases. Led to van Gulik's own stories with same characters. Extensive introduction. 9 illustrations. 237pp. 5⅜ × 8½.
23337-5 Pa. $6.95

THE MALLEUS MALEFICARUM OF KRAMER AND SPRENGER, translated by Montague Summers. Full text of most important witchhunter's "bible," used by both Catholics and Protestants. 278pp. 6⅝ × 10. 22802-9 Pa. $10.95

SPANISH STORIES/CUENTOS ESPAÑOLES: A Dual-Language Book, Angel Flores (ed.). Unique format offers 13 great stories in Spanish by Cervantes, Borges, others. Faithful English translations on facing pages. 352pp. 5⅜ × 8½.
25399-6 Pa. $8.95

THE CHICAGO WORLD'S FAIR OF 1893: A Photographic Record, Stanley Appelbaum (ed.). 128 rare photos show 200 buildings, Beaux-Arts architecture, Midway, original Ferris Wheel, Edison's kinetoscope, more. Architectural emphasis; full text. 116pp. 8¼ × 11. 23990-X Pa. $9.95

OLD QUEENS, N.Y., IN EARLY PHOTOGRAPHS, Vincent F. Seyfried and William Asadorian. Over 160 rare photographs of Maspeth, Jamaica, Jackson Heights, and other areas. Vintage views of DeWitt Clinton mansion, 1939 World's Fair and more. Captions. 192pp. 8⅞ × 11. 26358-4 Pa. $12.95

CAPTURED BY THE INDIANS: 15 Firsthand Accounts, 1750–1870, Frederick Drimmer. Astounding true historical accounts of grisly torture, bloody conflicts, relentless pursuits, miraculous escapes and more, by people who lived to tell the tale. 384pp. 5⅜ × 8½. 24901-8 Pa. $8.95

THE WORLD'S GREAT SPEECHES, Lewis Copeland and Lawrence W. Lamm (eds.). Vast collection of 278 speeches of Greeks to 1970. Powerful and effective models; unique look at history. 842pp. 5⅜ × 8½. 20468-5 Pa. $13.95

THE BOOK OF THE SWORD, Sir Richard F. Burton. Great Victorian scholar/adventurer's eloquent, erudite history of the "queen of weapons"—from prehistory to early Roman Empire. Evolution and development of early swords, variations (sabre, broadsword, cutlass, scimitar, etc.), much more. 336pp. 6⅛ × 9¼. 25434-8 Pa. $8.95

AUTOBIOGRAPHY: The Story of My Experiments with Truth, Mohandas K. Gandhi. Boyhood, legal studies, purification, the growth of the Satyagraha (nonviolent protest) movement. Critical, inspiring work of the man responsible for the freedom of India. 480pp. 5⅜ × 8½. (USO) 24593-4 Pa. $7.95

CELTIC MYTHS AND LEGENDS, T. W. Rolleston. Masterful retelling of Irish and Welsh stories and tales. Cuchulain, King Arthur, Deirdre, the Grail, many more. First paperback edition. 58 full-page illustrations. 512pp. 5⅜ × 8½. 26507-2 Pa. $9.95

THE PRINCIPLES OF PSYCHOLOGY, William James. Famous long course complete, unabridged. Stream of thought, time perception, memory, experimental methods; great work decades ahead of its time. 94 figures. 1,391pp. 5⅜ × 8½. 2-vol. set.
Vol. I: 20381-6 Pa. $12.95
Vol. II: 20382-4 Pa. $12.95

THE WORLD AS WILL AND REPRESENTATION, Arthur Schopenhauer. Definitive English translation of Schopenhauer's life work, correcting more than 1,000 errors, omissions in earlier translations. Translated by E. F. J. Payne. Total of 1,269pp. 5⅜ × 8½. 2-vol. set.
Vol. 1: 21761-2 Pa. $11.95
Vol. 2: 21762-0 Pa. $11.95

MAGIC AND MYSTERY IN TIBET, Madame Alexandra David-Neel. Experiences among lamas, magicians, sages, sorcerers, Bonpa wizards. A true psychic discovery. 32 illustrations. 321pp. 5⅜ × 8½. (USO) 22682-4 Pa. $8.95

THE EGYPTIAN BOOK OF THE DEAD, E. A. Wallis Budge. Complete reproduction of Ani's papyrus, finest ever found. Full hieroglyphic text, interlinear transliteration, word-for-word translation, smooth translation. 533pp. 6½ × 9¼. 21866-X Pa. $9.95

MATHEMATICS FOR THE NONMATHEMATICIAN, Morris Kline. Detailed, college-level treatment of mathematics in cultural and historical context, with numerous exercises. Recommended Reading Lists. Tables. Numerous figures. 641pp. 5⅜ × 8½. 24823-2 Pa. $11.95

THEORY OF WING SECTIONS: Including a Summary of Airfoil Data, Ira H. Abbott and A. E. von Doenhoff. Concise compilation of subsonic aerodynamic characteristics of NACA wing sections, plus description of theory. 350pp. of tables. 693pp. 5⅜ × 8½. 60586-8 Pa. $13.95

THE RIME OF THE ANCIENT MARINER, Gustave Doré, S. T. Coleridge. Doré's finest work; 34 plates capture moods, subtleties of poem. Flawless full-size reproductions printed on facing pages with authoritative text of poem. "Beautiful. Simply beautiful."—*Publisher's Weekly*. 77pp. 9¼ × 12. 22305-1 Pa. $5.95

NORTH AMERICAN INDIAN DESIGNS FOR ARTISTS AND CRAFTSPEOPLE, Eva Wilson. Over 360 authentic copyright-free designs adapted from Navajo blankets, Hopi pottery, Sioux buffalo hides, more. Geometrics, symbolic figures, plant and animal motifs, etc. 128pp. 8⅜ × 11. (EUK) 25341-4 Pa. $7.95

SCULPTURE: Principles and Practice, Louis Slobodkin. Step-by-step approach to clay, plaster, metals, stone; classical and modern. 253 drawings, photos. 255pp. 8⅛ × 11. 22960-2 Pa. $10.95

THE INFLUENCE OF SEA POWER UPON HISTORY, 1660–1783, A. T. Mahan. Influential classic of naval history and tactics still used as text in war colleges. First paperback edition. 4 maps. 24 battle plans. 640pp. 5⅜ × 8½.
25509-3 Pa. $12.95

THE STORY OF THE TITANIC AS TOLD BY ITS SURVIVORS, Jack Winocour (ed.). What it was really like. Panic, despair, shocking inefficiency, and a little heroism. More thrilling than any fictional account. 26 illustrations. 320pp. 5⅜ × 8½.
20610-6 Pa. $7.95

FAIRY AND FOLK TALES OF THE IRISH PEASANTRY, William Butler Yeats (ed.). Treasury of 64 tales from the twilight world of Celtic myth and legend: "The Soul Cages," "The Kildare Pooka," "King O'Toole and his Goose," many more. Introduction and Notes by W. B. Yeats. 352pp. 5⅜ × 8½.
26941-8 Pa. $8.95

BUDDHIST MAHAYANA TEXTS, E. B. Cowell and Others (eds.). Superb, accurate translations of basic documents in Mahayana Buddhism, highly important in history of religions. The Buddha-karita of Asvaghosha, Larger Sukhavativyuha, more. 448pp. 5⅜ × 8½. ,
25552-2 Pa. $9.95

ONE TWO THREE . . . INFINITY: Facts and Speculations of Science, George Gamow. Great physicist's fascinating, readable overview of contemporary science: number theory, relativity, fourth dimension, entropy, genes, atomic structure, much more. 128 illustrations. Index. 352pp. 5⅜ × 8½.
25664-2 Pa. $8.95

ENGINEERING IN HISTORY, Richard Shelton Kirby, et al. Broad, nontechnical survey of history's major technological advances: birth of Greek science, industrial revolution, electricity and applied science, 20th-century automation, much more. 181 illustrations. ". . . excellent . . ."—Isis. Bibliography. vii + 530pp. 5⅜ × 8¼.
26412-2 Pa. $14.95